Advances in Intelligent Systems and Computing

199

Editor-in-Chief

Prof. Janusz Kacprzyk
Systems Research Institute
Polish Academy of Sciences
ul. Newelska 6
01-447 Warsaw
Poland
E-mail: kacprzyk@ibspan.waw.pl

For further volumes:
http://www.springer.com/series/11156

Suresh Chandra Satapathy, Siba K. Udgata,
and Bhabendra Narayan Biswal (Eds.)

Proceedings of the International Conference on Frontiers of Intelligent Computing: Theory and Applications (FICTA)

Springer

Editors
Dr. Suresh Chandra Satapathy
Professor and Head
Dept. of Computer Science Engineering
Anil Neerukonda Institute of Technology
and Sciences
Sangivalasa
Vishakapatnam
India

Dr. Bhabendra Narayan Biswal
Director
Bhubaneswar Engineering College
Bhubaneswar
Odisha
India

Dr. Siba K. Udgata
AI Lab
Dept. of computer & Information Sciences
University of Hyderabad
Hyderabad
India

ISSN 2194-5357
e-ISSN 2194-5365
ISBN 978-3-642-35313-0
ISBN 978-3-642-35314-7 (eBook)
DOI 10.1007/978-3-642-35314-7
Springer Heidelberg New York Dordrecht London

Library of Congress Control Number: 2012953381

Printed on acid-free paper

Springer is part of Springer Science+Business Media (www.springer.com)

Preface

This AISC volume contains the papers presented at the First International Conference on Frontiers in Intelligent Computing: Theory and Applications (FICTA-2012) held during 22-23 December 2012 organized by Bhubaneswar Engineering College (BEC), Bhubaneswar, Odisa, India.

FICTA-2012 is a novel beginning of the prestigious international conference series that is aimed to bring researchers from academia and industry to report, deliberate and review the latest progresses in the cutting-edge research pertaining to intelligent computing and its applications to various engineering fields.

FICTA-2012 received 172 submissions and after a rigorous peer-review process with the help of our program committee members and external reviewers finally we accepted 86 papers with an acceptance ratio of 0.50.

The conference featured many distinguished keynote address by eminent speakers like Dr Ganesh Venayagamoorthy, Duke Energy Distinguished Professor, Director: Real-Time Power and Intelligent Systems Laboratory, Holcombe Department of Electrical and Computer Engineering, Clemson University, Clemson, Dr P N Suganthan, NTU Singapore, Dr JVR Murthy, JNTU Kakinada, India, Dr N R Pal, ISI Kolkota, Dr RavipudiVenketa Rao, SVNIT, Surat, India, Dr KPN Murthy, University of Hyderabad, India, Dr Kalyanmoy Dev, IIT Kanpur.

The conference also included a One day tutorial on "Evolutionary Computation and its engineering Applications". The tutorials were presented by Dr Swagatam Das, ISI, Kolkota, Dr B K Panigrahi, IIT Delhi and Dr S K Udgata, University of Hyderabad.

We take this opportunity to thank authors of all submitted papers for their hard work, adherence to the deadlines and patience with the review process. The quality of a referred volume depends mainly on the expertise and dedication of the reviewers. We are indebted to the program committee members and external reviewers who not only produced excellent reviews but also did these in short time frames.

We would also like to thank Bhubaneswar Engineering College (BEC), Bhubaneswar having coming forward to organize this first ever conference in the series. Our heartfelt thanks are due to Er Pravat Ranjan Mallick, Chairman, KGI, Bhubaneswar for the unstinted support to make the conference a grand success. Er Alok Ranjan Mallick, Vice-Chairman, KGI, Bhubaneswar and Chairman of BEC deserve kudos for the great support he has extended from the day one of the conceptualization of the idea of conducting the conference. We extend our thanks to him for coming ahead in supporting the conference financially and extending free accommodation and sightseeing trips to world famous Konark and Puri temple for all participants. Although BEC is just 5 years old but mind set of Er Alok Ranjan Mallick, the chairman of BEC, to make this as a world class institution has forced us to hold the first conference of FICTA series in the excellent campus of BEC. The green campus and excellent infrastructure along with very committed faculty and motivated students of BEC devoted their time to make the conference a memorable one. Each one of them deserves big thanks. We are confident that in future too we would like to organize many more international level conferences in this beautiful campus. We would also like to thank our sponsors for providing all the support and financial assistance.

We thank Prof P K Dash, SOA University, Bhubaneswar and Prof Ganapati Panda, Dy Director, IIT Bhubaneswar for providing valuable guidelines and inspirations to overcome various difficulties in the process of organizing this conference as Honorary General Chairs of this Conference. We extend our heart-felt thanks to Prof P N Suganthan, NTU Singapore for guiding us being the General chair of the conference. Dr B K Panigrahi, IIT Delhi and Dr Swagatam Das, ISI Kolkota deserves special thanks for being with us from the beginning to the end of this conference, without their support this conference could never have been successful. We would also like to thank the participants of this conference, who have considered the conference above all hardships. Finally, we would like to thank all the volunteers who spent tireless efforts in meeting the deadlines and arranging every detail to make sure that the conference can run smoothly. All the efforts are worth and would please us all, if the readers of this proceedings and participants of this conference found the papers and conference inspiring and enjoyable.

Our sincere thanks to all press print & electronic media for their excellent coverage of this conference.

December 2012

Volume Editors
Dr. Suresh Chandra Satapathy
Dr. Siba K. Udgata
Dr. Bhabendra Narayan Biswal

Organization

Organizing Committee

Chief Patron	Er Pravat Ranjan Mallick Chairman, KGI, Bhubaneswar
Patron	Er Alok Ranjan Mallick Vice-Chairman, KGI, Bhubaneswar Chairman, BEC, Bhubaneswar
Organizing Secretary	Prof B.N. Biswal Director (A&A), BEC, Bhubaneswar
Honorary Chairs	Dr P.K. Dask, SMIEEE, FNAE, Director (Research) SOA University, Bhubaneswar, India Dr Ganapati Panda, SMIEEE, FNAE Deputy Director, IIT, Bhubaneswar, India
General Chairs	Dr P.N. Suganthan, NTU, Singapore Dr Swagatam Das, ISI, Kolkota

Steering Committee Chair

Dr B.K. Panigrahi	IIT, Delhi, India

Program Chairs

Dr Suresh Chandra Satapathy	ANITS, Vishakapatnam, India
Dr S.K. Udgata	University of Hyderabad, India
Dr B.N. Biswal	Director (A&A), BEC, Bhubaneswar, India

International Advisory Committee/Technical Committee

P.K. Patra, India
J.V.R. Murthy, India
T.R. Dash, Kambodia
Kesab Chandra Satapathy, India
Maurice Clerc, France
Roderich Gross, England
Sangram Samal, India
K.K. Mohapatra, India
L. Perkin, USA
Sumanth Yenduri, USA
Carlos A. Coello Coello, Mexico
Dipankar Dasgupta, USA
Peng Shi, UK
Saman Halgamuge, Australia
Jeng-Shyang Pan, Taiwan
X.Z. Gao, Finland
Juan Luis Fernández Martínez, California
Oscar Castillo, Mexcico
Leandro Dos Santos Coelho, Brazil
Heitor Silvério Lopes, Brazil
Rafael Stubs Parpinelli, Brazil
S.S. Pattanaik, India
Gerardo Beni, USA
Namrata Khemka, USA
G.K. Venayagamoorthy, USA
K. Parsopoulos, Greece
Zong Woo Geem, USA
Lingfeng Wang, China
Athanasios V. Vasilakos, Athens
S.G. Ponnambalam, Malaysia
Pei-Chann Chang, Taiwan
Ying Tan, China
Chilukuri K. Mohan, USA
M.A. Abido, Saudi Arabia
Saeid Nahavandi, Australia
Almoataz Youssef Abdelaziz, Egypt
Hai Bin Duan, China
Delin Luo, China
M.K. Tiwari, India
A. Damodaram, India
Oscar Castillo, Mexcico
John MacIntyre, England
Frank Neumann
Rafael Stubs Parpinelli, Brazil
Jeng-Shyang Pan, Taiwan
P.K. Singh, India
Sachidananda Dehuri, India
P.S. Avadhani, India
G. Pradhan, India
Anupam Shukla, India
Dilip Pratihari, India
Amit Kumar, India
Srinivas Sethi, India
Lalitha Bhaskari, India
V. Suma, India
Ravipudi Venkata Rao, India

Organizing Committee

Prof. R.K. Behuria, Hod, Mech
Prof. P.K. Mohapatra, Hod, Civil
Prof. Debi Prasad Kar, Hod, ENTC
Prof. Manas Kumar Swain, Hod., CSE
Prof. P.M. Dash, Hod, EEE
Prof. S. Samal, Hod, Aeronautical Engg.
Prof. D. Panda, Hod, Sc.&H

Contents

Mixture Kernel Radial Basis Functions Neural Networks for Web Log Classification

Dash Ch. Sanjeev Kumar[1], Pandia Manoj Kumar[1], Dehuri Satchidananda[2], and Cho Sung-Bae[3]

[1] Department of Computer Science, Silicon Institute of Technology, Patia, Bhubaneswar-24, Odisha, India
[2] Department of Information and Communication Technology, Fakir Mohan University, Vyasa Vihar, Balasore-756019, Odisha, India
[3] Soft Computing Laboratory, Department of Computer Science, Yonsei University, 50 Yonsei-ro, Sudaemoon-gu, Seoul 120-749, South Korea.
{sanjeev_dash,manoj_pandia}@yahoo.com, satchi.lapa@gmail.com, sbcho@yonsei.ac.kr

Abstract. With the immense horizontal and vertical growth of the World Wide Web (WWW), it is becoming more popular for website owners to showcase their innovations, business, and concepts. Along side they are also interested in tracking and understanding the need of the users. Analyzing web access logs, one can understand the browsing behavior of users. However, web access logs are voluminous as well as complex. Therefore, a semi-automatic intelligent analyzer can be used to find out the browsing patterns of a user. Moreover, the pattern which is revealed from this deluge of web access logs must be interesting, useful, and understandable. A radial basis function neural networks (RBFNs) with mixture of kernels are used in this work for classification of web access logs. In this connection two RBFNs with different mixture of kernels are investigated on web access logs for classification. The collected data are used for training, validation, and testing of the models. The performances of these models are compared with RBFNs. It is concluded that mixture of appropriate kernels are an attractive alternative to RBFNs.

Keywords: Neural networks, Radial basis function neural networks, Classification, Web log, Mixture kernel.

1 Introduction

The web has influenced a lot to both users (visitors) as well as the web site owners. The web site owners are able to reach to all the targeted audience domestically and internationally. On the other side visitors are also availing those facilities. Over the last decades, the proliferation of information on the web has resulted in a large repository of web documents stored in multiple websites. These plethora and diversity of resources have promoted the need for developing a semi-automatic mining technique on the WWW, thereby giving rise to the term of web mining [14, 15].

S.C. Satapathy et al. (Eds.): Proc. of Int. Conf. on Front. of Intell. Comput., AISC 199, pp. 1–9.
DOI: 10.1007/978-3-642-35314-7_1

Every website contains multiple web pages, each of which has: 1) contents: which can be in any form, e.g., text, graphics, multimedia etc; 2) structure: containing links from one page to another; and 3) usage: users accessing the web pages. According to this the area of web mining can be categorized like Table 1.

Table 1. Web Mining Categorization

Type	Structure	Form	Object	Collection
Usage	Accessing	Click	Behavior	Logs
Content	Pages	Text	Index	Pages
Structure	Map	Hyperlinks	Map	Hyperlinks

Mining the contents of web pages is called as "Content Mining". Similarly, mining the links between web pages and the web access logs are respectively called "Structure" and "Web Usage" mining.

Web servers record and mount up data about user interactions whenever request for resources are received. Analyzing the web access logs of different web sites can help understand the user behavior and the web structure, thereby improving the design of this colossal collection of resources. To proceed towards a semi-automatic intelligent web analyzer, obviating the need for human intervention, we need to incorporate and embed computational or artificial intelligence into web analyzing tools. However, the problem of developing semi-automated tools in order to find, extract, filter, and evaluate the users desired information from unlabeled, distributed, and heterogeneous web access logs data is far from being solved. To handle these characteristics and overcome some of the limitations of existing methodologies, RBFN (a member of computational intelligence family) seems to be a good candidate. In this paper RBFN with mixture of kernels is used to design a classifier for web access logs classification. Li et al. [17] and Junjie et al. [18] works of RBFN for classification of web usage data are the source of motivation to carry out this work.

Over the last several decades multilayer perceptron (MLP) network was the popular network architectures and used in most of the applications. In MLP, the weighted sum of the inputs and bias term are passed to activation level through a transfer function to produce the output, and the units are arranged in a layered feed-forward topology called Feed Forward Neural Network (FFNN). In particular, MLP using back-propagation learning algorithm has been successfully applied to many applications. However, the training speed of MLP is typically much slower than those of feed-forward neural network comprising of single layer. Moreover, the problems such as local minima trapping, saturation, weight interference, initial weight dependence, and over-fitting make MLP training difficult [19]. Additionally, it is also very difficult to determine the parameters like number of neurons in a layer, and number of hidden layers in a network, thereby deciding a proper architecture is not trivial. These issues create the source of motivation to choose RBFNs over other alternatives to solve the problem focused in this paper.

Radial basis function neural network is based on supervised learning. RBFN networks are also good at modeling nonlinear data and can be trained in one stage rather than using an iterative process as in MLP. Radial basis function networks [1, 2, 3] have been studied in many disciplines like pattern recognition [4], medicine [5], multi-media applications [6], computational finance [7], software engineering [9], etc. It is emerged as a variant in late 1980's, but its root entrenched in much older pattern recognition, numerical analysis and other related fields [8]. The basic architectural framework of RBFN and its improvement by considering mixture of kernels are discussed in Section 3.4.

The rest of the paper is organized as follows. In Section 2, we cover briefly about the basics of RBFNs, as because this research is just improvement over RBFNs. Section 3 presents the framework of mixture kernels based RBFNs for classification. Sections 4 discuss the experimental study followed by conclusions and future work in Section 5.

2 Radial Basis Function Neural Networks

A RBFN is a multilayer and feed forward network which is often used for interpolation in multidimensional space. The basic architecture of a three-layered network is shown in Figure 2. A RBFN has three layers consisting of input, hidden, and output layers. The input layer receives input data. The hidden layer transforms the data from the input space to the hidden space using a non-liner function. The output layer, which is linear, yields the response to the network. Its training procedure is usually divided in to two stages. First, the centers and widths of the hidden layer are determined by various ways like clustering algorithms such as k-means algorithm [10], vector quantization [11], decision trees [12], or self-organizing future maps [13]. In this paper random initialization of centers and width is considered and then fine tuned over iterations. Second the weights connecting the hidden layer with the output are determined by singular value decomposition (SVD) or least mean square (LMS) algorithm. Here SVD is used for optimizing the weights. The number of basis functions controls the complexity and the generalization ability of RBF network.

The argument of the activation function of each hidden unit in an RBFN computes the Euclidean distance between an input vector and the center of the unit.

$$\Phi_i(x) = \exp\left(-\frac{\| x - \mu_i \|^2}{2\sigma_i^2} \right) \tag{1}$$

where ||...|| represents Euclidean norm, μ_i, σ_i and ϕ_i are center, width and the output of the ith hidden unit. The output layer, a set of summation units, supplies the response of the network.

The commonly used radial basis functions are enumerated in Table 2 as follows:

Table 2. Kernel functions used in RBFNs

Name of the Kernel	Mathematical Formula
Gaussian Functions	$\Phi(r) = \exp\left(-\frac{x^2}{2\sigma^2}\right)$ width parameter $\sigma > 0$
Generalized Multi-Quadric Functions	$\Phi(r) = \left(x^2 + \sigma^2\right)^{\beta}$ Parameter $\sigma > 0$, $1 > \beta > 0$
Generalized Inverse Multi Quadric Functions	$\Phi(x) = (x^2 + \sigma^2)^{-\alpha}$
Thin Plate Spline Function	$\Phi(x) = x^2 \ln(x)$
Cubic Function	$\Phi(x) = x^3$
Linear Function	$\Phi(x) = x$

3 The Framework of Mixture Kernel Based RBFNs

Figure 1 describes overall framework of our research. Subsection 3.1 - 3.3 describes the web usage mining and steps required for preprocessing the dataset. Subsection 3.4 describes RBFN with mixture kernels for classification of web access logs.

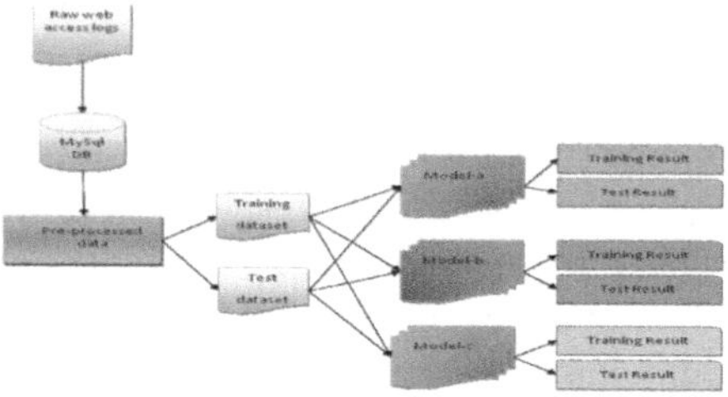

Fig. 1. Block diagram of web log classification

3.1 Web Usage Mining and Preprocessing

While visiting a website user navigates through multiple web pages. Each access to a web page is stored as a record in log files. These log files are managed by web servers. The format of log files varies from server to server. Apache web server maintains two different log file formats namely 1) Common Log Format, and 2) Combined Log Format.

The raw data from access log file are not in a state to be used for mining. These records must be pre-processed at first. Pre-processing [16] involves 1) Data cleaning, 2) User identification, and 3) Session identification.

For pre-processing we have followed the database approach where all the raw records are inserted into a table and the pre-processing tasks such as data cleaning, user identification and session identification are done using SQL queries. This approach gives better flexibility and faster execution.

First the raw web access log records are scanned from log files and inserted into a table which is designed in MySQL database. As we know query execution is easy and also the time taken for query execution is much less in comparison to file processing. Once all records are inserted successfully, the steps of preprocessing can be executed.

In this paper we have used the web access log data from www.silicon.ac.in. Silicon institute of technology is one of the premiere technical institutes in Odisha, India. We have collected the records between 22-Oct-2010 04:15:00 to 24-Oct-2010 04:05:48. The total number of records in the file is 12842.

3.2 Data Transformation

We have written a java program which reads each line from the log file and insert in to the table in MySQL database.

3.3 Data Cleaning

For data cleaning we need to remove all the records which contain the request for files like jpg, gif, css, js, etc. For this we execute the delete query.

Total number of records reduced to 1749 from 12842 after the data cleaning. Every unique page is assigned a PageID. The total unique web pages are found to be 97.

3.4 Mixture Kernels Based RBFNs

Suppose we take the region R to be a hypercube with sides of length h centered on the point x. Its volume is then given by

$$V=h^d . \tag{2}$$

We can find an expression for *K*, the number of points which fall within this region, by defining a kernel function *H*(*u*), also known as Parzen window given by

$$H(u)=\begin{Bmatrix} 1 & |u_j|<1/2 \\ 0 & \text{otherwise} \end{Bmatrix} \qquad j=1,\ldots,d \tag{3}$$

so that *H*(*u*) corresponds to a unit hypercube centered at the origin. Thus for all data points x^n, the quantity $H((x- x^{n)} /h)$ is equal to unit *y* if the point x^n falls inside a hypercube of side *h* centered on *x*, and is zero otherwise. The total number of points falling inside the hypercube is then simply

$$K=\sum_{n=1}^{N} H\left(\frac{x-x^n}{h}\right) \tag{4}$$

Mixture models are a divide-and-conquer learning method derived from the mixture estimation paradigm [21] that is heavily studied in artificial neural network research

[22]. They reduce the complexity by decomposing learning tasks and variance by combining multiple kernels.

A probability density function $p(x)$ can be modeled as a superposition of component densities as

$$p(x) = \sum_{j=1}^{M} P(j)\Psi_j(x) \tag{5}$$

Such a representation is called mixture distribution and the coefficients $P(j)$ are called the mixing parameters, where p is the prior probability of the component density ψ_j. The parameters of such a mixture model can be obtained using maximum likelihood [22]. These priors are chosen to satisfy the constraints.

$$\sum_{j=1}^{M} \psi(j) = 1 \tag{6}$$

$$0 \le \psi(j) \le 1 \tag{7}$$

Similarly, the component density functions $p(x \mid j)$ are normalized so that

$$\int p(x \mid j)dx = 1 \tag{8}$$

and hence can be regarded as class-condition densities. To generate a data point from the probability distribution [13], one of the component j is first selected at random with probability $\psi(j)$, and then a data point is generated from the corresponding component density p(x|j). Fatai [21] was done a comparative study of the application of Gaussian Mixture Model (GMM) and Radial Basis Function (RBF) in biometric recognition of voice.

Recall that RBF network contains three layers. In the hidden layer, the number of neurons is decided by leaning process. The two proposed models including RBFNs (model-a) have used nine neurons in the hidden layer. In model-b we have used combination of Gaussian, multi-quadric, and inverse multi-quadric functions.

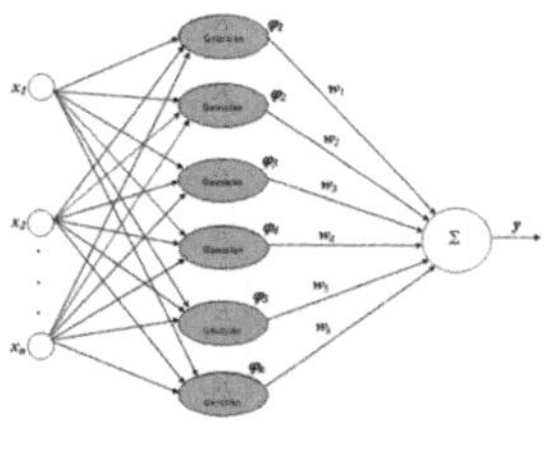

Fig. 2. (Model-a)

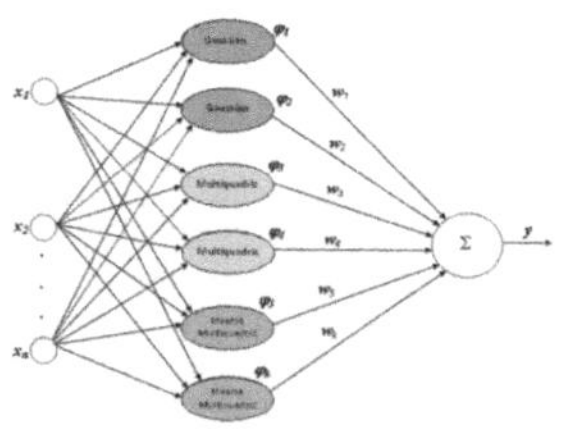

Fig. 3. (Model-b)

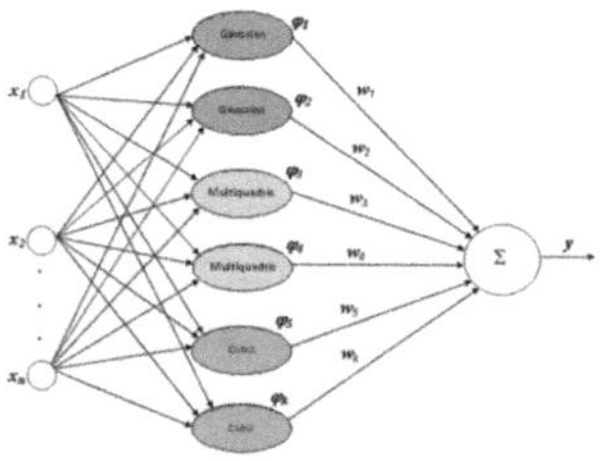

Fig. 4. (Model-c)

4 Experimental Study

4.1 Description of Dataset and Parameters

For classification purpose we have categorized the access time into two categories, i.e. from 00:00 hours to 12:00 hours as category 1 and 12:00 to 24:00 as category 2. Similarly we have grouped all the region of access into two, i.e. IP addresses accessing from within India as belongs to category 1 and IP addresses accessing from outside India as in category 2. To find the region from IP address we have used the website www.ip2location.com [18].

Our class attribute is frequent. For this we have considered two values i.e., 1 for not frequent and 2 for frequent.

This dataset includes 237 instances and each of which has 4 attributes, e.g., ID, request count, timeid, and locationid. The instances are classified into 2 classes (i.e., every instance either belongs to class 1 or 2). Class 1 has 129 instances, and class 2 contains 108. None of the attributes contain any missing values.

4.2 Results and Analysis

The results obtained from our experimental study are described in Table 3 according to different kernels like Gaussian functions, multi-quadric functions, inverse multi-quadric function, and cubic function. It is observed that model-b (Fig.3) [mixture of Gaussian, multi-quadric, and inverse multi-quadric RBF network] gives better result as compared to model-a (Fig.2) [RBFN with Gaussian] and model-c (Fig.4) [mixture of Gaussian, multi-quadric, and cubic] for web log dataset.

Table 3. Classification accuracy obtained from simulation of model-a, model-b, and model-c

Name of the Kernel	Hidden Units	Training	Testing	Average Accuracy
RBFN (with Gaussian)	9	83.7209	51.9380	67.82945
Mixture RBFN (Gaussian, multi-quadric, and inverse multi-quadric)	9	75.1938	76.7442	75.969
Mixture RBFN (Gaussian, multi-quadric, and cubic)	9	74.4186	76.7442	75.5814

5 Conclusion and Future Work

In this paper, we have compared RBFNs with two mixture models of RBFNs for classification on web log data. The average accuracy of second model (i.e., model-b)

gives better result than the first and third model. From the analysis we conclude that the web pages of www.silicon.ac.in are accessed most frequently at daytime inside India. As we have applied the classification on the attributes region and time, similarly this can be extended to consider some more attributes like user agent and referrer. Here we have considered only Gaussian, multi-quadric, inverse multi-quadric, and cubic kernel. By evolutionary technique this can be further extended to use other combination of kernels for higher accuracy.

References

1. Powell, M.J.D.: Radial Basis Functions for Multi-variable Interpolation: A Review. In: IMA Conference on Algorithms for the Approximations of Functions and Data, RMOS Shrivenham, UK (1985)
2. Broomhead, D.S., Lowe, D.: Multivariable Functional Interpolation and Adaptive Networks. Complex Systems 2, 321–355 (1988)
3. Buhmann, M.D.: Radial Basis Function Networks. In: Encyclopedia of Machine Learning, pp. 823–827 (2010)
4. Theodoridis, S., Koutroumbas, K.: Pattern Recognition. Academic Press, San Diego (2008)
5. Subashini, T.S., Ramalingam, V., Palanivel, S.: Breast Mass Classification Based on Cytological Patterns Using RBFNN and SVM. Expert Systems with Applications 36(3), 5284–5290 (2009)
6. Dhanalakshmi, P., Palanivel, S., Ramalingam, V.: Classification of Audio Signals Using SVM and RBFNN. Expert Systems with Applications 36(3), part 2, 6069–6075 (2009)
7. Sheta, A.F., De Jong, K.: Time Series Forecasting Using GA Tuned Radial Basis Functions. Information Sciences 133, 221–228 (2001)
8. Park, J., Sandberg, J.W.: Universal Approximation Using Radial Basis Function Networks. Neural Computation 3, 246–257 (1991)
9. Idri, A., Zakrani, A., Zahi, A.: Design of Radial Basis function Neural Networks for Software Effort Estimation. International Journal of Computer Science 7(4), 11–17 (2010)
10. Moody, J., Darken, C.J.: Fast Learning Networks of Locally-Tuned Processing Units. Neural Computation 6(4), 281–294 (1989)
11. Falcao, A., Langlois, O.T., Wichert, A.: Flexible Kernels for RBF Networks. Neurocomputing 69, 2356–2359 (2006)
12. Ghodsi, A., Schuurmans, D.: Automatic Basis Selection Techniques for RBF Networks. Neural Networks 16, 809–816 (2003)
13. Bishop, C.M.: Neural Networks for Pattern Recognition. Oxford University Press, New York (1995)
14. Hu, C., Zong, X., Lee, C.W., Yeh, J.H.: World Wide Web Usage Mining Systems and Technologies. Journal of Systemics, Cybernetics and Informatics 1(4), 53–59 (2003)
15. Srivastava, J., Cooley, R., Deshpande, M., Tan, P.N.: Web Usage Mining: Discovery and Applications of Usage Patterns from Web Data. SIGKDD Explorations 1(2), 1–12 (2000)
16. Li, Z., He, P., Lei, M.: Applying RBF Network to Web Classification Mining. Journal of Communication and Computer 2(9) (2005) ISSN 1548-7709
17. Junjie, C., Rongbing, H.: Research of Web Classification Mining based on RBF Neural Network. In: Proceedings of Control, Automation, Robotics and Vision Conference, vol. 2, pp. 1365–1367 (2004)

18. IP to location mapping, http://www.ip2location.com
19. Dehuri, S., Cho, S.B.: A Comprehensive Survey on Functional Link Neural Networks and an Adaptive PSO-BP Learning for CFLNN. Neural Computing and Applications 19(2), 187–205 (2010)
20. Anifowose, F.A.: A Comparative Study of Gaussian Mixture Model and Radial Basis Function for Voice Recognition. International Journal of Advanced Computer Science and Applications 1(3), 1–9 (2010)
21. Duda, R.O., Hart, P.E.: Pattern Classification and Scene Analysis. Wiley, New York (1973)
22. Jordan, M.I., Jacobs, R.A.: Hierarchical Mixtures of Experts and the EM Algorithm. Neural Computation 6, 181–214 (1994)

A Neural Network Based Image Watermarking Technique Using Spiral Encoding of DCT Coefficients

Jhuma Dutta[1,*], Subhadip Basu[2], Debotosh Bhattacharjee[2], and Mita Nasipuri[2]

[1] Computer Sc. & Engg.Dept., Jalpaiguri Govt. Engg. college, Jalpaiguri-735102, India
jhumadutta81@gmail.com
[2] Computer Sc. & Engg.Dept., Jadavpur University, Kolkata-700032, India

Abstract. Secret data is more prone to unauthorized access when transmitted over the network. Hence data must be protected using some digital media protection scheme. Digital watermarking is now very popular in this field. Here we have proposed a new watermarking scheme to protect some secret messages by putting into a cover image during network transmission. The scheme is based on feed-forward back propagation neural network. The work is done in the frequency domain by considering 8×8 block of the image at a time. The neural network will be trained with spiral scan values of each block. Once the network has been trained it can be used to embed some secret message into any cover image and from the embedded image the message can again be extracted using the trained network. The proposed solution is highly effective and does not produce any visual deterioration of the cover image.

Keywords: Digital Watermarking, Frequency Domain, Neural Network, Spiral Scan, Zigzag Scan.

1 Introduction

Digital Watermarking is a method for hiding information. Using digital watermarking information like number, text or some digital media such as image, audio or video can be hidden into another digital media by manipulating the content of the digital data. As digital protection is becoming popular day by day, digital watermarking provides a good platform to extend and develop new schemes. There are a number of possible applications for digital watermarking technologies and this number is increasing rapidly. For example, in the field of data security, watermarks may be used for certification, authentication, and conditional access. Watermarking can be used to protect as well as detect illegal use of digital information in the area of copyright protection.

Since the inception of watermarking technology, it has evolved through different concepts proposed by different persons at different times. The main idea is to make it more secure, reliable and robust. Ahmidi *et al.* [1] focused on visually meaningful color image watermarks based on Discrete Cosine Transformation

* Corresponding author.

S.C. Satapathy et al. (Eds.): Proc. of Int. Conf. on Front. of Intell. Comput., AISC 199, pp. 11–18.
DOI: 10.1007/978-3-642-35314-7_2

(DCT). They used the sensitivity of human eyes to adaptively embed a watermark in a color image. Song Huang *et al.* [2] in 2009 proposed a watermarking technique based on image features and neural network. They considered the watermark as the fusion of a binary copyright symbol and image feature as got by analyzed the image fractal dimension. Fengsen Deng *et al.* [3] also presented an algorithm for embedding a watermark into a still host image in the DCT domain. They embedded a binary image into the dc components of the DCT coefficients of 8 x 8 blocks. They incorporated the feature of texture masking and luminance masking of the human visual system into watermarking. Summrina *et al.* [4] proposed a watermarking method based on Full Counter Propagation Neural Network to train multiple grey or color cover images to produce a desired watermark. Bibi Isac and V. Santhi give a review of different watermarking techniques on digital image and video using neural network [5]. They explained the watermarking works using different neural networks as Cellular Neural Network, Full Counter Propagation Neural Network, and RBF neural network in DWT domain. Er. Ashish Bansal *et al.* [6] implemented a technique based on back propagation neural network to train a given cover image to produce a desired watermark. Here the entire trained neural network weights have been successfully hidden within the cover image itself. Chang C. –C *et al.* [7] described different intelligent watermarking techniques using neural network. They have used a back-propagation neural network to design the watermarking scheme. To train the network they have applied zigzag scan of the DCT coefficients. As the performance goal depends on image this scheme is not able to give a satisfactory result for training the network using the image as shown in Fig. 1. However, in the current work we have overcome this problem by considering spiral scan technique instead of popularly used zigzag scan method.

Although a lot of work has already been done on this technology still we can enhance it by making it more reliable, secure, and robust. The work is done in frequency domain. For embedding and for extracting the watermark file we have used a feed forwarded back-propagation neural network and the network has been trained using spiral encoding of the DCT coefficient. Once the network has been trained it can be used for watermarking irrespective of the cover image.

Relevant theory and flow charts are discussed in Section-2. Section-3 gives an overview of our work. Experimental results are discussed in Section-4 by analyzing histograms, PSNR value and visual fidelity, and finally, conclusions are drawn in Section-5 of the current paper.

(a) Original Hill image

Fig. 1. Source image and the network performances using zigzag and spiral encoding resp

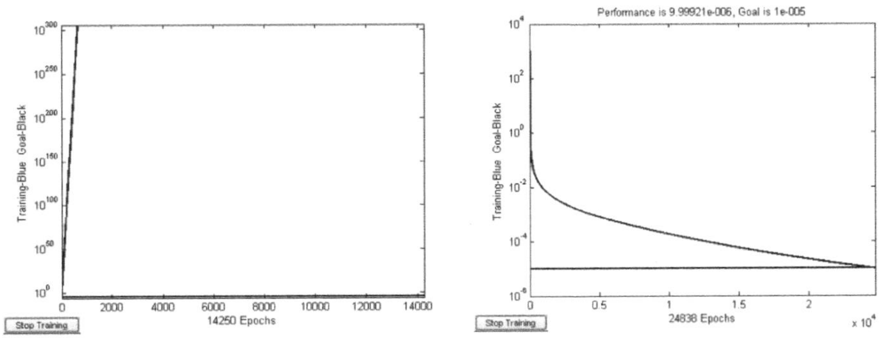

(b) Network using zigzag scan (c) Network using spiral scan

Fig. 1. *(continued)*

2 Theory and Flow Charts

In this section we have discussed the basic concepts about Zigzag Scan and Spiral Scan. This section also includes flowcharts for watermark embedding and extracting processes.

2.1 Zigzag Scan

It is a specific sequential ordering or traversal of the DCT coefficient. The scan puts the high frequency components together. The sequence of traversal of the image pixels is shown in Fig. 2 (a). This is one of the most commonly used scanning methods. But in our work while training the neural network, Zigzag scan failed to provide satisfactory result whereas Spiral scan (discussed in section 2.2) went really well.

2.2 Spiral Scan

This traversal technique gives another pattern of sequential ordering of the DCT coefficients. For spiral scan we can move either clockwise or anti clock wise direction. Here we have used clock wise E (East), S (South), W (West), N (North) traversal, as shown in Fig. 2(b).

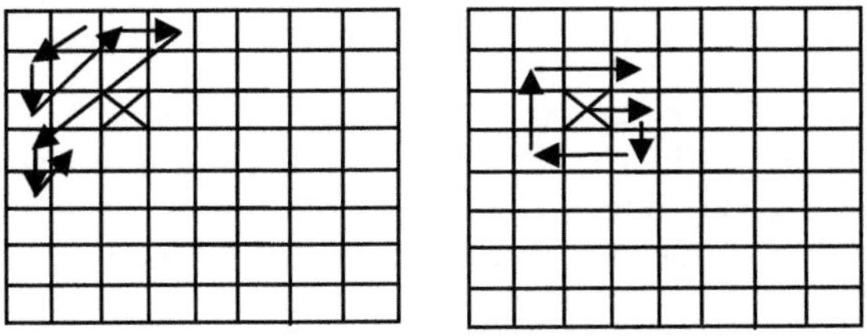

(a) Zigzag scan (b) Spiral scan

Fig. 2. Sequential ordering of the DCT coefficients

2.3 Flow Charts

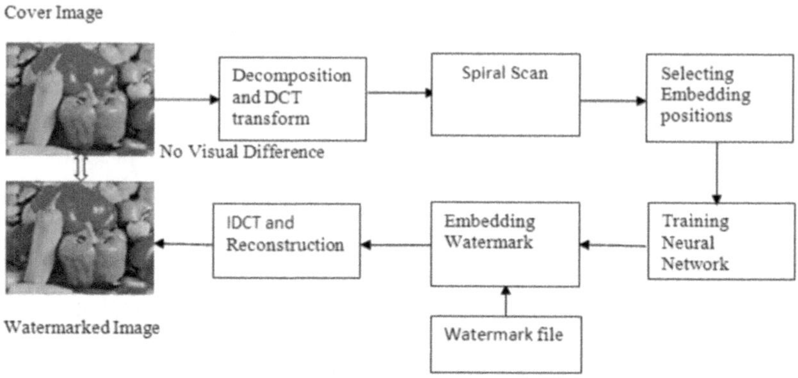

Fig. 3. The flow chart of the embedding phase

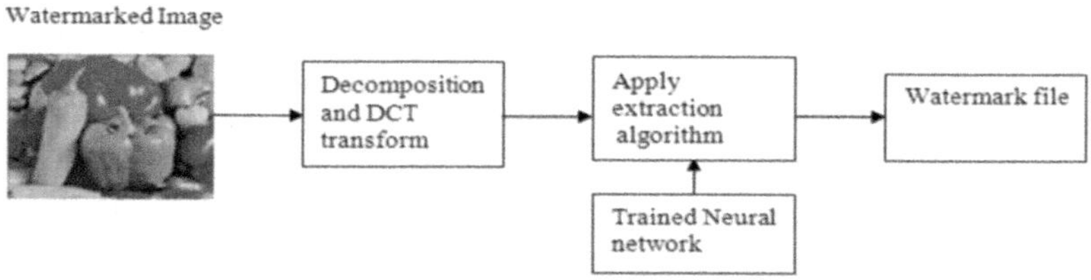

Fig. 4. The flow chart of the extracting phase

3 Present Work

Our work is based on data hiding into a cover image and then extracts the hidden data from the embedded image[1]. The work is done in the frequency domain. We have used spiral encoding to train the network. Once the network has been trained that can be used for watermarking to any images. Two flowcharts, one for the embedding phase and the other for the extracting phase, are shown in Figs. 3-4 respectively. The overall work contains the following three parts:

- Create an artificial neural network and then train the network.
- Embed some watermark into a cover image[2] using the trained network.
- Extract the watermark from the watermarked image using the trained network.

Instead of considering the whole image at a time the image is first divided into 8×8 blocks. At a time one block is taken then applied DCT to convert the image from spatial domain to frequency domain. To train the network we concentrate on the DC components of DCT coefficients. To get a better result we have make a spiral scan over a point on DCT block. Here we have considered $(3, 3)$ position of each DCT

[1] The watermarked image
[2] The image hides the watermark

block. Spiral scan values of the first block makes first column of the input matrix and the value of the position $(3, 3)$ makes first input value to the target of the network. The spiral scan values of the second block makes second column of the input matrix and the value of the position$(3, 3)$ is the second value of the target. In this way the process is going on until all the blocks have been considered. At the end we get the input and the target of the network. Now the network will be trained. Once the network has been trained it will be used for embedding and extracting the watermark.

4 Results and Discussion

In this section we have discussed the experimental results by considering different images in terms of their visual fidelity, histogram and PSNR analysis. Fig. 5 shows the original Hill image and the watermarked Hill images. In this figure Hill is used both for a cover image and a source image[3]. We get two different watermarked images, one by using the network trained by Hill itself, in this case no such visual deterioration is found and the other by using network trained by another source image Lena and some visual deterioration is found all though we can successfully extract the watermark from both the watermarked images. Fig. 6 and Fig. 7 represent experimental results by considering another cover images Lena and Peppers. For this case we also get same type of results.

(a) Hill (b) Watermarked Hill (using network trained by Hill) (c) Watermarked Hill (using network trained by Lena)

Fig. 5. Comparison of visual fidelity in Hill and Watermarked Hill

(a) Lena (b) Watermarked Lena (using network trained by Lena) (c) Watermarked Lena (using network trained by Peppers)

Fig. 6. Comparison of visual fidelity in Lena and Watermarked Lena

[3] The image used to create and train the neural network.

(a) Peppers (b) Watermarked Peppers (using network trained by Peppers) (c) Watermarked Peppers (using network trained by Lena)

Fig. 7. Comparisons of visual fidelity in Peppers and Watermarked peppers

4.1 Histogram Analysis

Fig. 8, Fig.9 and Fig. 10 shows the histograms for cover images Hill, Lena and Peppers and their watermarked images. In Fig. 8 we have considered the gray level value 110 and get the pixel count value 15for the cover image (left). The watermarked file is embedded within this cover image but we also get the same pixel count in that level for the watermarked image (middle) when the network is trained by the image itself. And we get the value 22 when the network is trained by some other source image Lena. The changes in their Stand Deviations and Percentiles are very minimal. So we get considerable changes in the watermarked images. From the Histograms in Fig. 9 and Fig. 10 we also get some considerable changes by considering another cover images and source images. By analyzing the Histograms we can say that the noise addition in the watermarked images is very small. So our technique gives a satisfactorily results for watermarking.

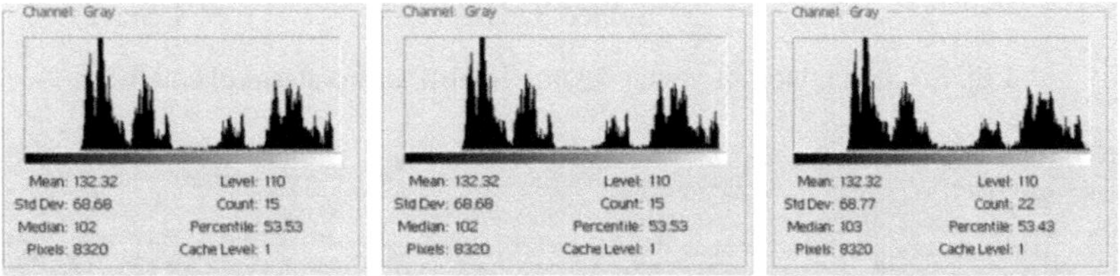

Fig. 8. Comparisons of Histogram of Hill and Watermarked Hill

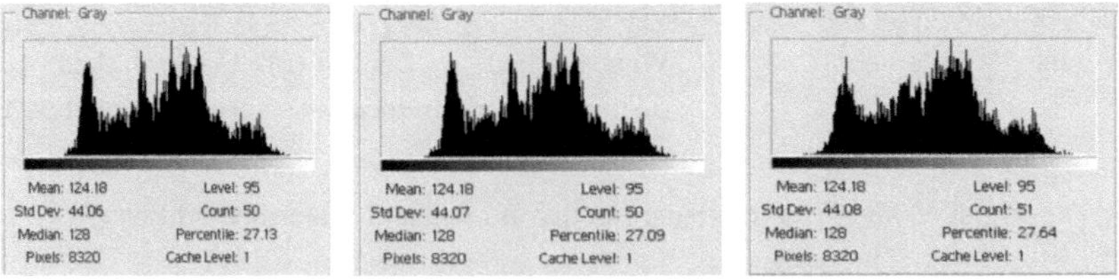

Fig. 9. Comparisons of Histogram of Lena and Watermarked Lena

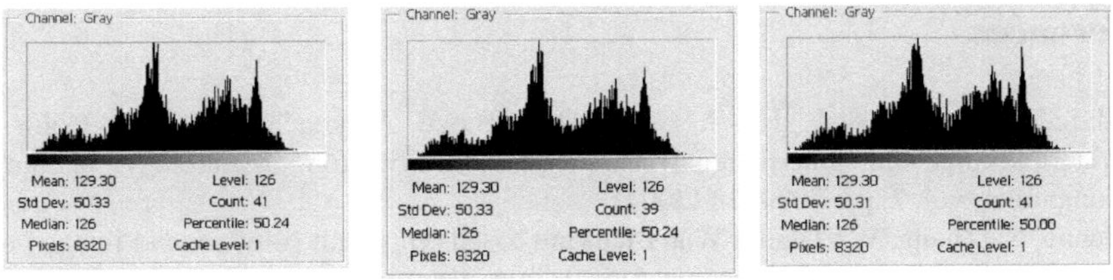

Fig. 10. Comparisons of histogram of Peppers and Watermarked peppers

4.2 PSNR Analysis

Peak-Signal-to-Noise-Ratio is a commonly used method to measure the imperceptibility of the watermarked image. Higher PSNR value implies better similarity. We have prepared a table (Table 1) depending on the calculated PSNR value of the source images and the watermarked images for the network trained with both, same cover image or different cover image. Result with same cover image, Hill, Lena, Peppers is consistently high, indicates the watermarked images look as similar as the original one. Result with different cover image is also at par i.e. almost similar to the original one.

Table 1. The PSNR value of the cover image and the Watermarked Image

Cover Image	Watermarked Image	PSNR
Hill	Watermarked Hill by Hill	55.8760
Hill	Watermarked Hill by Lena	37.3876
Lena	Watermarked Lena by Lena	55.4062
Lena	Watermarked Lena by Peppers	33.4692
Peppers	Watermarked Peppers by Peppers	55.2941
Peppers	Watermarked Peppers by Lena	33.4111

5 Conclusions

In this paper we present a technique to embed a watermark within a cover image using spiral encoding of DCT efficient in addition of feed forward back propagation neural network. This is a new approach as we trained the network by spiral encoding in the frequency domain. The technique gives a more secured method of watermarking, than the conventional zigzag scan technique. The developed methodology also has an advantage that once the neural network has been created and trained that can be used for watermarking irrespective of cover image.

References

1. Ahmidi, N., Safabaksh, R.: A Novel DCT Based Approach for secure Color Image Watermarking. In: International Conference Information Technology: Coding and Computing, vol. 2, pp. 709–713 (2004)
2. Huang, S., Zhang, W.: Digital Watermarking based on Neural Network and Image Features. In: Second International Conference on Information and Computing Science, vol. 2, pp. 238–240 (2009)
3. Deng, F., Wang, B.: A Novel Technique for Robust Image Watermarking in the DCT Domain. In: International Conf. of Neural Networks on Signal Processing, vol. 2, pp. 1525–1528 (2003)
4. Wajid, S.K., Jaffar, M.A., Rasul, W., Mirza, A.M.: Robust and Imperceptible Image Watermarking using Full Counter Propagation Neural Networks. In: International Conference on Machine Learning and Computing, vol. 3, pp. 385–391. IACSIT Press, Singapore (2011)
5. Isac, B., Santhi, V.: A Study on Digital Image and Video Watermarking Schemes using Neural Networks. International Journal of Computer Applications 12(9) (January 2011)
6. Bansal, A., Singh Bhadauria, S.: Watermarking using Neural Network and Hiding the Trained Network within the cover Image. Journal of Theoretical and Applied Information Technology, 663–670 (2005-2008)
7. Chang, C.-C., Lin, I.-C.: Robust Image Watermarking Systems Using Neural Networks. Series on Innovative Intelligence, vol. 7, pp. 402–404 (2004)

Evolutionary Neuro-Fuzzy System for Protein Secondary Structure Prediction

Andey Krishnaji[1] and Allam Appa Rao[2]

[1] Assistant Professor, Dept. of Computer Applications, Swarnandhra College of Engineering & Technology, Narasapur, Andhra Pradesh, India 534 280
[2] Former Vice-Chancellor, JNTUK, Director, CRRao Advanced Institute for Mathematics, Statistics & Computer Science, University of Hyderabad Campus, Gachibowli, Hyderabad, India 500 046
{krishnaji.scet,allamapparao}@gmail.com

Abstract. Protein secondary structure prediction is an essential step for the understanding of both the mechanisms of folding and the biological function of proteins. Experimental evidences show that the native conformation of a protein is coded within its primary structure. This work investigates the benefits of combining genetic algorithms, fuzzy logic, and neural networks into a hybrid Evolutionary Neuro-Fuzzy System, especially for predicting a protein's secondary structure directly from its primary structure. The proposed system will include more biological information such as protein structural class, solvent accessibility, hydrophobicity and physicochemical properties of amino acid residues to improve accuracy of protein secondary structure prediction. The proposed system will experiment on three-class secondary structure prediction of proteins, that is, alpha helix, beta sheet or coil. The experimental results indicate that the proposed method has the advantages of high precision, good generalization, and comprehensibility. The method also exhibits the property of rapid convergence in fuzzy rule generation.

Keywords: Artificial Neural Networks, Fuzzy Logic, Genetic Algorithms, Protein, Secondary Structure.

1 Introduction

One of the greatest mysteries on the earth is life. Proteins are essential biochemical compounds for the life to exist on the earth. In order to understand both the mechanisms of folding and the biological activity of proteins, the knowledge of protein secondary structure is essential. Although X-ray diffraction has been accurate and successful in understanding the three dimensional structure of many crystallized proteins, it is quite time-consuming and expensive. Experimental evidences show that the native conformation of a protein is coded within its primary sequence, i.e, amino acid sequence. So, many methods have been developed to predict the secondary structure of proteins from the sequence data. Protein secondary structure prediction is an intermediate step in the prediction of 3D structure from amino acid sequence [2][3][7].

S.C. Satapathy et al. (Eds.): Proc. of Int. Conf. on Front. of Intell. Comput., AISC 199, pp. 19–26.
DOI: 10.1007/978-3-642-35314-7_3

1.1 Levels of Protein Structure

There are four levels of protein structure. They are:

Primary Structure: Primary structure refers to the sequence of the different amino acids of the peptide or protein. The primary structure is held together by covalent or peptide bonds, which are made during the process of protein biosynthesis or translation. The two ends of the polypeptide chain are referred to as the carboxyl terminus (C-terminus) and the amino terminus (N-terminus) based on the nature of the free group on each extremity. Counting of residues always starts at the N-terminal end (NH_2-group), which is the end where the amino group is involved in a peptide bond. The primary structure of a protein is determined by the gene corresponding to the protein.

Secondary Structure: Secondary structure refers to highly regular local sub-structures. These secondary structures are defined by patterns of hydrogen bonds between the main-chain peptide groups. There are three types of local segments:

- **Helices:** Where residues seem to be following the shape of a spring. The most common are the so-called alpha helices.
- **Extended or Beta-strands:** Where residues are in line and successive residues turn their back to each other.
- **Random coils:** When the amino-acid chain is neither helical nor extended.

Tertiary Structure: Tertiary structure refers to three-dimensional structure of a single protein molecule. The alpha-helices and beta-sheets are folded into a compact globule. The folding is driven by the *non-specific* hydrophobic interactions (the burial of hydrophobic residues from water), but the structure is stable only when the parts of a protein domain are locked into place by *specific* tertiary interactions, such as salt bridges, hydrogen bonds, and the tight packing of side chains and disulfide bonds. The disulfide bonds are extremely rare in cytosolic proteins, since the cytosol is generally a reducing environment.

Quaternary Structure: Quaternary structure is a larger assembly of several protein molecules or polypeptide chains, usually called subunits in this context. The quaternary structure is stabilized by the same non-covalent interactions and disulfide bonds as the tertiary structure. Many proteins do not have the quaternary structure and function as monomers.

1.2 Neural Networks, Fuzzy Logic, and Genetic Algorithms (GA)

Neural Networks are information processing systems. They can be thought of as black box devices that accept inputs and produce outputs. Neural Networks map input vectors onto output vectors. Fuzzy Logic provides a general concept for description and measurement. Fuzzy logic systems can be used to encode human reasoning into a program to make decisions or control machinery. More information on neural networks and fuzzy logic can be found in [1][5][7][8][9].

Genetic Algorithms are search algorithms that are based the mechanics of natural selection and natural genetics. Genetic algorithms consist of three fundamental operations: reproduction, crossover and mutation. More information on genetic algorithms can be found in [8][10].

2 Problem Statement

Protein structure prediction is the prediction of the three-dimensional structure of a protein from its amino acid sequence. The primary sequence of the protein contains the full information to determine the three dimensional structure. The primary sequence of a protein can be represented as

$$\{A^{n1},R^{n2},N^{n3},D^{n4},C^{n5},Q^{n6},E^{n7},G^{n8},H^{n9},I^{n10},L^{n11},K^{n12},M^{n13},F^{n14},P^{n15},S^{n16},T^{n17},W^{n18},Y^{n19},V^{n20}\}$$

where the letters are the one letter codes of the amino acid residue (total 20 possible amino acids), and n1, n2, n3, …,n20 represent the number of times the corresponding amino acid code repeats in the protein sequence and n=(n1+n2+…+n20) is the length of the protein to be predicted. The secondary structure of the sequence having length n is $\{L^{m1},H^{m2},E^{m3}\}$, where H, L, E are different secondary structure classes and m1, m2, m3 represent the number of times the corresponding secondary structural class repeats in the secondary structure of the protein. So, the problem of secondary structure prediction can be represented as a mapping problem as follows:

$$\{A^{n1}R^{n2}N^{n3}D^{n4}C^{n5}Q^{n6}E^{n7}G^{n8}H^{n9}I^{n10}L^{n11}K^{n12}M^{n13}F^{n14}P^{n15}S^{n16}T^{n17}W^{n18}Y^{n19}V^{n20}\} \rightarrow \{L^{m1}H^{m2}E^{m3}\}$$

3 Hypothesis

In order to develop and implement a better protein secondary structure prediction system, the hypothesis of this work can be stated as follows:

"Constructing and designing advanced well organized artificial neural networks architecture combined with fuzzy logic and genetic algorithms to extract more information from neighboring amino acids can increase accuracy of secondary structure prediction of proteins".

4 Methodology

This section briefly describes the methodological framework used in developing and implementing a method to achieve a better prediction method for the protein secondary structure from its primary sequence (amino acid sequences). Due to the complex and dynamic nature of biological data, the application of conventional methods of machine learning approaches including neural networks without augmentation does not achieve good performance. This work proposes to use a hybrid computational intelligence technique, which combines artificial neural network approach with Fuzzy Logic and Genetic Algorithms into Evolutionary Neuro-Fuzzy System to include more biological information to achieve a better and more accurate prediction method for protein secondary structure. The architecture of the proposed Evolutionary Neuro-Fuzzy System shown in Fig. (1):

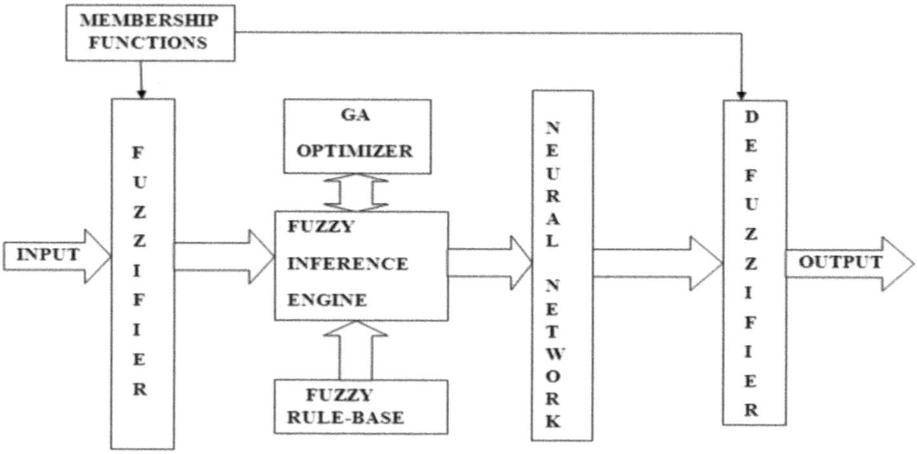

Fig. 1. Architecture of Evolutionary Neuro-Fuzzy System

4.1 Description of Evolutionary Neuro-Fuzzy System

It uses fuzzy data, fuzzy rules, and fuzzy inference. The fuzzy rules and the membership functions make up the system knowledge base [4]. It can handle different types of production rules depending up on the type of the antecedent and the consequent part in the rule: crisp to crisp, crisp to fuzzy, fuzzy to crisp, or fuzzy to fuzzy. The membership functions transform the approximate measurements of protein features into membership values. Fuzzy inference engine activates all the satisfied rules at every cycle and maps the primary fuzzified features to other secondary fuzzy features by using the fuzzy production rules. The membership values are supplied to the input of a pre-trained neural net for classification [1][6].GA Optimizer is a genetic algorithm based optimizer which optimizes the parameters of the rule-base.

4.2 Major Operational Steps of GA Optimizer

(i) Initialize the population
(ii) Calculate the fitness for each individual in the population
(iii) Reproduce selected individuals to form a new population
(iv) Perform evolutionary operations, such as crossover and mutation on the population
(v) Loop to step (ii) until the required condition is met.

4.3 Feature Selection and Normalization

There are 20 amino acids which are named as: alanine(A), aspartic acid(D), phenylalanine(F), histidine(H), lycine(K), methionine(M), Proline(P), ARginine(R), threonine(T), tryptophan(W), cysteine(C), glutamic acid(E), glicine(G), Isoleucine(I), leucine(L), asparagiNe(N), glutamine(Q), serine(S), valine(V), tYrosine(Y). These are called magic 20. The set of twenty amino acids can be represented by X={A, C, D, E, F, G, H, I, K, L, M, N, P, Q, R, S, T, V, W, Y}. The Amino Acid Index Datsbase (AAindex) contains a number of physicochemical properties of amino acids[5]. The most appropriate features such as hydrophobicity and other physicochemical properties for our work have been selected carefully from AAindex databse. Now an amino acid normalization function is defined as follows:

$$\mathbf{y = f(x)}$$

where x is a member of X and the value of y will be in the closed interval [0,1].

4.4 Data Collection

Protein Sequence data are very much essential for this work for training and testing the proposed Evolutionary Neuro-Fuzzy System. Protein sequence files are openly available in the protein data bank (PDB). The PDB website, *http://www.rcsb.org/pdb,* is used as the main source of data for this work. The data stored in the .pdb files are basically the protein primary structure sequences and the three dimensional coordinates of all the atoms of the amino acid molecules i.e. the residues in the sequence. The format in which the secondary structure is given in these files is not suitable for protein secondary structure prediction. For this reason, the PDB data are transformed into secondary structural data, geometrical features and solvent exposure of proteins. In order to transform PDB sequence data into the suitable and compatible input to proposed Evolutionary Neuro-Fuzzy System, Amino Acid Encoding Normalization methods are used. The data set entries that match the following criteria are include: (a) protein sequences with a length of greater than 80 amino acids (b) protein sequences that have no breaks (c) protein sequences of those proteins whose structures are determined by X-ray diffraction method.

5 Results and Discussion

The data set in our experiment is extracted as described in Data Collection section (sec. 4). The experimental details of Evolutionary Neuro-Fuzzy System are shown in the following screen shots. They show the results when the program runs on the chosen protein data. Figure (2) shows the start up window which contains two main buttons: one for moving to prediction interface and the other to move to prediction analysis. Figure (3) shows a window which allows user to browse the sequences for which secondary structures can be assigned. Figure (4) shows a window which displays tetra peptide averages and corresponding plots. The experimental results indicate that the proposed system has the advantages of high precision, good generalization, and comprehensibility. The method also exhibits the property of rapid convergence in fuzzy rule generation.

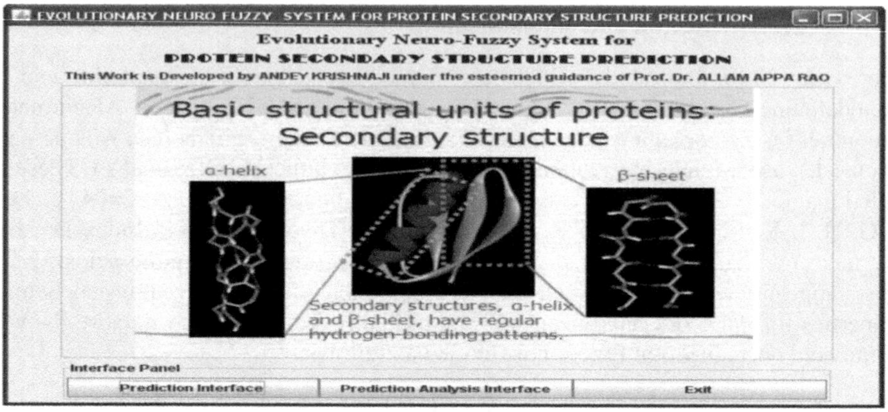

Fig. 2. Start-up window of Evolutionary Neuro-Fuzzy System

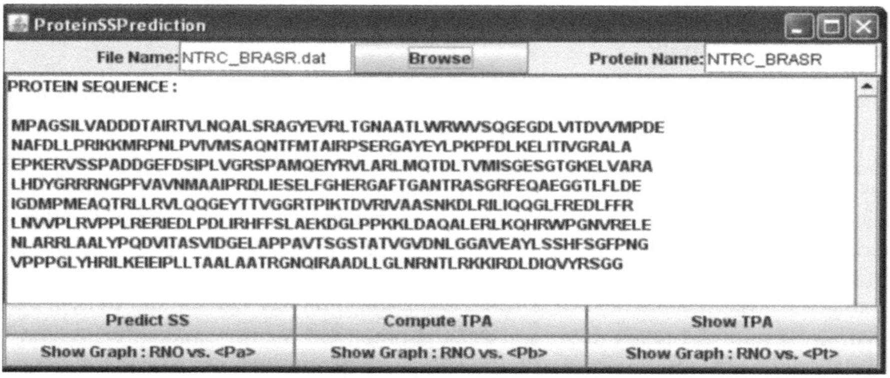

Fig. 3. Select a protein whose secondary structure is predicted

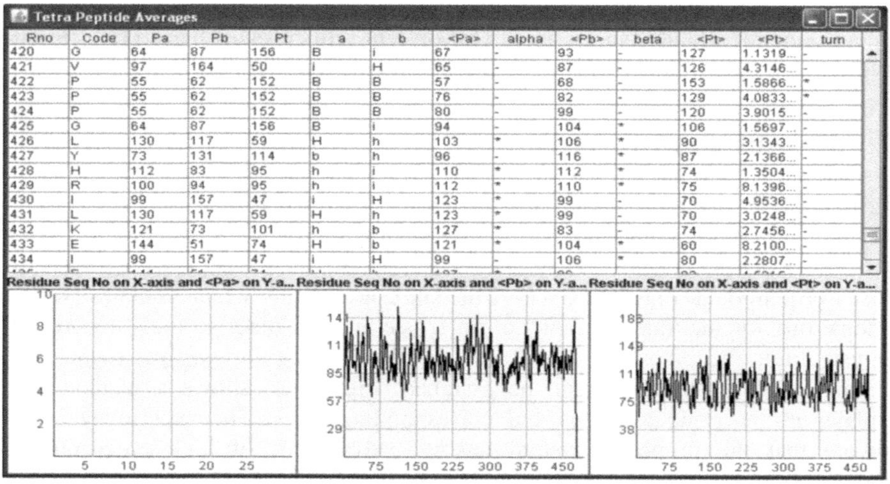

Fig. 4. Compute tetra peptide averages(TPAs)

Secondary Structure Assignment: The following screen shot (fig.5) shows how the system assigns secondary structure classes to amino acid residues in the given protein sequence. The experimental results indicate that the proposed Evolutionary Neuro-Fuzzy System has the advantages of high precision, good generalization, and comprehensibility. The method also exhibits the property of rapid convergence in fuzzy rule generation.

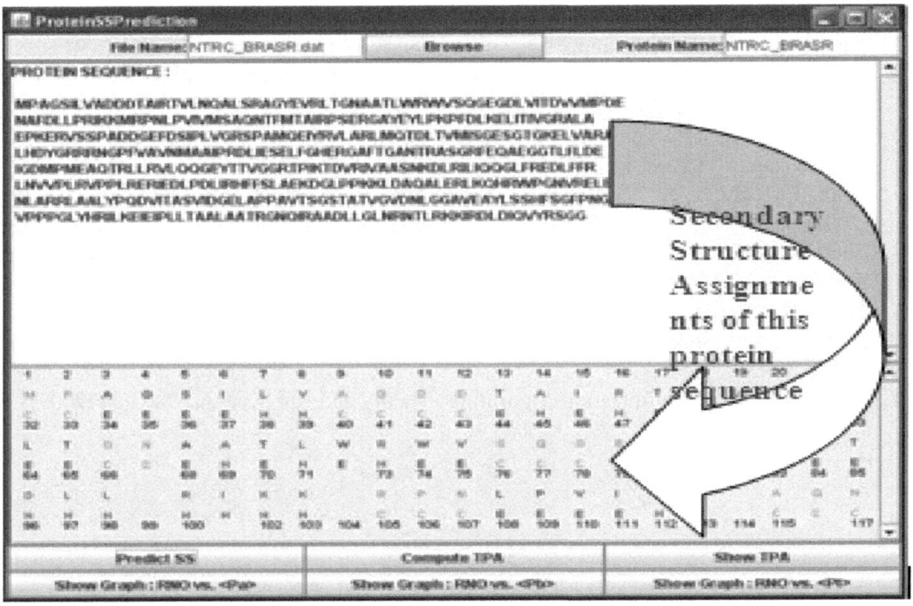

Fig. 4. Assigning Secondary Structure to the target Protein Sequence

6 Conclusion

This paper presents a technique called Evolutionary Neuro-Fuzzy System for protein secondary structure prediction. The main characteristic of this technique is to combine the best properties of both neural networks and fuzzy logic into Evolutionary Neuro-Fuzzy System. This paper also reports an experiment on 3-class secondary structure prediction of proteins using this technique. The experimental results indicate that the proposed system has the advantages of high precision, good generalization, and comprehensibility. The method also exhibits the property of rapid convergence in fuzzy rule generation.

References

1. Jang, J.-S.R.: ANFIS: Adaptive-Network-Based Fuzzy inference system. IEEE Transactions on Systems, Man and Cybernetics 23(0018-9472), 665–685 (1993)
2. Baker, D., Sali, A.: Protein Structure Prediction and Structural genomics. Science 294(5540), 93–96 (2001)
3. Mount, D.W.: Bioinformatics: Sequence and Genome Analysis. Gold Spring Harbor Laboratory Press
4. Sugeno, T., Yasukawa, M.: A fuzzy logic based approach to qualitative modeling. IEEE Transactions on Fuzzy Systems 1(1), 7–31 (1993)
5. Kawashima, S., Kenehisa, M.: AAIndex: Amino acid index database. Nucleic Acids Research 28, 374 (2000)
6. Kasabov, N.K.: Foundations of Neural Networks, Fuzzy Systems, and Knowledge Engineering. MIT Press (1998)
7. Baldi, P., Brunak, S.: Bioinformatics: The machine learning approach. The MIT Press (2001)
8. Eberhart, R.C., Shi, Y.: Computational Intelligence: Concepts & Implementation. Morgan Kalfman Publishers (2007)
9. Takagi, H.: Introduction to Fuzzy Systems. Neural Networks, and Genetic Algorithms
10. Fausette, L.: Fundamentals of Neural Networks: Architectures, Algorithms, and Applications
11. Weise, T.: Global Optimization Algorithms: Theory and Applications, 2nd edn (2009)

A New Approach to Time–Time Transform and Pattern Recognition of Non-stationary Signal Using Fuzzy Wavelet Neural Network

Birendra Biswal[1] and A. Jaya Prakash[2]

[1] Electronics and Communication Dept, GMR Institute of Technology
[2] Student of Electronics and Communication, GMR Institute of Technology

Abstract. This paper discusses new approaches in time- time transform and Nonstationary power signals classification using fuzzy wavelet neural networks. The time-time representation is derived from the S-transform, a method of representation of a real time series as a set of complex, time-localized spectra. When integrated over time, the S-transform becomes the Fourier transform of the primary time series. Similarly, when summed over the primary time variable, the TT-transform reverts to the primary time series. TT-transform points to the possibility of filtering and signal to noise improvements in the time domain. In our research work visual localization, detection and classification of Nonstationary power signals problem using TT-transform and automatic Nonstationary power signal classification using FWNN (Fuzzy wavelet Neural Network) have been considered. Time-time analysis and Feature extraction from the Nonstationary power signals is done by TT-transform. In the proposed work pattern recognition of various Nonstationary power signals have been considered using particle swarm optimization technique. This paper also emphasizes the robustness of TT-transform towards noise. The average classification accuracy of the noisy signals due to disturbances in the power network is of the order 92.1.

Keywords: Nonstationary power signals, FWFNN, S-Transform, TT-Transform, particle swarm optimization.

1 Introduction

In electrical power networks, the voltage and current signals exhibit fluctuations in amplitude, phase, and frequency due to the operation of solid-state devices that are prolifically used for power control. TT-Transform provides a unified framework for processing distorted power signals. In this paper we present a two dimensional time-time representation of different Nonstationary power signals, based upon the S-transform. This is termed the TT-transform. The extracted features (Standard deviation, &Normalized values) have been taken for non- stationary power signal classification using fuzzy wavelet neural networks (FWNN). Fig-1 illustrates the use of popular TT-transform algorithm to extract the features of the Nonstationary power signals. The extracted features are applied to the FWNN classifier for automatic classification.

S.C. Satapathy et al. (Eds.): Proc. of Int. Conf. on Front. of Intell. Comput., AISC 199, pp. 27–34.
DOI: 10.1007/978-3-642-35314-7_4

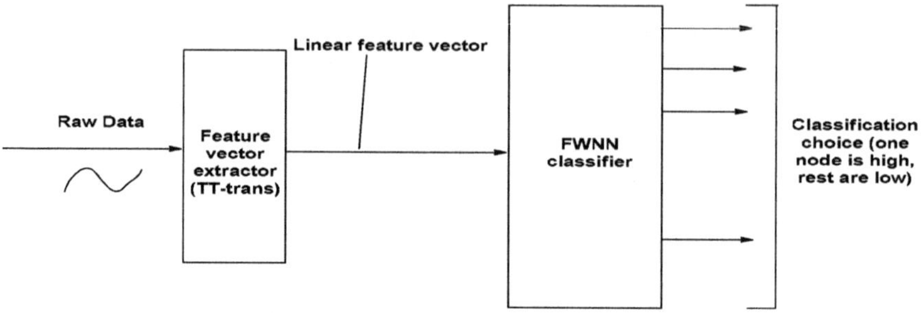

Fig. 1. Feature vector approach to classification

2 The S-Transform

The S-Transform (R.G. Stockwell, L. Mansinha, and R.P. Lowe) is a time frequency spectral localization method [1] with a Gaussian window whose height and width vary with frequency, the expression of the S-Transform is given by

$$S(\tau, f) = \int_{-\infty}^{\infty} h(t)\left\{\frac{|f|}{\sqrt{2\pi}} \times \exp\left[\frac{-f^2(\tau-t)^2}{2}\right]\exp(-2\pi i f t)\right\}dt. \quad (1)$$

Here the window is a scaled Gaussian whose midpoint is τ =t, At any t and f, the S-transform may be considered as a set of localized Fourier coefficients, obtained by considering only the portion of the primary function lying within a few wavelengths on either side of τ = t.

3 The TT-Transform

We define a second time –time distribution, the TT-transform (C. R. Pinnegar and L. Mansinha), obtained from the inverse Fourier transform of the S-transform

$$TT(t,\tau)] = \int_{-\infty}^{\infty} S(t, f)\exp(+2\pi i f \tau)df. \quad (2)$$

If TT (t, τ) is considered at all τ but a specific t, the result is a time-local function, conceptually similar to a windowed function From (3)

$$\int_{-\infty}^{\infty} TT(t,\tau)dt = h(\tau). \quad (3)$$

So, like the S-transform, the TT-transform is invertible [2].

In our research work we have extracted various features like Energy, Standard deviation, Autocorrelation, mean, variance and normalized values from the Nonstationary power signals. During feature extraction it is found that the standard deviation and normalized values are the most distinguished features. Therefore standard deviation and normalized values have been taken for classification of various Nonstationary power signals.

4 Simulation Results of Nonstationary Power Signals Using TT-Transform

In our study we have discussed different types of Nonstationary power signal problems such as Voltage sag, Voltage flicker, Voltage spikes, etc with MATLAB software. The TT output shows the plot of the normalized TT- contour and the absolute value of a given magnitude of change in the time-time co-ordinate system. The following case studies are presented in this paper: From the above simulation results it is quite clear that the TT-transform doe's excellent detection and visual classification of Nonstationary power signals because of superior time-time resolution characteristic. Which are given below

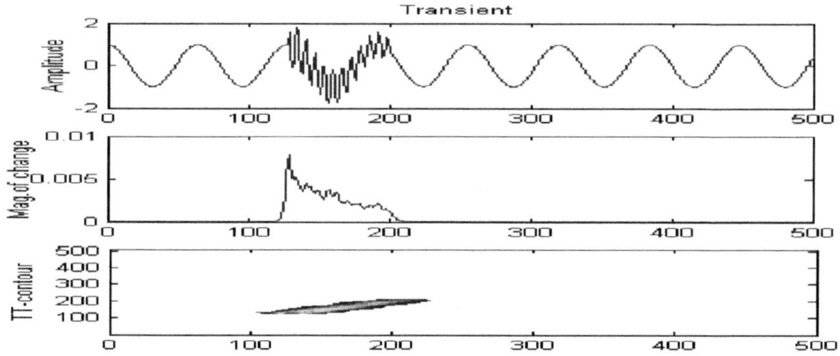

Fig. 2. Transient voltage waveform and TT-transform result

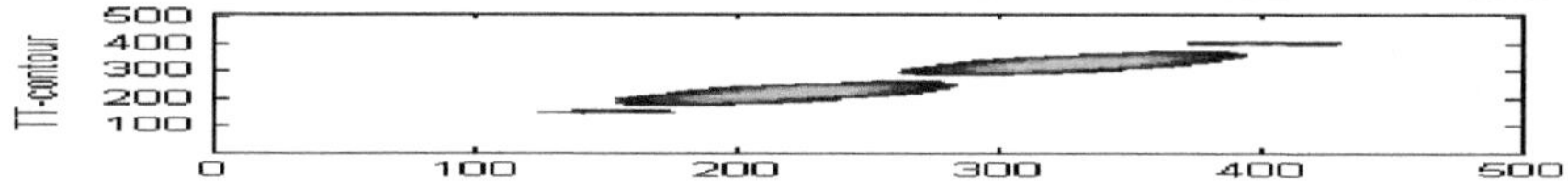

Fig. 3. Voltage flicker waveform and TT-transform result

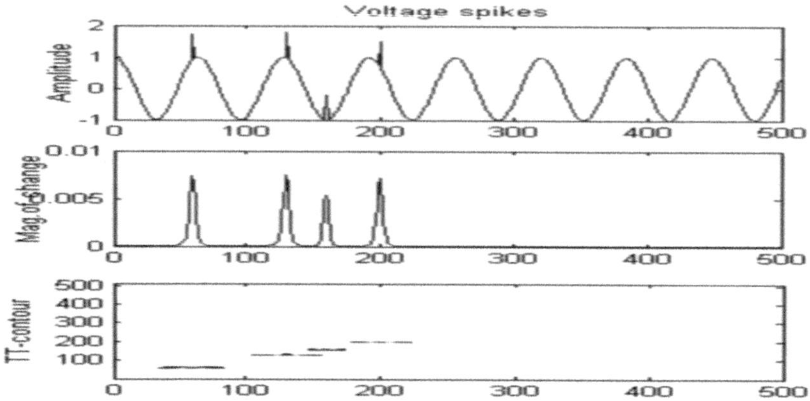

Fig. 4. Voltage spike waveform and TT-transform result

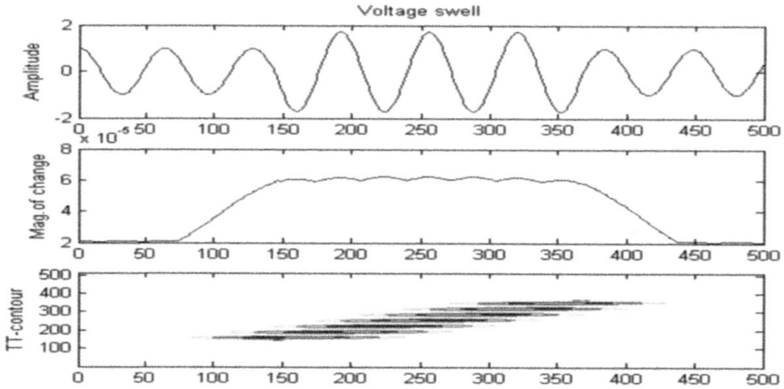

Fig. 5. Voltage swell waveform and TT-transform result

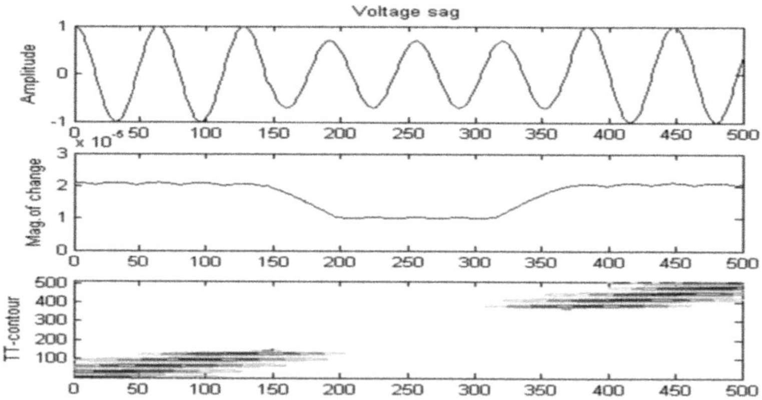

Fig. 6. Voltage sag waveform and TT-transform result

5 The Proposed Fuzzy Wavelet Neural Network

The fundamental idea behind wavelets is to analyze according to scale. Wavelets are functions that satisfy certain mathematical requirements and are used in representing data or other functions. If we look at a signal with a large "window," we would notice gross features. Similarly, we know the Gaussian function is local in both time and frequency domains and is a $C^{\infty}(R)$ function. Therefore, any derivative of the Gaussian function can be used as the basis for a wavelet transform. In our model we have taken first derivative Gaussian wavelet function for pattern recognition of Nonstationary power signals. The Gaussian first derivative wavelet, is defined as:

$$\psi(x) = -x\exp\left(-\frac{x^2}{2}\right)$$

Wavelets are well suited for approximating data with sharp discontinuities.

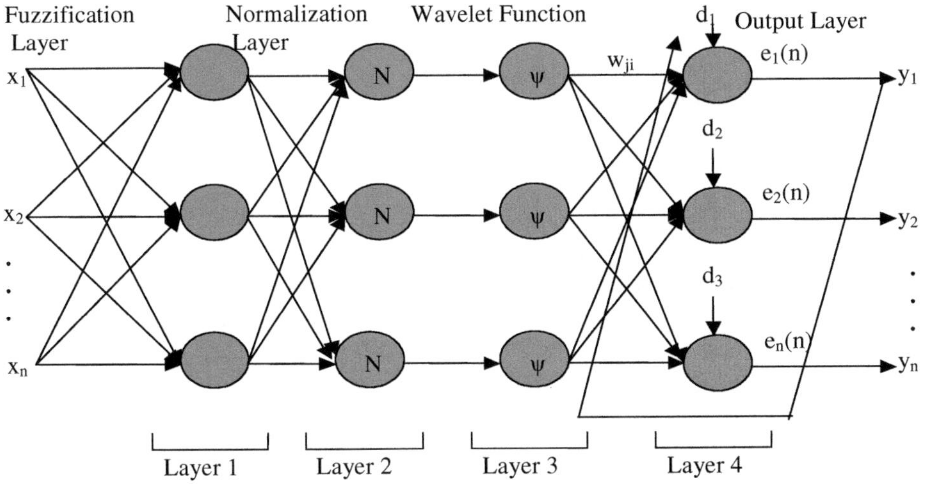

Fig. 7. Fuzzy wavelet neural network model

6 Particle Swarm Optimization

PSO (Fun Ye and Ching-Yi Chen, Dw van der Merwe and AP Engelbrecht) uses a number of agents (particles) [4] that constitute a swarm moving around in the search space looking for the best solution. Each particle is treated as a point in a D-dimensional space, which adjusts its “flying” according to its own flying experience as well as the flying experience of other particles. The position with the highest fitness value in the entire run is called the global best (gbest The PSO concept consists of changing the velocity (or accelerating) of each particle towards its pbest and the gbest position at each time step.

After finding the two best values, the particle updates its velocity and positions with following equation (4) and (5).

$$\boldsymbol{v}_i^{k+1} = \boldsymbol{w}\boldsymbol{v}_i^k + \boldsymbol{c1} * \boldsymbol{rand1} * (\boldsymbol{pbest} - \boldsymbol{s}_i^k) + \boldsymbol{c2} * \boldsymbol{rand2} * (\boldsymbol{gbest} - \boldsymbol{s}_i^k) \quad (4)$$

$$\boldsymbol{s}_i^{k+1} = \boldsymbol{s}_i^k + \boldsymbol{v}_i^{k+1} \quad (5)$$

Where $\boldsymbol{v}_i^k$: velocity of centre points $\quad w$: $weighting\ function$
ran : $random\ no.$ $Between$ 0 & 1 $\quad c1, c2$: $weighting\ factors$
$\boldsymbol{s}_i^k$: $current\ position\ of\ agent\ i\ at\ iteration\ k$
$pbest$: $pbest\ of\ centre\ i$
$gbest$: $gbest\ of\ the\ group$

The above weighting function is given by

$$w = w_{max} - [(w_{max} - w_{min})/iter_{max}] * iter \quad (6)$$

Where w_{max} : $initial\ weight$
w_{min} : $final\ weight$
$iter_{max}$: $maximum\ iteration\ number$
$iter$: $current\ iteration\ number$

Then the particle flies towards a new position according to equation (5). The inertia weight ‘w’ is employed to control the impact of the previous history of velocities on the current velocity, thus to influence the trade-off between global (wide-ranging) and local (nearby) exploration abilities of the "flying points". A larger inertia weight w facilitates global exploration (searching new areas) while a smaller inertia weight tends to facilitate local exploration to fine-tune the current search area

7 Discussions

As can be seen from the Tables, the proposed method is doing very well in classifying these ten types of disturbances using the concept of PSO. Even after denoising, the variation of the classification accuracy of the fuzzy wavelet neural network is negligible. This is the result of the TT-transforms robustness towards noise. An interesting point to note is that the classification accuracy of harmonics actually increases with the presence of noise. The mean square error plot in dB and the classification accuracy table is given below for 1000 & 2000 iteration respectively

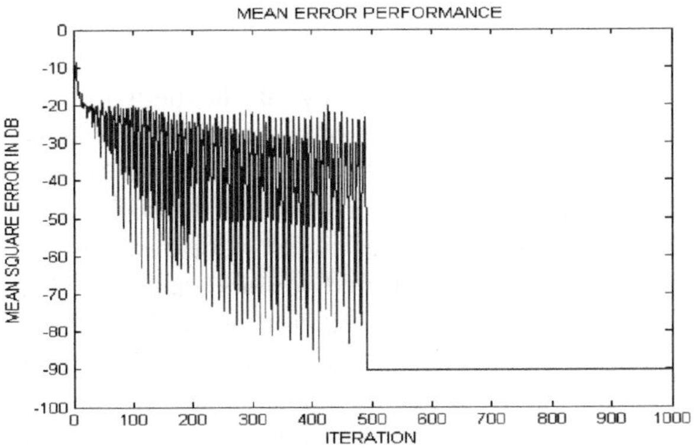

Fig. 8. Mean square error (dB) plot without noise for 1000 iteration

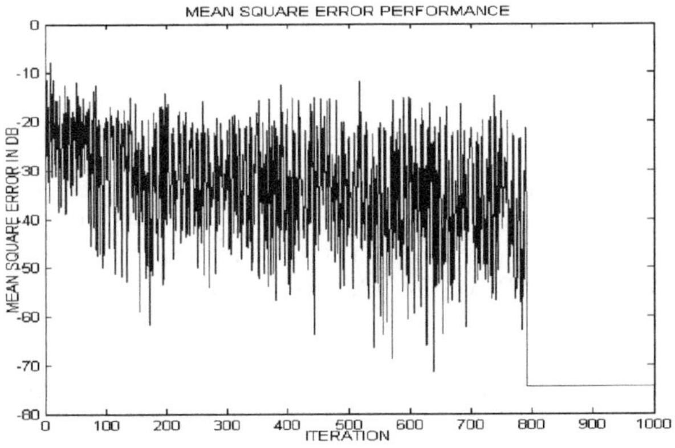

Fig. 9. 30 dB de-noised. Mean square error (dB) plots for 1000iteration

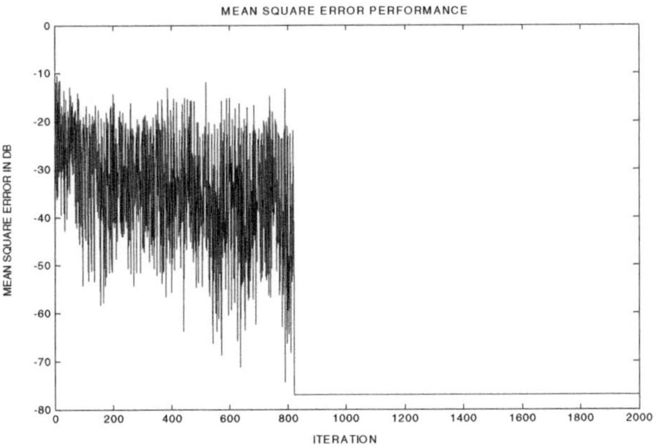

Fig. 10. 30 dB de-noised. Mean square error (dB) plots for 2000 iteration

8 Conclusion

It is shown that the classification accuracy of the network is very high and is practically invariant even in the presence of noise. The network is quick to train, as only 2000 iteration are needed to give 92.1% classification accuracy. The time-time transform is used in this paper as a powerful analysis tool for detection, localization and classification of Nonstationary power signal waveforms. TT-Transform provides a unified framework for processing distorted power signals. The learning curve of the FWNN model appears to have some oscillation at the beginning of learning. This situation reflects the structural change during the early stages of learning. With progressive iterations our model has rapid convergence. The proposed FWNN model requires only one adjustable parameter (weight matrix) and hence converges more quickly than other alternative models. It is successful in achieving faster learning and higher pattern recognition accuracy with fewer parameters in many problems. Thus the FWNN model with feature vector of TT-Transform has a powerful ability to pattern recognition of power signal disturbances using PSO.

References

[1] Stockwell, R.G., Mansinha, L., Lowe, R.P.: Localization of the complex Spectrum: The S-Transform. IEEE Transaction on Signal Processing 44(4), 998–1001 (1996)

[2] Pinnegar, C.R., Mansinha, L.: A method of time-time analysis: The TT-transform. Elsevier Science on Digital Signal Processing 13, 588–603 (2003)

[3] Chu, P.C.: The S-transform for obtaining localized spectral. Mar. Technol. Soc. J. 29, 28–38

[4] Chen, C.-Y., Ye, F.: Particle Swarm Optimization Algorithm and its Application to Clustering Analysis. In: IEEE ICNSC 2004, Taiwan, R.O.C, pp. 789–794 (2004)

[5] Ye, F., Chen, C.-Y.: Alternative KPSO-Clustering Algorithm. Tamkang Journal of science and Engineering 8(2), 165–174 (2005)

[6] van der Merwe, D.W., Engelbrecht, A.P.: Data Clustering using Particle Swarm Optimization. Department of Computer Science University of Pretoria

A Non-Fuzzy Self-Tuning Scheme of PD-Type FLC for Overhead Crane Control

A.K. Pal[1], R.K. Mudi[2], and R.R. De Maity[3]

[1] Dept. of A.E.I.E, HIT, Kolkata, India
arabindakumarpal@gmail.com
[2] Dept. of I.E.E, Jadavpur University, Kolkata, India
rkmudi@yahoo.com
[3] Dept. of E.I.E, DBCREC, Durgapur, India
ritu_maity_8@yahoo.co.in

Abstract. A non-fuzzy self-tuning scheme is proposed for Fuzzy PD controller in this paper. To eliminate the design complexity, output scaling factor (SF) of the proposed fuzzy controller is updated according to the process trend by a gain modification factor, which is determined by the normalized change of error of the system and its number of fuzzy partitions. The proposed non-fuzzy self-tuning fuzzy PD controller (NFST-FPDC) is demonstrated on a laboratory scale overhead crane. Moving a suspended load along a pre-specified path is not an easy task when strict specifications on the swing angle and transfer time need to be satisfied. In this study, twin NFST-FPDC are designed to control the trolley position of the crane and swing angle of the load. The proposed non-fuzzy gain tuning scheme guarantees a fast and precise load transfer and the swing suppression during load movement, despite of model uncertainties.

Keywords: Fuzzy control, Crane position control, Swing angle, Self-tuning.

1 Introduction

Overhead cranes are used in different industries for the loading and unloading of raw materials, freight and heavy equipments [1]. Control of overhead cranes, particularly the swing of trolley has become the requirements as a core technology for automated crane system. The purpose of crane control is to reduce the pendulum type motion of the loads while moving the trolley to the desired position as fast as possible. Thus, the need for faster cargo handling requires the precise control of crane motion so that its dynamic performance is improved [2- 3]. Various attempts have been made to solve the problem of swing of load [4- 5]. Most of them focus the control on suppression of load swing without considering the position error in crane motion. Besides, several authors have considered optimization techniques to control the cranes. They have used minimal time control technique to minimize the load swing. Since the swing of load depends on the motion and acceleration of the trolley, minimizing the cycle time and minimizing the load swing are partially conflicting requirements.

S.C. Satapathy et al. (Eds.): Proc. of Int. Conf. on Front. of Intell. Comput., AISC 199, pp. 35–42.
DOI: 10.1007/978-3-642-35314-7_5

The aim of fuzzy techniques is to get ahead of the limits of conventional techniques. A number of approaches have been proposed to implement hybrid control structures to control the nonlinear systems. Among the various types of hybrid controllers, PI-type fuzzy logic controllers (FLCs) are most common and practical [6] followed by the PD-type FLCs. But like conventional PI-controllers [7], performance of PI-type FLCs for higher order systems, systems with integrating elements or large dead time, and also for nonlinear systems may be very poor due to large overshoot and excessive oscillation. PD-type FLCs are suitable for a limited class of systems [8], like integrating, non-minimum and non-linear systems.

Practical processes are usually nonlinear in nature and associated with dead time, and their parameters may change with time and ambient conditions. Conventional FLCs with fixed values of SFs and simple MFs are not expected to provide good control performance. Mudi *et al* [9-14] proposed robust self-tuning schemes based on fuzzy rules, where the output SF of FLC is modified on-line by a gain updating factor, which is further multiplied by a fixed factor chosen empirically.

Instead of expert's defined fuzzy rules, in this paper we propose a non-fuzzy self-tuning scheme for fuzzy PD-type controller (NFST-FPDC). In the proposed NFST-FPDC, its output SF is continuously modified by a single deterministic rule defined on the normalized change of error, *i.e.*, Δe_N, and the number of its linguistic values of MFs. Observe that the proposed heuristic rule acts on the instantaneous speed of response of the process under control. Thus, the on-line adjusted output SF of the proposed NFST-FPDC is expected to improve the close-loop performance, since it incorporates the dynamics of the process.

In this study, we attempt to provide a practical solution for the anti-swing and precise position control of an overhead crane. The position of trolley, swing angle of load and their differentiations are applied to derive the proper control input of the trolley crane. Two PD-type fuzzy logic controllers are used to deal separately with the feedback signals, swing angle and trolley position, and their differentiations [15-17]. The main advantage of this separated approach is to greatly reduce the computational complexity of the crane control system. The total number of fuzzy rules for the complete control system is therefore less than the number of rules used by conventional fuzzy system. Besides, when designing the proposed fuzzy controllers, no mathematical model of the crane system is required in advance.

2 Laboratory Based Overhead Crane Set-up

A laboratory scale crane setup (FEEDBACK, UK) shown in Fig. 1 and Fig. 3 consists of a cart moving along the 1m length track and a load is attached with the cart through shaft. The cart can move back and forth causing the load to swing. The movement of the cart is caused by pulling the belt in two directions by the DC motor attached at the end of the rail. By applying a voltage to the motor we control the force with which the cart is pulled. The value of the force depends on the value of the control voltage. Two variables that are read using optical encoders, installed on the cart, are the load angle and the cart position on the rail. The controller's task will be to change the DC motor voltage depending on these two variables, in such a way that the desired crane control task is fulfilled. Initially the control signal is set to -2.5v to 2.5v and the generated force is of around -20N to +20N. The cart position is physically bounded by the rail length and is equal to -0.5m to +0.5m.

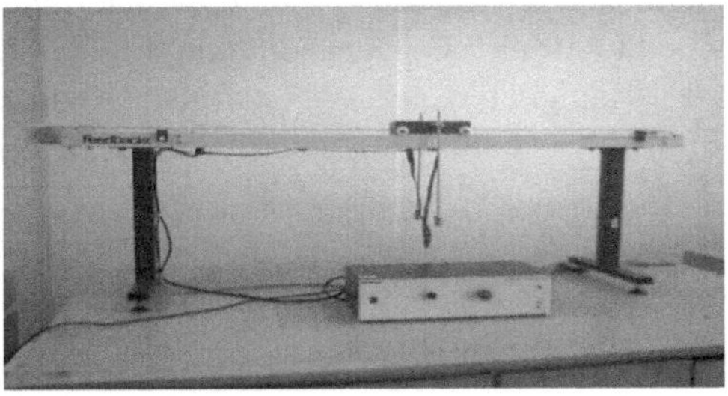

Fig. 1. Overhead crane set-up

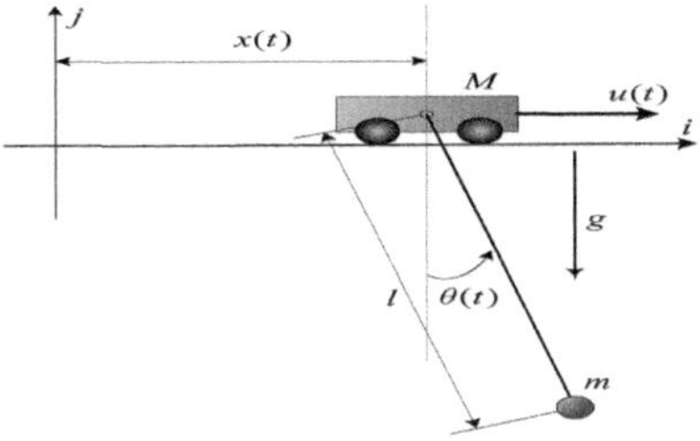

Fig. 2. Model of the overhead crane

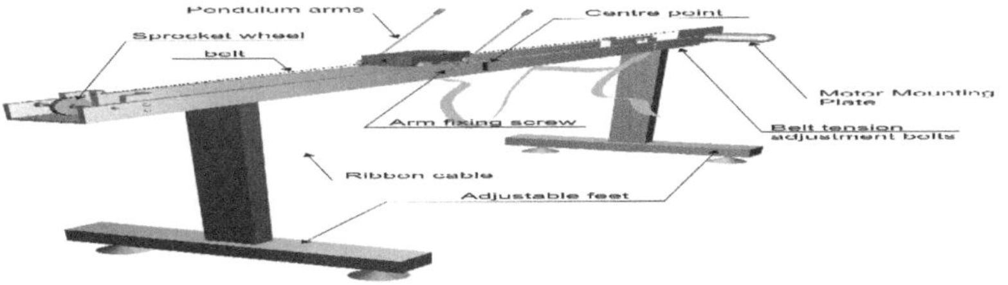

Fig. 3. Mechanical unit of the overhead crane system

Fig. 2 shows the schematic of overhead crane traveling on a rail, where *x(t), θ(t),* and *u(t)* are the cart position, load swing angle, and cart driving force respectively. The cart mass, load mass, load arm length, and gravity are represented by *M, m, l,* and *g*respectively. In this paper, the stiffness and mass of the rope are neglected and the load is considered as a point mass. The proposed scheme is focused on anti sway tracking control of an indoor overhead crane; therefore, the hoisting motion and the effects of wind disturbance are not considered. Then, the equations of motion of the overhead crane system without uncertainty [4] are obtained through the following equations:

$$(M+m)x''+(ml\cos\theta)\theta''-(ml\sin\theta)\theta'^2 = u \quad (1)$$

$$ml^2\theta''+(ml\cos\theta)x''-mgl\sin\theta = 0 \quad (2)$$

The main difficulty in controlling the overhead crane system basically lies in the handling of the coupled nature between the sway angle and trolley movement. The dynamic model obtained is nonlinear in nature, that means the cart position and its derivative or swing angle and its derivative is a nonlinear function.

3 Proposed Controller Design

The output scaling factor should be considered a very important parameter of the FLC since its function is similar to that of the controller gain. Moreover, it is directly related to the stability of the control system. So the output SF should be determined very carefully for the successful implementation of a FLC. Depending on the error (e) and change of error (Δe) of the controlled variable, an expert operator always tries to modify the controller gain, i.e., output SF, to enhance the system performance [9-14].

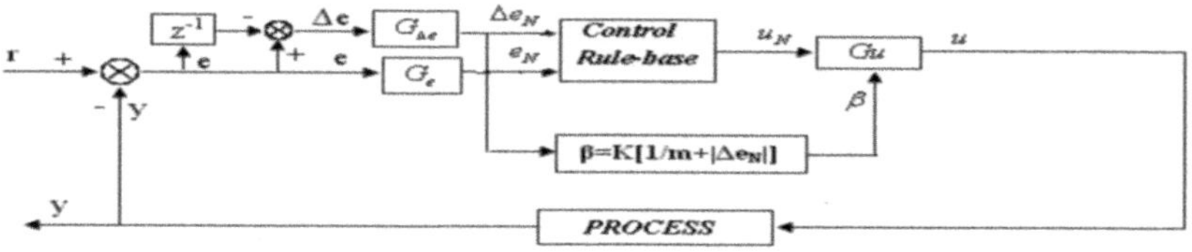

Fig. 4. Diagram of NFST-FPDC

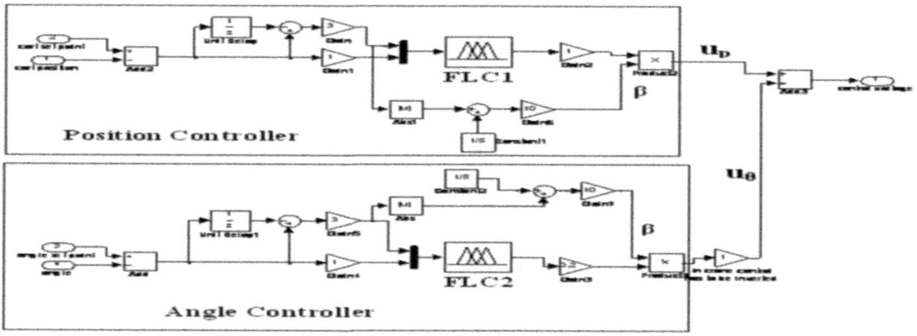

Fig. 5. Diagram of twin NFST-FPDC for overhead crane control

Following such an operator's policy, here, we suggest a simple self-tuning scheme of Fuzzy PD Controller (FPDC), where an online gain modifier (β) is determined from the relation, $\beta = K[1/m + |\Delta e_N|]$ as shown in Fig. 4. Here, β is the on-line adjustable parameter for the output SF G_u, m is the number of fuzzy partitions of Δe_N (*i.e.*, m= 5), and K is a positive constant that will bring appropriate variation in β. The difference between conventional (FPDC) and proposed (NFST-FPDC) controllers is that the FPDC uses only G_u to generate its output (*i.e.*, $u = G_u u_N$), whereas the output of NFST-FPDC is obtained by using the effective SF, *i.e.*, βG_u, as shown in Fig. 4. Unlike fuzzy based self-tuning scheme, which requires expert's defined fuzzy self-tuning rules, here β is computed on-line by a single model independent non-fuzzy relation.

In place of single controller as shown in Fig. 4, in crane control we use dual controllers, one for position and another for angle control. The feedback signals from the crane act as the input variables of FPDC and NFST-FPDC as shown in the Fig. 5.

The same NFST-FPDC shown in Fig. 5 can be used as FPDC by eliminating the gain modifier β. There are two similar fuzzy logic controllers, which work separately with cart position and sway angle. The position controller and angle controller, which deal separately with the cart position and swing angle, drive the overhead crane. In this design, the position error (e) and change of position error (Δe) are selected as the input linguistic variables of fuzzy position controller. The input linguistic variables of fuzzy angle controller are selected as the swing angle error (e_θ) and its derivative Δe_θ.

Control surfaces (e and Δe versus u) of FPDC and NFST-FPDC are depicted in Fig. 6 and Fig. 7 respectively. After a careful inspection of the two surfaces it can be realized that the control surface of the proposed NFST-FPDC is more non-linear in nature but smooth than that of FPDC. In practical implementation the smoothness of the control surface is highly desirable for the limited speed of the actuator and to avoid the chattering problem.

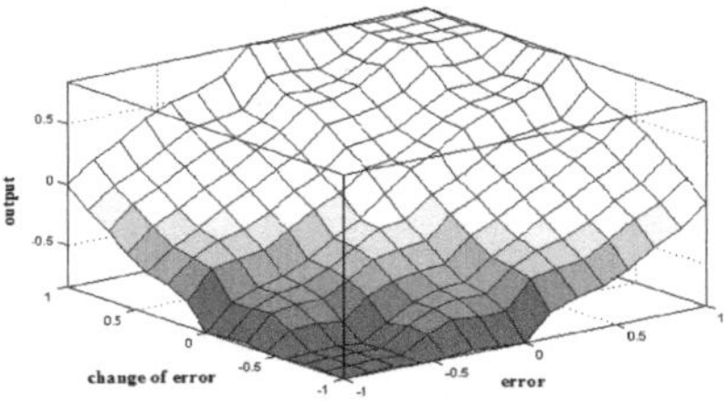

Fig. 6. Control surface of FPDC

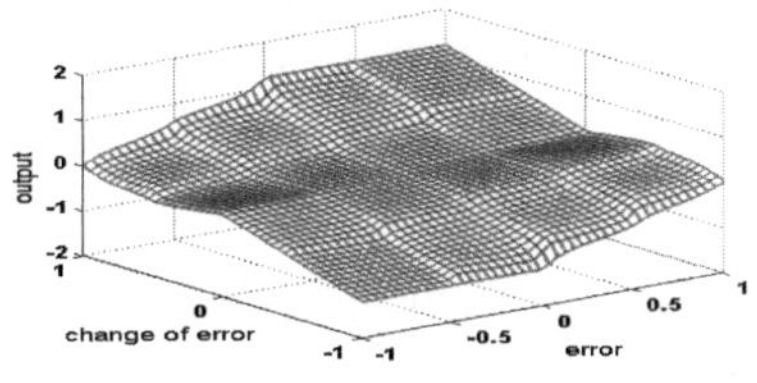

Fig. 7. Control surface of NFST-FPDC

In our design for crane swing control, the left swing of the load is considered as positive swing, while the right swing of the load is negative swing. The output of the FPDC for position and swing angle control are u_P and u_θ respectively. Thus, the actual control action to drive the cart is defined as: $u=u_P+u_\theta$. For the overhead crane control using NFST-FPDC, we incorporate a self-tuning scheme through an online gain modifier β determined by the relation $\beta = K[1/m + |\Delta e_N|]$ as shown in Fig. 5. The controller output u of FPDC and NFST-FPDC is used to drive the DC motor of the overhead crane system. Fig. 8 shows the MFs of e, Δe and u_P, whereas the MFs of e_θ, Δe_θ and u_θ are represented by Fig. 9. Error (e) due to position and error (e_θ) due to angle are obtained respectively from the cart position encoder and swing angle encoder. The ranges of input-output variables for position controller are [-1, +1] and [-20°, +20°] for angle controller.

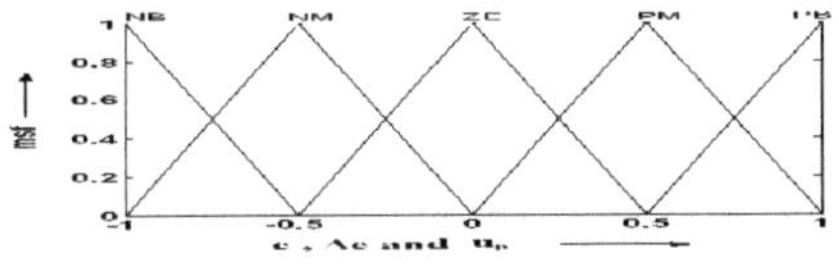

Fig. 8. MFs of e, Δe and u_p

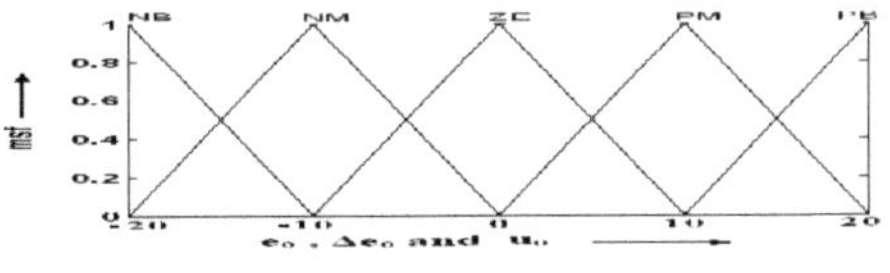

Fig. 9. MFs of e_θ, Δe_θ and u_θ

Table 1. Fuzzy rules for computation of position controller output

Δe \ e	NB	NM	ZE	PM	PB
NB	NB	NB	NB	NM	ZE
NM	NB	NB	NM	ZE	PM
ZE	NB	NM	ZE	PM	PB
PM	NM	ZE	PM	PB	PB
PB	ZE	PM	PB	PB	PB

Table 2. Fuzzy rules for computation of angle controller output

Δe_θ / e_θ	NB	NM	ZE	PM	PB
NB	PB	PB	PB	PM	ZE
NM	PB	PB	PM	ZE	NM
ZE	PB	PM	ZE	NM	NB
PM	PM	ZE	NM	NB	NB
PB	ZE	NM	NB	NB	NB

Each of the position and angle controllers consists of only 25 fuzzy rules as shown in Table 1 and Table 2 respectively. The proposed dual controller structure for crane control divides the input antecedents of fuzzy rules into two parts. Hence, both position controller and angle controller have only $i/2$ fuzzy antecedents, where 'i' is the number of input linguistic variables, here $i=4$. If each input variable has 'n' linguistic terms, here $n=5$, then the possible control rules required for our scheme is $2{*}n^{i/2}$ =50. Thus the total number of rules for FPDC and NFST-FPDC in the crane control scheme are greatly reduced compared to traditional fuzzy control schemes, which may need n^i, *i.e.*, $5^4=625$ rules.

4 Results

The proposed self-tuning scheme is tested on an overhead crane (Fig. 1) with sinusoidal and step input with amplitude of 0.3m. The controller output u of FPDC and NFST-FPDC separately applied to the overhead crane to control the crane position as well as swing angle of the load attached. The NFST-FPDC outperforms the FPDC as shown in Fig. 10 to Fig. 17 under various inputs. Real-time experiments on the overhead crane illustrate the advantages of proposed non-fuzzy self-tuning scheme. From Fig. 11 and Fig. 15, we observed negligible deviation in trolley position from the set point value. From Table 3, we find that the different performance parameters such as IAE, ITAE, and ISE are reduced by a large percentage when controlled by NFST-FPDC compared to FPDC. Figs. 13 and 17 shows that the load swing is minimum, especially in case of step input the swing angle approaches to almost zero for our proposed scheme. We also study the system with conventional controller and found that the load sway is not smooth for different inputs, which is one of the most desirable parameters for overhead crane control in industry. Thus, the above study reveals that the proposed self-tuning scheme for fuzzy controller can fix the over-head crane in its desired position with negligible sway angle.

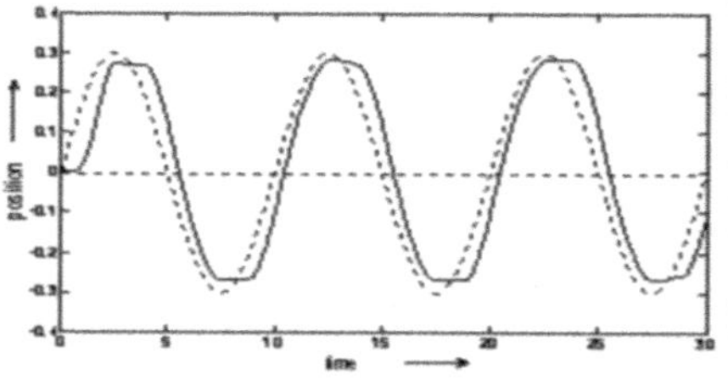

Fig. 10. Crane position control for sine input using FPDC (dotted lines – reference crane position and solid lines – actual crane position)

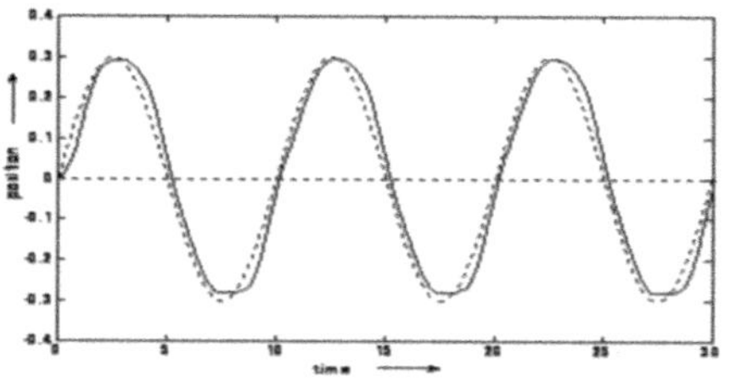

Fig. 11. Crane position control for sine input using NFST-FPDC (dotted lines – reference crane position and solid lines – actual crane position)

Table 3. Performance analysis for the overhead crane control

Reference Input	FLC	IAE	ITAE	ISE
Sine (amplitude 0.3)	FPDC	32.6416	799.2207	2.8917
	NFST-FPDC	9.8239	147.4995	0.4209
Step (amplitude 0.3)	FPDC	7.5129	75.5667	0.6952
	NFST-FPDC	2.1929	3.9660	0.3458

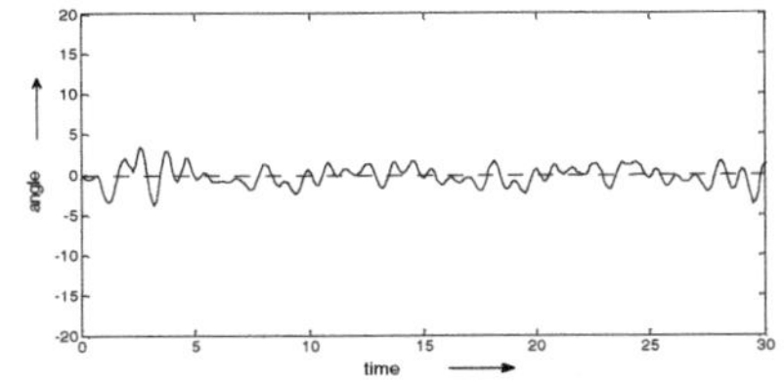

Fig. 12. Overhead crane swing angle control for sine input with FPDC

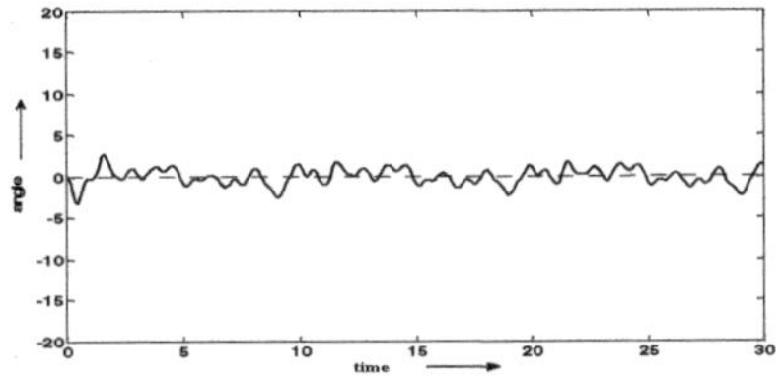

Fig. 13. Overhead crane swing angle control for sine input with NFST-FPDC

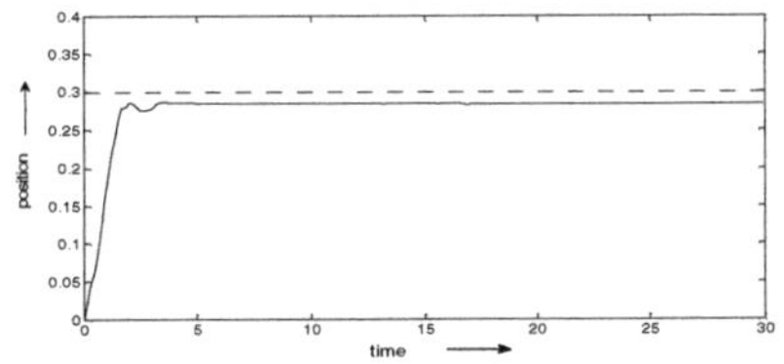

Fig. 14. Overhead crane position control for step input with FPDC (dotted lines – reference crane position and solid lines – actual crane position)

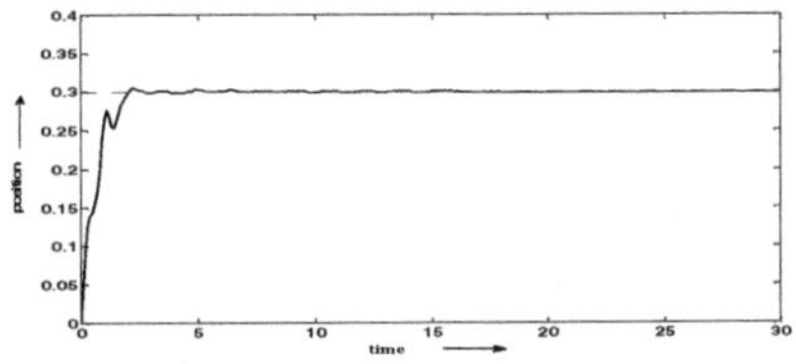

Fig. 15. Overhead crane position control for step input with NFST-FPDC (dotted lines – reference crane position and solid lines – actual position)

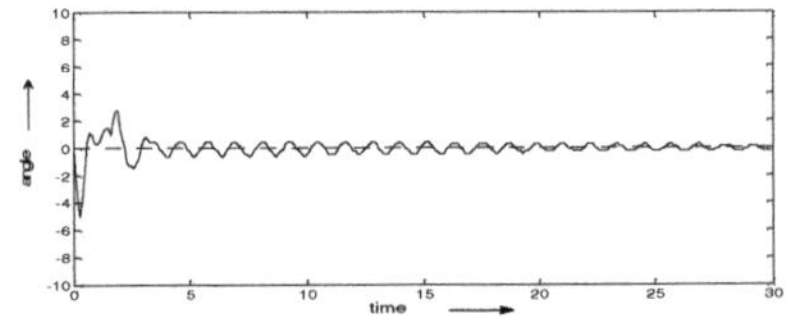

Fig. 16. Overhead crane swing angle control for step input with FPDC

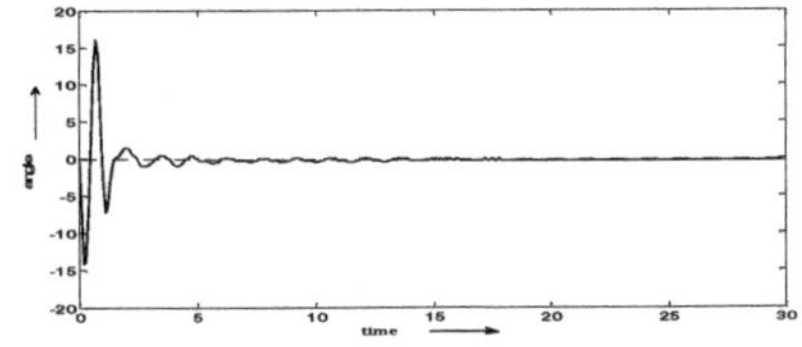

Fig. 17. Overhead crane swing angle control for step input with NFST-FPDC

5 Conclusion

In this paper, we proposed a simple self-tuning scheme for PD-type FLCs. Here, the controller gain (output SF) has been updated on-line through a gain modifying parameter β defined on the change of error (Δe) and its number of fuzzy partitions. Our proposed NFST-FPDC exhibited effective and improved performance compared to its conventional fuzzy counterpart. The proposed twin control scheme for overhead crane reduces the computational complexity and is very easy to understand. By applying the proposed self-tuning method and dual control scheme the load swing angle of the crane comes to a minimum. Experimental results verified that the proposed NFST-FPDC not only positioned the trolley in the desired location, it also significantly reduced the load swing during movement.

References

1. Hong, K.S., Ngo, Q.H.: Port Automation: modeling and control of container cranes. In: Inter. Conf. on Instrumentation, Control and Automation, pp. 19–26 (October 2009)
2. Hamalainen, J.J., Marttinen, A., Baharova, L., et al.: Optimal path planning for a trolley crane: fast and smooth transfer of load. In: IEE Proc. Control Theory and Applications, vol. 142(1), pp. 51–57 (1995)
3. Li, C., Lee, C.Y.: Fuzzy motion control of an auto-warehousing crane system. IEEE Trans. on Ind. Electron. 48(5), 983–994 (2001)
4. Park, M.S., Chwa, D., Hong, S.K.: Antisway tracking control of overhead cranes with system uncertainity and actuator nonlinearity using an adaptive fuzzy sliding mode control. IEEE Trans. on Industrial Electronics 55(11), 3972–3984 (2008)
5. Sorensen, K.L., Singhose, W., Dickerson, S.: A controller enabling precise positioning and sway reduction in bridge and grany cranes. Control Engineering Practice 15, 825–837 (2007)
6. Lee, C.C.: Fuzzy logic in control systems: Fuzzy logic controller—Parts I, II. IEEE Trans. on Syst., Man, Cybern. 20, 404–435 (1990)
7. Shinskey, F.G.: Process Control Systems—Application, Design, and Tuning. McGraw-Hill, New York (1998)
8. Malki, H.A., Li, H., Chen, G.: New design and stability analysis of fuzzy proportional-derivative control systems. IEEE Trans. on Fuzzy Systems 2, 245–254 (1994)
9. Mudi, R.K., Pal, N.R.: A Self-Tuning Fuzzy PD Controller. IETE Journal of Research (Special Issue on Fuzzy Systems) 44(4&5), 177–189 (1998)
10. Mudi, R.K., Pal, N.R.: A robust self-tuning scheme for PI and PD type fuzzy control- lers. IEEE Trans. on Fuzzy Systems 7(1), 2–16 (1999)
11. Mudi, R.K., Pal, N.R.: A self-tuning fuzzy PI controllers. Fuzzy Sets and Systems 115, 327–338 (2000)
12. Pal, N.R., Mudi, R.K., Pal, K., Patranabis, D.: Rule Extraction through Exploratory Data Analysis for Self-Tuning Fuzzy Controller. Int. J. of Fuzzy Systems 6(2), 71–80 (2004)
13. Pal, A.K., Mudi, R.K.: Self-Tuning Fuzzy PI controller and its application to HVAC system. IJCC (US) 6(1), 25–30 (2008)
14. Pal, A.K., Mudi, R.K.: Development of a Self-Tuning Fuzzy Controller through Relay Feedback Approach. In: Das, V.V. (ed.) CIIT 2011. CCIS, vol. 250, pp. 424–426. Springer, Heidelberg (2011)
15. Chang, C., Hsu, S., Chiang, K.: A practical fuzzy controllers scheme of overhead crane. Journal of Control Theory and Applications 3, 266–270 (2005)
16. Liu, D., Yi, J., Zhao, D., Wang, W.: Adaptive sliding mode fuzzy control for a two dimensional overhead crane. Mechatronics 15, 505–522 (2005)
17. Yang, J.H., Yang, K.S.: Adaptive coupling control for overhead crane systems. Mechatronics 17(2/3), 143–152 (2007)

ANN and Data Mining Approaches to Select Student Category in ITS

Aniruddha Dey[1], R.B. Mishra[2], and Kanai Chandra Pal[3]

[1] Department of Computer Science & Engineering, Jadavpur University. Kolkata-700032. India
[2] Department of Computer Engineering, IIT-BHU. Varanasi-221005, India
[3] Department of Printing Engineering, Jadavpur University. Kolkata-700032. India
anidey007@gmail.com, ravibm@bhu.ac.in, kanaipal@yahoo.com

Abstract. Data mining Methods have widely used to classification and categorization problems. It requires the categorization of the student on the basics of their performance. In this work an application of the data mining technique such as: Decision Tree (DT), Classification and Regression Trees (C&RT algorithm) have been used in the data set for the categorizing the student as high, medium and low. It's important to use (ANN) because this is a method by which students are categorized through cognitive input and behavioral input. We have used a data mining method Classification and Regression Trees (C&RT) to categorize the students in different category based on their cognitive and behavioral parameter.

Keywords: Data mining, ANN, E-learning, ITS, Decision tree.

1 Introduction

Web usage mining is a special form of data mining to address behavior in website. Web usage mining shall be proposed as central, non-instructional and objective evaluation technique for the analysis of the usage of web based learning and training system and in particular for the instruction with educational content in these system. The capability of classify the learners' performance is an important task in ITS which can save learners' time and effort [1]. The object of the data mining process is to discover new, interesting, and useful knowledge using variety of techniques such as predication, classification, clustering association rule mining and sequential pattern discovery from large data collection[3].Different using ANN and data mining methods focus on the issue of the adaptive learning, analysis the student's solution and finding optimal learning path in e-learning[4].Another study that use association rule has been used to improve the e-learning course. A variety of interactive learning and training activity are integrated in highly interactive environment in this content [5]. The learning behavior of learner and their interaction with content in particular are in central for both design and evaluation activities [6]. The behavior of learner in learning and training technology system is needed to be analyzed and evaluated in order to show the effectiveness. To improve instructional design we look at technique to determine the goal of learning sessions, the detail interaction with content and the changing of learning behavior over time [9].

S.C. Satapathy et al. (Eds.): Proc. of Int. Conf. on Front. of Intell. Comput., AISC 199, pp. 43–50.
DOI: 10.1007/978-3-642-35314-7_6

2 Problem Description

In this section we address the issue of learning strategy and learning performance using data mining method and ANN. Five subject taken as a courseware to see the case study. The problem is also described by collecting information six cognitive parameters and three behavioral parameters. Three cognitive parameters and three behavioral parameters are important for the input data. Data Mining (DM) helps to extract and analyze the meaningful relationship between various Cognitive and Behavioral parameter and it also provides relative importance of various parameters based on 307 records. According to experiment using C&RT there are two types of rule different category (High, Medium, and Low). According to C&RT the overall accuracy of the C&RT algorithm is 100%.

3 Performance Computation

A. Cognitive Computation Model

Cognitive computational model have been develop by many researchers, taking into account the mental states, psychological parameter and cognitive parameter to display human behavior in general and specific tasks in particular.*BDI(Belief, Desire and Intension)* theory has been developed to model the mental states in cognitive tasks.

a. Performance:-The performance is calculated on the basis of numbers of questions that are answered in good category, numbers of question that are answered in fair best category, number of questions that are answered in defective category by the student. Performance denoted by P_{ist}^{js}, which is denoted by follows:

$$P_{ist}^{js} = \frac{Q_{ist}^{js,g} * w_g + Q_{ist}^{js,avg} * w_{avg} - Q_{ist}^{js,b} * w_b}{w_g + w_{avg} + w_b} \tag{1}$$

Where P_{ist}^{js} is the performance of ith student for jth subject without weight.

$Q_{ist}^{js,g}$ is the ith type of student provided the good answer for j type subject question.

$Q_{ist}^{js,avg}$ is the ith type of student provided the avg answer for j type subject question.

$Q_{ist}^{js,b}$ is the ith type of student provided the bad answer for j type subject question.

w_g is weight assigned for the student provided the good answer.

w_{avg} is weight assign for the student provided the average answer.

w_b is the weight assign for the student provided the bad answer.

In our problem we assume that w_g=0.8, w_{avg} =0.3 and $w_{b;}$ =0.9. The higher value of w_b gives negative impact through formula.

b. Performance (with Difficulty weight):- P_{ist}^{js} is multiplied with the difficulty weight w_d . w_d is related with level of difficulty of question. Performance (with difficulty weight) is defined as follows:

$$p_{ist,w}^{js} = p_{ist}^{js} * w_d \tag{2}$$

Where $p_{ist,w}^{js}$ is the performance with difficulty weight.

w_d is the difficulty weight assign according to the question difficulty.

c. Capability:- It is computed on the basis of performance and difficulty weight, performance with difficulty and total number of questions to be answer. The capability shows the how much question can be handled by a particular student. Capability is defined as follows:

$$(Capability)_{ist} = \frac{\sum_j p_{ist,w}^{js}}{Tota\ln oofQuesion * \sum w_d} \tag{3}$$

$$X_{ist}^{ss} = \frac{Q_{ist}^{js,g} * w_{ss,g} + Q_{ist}^{js,avg} * w_{ss,avg} - Q_{ist}^{js,b} * w_{ss,b}}{w_{ss,g} + w_{ss,avg} + w_{ss,b}} \tag{4}$$

Where x_{ist}^{ss} is a parameter for calculate the desire.

$w_{ss,g}$ is preferential weight assign for student provided the good answer.

$w_{ss,avg}$ is the preferential weight assign for student provided the average answer.

$w_{ss,b}$ is the preferential weight assign for student provided the bad answer.

In our problem we assume that $w_{ss,g}$=0.8, $w_{ss,avg}$ =0.3 and $w_{ss,b}$ =0.9. The higher value of $w_{ss,b}$ gives negative impact through formula.

d. Desires:- It denote states that student wish to answer question which is based upon the total factor X and difficulty weight. Desire is defined as follows:

$$(Desire)_{ist} = \frac{\sum_5 X_{ist}^{ss} * w_d}{Total\ no\ of\ Quesion * \sum_5 w_d} \tag{5}$$

e. Preference:- A student preference play an active role in social practical reasoning, where a question is to be selected in order for given intension to be fulfilled. Preference is defined as follows:

$$(preference)_{ist} = \frac{\sum w_d * \left(\frac{\left(Q_{ist}^{js,g} * w_{p,g} + Q_{ist}^{js,Avg} * w_{p,avg} - Q_{ist}^{js,b} * w_{p,b}\right)}{w_{p,g} + w_{p,avg} + w_{p,b}}\right)}{Total\ no\ of\ Question} \tag{6}$$

In our problem we assume that $w_{p,g}$=0.8 , $w_{p,avg}$ =0.5 and $w_{p,b}$ =0.7. The higher value of $w_{p,b}$ gives negative impact through formula.

f. Intention:- Intentions are viewed as those questions that a student has committed to achieve. The intention computed on the basis of choice (desire) and preference. Intention is defined as follows:

$$(Intention)_{ist} = (Choice)_{ist} * (preference)_{ist} \tag{7}$$

g. Commitment:- It means the acts of binding yourself (intellectually or emotionally) to a course of action. Commitment is defined as follows:

$$(Commitment)_{ist} = (Desire)_{ist} * (Capability)_{ist} \tag{8}$$

h. Cognitive Index Factor (C.I.F):- The Cognitive Index Factor is computed on the basis of commitment and capability of a Student. Cognitive Index Factor is defined as follows:

$$(C.I.F)_{ist} = (Commitment)_{ist} * (Capability)_{ist} \tag{9}$$

B. Behavioral Computation Model

a. Pleasure:- Generally the word "pleasure" stands for a feeling of happy satisfaction and enjoyment. In this work is concerned with the student's behavioral activity.

b. Fatigue- Fatigue is extreme tiredness which can be observed on a student's face when he is being asked question repeatedly but not being able to answer them.

c. Distortion- Meaning of distortion is pulling or twist out of shape. In this work the word distortion is used to define one kind of behavioral parameter.

The Behavioral Expression Index Factor helps to select a student for a particular question. It is computed on the basis of pleasure, fatigue, distortion, which is defined as follows:

$$(\text{BEIF})_{ist} = \frac{\sum Q^{js}_{ist,ple} * w_{ple} - \sum Q^{js}_{ist,fatig} * w_{fatig} - \sum Q^{js}_{ist,Dist} * w_{dist}}{w_{ple} + w_{fatig} + w_{dist}} \tag{10}$$

Where , $W_{dist} > W_{fatig}$.

$BEIF^{js}_{ist}$ is the behavioral expression of ith student for j^{th} subject with weight.

$Q^{js}_{ist,ple}$ is the ith type of student answer j type subject question with pleasure.

$Q^{js}_{ist,fatig}$ is the ith type of student answer j type subject question with fatigue .

$Q^{js}_{ist,dist}$ is the ith type of student answer j type subject question with distortion.

w_{ple} is weight assign for the student provided answer with pleasure.

w_{fatig} is weight assign for the student provided answer with fatigue.

w_{dist} is the weight assign for the student provided answer with distortion.

Student Index Factor (SIF):-The Student Index Factor helps to select a student for a particular question. It is computed on the basis of Cognitive Index Factor and Behavioral Expression index factor. α is a negotiation index factor. α is a factor which is control SIF. It is in $0< \alpha <1$. Where α=0.5. SIF is defined as follows:

$$\boxed{(\text{SIF})_{ist} = \alpha * (\text{CIF})_{ist} + (1-\alpha) * (\text{BEIF})_{ist}} \tag{11}$$

4 Experimentation

Data mining is usually defined as searching, analyzing and sifting through large amounts of data to find relationships, patterns, or any significant statistical correlations. With the advent of computers, large databases and the internet, it is easier than ever to collect millions, billions and even trillions of pieces of data that can then be systematically analyzed to help look for relationships and to seek solutions to difficult problems[3].

A. Decision Tree Methods

Decision Tree method is an analytical approach to making decisions, especially those that have the potential to be risky or costly We have made an attempt to drive the importance of attributes in negotiation using data mining methods. The decision tree method is a visual, easy-to-understand alternative to the numerical charts and statistical probabilities in other decision analysis methods, such as spreadsheets. Decision trees also are adaptable, meaning they can be modified as new decisions present themselves or as new information becomes available and changes the scenarios. We proposed C&RT algorithm using the Decision Tree experimental strategies is shown in Table 1.

Table 1. Rule Set for categorization of participant by Decision Tree(C&RT algorithm)

Rules for H - contains 3 rule(s)
Rule 1 for H (243; 0.926)
if DISTORTION = H then H
Rule 2 for H (81; 0.988)
if DISTORTION = M and FATIGUE = H then H
Rule 3 for H (27; 0.926)
if DISTORTION = M and FATIGUE = M and PLEASURE in ["H"] then H
Rules for M - contains 3 rule(s) Rule 1 for M (243; 0.992)
if DISTORTION = L then M
Rule 2 for M (81; 1.0)
if DISTORTION = M and FATIGUE = L then M
Rule 3 for M (54; 0.852)

B. ANN Method

ANN model implemented using quick method, dynamic method and RBFN. Quick method produce smaller hidden layer that are faster to train and generalize better Dynamic method creates initial topology but modifies the topology by adding and/or removing hidden units as training process. The RBFN uses techniques similar to k-means clustering to partition the data based on values of the target The comparative view of these three method shows that dynamic method provided highest corrected case and in term of accuracy it dominated other two methods. This method produce smaller hidden layer that are faster to train and generalize better [10]. In quick method 6 input parameters and all parameter have categorical values. Among these 6 parameters each parameters has 3 stages. Therefore 3 neurons are for each categorical variable having three stages. Therefore, total number of neuron in input layer of ANN is (3*6=18) neurons. All combined ANN methods review are shown in Table 2.

Table 2. Comparative review of all combined ANN methods

Parameter	Value					
Method	Quick		Dynamic		RBFN	
Input data	307training data sets		307 training data sets		307Training Data Sets	
Correct cases	294	95.77%	302	98.37%	287	93.49%
Wrong cases:	13	4.23%	5	1.63%	20	6.51%
Input layer	18 neurons		18 neurons		18 neurons	
Hidden layer	1:3 neurons		1:6neurons,2:6 neurons		1: 20 neurons	
Output layer	3 neurons		3 neurons		3 neurons	

C. Sensitivity Analysis(SA)

Sensitivity analysis is an approach that analyzes and reflects the sensitivity degree of how the outcome of the model can be apportioned and altered to different circumstances of variation. Sensitivity analysis is performed to reduce network complexity by deleting the variables that have no or less influence on network training and to understand the degree of influence of each variable to network training. The greater the sensitivity degree, the larger the impact it has to the outcomes of artificial neural networks. For sensitivity analysis, feature selection node helps to identify the fields that are most important in a certain outcome. It may end up with a quicker, more efficient method, one that uses fewer predictors, executes more quickly, and may be easier to understand. For sensitivity analysis, feature selection node helps to identify the fields that are most important in a certain outcome. Sensitivity Analysis of ANN methods review are shown in Table 3.

Table 3. Comparative view of Sensitivity Analysis of ANN Method

SENSITIVITY ANALYSIS BY QUICK METHOD		SENSITIVITY ANALYSIS BY DYNAMIC METHOD		SENSITIVITY ANALYSIS BY RBFN METHOD	
Inputs	Relative Importance	Inputs	Relative Importance	Inputs	Relative Importance
DISTORTION	0.61995	DISTORTION	0.625584	PLEASURE	0.53397
FATIGUE	0.26017	FATIGUE	0.248239	PERFOMANCE	0.310547
PLEASURE	0.0893055	PLEASURE	0.0828965	FATIGUE	0.0998198
PERFOMANCE	0.0227612	PERFOMANCE	0.0579524	DISTORTION	0.0992947
COMMITMENT	0.011368	CAPABILITY	0.0335146	COMMITMENT	0.0965612
CAPABILITY	0.0112035	COMMITMENT	0.0321814	CAPABILITY	0.0871837

D. Feature Selection through DM

In this model we use feature selection node of Clementine 11.1 software. The feature selection node helps to identify the fields that are most important in predicting a certain outcome. From a set of hundreds or even thousands of predictors, the feature selection node screens, ranks, and selects the predictors that may be most important. Ultimately it may end up with quicker more efficient model one that uses few predictor, execute more quickly, and may be easier to understand. Sensitivity Analysis of ANN methods review are shown in Table 4.

Table 4. Important parameter through Feature Selection

Si no		Rank	Field	Type	Importance	Value
1	true	1	DISTORTION	set	Important	1.0
2	true	2	FATIGUE	set	Important	1.0
3	false	3	PLEASURE	set	Marginal	0.944
4	false	4	PERFOMANCE	set	Unimportant	0.527
5	false	5	COMMITMENT	set	Unimportant	0.351
6	false	6	CAPABILITY	set	Unimportant	0.327

5 Implementation

In this study we developed a web based learning system to implement proposed method above. Here we are designed .net in windows environment. Apache tomcat is the web server we use to this implementation. The best student SIF is found 49.375 from snapshot of figure 1. SIF is depending upon the BEIF and here CIF is negligible.

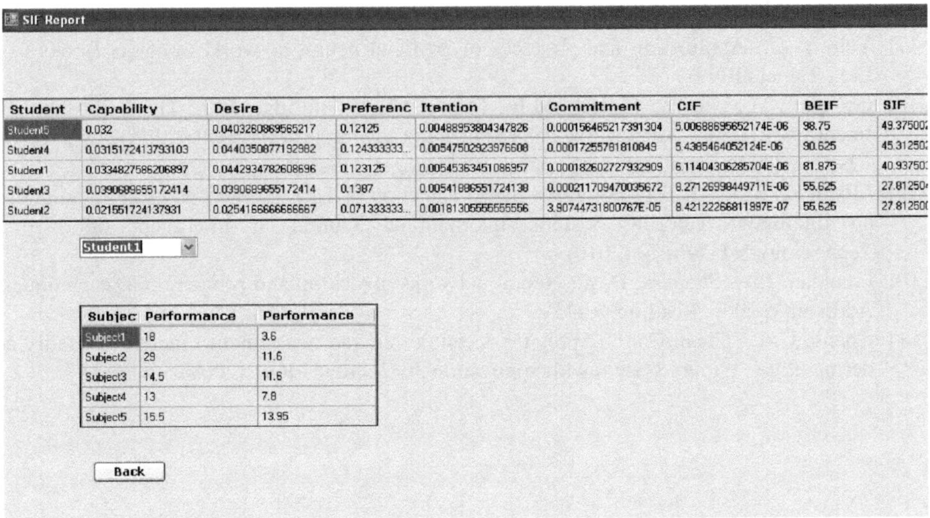

Fig. 1. Snapshot of the proposed e-learning System (test result)

6 Conclusion

In this work we have used data mining approach to determine the rules for categorization of student, importance of the input and the accuracy of classification and categorization of student. Using importance cognitive and behavioral input we accurately classify and categorize the student. We have been proposed combined ANN model (Quick, Dynamic and RBFN) to decide importance of input question type responsible for category of student. A comparative review of all combined ANN methods has been represented. Here find that student Behavioural parameter is much important rather cognitive parameter and figure 1 is also shown that.

References

[1] Grandison, T., Sloman, M.: A survey of trust in internet applications. IEEE Communications Surveys and Tutorials 4(4), 2–16 (2000)

[2] Romero, C., Ventura, S.: Educational Data Mining: a Survey from 1995 to 2005. Expert Systems with Applications 33(1), 135–146 (2007)

[3] Merceron, A., Yacef, K.: Mining Student Data Captured from a Web-Based Tutoring Tool: Initial Exploration and Results. Journal of Interactive Learning Research (JILR) 15(4), 319–346 (2004)

[4] Pahl, C.A.: Conceptual Architecture for Interactive Educational Multimedia. In: Ma, J. (ed.) Web-based Intelligent e-Learning Systems: Technologies and Applications. Idea Group Inc., Hershey (2005)

[5] He, M., Jennings, N.R., Leung, H.: On agent-mediated electronic commerce. IEEE Transaction on Knowledge and Data Engineering 15(4), 983–1003 (2003), doi:10.1109/TKDE.2003.1209014

[6] Leung, E.W.C., Li, Q.: An experimental study of a personalized learning environment through open-source software tools. IEEE Transaction on Education 50(4) (2007), doi:10.1109 /TE.2007.904571

[7] Yeh, Y.-C.: Application and practices of artificial neural network. Scholars Books Co. Ltd., Taipai (1999)

[8] Georgeff, M., Pell, B., Pollack, M.E., Tambe, M., Wooldridge, M.J.: The Belief-Desire-Intention Model of Agency. In: Papadimitriou, C., Singh, M.P., Müller, J.P. (eds.) ATAL 1998. LNCS (LNAI), vol. 1555, pp. 1–10. Springer, Heidelberg (1999)

[9] Misra, K., Misra, R.B.: Multiagent Based Selection of Tutor-Subject-Student Paradigm in an Intelligent Tutoring System. International Journal of Intelligent Information Technology 5(1), 46–70 (2010)

[10] Freeman, J.A., Skapura, D.M.: Neural networks algorithm'and programming techniques. Addison-Wesley, Reading (1992)

[11] Chang, C.-L., Chen, C.-H.: Applying decision tree and neural net to increases quality of dermatology. Expert System with application 36(2), 4035–4041 (2009)

Fuzzy PI Controller with Dynamic Set Point Weighting

Pubali Mitra[1], Chanchal Dey[2], and Rajani K. Mudi[3]

[1] Department of Instrumentation & Control Engineering, CIEM, India
pubali.mitra.cu@gmail.com
[2] Department of Applied Physics, University of Calcutta, India
cdaphy@caluniv.ac.in
[3] Department of Instrumentation & Electronics Engineering, Jadavpur University, India
rkmudi@iee.jusl.ac.in

Abstract. Fuzzy PI controllers usually fail to provide satisfactory performance for high-order as well as integrating processes with dead time. To lower the overshoot during set point response fixed set point weighting technique is used. But, for achieving improved responses during set point change as well as load variation simultaneously, here, dynamic set point weighting technique is proposed for fuzzy PI controller. Significant performance enhancement is found for second and third-order processes with dead time compared to conventional fuzzy PI controller. Adequate robustness of the proposed controller is also observed against the variation in process dead time.

Keywords: Fuzzy PI control, fixed set point weighting, dynamic set point weighting.

1 Introduction

Fuzzy logic controllers (FLCs) are becoming quite popular in process industries [1] as they can be easily designed to cope with considerable amount of process nonlinearity [2]. Among the various forms, PI type FLCs (FPICs) is the most popular [3, 4] due to their offset eliminating property. But, similar to the conventional PI controllers, performance of FPICs is not acceptable for integrating and higher-order processes with dead time. Mostly, they produce significant amount of oscillations during transient phase of the responses [5, 6]. FLCs can be successfully used for complex and nonlinear processes [7] and they are more robust [8] compared to conventional controllers. A conventional FLC has a fixed set of control rules, usually derived from expert's knowledge. Selection of the suitable values of membership functions (MFs) and scaling factors (SFs) are very crucial for successful design of FLCs. Till today; there is no general guideline for the selection of MFs and SFs for designing an FLC. Most of the cases, it is done by trial and error. Researchers have made various attempts for online tuning of SFs [4, 5] as well as MFs [9] for performance enhancement of the conventional FPICs.

To obtain an improved transient response set point filtering and set point weighting methods [10-12] are widely accepted for conventional PID controllers. In such cases,

S.C. Satapathy et al. (Eds.): Proc. of Int. Conf. on Front. of Intell. Comput., AISC 199, pp. 51–58.
DOI: 10.1007/978-3-642-35314-7_7

smaller overshoot is achieved at the cost of increased rise time but no improvement is found during load rejection. Instead of a fixed set point weighting, variable set point weighting technique is suggested in [13] for reducing the process rise time. Authors in [14] proposed separate set point weighting factors for proportional and derivative terms towards performance improvement of first-order time delay processes. Fuzzy logic based set point weight tuning is proposed in [15], which offers commendable performance during transient phase of the process response. Dynamic set point weighting (DSW) technique is suggested for conventional PI and PID controllers in [16] and [17] respectively. It offers simultaneous performance improvement during set point change as well as load variation. Here, we have extended the methodology of DSW in fuzzy logic based PI controller. It is expected that this mechanism would provide smaller oscillation during set point change and a faster recovery during load rejection.

In case of conventional set point weighting [11], the weighted set point for the proportional component of PI/PID controller is obtained by multiplying the actual set point with a fixed weighting factor where the weighting factor (β) has a value in between 0 and 1(i.e., $0<\beta\leq 1$). Whereas, in dynamic set point weighting [16, 17], instead of a fixed weighting factor (β), a continuously varying weighting factor (β_d) is used to get the dynamically weighted set point for the proportional component of PI/PID controller at each sampling instant. The dynamic weighting factor is calculated online through a simple expression based on the instantaneous normalized change of error (Δe_r) of the controlled variable and the normalized dead time (θ) of the process. In the proposed dynamic set point weighted fuzzy PI controller (DSWFPIC) both the inputs to fuzzy controller i.e., error (e) and change of error (Δe) are computed based on dynamic weighted set value. The input and output SFs are chosen through trial and error and the same SFs are used for performance comparison among the conventional FPIC and the proposed DSWFPIC.

The performance of the proposed DSWFPIC is verified for second and third-order linear processes with dead time along with a second-order integrating process with dead time. For having a clear comparison, a set of performance indices are evaluated for both the FPIC and DSWFPIC with same set of SFs and MFs. Robustness of the proposed controller is also verified by increasing the process dead time while keeping the controller settings (corresponding to the nominal dead time) unchanged. From the performance analysis it is found that the proposed DSWFPIC is quite capable of providing a considerable performance improvement compared to the conventional FPIC during set point change as well as load variation.

2 Controller Design

Block diagram of the proposed DSWFPIC is shown in Fig. 1(a). All the input (e_N, Δe_N) and output (Δu_N) variables of DSWFPIC are defined over -1 to +1 by 7 MFs (symmetric triangles with equal base and 50% overlap) as shown in Fig. 1(b). The blocks F and DF represent 'Fuzzification' and 'Defuzzfication' modules respectively.

In DSWFPIC, dynamic set point weighting (DSW) mechanism is incorporated with FPIC as shown by the dotted boundary in Fig. 1(a). The control rule base (CRB) of FPIC consists of 49 rules as depicted in Table 1. The relationships between SFs (G_e, $G_{\Delta e}$, and $G_{\Delta u}$) and the input and output variables (e, Δe, and Δu) are given by equations 1-3. Control action at k^{th} sampling instant is obtained by equation 4. The nature of control surface is shown in Fig. 1(c) which is highly nonlinear in nature.

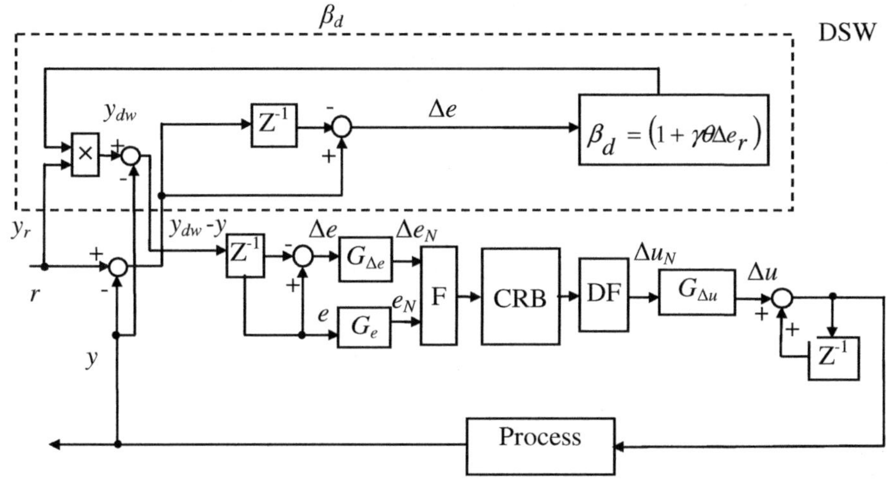

Fig. 1(a). Block diagram of the proposed DSWFPIC

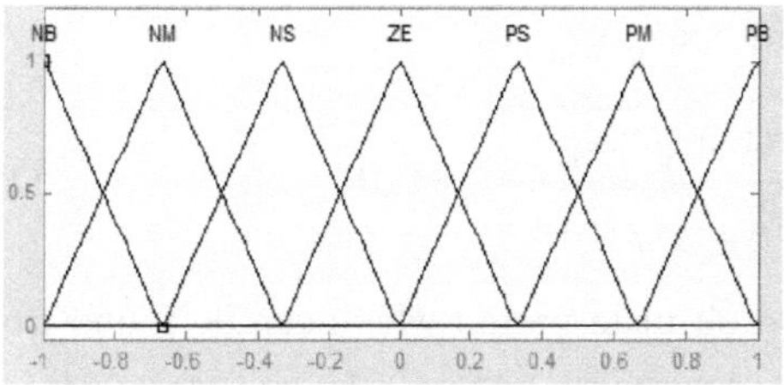

Fig. 1(b). MFs for input and output variables (e_N, Δe_N, and Δu_N)

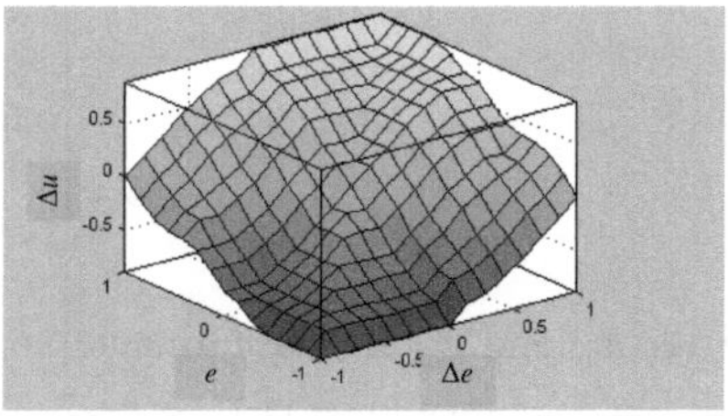

Fig. 1(c). Control surface

Table 1. Control rule base (CRB)

Δe/e	**NB**	**NM**	**NS**	**ZE**	**PS**	**PM**	**PB**
NB	NB	NB	NB	NM	NS	NS	ZE
NM	NB	NM	NM	NM	NS	ZE	PS
NS	NB	NM	NS	NS	ZE	PS	PM
ZE	NB	NM	NS	ZE	PS	PM	PB
PS	NM	NS	ZE	PS	PS	PM	PB
PM	NS	ZE	PS	PM	PM	PM	PB
PB	ZE	PS	PS	PM	PB	PB	PB

$$e_N = G_e e \ , \tag{1}$$

$$\Delta e_N = G_{\Delta e} \Delta e \ , \tag{2}$$

$$\Delta u = G_{\Delta u} \Delta u_N \ , \tag{3}$$

$$u(k) = u(k-1) + \Delta u(k) \ . \tag{4}$$

Discrete form of fixed set point weighted PI controller [11] at k^{th} sampling instant is given by

$$u(k) = k_c\left[\{\beta y_r - y(k)\} + \frac{\Delta t}{T_i}\sum_{i=0}^{k} e(i)\right] . \tag{5}$$

In equation (5), β is the fixed weighting factor, $y(k)$ is the process output at k^{th} instant, k_c is the proportional gain, T_i is the integral time, and Δt is the sampling interval. In case of dynamic set point weighting [16, 17] the weighted set point (y_{dw}) is modified at each sampling instant by the dynamic weighting factor β_d according to the following relation:

$$y_{dw}(k) = \beta_d(k) \times y_r . \tag{6}$$

where the dynamic weighting factor $\beta_d(k)$ is calculated online by equation (7).

$$\beta_d(k) = [1 + \gamma \times \theta \times \Delta e_r(k)] . \tag{7}$$

Here, γ is an adjustable tuning parameter, θ is the normalized dead time of the process, and $\Delta e_r(k)$ is the normalized change of error at k^{th} instant, i.e.,

$$\Delta e_r(k) = \frac{e(k) - e(k-1)}{y_r} = \frac{\Delta e(k)}{y_r} . \tag{8}$$

The tuning parameter γ and the normalized dead time θ can be obtained from the following relations:

$$\gamma = 1.5 \times k_u \times t_u , \tag{9}$$

$$\theta = \frac{\pi - 2\arctan(\sqrt{k_u k_p - 1})}{2.72\sqrt{k_u k_p - 1}} + 0.10 . \tag{10}$$

Here, k_u and t_u are the ultimate gain and ultimate period which can be found from the relay feedback test [18] of the concerned process and k_p is the open loop process gain.

So, the discrete form of dynamic set point weighted PI controller at k^{th} sampling instant is given by the following expression:

$$u(k) = k_c\left[\{y_{dw}(k) - y(k)\} + \frac{\Delta t}{T_i}\sum_{i=0}^{k} e(i)\right] . \tag{11}$$

Here, in case of the proposed DSWFPIC, both error $(e(k))$ and change of error $(\Delta e(k))$ are calculated from the dynamic weighted set value at each sampling instant by the following relations:

$$e(k) = y_{dw}(k) - y(k) . \tag{12}$$

$$\Delta e(k) = e(k) - e(k-1) . \tag{13}$$

3 Results

Performance of the proposed DSWFPIC is compared through simulation study with conventional FPIC for a difficult system like second-order marginally stable process with dead time. In addition, second-order and third-order linear processes with dead time are also considered for performance evaluation. To have a clear comparison between FPIC and the proposed DSWFPIC a number of performance indices – *%OS* (percentage overshoot), t_r (rise time), t_s (settling time), *IAE* (integral absolute error), and *ITAE* (integral time absolute error) are evaluated. To verify the robustness of the proposed DSWFPIC, process dead time is increased by 20% in each case from its nominal value keeping the controller settings unchanged. Mamdani type inferencing is used with centroid method for defuzzification.

3.1 Second-Order Marginally Stable Process

The transfer function of a second-order marginally stable process is given by-

$$G_p(s) = \frac{k_p e^{-Ls}}{s(\tau s + 1)} . \tag{14}$$

Here, we consider open loop process gain $k_p = 1$ and time constant $\tau = 1$s. Responses and corresponding control actions for dead time $L = 0.3$s are shown in Fig. 2(a). With a +20% perturbation in L, responses and control actions are depicted in Fig. 2(b). List of performance indices for both the nominal and perturbed dead time are given in Table 2(a) and Table 2(b) respectively. Results (Fig. 2 and Table 2) demonstrate a remarkably improved performance of DSWFPIC over FPIC during set point change as well as load disturbance.

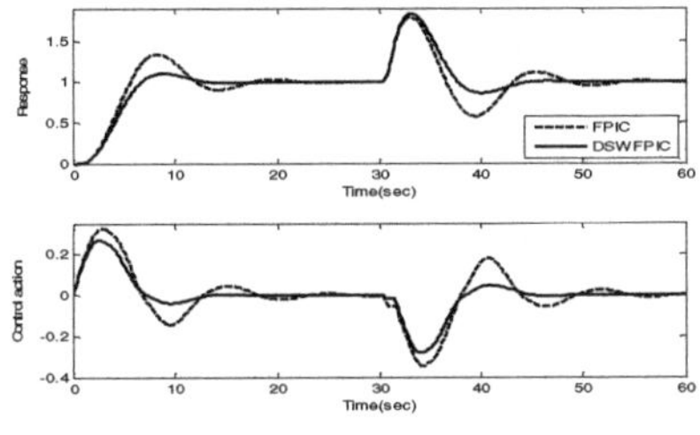

Fig. 2(a). Responses and control actions for $G_p = \frac{e^{-0.3s}}{s(s+1)}$

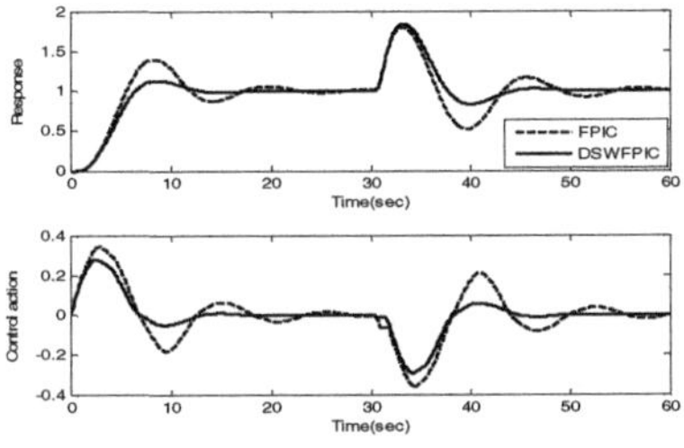

Fig. 2(b). Responses and control actions for $G_p = \frac{e^{-0.36s}}{s(s+1)}$

Table 2(a). Performance analysis for $G_p = \frac{e^{-0.3s}}{s(s+1)}$

	$G_{\Delta e}$	G_e	$G_{\Delta u}$	*%OS*	t_r(s)	t_s(s)	*IAE*	*ITAE*
FPIC	32	1	0.02	34.0	5.5	21.5	10.91	224.4
DSWFPIC				10.4	6.6	13.7	8.59	155.6

Table 2(b). Performance analysis for $G_p = \frac{e^{-0.36s}}{s(s+1)}$

	$G_{\Delta e}$	G_e	$G_{\Delta u}$	*%OS*	t_r(s)	t_s(s)	*IAE*	*ITAE*
FPIC	32	1	0.02	39.7	5.5	24	12.19	264.3
DSWFPIC				13.3	6.5	13.6	8.94	165.5

3.2 Second-Order Linear Process

Transfer function of the second-order linear process is given by

$$G_p(s) = \frac{k_p e^{-Ls}}{(\tau s + 1)^2} \quad . \tag{15}$$

Here, we consider open loop process gain $k_p = 1$ and time constant $\tau = 1$s. Figures 3(a) and 3(b) show the responses and corresponding control actions with dead time $L = 0.2$s and with its +20% perturbation (i.e., $L = 0.24$s) for FPIC and DSWFPIC under both set point change and load variaton. Corresponding performance indices are listed in Table 3(a) and Table 3(b) respectively.

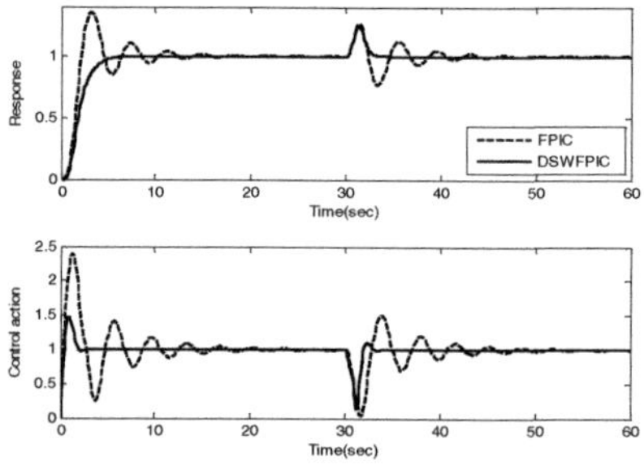

Fig. 3(a). Responses and control actions for $G_p = \frac{e^{-0.2s}}{(s+1)^2}$

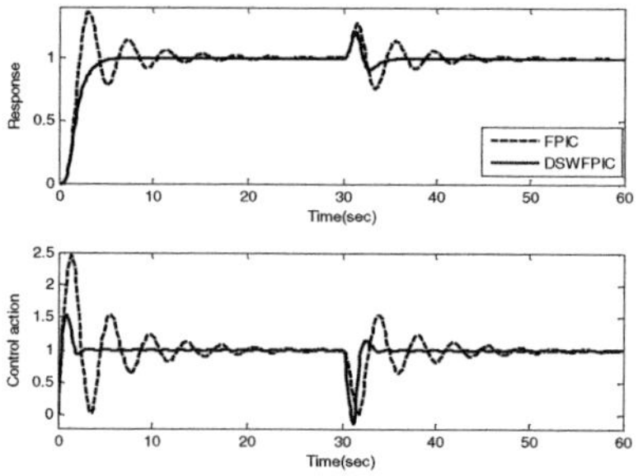

Fig. 3(b). Responses and control actions for $G_p = \frac{e^{-0.24s}}{(s+1)^2}$

Table 3(a). Performance analysis for $G_p = \frac{e^{-0.2s}}{(s+1)^2}$

	$G_{\Delta e}$	G_e	$G_{\Delta u}$	%OS	t_r(s)	t_s(s)	IAE	ITAE
FPIC	10	1	0.3	35.7	2.1	17	3.61	44.6
DSWFPIC				0.00	11.6	7.1	2.418	13.17

Table 3(b). Performance analysis for $G_p = \frac{e^{-0.24s}}{(s+1)^2}$

	$G_{\Delta e}$	G_e	$G_{\Delta u}$	%OS	t_r(s)	t_s(s)	IAE	ITAE
FPIC	10	1	0.3	36.3	2.2	20.5	4.14	54.74
DSWFPIC				0.0	11.5	6.6	2.46	14.92

3.3 Third-Order Linear Process

Transfer function of third-order linear process is given by

$$G_p(s) = \frac{k_p e^{-Ls}}{(\tau s + 1)^3} \quad . \tag{16}$$

We consider the open loop process gain $k_p = 1$ and time constant $\tau = 1$s. Dead time L is considered as 0.4s. Responses and control actions are shown in Figs. 4(a) and 4(b) for nominal and with a +20% perturbation in dead time ($L = 0.48$s). Performance indices are given in Tables 4(a) and 4(b) respectively. Like previous two examples, here also DSWFPIC exhibits noticeable improvement over FPIC.

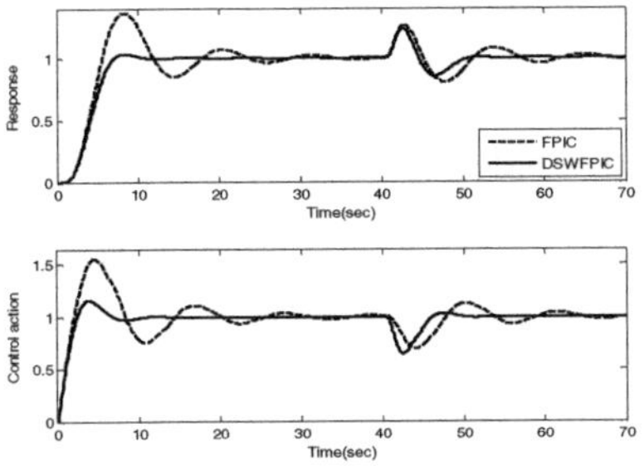

Fig. 4(a). Responses and control actions for $G_p = \frac{e^{-0.4s}}{(s+1)^3}$

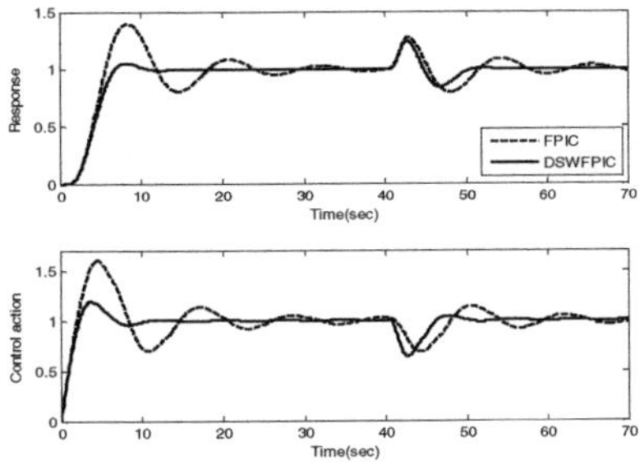

Fig. 4(b). Responses and control actions for $G_p = \frac{e^{-0.48s}}{(s+1)^3}$

Table 4(a). Performance analysis for $G_p = \frac{e^{-0.4s}}{(s+1)^3}$

	$G_{\Delta e}$	G_e	$G_{\Delta u}$	$\%OS$	t_r(s)	t_s(s)	IAE	$ITAE$
FPIC	10	1	0.06	36.1	5.6	34.7	8.46	140.2
DSWFPIC				3.5	7.1	12.2	5.37	57.62

Table 4(b). Performance analysis for $G_p = \frac{e^{-0.48s}}{(s+1)^3}$

	$G_{\Delta e}$	G_e	$G_{\Delta u}$	$\%OS$	t_r(s)	t_s(s)	IAE	$ITAE$
FPIC	10	1	0.06	39.5	5.7	36.4	9.23	159.6
DSWFPIC				5.2	6.9	13.1	5.52	61.63

In summary, responses and performance indices obtained from simulation study clearly indicates that the proposed DSWFPIC is capable of providing a significantly improved close-loop performance compared to FPIC under set point change as well as load variation. On increasing the dead time by 20% with the same controller settings it still maintains the same level of performance.

4 Conclusion

Here, we proposed a dynamic set point weighting based fuzzy PI controller where the set value get adjusted by dynamic weighting factor at each sampling instant depending on the process operating condition. The distinct feature of the proposed controller is that it can provide considerably improved set point response as well as load regulation simultaneously. Robustness feature of the proposed technique is observed against the variation of process dead time. Similar to the linear process the proposed controller may also be applied for performance enhancement of nonlinear processes. The mechanism for online computation of dynamic weighting factor is quite simple and straight forward and hence this technique can be implemented in practical applications and all such possibilities are in our future scope.

References

1. Sugeno, M.: Industrial Applications of Fuzzy Control. Elsevier, Netherlands (1985)
2. Driankov, D., Hellendron, H., Reinfrank, M.: An Introduction to Fuzzy Control. Springer, New York (1993)
3. Pal, A.K., Mudi, R.K.: Self-tuning fuzzy PI controller and its application on HVAC systems. Int. J. Comp. Cogn. 6(1), 25–30 (2008)
4. Mudi, R.K., Pal, N.R.: A robust self-tuning scheme for PI and PD type fuzzy controllers. IEEE Trans. Fuzzy Sys. 7(1), 2–16 (1997)
5. Chopra, S., Mitra, R., Kumar, V.: Auto tuning of fuzzy PI controller using fuzzy logic. Int. J. Comp. Cogn. 6(1), 12–18 (2008)
6. Lee, J.: On methods for improving performance of PI-type fuzzy logic controllers. IEEE Trans. Fuzzy Sys. 1(4), 298–301 (1993)
7. Zhao, Y., Collins, E.G.: Fuzzy PI control design for an industrial weigh feeder. IEEE Trans. Fuzzy Sys. 11(3), 311–319 (2003)
8. Palm, R.: Sliding mode fuzzy control. In: Int. Conf Fuzz IEEE-1992, SanDiego, pp. 519–526 (1992)
9. Dey, C., Mudi, R.K.: Design of a PI-type fuzzy controller with on-line membership function tuning. In: 12th Int. Conf. Neural Information Processing ICONIP-2005, Taipei (2005)
10. Khan, B.Z., Lehman, B.: Set-point PI controllers for systems with large normalized deadtime. IEEE Trans. Control Sys. Tech. 4(4), 459–466 (1996)
11. Hang, C.C., Astrom, K.J., Ho, W.K.: Refinements of Zeigler-Nichols tuning formula. IEE Proc.-D. 138(2), 111–118 (1991)
12. Rangaiah, G.P., Krishnaswamy, P.R.: Set-point weighting for simplified model predictive control. Chem. Eng. J. 50(3), 159–163 (1992)
13. Hang, C.C., Cao, L.: Improvement of transient response by means of variable set-point weighting. IEEE Trans. Ind. Elect. 43(4), 477–484 (1996)
14. Prashanti, G., Chidambaram, M.: Set-point weighted PID controllers for unstable systems. J. Franklin Institute 337(2-3), 201–215 (2000)
15. Visioli, A.: Fuzzy logic based set-point weight tuning of PID controllers. IEEE Trans. Sys. Man Cyb. 29(6), 587–592 (1999)
16. Mudi, R.K., Dey, C.: Performance improvement of PI controllers through dynamic set-point weighting. ISA Trans. 50(2), 220–230 (2011)
17. Dey, C., Mudi, R.K., Lee, T.T.: Dynamic set-point weighted PID controller. Cont. Int. Sys. 37(4), 212–219 (2010)
18. Seborg, D.E., Edgar, T.F., Mellichamp, D.A.: Process dynamics and control. Wiley, New York (2004)

A Particle Swarm Optimized Functional Link Artificial Neural Network (PSO-FLANN) in Software Cost Estimation

Tirimula Rao Benala[1], Korada Chinnababu[1], Rajib Mall[2], and Satchidananda Dehuri[3]

[1] Anil Neerukonda Institute of Technology and Sciences
Sangivalasa-531162, Visakhapatnam, Andhra Pradesh, India
tirimula@gmail.com

[2] Department of Computer Science and Engineering
Indian Institute of Technology, Kharagpur
rajib@cse.iitkgp.ernet.in

[3] Department of Systems Engineering
Ajou University, San 5, Woncheon-dong, Yeongtong-gu, Suwon 443-749, Republic of Korea
satchi.lapa@gmail.com

Abstract. We use particle swarm optimization (PSO) to train the functional link artificial neural network (FLANN) for software effort prediction. The combined framework is known as PSO-FLANN. This framework exploits the global classification capability of PSO and FLANN's complex nonlinear mapping between its input and output pattern space by using functional expansion. The Chebyshev polynomial has been used as choice of expansion in FLANN to exhaustively study the performance in three real time datasets. The simulation results show that it not only deals efficiently with noisy data but achieves improved accuracy in prediction.

Keywords: Software cost estimation, Particle Swarm optimization, Functional Link Artificial Neural Networks.

1 Introduction

Successful software project development primarily relies on accurate effort prediction at an early stage of development. Hence, software effort prediction models are important for software practitioners and project managers. However, constructing an accurate model for predicting the effort is a challenging task for most of the software systems. The importance of accurate estimation has led to extensive research efforts to software cost estimation methods. In this paper, we are concerned with cost estimation models that are based on Particle swarm optimized Functional link artificial neural networks (PSO-FLANN). PSO-FLANN, is a typical three layer feed forward neural network. It consists of input layer, hidden layer and output layer. However in FLANN the weight vector is evolved by PSO during training of the network. The FLANN architecture for predicting software development effort is a single-layer feed forward neural network consisting of one input layer and an output

S.C. Satapathy et al. (Eds.): Proc. of Int. Conf. on Front. of Intell. Comput., AISC 199, pp. 59–66.
DOI: 10.1007/978-3-642-35314-7_8

layer. The FLANN generates output (effort) by expanding the initial inputs (cost drivers) and then processing in the final output layer. Each input neuron corresponds to a component of an input vector. The output layer consists of one output neuron that computes the software development effort as a linear weighted sum of the outputs of the input layer [14,15]. The large and non-normal data sets leads FLANN methods to low prediction accuracy and high computational complexity.

The paper is organized as follows. Functional Link Neural networks, Particle swarm optimization, and cost estimation fundamentals are briefly reviewed in Section 2. The proposed PSO-FLANN approach is described in Section 3. In Section 4, numerical examples from Cocomo81 (Coc81), Nasa93, Maxwell dataset is used to illustrate the performance. This paper is concluded with a future works in Section 5.

2 Background

In this section we provide an overview of a few basic concepts, based on which our work has been developed.

2.1 Software Cost Estimation

Accurate software cost estimation is an indispensible process for effective software management. There are two fundamental approaches to software cost estimation: top-down and bottom-up. In a top-down approach, the overall estimate for the project is first determined based on some models and then the estimates for different tasks are determined. Such a model typically requires, as input, the size estimate (in LOC or function points) of the overall system. The bottom-up approach, on the other hand, first defines the various activities that need to be executed for the project. As these activities are typically at a sufficiently low level of granularity, effort estimate of each can be done from past experience. The estimate for the total project is then obtained from the estimates of these activities. According to Oliveira [10], the main risk factors for software projects are the schedule and effort (cost) to finish it; the particularities of software project requisites, project team, and the employed technology make the process of cost estimation too hard. Due to these and other peculiarities of each project, it is known in practice that the accurate measurement of the cost and development time of software is only possible when the project is finished [1, 2,6,10]. However, it is necessary to perform estimations before the project begins. We investigate a novel technique aimed to predict (estimate) the software development cost; the proposed technique is based on PSO and FLANN.

2.2 Architecture of FLANN

A FLANN network can be used not only for functional approximation but also for decreasing the computational complexity. Further, the FLANN network is much faster than other network. The primary reason for this is that the learning process in FLANN network has two stages and both stages can be made efficient by appropriate learning algorithms. The use of on FLANN to estimate software development effort requires the determination of its architecture parameters according to the characteristics of datasets [15]. The architecture of FLANN is shown in figure 1.

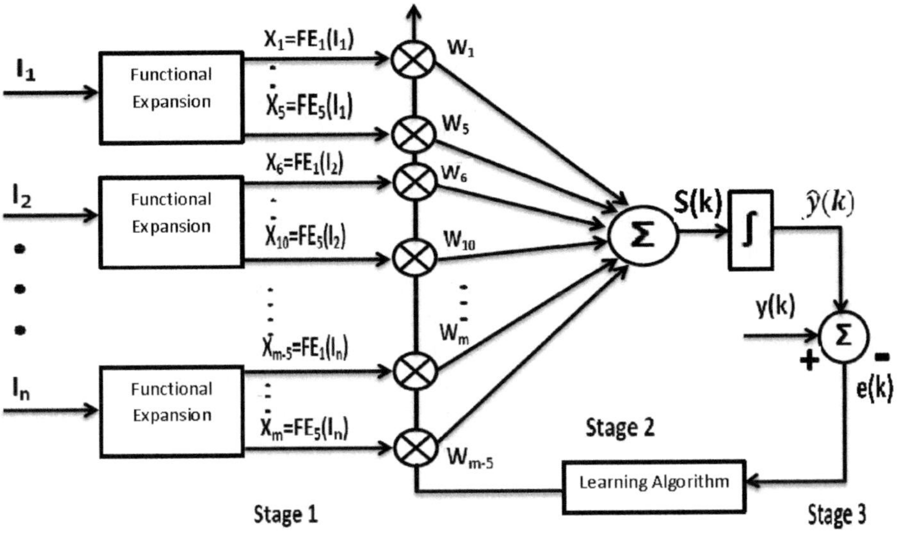

Fig. 1. FLANN Architecture

2.3 Particle Swarm Optimization

Particle swarm optimization (PSO) was originally designed and introduced by Ebarhart and Kennedy [7]. In the standard PSO algorithm [3,7], at iteration t, the velocity and position can be updated using eqns. (1) and (2) respectively.

$$\overrightarrow{v_k}(t+1) = w \otimes \overrightarrow{v_k}(t) + \overrightarrow{c_1} \otimes \overrightarrow{r_1}(t) \otimes \left(\overrightarrow{p_k}(t) - \overrightarrow{x_k}(t)\right) + \overrightarrow{c_2} \otimes \overrightarrow{r_2}(t) \otimes \left(\overrightarrow{p_g}(t) - \overrightarrow{x_k}(t)\right) \quad (1)$$

$$\overrightarrow{x_k}(t+1) = \overrightarrow{x_k}(t) + \overrightarrow{v_k}(t+1) \quad (2)$$

The symbol $\otimes$ denotes point by point vector multiplication. The inertia momentum factor w, $(0 \leq w \leq 1)$, self-confidence factor c_1 and swarm confidence factor c_2 are non-negative real constants. Randomness (useful for good state space exploitation) is introduced via the vector of random numbers $\overrightarrow{r_1}$ and $\overrightarrow{r_2}$. These are usually selected as a uniform random number in the range [0, 1]. The steps of velocity update, position update and fitness computations are repeated until a desired convergence criterion is met. The stopping criteria is usually that the maximum change in the best fitness should be smaller than the specified tolerance for a specified number of iterations, I, as shown in Eq. (3). Alternatively, the algorithm can be terminated when the velocity updates are close to zero over a number of iterations.

$$\left| f\left(\overrightarrow{p_g}(t)\right) - f(\overrightarrow{p_g}(t-1)) \right| \leq \epsilon, \qquad t = 2,3,\ldots\ldots,I \quad (3)$$

3 PSO-FLANN Framework

PSO-FLANN [3] is a typical three layer feed forward neural network. It consists of input layer, hidden layer and output layer. The only difference with FLANN is that the weight vector is evolved by the PSO during the training of the network. The nodes between input and hidden layers are connected without weight vector, but the nodes between hidden and output layer are connected by weights. In PSO-FLANN architecture there are n input nodes (i.e. equal to number of cost drivers of the dataset), and m nodes in the hidden layer, where m is the number of functionally expanded node and one output layer in the output neuron. The connection between hidden layer and output layer is assigned with the weight vector. In this work we have used Chebyshev polynomial for mapping the input features from one form to another form of higher dimension. In the following, we present the algorithm in pseudocode form.

PSO-FLANN Algorithm

```
1.  DIVISION OF DATASET
      Divide the dataset into two parts: Training and Testing.
2.  MAPPING OF INPUT PATTERNS
      Map each pattern from lower dimension to higher dimension, i.e. expand
      each feature value according to the predefined set of functions.
3.  RANDOM INITIALIZATION
      Initialize each particle randomly with small values from the domain [-1, 1].
4.  WHILE (THE TERMINATION CRITERIA NOT MET)
      FOR entire swarm
        FOR Each particle in the swarm
          FOR Each sample of training sample
                Calculate the weighted Sum and feed as an input to the node of the
                Ouput layer.
                Calculate the error and accumulate it.
          END
          Fitness of the particle is equal to the accumulated error.
          If fitness value is better than the best fitness value in history,
          Set the current value as new personal best,
      END
         Choose the particle with best fitness value of all particles as global best
      END
  END
  MUTATION
5.  WHILE END
```

3.1 Performance Evaluation Metrics

To evaluate the accuracy of our proposed method, three performance metrics are considered: Mean Magnitude of Relative Error (MMRE), Median Magnitude of relative error (MdMRE),and PRED (0.25), because these measures are widely referred to in literature [13].

The MMRE is defined as:

$$\text{MMRE}=\frac{1}{n}\sum_{i=1}^{n} MRE$$

$$MRE = |\frac{(E_i-\hat{E}_i)}{E_i}|$$

Where n denotes the total number of projects, E_i denotes the actual cost of ith project, and $\hat{E}_i$ denotes the estimated cost of the ith project. Small MMRE value indicates the low level of estimation error. However this metric is unbalanced and penalizes overestimation more than underestimation. The MdMRE is the median of all the MREs.

$$\text{MdMRE} = \text{Median (MRE)}$$

It exhibits similar pattern to MMRE but it is more likely to select the true model especially in the underestimation case since it is less sensitive to extreme outlier [4]. The PRED (0.25) is the percentage of prediction that fall within 25 percent of actual cost

$$PRED(q) = \frac{k}{n}$$

Where n denotes the total number of projects and k denotes the number of projects whose MRE is less than or equal to q. Normally, q is set to be 0.25. The PRED(0.25) identifies cost estimations that are generally accurate, while MMRE is biased and not always reliable as a performance metric. However, MMRE has been de facto standard in the software cost estimation literature.

4 Experimental Results

In this section, three real world software engineering datasets, namely, Cocomo81 (Coc81), Nasa93, Maxwell [8] are utilized for empirical evaluation of our methods.

4.1 Dataset Preparation

Before the experiments, all types of features are normalized into the interval [0, 1] in order to eliminate their different influences. The three real datasets are randomly split into three nearly equal sized sub-sets for training and testing. The training set is treated as the targets for the optimization of feature weights and project subsets. The testing set is exclusively used to evaluate the optimized FLANN models.

4.2 Cost Estimation Models

Out of the three functional expansions namely, C-FLANN, P-FLANN and L-FLANN, C-FLANN based model was considered in our experiments. Hereafter, in this paper C-FLANN will be annotated as FLANN. The proposed models using PSO as learning algorithm in place of back propagation for FLANN will be hereafter known as PSO-FLANN. For a comprehensive evaluation of the proposed models, For comparison, other popular estimation models including Step wise regression (SWR) [11], Functional Link Artificial Neural Networks (FLANN) [15], classification and regression trees (CART) [12], are also included in the experiments.

4.3 Experitmental Setup

For the purpose of validation, we adopt three-fold cross validation [5,9] to evaluate the accuracy of the methods. In this scheme all the three datasets are randomly divided into three nearly equal sized subsets. At each time one of three subsets is used as the test sets which are exclusively used to evaluate the estimation performance, and other two subsets treated as the Validation data set and training data set exclusively used to optimize the cost drivers. This process is repeated three times. Then the average training error and testing error across all three trials are computed. The advantage of this scheme is that it becomes nearly independent of how the data is split since each data point is assigned into a test set, a training set and a validation set respectively. First, the performance of PSO-FLANN is investigated. The best variants on training set are selected as the candidate for comparisons. Subsequently, the optimizations of machines learning methods are conducted on the training dataset by searching through their parameter spaces. Finally, the training and testing results of the best variants of all estimation methods are summarized and compared. The experiments results and their analysis are presented in next section.

4.4 Experimental Results

Tables 1 to 3 summarize the performance results of all the methods applied on three real time datasets. The second annotated column in each table shows performance of various methods with respect to performance metrics MMRE. Similarly, the third column and fourth column of each table summarizes the results with respect to performance metrics MdMRE and PRED(0.25) respectively. With these values it can be inferred that the testing results in the proposed methods outperform the testing results in traditional methods. Thus our results indicate that hybrid combination of PSO and FLANN improves the accuracy very efficiently when compared to SWR, FLANN and CART.

Table 1. Results on Coc81 Dataset

Methods	*MMRE*		*MdMRE*		*PRED(0.25)*	
	Training	**Testing**	**Training**	**Testing**	**Training**	**Testing**
PSO-FLANN	**0.43**	**0.37**	**0.48**	**0.42**	**0.39**	**0.52**
FLANN	**0.45**	**0.38**	**0.49**	**0.47**	**0.35**	**0.49**
SWR	**0.34**	**0.35**	**0.42**	**0.44**	**0.52**	**0.50**
CART	**1.28**	**1.12**	**0.62**	**0.58**	**0.17**	**0.19**

Table 2. Results on Nasa93 Dataset

Methods	*MMRE*		*MdMRE*		*PRED(0.25)*	
	Training	**Testing**	**Training**	**Testing**	**Training**	**Testing**
PSO-FLANN	**0.49**	**0.34**	**0.44**	**0.45**	**0.39**	**0.50**
FLANN	**0.42**	**0.49**	**0.46**	**0.48**	**0.38**	**0.48**
SWR	**0.39**	**0.34**	**0.47**	**0.49**	**0.44**	**0.44**
CART	**1.34**	**1.28**	**0.85**	**0.66**	**0.23**	**0.30**

Table 3. Results on Maxwell Dataset

Methods	*MMRE*		*MdMRE*		*PRED(0.25)*	
	Training	**Testing**	**Training**	**Testing**	**Training**	**Testing**
PSO-FLANN	**0.55**	**0.38**	**0.49**	**0.42**	**0.32**	**0.48**
FLANN	**0.48**	**0.42**	**0.39**	**0.40**	**0.45**	**0.28**
SWR	**0.42**	**0.42**	**0.47**	**0.39**	**0.39**	**0.45**
CART	**0.92**	**0.91**	**0.46**	**0.39**	**0.29**	**0.31**

5 Conclusion and Future Work

We have done our research in the direction of software cost estimation by hybrid system using PSO and FLANN. We extend connotations to our work with Artificial Bee Colony (ABC), Differential Evolution (DE), Artificial Immune System (AIS), Bacterial foraging optimization algorithm, Neuro Fuzzy, Neuro Genetic, Simulated Annealing and fuzzy logic. We have evaluated the performance of PSO-FLANN for cost prediction. It provides better prediction accuracy compared to FLANN. The experimental results show that our method gives promising results as compared to conventional FLANN and outperform the comparative techniques such as FLANN, SWR and CART. Motivation is therefore exploring the scope of application of soft computing in the field of Software Cost Estimation..

References

1. Araújo, R., de, A., Oliveira, A.L.I., Soares, S.: A shift-invariant morphological system for software development cost estimation. Expert Systems with Applications 38, 4162–4168 (2011)
2. Braga, P.L., Oliveira, A.L.I., Ribeiro, G.H.T., Meira, S.R.L.: Software effort estimation using machine learning techniques with robust confidence intervals. In: IEEE International Conference on Tools with Artificial Intelligence (ICTAI) (2007)
3. Dehuri, S., Roy, R., Cho, S.-B., Ghosh, A.: An Improved Swarm Optimized functional link artificial neural network (ISO-FLANN) for Clasification. J. Syst. Software 85(6) (2012)
4. Foss, T., Stensrud, E., Kitchenham, B., Myrtveit, I.: A simulation study of themodel evaluation criterion MMRE. IEEE Transactions on Software Engineering 29(11) (2003)
5. Huang, S.J., Chiu, N.H.: Optimization of analogy weights by genetic algorithm for software effort estimation. Information and Software Technology 48, 1034–1045 (2006)
6. Keung, J.W.: Theoretical Maximum Prediction Accuracy for Analogy-Based Software Cost Estimation. In: 15th Asia-Pacific Software Engineering Conference, pp. 495–502 (2008), http://ieeexplore.ieee.org/lpdocsepic03/wrapper.htm?arnumber=4724583
7. Kennedy, J., Eberhart, R.C.: Particle swarm optimization. In: Proceedings of the IEEE International Conference on Neural Networks, Perth, Australia, pp. 1942–1948 (1995)
8. Menzies, T.: The PROMISE Repository Of Software Engineering Databases. School of Information Technology and Engineering, University of Ottawa, Canada (2006), http://promise.site.uottawa.ca/SERepository
9. Mendes, E., Watson, I., Triggs, C., Mosley, N., Counsell, S.: A Comparative Study of Cost Estimation Models for Web Hypermedia Applications. Empirical Software Engineering 8, 163–196 (2003)
10. Oliveira, A.L.I.: Estimation of software project effort with support vector regression. Neurocomputing 69(13-15), 1749–1753 (2006)
11. Shepperd, M., Kadoda, G.: Comparing Software Prediction Techniques using Simulation. IEEE Transaction on Software Engineering 27(11), 1014–1022 (2001)
12. Stensrud, E.: Alternative Approaches to Software Prediction of ERP Projects. Information and Software Technology 43(7), 413–423 (2001)
13. Stensrud, E., Foss, T., Kitchenham, B.A., Myrtveit, I.: An empirical validation of the relationship between the magnitude of relative error and project size. In: Proceedings of the IEEE 8th Metrics Symposium, pp. 3–12 (2002)
14. Tirimula Rao, B., Sameet, B., Kiran Swathi, G., Vikram Gupta, K., Raviteja, C., Sumana, S.: A Novel Neural Network approach for Software Cost Estimation Using Functional Link Artificial Neural Networks. International Journal of Computer Science and Network Security (IJCSNS) 9(6), 126–131 (2009)
15. Tirimula Rao, B., Dehuri, S., Mall, R.: Functional Link Artificial Neural Networks for Software Cost Estimation. International Journal of Applied Evolutionary Computation (IJAEC) 3(2), 62–82 (2012)

An Augmented Self-tuning Fuzzy Logic Controller with Its Real-Time Implementation

Rajani K. Mudi, D. Simhachalam, and Arindam Mondal

Dept. of Inst. & Electronics Engg, Jadavpur University, Kolkata, India
{Rkmudi,arinda_robotics}@yahoo.com, chalamju10@gmail.com

Abstract. A self-tuning fuzzy PI controller (STFPIC1) is reported elsewhere whose output scaling factor (SF) is continuously adjusted by a fuzzy gain modifying factor α, which is further multiplied by an empirically chosen constant, irrespective of the type of process. Instead of such a fixed value, here, we propose to augment the gain modifier α based on the process dynamics, which is closely related to its *critical point*, *i.e.*, ultimate gain and ultimate period. The critical point is obtained by conducting relay-feedback experiment. Thus, the overall output SF of the proposed augmented self-tuning fuzzy PI controller (STFPIC2) will be more rational, as its design considers the dynamics of the process. Performance of the proposed controller is studied for a number of high-order dead-time processes under both set-point change and load disturbance. Performance comparisons with other FLCs are provided in terms of various performance indices. Effectiveness of the proposed STFPIC2 is tested through real-time implementation on a practical temperature control system.

Keywords: Self-tuning fuzzy control, Critical point, Relay-feedback.

1 Introduction

In process control applications, fuzzy logic controllers (FLCs) are being popular due to their inherent capabilities of handling linear as well as highly non-linear systems. They have been successfully used, and proved to be superior to the conventional non-fuzzy controllers for a number of difficult processes [1]. Even they are found to be less sensitive to parametric variations than conventional controllers [2]. Usually two types of FLC structures have been considered, *i.e.*, PI-type FLC (FPIC) and PD-type FLC (FPDC). PI-type FLCs are most common and practical. However, the performance of PI-type FLCs is known to be quite satisfactory for linear systems. But like conventional PI-controllers, performance of PI-type FLCs for non-linear systems, systems with integrating element, and also for higher-order systems found to be poor due to large overshoot and excessive oscillation [3-5].

Control policy of a conventional FLC is defined by a number of fuzzy *if-then* experts' rules defined on error (e) and change of error (Δe) of the controlled variable. The membership functions (MFs) of the input and output linguistic variables are usually defined on a common normalized domain. While designing an efficient FLC proper selection of its input and output scaling factors (SFs) is a very important task,

S.C. Satapathy et al. (Eds.): Proc. of Int. Conf. on Front. of Intell. Comput., AISC 199, pp. 67–75.
DOI: 10.1007/978-3-642-35314-7_9

which in many cases are done through trial or based on experimental data [6, 7]. A conventional FLC with a limited number linguistic rules and simple MFs, may not fulfill the desired performance specification for practical systems, which are nonlinear and high-order systems. To rise above such limitations many research works have been reported where either the input-output SFs or the definitions of MFs and sometimes the control rules are tuned to improve the close-loop performance [6-13].

The self-tuning fuzzy PI-type controller (STFPIC1) reported in [8] is tuned by dynamically adjusting its output SF in each sampling instant by a gain updating factor (α), which is further augmented by a fixed multiplicative factor (K) chosen empirically. Here, we propose a critical point based self-tuning fuzzy PI-type controller (STFPIC2), which is similar to that of STFPIC1. However, unlike a fixed value of K in STFPIC1, in the proposed STFPIC2 the value of K is determined from the information of the critical point of the concerned process, and is obtained through a heuristic relation of ultimate gain (K_c) and ultimate period (P_c) of the process under control. Thus, the adjustable output SF of the proposed STFPIC2 seems to be more rational, since it incorporates the dynamics of the process under control, which is directly related to its *critical point*. The value of α is determined by fuzzy rules defined on e and Δe, and derived from the knowledge of process control engineering. In this context, it is to be mentioned that such knowledge and information have been embedded while developing improved auto-tuning PI/PID controllers [14-16]. The performance of STFPIC2 is tested by simulation experiments on a wide variety of linear and nonlinear high-order processes with different values of dead time. Results in each case show a significantly improved performance of the proposed STFPIC2 compared to its conventional fuzzy (FPIC), and better than or comparable with that of STFPIC1 [8]. Usefulness of the proposed STFPIC2 is justified through real-time implementation on a laboratory scale practical temperature control system.

2 Design of the Proposed Controller – STFPIC2

Figure 1 shows the simplified block diagram of STFPIC2. The output SF of the controller is modified by a self-tuning mechanism, shown by the dotted boundary Detailed design considerations of STFPIC2 are available in [8]. However, to make this study self-contained, various design aspects of the STFPIC2 are briefly discussed below.

MFs for inputs, (*i.e.*, e_N and Δe_N) and output, (*i.e.*, Δu_N) of the controller (shown in Fig. 2a) are defined on the common normalized domain [-1, 1], whereas the MFs for α (shown in Fig. 2b) is defined on [0, 1]. Except at the two extreme ends, symmetric triangular MFs are used. The relationships between the SFs (G_e , $G_{\Delta e}$ and G_u), and the input and output variables of the STFPIC2 are as follows:

$$e_N = G_e \times e, \tag{1}$$

$$\Delta e_N = G_{\Delta e} \times \Delta e, \tag{2}$$

$$\text{and } \Delta u = (K \alpha G_u) \times \Delta u_N \tag{3}$$

Unlike fuzzy PI controllers (FPIC), which uses only G_u to generate the incremental output (*i.e.*, $\Delta u = G_u \times \Delta u_N$), the output ($\Delta u$) for STFPIC2 is obtained by using the effective SF, *i.e.*, $K\alpha G_u$ as shown in Fig. 1, where K is a process specific constant. The value of α is computed on-line using a model independent fuzzy rule-base defined on e and Δe.

A PI-type FLC in its velocity form can be described by

$$u(k) = u(k\text{-}1) + \Delta u(k)\,. \tag{4}$$

In Eqn. (4) k is the sampling instance and Δu is the incremental change in controller output, which is determined by the rules of the form, R_{PI} : If e is E and Δe is ΔE then Δu is ΔU. The rule-base for computing Δu is shown in Fig. 3a, which is derived following the principle of sliding mode control [17, 18]. The gain updating factor (α) is calculated using fuzzy rules of the form: R_α: If e is E and Δe is ΔE then α is α. The rule-base in Fig. 3b is used for the computation of α, which is designed according to the controller rule-base in Fig. 3a with a view to incorporating an operator's strategy. The logic behind using such a rule-base in Fig. 3b has been elaborated in [8]. Next we explain how the value of K is determined.

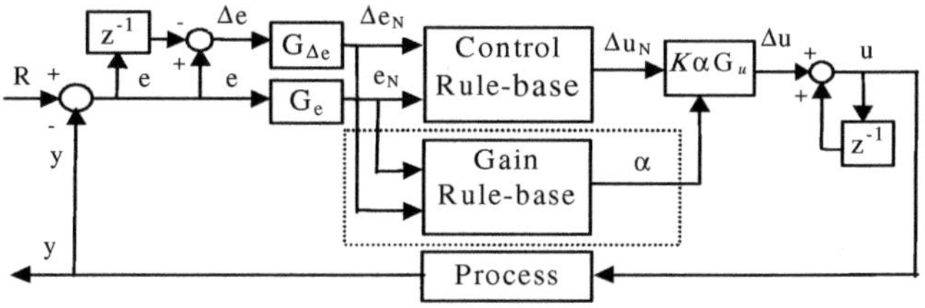

Fig. 1. Block Diagram of the proposed STFPIC2

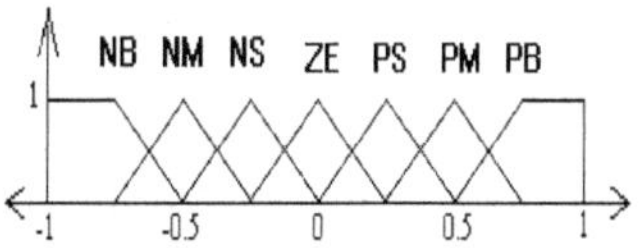

Fig. 2a. MFs of e_N, Δe_N and Δu_N

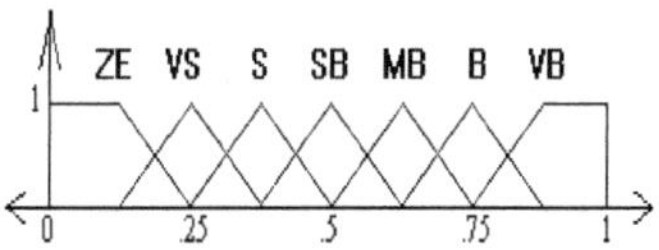

Fig. 2b. MFs of α

Δe/e	NB	NM	NS	ZE	PS	PM	PB
NB	NB	NB	NB	NM	NS	NS	ZE
NM	NB	NM	NM	NM	NS	ZE	PS
NS	NB	NM	NS	NS	ZE	PS	PM
ZE	NB	NM	NS	ZE	PS	PM	PB
PS	NM	NS	ZE	PS	PS	PM	PB
PM	NS	ZE	PS	PM	PM	PM	PB

Fig. 3a. Fuzzy Rules for Computation of Δu

Δe/e	NB	NM	NS	ZE	PS	PM	PB
NB	VB	VB	VB	B	SB	S	ZE
NM	VB	VB	B	B	MB	S	VS
NS	VB	MB	B	VB	VS	S	VS
ZE	S	SB	MB	ZE	MB	SB	S
PS	VS	S	VS	VB	B	MB	VB
PM	VS	S	MB	B	B	VB	VB

Fig. 3b. Fuzzy Rules for Computation of α

2.1 Computation of *K*

In case of STFPIC1 [8], the value of *K* is empirically chosen as 3, irrespective of the type of process to be controlled, though it can directly influence the close-loop performance. But in STFPIC2, the value of *K* depends on the process dynamics. For computation of *K*, relay-feedback test is performed on the given process to estimate the corresponding critical point, *i.e.*, ultimate gain (K_c) and ultimate period (P_c). Åström and Hägglund [19] have developed an attractive and simple experimental relay feedback method to determine K_c and P_c. The conventional relay feedback consists of a relay of height '*d*' in the feedback loop as shown in Fig. 4. The loop starts to oscillate around the set point. The time period of the output (*y*) oscillation is the critical period P_c. If the maximum amplitude of the process response is recorded as '*a*', then the critical gain K_c is approximately determined by the following relation:

$$K_c = \frac{4d}{\pi a} \tag{5}$$

Now, the following heuristic relation is developed to find the value of *K* for a given process :

$$K = xe^{1/x} , \tag{6}$$

Where, $x = \dfrac{K_c P_c}{K_c + P_c}$.

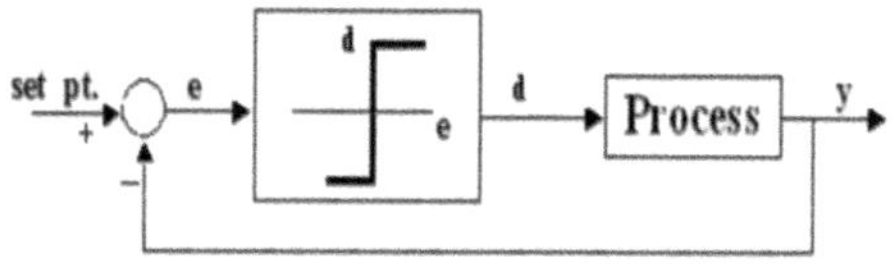

Fig. 4. Relay-feedback experiment

3 Results

The performance of STFPIC2 is compared with STFPIC1 [8] and conventional fuzzy PI controller (FPIC) for various second-order linear and nonlinear processes with different values of dead time. Control performance is evaluated in terms of peak overshoot (%OS), settling time (t_s), integral absolute error (IAE), and integral time absolute error (ITAE). To establish the robustness of the proposed scheme we use the same MFs (Fig. 2) and same rule-bases (Fig. 3) for all the processes. We have used Mamdani type inferencing and Height method of defuzzification [18]. Effectiveness of the proposed STFPIC2 is also tested on a practical temperature control system.

3.1 Simulation Results

Most of the practical processes can be fairly approximated by a second-order dead time (*L*) model. Here, we report the results for the following three process models.

$$\textit{Linear}: \quad \ddot{y} + \dot{y} + 0.2y = u(t-L) \tag{7}$$

$$\textit{Nonlinear}: \quad \ddot{y} + \dot{y} + 0.25y^2 = u(t-L) \tag{8}$$

$$\textit{Marginally stable}: \quad \ddot{y} + \dot{y} = u(t-L) \tag{9}$$

We consider different values of dead time (*i.e.*, L = 0.2, 0.5. 0.6, and 0.8) for the linear process of (7). Responses of this process with L = 0.2, 0.6, and 0.8 due to both set point change and load disturbance under various FLCs (*i.e.*, STFPIC1, STFPIC2, and FPIC) are shown in Fig. 5. Detailed performance analysis in terms of various performance indices are provided in Table 1. From the results (Fig. 5 and Table 1) it clearly reveals that the STFPIC1 with fixed value of K (= 3) fails to provide acceptable performance when the process is subjected to a large change in dead time. But in the same situation, our proposed STFPIC2 maintains satisfactory performance; obviously, this has been possible due to process dependent K, which justifies the effectiveness of our proposed scheme.

Similar to the linear process, we have also considered a wide variation in L (*i.e.*, L = 0.2, 0.6, and 0.8) for the nonlinear process in (8). Response characteristics are depicted in Fig. 6 for different values of L. Table 2 provides the detailed performance comparison. Like previous results, STFPIC2 exhibits better performance compared to STFPIC1.

Responses of the marginally stable process in (9) are shown in Fig. 7, and various performance indices are recorded in Table 3. In this case, we see that the performance of STFPIC2 is comparable with that of STFPIC1, possibly due to small variation in K compared to two previous cases.

3.2 Experimental Results

We have also successfully tested the performance of STFPIC2 on a laboratory scale temperature control system using Process Control Trainer 37-100, FEEDBACK, UK as shown in Fig. 8. Responses under set point change and load disturbance are shown in Fig. 9. Fig. 9(a) shows the nominal responses, whereas Fig. 9(b) shows the responses with additional forward path delay introduced in the control algorithm. Performance analysis in terms integral criteria is provided in Table 4. For this practical process also, we observe comparable results (Fig. 9 and Table 4) of STFPIC1 and STFPIC2, which justifies the usefulness of our proposed scheme.

Table 1. Performance comparison for the second-order linear process in (7)

L	K	FLC	%OS	t_s(s)	IAE	ITAE
0.2	3.12	FPIC	36.9	29.8	10.7	284.7
		STFPIC1	20.2	27.1	8.8	228.2
		STFPIC2	20.4	30.9	8.6	224.6
0.5	2.82	FPIC	47.6	48.2	17.73	860.8
		STFPIC1	29.0	53.4	14.58	693.4
		STFPIC2	25.8	39.6	11.52	481.8
0.6	2.80	FPIC	51.6	60.5	20.77	1070
		STFPIC1	32.4	69.5	20.28	1144
		STFPIC2	28.8	46.3	13.71	629.4
0.8	2.72	FPIC	60.0	111.2	33.14	2490
		STFPIC1	-	-	-	-
		STFPIC2	35.4	75.8	22.94	1622

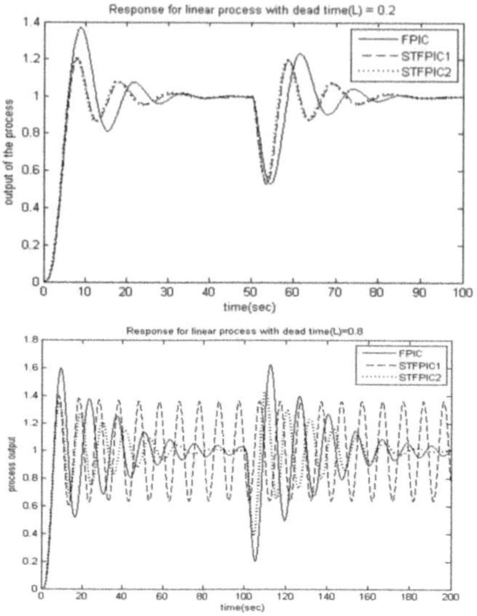

Fig. 5. Responses of the second-order linear process in (7) with different values of *L*.

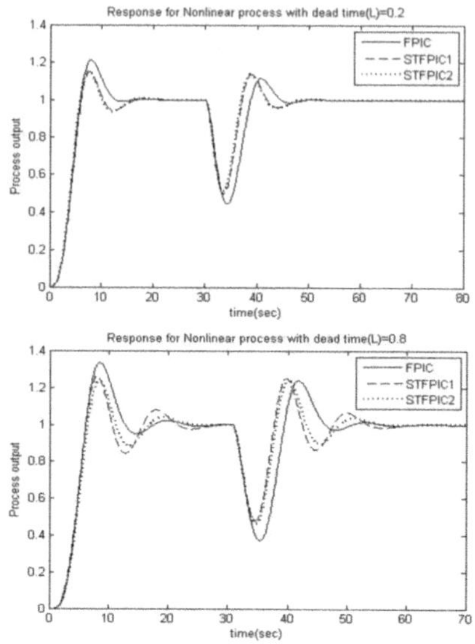

Fig. 6. Responses of the second-order nonlinear process in (8) with different values of *L*.

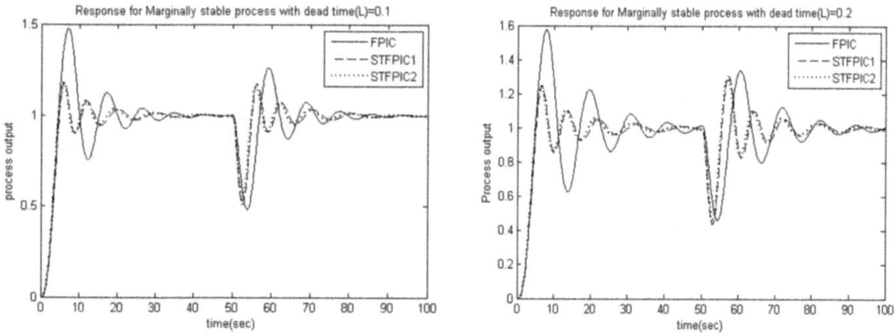

Fig. 7. Responses of the marginally stable process in (9) with different values of (*L*).

Table 2. Performance comparison for the second-order nonlinear process in (8)

L	*K*	FLC	%OS	t_s(s)	IAE	ITAE
0.2	3.18	FPIC	21.29	12.2	7.83	128
		STFPIC1	15.04	15.2	7.04	104
		STFPIC2	15.39	14.8	6.90	101
0.6	2.78	FPIC	29.17	15.8	8.52	136
		STFPIC1	21.85	20.3	8.22	131
		STFPIC2	20.52	20.3	8.21	130
0.8	2.71	FPIC	33.43	20.8	10.1	185
		STFPIC1	26.25	25.1	9.44	167
		STFPIC2	23.61	21.7	9.24	158

Table 3. Performance comparison for the marginally stable process in (9)

L	*K*	FLC	%OS	t_s(s)	IAE	ITAE
0.1	3.17	FPIC	47.7	31.7	10.0	268.2
		STFPIC1	17.8	25.9	6.57	161.5
		STFPIC2	17.6	24.9	6.34	152.2
0.2	3.15	FPIC	57.5	46.6	14.4	431.8
		STFPIC1	24.6	34.2	8.96	264.3
		STFPIC2	24.3	32.8	8.60	252.6
0.3	3.05	FPIC	66.9	72.1	21.7	1004
		STFPIC1	30.7	50.4	12.3	601.6
		STFPIC2	30.5	49.9	12.1	594.9

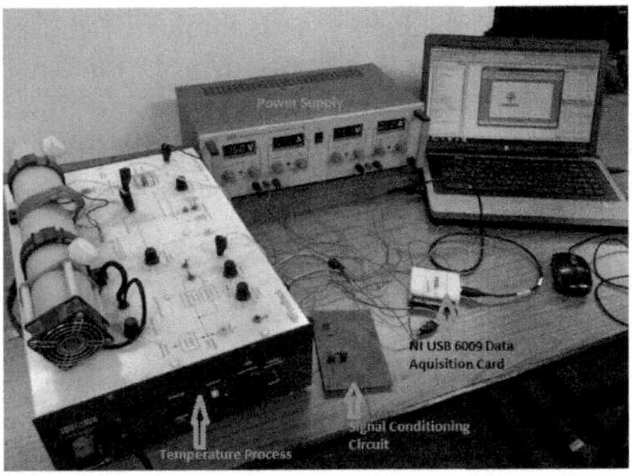

Fig. 8. Experimental setup of temperature control system

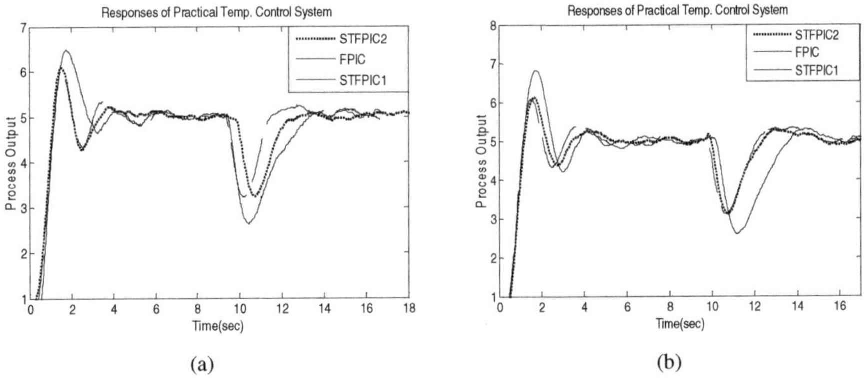

Fig. 9. Responses of the temperature control system; (a) nominal, (b) with extra forward path delay.

Table 4. Performance comparison for the practical temperature control system

	Nominal		With extra delay	
FLCs	**IAE**	**ITAE**	**IAE**	**ITAE**
FPIC	14.0	103.16	14.43	91.97
STFPIC1	9.5	53.3	10.8	68.5
STFPIC2	9.4	50.9	10.7	64.6

4 Conclusion

We proposed a self-tuning fuzzy PI controller whose output scaling factor gets real-time adjustment by a fuzzy gain updating parameter, which has been further augmented by a *critical point* based multiplicative factor. Relay-feedback approach has been adopted to find the critical point. The output SF of the proposed controller thus became more rational, as it embedded the dynamics of the process under control. Performance of the proposed controller has been tested under both set-point change and load disturbance for second-order processes as well as practical temperature control system with a large change in dead time. Detailed performance comparisons with other fuzzy logic controllers justified the effectiveness of the proposed controller.

References

[1] Sugeno, M.: Industrial Applications of Fuzzy Control. Elsevier Sc, Amsterdam (1985)
[2] Harris, C.J., Moore, C.G., Brown, M.: Intelligent Control - Aspects of Fuzzy Logic and Neural Nets. World Scientific, Singapore (1993)
[3] Ying, H., Siler, W., Buckley, J.J.: Fuzzy Control Theory: A Nonlinear Case. Automatica 26, 513–520 (1990)
[4] Boverie, S.: Fuzzy Logic Control for High-order Systems. In: Proc. 2nd IEEE Int. Conf. on Fuzzy Systems, pp. 117–121 (1993)
[5] Lee, J.: On Methods for Improving Performance of PI-Type Fuzzy Logic Controllers. IEEE Trans. on Fuzzy Syst. 1, 298–301 (1993)
[6] Nomura, H., Hayashi, I., Wakami, N.: A Self-Tuning Method of Fuzzy Control by Decent Method. In: Proc. IFSA 1991, pp. 155–158 (1991)
[7] Chung, H.Y., Chen, B.C., Lin, J.J.: A PI-type Fuzzy Controller with Self-tuning Scaling Factors. Fuzzy Sets and Syst. 93, 23–28 (1998)
[8] Mudi, R.K., Pal, N.R.: A Robust Self-Tuning Scheme for PI and PD Type Fuzzy Controllers. IEEE Trans. on Fuzzy Systems 7, 2–16 (1999)
[9] Palm, R.: Scaling of Fuzzy Controller Using the Cross-Correlation. IEEE Trans. on Fuzzy Syst. 3, 116–123 (1995)
[10] Mudi, R.K., Pal, N.R.: A Self-Tuning Fuzzy PI Controller. Fuzzy Sets and Systems 115, 327–338 (2000)
[11] Li, H.X., Gatland, H.B.: Conventional Fuzzy Control and Its Enhancement. IEEE Trans. on Syst., Man, Cybern. 26, 791–797 (1996)
[12] Mudi, R.K., Pal, N.R.: A Self-Tuning Fuzzy PD Controller. IETE Journal of Research 44, 177–189 (1998)

[13] Pal, A.K., Mudi, R.K.: Self-Tuning Fuzzy PI controller and its application to HVAC system. Int. Journal of Computational Cognition 6, 25–30 (2008)
[14] Mudi, R.K., Dey, C., Lee, T.T.: An improved auto-tuning scheme for PI controllers. ISA Transactions 47, 45–52 (2008)
[15] Dey, C., Mudi, R.K.: An improved auto-tuning scheme for PID controllers. ISA Tranactions 48, 396–409 (2009)
[16] Mudi, R.K., Dey, C.: Performance improvement of PI controllers through dynamic set-point weighting. ISA Transactions 50, 220–230 (2011)
[17] Palm, R.: Sliding Mode Fuzzy Control. In: Proc. 1st IEEE Int. Conf. on Fuzzy Systems, pp. 519–526 (1992)
[18] Dirankov, D., Hellendoorn, H., Reinfrank, M.: An Introduction to Fuzzy Control. Springer, NY (1993)
[19] Åström, K.J., Hägglund, T.: Automatic Tuning of Simple Regulators with Specifications on Phase and Amplitude Margins. Automatica 20, 645–651 (1984)

Speaker Invariant and Noise Robust Speech Recognition Using Enhanced Auditory and VTL Based Features

S.D. Umarani[1], R.S.D. Wahidabanu[1], and P. Raviram[2]

[1] Government College of Engineering, Salem - 636011, India
[2] Department of CSE
Mahendra Engineering College, Tiruchengode- 637503, India
{umaraviram,drwahidabanu,drpraviram}@gmail.com

Abstract. This paper focuses on design and implementation of a noise-resilient and speaker independent speech recognition system for isolated word recognition. In this work auditory transform (AT) based features called as Cochlear Filter Cepstral Coefficients (CFCCs) has been used for feature extraction and its robustness against noise and variation in vocal track length (VTL) performance has been enhanced by the application of wavelet based denoising algorithm and invariant-integration method respectively. The resultant features are called as enhanced CFCC Invariant-Integration Features (ECFCCIIFs). To accomplish the objective of this paper, feature-finding neural network (FFNN) is used as classifier for the recognition of isolated words. Results are compared with the results obtained by the standard CFCC features and it is observed that, at both matching and mismatching conditions the ECFCCIIFs features remains high recognition rate under low Signal-to-noise ratios (SNRs) and their performance are more effective under high SNRs too.

Keywords: Denoising, Invariant integration, CFCC, FFNN, SNR, Auditory, VTL.

1 Introduction

In recent years performance of the Automatic Speech Recognition (ASR) systems has extremely improved. Although many technological improvements have been done in ASR, the performance degrades when the training and testing environments are differing. These environments are speaker variation, channel distortion, reverberation, noise etc [1]. As human hearing system is robust to these environmental variations, a human auditory-system based feature extraction algorithm (AT) has been developed by Qi Li [2]. In this paper, to improve the noise robustness of AT features a wavelet based thresholding algorithm called adaptive wavelet thresholding [3] has been applied. The effects of inter-speaker variability originating from different vocal tract lengths (VTLs), reflects as translations in the subband- index space of time frequency representation of a speech signal. A method "invariant integration", integrates regular nonlinear functions of the features over the transformation group for which invariance should be achieved [4], is applied in this paper to make the CFCC based speech recognition as robust to VTL changes.

S.C. Satapathy et al. (Eds.): Proc. of Int. Conf. on Front. of Intell. Comput., AISC 199, pp. 77–82.
DOI: 10.1007/978-3-642-35314-7_10

2 CFCC with Adaptive Wavelet Thresholding and Invariant Integration

The proposed method of enhanced auditory transform based feature extraction is shown in fig. 1. The computation process is discussed bellow.

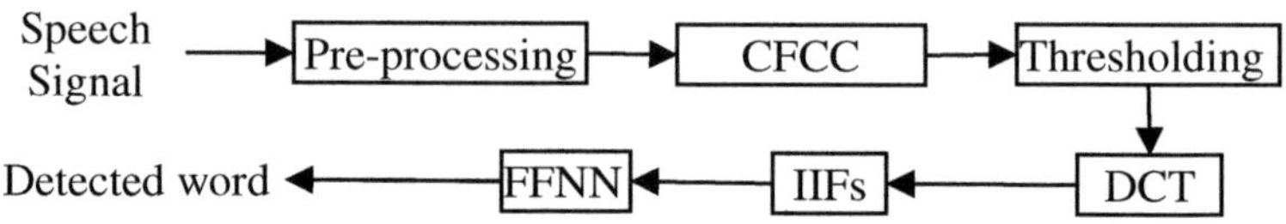

Fig. 1. Proposed method of computation of ECFCCIIFs

Step 1: Preprocessing. As preprocessing increases the recognition rates considerably, initially preprocessing is done, which is a process of sampling followed by Exerting pre-emphasis, framing and window adding processing to the original speech signal.

Step 2: CFCC feature extraction. The CFCC feature extraction system consists of the modules that conceptually replicate the human hearing system namely auditory transform (AT), Hair cell windows and Non linearity. The auditory transform is the forward transform and has been implemented as a filter bank. Transform of preprocessed speech signal $\mathbf{f(n)}$ with respect to a cochlear filter $\boldsymbol{\psi}(\mathbf{n})$, gives the basilar membrane impulse response in the cochlea and is defined as,

$$T(a,b) = \sum_{n=0}^{N} f(n) \frac{1}{\sqrt{|a|}} \psi\left(\frac{n-b}{a}\right) . \tag{1}$$

The cochlear filter, as the most important part of the transform, is defined as,

$$\psi_{a,b}(n) = \frac{1}{\sqrt{|a|}} \psi \frac{1}{\sqrt{|a|}} \psi \left(\frac{n-b}{a}\right)^{\alpha} exp\left[-2\pi f_L \beta \left(\frac{n-b}{a}\right)\right] cos\left[2\pi f_L \left(\frac{n-b}{a}\right) + \theta\right] u(n-b) . \tag{2}$$

Based on the values of α and β, shape and width of the cochlear filter varies in the frequency domain. In human auditory system hair cells amplifies the vibration of the basilar membrane in a non-linear manner and is implemented as,

$$h(a,b) = T(a,b)^2; \forall T(a,b) . \tag{3}$$

Now the non-linearly amplified signal has been represented as nerve spike count density and is given by,

$$S(i,j) = \frac{1}{d} \sum_{b=l}^{l+d-1} h(a,b) , l = 1, L, 2L, \ldots ; \forall i,j . \tag{4}$$

Where, d is the window length, T_i is the period of the central frequency of the ith band, and L is the window shift duration. Furthermore, the scales of loudness function is applied as,

$$f(i,j) = S(i,j)^{1/3} . \quad (5)$$

where, i is band number and j is the index of frequency band. This operation implements cubic root nonlinearity from the physical energy to the perceived loudness. Thus CFCC features are obtained.

Step 3: Denoising by thresholding. In order to enhance the ability to resist the noises of different environments, ZHANG Jie et al. [3] has introduced an adaptive wavelet thresholding approach. When the background noise is white noise, including color noise with flatting spectrum amplitude, the thresholding function is,

$$\lambda = \delta \sqrt{2 \log_{10}^{N}} \; g(i) . \quad (6)$$

where, g(i) is the inverse proportion function of variable i, is given by,

$$g(i) = \frac{1}{\left[2^{\frac{i-1}{2}} \ln(i+1) \right]} . \quad (7)$$

When the background noise is colored noise, the threshold function is,

$$\lambda' = \delta \sqrt{2 \log_{10}^{N}} \; \varphi(\gamma) . \quad (8)$$

Where, γ denote the flatness and φ(γ) is the correction function as,

$$\varphi(\gamma) = \frac{2}{\left[\log_{10}^{0.05\gamma} \right]} . \quad (9)$$

These threshold functions adjust adaptively so that it can adapt all kinds of noisy environment and the noise present in the transformed signal get removed.

Step 4: Discrete Cosine Transform. In final step, the inverse DFT is applied as,

$$y(m,n) = \frac{2}{MN} C(m)C(n) \sum_{i=0}^{M-1} \sum_{j=0}^{N-1} f(i,j) \cos\frac{(2i+1)m\pi}{2M} \cos\frac{(2j+1)n\pi}{2N} . \quad (10)$$

The result is called the Enhanced Cochlear Frequency Cepstral Coefficients (ECFCC). where, C(m)C(n)=1/√2, for m, n=0 and C(m)C(n)=0, otherwise. i is band number and j is the index of frequency band and M is the number of frequency bands and N is the number of indexes of each frequency band.

Step 5: Invariant Integration Feature (IIF). Let $\tilde{\boldsymbol{y}}_j(\boldsymbol{n},\boldsymbol{\tau})$ denote the magnitudes of the obtained subband coefficients of ECFCC value at the final frame rate, at time instance $\boldsymbol{n}$, with $\boldsymbol{j}$ being the subband index and $\boldsymbol{\tau}$ being the time interval. Furthermore, let $\boldsymbol{c} = (\boldsymbol{c_1}, \boldsymbol{c_2}, \ldots, \boldsymbol{c_j})$ denote the center frequencies of the filters. Now, for a given subband index $\boldsymbol{i} \in \mathbb{N}$ and a cycle number $\boldsymbol{p} \in \mathbb{R}^+$,

$$\tau_i(p) := \frac{p}{c_i} . \tag{11}$$

defines the time interval for each subband [5], where, $\boldsymbol{\alpha_T}$ is the ratio between the vocal track length of speaker A and B. Now, let $\boldsymbol{p} = (\boldsymbol{p_1}, \boldsymbol{p_2}, \ldots, \boldsymbol{p_M})$ contain cycle numbers. A monomial $\tilde{\boldsymbol{m}}$ can be defined as,

$$\tilde{m}(n; \omega, \boldsymbol{j}, \boldsymbol{l}, \boldsymbol{m}, \boldsymbol{p}) := \left[\prod_{i=1}^{M} \tilde{y}_{j_i+\omega}^{l_i}(n, m_i, \tau_{j_i+\omega}(p_i))\right]^{1/\gamma(l)} . \tag{12}$$

Now, an Enhanced CFCC Invariant Integration Feature (ECFCCIIF) $A_{\tilde{m}}(n)$ is then computed as,

$$A_{\tilde{m}}(n) := \frac{1}{2W+1} \sum_{w=-W}^{W} \tilde{m}(n; w, \mathbf{j}, \mathbf{l}, \mathbf{m}, \mathbf{p}) . \tag{13}$$

yields features that are robust against noise and variation in VTLs.

3 Training/Recognition of Isolated Word Using FFNN

The FFNN [6] has high recognition rate than the classical HMM and DTW recognizers and yields similar recognition rates. Its architecture is depicted in Fig. 2. The enhanced features, ECFCCIIFs are the inputs to FFNN. From these, the most important features are selected by feature detector layer using the substitution rule. The resulting activity vectors are classified in a linear, optimal (least mean square) manner. Although the classification is performed by a linear neural network, the whole classification process is highly nonlinear due to the second order characteristics of the extracted features.

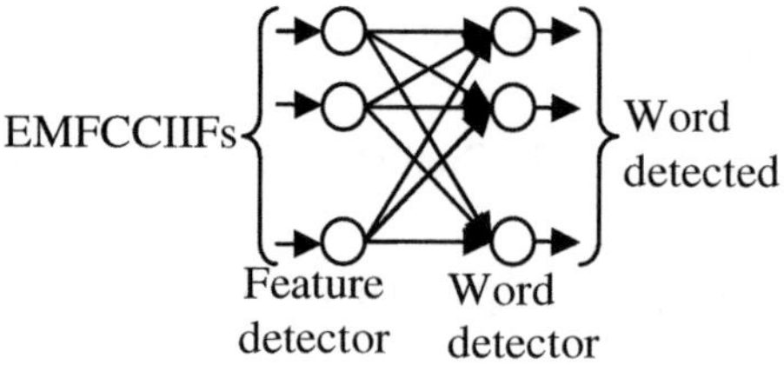

Fig. 2. System structure of the Feature Finding Neural

4 Experimental Evaluation

The ECFCCIIFs/FFNN system was evaluated with the training data set recorded under a clean condition and the testing data sets mixed with white and color noise at various noise levels. In this work the Speech Separation Challenge database [7] has been used for training and testing. During experiment, the speech sampling frequency is 11.025 kHz, frame length is 256 points and the frame shift is 128. The filterbank used for the computation of the CFCCs had 32 filters. In the Invariant Integration methods, 30 IIFs of order one has been used. In FFNN, the substitution rule starts with 32 randomly taken features and each iteration low-relevance features are replaced with features with higher relevance.

5 Results and Discussions

The ECFCCIIFs has been implemented. Performance comparison of the CFCC and ECFCCIIFs algorithm for six different noise conditions and for matching and mismatching VTLs is tabulated in Table 1. As expected, the cochlear filter cepstral coefficient with adaptive wavelet thresholding and invariant-integration has been yielded higher accuracies in all scenarios than the standard cochlear filter cepstral coefficient. This is plotted in Fig. 3.

Table 1. The comparison of performances (%)

SNR	Match (MF-MF)		Mismatch(M-F/F-M)	
	CFCC	ECFCCIIFs	CFCC	ECFCCIIFs
-10dB	20.31	37.76	12.82	35.71
- 5dB	32.47	61.2	23.28	44.9
0dB	58.75	69.01	52.16	73.78
5dB	82.54	91.21	73.78	86.75
10dB	94.26	97.11	80.73	90.78
Clean	98.06	99.01	85.57	94.19

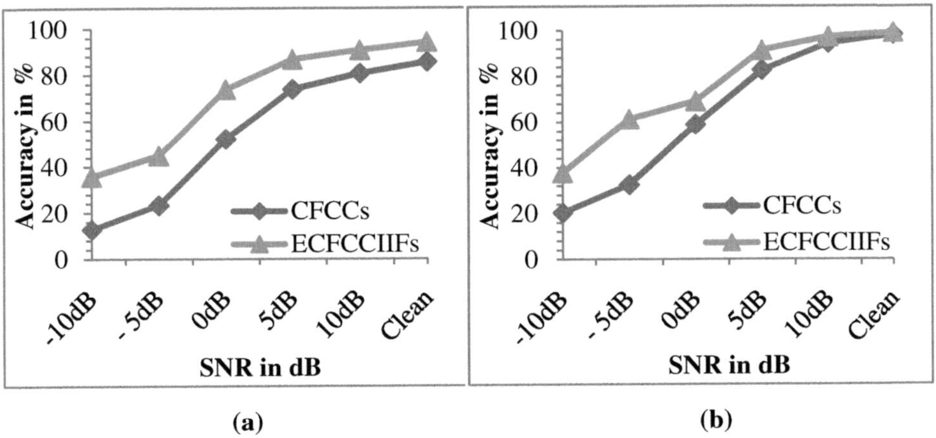

Fig. 3. Comparison of Accuracy for (*a*) Mismatching Condition and (*b*) Matching Condition

6 Conclusions

A noise-resilient and speaker independent speech recognition system for isolated word recognition has been designed and implemented. Noise robust performance of CFCC is enhanced by the application of adaptive wavelet thresholding and also the enhanced features are made as robust to variation in VTL by the invariant-integration. Classifier called feature-finding neural network (FFNN) is used for the recognition of isolated words. Results are compared with the results obtained by the standard CFCC features. Through experiments it is observed that under mismatched conditions, the combined ECFCC and IIFs features remains high recognition rate under low SNRs and their performance are more effective under high SNRs too.

References

1. Acero, A., Stern, R.M.: Environmental robustness in automatic speech recognition. In: International Conference on Acoustics, Speech, and Signal Processing (ICASSP 1990), vol. 2, pp. 849–852. IEEE Press, Albuquerque (1990)
2. Li, Q.: An auditory-based transform for audio signal processing. In: Proceedings of IEEE Workshop on Applications of Signal Processing to Audio and Acoustics, New Paltz, New York (2009)
3. Zhang, J., Li, G.-L., Zheng, Y.-Z., Liu, X.-Y.: A Novel Noise-robust Speech Recognition System Based on Adaptively Enhanced Bark Wavelet MFCC. In: Sixth International Conference on Fuzzy Systems and Knowledge Discovery (FSKD 2009), Tianjin, pp. 443–447 (2009)
4. Muller, F., Mertins, A.: Invariant-integration method for robust feature extraction in speaker-independent speech recognition. In: Int. Conf. Spoken Language Processing (Interspeech 2009-ICSLP), Brighton, pp. 2975–2978 (2009)
5. Muller, F., Mertins, A.: On Using the Auditory Image Model and Invariant-Integration for Noise Robust Automatic Speech Recognition. In: Proc. Int. Conf. Audio, Speech, and Signal Processing, Kyoto, Japan, pp. 4905–4908 (2012)
6. Gramss, T., Strube, H.W.: Recognition of isolated words based on psychoacoustics and neurobiology. Speech Communication 9, 35–40 (1990)
7. Cooke, M., Lee, T.-W.: Speech separation challenge, `http://www.interspeech2006.org`

A Noble Fuzzy Self-Tuning Scheme for Conventional PI Controller

Rajani K. Mudi[1] and Ritu Rani De Maity[2]

[1] Dept of I.E.E., Jadavpur University, Salt-lake Campus, Kolkata, India
rkmudi@yahoo.com
[2] Dept of E.I.E, Dr. B.C. Roy Engg College, Durgapur, India
ritu_maity_8@yahoo.co.in

Abstract. In this study, we propose an improved auto-tuning PI controller. The widely practiced Ziegler-Nichols tuned PI controller (ZNPIC) exhibits poor performance for nonlinear and high-order systems. In order to have an overall improved transient response, the proportional constant of the proposed PI controller is parameterized by a real-time nonlinear updating factor α depending on the process trend. The value of α is determined through a fuzzy inference engine with 49 if-then rules defined on the process error and change of error. Performance of the proposed fuzzy self-tuning ZNPIC, termed as FST-ZNPIC, is tested for second-order linear and nonlinear dead time processes including a marginally stable system under both set point change and load disturbance. Simulation results reveal our FST-ZNPIC provides significantly improved transient responses compared to other conventional and fuzzy self-tuning controllers. Robustness of FST-ZNPIC is observed with a considerable change in process dead time.

Keywords: Ziegler-Nichols tuning, Auto-tuning, Fuzzy self-tuning.

1 Introduction

PI and PID controllers are normally used in process industries due to their simple design and tuning methods [1, 2]. In process industries, PI controllers are more preferable than PID controllers due to presence of measurement noise. However, PI controllers provide poor performance for high-order and nonlinear processes. To overcome such limitations of PI controllers a lot of research works have been done towards developing effective tuning rules using various adaptive methods [3-6]. Performances of conventional PI/PID controllers are improved incorporating different adaptive schemes for auto-tuning and self-tuning controllers either by simple heuristic rules [7-11] or using fuzzy inference systems [12, 14].

Here, we propose a fuzzy self-tuning scheme for continuous modification of the proportional constant k_p of a ZNPIC [15] with a view to achieving improved close-loop performance. The proportional gain k_p is adjusted online by a fuzzy gain updating factor α. The value of α is determined using fuzzy rules defined on the error

S.C. Satapathy et al. (Eds.): Proc. of Int. Conf. on Front. of Intell. Comput., AISC 199, pp. 83–91.
DOI: 10.1007/978-3-642-35314-7_11

(e) and change of error (Δe) of the controlled variable. Here, we use the same fuzzy rule-base suggested for online gain modification of self-tuning fuzzy PI controller (STFPIC) developed in [16]. Performance of the proposed FST-ZNPIC, is observed for different second-order dead time processes due to both set point change and load disturbance. Simulation results in terms of various performance indices show a remarkably improved performance of FST-ZNPIC compared to other conventional and fuzzy self-tuning controllers. Robustness of the proposed controller is tested by applying considerable perturbations in the process dead time.

2 The Proposed Fuzzy Self-tuning Controller – FST-ZNPIC

Block diagram of the proposed FST-ZNPIC is shown in Fig. 1. The proportional constant k_p of a ZNPIC, in its velocity form [3], is continuously modified by a fuzzy updating factor α, which is determined using a fuzzy rule-base defined on e and Δe [16]. Next we present the concept of tuning strategy and details of the proposed FST-ZNPIC.

2.1 Design of FST-ZNPIC

The discrete form of a PI controller can be expressed by

$$u(k) = k_p[e(k) + \frac{\Delta t}{T_i}\sum_{i=0}^{k} e(i)] \tag{1}$$

where $e(k) = [r - y(k)]$, r is the set point and $y(k)$ is the process output at k^{th} instant, Δt is the sampling interval, k_p and T_i are the proportional constant and integral time, respectively. In case of a ZNPIC, the k_p and T_i are obtained according to the ZN ultimate cycle tuning rules [15]:

$$k_p=0.45k_u\,, \tag{2}$$

$$\text{and } T_i=0.833t_u \tag{3}$$

where k_u and t_u are the ultimate gain and ultimate period respectively. Here, we use the velocity form of ZNPIC as revealed by Fig. 1. Therefore, the controller output is obtained with the following rule:

$$u(k) = u(k\text{-}1) + \Delta u(k)\,, \tag{4}$$

$$\text{where, } \Delta u(k) = k_p\,[\Delta e(k) + (\Delta t/T_i)\,e(k)] \tag{5}$$

is the incremental change in controller output. The proposed FST-ZNPIC is a gain adaptive controller, which gets online adjustment of its proportional constant k_p through a fuzzy self-tuning scheme.

2.2 Fuzzy Self-tuning Scheme

In FST-ZNPIC the ZN-tuned proportional constant k_p as obtained by Eqn. (2) is proposed to modify through α by the following empirical relation:

$$k_{pfst} = k_p(1 - K\alpha). \quad (6)$$

Here, k_{pfst} is the modified proportional constant for FST-ZNPIC. In (6) K is a constant, which will make required variation in k_{pfst} to achieve the desired close-loop performance. Observe that, unlike ZNPIC, the value of k_{pfst} is not fixed while the controller is in operation. k_{pfst} is being modified at each sampling time by α depending on the trend of the controlled process. Using a suitable value of K the desired variation in k_{pfst} is to be achieved. The functional relation of α is as follows:

$$\alpha = f(e(k), \Delta e(k)). \quad (7)$$

Where $e(k)$ is the error and $\Delta e(k)$ is the change of error at the kth instant. The following subsection includes the description of the fuzzy self-tuning scheme for k_p.

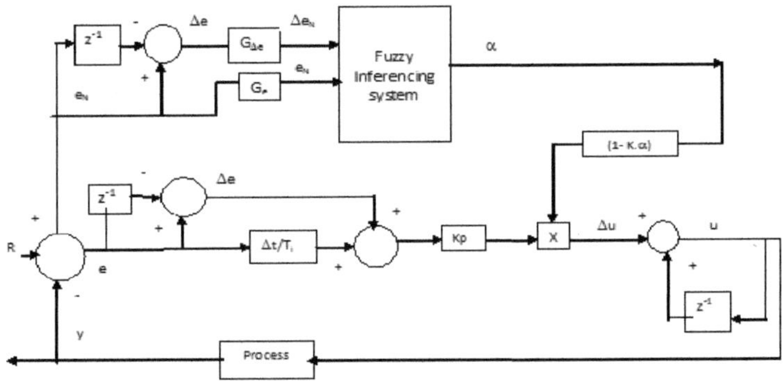

Fig. 1. Block diagram of the proposed FST-ZNPIC

2.3 Membership Function

The input membership functions (MFs) for e and Δe are defined in the common interval [-1, 1], whereas the MFs for α is defined in [0,1] as shown in Fig. 2. We select triangular MFs with equal base width and 50% overlap except at the two extreme ends.

2.4 Scaling Factor

The actual input variables e and Δe are mapped into the common interval [1, -1] by input scaling factors (SFs) G_e and $G_{\Delta e}$. Selection of the appropriate values of SFs are done on the basis of knowledge about the process under control and sometimes by trial and error. Following are the relationships of SFs and input variables:

$$e_N = G_e . e \tag{8}$$

$$\Delta e_N = G_{\Delta e} . \Delta e \tag{9}$$

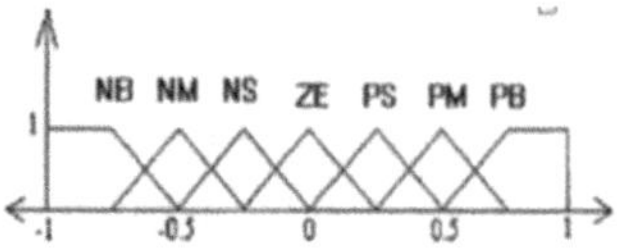

(a)

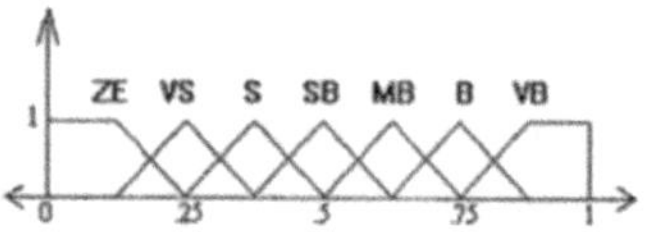

(b)

Fig. 2. Membership functions of (a) e and Δe, and (b) gain updating factor (α)

Δe/e	**NB**	**NM**	**NS**	**ZE**	**PS**	**PM**	**PB**
NB	VB	VB	VB	B	SB	S	ZE
NM	VB	VB	B	B	MB	S	VS
NS	VB	MB	B	VB	VS	S	VS
ZE	S	SB	MB	ZE	MB	SB	S
PS	VS	S	VS	VB	B	MB	VB
PM	VS	S	MB	B	B	VB	VB
PB	ZE	S	SB	B	VB	VB	VB

Fig. 3. Fuzzy Rules for Computation of α

2.5 Rule Base

Online computation of α is done on the basis of a rule-base defined in terms of e and Δe as shown in Fig. 3. The rule-base in Fig. 3 for the gain updating factor α is developed from the knowledge of process dynamics keeping in mind an improved overall close-loop performance. The rules are in the form: R_α: *if* e is E and Δe is ΔE then α is $\boldsymbol{\alpha}$

2.6 Tuning Strategy

The main objective of the auto-tuning scheme is to bring the system under control in steady state as quickly as possible with minimum oscillation. The value of k_{pfst} is updated according to (6) with the rule-base of Fig. 3. The rule-base in Fig. 3 is so designed that when the process moves towards its set point (*i.e.,* α is –ve) k_{pfst} is increased from its nominal value k_p , on the other hand, when the process moves away from the set point (*i.e.,* α is +ve) k_{pfst} is decreased. Such nonlinear gain variation is expected to provide a quick recovery of the process during set point change as well as load disturbance.

3 Simulation Results

The performance of the proposed controller is tested on a number of linear and nonlinear second-order processes with dead time (L), including a marginally stable system with set point change and load disturbance. A comparative performance analysis is made with the proposed FST-ZNPIC, ZNPIC [15] and STFPIC [16]. The performance parameters include the overshoot ($\%OS$), settling time (t_s), rise time (t_r), and integral absolute error (IAE). In case of of FST-ZNPIC, the value selected for the constant K is 0.5 for all the examples. Next, we present the results for different processes.

3.1 Second-Order Linear System

The transfer function of the second-order system is

$$G_P(s) = e^{-LS}/(s+1)^2 \tag{10}$$

For this process, we consider three different values of dead time (L) 0.1, 0.2 and 0.3. Fig. 4 shows the transient responses under set point change and load disturbance for FST-ZNPIC, ZNPIC and STFPIC. The proposed controller is tuned for L= 0.2. Results recorded in Table 1 show that even the controller is tuned at L=0.2 it shows outstanding performance for 50% change in dead time, L=0.1 and L=0.3. This proves the robustness of the proposed controller. Table 1 exhibits that FST-ZNPIC significantly reduces the percentage overshoot compared to ZNPIC and ZNPID. The performance of FST-ZNPIC (with 49 rules) is either better or comparable to STFPIC, although STFPIC is designed with 98 rules, *i.e.*, double of FST-ZNPIC.

3.2 Second-Order Nonlinear Process

The proposed controller is used in the second order nonlinear process whose transfer function is

$$\ddot{y} + \dot{y} + 0.2y^2 = u(t - L) \tag{11}$$

Fig. 5 and Table 2 provide simulation results of the process with dead time under set point change and load disturbance. Keeping the controller setting at L=0.2, the results are taken for dead time L=0.1 and L=0.3. Table 2 shows that in comparison to ZNPIC, ZNPID, and STFPIC, $\%OS$, ts and even IAE are significantly reduced. Hence, for this process in (11), FST-ZNPIC is capable of providing an outfitting performance compared to other controllers.

3.3 Second-Order Marginally Stable Process

To establish the effectiveness of FST-ZNPIC, we consider the following second-order marginally stable process:

$$G_P(s) = e^{-LS}/[s(s+1)] \tag{12}$$

Table 3 shows ZNPIC produces large overshoot and oscillation with *L*=0.3. Whereas, FST-ZNPIC reduces overshoot as well as the oscillation. Keeping the controller setting at *L*=0.3 when applied to the process at *L*=0.2 and *L*=0.4, it results satisfactory performance. Fig 6 shows the transient responses of (12) under set point change and load disturbance. Table 3 shows that FST-ZNPIC provides almost similar performance like STFPIC.

Table 1. Performance comparison of the second-order linear process in (10)

L	*Controller*	*%OS*	t_s	t_r	*IAE*
0.1	ZNPID	37.42	3.7	0.5	1.43
	ZNPIC	40.25	6.8	0.7	2.13
	STFPIC	2.49	7.1	3.2	4.39
	FST-ZNPIC	0.0	7.4	1.2	2.46
0.2	ZNPID	54.44	4.8	0.5	1.64
	ZNPIC	57.4	12.9	0.6	3.2
	STFPIC	3.26	7.1	3.0	4.51
	FST-ZNPIC	7.01	8.0	1.0	2.6
0.3	ZNPID	72.6	8.4	0.5	1.94
	ZNPIC	71.43	31.5	0.6	6.75
	STFPIC	4.2	6.9	2.9	4.73
	FST-ZNPIC	14.05	13.2	1.0	3.29

Table 2. Performance comparison of the second-order nonlinear process in (11)

L	*Controller*	*%OS*	t_s	t_r	*IAE*
0.1	ZNPID	50.02	9.5	1.0	3.04
	ZNPIC	46.85	14.4	1.2	3.98
	STFPIC	19.1	15.1	3.2	7.99
	FST-ZNPIC	13.16	10.3	1.9	4.10
0.2	ZNPID	58.6	11.6	0.9	3.49
	ZNPIC	55.3	17.7	1.1	4.9
	STFPIC	21.22	15.0	3.1	8.43
	FST-ZNPIC	18.2	13.9	1.8	4.53
0.3	ZNPID	65.33	14.0	0.9	4.04
	ZNPIC	61.6	23.6	1.0	6.08
	STFPIC	23.41	20.0	3.1	8.96
	FST-ZNPIC	21.87	16.7	1.8	5.07

Table 3. Performance comparison of the marginally stable process in (12)

L	*Controller*	*%OS*	t_s	t_r	*IAE*
0.2	ZNPID	68.65	19.0	1.2	5.4
	ZNPIC	68.99	21.4	1.3	6.93
	STFPIC	22.25	29.5	2.8	7.18
	FST-ZNPIC	34.28	16.7	2.1	6.96
0.3	ZNPID	74.42	19.6	1.1	6.13
	ZNPIC	75.61	28.7	1.3	8.44
	STFPIC	27.78	40.2	3.0	9.51
	FST-ZNPIC	38.13	20.9	2.1	7.82
0.4	ZNPIC	89.55	37.6	1.3	11.75
	STFPIC	33.77	54.1	3.0	12.2
	FST-ZNPIC	47.2	29.3	2.0	9.51

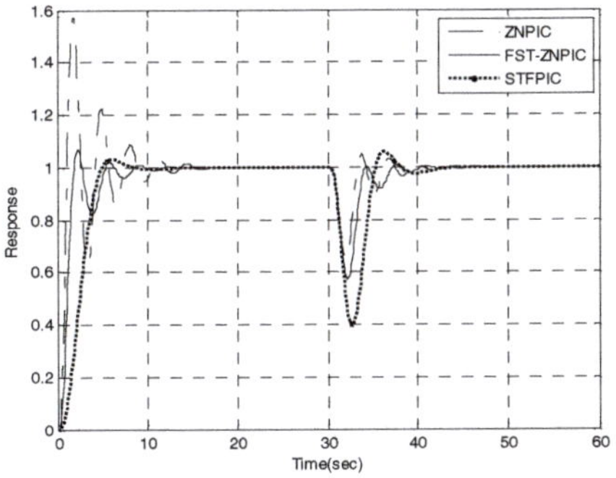

Fig. 4a. Responses of (10) for *L*=0.2

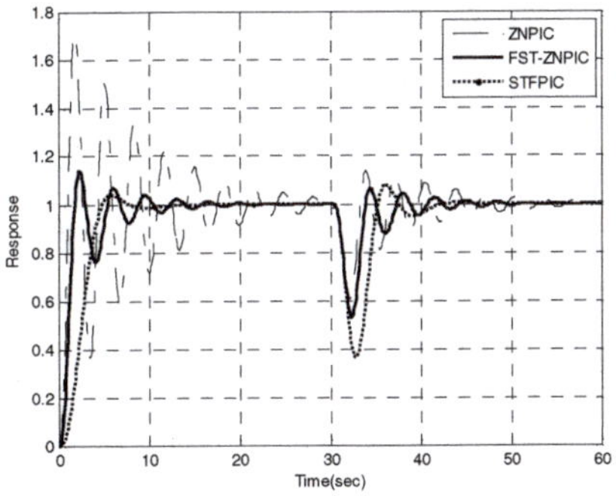

Fig. 4b. Responses of (10) for *L*=0.3

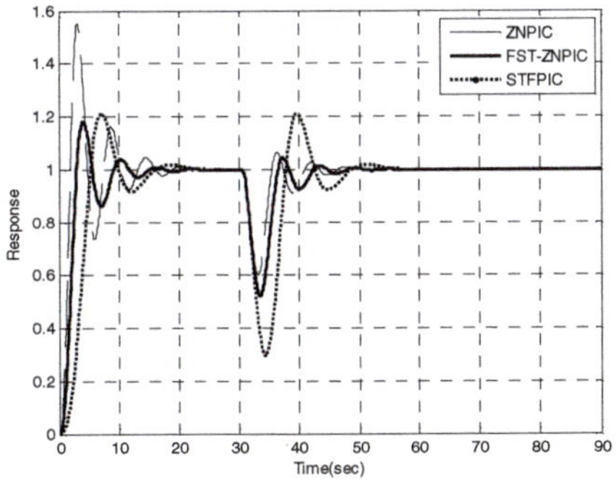

Fig. 5a. Responses of (11) for *L*=0.2

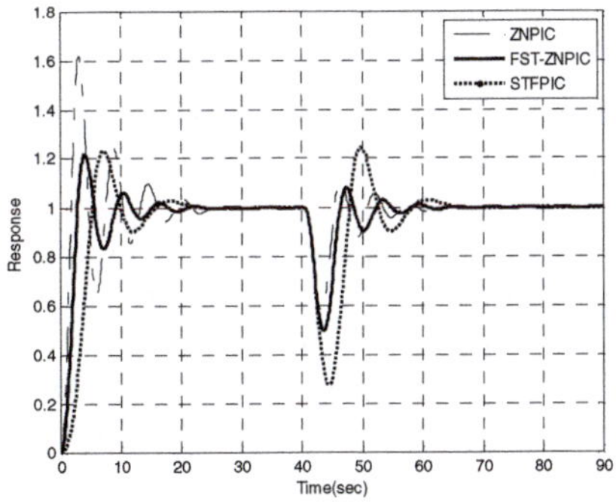

Fig. 5b. Responses of (11) for *L*=0.3

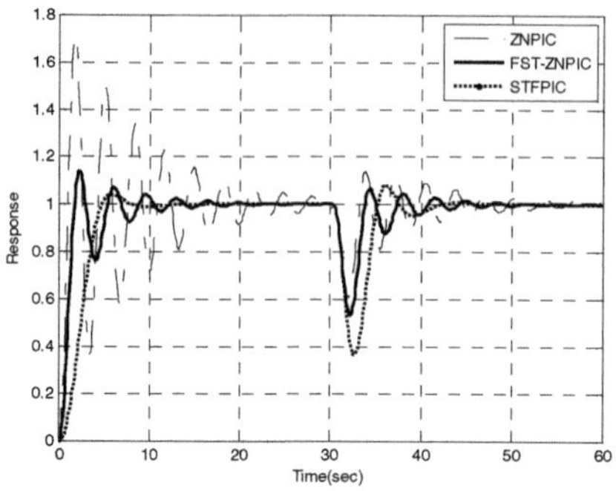

Fig. 6a. Responses of (12) for L=0.3

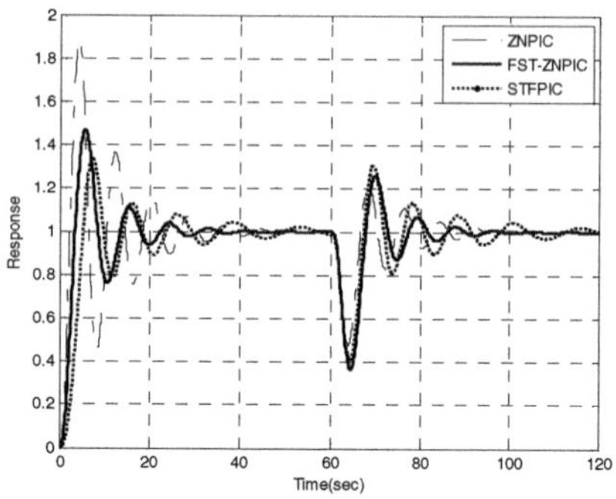

Fig. 6b. Responses of (12) for L=0.4

4 Conclusion

Here, we proposed a fuzzy logic based gain adaptive PI controller. The proposed scheme is simple and model-independent. The nonlinear gain adaptive factor α continuously modifies the proportional constant of the proposed FST-ZNPIC. Simulation results for different process with varying dead time established the robustness and high performance quality of the proposed controller.

References

[1] Shinsky, F.G.: Process control systems — application, design, and tuning. McGraw-Hill, New York (1998)

[2] Astrom, K.J., Hang, C.C., Person, P., Ho, W.K.: Towards intelligent PID control. Automatica 28(1), 1–9 (1992)

[3] Panda, S.K., Lim, J.M.S., Dash, P.K., Lock, K.S.: Gain scheduled PI speed controller for PMSM Drive. In: Proc. IEEE Industrial Electronics Society International Conference, IECON 1997, vol. 2, pp. 925–930 (1997)

[4] Seborg, D.E., Edgar, T.F.: Adaptive control strategies for process control: A survey. AICHE J. 32(6), 881–913 (1986)

[5] Kristiansson, B., Lennartson, B.: Robust and optimal tuning of PI and PID controllers. IEE Proc. Control Theory Appl. 149(1), 17–25 (2002)

[6] Tursini, M., Parasiliti, F., Zhang, D.: Real time gain tuning of PI controllers for high performance PMSM drives. IEEE Trans. Industry Appl. 38(4), 1018–1026 (2002)

[7] Mudi, R.K., Dey, C., Lee, T.T.: An improved auto-tuning scheme for PI controllers. ISA Trans. Actions 47(1), 45–52 (2008)

[8] Dey, C., Mudi, R.K.: An improved auto-tuning scheme for PID controllers. ISA Transactions 48(4), 396–409 (2009)

[9] Hang, C.C., Cao, L.: Improvement of transient response by means of variable set-point weighting. IEEE Trans. Industrial Electronics 43(4), 477–484 (1996)

[10] Dey, C., Mudi, R.K., Lee, T.T.: Dynamic set-point weighted PID controller. Control Intelligent Sys. 37(4), 212–219 (2010)

[11] Mudi, R.K., Dey, C.: Performance improvement of PI controllers through dynamic set-point weighting. ISA Transactions 50(2), 220–230 (2011)
[12] Visioli, A.: Fuzzy logic based set-point weight tuning of PID controllers. IEEE Trans. Sys. Man Cyb. 29(6), 587–592 (1999)
[13] He, S.Z., Tan, S., Xu, F.L., Wang, P.Z.: Fuzzy Self-Tuning of PID Controller. Fuzzy Sets and Syst. 56(1), 37–46 (1993)
[14] Maeda, M., Murakami, S.: A Self-Tuning Fuzzy Controller. Fuzzy Sets and Syst. 51(1), 29–40 (1992)
[15] Ziegler, J.G., Nichols, N.B.: Optimum settings for automatic controllers. ASME Trans. 64, 759–768 (1942)
[16] Mudi, R.K., Pal, N.R.: A robust self-tuning scheme for PI and PD type fuzzy controllers. IEEE Trans. Fuzzy Syst. 7(1), 2–16 (1999)

An Efficient Similarity Search Approach to Incremental Multidimensional Data in Presence of Obstacles

Shelley Gupta[1], Avinash Dwivedi[1], R.K. Issac[2], and Sachin Kumar Agrawal[3]

[1] Department of Computer Science and Engineering,
Krishna Engineering College, Ghaziabad, Uttar Pradesh, India
shelley.g17@gmail.com, avinash_dwivedi@rediff.com
[2] Faculty of Engineering
SHIATS, Uttar Pradesh, India
rkisaac@rediffmail.com
[3] Department of Information Technology
TCS, Banglore, India
agrawalsachinkr@gmail.com

Abstract. In data mining field similarity search has always been a crucial task. A similarity search finds the data points from the same data set space that matches the given query sequence exactly or differs slightly, and is done for whole sequence matching or partial sequence matching. In data sets the existence of obstacle information greatly affects the performance of similarity search in terms of efficiency and effectiveness. Thus, in this paper we present an efficient approach to similarity search based on dynamic selection of input features or attributes in presence of obstacles in respect to better running time and accuracy, with the incremental multidimensional data set. The results show that performance of the similarity search is highly dependent on data size. Thus, our approach can improve the data analysis of financial market, engineering and scientific databases, and telecom industry, providing better performance of classification, clustering, machine learning, and medical diagnosis.

Keywords: Similarity search, Obstacles, Multidimensional Data, Efficiency, Accuracy.

1 Introduction

Data mining serves as an important tool in predicting the behavior of the new data sets in wide range of applications. Similarity search has been a crucial task in data analysis and also one of the most studied area in data mining field. Most of the earlier approaches [1, 5, 10] models the nearest neighbor search over a dynamic set of features, dealing subsequence matching and the negative impact of high dissimilarity in few dimensions and also considers the distance between the data point and the query point Q more accurately dealing with both the subset of dimensions a data point is close to a query point Q and the average distance on this subset in presence or absence of obstacles. But most of these approaches are good for small data set, as

S.C. Satapathy et al. (Eds.): Proc. of Int. Conf. on Front. of Intell. Comput., AISC 199, pp. 93–101.
DOI: 10.1007/978-3-642-35314-7_12

considers the entire data set in the approach therefore affecting the performance of similarity search problems. Thus, in this paper we aim at improving the efficiency and effectiveness of similarity search algorithms in presence of obstacles dealing with the data points only in the similarity range of the query point, eliminating the tedious calculations involving the whole large data set. Thus,improving the performance of the similarity search approach with the dramatically increasing data set with respect to running time and accuracy.

2 Related Work and Motivation

In most of the earlier approaches, similarity search is based on fixed set of features (dimensions), focusing on full similarities (full data space of the data set) as they determine the similarity between the data point and query point aggregating the difference between each dimension of the two data point using similarity functions such as Euclidean distance, leaving uncovered partial similarities and also affected by few dimensions with high dissimilarity [2, 5].

Some algorithms targeting partial similarities [4, 2, 3] consider fixed subset of dimensions as the limitation for input parameters. Tung etc. [5] covers partial similarity as well as the effect of the dimensions with high dissimilarities, but have drawbacks as choosing the value for δ does not always work well as well as it does not take into account the average distance among the dimensions the data point is closest to a query point.

In high dimensional space the similarity functions such as Euclidean distance may not work well as dimensionality goes higher and the data are usually sparse [6]. Many approaches also suffer from the "curse of dimensionality" [7, 9, 8]. OPanKNN [1] coversmost of the above features but have drawbacks of considering the entire data set in most of the steps, thus increasing the running time with the increment of the data set . Thus, OPanKNN [1] suffers from drawbacks such as:

1. Forming S_l by sorting theentire data set *ids* based on δ_{il} on each dimension for determining KS_l i.e. K nearest neighbors to the Query point on each dimension, thus consuming more time as most of the sorting algorithm has time complexity O(nlogn)
2. Generate F_i for the entire data set whereas F_i is needed only in PDO (X_i, Q) calculation which requires only KS_l *ids*.

From the above we can say that it involves tedious calculations involving the entire data set, thus deteriorating the performance. The above algorithm is good for small data set.

Thus, our motive is to design an efficient similarity search approach in terms of time and accuracy of results, irrespective of multidimensional data size by reducing the tedious execution of OPanKnn [1] set making use of Shi [11] with changes.

3 Efficient Similarity Search Problem Formulation in Presence of Obstacles

Let the data set be of n instances with d dimensionality, thus $X = \{X_1, X_2, \ldots, X_n\}$ and $X_i = \{x_{i1}, x_{i2}, \ldots, x_{id}\}$ with i as the *id* of data point X_i. Let the query point be $Q = [q_1, q_2, \ldots, q_d]$, along with $O = \{O_1, O_2, \ldots, O_m\}$ as a set of m obstacle points in the same data space and D_l be the dimension, $l = 1, 2, \ldots, d$.

The data set is normalized in the range of [0, 1].

3.1 Algorithm

In our approach fig. 1 , applying Shi [11] concept with modified OPanKNN [1] we make the value range on each dimension into V_{lmin} and V_{max}, with $V_{lmin} = 0$ and $V_{max} = 1$, as data set is normalized in the range of [0, 1]. Then, divide it evenly in $\lfloor n/k \rfloor$ segments. Let S_{lj}be the segment, q_l belongs on each dimension and performs segment mergence process [11] from S_{lj}, the resulting S_{lj} will contain at least k data points. Thus, the sorting will be done on this set of data points rather than on the entire data set. For each i Є X, calculate $\Delta i\ [\delta_{i1}, \delta_{i2}, \ldots \delta_{id}]$ in which $\delta_{il} = |x_{il} - q_l|$. Now on each dimension, for each X_i Є S_{lj}, sort the ids in S_{lj} instead of in whole data set, based on δ_{il} , then construct S_l and KS_l. Calculate GS = {i} in which $GS = U_l\ KS_l$.

On each dimension D_l,sort the set of m obstacle points $O_{1l}, O_{2l} \ldots O_{ml}$, dimension-wise in ascending order, $O1l'$, $O_{2l'}, \ldots, O_{ml'}$. The value range of D_l, [0, 1], divided into m+1 segments: $[O_{0l'}, O_{1l,}), [O_{1l'}, O_{2l'}) \ldots [O_{ml'}, O_{m+1l'}]$ represented by $Z_{l0}, Z_{l1}, \ldots, Z_{lm}$, $O_{0l'} = 0$, $O_{m+1l'} = 1$.Suppose q_l belongs to the obstacle zone Z_{lk}.

For each n data points in DS, calculate $\Delta i\ [\delta_{i1}, \delta_{i2}, \ldots \delta_{id}]$ in which $\delta_{il} = |x_{il} - q_l|$. For the data point i Є GS,generate $Fi = [f_{i1}, f_{i2}, \ldots f_{id}]$ in which f_{il}, $l = 1, 2, \ldots, d$ based modified formulae: if i Є GS then $f_{il} = 0$, if i Є KS_l and x_{il} Є Z_{lk} then $f_{il} = 1$, and if i Є KS_l and x_{il} ¬Є Z_{lk} then $f_{il} = \min((|\delta_{il}| / \min(|q_l - O_{kl}|, |q_l - O_{k+1l}|), 1/|\delta_{il}|)$. Thus, Fiis calculated only for the data points or *ids* Є GS, on each dimension (maximum of dK or minimum of K data points), rather than for the data set of n instances . Calculate EPDO (Xi, Q) [1], such that EPDO (Xi, Q) = $(\Sigma\ \delta_{il} * f_{il}) / (\Sigma f_{il})^2$ $l = 1,2\ldots,d$, for each i Є GS, then based on increasing values of EPDO (Xi, Q) sort GS = {i}. Let set EOKSS contain first k ids Є GS. Finally return EOKSS as the K similar search to the query point.

*Note:*In segment mergence process [11] the data point value on each dimension with value equal to q_l is added N_l, i.e. N_l be the number of *ids* of data points less than or equal to q_l in our approach.

Lemma 1: The resulting Slj after segment mergence process will contain at least K data points.

Proof: Explained by means of three cases:Let the value ql lie in the segment Slj

Case1. In fig. 2 ql lies in extreme left segment, therefore no segment on left for segment mergence present:N_l< K, therefore a segment on the left of Slj which is not merged is required, butsuch a segment not present. N_l> K,therefore no segment

Algorithm EOKSS (DS: Dataset, Q: Query Point, d: dimensionality of DS, K: number of similar data points required, m: number of obstacle points)

```
Begin
  For l=1,2,.......,d                // On each dimension D_l //
        Suppose the value range is [V_lmin, V_lmax], dividing it evenly by k, resulting in segments S_l1, S_l2,
        ......., S_l|_n/k_|.
  For l= 1, 2, ....., d
        Let the value q_l belongs to the segment S_lj.
        Perform Segment Mergence Process from S_lj.
        The resulting S_lj will contain at least K data points;
  For l = 1 to d                     // On each dimension D_l //
        (
        For each X_i Є S_lj, sort the ids in S_lj instead of in whole data set, based on δ_il .
        // On each dimension D_l //
         Construct S_l           // sorted list of ids based on δ_il //
         Construct KS_l          // KS_l  =  first K ids of S_l.
        )
  For  l = 1 to d                   // On each dimension  D_l//
        Calculate GS = {i} in which GS = U_l KS_l .
  For  l = 1 to d                   // On each dimension  D_l//
        (
         Sort the set of m obstacle points O_11, O_21 ........O_ml, dimension-wise in ascending order,  O_1l',
        O_2l', ......., O_ml'. The value range of D_l, [0, 1], divided into m+1 segments: [O_0l', O_1l,), [O_1l',
        O_2l') ................. [O_ml', O_m+1l']   represented by Z_l0, Z_l1 ,......,  Z_lm, O_0l'  = 0, O_m+1l' = 1.
        Let q_l belongs to the obstacle zone Z_lk
        )
  For each i Є DS, calculate Δi [δ_i1, δ_i2, ......δ_id] in which  δ_il = |x_il – q_l|;
  For the data point i Є GS,generate Fi = [f_i1, f_i2,.......f_id] in which f_il, l = 1, 2, ..., d based on formulae:
                    if i Є GS
                         f_il = 0
                  if  i Є KS_l and x_il Є Z_lk
                         f_il =  1
                    if i Є KS_l and x_il ¬ЄZ_lk
                         f_il =  min( (| δ_il |  /  min  (|q_l - O_kl|, |q_l - O_k+1l|),  1/| δ_il |)
  For each   i Є GS
        Calculate EPDO (Xi, Q), such that EPDO (Xi, Q) = (Σ  δ_il*f_il)  / (Σ f_il)^2   l = 1,2.......,d
  Sort GS = {i} based on increasing values of EPDO (Xi, Q).
  Let set EOKSS contain first k ids Є GS. Return EOKSS.
  End
```

Fig. 1. The Efficient EOKSS Algorithm [1, 11]

mergence required. $N_l = K$, therefore no segment mergence required. For all the above cases, if N_r is > or = K, number of data points will be greater than or equal to K. If N_r is < K, then segment mergence is done in the right. Thus, from the above cases the combination of data points in N_l an N_r, S_{lj} will contain at leastK data points.

Case 2. Fig. 3 ql lies in extreme right segment, therefore no segment on right for segment mergence present: Same as case 1, Thus, from the above cases the combination of data points in N_l an Nr, S_{lj} will contain at least K data points.

Case 3. Fig. 4 ql lies in the segment, with segment for mergence on the both the right and left side:

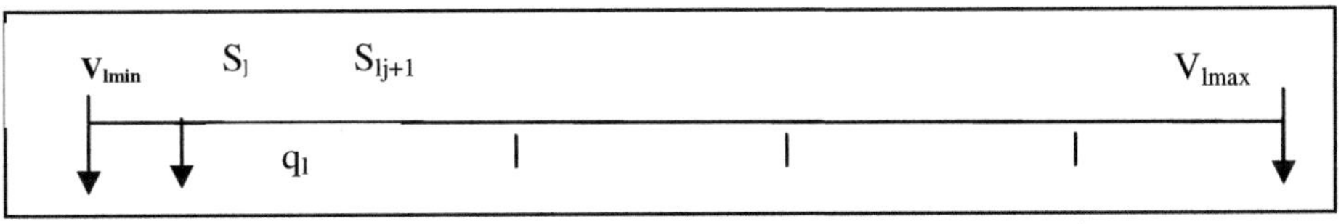

Fig. 2.

If N_l is > or = K, number of data points greater than or equal to K. If N_l is < K, then segment mergence is done in left . If N_r is > or = K, number of data points greater than or equal to K. If N_r is < K, then segment mergence is done on the right.

Thus, from the above cases the combination of data points in N_l and Nr, Slj will contain at least K data points.

Thus from the above three cases we can say that the resulting S_{lj} after segment mergence process will contain at least K data points, on each dimension.

4 Experiments

We use the real data sets from real world applications obtained from UCI Machine Learning Repository [12], for conducting comprehensive experiments, applying the strategy similar to the one described in [5].We executed our experiments on Intel(R) Core(TM)2 Quad CPU Q9650 @3.00 GHz 3 GHz, 4.00 GB (3.25 GB Usable)of RAM and 32 bit operating system.

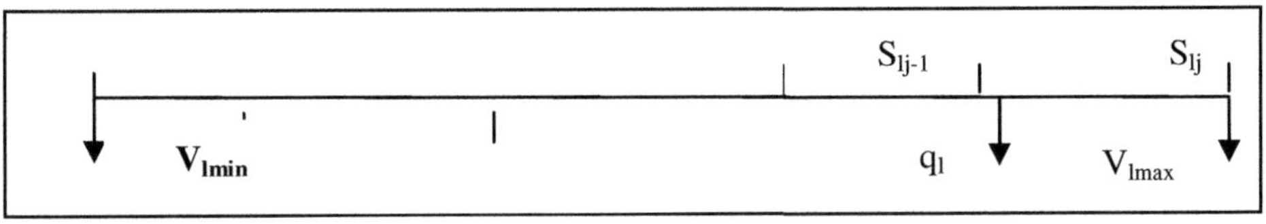

Fig. 3.

4.1 Experiments to Indicate the Scalability of EOKSS over K, Data Size, Dimensions and Obstacles.

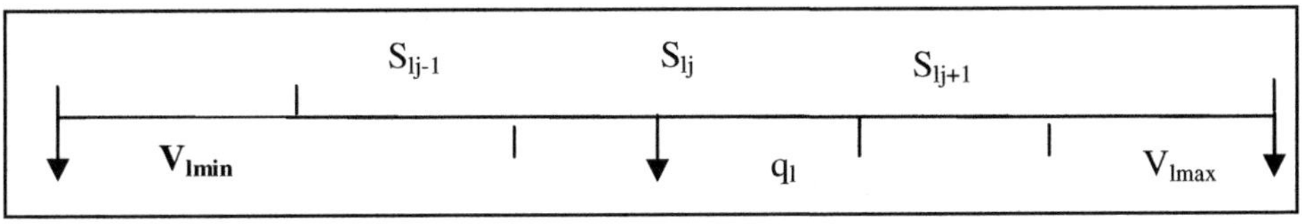

Fig. 4.

In this section, the experiments are conducted on real data set Wine quality - white wine samples, the wine quality graded between 0 (very bad) and 10 (very excellent), with numbers of instances 4898 and number of attributes: 11 + output attribute.

In figure 5, each group has K increasing from 100 to 240 over fixed number of dimensions (9), obstacles (12) and data size (2000, 3000 and 4000) for each group, showing the running time of 3 group of data sets using one query point.

Similarly, figure 6, 7 and 8 shows the experimental results for increasing values of data sizes, dimensions and obstacles respectively, keeping other parameters constant.

Thus, the below given figures 5, 6, 7 and 8 clearly indicates that the proposed approach is scalable over the number of similarity search required (K), data size, dimensions and obstacles.

4.2 Experiments for Comparative Study of EOKSS vs. OPanKNN

Timing Analysis

In figure 9, the data size increasing from 500 to 800, with dimensionality and K set as 9 and 100 respectively. From this figure we see that EKSS improves the performance with respect to running time compared to OPanKNN with greater increments in the data size.

Accuracy

The proposed approach, EOKSS effectiveness or accuracy has been evaluated from a quantitative view using class striping technique [5]. The wine data set contains 178 instances of data points with 13 dimensions and 3 classes. The class of the data points is indicated with an extra dimension in data set record. The class tag is stripped from the data points and the proposed approach, EOKSS and the OPanKNN algorithm are used to find the similar data points to the query point with in the same data space. Predicted class of the retrieved similar data points to the query point is considered same as that of the respective query point. Actual class of the retrieved similar data points to the query point is the class of the data point in the data set mentioned.

The answer is correct if the predicted class and the actual class of the retrieved similar data points are same and also if the retrieved answer has no obstacle object in between the retrieved data points and the query point. Otherwise, the answer is incorrect. Statistically, the similarity search method is more accurate or effective if the searched similar data points are more correct [5].

We executed 45 different query points and 2 obstacle points, selected randomly from the wine data set using K as 8. Firstly, we construct the confusion matrix and then calculate the accuracy rate and error rate for the retrieved similar data points to the query point. In confusion matrix columns and rows represents predicted and actual class respectively. QC represents the class of query point and ¬QC the class not same as that of query point class. Accuracy rate is evaluated by dividing the correctanswers by the number of query points times K. Error rate is evaluated by dividing the incorrect answers by the number of query points times K.

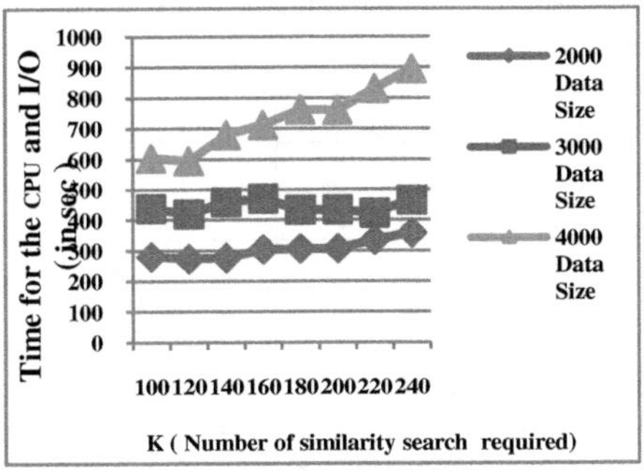

Fig. 5. Running Time with the increasing K

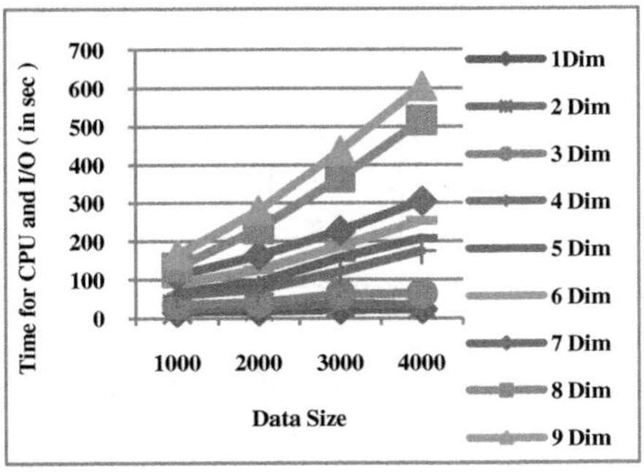

Fig. 6. Running Time with increasing Data Size

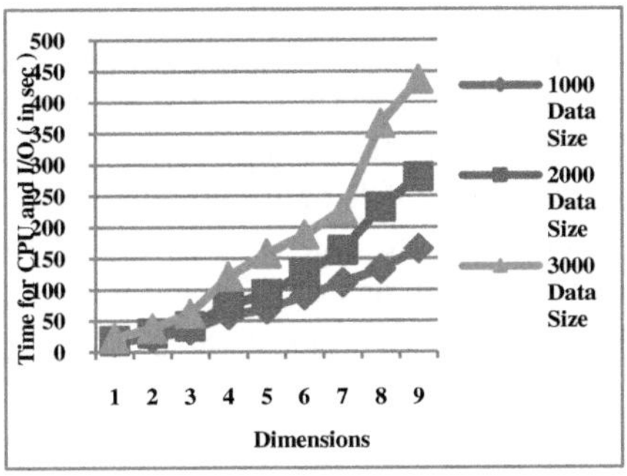

Fig. 7. Running time with increasing dimensions

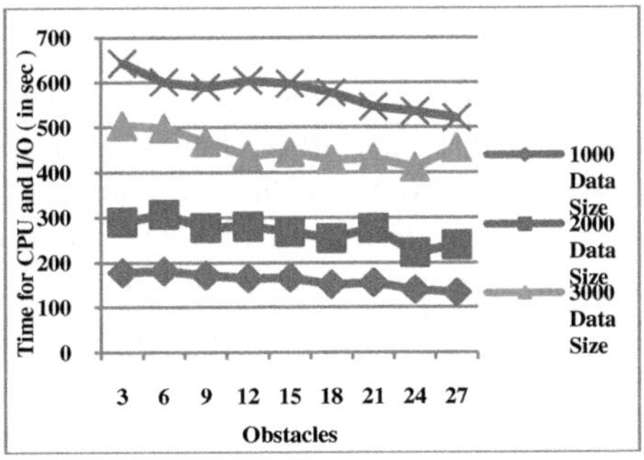

Fig. 8. Running Time with increasing Obstacles

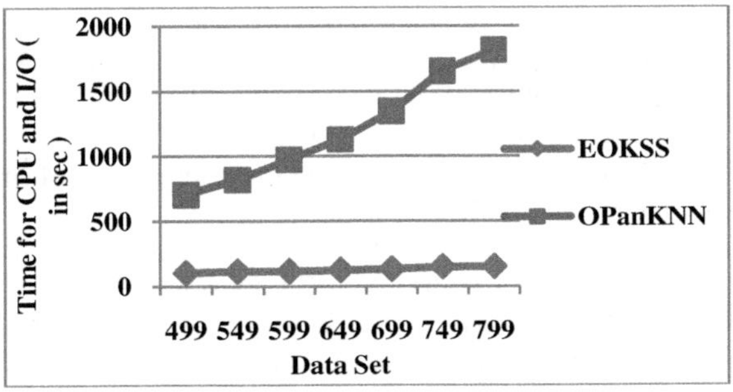

Fig. 9. Comparative study of EOKSSS and OPANKNN in terms of Running Time on one Query Point, with K = 100, dimensionality = 9 and obstacles = 12

Table 1. Confusion matrix for proposed approach (EOKSS)

Classes	QC	¬QC	Total
QC	344	-	344
¬QC	16	-	16
Total	360	-	360

Table 2. Confusion matrix for OPanKNN

Classes	QC	¬QC	Total
QC	333	-	333
¬QC	27	-	27
Total	360	-	360

The accuracy rate of proposed approach (EOKSS) is 95.55% which is higher than the accuracy rate of OPanKNN 92.5% Whereas, the error rate of proposed approach (EOKSS) is 4.44% which is lower than the error rate of OPanKNN 7.5% . Thus, the proposed approach is more accurate and effective.

5 Conclusion

In this paper we proposed an approach that requires lesser time for searching similarity in the multidimensional data analysis, with the increasing data size. Within our approach, binary array is calculated based on the *ids* of the data points in GS, i.e. the set of data points within the maximum and minimum of K and dK ids nearest to Query point as compared to whole large data set. Also the greater accuracy of the proposed approach is achieved. Thus, this improved approach can be significant in fields of data clustering, bioinformatics, telecom, retail, signal processing, financial market analysis and pattern recognition.

References

1. Shi, Y., Zhang, L., Zhu, L.: An approach to nearest neighboring search for multi-dimensional data. International Journal of Future Generation Communication and Networking, 4(1) (March 2011)
2. Hinneburg, A., Aggarwal, C.C., Keim, D.A.: What is the nearest neighbor in high dimensional spaces? The VLDB Journal, 506–515 (2000)
3. Aggarwal, C.C., Hinneburg, A., Keim, D.A.: On the Surprising Behavior of Distance Metrics in High Dimensional Space. In: Van den Bussche, J., Vianu, V. (eds.) ICDT 2001. LNCS, vol. 1973, pp. 420–434. Springer, Heidelberg (2000)
4. Aggarwal, C.C.: Towards meaningful high-dimensional nearest neighbor search by human-computer interaction. In: ICDE (2002)
5. Tung, A.K.H., Zhang, R., Koudas, N., Ooi, B.C.: Similarity Search: A matching based approach. In: VLDB 2006, pp. 631–642. VLDB Endowment (2006)
6. Beyer, K., Goldstein, J., Ramakrishnan, R., Shaft, U.: When is "nearest neighbor" meaningful? In: International Conference on Database Theory 1999, Jerusalem, Israel, pp. 217–235 (1999)
7. White, D.A., Jain, R.: Similarity Indexing with the SS-tree. In: Proceedings of the 12th Intl. Conf. on Data Engineering, New Orleans, Louisiana, pp. 516–523 (February 1996)

8. Berchtold, D.A., Keim, S., Kriegel, H.P.: The X-tree: An index structure for high-dimensional data. In: VLDB 1996, Bombay, India, pp. 28–39 (1996)
9. Weber, R., Schek, H.J., Blott, S.: A quantitative analysis and performance study for similarity-search methods in high-dimensional spaces. In: Proc. 24th Int. Conf. Very Large Data Bases, VLDB, August 24-27, pp. 194–205 (1998)
10. Shi, Y., Zhang, L.: A dimension-wise approach to similarity search problems. In: The 4th International Conference on Data Mining, DMIN 2008 (2008)
11. Shi, Y.: A scalable approach to multi-dimensional data analysis. the International Journal of Bio-Science and Bio- Technology 2(4) (March 2010)
12. Bay, S.D.: The UCI KDD Archive. University of California, Irvine, Department of Information and Computer Science, `http://kdd.ics.uci.edu`

Alleviating Sparsity and Scalability Issues in Collaborative Filtering Based Recommender Systems

Akshi Kumar and Abhilasha Sharma

Dept. of Computer Engineering, Delhi Technological University
akshi.kumar@gmail.com, abhilasha_sharma87@yahoo.com

Abstract. Commercial recommender systems in general are used to evaluate very large product sets. In a user – item rating database, though users are very active, there are a few rating of the total number of items available. The user-item matrix is thus extremely sparse. Since a collaborative filtering algorithm is mainly based on similarity measures computed over the co-rated set of items, the large levels of sparsity can lead to less accuracy and can challenge the predictions or recommendations of the collaborative filtering (CF) systems. Further, a CF algorithm is assumed to be efficient if it is able to filter items that are interesting to users. But, they require computations that are very expensive and grow non-linearly with the number of users and items in a database. In general, the whole ratings database is searched in collaborative filtering and thus it suffers from poor scalability when more and more users and items are added into the database. Instigated by these challenges, we investigate two collaborative filtering algorithms, firstly an algorithm based on weighted slope one scheme and item clustering & secondly an algorithm based on item classification & item clustering, which deal with the sparsity and scalability issues simultaneously. Experiments were carried to determine which is better in terms of simplicity and accuracy among the two methods.

Keywords: Recommender System, Collaborative Filtering, Sparsity, Scalability.

1 Introduction

With the advent and proliferation of the Internet and e-commerce, it is evident that the complexity of finding relevant information on the Web has become increasingly intricate and crucial. In fact, "information overload" on the Web is a well recognized problem [1], where users find it increasingly difficult to locate the right information at the right time. Moreover, the World Wide Web have given us a world of endless possibilities- like items to consume, movies to watch, music to listen, conversations to participate in etc. Amidst all this range of endless options, a consumer faces the task of what to choose which might interest him. Recommender system [2, 3] comes to the rescue for such a consumer. These systems aim to mediate, support, or automate the everyday process of sharing recommendations [4]. One of the most successful technologies in recommender systems research and practice is collaborative filtering.

S.C. Satapathy et al. (Eds.): Proc. of Int. Conf. on Front. of Intell. Comput., AISC 199, pp. 103–112.
DOI: 10.1007/978-3-642-35314-7_13

Collaborative filtering (CF) [5, 6, 7, 8, 9] (also known as social information filtering) is based on the basic principle of finding a subset of users who have similar tastes and preferences to that of the active user, and offering recommendations based on the subset of users. Research has substantiated some prominent problems existent in collaborative filtering, namely the cold-start problem, issues of data sparsity and scalability, shilling attacks, synonymy, amongst others.

In response to the identified need for improved users' experience by personalizing what they see and resolving the data sparsity & scalability issues simultaneously, we propose & validate two techniques, namely, Weighted Slope One Scheme and Item Classification to determine vacant ratings in the given sparse data set and further combine each of them with k-means clustering to deal with scalability issue. We call them Algorithm 1 (Weighted Slope One Scheme+ Item Clustering) & Algorithm 2 (Item Classification + Item Clustering) throughout this paper. The weighted slope one scheme is based on popularity differential of items i.e. in a pair wise fashion it is determined which item is liked better than other. This information is further used to predict rating for a user given their rating for other items. The item classification technique classifies items (here movies) into groups based on some attribute. The attribute chosen here is genre. Further in each group user based collaborative filtering is used to predict ratings. Since a movie could belong to various genres in that case mean of values predicted from each group is used to get final rating for that movie. Item clustering (K-means) is then performed on the dense data set obtained from each of the techniques. Using the item clustering, items are clustered into groups, i.e., 48 clusters are created. Clustering results in similar items grouped together. The item for which rating needs to be predicted is matched with the centroid of each of the clusters. Then neighbours in the nearest matching centroids is considered as neighbours for the item in question. Once neighbours are identified weighted average weighted by similarity is used to determine rating. We compare the algorithms on Movielens data set. The data set is reduced to ease computation problems. It is observed that algorithm 2 outperforms algorithm 1 as far as accuracy is concerned and the computation time required for creating dense data set is also lesser. The system is easier to implement, provides us with a way to decipher two prominent issues in one go and can be deployed in any kind of recommender systems.

2 Background Work

Recommender systems are systems that provide recommendations to customers based on their past purchases, tastes, and preferences [10]. Examples of such applications include recommending books, CDs and other products at Amazon.com [11], movies by MovieLens [12]. The commonly accepted formulation of the recommendation problem was first stated in [5, 6] and this problem has been studied extensively since then. Moreover, recommender systems are usually classified into Content-based, Collaborative and Hybrid based on how recommendations are made [8]. However, despite all these advances, the current generation of recommender systems still requires further improvements to make recommendation methods more effective and applicable to an even broader range of real-life applications. The term "collaborative filtering" was introduced in the context of the first commercial recommender system, called Tapestry [7], which was designed to recommend documents drawn from newsgroups to a collection of users. Seminal collaborative filtering systems included GroupLens [5], the Bellcore Video Recommender [2], and Firefly [6]. However, collaborative systems have their own limitations, namely, the First Rater Problem, Sparsity & Scalability issues [8, 9, 10 and 13]. The paper [4] considers several challenges for recommender systems.

Commercial recommender systems in general are used to evaluate very large product sets. In a user – item rating database, though users are very active, there are a few rating of the total number of items available. The user-item matrix is thus extremely sparse. Since a collaborative filtering algorithm is mainly based on similarity measures computed over the co-rated set of items, thus large levels of data sparsity can lead to less accuracy and can challenge the predictions or recommendations of the CF systems. Scalability is another major issue with the size of the data set being enormously huge. It becomes difficult to search the entire rating database and there is poor scalability when more and more users and items are added into the database. Keeping this in mind, we have investigated two collaborative filtering algorithms, which deal with the above two issues simultaneously and determine their performance in terms of simplicity and accuracy.

3 CF Algorithms to Resolve Sparsity and Scalability Simultaneously

Two techniques Weighted Slope One Scheme and Item Classification have been used to determine vacant ratings in the given sparse data set. Further to deal with scalability issue, K-means Clustering have been used that groups items in both cases for producing the final recommendations.

3.1 Sparsity Reduction (Pre – prediction)

For reducing the sparseness of data, the techniques adopted are:

Weighted Slope One Scheme. Slope One is a family of algorithms used for collaborative filtering. It was introduced in a 2005 by Daniel Lemire and Anna Maclachlan [14]. These are very simple to implement and their accuracy is often at par with more complicated and computationally expensive algorithms. The slope one predictors was first introduced for online rating. The basic idea is to answer the question how a user would rate a give item, given other users' ratings [15]. Slope One algorithms work on the intuitive principle of a "**popularity differential**" between items for users. In a pair wise fashion, we determine how much better one item is liked than another. One way to measure this differential is simply to subtract the average rating of the two items. In turn, this difference can be used to predict another user's rating of one of those items, given their rating of the other. The slope one method uses a simpler form of regression f(x) = x + b, hence the name "slope one".

Notation. The following notation is used to describe the scheme. The ratings from a given user, is called an *evaluation*, is represented as an incomplete array u, where u_i is the rating that this user gives to item i. The subset of the set of items consisting of all those items which are rated in u is S(u). The set of all evaluations in the training set is χ. The number of elements in a set *S* is *card (S).* The average of ratings in an evaluation *u* is denoted *u*. The set $S_i(\chi)$ is the set of all evaluations $u \in \chi$ such that they contain item *i* ($i \in S(u)$). Given two evaluations *u*, *v*, we define the scalar product $\langle u,v \rangle$ as $\Sigma_{i \in S(u) \cap S(v)} u_i v_i$. Predictions, P(u), represent a vector where each component is the prediction corresponding to one item: predictions depend implicitly on the training set χ [14].

Expressing Slope one scheme using above notation. Formally, given two evaluation arrays v_i and w_i with $i = 1, \ldots, n$, we search for the best predictor of the form $f(x) = x+b$ to predict *w* from *v* by minimizing $\Sigma_i (v_i + b - w_i)^2$. Deriving with respect to *b* and setting the derivative to zero, we get $b = (\Sigma_i wi - vi)/n$. In other words, the constant *b* must be chosen to be the average difference between the two arrays. This result motivates the following scheme. Given a training set c, and any two items *j* and *i* with ratings u_j and u_i respectively in some user evaluation u (annotated as $u \in S_{j,i}(\chi)$), we consider the average deviation of item *i* with respect to item *j* as:

$$\text{dev}_{j,i} = \sum_{u \in S_{j,i}(\chi)} \frac{u_j - u_i}{card\left(S_{j,i}(\chi)\right)}..$$

The symmetric matrix defined by $\text{dev}_{j,i}$ can be computed once and updated quickly when new data is entered. Given that $\text{dev}_{j,i} + u_i$ is a prediction for u_j given u_i, a reasonable predictor might be the average of all such predictions

$$P(u)_j = \frac{1}{card(R_j)} \sum_{i \in R_j} \left(\text{dev}_{j,i} + u_i\right)$$

where $R_j = \{i | i \in S(u), i \neq j, card(S_{j,i}(\chi)) > 0\}$ is the set of all relevant items.

One of the drawbacks of slope one is that the number of ratings observed is not taken into consideration. Intuitively, to predict user *A*'s rating of item *L* given user *A*'s rating of items *J* and *K*, if 2000 users rated the pair of items *J* and *L* whereas only 20 users rated the pair of items *K* and *L*, then user *A*'s rating of item *J* is likely to be a far better predictor for item *L* than user *A*'s rating of item *K* is. Thus, we define the weighted slope one prediction as the following weighted average

$$p^{wS1}(u)_j = \frac{\sum_{i \in S(u) - \{j\}} \left(\text{dev}_{j,i} + u_i\right) c_{j,i}}{\sum_{i \in S(u) - \{j\}} c_{j,i}}$$

where $c_{j,i} = card(S_{j,i}(\chi))$

Item Classification. This approach classifies the items to pre-produce the ratings, where necessary. Items may be categorized or clustered based on the attributes of the items. For Example, in the context of movies, every movie can be classified according to the "genre" attribute of each item. In our work , we have used movies as the items and the various genre that they could belong are Action, Adventure, Animation, Children's, Comedy, Crime, Documentary, Drama, Fantasy, Film-Noir, Horror, Musical, Mystery, Romance, Sci-Fi, Thriller, War, Western. So in all we have divided the collection of all movies in 18 groups. Once the items have been classified according to the attribute content, then in the sub-matrix formed user based collaborative filtering is used to fill the vacant rating. An item can belong to one or more genre. In such case the mean of the value from each of the sub-matrix formed is used as the value for that item.

3.2 Improving Scalability (Using Item Clustering)

Scalability is improved using K-means clustering [16]. Firstly, the items are clustered into groups and then the target item for which recommendation needs to be computed is matched with the centroid of each of the clusters formed. The nearest similarity clusters are identified and then nearest similarity items within these clusters are identified. These neighbors' are further used to predict the value for the target item. Scalability is better because the entire rating database is not searched rather specific clusters are searched to identify similar items.

K-means Clustering. The algorithm first takes K items as the centers of K unique clusters. Each of the remaining items is then compared to the closest center. In the following passes, the cluster centers are re-computed based on cluster centers formed in the previous pass and the cluster membership is re-evaluated. The pearson's correlation, as following formula, is used to measure the linear correlation between two vectors of ratings as the target item t and the remaining item r.

$$sim(t,r) = \frac{\sum_{i=1}^{m}(R_{it} - A_t)(R_{ir} - A_r)}{\sqrt{\sum_{i=1}^{m}(R_{it} - A_t)^2 \sum_{i=1}^{m}(R_{ir} - A_r)^2}}$$

where, R_{it} is the rating of the target item t by user i, R_{ir} is the rating of the remaining item r by user i, A_t is the average rating of the target item t for all the co-rated users, A_r is the average rating of the remaining item r for all the co-rated users, and m is the number of all rating users to the item t and item r. K-means clustering used above groups items with maximum similarity.

3.3 Collaborative Filtering Combining Sparsity Reduction Techniques and Item Clustering

We combine the sparsity reduction technique and item clustering for more scalable and accurate recommendations. The sparsity reduction technique help to determine vacant ratings in the entire data set thus providing dense data set. Item clustering using K-means is used to further cluster similar items .The scalability is improved using item clustering because similar items can be found easily by selecting nearest matching cluster centroids with the target item [17].

Once the items are clustered, the item centers are obtained. This center is represented as an average rating over all items in the cluster. The target item neighbors are chosen in some of the item center clustering. Pearson's correlation is used to compute similarity between target item and the item centers. Once the similarity is calculated between the target item and the item centers, the items in the most similar centers are chosen as the candidates. As the target item nearest clustering centers are chosen, the similarity is calculated between the target item and items in the selected clustering centers. The top K most similar items based on cosine measure are selected. Cosine measure looks at the angle between two vectors of ratings as the target item t and the remaining item r. The following formula is used:

$$\text{sim}(t, r) = \frac{\sum_{i=1}^{m} R_{it} R_{ir}}{\sqrt{\sum_{i=1}^{m} R_{it}^{2} \sum_{i=1}^{m} R_{ir}^{2}}}$$

where, R_{it} is the rating of the target item t by user i, R_{ir} is the rating of the remaining item r by user i, and m is the number of all rating users to the item t and item r. Once the membership of items is computed, we calculate the weighted average of neighbors' ratings, weighted by their similarity to the target item. The rating of the target user u to the target item t is as following:

$$P_{ut} = \frac{\sum_{i=1}^{c} R_{ui} \times \text{sim}(t, i)}{\sum_{i=1}^{c} \text{sim}(t, i)}$$

where, R_{ui} is the rating of the target user u to the neighbour item i, sim(t, i) is the similarity of the target item t and the neighbour it user i for all the co-rated items, and m is the number of all rating users to the item t and item r.

4 Experimental Analysis

Data is collected from Grouplens website [18].We have used Movielens collaborative filtering data set to evaluate the performance of the two algorithms. The data consists of 1, 00, 000 ratings from 943 users who rated for 1682 items (movies). The programming language used for source code is Python 2.7. The data set from grouplens website was too huge and for dealing with sparsity, 1486126 unknown ratings required to be calculated which needed long time for computation. So we decided to implement our algorithms on a reduced data set. We consider the data set having only those items which are rated by 60 or more users. This reduced the original data set to a size having 309 users and 536 items. Total known ratings in this data set is 53536. 5% of the users are randomly selected to be the test users. From each user in the test set, ratings for 5 items were withheld and predictions were computed for those 5 items using both algorithm 1(weighted slope one scheme and item clustering) and algorithm 2 (item classification and item clustering). For item classification technique, we need to consider the genre of movies. On reducing the data set, no movie from among the selected ones belonged to genre 'unknown'. So finally for our dataset, 18 genres were considered as mentioned earlier. First the dense data set is created using each of the technique described in previous section to reduce sparseness. Then item clustering is done on both the dense datasets produced. K-means clustering is used to group similar items. The number of clusters created is 48.

Metric used to evaluate the accuracy of algorithms is MAE (Mean Absolute Error).It compares the deviation of the predicted ratings from the respective actual user ratings. The size of the neighborhood has a significant effect on the prediction quality. We varied the number of neighbors and compute the MAE. The conclusion as depicted from the fig.1, which includes the Mean Absolute Errors for the two algorithms as observed in relation to the different number of neighbors, is that Item Classification technique is better. The predicted values for each of the target item for different users (5% of total users), along with their actual rating and absolute difference between predicted and actual ratings for each of the collaborative filtering algorithms were calculated.

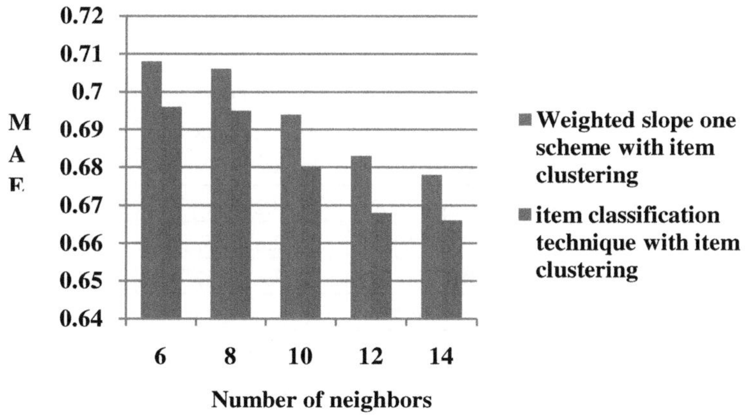

Fig. 1. Comparative results for the two algorithms

The sample results for 6 neighbors using Algorithm 1 (Weighted Slope One Scheme+ Item Clustering) & Algorithm 2 (Item Classification + Item Clustering) are shown below in table 1 & table 2 respectively:

Table 1. Using Weighted Slope one scheme and Item clustering (**MAE= 0.708669**)

Predicted rating	Actual rating	Predicted rating	Actual rating	Predicted rating	Actual rating	Predicted rating	Actual rating	Predicted rating	Actual rating
3.771887	4	4.252483	4	3.703456	5	3.478749	4	3.527947	4
3.978097	4	4.538225	5	4.109109	5	3.70761	2	3.519338	4
3.131967	3	3.714024	3	3.540758	5	2.948917	2	2.817101	3
3.552932	5	4.024951	2	3.854411	5	4.149563	3	3.611031	3
4.301203	4	4.406705	5	4.423963	5	3.79972	4	3.649225	3
3.72715	4	4.436714	5	4.071103	5	3.666714	3	3.833281	4
3.625225	4	4.038809	4	3.552864	5	3.399332	4	3.308322	4
3.744442	4	3.913577	2	4.06821	3	3.556006	3	3.506115	4
3.594493	2	3.815497	3	3.958245	5	3.492522	3	3.574611	4
4.017775	4	3.698761	5	3.998824	5	4.111313	3	4.058118	4
3.857931	2	4.295343	4	4.271942	5	3.527524	3	3.630347	5
3.088844	3	3.147182	3	2.571817	4	2.99972	2	3.166497	3
4.693009	5	4.488164	5	4.584289	5	4.241981	4	4.075443	4
3.728595	5	3.936152	3	3.524587	5	3.51772	2	3.379486	4
3.305931	4	3.716047	3	3.816923	4	3.133352	3	3.074971	3

Table 2. Using Item Classification Technique and Item clustering (**MAE= 0.696975**)

Predicted rating	Actual rating	Predicted rating	Actual rating	Predicted rating	Actual rating	Predicted rating	Actual rating	Predicted rating	Actual rating
4.447092	4	4.054278	4	3.885695	5	2.878094	4	3.579958	4
4.679484	4	4.841588	5	4.157742	5	3.455908	2	3.669238	4
3.5004	3	2.62092	3	3.628885	5	2.692162	2	2.844484	3
3.898149	5	2.84753	2	3.676934	5	3.022588	3	3.502278	3
4.420494	4	3.925422	5	4.268989	5	3.41981	4	2.901473	3
3.877591	4	3.510301	5	4.311304	5	3.2769	3	3.379595	4
3.24533	4	3.819959	4	3.543191	5	2.832477	4	3.19495	4
3.160002	4	2.959836	2	4.152219	3	3.672887	3	3.71173	4
3.822456	2	3.416087	3	3.996633	5	2.806284	3	3.071239	4
4.267351	4	4.123496	5	4.333419	5	3.39399	3	3.702737	4
3.644186	2	4.050412	4	4.335038	5	3.548201	3	3.762214	5
3.499133	3	3.139822	3	2.551045	4	2.662468	2	2.382305	3
4.337399	5	3.722277	5	4.521594	5	3.828206	4	3.693621	4
3.650867	5	3.949098	3	3.464083	5	3.173247	2	3.141476	4
3.815945	4	3.653458	3	3.864616	4	2.807937	3	3.065456	3

Limitations. Following are some limitations of the proposed system:

- It might be the case that there exist data for which attribute identification is troublesome.
- The proposed algorithms do not solve the cold start problem.
- The system does not work well where the attributes have synonymous names, for eg *children movie* and *children films* refer to same thing, but the algorithm does not have any provision to shield itself from effects of synonymy.
- The system does not provide any explanation for the predicted recommendations which is crucial for building user trust.

5 Conclusion

We proposed and compared two collaborative filtering recommendation algorithms which help to alleviate the issues of sparsity and scalability. The first algorithm combines weighted slope one scheme and item clustering and the other algorithm combines item classification technique and item clustering. Item clustering helps to meet the real time requirement of recommender system by reducing the search effort needed to find neighbours of the target item. We demonstrated software which can be used to make predictions based on the two collaborative filtering algorithms. We have shown a case study of these two algorithms on Movielens data set. The data set was reduced to ease computation problems. Algorithm 2(Item Classification + Item Clustering) outperforms algorithm 1 (Weighted Slope One Scheme+ Item Clustering) as far as accuracy is concerned and the computation time required for creating dense data set is also lesser. This collaborative filtering algorithm can be deployed in any kind of recommender systems. The model proposed for the collaborative filtering algorithms is executed on a reduced data set from movielens. But this work can be easily extended to other datasets like Jester, datasets related to e-commerce. Further, we have incorporated two sparsity reduction techniques in our algorithms but there exists other sparsity reducing techniques like case based reasoning, content based predictor (like TAN-ELR : tree augmented naïve Bayes optimized by extended logistic regression) ,extended BMI (Bayesian Multiple Imputation) etc. which can be used as well. The technique used to overcome scalability issue in our work is k-means clustering which is the simplest one. Other clustering techniques that can be tried are Fuzzy c-means clustering, clustering using genetic algorithms, hierarchical clustering etc. A work could be carried out to see how well these techniques can fit in our algorithm and provide reasonably better results.

References

[1] Carlson, C.N.: Information overload, retrieval strategies and Internet user empowerment. In: Proceedings of the Good, the Bad and the Irrelevant, pp. 169–173. University of Art and Design, Helsinki (2003)

[2] Hill, W., Stead, L., Rosenstein, M., Furnas, G.: Recommending and evaluating choices in a virtual community of use. In: Proceedings of the ACM Conference on Human Factors in Computing System, pp. 194–201 (1995)
[3] Resnick, P., Varian, H.R.: Recommender Systems. Guest Editor's Introduction to the Special Section. Communications of the ACM 40(3), 56–58 (1997)
[4] Terveen, L.G., Hill, W.: Beyond Recommender Systems: Helping People Help Each Other. In: Carroll, J. (ed.) HCI in the New Millennium. Addison Wesley (2001)
[5] Resnick, P., Iakovou, N., Sushak, M., Bergstrom, P., Riedl, J.: GroupLens: An open architecture for collaborative filtering of netnews. In: Proceedings of the Computer Supported Cooperative Work Conference (1994)
[6] Shardanand, U., Maes, P.: Social information filtering: Algorithms for automating 'word of mouth'. In: Proceedings of the Conference on Human Factors in Computing Systems (1995)
[7] Goldberg, D., Nichols, D., Oki, B.M., Terry, D.: Using collaborative filtering to weave an information tapestry. Communications of the ACM 35(12), 61–70 (1992)
[8] Balabanovic, M., Shoham, Y.: Fab: Content-Based, Collaborative Recommendation. In: Resnick, Varian (eds.), pp. 66–72
[9] Lee, W.S.: Collaborative learning for recommender systems. In: Proceedings of the International Conference on Machine Learning (2001)
[10] Schafer, J.B., Konston, J.A., Riedl, J.: Recommender systems in e-Commerce. In: Proceedings of ACM Conference on e-commerce, pp. 158–166 (1999)
[11] Linden, G., Smith, B., York, J.: Amazon.com Recommendations: Item-to-Item Collaborative Filtering. IEEE Internet Computing 7(1), 76–80 (2003)
[12] Miller, B.N., Albert, I., Lam, S.K., Konstan, J.A., Riedl, J.: MovieLens Unplugged: Experiences with an Occasionally Connected Recommender System. In: Proceedings of the International Conference on Intelligent User Interfaces, pp. 263–268 (2003)
[13] Good, N., Schafer, J.B., Konstan, J., Borchers, A., Sarwar, B., Herlocker, J., Riedl, J.: Combining Collaborative Filtering with Personal Agents for Better Recommendations. In: Proceedings of American Association of Artificial Intelligence, pp. 439–446 (1999)
[14] Lemire, D., Maclachlan, A.: Slope One Predictors for Online Rating-Based Collaborative Filtering. In: SIAM Data Mining (2005)
[15] Wang, P., Ye, H.W.: A Personalized Recommendation Algorithm Combining Slope One Scheme and User Based Collaborative Filtering. In: Proceedings of the 2009 International Conference on Industrial and Information Systems, IIS 2009 (2009)
[16] Tan, P.-N., Steinbach, M., Kumar, V.: Introduction to Data Mining. Pearson Education
[17] Gong, S.J.: A Collaborative Filtering Recommendation Algorithm Based on User Clustering and Item clustering. Journal of Software 5(7), 745–752 (2010)
[18] `http://www.grouplens.org`

Handling Unlabeled Data in Gene Regulatory Network

Sasmita Rout, Tripti Swarnkar, Saswati Mahapatra, and Debabrata Senapati

Department of Computer Applications, ITER,
SOA University, Bhubaneswar, India
{rout_mca_sasmita,s_aswati}@yahoo.co.in,
tripti_sarap@yahoo.com,
debabratasenapati@gmail.com

Abstract. A gene is treated as a unit of heredity in a living organism. It resides on a stretch of DNA. Gene Regulatory Network (GRN) is a network of transcription dependency among genes of an organism. A GRN can be inferred from microarray data either by unsupervised or by supervised approach. It has been observed that supervised methods yields more accurate result as compared to unsupervised methods. Supervised methods require both positive and negative data for training. In Biological literature only positive example is available as Biologist are unable to state whether two genes are not interacting. A common adopted solution is to consider a random subset of unlabeled example as negative. Random selection may degrade the performance of the classifier. It is usually expected that, when labeled data are limited, the learning performance can be improved by exploiting unlabeled data. In this paper we propose a novel approach to filter out reliable and strong negative data from unlabeled data, so that a supervised model can be trained properly. We tested this method for predicting regulation in E. Coli and observed better result as compared to other unsupervised and supervised methods. This method is based on the principle of dividing the whole domain into gene clusters and then finds the best informative cluster for further classification.

Keywords: Gene, Gene Regulatory Network, Unlabeled data, SVM, K Means, Cluster, Transcription Factor.

1 Introduction

A gene is a unit of heredity of a living organism which resides on a stretch of DNA. All living organism depend on genes, as they specify all proteins and functional RNA chains. In other way a gene is "a locatable region of genomic sequence, corresponding to a unit of inheritance, which is associated with regulatory regions, transcribed regions, and other functional sequence regions". Gene regulatory networks (GRN) [1] explicitly represent the causality of developmental processes. They explain exactly how genomic sequence encodes the regulation

S.C. Satapathy et al. (Eds.): Proc. of Int. Conf. on Front. of Intell. Comput., AISC 199, pp. 113–120.
DOI: 10.1007/978-3-642-35314-7_14

of expression of the sets of genes that progressively generate developmental patterns and execute the construction of multiple states of differentiation. These are inhomogeneous compositions of different kinds of sub circuits, each performing a specific kind of function. This concept is important, because it holds the key to network design principles. Better understanding of the complexity of interdependencies among gene up and down regulation helps in inferring GRN. Different model architectures to reverse engineer gene regulatory networks from gene expression data have been proposed in literature [2]. These models represent biological regulations as a network genes, proteins etc and edges represents the presence of interaction activities between such network components. Four main network models based on unsupervised method can be distinguished: such as information theory models, Boolean network models, Differential and difference equation model and Bayesian models. Information theory model correlates two genes by means of a correlation coefficient and a threshold. Two genes are predicted to interact if the correlation coefficient of their expression levels is above a threshold. For example, TD-ARACNE [3], ARACNE [4] etc. infer the network structure. Boolean network model uses a binary variable to represent the state of a gene activity and a directed graph; here edges are represented by boolean functions to represent the interactions between genes. For example REVEAL [5] infers boolean network model from gene expression data. Differential and difference equation [6] describes gene expression changes as a function of the expression level of other genes. Bayesian model makes use of Bayes rules and consider gene expressions as random variables. The major advantage is that the Bayesian framework allows combining different types of data and prior knowledge in gene networks inference [7]. Just like unsupervised method, recently different supervised methods are also used to find the gene regulatory network. But in this approach unlike unsupervised method, it requires not only gene expression data but also a list of known regulation relationship. The following table lists some of the supervised and unsupervised methods. The basic principle to predict new regulations is: if a gene X having expression profile ep(X) is known to regulate a gene Y with expression profile ep(Y), then all other couples of genes A and B, having respectively expression profiles similar to ep(X) and ep(Y) are likely to interact. Expression profiles are taken as the feature vectors in the machine learning algorithm, while the result is a binary variable representing whether two genes interact or not.

Table 1. Methods under Unsupervised and Supervised approach

Unsupervised Approach	Supervised Approach
Information Theory Model	Decision Tree
Boolean Networks	SVM
Ordinary Differential Equation	Neural Network

It has been observed that supervised method give more accurate result as compared to unsupervised methods. Supervised methods require both genes and their complete linkage for their training. But in Biology literature only positive data is available as Biologist only able to tell which are interacting, i.e. Biological databases lists only interacting genes, it does not provide any genes information regarding non-interacting genes, which is a great challenge in finding gene regulatory network through supervised approach.

2 Related Work

2.1 Gene Regulatory Networks

Selection of Reliable Negatives: In [8] the authors tried to predict non-coding RNA genes, where the first set of negative examples is built by maximizing the distances of negative sample points to the known positive sample points by using a distance metric built upon the RNA sequence. Such a negative set is iteratively refined by using a binary classifier based on current positive and negative examples until no further additional negative examples can be found. In [9] they proposed a method applied to gene regulatory network, which selects a reliable set of negatives by exploiting the known network topology.

Probability Estimate Correction: PosOnly method: In paper [10], the conditional probabilities produced by a model trained on the labeled and unlabeled examples differ by only a constant factor from the conditional probabilities produced by a model trained on fully labeled positive and negative examples. Such result can be used to learn a probabilistic binary classifier, such as SVM (Support Vector Machine) with Platt scaling [11], using only positive and unlabeled data.

PSEUDO-RANDOM Method: In paper [9], a gene interaction network is modeled as a directed graph $< G, E >$ where G represents the genes, and E represents the set of directed interactions between genes. Let P be the known gene-gene interactions in E, then Q = E - P the unknown regulatory links, and N=Complement(E) the edges not contained in E. The unknown gene regulatory connections Q can be inferred by a machine learning scheme trained with the set of known regulatory connections. Precisely, P is the set of known positive examples, N is the set of all unknown negative examples and Q is the set of unknown positive examples. A selection of reliable negatives approach selects, from the unlabeled set $N \cup S$ of unknown connections, a subset of reliable negative examples $S \cong N$ and $S \cap Q$ which should be as much as possible composed of negative examples, i.e. and . Such negative examples are used to improve the training phase of a classifier. The PSEUDO-RANDOM method is built over the assumption that a regulatory network has no or few cycles and that it has a tree like structure. For complex eukaryote organisms such an assumption may not be true as many complex cell functions are based on homeostasis and feedback loops.

In contrast, for simpler including Escherichia coli and Saccharomyces cerevisiae, such an assumption may be correct: there are unsupervised approaches, such as ARACNE, that prune the final network by removing 3-arc cycles [3]. This leads to an heuristic that selects as candidate negatives those given by the union of the transitive closure of the known network and its transpose.
S = TC(P) ∪ Transpose (TC(P)) ∪ Transpose(P)

SIRENE: SIRENE (Supervised Inference of Regulatory Networks) [12] is a method to infer gene regulatory networks on a genome scale from a compendium of gene expression data. SIRENE differs from other approaches in that it requires not only gene expression data, but also a list of known regulation relationships both interacting and non-interacting. The authors used Support Vector Machine algorithm for predicting gene regulatory network.

2.2 Text Mining

In traditional text classification, a classifier is built using labeled training documents of every class. In paper [13], Given a set P of documents of a particular class (called positive class) and a set U of unlabeled documents that contains documents from class P and also other types of documents , called negative class documents, the authors build a classifier to classify the documents in U into documents from P and documents not from P. The key feature of the problem is that there is no labeled negative document, which makes traditional text classification techniques inapplicable. In this paper, the author proposed an effective technique to solve the problem. It combines the Rocchio method and the SVM technique for classifier building. Experimental results show that this method outperforms existing methods significantly.

3 Proposed Model

This is a general method for extracting strong reliable negative data for training the supervised model. As it has been already discussed that, GRN can be inferred from microarray data either by unsupervised or by supervised approach. It has been observed that supervised methods yields more accurate result as compared to unsupervised methods. Supervised methods require both positive and negative data for training. In Biological literature only positive example is available as Biologist are unable to state whether two genes are not interacting. A common adopted solution is to consider a random subset of unlabeled example as negative. Random selection may degrade the performance of the classifier. It is usually expected that, when labeled data are limited, the learning performance can be improved by exploiting unlabeled data. As shown in figure 2, p is the set of known interactions and U is unknown (both interacting and non-interacting). Traditionally, while training a supervised model, a random subset of U is taken for negative data, which used to degrade the performance of the classifier as while doing random selection some positive example from Q might be taken as negative.

3.1 Data

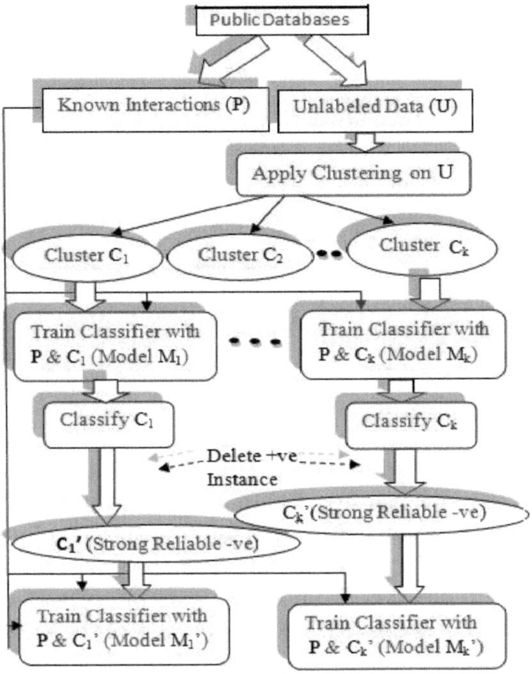

Fig. 1.

In our experiment, we used the expression and regulation data of E. Coli, which is publicly available in [14]. The expression data consist of a compendium of 445 E.coli microarray expressions profiles for 4345 genes. The microarrays were collected under different experimental conditions such as growth phases, antibiotics, different media, numerous genetic perturbations and varying oxygen concentrations. The regulation data consist of 3293 experimentally confirmed regulations between 154 TF and 1164 genes, extracted from the RegulonDB database [15].

3.2 Algorithm

Step 1 Consider the available interacting genes as true positive (P) and unlabeled genes as U
Step 2 Apply K-Means on U to build k number of clusters $(C_1, C_2, ...C_k)$
Step 3 for $i = 1$ to k do
Step 3.1 Train model M_i with P and C_i
Step 3.2 Classify C_i itself with model M_i

Step 3.3 P=Performance of M_i
Step 3.4 Delete Positive examples from C_i if any
Step 3.5 Train classifier M_i* with P and the remaining instances of C_i i.e. C_i*
Step 3.6 P*=Performance of M_i*
Step 3.7 Compare P and P*

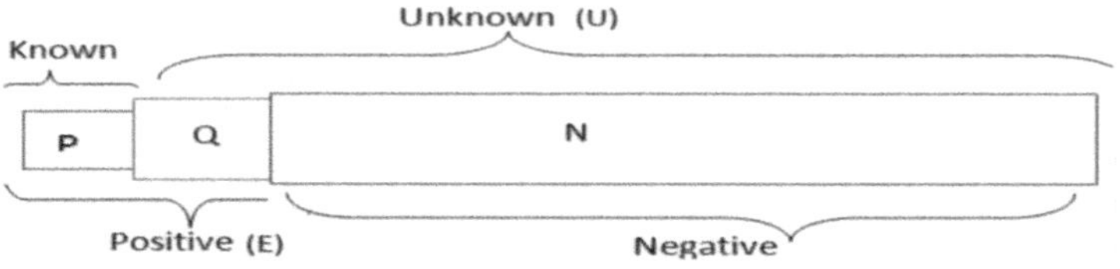

Fig. 2.

3.3 Experimental Result

The experiment is performed on those Transcription Factors (TF) having more than 50 interactions, such as crp (900), fis (1166), fnr (1218), himD (1451), rpoD (2307) etc. where each TF is associated with an unique number. We run the algorithm for each TF and observed that the performance of the classifier after removing the supposed to be positive example is better than the classifier taken earlier. It has been observed that irrespective of the number of cluster in K-means , the correct rate of almost all cluster (after removing the +ve instances) are better than the earlier model which has been shown in figure 3a. Figure 3b shows the classifiers in the ROC space. The classifiers performances are measured for both k=10 and k=15. And it has been observed that the performance is good irrespective of the number of cluster. But we have shown the results only for k=10. We have taken SVM [16] as the classifier for each cluster. The correct rate of different TF is shown in Figure 4. a and b.

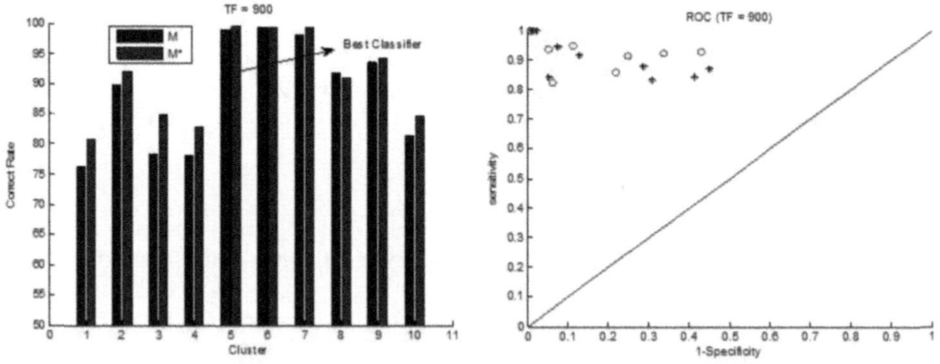

Fig. 3. a,b

	TF = 900		TF = 1166		TF = 1218		TF=1451		TF=2307		TF = 98	
Perf	P	P*	P	P*	P	P*	P	P*	P	P*	P	P*
Clu1	0.7625	0.8076	0.9003	0.9153	0.8750	0.8925	0.9988	1.0000	0.8324	0.8612	1.0000	1.0000
Clu2	0.8967	0.9210	0.9594	0.9668	0.9029	0.9070	0.8743	0.8891	0.9045	0.9049	0.9258	0.9361
Clu3	0.7825	0.8480	0.9441	0.9360	0.9750	0.9832	0.9040	0.9317	0.6414	0.7584	0.9174	0.9352
Clu4	0.7793	0.8274	0.6889	0.7561	1.0000	1.0000	0.9197	0.9476	0.9819	0.9842	0.9331	0.9321
Clu5	0.9895	0.9947	0.9999	1.0000	0.8726	0.9398	0.8770	0.8787	0.7871	0.8665	0.9755	0.9691
Clu6	0.9941	0.9941	0.9916	0.9916	0.9261	0.9650	0.8854	0.8947	0.9368	0.9550	0.9437	0.9781
Clu7	0.9800	0.9933	0.9548	0.9817	0.9825	0.9941	0.9769	0.9846	0.6897	0.7945	0.9322	0.9153
Clu8	0.9350	0.9101	0.9872	1.0000	0.9828	0.9828	0.9928	1.0000	0.7797	0.8219	0.8758	0.9170
Clu9	0.9350	0.9424	0.8313	0.8410	0.9044	0.8956	0.7865	0.9162	0.9974	0.9975	1.0000	1.0000
Clu10	0.8133	0.8464	0.9777	0.9824	0.9104	0.9308	0.9558	0.9609	0.9370	0.9604	0.9904	0.9904

	TF = 1450		TF = 1473		TF=1671		TF=1863		TF = 2310		TF = 2311	
Perf	P	P*	P	P*	P	P*	P	P*	P	P*	P	P*
Clu1	1.0000	1.0000	0.9928	0.9928	0.9324	0.9726	1.0000	1.0000	0.9538	0.9535	0.8905	0.8955
Clu2	0.9489	.9830	0.9617	0.9713	0.9513	0.9162	0.9375	0.9509	1.0000	1.0000	0.9294	0.9477
Clu3	0.9667	0.9497	0.9167	0.9379	0.9839	1.0000	0.9600	0.9882	1.0000	1.0000	0.9330	0.9258
Clu4	0.8509	0.8825	0.9911	1.0000	1.0000	1.0000	1.0000	1.0000	0.8997	0.9014	0.9302	0.9634
Clu5	0.9029	0.9104	0.8606	0.8909	0.9877	0.9939	0.9544	0.9514	0.9916	0.9914	0.9237	0.9395
Clu6	0.9839	0.9919	0.9899	1.0000	1.0000	0.9881	0.9740	1.0000	0.9237	0.9652	0.9943	0.9942
Clu7	0.8602	0.8703	0.8510	0.9241	0.8702	0.8898	0.9932	0.9932	0.9948	1.0000	0.9956	1.0000
Clu8	0.8681	0.9119	0.9185	0.9167	0.9112	0.9275	0.9600	0.9726	0.9024	0.9271	0.8267	0.8197
Clu9	0.8710	0.8337	0.8687	0.9055	0.9498	0.9766	0.9010	0.9016	0.9454	0.9496	0.9640	0.9628
Clu10	0.9038	0.9114	0.8701	0.8551	0.9353	0.9847	1.0000	1.0000	0.9655	0.9706	0.9360	0.9580

Fig. 4. a, b

4 Conclusion

Supervised methods always need a complete set of known regulatory networks i.e. gene expression data and list of known regulation relationship both interacting and non-interacting. But In Biology literature only positive examples are available, as Biologists do not have idea about the genes which are not interacting. That means only positive examples are available. So a common adopted solution is to consider all or a random subset of unlabeled example as negative, for the training of a supervised model. But the random selection of false negatives could affect the performance of the classifier, as it learns wrongly potentially positive examples as negatives. Hence learning from positive and unlabeled data is a hot topic. So instead of selecting a random subset from unlabeled data, the subset of instances can be further processed to delete the potentially positive example through clustering and classification. The instances left behind in the clusters are the strong and reliable negative instances, which can be used for training a supervised model. As supervised approach yields better result and can help in finding the functions of unknown genes, identifying pathways, finding potential target and managing patient's health based on genomic sequence.

Acknowledgments. To complete this work, we have got valuable suggestions and guidance from different experts of this field. We also thankful to all the authors of the references we have taken.

References

1. Davidson, E., Levine, M.: Gene Regulatory Network. PNAS 102(14), 4935 (2005)
2. Hecker, M., Lambeck, S., Toepfer, S., van Someren, E., Guthke, R.: Gene regulatory network inference: Data integration in dynamic models-A review. Bio Systems (2008)
3. Zoppoli, P., Morganella, S., Ceccarelli, M.: TimeDelay-ARACNE: Reverse engineering of gene networks from time-course data by an information theoretic approach. BMC Bioinformatics (2010)
4. Margolin, A.A., Nemenman, I., Basso, K., Wiggins, C., Stolovitzky, G., Dalla Favera, R., Califano, A.: ARACNE: An Algorithm for the Reconstruction of Gene Regulatory Networks in a Mammalian Cellular Context. BMC Bioinformatics (2006)
5. Liang, S., Fuhrman, S., Somogyi, R.: Reveal, A General Reverse Engineering Algorithm for Inference of Genetic Network Architectures. In: Pac. Symp. Biocomput., pp. 18–29 (1998)
6. de Jong, H.: Modeling and simulation of genetic regulatory systems: a literature review. J. Comput. Biol. (2002)
7. Werhli, A.V., Husmeier, D.: Reconstructing gene regulatory networks with Bayesian networks by combining expression data with multiple sources of prior knowledge. Stat. Appl. Genet. Mol. Biol. (2007)
8. Wang, C., Ding, C., Meraz, R.F., Holbrook, S.R.: PSoL: a positive sample only learning algorithm for finding non-coding RNA genes. Bioinformatics, 2590–2596 (2006)
9. Ceccarelli, M., Cerulo, L.: Selection of negative examples in learning gene regulatory networks. In: IEEE International Conference on Bioinformatics and Biomedicine Workshop, BIBMW 2009, pp. 56–61 (2009)
10. Elkan, C., Noto, K.: Learning classifiers from only positive and unlabeled data. In: Proceeding of the 14th ACM SIGKDD International Conference on Knowledge Discovery and Data Mining, KDD 2008, pp. 213–220. ACM, New York (2008)
11. Lin, H.T., Lin, C.J., Weng, R.C.: A note on Platt's probabilistic outputs for support vector machines. Mach. Learn., 267–276 (2007)
12. Mordelet, F., Vert, J.P.: SIRENE: supervised inference of regulatory networks. Bioinformatics, 76–82 (2008)
13. Li, X., Liu, B.: Learning to Classify Texts Using Positive and Unlabeled Data. In: Proceedings of the Eighteenth International Joint Conference on Artificial Intelligence, IJCAI 2003, Acapulco, Mexico, August 9-15, pp. 587–594 (2003)
14. Faith, J.J., et al.: Large-scale mapping and validation of Escherichia coli transcriptional regulation from a compendium of expression profiles. PLoS Biol. (2007)
15. Salgado, H., et al.: Regulondb (version 5.0): Escherichia coli k-12 transcriptional regulatory network, operon organization, and rowth conditions. Nucleic Acids Res. 34(Database issue), D394–D397 (2006)
16. Vapnik, V.N.: Statistical Learning Theory. Wiley, New York (1998)

Naive Credal Classifier for Uncertain Data Classification

S. Sai Satyanarayana Reddy[1], G.V. Suresh[1], T. Raghunadha Reddy[2], and B. Vishnu Vardhan[3]

[1] LBRCE, Myalvaram
{saisn90,vijaysuresh.g}@gmail.com
[2] SIET, Narsapur
raghu.sas@gmail.com
[3] JNTUCEJ, Jagitiyala
vishnubulusu@yahoo.com

Abstract. Data uncertainty due to various causes, including imprecise measurement, network latency, out-dated sources and sampling errors, is common in real-world applications. Data Analysis applications are typical in collecting and accumulating large amounts of uncertain data. This attracted more and more database community to analyze and resolve the uncertainty incured in the large data sets. We, in this article, present a naive classifier, which is a Set-Valued counterpart of Naive Bayes that is extended to a general and flexible treatment of incomplete data, yielding to a new classifier called *Naïve Credal Classifier*. Naïve Credal Classifier *is* an application on closed and convex sets of probability distributions called Credal sets, of uncertainty measures. The Naïve Credal Classifier extends the discrete Naive Bayes classifier to imprecise probabilities and also models both prior ignorance and ignorance about the likelihood by sets of probability distributions. This is a new means to deal with uncertain data sets that departs significantly from most established conventional classification methods. Experimental results show that proposed model exhibits reasonable accuracy performance in classification on uncertain data.

Keywords: Uncertain Data, Naive Bayesian Classifier, Credal Classifier, Data Mining.

1 Introduction

Data mining and knowledge discovery techniques are widely used in various applications in business, government, and science. Data is frequently coupled with uncertainty because of inaccuracy in measurement, sampling discrepancy, collecting data from outdated data sources, or other errors (i.e. limitations of the observation equipment, limited resources to collect, store, transform, analyze, or understand data) [1] [2]. Uncertainty may be present in various application include banking, bioinformatics, environmental modeling, epidemiology, finance, marketing, medical diagnosis, and meteorological data analysis [1] [2] [3]. Moreover, sensors are often

S.C. Satapathy et al. (Eds.): Proc. of Int. Conf. on Front. of Intell. Comput., AISC 199, pp. 121–132.
DOI: 10.1007/978-3-642-35314-7_15

used to collect data in applications such as environment surveillance, security, and manufacturing systems will introduce dynamic errors like measurement inaccuracies, sampling frequency of the sensors, deviation caused by a rapid change of the measured property over time (e.g., drift, noise), wireless transmission errors, or network latencies. There is also uncertainty in survey data (e.g., number '1' vs. uppercase letter 'I' vs. lowercase letter 'L') and uncertainty due to data granularity (e.g., city, province) in taxonomy. Disguised or missing data also introduce uncertainty. Data uncertainty can be categorized into two types, namely existential uncertainty and. value uncertainty [2] [3]. In the first type it is uncertain whether the object or data tuple exists or not. For example, a tuple in a relational database could be associated with a probability value that indicates the confidence of its presence. In value uncertainty, a data item is modeled as a closed region which bounds its possible values, together with a probability density function of its value. All these scenarios lead to huge amounts of uncertain data in various real-life situations.

2 Research Background

2.1 Credal Set Concepts

Let Xdenotes a generic variable, taking values in a finite set Ξ:= {x(1). . . x(n)}. A probability mass function over X, which is a nonnegative real map over Ξ normalized to one, will be denoted by P(X). A credal set over X, which is a convex set of probability mass functions over X, will be denoted by K(X).The extreme points of K(X) are denoted as ext [K(X)][1]. Here we only consider CSs with a finite number of extreme points, i.e., such that |ext [K(X)]| < +∞. Geometrically, a CS is therefore a polytope in the probability simplex, and can be equivalently specified through an explicit enumeration of its extreme points (V-representation) and a finite set of linear constraints (H-representation).Unlike the V-representation, which is clearly uniquely defined; different H-representations can specify the same CS. The notation $\overline{K}(X)$ is used for the vacuous CS, i.e., the (convex) set of all the probability mass functions over X. It is easy to note that |ext [$\overline{K}(X)$]| = |$\mathcal{X}$|.

2.2 Lower Probabilities

A conjugate pair of *lower/upper probability* operators [1] is defined as a pair $(\underline{P}, \overline{P})$ of nonnegative real maps over the power set $2^{\mathcal{X}}$, such that: (i) $\underline{P}(\theta) = 0$; (ii) the operators are respectively super- and sub-additive, i.e.,

$$\underline{P}(A \cup B) \geq \underline{P}(A) + \underline{P}(B)$$
$$\overline{P}(A \cup B) \leq \overline{P}(A) + \overline{P}(B)$$

for each A , B $\in 2^{\mathcal{X}}$; (iii) the following conjugacy relation holds for each $A \in 2^{\mathcal{X}}$

$$\overline{P}(A) = 1 - \underline{P}(\mathcal{X} \setminus A) \tag{1}$$

According to (1), the operator $\overline{P}$ is completely determined by its conjugate $\underline{P}$ (and vice versa).

2.3 Credal Classification

A classifier is an algorithm that allocates new objects to one out of a finite set of previously defined groups (or classes) on the basis of observations on several characteristics of the objects, called attributes or features [4][5][6]. A credal classifier is defined as a function that maps an object into a set of classes. Credal classification assumes that the knowledge hidden in data does not always allow the classes to be completely ranked: in this case, only the classes that are dominated in the partial order can be discarded from consideration, producing the set of *undominated* classes as output. Credal classification is closely related to the theory of *imprecise probabilities* [7][8][9] in that it does not require that probability values be precise: they can be intervals or, more generally, uncertainty can be modelled by a set of distributions. The classifier starts in condition of prior ignorance. Every new instance is first classified and only then stored in the knowledge base together with the actual class, which is unknown at the classification stage. The classifier's knowledge grows incrementally, so that its predictions become more reliable as more units are collected. A credal classifier naturally shows this behaviour. Initially it will produce all the classes with more units, the average output set size will decrease approaching one in the limit [10]. When we do not assume anything about the mechanism generating the missingness, missing data give rise to a set of possible distributions modelling the domain. From this, it is immediate to obtain a credal classifier that is robust to every possible mechanism of missingness, i.e. to all the possible replacements of missing data with known values.

2.4 Naive Credal Classification

Credal sets can be applied to the naive classifier by substituting the conditional distributions $P[C], P[A_i \mid C](i = 1, ..., n, C \in \Omega_C)$ with the credal sets $\wp_C, \wp_{A_i}^C \ (i = 1, ..., n, C \in \Omega_C)$ [14][18] then, the focus is on the way the classification should be realized. In the sequel first it is analyzed the straight extension of the point probability procedure to credal sets. This implies to compute $P[C \mid A_1, ...A_n] \forall C \in \Omega_c$ and compare them.

2.5 Interval Computation

Consider the minimization, the maximization is analogous. Furthermore, consider the computation of the posterior probability for a fixed value c of *C*. The objective function call be rewritten applying the definition [13][14]

$$\min_{P[C,A_1,\ldots,A_n]\in\wp} P[C \mid A_1 = a_1,\ldots,A_n = a_n] \tag{2}$$

$$\max_{P[C,A_1,\ldots,A_n]\in\wp} P[C \mid A_1 = a_1,\ldots,A_n = a_n] \tag{3}$$

$$P[c \mid a_1,\ldots\ldots,a_n] =$$

$$\frac{P[c,a_1,\ldots\ldots,a_n]}{\Sigma_{C\in\Omega_C} P[C,a_1,\ldots\ldots,a_n]} = \tag{4}$$

$$\left(\frac{\Sigma_{C\in\Omega_C} P[C,a_1,\ldots\ldots,a_n]}{P[c,a_1,\ldots\ldots,a_n]}\right)^{-1} = \tag{5}$$

$$\left(1+\frac{\Sigma_{C\in\Omega_C} P[C,a_1,\ldots\ldots,a_n]}{P[c,a_1,\ldots\ldots,a_n]}\right)^{-1} \tag{6}$$

where it is assumed $P[c,a_1,\ldots\ldots,a_n] \neq 0$ for the passage from Eq. (4) to Eq. (5) notice that $P[c,a_1,\ldots\ldots,a_n] = 0$ it is clear from Eq. (4) that $P[c,a_1,\ldots\ldots,a_n] = 0$ (the case when also $P[a_1,\ldots\ldots,a_n] = 0$ is not considered because it would imply an undefined conditional probability $P[c \mid a_1,\ldots\ldots,a_n]$.The naive classifier has a single root node *(C)* which corresponds to the classification variable and its children nodes are the attribute variables [18]. Each node X hi the network has the usual conditional distributions $P[X \mid P_a(X)]$ of the node given the state of the Parent *Pa* (X*)* which are estimated from data or from other types of knowledge. Given a particular instance $A_1 = a_1,\ldots\ldots,A_n = a_n$ of the attribute variables the classification is made by computing the posterior probability $P[C \mid a_1,\ldots.a_n]$ for each value of ***C*** and by picking up the value with Maximum probability, $c^* = \arg\max_c P[C \mid a_1,\ldots\ldots,a_n]$.

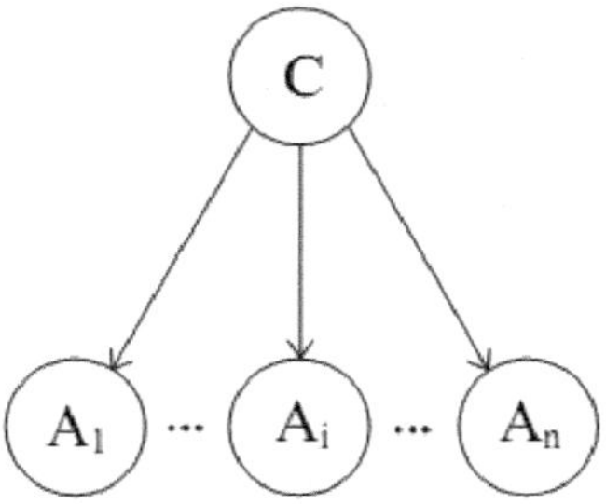

Fig. 1. Naive Classifier

Finally, Eq. (6) is written according to the topology of the graph.

$$P[c \mid a1, \ldots., an]$$

$$\left(1 + \frac{\sum_{C \in \Omega_{C \setminus \{c\}}} P[C] \prod_{i=1}^{n} P[a_i | C]}{P[c] \prod_{i=1}^{n} P[a_i | c]} \right)^{-1} \tag{7}$$

Now the mininmization problem is written by substituting **its** objective function with the right member of Eq (7),

$$\min_{P[C] \in \wp C} \; \min_{P[A_i | C] \in \wp_{A_i}^{C}, C \in \Omega_C, 1, \ldots., n}$$

$$\left(1 + \frac{\sum_{C \in \Omega_{C \setminus \{c\}}} P[C] \prod_{i=1}^{n} P[a_i | C]}{P[c] \prod_{i=1}^{n} P[a_i | c]} \right)^{-1} \tag{8}$$

Notice that since the objective is now expressed by means of the local conditional distributions, also the minimization is taken over the possible conditional distributions in the credal sets local to the nodes. Let us now focus on the inner minimization only. The goal is the *maximization of* the fractional function inside parentheses since this is Equivalent minimizing the reciprocal. First of all, notice that it is possible to minimize the denominator and to maximize the numerator separately, in fact they do not share any term. Second, notice that the quantities $P[a_i \mid C](C \in \Omega_C, i = 1, \ldots., n)$ are in dependent one another' because they are defined by means of disjoint sets of constraints. This means that the choice of conditional distribution $P[a_i \mid C]$ in $\wp_{A_i}^{C}$ can be made without taking into account the choice of any other conditional distribution. Consider the denominator $P[c]$ is a non-negative number; therefore, the denominator is minimized when the product $\prod_{i=1}^{n} P[a_i \mid c]$ is minimized. This is obtained by setting any $P[a_i \mid c]$ to its minimum *i.e.* $\prod_{i=1}^{n} \underline{P}[a_i \mid c]$. An analogous argument holds for the numerator $P[C]$ is non-negative $\forall C \in \Omega_C$ and the sum is made by terms that can be optimized separately. Hence, the numerator is maximized when the product of the conditional distributions is set to $\prod_{i=1}^{n} \overline{P}[a_i \mid c] \forall C \in \Omega_{C \setminus \{c\}}$ then it becomes

$$\min_{P[C] \in \wp_C} \left(1 + \frac{\sum_{C \in \Omega_{C \setminus \{c\}}} P[C] \prod_{i=1}^{n} \overline{P}[a_i \mid C]}{P[c] \prod_{i=1}^{n} \underline{P}[a_i \mid c]} \right)^{-1} \tag{9}$$

For any given node X denote with $\overline{\wp}_X^{Pa(X)}$ the finite sub set of $\wp_X^{Pa(X)}$ made by its extreme points then the Eq.9 can written as

$$\min_{P[C] \in \overline{\wp}_C} \left(1 + \frac{\sum_{C \in \Omega_{C \setminus \{c\}}} P[C] \prod_{i=1}^{n} \overline{P}[a_i \mid C]}{P[c] \prod_{i=1}^{n} \underline{P}[a_i \mid c]} \right)^{-1} \tag{10}$$

which is simply solved by enumerating the extreme distributions in $\overline{\wp}_C$ .of course, it is also necessary to know the extremes of $P[a_i \mid C] \forall C \in \Omega_C, i = 1, \ldots\ldots, n)$,in order to apply formula (10). This is obtained by solving problems $\min_{P[A_i \overline{\wp}_{A_i}^{c}} P[a_i \mid c]$ and $\max_{P[A_i \mid C] \in \overline{\wp}_{A_i}^{C}} P[a_i \mid C] \forall C \in \Omega_C \setminus \{c\}, \forall i = 1, \ldots\ldots, n$ which can be again done enumerating the extreme distributions in the respective feasible sets. We can also write the Eq(10) for maximizing

$$\max_{P[C] \in \overline{\wp}_C} \left(1 + \frac{\sum_{C \in \Omega_{C \setminus \{c\}}} P[C] \prod_{i=1}^{n} \underline{P}[a_i \mid C]}{P[c] \prod_{i=1}^{n} \overline{P}[a_i \mid c]} \right)^{-1} \tag{11}$$

2.6 Credal Dominance

So far, the analysis has focused on the computation of the ignorance intervals for the states of **C**. It is useful to observe that the potentially available information with credal sets is greater than that provided by intervals; the credal set for $P[C \mid a_1, \ldots\ldots, a_n]$, say $\wp_C^{a_1, \ldots, a_n}$ generally conveys greater knowledge than that given by the separate intervals for the posterior probabilities of **C.** In fact, the credal set can also represent possible constraints between the above probabilities, which disappear with the interval view [14][15].

Let *X* be a discrete variable and x_1, x_2 two states in the domain of X. Consider the distribution $P[X \mid E] \in \wp_X^E$ where E represents what is known and $\wp_X^E$ is a non-empty set of distributions. The state x_1 is said to be credal-dominant as compared to x_2, $x_1 \succeq x_2$ if for any distribution $P[X \mid E] \in \wp_X^E$, $P[X = x_1 \mid E] \geq P[X = x_2 \mid E]$.

2.7 Classification Procedure

Let us denote the classification variable by *C*, taking values in the nonempty and finite set *w*, where the possible classes are denoted k by lower-case letters. We measure k features ($A_1,\ldots,A_k$) taking generic values ($a_1, \ldots, a_k$) = a from the sets $\mathcal{A}_1,\ldots,\mathcal{A}_k$, which are assumed to be nonempty and finite. Let E [U(c)|a,n,t] denote the expected utility with respect to from choosing class c, given a, the previous data n and a vector t of hyperparameters[20]. Since t belongs to a region, there are many such utilities for every class c, so that we cannot always compare two classes: generally, we have a partial order on the classes that only allows us to discard the dominated ones. The partial order depends on the chosen dominance criterion. We use credal dominance, defined below. We say that class c' credal-dominates class c'' if and only if $E[U(c')a,n,t] > E[U(c'')a,n,t]$ for all values of t in the imprecise model.

3 Related Work

3.1 Naïve Credal Classifier (NCC)

The naive credal classifier [12][13][14] addresses this problem by specifying a set of priors through the IDM. More in particular, it allows the marginal prior probability of each class to vary between 0 and 1 and the conditional prior probability of each value of the features given each class to identically vary between 0 and 1. Overall, NCC adopts a set of joint priors over features and classes; in fact, it considers all the Dirichlet priors which are admissible by NBC (*Naïve Bayes Classifier*)[14]. The set of priors is then turned into a set of posteriors by the generalized Bayes' rule, once the likelihood has been computed from the data **d;** at this point, NCC is trained. Assuming to have **k** features, NCC classifies an instance when the values $f = \{f_1, \ldots, f_k\}$ of the features are known. NCC identifies the non-dominated classes; this is achieved by running pair-wise comparison between the classes. Let us consider an example with two classes c_1 and c_2; if the set of posteriors contains a distribution such that $P(c_1 \mid d, f) > P(c_2 \mid d, f)$ but also a distribution such that

$P(c_2 \mid d, f) > P(c_1 \mid d, f)$, both c_1 and c_2 are non dominated; instead, if $P(c_1 \mid d, f) > P(c_2 \mid d, f)$ on every posterior, c_1 dominates c_2 .Therefore there can be one or multiple non-dominated classes, yielding respectively determinate or indeterminate classifications. When there is a single non-dominated class, this corresponds to the class identified by NBC; if instead there are more non-dominated classes, they contain the class returned by NBC and the classification issued by NBC is prior-dependent [18]. On prior-dependent instances, NCC returns a less informative but more robust answers than NBC. Indeterminate classifications are less frequent on large data sets, because the choice of the prior becomes less influential as more data are available; yet, prior-dependent instances can be present even on large data sets[19].

3.2 An Example

A company wants to assess the risk it incurs in getting selling the bikes to the customers who live in different cities. The risk(R) can be classified as *Low Medium or High* and is related to two attributes of the customers age (A) defined over (*young, middle-aged, old*) and city(C) over (*HYD BAN CHN*) in which they live. Concerning the Customer's Age it is supposed that middle-aged persons have better behaviour, than young and old people concerning the risks of the cities that are ranked as HYD<BAN<CHN.We will calculate the prior probability intervals for the risk classes in Table1.*(all the Probabilities are artificial)*

Table 1. Prior probability intervals, $[\underline{P}[R], \overline{P}[R]]$

Risk Class	$[\underline{P}[R]$	$[\,P[R]$
Low	0.77	0.85
Medium	0.10	0.15
High	0.05	0.08

The Table 1 infers the fact that most of the customers are known to be low-risk with a percentage variation from 75% to 85% and 10% to 15% medium Risk and 5%-8% high risk. We can calculate the conditional probabilities for the age and city for given risk class Table 2 Table 3 gives the $[\underline{P}[A \mid R],\ \overline{P}[A \mid R]]$ and $[\underline{P}[T \mid R],\ \overline{P}[T \mid R]]$

Table 2. Conditional Probability intervals, $[\underline{P}[A \mid R], \overline{P}[A \mid R]]$

	R		
Age	Low	Medium	High
Low	[0.15,0.22]	[0.27,0.32]	[0.60,0.70]
Medium	[0.50,0.55]	[0.33,0.38]	[0.50,0.15]
High	[0.28,0.34]	[0.34,0.38]	[0.20,0.30]

Table 3. Conditional Probability intervals, $[\underline{P}[T \mid R], \overline{P}[T \mid R]]$

	R		
City	Low	Medium	High
HYD	[0.70,0.72]	[0.15,0.20]	[0.02,0.06]
BAN	[0.18,0.20]	[0.60,0.65]	[0.22,0.28]
CHN	[0.08,0.10]	[0.20,0.25]	[0.66,0.72]

Let us consider the case of an old person living in HYD; We compute intervals for $P[R, A = old, T = HYD]$,recalling that $\underline{P}[R, A = old, T = HYD] = \underline{P}[R]\ \underline{P}[A = old \mid R]\ \underline{P}[T = HYD \mid R]$ and $\overline{P}[R, A = old, T = HYD] = \overline{P}[R]\ \overline{P}[A = old \mid R]\ \overline{P}[T = HYD \mid R]$.The intervals are shown by means of line segments for instance, $P[R = low, A = old, T = VE]$ belongs to the interval [0151,0.208]. By applying interval dominance to such intervals, we obtain a total order on the states of R because the intervals do not overlap. In this case the customer is classified as low risk.

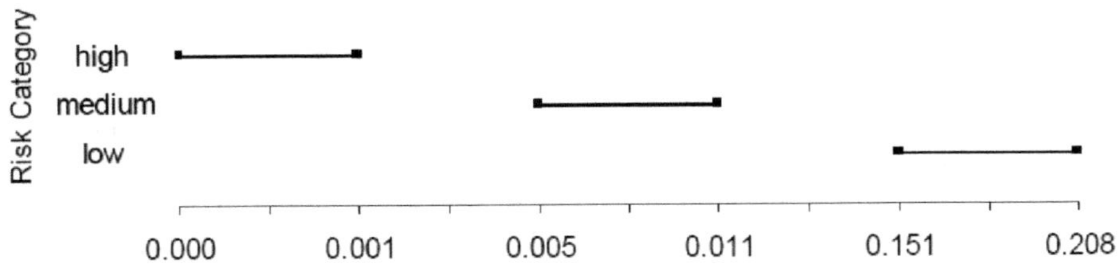

Fig. 2. Joint Probability intervals for risk Category age and city $\underline{P}[R, A = old, T = HYD]$ $\overline{P}[R, A = old, T = HYD]$

We have to compute posterior probability for risk category by using Eq (9) and Eq (10) then *P[R=low|A=old, T=VE]* lies in the interval [0.922, 0.975].Let us now compute young people in *CHN P[R, A=young, T=CHN]*.Now interval dominance only implies a partial order of states of R because the intervals for the states low and medium overlap but it is still possible to obtain a single credal-dominant state ,i.e., *high risk.* The probability *P[R=high |A=young, T=MI]* lies in the interval [0.435, 0.693].

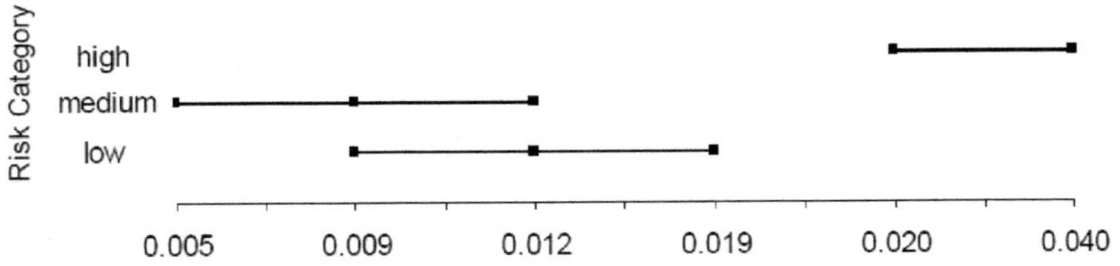

Fig. 3. Probability intervals age and city $\underline{P}[R, A = young, T = CHN]$ $\overline{P}[R, A = young, T = CHN]$

Next we compute for young people in in BAN, the intervals are shown below in Fig 4.this time a single Credal-dominant state is not available because state high is credal-dominated and the intervals for the other two states overlap. The result of the classification is the set*{low, medium}*

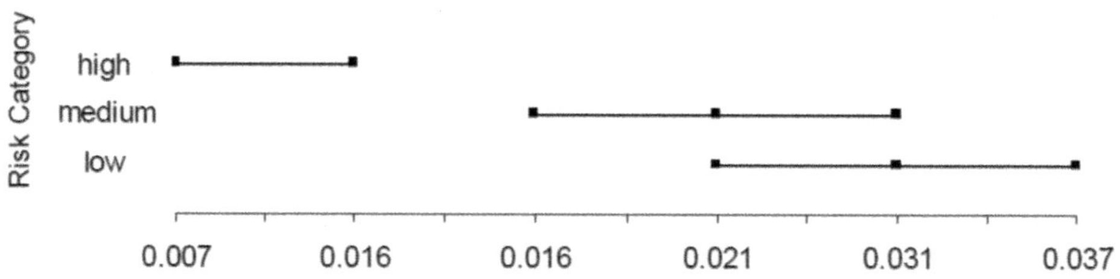

Fig. 4. Joint Probability intervals age and city $\underline{P}[R, A = young, T = BAN]$ $\overline{P}[R, A = young, T = BAN]$

4 Experiment and Results

We focus on the spect data set, which is made of 2 classes and 267 instances[19][20]. We ran 10-folds cross-validation for the NCC. The NCC produced a precise classification (i.e., a single class) for about the 59.55% of the 155 instances, with an accuracy C_1=52.01%. In the remaining 40.45% (S) of instances, it produced an average of Z=2.36 classes out of the possible 4. This set of classes contained the actual class with probability 0.82 (Cs). The most relevant output here is S: the NCC states that on about 40% of the instances, the available knowledge is not sufficient to produce a single class, but only a set of possible alternative classes. We can appreciate the behaviour of the NCC by also noting that the NCC isolates a subset of instances on which robust predictions are possible (C_1). Also, instead of predicting at random on the remaining instances, the NCC produces a set of classes with a high probability (Cs) of including the actual class: in other words, we can be confident that the discarded classes have low chance of being true.

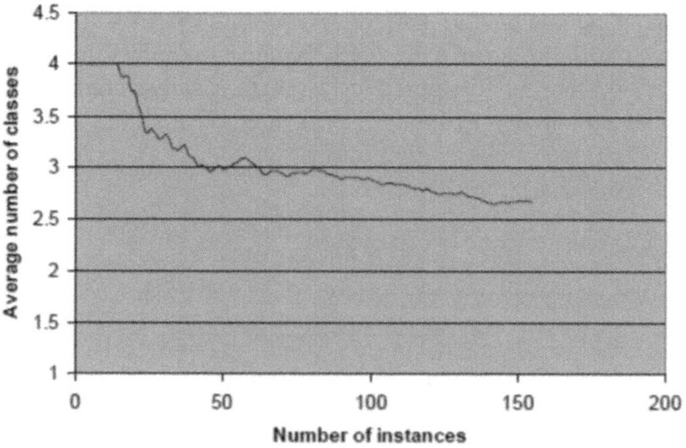

Fig. 5. The average number of classes produced by the NCC as a function of the number of instances

5 Experiment and Results

This paper proposes credal classification as a generalization of standard classification and realizes credal classification by extending the naive Bayes classifier to credal sets. It derives the related procedures for classification and for the computation of posterior lower and upper probabilities. By analyzing the computational complexity of the procedures, it shows that Naive Credal Classification is a well-solvable task. Naive Credal Classifier is efficient to deal with complete and uncertain data which are a pervasive problem in applied statistical inference. Our proposal deals with the combination of imprecise due to uncertain data with imprecise Dirichlet model. This approach will enable the NCC to be inferred from incomplete data sets in a simple and sound way. We plan to explore more classification approaches for various uncertainty models and find more efficient training algorithms in the future.

References

[1] Leung, C.K.-S.: Mining uncertain data. In: WIREs Data Mining and Knowledge Discovery, vol. 1. John Wiley & Sons, Inc. (July/August 2011)
[2] Suresh, G.V., Shaik, S., Reddy, E.V., Shaik, U.A.: Gaussian Process Model for Uncertain Data Classification. International Journal of Computer Science and Information Security (IJCSIS) 8(9), 111–115 (2010)
[3] Chau, M., Cheng, R., Kao, B.: Uncertain Data Mining: A New Research Direction. In: Proceedings of the Workshop on the Sciences of the Artificial, Hualien, Taiwan, December 7-8 (2005)
[4] Aggarwal, C.C.: Managing and Mining Uncertain Data. Kluwer Academic Publishers, Boston
[5] Aggarwal, C.C.: A Survey of Uncertain Data Algorithms and Applications. IEEE Transactions on Knowledge and Data Engineering 21(5) (2009)
[6] Duda, R.O., Hart, P.E.: Pattern classification and scene analysis. Wiley, New York (1973)
[7] Duda, R.O., Hart, P.E., Stork, D.G.: Pattern classification, 2nd edn. Wiley (2001)
[8] Zaffalon, M.: A credal approach to naive classification. In: de Cooman, G., Cozman, F., Moral, S., Walley, P. (eds.) ISIPTA 1999, pp. 405–414. Univ. of Gent, Belgium (1999); The Imprecise Probabilities Project
[9] Zaffalon, M.: Statistical inference of the naive credal classifier. In: de Cooman, G., Fine, T., Seidenfeld, T. (eds.) ISIPTA 2001, pp. 384–393. Shaker Publishing, The Netherlands (2001)
[10] Walley, P.: Statistical Reasoning with Imprecise Probabilities. Chapman and Hall, New York (1991)
[11] Walley, P.: Inferences from multinomial data: learning about a bag of marbles. J. R. Statist. Soc. B 58(1), 3–57 (1996)
[12] Zaffalon, M., Wesnes, K., Petrini, O.: Reliable diagnoses of dementia by the naive credal classifier inferred from incomplete cognitive data. In: IDSIA, Galleria 2, 6928 Manno (Lugano), Switzerland
[13] Zaffalon, M.: The naive credal classifier. Journal of Statistical Planning and Inference 105(1), 5–21 (2002)

[14] Zaffalon, M.: Statistical inference of the naive credal classifier. In: de Cooman, G., Fine, T.L. (eds.) Proceedings of the Second International Symposium on Imprecise Probabilities and Their Applications, ISIPTA 2001, pp. 384–393. Shaker, The Netherlands (2001)
[15] Zaffalon, M., Wesnes, K., Petrini, O.: Reliable diagnoses of dementia by the naive credal classifier inferred from incomplete cognitive data. Artificial Intelligence in Medicine 29(1-2), 61–79 (2003)
[16] Pearl, J.: Probabilistic Reasoning in Intelligent Systems: Networks of Plausible Inference. Morgan Kaufmann, San Mateo
[17] Wolpert, D.H., Wolf, D.R.: Estimating functions of probability distributions from a finite set of samples. Physical Review E 52(6), 6841–6854 (1995)
[18] Zaffalon, M., Hutter, M.: Robust inference of trees, Tech. Rep. IDSIA-11-03, IDSIA (2003)
[19] Murphy, P.M., Aha, D.W.: UCI repository of machine learning databases (1995), http://www.sgi.com/Technology/mlc/db/
[20] Kohavi, R., John, G., Long, R., Manley, D., Pfleger, K.: MLC++: a machine learning library in C++. In: Tools with Artificial Intelligence, pp. 740–743. IEEE Computer Society Press (1994)

An Effective Analysis of Server Log for Website Evaluation

Saritha Vemulapalli[1] and Shashi M.[2]

[1] Department of Computer Science & Engineering,
C M R Institute of Technology, Bangalore, India
saritha_vemulapalli@yahoo.com
[2] Department of CS & SE,
Andhra University College of Engg (A), Vizag, A.P, India
smogalla2000@yahoo.com

Abstract. The Web constitutes huge, distributed and dynamically growing hyper medium, supporting access to data and services. In corporate business due to strong market competition more organizations rely on web to conduct business, website design & management becoming critical issue in web based applications. One of the vital goals of organizations is having attractive & well organized website. Website managers are responsible to take decisions about contents & hyperlink structure in order to capture the attention of visitor's. Visitor's interactions with website are stored in server logs and serves as huge electronic survey of website. In this paper server logs are analyzed using the web log analyzer program to get general statistics about hit's, visitor's, visit's, browsers, operating systems, referring sites, spider URL's, eminent & delicate pages and statistics about error pages, broken links. Obtained results can be useful to website manager to evaluate website, helps in improving the effectiveness of website.

Keywords: Data mining, Web log analysis, Web usage mining, Web usage analysis, preprocessing, Website design & management.

1 Introduction

The Web constitutes a huge, distributed and dynamically growing hyper medium, supporting access to data and services. Effective presence of website is the key to success in global market. One of the vital goals of an enterprises and organizations is having an attractive & well organized website in terms of both content and structure, in case of content based websites such as universities, e-education, e-commerce & newspapers. Usage of an automated tool becomes necessary in order to search, extract, filter, and judge the required information. As a result, in recent time web usage mining has attracted lot of attention [1]. Web based applications generate and collect large volumes of data in their day-to-day activities. Website visitor's actions can be collected in server logs in an unstructured format and later this information can be used for user behaviour analysis. Large quantities of such data are typically

S.C. Satapathy et al. (Eds.): Proc. of Int. Conf. on Front. of Intell. Comput., AISC 199, pp. 133–141.
DOI: 10.1007/978-3-642-35314-7_16

generated by e-commerce web servers. Web mining is the application of data mining which deals with the extraction of interesting knowledge from the web documents and services which are expressed in the form of textual, linkage or usage information [2]. Web mining is divided into web content mining, web structure mining and web usage mining. The process of discovering useful knowledge from the raw information (text, image, audio or video data) available in web pages is web content mining. Analyzing the link between pages of a website using web topology is web structure mining. Cooley et al. [3] introduced the term web usage mining in 1997 and is defined as process of extracting useful information from server logs (i.e. user's history) to improve web services and performance. Source data mainly consist of the (textual) logs stores click stream data, as a result of user's interactions with a website and are represented in standard formats. Obtained user access patterns will be utilized in variety of applications, for example, to keep track of previously accessed pages of a user, to identify the typical behavior of the user [4], making clusters of users with similar access patterns and by adding navigational links [5], reorganization of a website to facilitate clients access to the desired pages more easily and with the minimum delay [6]. In addition to website evaluation, common access behaviours of the users can be used to improve the actual design and for making other modifications to a website [7]. Moreover, usage patterns can be used for business intelligence in order to improve sales and advertisement.

In this paper web log analyzer program is used to analyze the server logs of www.vnrvjiet.ac.in to get general statistics about hit's, visitor's, visit's, browsers, operating systems, referrer sites, spider URL's, eminent & delicate pages and error statistics such as client & server errors, corrupted & broken links, which helps the website manager to improve the effectiveness of the website.

The paper is organized as follows. Section 2 covers overview of web usage mining process and format of web log data. Implementation issues of web log analyzer program are presented in section 3. Section 4 covers experimental results. Conclusion & future enhancements are presented in section 5.

2 Web Usage Mining Process

Web usage mining is the discovery of user access patterns from server logs. The web usage mining process consists of data collection, data preprocessing, pattern discovery & analysis and visualization [8].

Web is interconnection between web documents and these documents are delivered by hypertext transfer protocol. The data collected from server side, client side, proxy servers, topology of website, web page contents, user registration or profile information can be used for mining process. Server logs are the primary source of data for web usage mining that are collected when users access web servers, represented in standard formats (e.g. Common Log Format [9] and Extended Common Log Format [10]). The raw information contained in a web server log file doesn't represent a structured, complete, reliable & consistent data. The quality of data can be improved with preprocessing techniques, such as data cleaning, user identification, sessionization, session reconstruction and data structurization [11]. Statistical & data mining techniques can be applied to the preprocessed web log data, in order to discover statistics & useful hidden patterns and are represented in visualization techniques such as graphs and reports.

2.1 Common Log Format (CLF)

Each entry of log file represented in the common log format has the following syntax. [Host/IP Rfcname Userid [DD/MMM/YYYY: HH:MM:SS -0000] "Method /Path HTTP/version" Code Bytes]

The "-" shown in a field indicates missing data.

- Host/IP is the IP address of the client (remote host), which made the request to the web server.
- Rfcname returns user's authentication. It operates by verifying specific TCP/IP connections and returns the user identifier of the process who owns the connection.
- Userid is the user id of the person requesting for the document.
- [DD/MMM/YYYY: HH:MM:SS -0000] is the date, time, and time zone when the server completed processing of the request.
- "Method /Path HTTP/version" is the request line from the client. Method is the request method, /path is the requested resource, and HTTP/version is the HTTP protocol.
- Code is the HTTP status code returned to the client.
- Bytes are the size of the object returned to the client, measured in bytes.

2.2 Extended Common Log Format (ECLF)

It's an extension to CLF, having some additional information about user_agent, cookie and referrer. User_agent is the visitor's browser & O.S version. Cookie is a persistent token, which defines the cookie sent to a visitor. Referrer defines the URL from where the visitor came from. Each entry of log file, represented in the ECLF has the following syntax. Fig. 1 shows log file represented in ECLF.
[s-computername s-ip s-port c-ip rfcname cs-userid date time cs-method cs-uri-stem cs-uri-query cs-version sc-status time-taken sc-bytes cs(user-agent) cs(cookie) cs(referrer)]

```
74.110.62.155 – user1 [10/Mar/2011:13:55:34 -0700] "GET /www.vnrvjiet.ac.in/home.html
HTTP/1.0" 200 2326 Mozilla/4.0+(compatible;+MSIE+6.0;+Windows+98) – –
74.110.62.155 – user1 [10/Mar/2011:13:55:36-0700] "GET /VNRInfrastructure/index3.html
HTTP/1.0" 200 2326 Mozilla/4.0+(compatible;+MSIE+6.0;+Windows+98)
ASPSESSIONIDASRDQQCA=NDBJHEFADELJLEAGJOPIEFBP
www.vnrvjiet.ac.in/home.html
```

Fig. 1. Example of Typical Extended Common Log Format Web Server Log

2.3 Http Protocol Status Codes

Http status codes returned by the server are classified into five classes [12]. i) Continue (100 series) ii) Success (200 Series) iii) Redirect (300 Series) iv) Failure (400 Series) v) Server Error (500 Series)

A status code of 100 series means that server has received the request, continuing process. A status code of 200 series means that the transaction was successful. A status code of 300 series means that the transaction was redirected. A status code of 400 series means that the transaction was failed due to error at client side. The most common failure codes are 401 (failed authentication), 403 (Forbidden request), and 404 (file not found). A status code of 500 series means that the transaction was failed due to server side error. The most common failure codes are 503 (Out of Resource).

3 Implementation of Web Log Analyzer Program

The Web Log Analyzer Program consists of components such as data collection, pre-processing, pattern discovery & analysis, and visualization is shown in Fig. 2. The implementation details of various components are explained below.

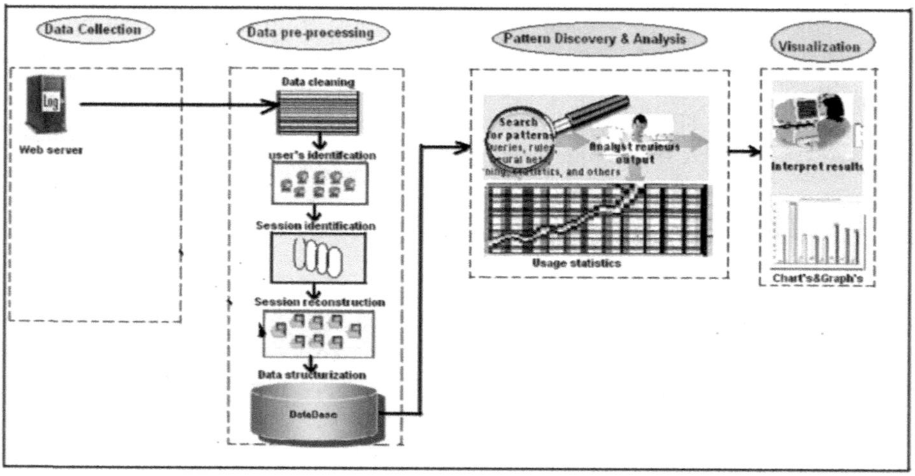

Fig. 2. Components of Web Log Analyzer Program

3.1 Data Collection

The web log analyzer program uses server logs of www.vnrvjiet.ac.in, represented in extended common log format (ECLF).

3.2 Data Preprocessing

The raw data contained in a server log file doesn't represent structured, complete, reliable & consistent information. As the web server logs aren't designed for data mining, preprocessing should be carried out in order to get reliable and accurate information. Low-quality of the data will produce low-quality mining results. The quality of the data can be improved with data preprocessing techniques, thereby helping to enhance the accuracy and efficiency of the subsequent mining process. Nearly 80% of mining efforts are required to improve the quality of data [13].

Data Cleaning. The process of removing the irrelevant entry's in pattern discovery. Irrelevant information includes the following [14]:

i) Removing all the attributes with no data at all and are not essential for the analysis.
ii) Removing the log entry's represents image, sound, video, flash animations, frames, pop-up pages, script's and style sheet files.
iii) Removing the access records generated by automatic search engine agents such as crawler, spider, robot, etc.
iv) Removing the access records requested by proxy servers.
v) Removing Log entries that have status of either "error" or "failure".

User Identification. The process of identifying the distinct user's, interacting with a website using the web browser. Users can be identified based on following factors [14]:

i) Different IP address is assumed as new user.
ii) The same IP address, but with different operating system or browser software is assumed as new user.
iii) The same IP address, operating system & browser software, but with different version is assumed as new user.

User's Session Identification. The users will visit the web site more than once. Session identification is the sequence of activities of a single user during a single visit at a defined duration [15]. Since HTTP protocol is stateless and connectionless discovering the user sessions from server log is a complex task.

Session identification can be done with the following rules [14]:

i) When there is a new user, a new session begins.
ii) When the time gap between consecutive requests made by the same user exceeds threshold $\Delta t=10$ minutes and if the referrer is "-", a new session begins.
iii) If the URL in the referrer field has never been accessed before in a current session, a new session begins.

Path Completion. Cache causes some important page requests are not recorded in server log, causing the problem of incomplete path. It is the process of reconstruction of user's navigation path, by appending missed page requests within the identified sessions.

The following rules are used for path completion [14]:

i) If the URL in the referrer field of the page request made is not equivalent to URL of last page user has requested & if the URL in the referrer field is in the user's history of the identified user's session, it is assumed that user uses "back" button. Missed page references that are inferred through this rule are added to the user's session file.

Data Structurization. The web log analyzer program translates the log file in to relational database for input to the pattern discovery & analysis phase. Different tables are designed in the relational database for each object, identified in various stages of preprocessing process.

3.3 Pattern Discovery and Analysis

The process of applying statistical and data mining techniques on the preprocessed web log data, in order to discover useful hidden patterns. This paper concentrates on statistical analysis; web log analyzer program analyzes the server logs of www.vnrvjiet.ac.in to get general statistics about hit's such as total number of hits, successful hit's, spider hits, visitor's such as number of visitor's, visitor's who visited once, repeat visitor's, average visit's per visitor, visit's such as number of session's, average visit duration, browsers, operating systems, top 10 viewed pages, least viewed pages, referrer sites, spider URL's. Error statistics such as server errors, client errors, and page not found errors. Discovered knowledge can be potentially useful in website design & management and providing support for marketing decisions.

3.4 Visualization and Result Presentation

The web log analyzer program generates different types of charts & reports, which represents usage statistics for an easier interpretation of the results.

4 Experimental Results

The web log analyzer program was developed based on IIS web server log represented in ECLF, using java programming language. The server log file of www.vnrvjiet.ac.in of 15th Nov 2010, having 10,375 records is selected for analysis. The results of preprocessing are shown in Table1. After cleaning the No. of records reduces down to 1,220 (12% of original records), 235 unique visitor's & 589 visitor's sessions are identified.

Table 1. The results of data preprocessing

Records in logfile	Records after cleaning	No of unique Visitors	Sessions
10,375	1,220	235	589

General statistics about hit's summary is shown in Table 2. Visitor's & visit's summary are shown in Table 3 & Table 4. 404 error (page not found) URL's and 500 error (internal server error) URL's are shown in Table 5 & Table 6. Eminent pages and delicate pages are shown in Table 7 & Table 8. Referring sites and spiders URL's are shown in Table 9 &Table 10. Browser & Operating system statistics are shown in Fig. 3 & Fig. 4.

Table 2. Hit's Summary

Category	hits
Total No. of Hits	1523
Successful Hits	1460
Spider Hits	293

Table 3. Visitor's Summary

Category	hits
Number of Unique Visitor's	235
Visitor's who visited once	175
Repeat Visitor's	60
Average Visit's per Visitor	2

Table 4. Visit's Summary

Category	hits
No. of Visit's	589
Avg.visit duration	1.6 min

Table 5. 404 Error Statistics

URL
/Annexure%20III.pdf
/adroit/home.html
/alumini_generalinformation.asp
/convergence2k8/imageprocessing.html
/vglug/alternative.html

Table 6. 500 Error Statistics

URL
/btech_mechnicalinfrastructure.asp

Table 7. Eminent Pages

URL
/Index.asp
/contact.asp
/btech_cse.asp
/btech_ece.asp
/placements_selected.asp
/place2009-2010.asp

Table 8. Delicate Pages

URL
/ADROIT/Adroit_mirror/ Index.html
/ADROIT/events.html
/ADROIT/pptresults.html
/ADROIT/shortcuts.html
/Careercounsellingandguidance/ dsc01743.html
/Convergence2k10pics/dsc_ 6968.html

Table 9. Referrer URL'S

REFERRER URL'S
http://adroit2k9.blogspot.com
http://khup.com
http://search.yahoo.com
http://www.facebook.com
http://www.google.co.in
http://www.google.com
http://www.vignanajyothi.com
http://www.way2college.com

Table 10. Spider URL'S

SPIDER URL'S
http://help.soso.com/webspider. htm l
http://search.msn.com/msnbot. html
http://www.bing.com/bingbot. html
http://www.exabot.com/go/robot. html
http://www.google.com/bot.html

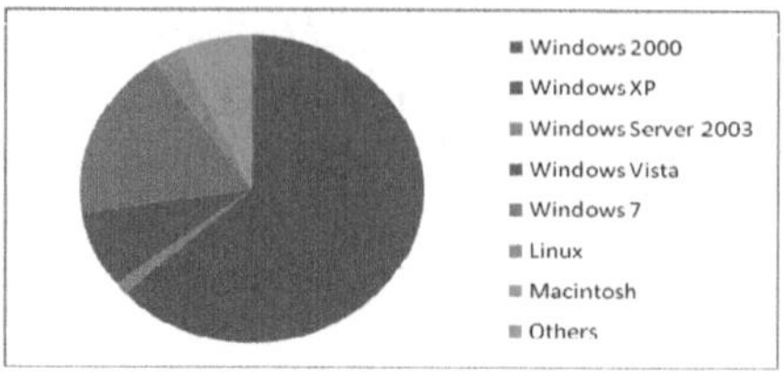

Fig. 3. Browser Statistics

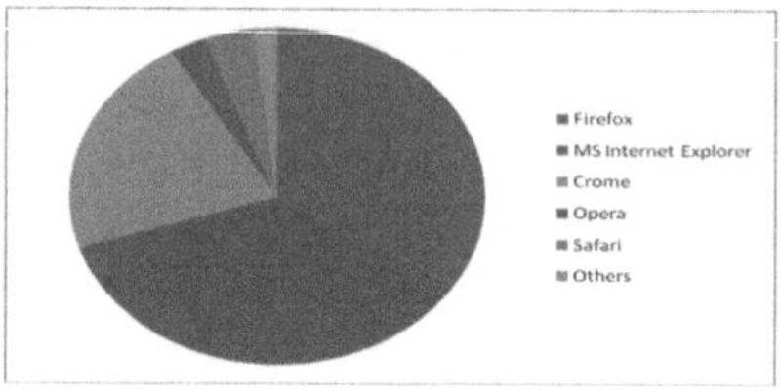

Fig. 4. Operating System Statistics

5 Conclusion and Future Enhancements

The attractiveness of a website in terms of both content & structure is critical for web based applications. Server logs of www.vnrvjiet.ac.in are analyzed using the web log analyzer program to get general statistics about hit's, visitor's, visit's, browsers, O.S, referring sites, spider URL's, eminent & delicate pages and corrupted & broken links. The obtained results can be used by website manager to increase the effectiveness of the website. This can be enhanced in future in order to find association among pages & to relate pages that are most often occur together in a single session by applying association rule generation & clustering algorithms. Such rules can also be helpful to web site managers to restructure the website.

References

1. Cooley, R., Mobasher, B., Srivastava, J.: Web mining: information and pattern discovery on the World Wide Web. In: International Conference on Tools with Artificial Intelligence, pp. 558–567. IEEE, Newport Beach (1997)
2. Cooley, R., Mobasher, B., Srivastava, J.: Data preparation for mining world wide web browsing patterns. Journal of Knowledge & Information System, 1–27 (1999)
3. Cooley, R., Mobasher, B., Srivastava, J.: Grouping Web page references into transactions for mining World Wide Web browsing patterns. In: Knowledge and Data Engineering Workshop, pp. 2–9. IEEE, Newport Beach (1997)
4. Srivastava, J., Cooley, R., Deshpande, M., Tan, P.-N.: Web usage mining: Discovery and applications of usage patterns from Web data. SIGKDD Explorations 1, 12–23 (2000)
5. Masseglia, F., Poncelet, P., Teisseire, M.: Using data mining techniques on Web access logs to dynamically improve Hypertext structure. ACM SigWeb Letters 8(3), 13–19 (1999)
6. Pirolli, P., Pitkow, J., Rao, R.: Silk from a sow's ear: Extracting usable structure from the web. In: Human Factors in Computing Systems: Common Ground, CHI 1996, Vancouver, Canada, New York (1996)
7. Bosnjak, S., Maric, M., Bosnjak, Z.: The Role of Web Usage Mining in Web Applications. Evaluation Management Information Systems 5(1), 031–036 (2010)
8. Pabarskaite, Z., Raudys, A.: A process of knowledge discovery from web log data: Systematization and critical review. Journal of Intelligent Informatin Systems 28(1), 79–104 (2007)

9. Configuration file of W3C httpd (1995),
`http://www.w3.org/Daemon/User/Config/`
10. W3C Extended Log File Format (1996),
`http://www.w3.org/TR/WD-logfile.html`
11. Cooley, R., Mobasher, B., Srivastava, J.: Data Preparation for Mining World Wide Web Browsing Patterns. J. Knowledge and Information Systems 1(1), 5–32 (1999)
12. Hypertext Transfer Protocol Overview (1995),
`http://www.w3.org/Protocol/rfc2616/rfc216sec1.html`
13. Frieder, O., Grossman, D.A.: Information Retrieval: Algorithms and Heuristics, 2nd edn. The Information Retrieval Series (2004)
14. Vemulapalli, S., Shashi, M.: Design and Implementation of an Effective Web Server Log Preprocessing System. In: Satapathy, S.C., Avadhani, P.S., Abraham, A. (eds.) InConINDIA 2012. AISC, vol. 132, pp. 897–905. Springer, Heidelberg (2012)
15. Spiliopoulou, M.: Managing Interesting Rules in Sequence Mining. In: Żytkow, J.M., Rauch, J. (eds.) PKDD 1999. LNCS (LNAI), vol. 1704, pp. 554–560. Springer, Heidelberg (1999)

Design of New Volterra Filter for Mammogram Enhancement

Ashutosh Pandey, Anurag Yadav, and Vikrant Bhateja

Deptt. of Electronics and Communication Engineering, Shri Ramswaroop Memorial Group of Professional Colleges, Faizabad Road, Lucknow-227105, (U.P.), India
{ashutosh91.p,anurag.yadav.ece,bhateja.vikrant}@gmail.com

Abstract. Non-linear filters are generally preferred for image enhancement applications as they provide better filtering results not only by suppressing background noise but also preserving the edges. This paper introduces a new technique for enhancement of digital mammograms using a Volterra filter. The proposed Volterra filter design is obtained by truncation of Volterra series to the first non-linear terms. Truncation of Volterra series leads to a simpler and effective representation without having prior knowledge of higher order statistics. The weight indices of the proposed filter are optimally selected in a manner to provide better enhancement of lesions in the mammograms in comparison to other techniques.

Keywords: Isotropic, Mammograms, MIAS Database, Symmetric, Volterra Filter.

1 Introduction

Mammography is one of the most effective method for early breast cancer detection before physical symptoms develop. Being a low-dose X-ray examination, it poses minimal risks from radiation exposure. However, mammographic screening still poses some limitations like: lower detection rates as the detected tumor may have poor prognosis and detection of false positives which may lead to unnecessary biopsies [1]. Mammographic images are generally noisy due to ill performance of X-ray hardware system and also contain poor contrast. Many enhancement techniques have been evolved for the enhancement of the mammograms. H. Tang *et al.* [2] proposed a technique based on fuzzy domain transformation, which satisfactorily removed the adverse effects of noise but proved to be less flexible in approach, as more than one parameter has to be varied for different types of images. Yessi Jusman, *et al.* [3] in their work performed contrast enhancement by adaptive histogram equalization. But, their technique gave satisfactory results only with mammograms affected by low noise density. J. Zheng, *et al.* [4] used skeletonization and mesh formation to increase the resolution of the mammograms, but the fine details of the tumors were not visible. S. Al-Kindi and G. Al-Kindi [5] utilized a combination of histogram equalization and Cannys' edge detection to enhance the contrast of sonograms and mammograms. However, the overall quality of the finally transformed image was dependent on the

S.C. Satapathy et al. (Eds.): Proc. of Int. Conf. on Front. of Intell. Comput., AISC 199, pp. 143–151.
DOI: 10.1007/978-3-642-35314-7_17

iteration decision factor. The edges and boundaries of the tumor were also eroded. Linear filters [6] generally exhibit simplicity in design, analysis and synthesis, but their performance limits in applications related to saturation type non-linear systems. These filters do not give impressive results for images coupled with signal-dependent or multiplicative noises as well as for those with Non-Gaussian statistics. Noises removed by the linear filters often leads to image blurring as edges could not be preserved. To overcome this drawback non-linear filters came into existence [6-7]. These filters are created by models which utilize Volterra filters [6], order statistics filters [8], and morphological filters [9]. The use of non-linear filters provide better image filtering results not only by suppressing effect of noises but also by preserving edges of the image. The only demerit possessed by non-linear filters is that it requires a large number of coefficients for designing. Hence, many techniques are developed to reduce the number of independent weighted coefficients of the non-linear filters [10-11]. The quadratic filters [10] are the simplest form of the non-linear filters. For effective quadratic filtering, methodical design algorithms are demanded but very few results have been notified in the realization of non-linear filters due to the fact that the quadratic filters do not possess simpler characterization as required for linear filters in frequency domain [11]. This paper introduces a new Volterra filter design for enhancement of digital mammograms. The proposed filter operates over a 3x3 mask of the mammographic image and enhances the contrast along with due suppression of background noises while preserving finer details of the lesions. The visual quality of the mammograms enhanced using the proposed filter is estimated using Contrast Improvement Index (*CII*) and Peak Signal-to-Noise Ratio (*PSNR*). The remaining part of paper is organized as follows: Section 2 describes design methodology of the proposed Volterra filter; the parameters used for quality assessment are given under Section 3. Section 4 details the obtained simulation results, their analysis and comparison, whereas the conclusions drawn are in Section 5.

2 Proposed Volterra Filter Design

2.1 Generalized Form of Volterra Filter

A new version of Volterra filter is introduced in this work for enhancement of digital mammograms. This can be considered as a discrete time invariant non-linear filter equipped with memory. It can be represented by means of a discrete Volterra series expansion [12] as given in eq. (1).

$$y(n) = \beta_0 + \sum_{L=1}^{\infty} \beta_L[x(n)] \tag{1}$$

where: $y(n)$ and $x(n)$ are the output and input images respectively; n represents the pixel gray level value at a particular location (i, j). β_L represents the Lth-order Volterra filter. The constant β_0 is an offset term used only in case of adaptive structures. Its value can be neglected in the present work.

By means of a second truncation on the upper limit of (1) yields a Volterra filter (of order L=2). Hence, the input–output relationship of the second order Volterra filter proposed in this work is given as:

$$y(n) = \underbrace{\sum_i \theta(i) x^{2\gamma(i)}(n-i)}_{y_{linear}} + \underbrace{\sum_i \sum_j \phi(i,j) x^{\lambda(i)}(n-i) x^{\lambda(j)}(n-j)}_{y_{quadratic}} \quad (2)$$

where: $\theta(i)$ and $\phi(i, j)$ represent the linear and quadratic filter coefficients. γ is the weight index for the linear term in (2). In this case, the input pixels of the linear filter are raised to the power of 2γ. Similarly, each of the collected and grouped input pixels for the quadratic filter in (2) is raised to the power of λ.

2.2 Determination of Filter Coefficients

A 3x3 Volterra filter used in this work comprises of linear coefficients represented as a 3x3 matrix and the quadratic filter coefficients represented with a 9x9 matrix. Here, the complexity of the proposed filter comes into play as its realization involves computation of a linear component consisting of 9 coefficients and quadratic component of 81 independent coefficients. The design complexity of this filter can be therefore simplified by utilizing the properties of symmetric and isotropic kernels. Symmetry and isotropy are two common properties of the Volterra kernels necessary for preservation of input gray levels and finer details of the image. This can be explained as under:

(a) In order to preserve an input pixel which remains untransformed at the output during an image transformation process, the sum of linear coefficients must be made equal to unity and that of quadratic coefficients to zero. Further, to ensure a linear response against a uniform variation of luminance on input pixels, the sum of the coefficients available on the rows and columns of the matrix must be made zero.

$$\sum_i \theta(i) = 1 \quad (3)$$

$$\sum_i \sum_j \phi(i,j) = 0 \quad (4)$$

(b) The filter operator must be isotropic in order to produce a response independent of features like edges or textures of the input image. The isotropic conditions leads to simplification in the filter design procedures not only by reduction in number of independent coefficients but also by simplification of their expression.

$$\theta(i) = \theta(N-1-i),\ i = 0, 1, \ldots, N-1 \quad (5)$$

$$\phi(i,j) = \phi(N-1-i, N-1-j),\ i, j = 0, 1, \ldots, N-1 \quad (6)$$

By utilizing kernel symmetry, 81 independent coefficients of the quadratic component get reduced to 45, in which each of them is involved in one of the 13 independent responses. By applying isotropic property, only 11 independent coefficients and 6 impulse responses are left.

2.3 Implementation of Proposed Filter

For implementation of the proposed filter, a kernel of size 3x3 is used; kernel of sizes above 3x3 poses limitations in hardware implementation. Hence, for a 3x3 filter kernel, (3) reduces to:

$$4\theta_1 + 4\theta_2 + \theta_0 = 1 \tag{7}$$

Optimal selection of values of the filter coefficients θ_0, θ_1 and θ_2 in (7) will be made in such a manner that they satisfy the above equation. Similarly, values of 11 independent coefficients of the quadratic filter ($\phi_0 - \phi_{10}$) will be optimally selected to satisfy equation (8) [obtained from (4) for a 3x3 kernel].

$$4\phi_1 + 16\phi_3 + 8\phi_7 + 8\phi_4 + 16\phi_{10} + 4\phi_8 + 4\phi_2 + 8\phi_6 + 8\phi_5 + 4\phi_9 + \phi_0 = 0 \tag{8}$$

With the determination of filter coefficients for both linear and quadratic components using a 3x3 kernel the generalized version of the proposed Volterra filter defined in (2) can be expressed in the following form:

$$y(n) = y_{linear} + y_{quadratic} \tag{9}$$

where: the linear and quadratic components of (9) can be stated as:

$$y_{linear} = \theta_0 x_5^{2a} + \theta_1(x_1^{2b} + x_3^{2b} + x_7^{2b} + x_9^{2b}) + \theta_2(x_2^{2c} + x_4^{2c} + x_6^{2c} + x_8^{2c}) \tag{10}$$

$$y_{quadratic} = \phi_7(x_1^b x_3^b + x_1^b x_7^b + x_3^b x_9^b + x_7^b x_9^b) + \phi_8(x_1^b x_9^b + x_3^b x_7^b)$$

$$+\phi_9(x_2^c x_8^c + x_4^c x_6^c) + \phi_{10}(x_1^b x_6^c + x_1^b x_8^c + x_2^c x_7^b + x_2^c x_9^b + x_3^b x_4^c + x_3^b x_8^c + x_4^c x_9^b + x_6^c x_7^b) \tag{11}$$

The quadratic component (11) of $y(n)$ consists of those terms where the distance between any two pixels (in a 3x3 kernel) is of two units. Imposition of this condition for framing the quadratic component (11) would lead to cancelation of input pixels existing in isolation or in adjacent pairs. This will have no effect on pair of pixels having inter-pixel distance of two units. This filter would be therefore capable in discrimination of texture patterns formed by adjacent or isolated pixels. The weight indices γ and λ of (2) are substituted with a, b and c which are the powers on the pixels as shown in (10) and (11) above. The values for these powers will be determined experimentally. The above design constraints leads to development of filter suitable for enhancement coupled with significant suppression of background noise.

3 Parameters Used for Evaluation of Filter Performance

The performance evaluation of an enhancement filter for mammograms can be ascertained by its capability to improve the difference between the mean gray levels lying in the image foreground with respect to the background. The objective evaluation of the proposed filter and its performance comparison with other mammogram enhancement techniques has been made using Contrast Improvement Index (*CII*) and Peak Signal-to-Noise Ratio (*PSNR*) [13].

A. *CII*

The contrast '*C*' of an object is given by:

$$C = \frac{m_f - m_b}{m_f + m_b} \quad (12)$$

where: m_f is the mean grey-level value of the foreground and m_b is the mean gray level value of background. The quantitative measure of contrast enhancement is defined as 'Contrast Improvement Index (*CII*)'.

$$CII = \frac{C_E}{C_O} \quad (13)$$

where: C_E and C_O are the contrasts of the region of interest in the enhanced and original images respectively. As *CII* does not contain enough information to quantize the background, the other parameter used is *PSNR*.

B. *PSNR*

These parameters are used to quantify the degree of noise suppression provided by an enhancement technique. Mathematical form of *PSNR* can be stated as:-

$$PSNR = \frac{m_f^m - m_b}{\sigma} \quad (14)$$

where: σ is the standard deviation, which gives the measurement of the level of noise in the background, m_f^m is the maximum grey level value of the foreground region. Higher the values of *CII* and *PSNR* more promising is contrast enhancement filter.

4 Results And Discussions

4.1 Simulation Results

The digital mammograms used in this work for simulation are taken from the Mammographic Image Analysis Society (MIAS) database [14]. MIAS is a UK based organization involved in mammogram related researches, and has generated a database of 322 digital mammograms. The three mammograms chosen as test images for this work are: mdb184 (with spiculated mass), mdb028 (with circumscribed mass) and mdb271 (with ill-defined mass). Digital mammograms (of size 1024x1024) obtained from the database are initially normalized before further processing. In these test images, the region containing the tumor is treated as foreground and the remaining area as background. As explained under section 2.3, for a 3x3 kernel, the values of the linear and quadratic filter coefficients are experimentally determined. These are those set of coefficient which satisfy the equations (7) and (8). The experimentally determined values are optimized using *CII* as the performance parameter. In this work, the optimal values of linear and quadratic filter coefficients used for implementation are: θ_0=0.2, θ_1=θ_2=0.1, ϕ_7=-2€, ϕ_8=-4€, ϕ_9=4€ and ϕ_{10}=€. Significantly good enhancement results are obtained for €=0.1. In a similar fashion, the values of the weight indices are taken as: a=8μ, b=c=μ; where value of μ varies between 2.8 to 3.4.

Once the optimal values of filter coefficients as well as weight indices are determined, the input test mammograms are then processed with the proposed Volterra filter given in (9)-(11). For the sake of comparison, the enhancement of the test mammograms is also performed using conventional enhancement techniques like: Unsharp Masking (UM) [15] & CLAHE [16] as well as recently developed enhancement approaches, which includes: Quadratic Filter (QF) [17] & Alpha Weighted Quadratic Filter (AWQF) [18]. The original as well as the enhanced mammograms for the three test images are shown in fig. 1(a)-(f).

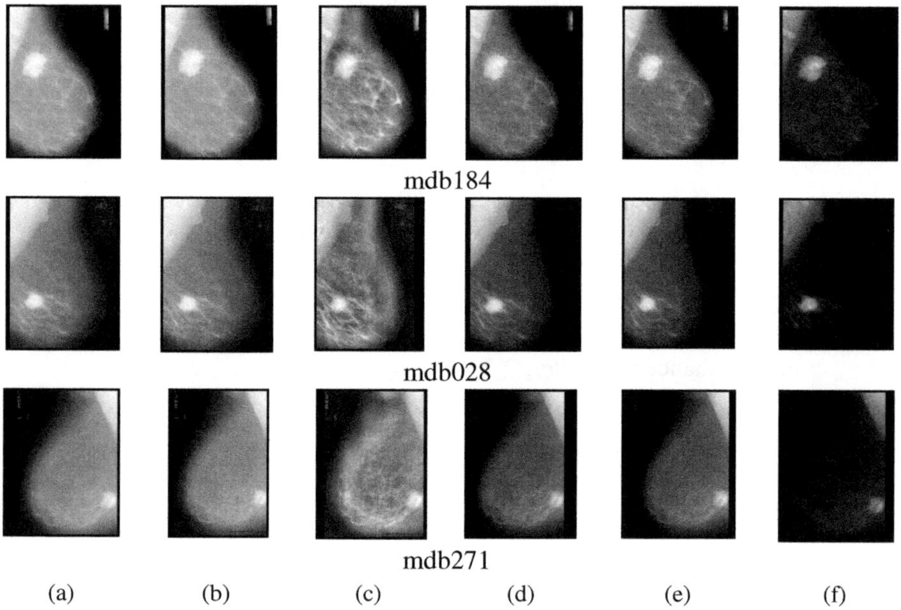

Fig. 1. (a) Original Mammograms. Mammograms processed with different enhancement techniques: (b) UM [15] (c) CLAHE [16] (d) QF [17] (e) AWQF [18] (f) Proposed Volterra Filter.

From the enhancement results given in fig. 1, it can be visualized that the targeted tumor region in the mammogram is very clearly visible using the proposed Volterra filter. In the results obtained by other mammogram enhancement techniques, the white tumor region is camouflaged with the background tissues. These techniques are not able to provide reasonable suppression of the background. The values of *CII* and *PSNR* are computed for the enhanced images and are tabulated under table 1 to table 3.

Table 1. Performance Comparison of Different Enhancement Techniques on mammogram (mdb184) containing a spiculated mass

	CII	*PSNR*
UM [15]	1.051698	1.018339
CLAHE [16]	0.957375	1.290735
QF [17]	1.813931	1.423190
AWQF [18]	1.628538	1.548328
Proposed Volterra Filter	4.564484	2.859962

Table 2. Performance Comparison of Different Enhacement Techniques on mammogram (mdb028) containing a circumscribed mass

	CII	*PSNR*
UM [15]	1.019045	0.998189
CLAHE [16]	0.783021	0.844278
QF [17]	1.779677	1.150828
AWQF [18]	1.772233	1.229548
Proposed Volterra Filter	3.615706	2.082246

Table 3. Performance Comparison of Different Enhacement Techniques on mammogram (mdb271) containing an ill-defined mass

	CII	*PSNR*
UM [15]	1.009581	1.102314
CLAHE [16]	0.962227	1.008748
QF [17]	1.977517	1.185189
AWQF [18]	1.668540	1.513928
Proposed Volterra Filter	4.558999	2.365767

4.2 Comparison

Higher values of *CII* depict a higher degree of enhancement obtained by an enhancement technique. From the *CII* results given under table 1 to table 3, it can be ascertained that QF [17] and AWQF [18] enhances contrast of the tumor region better than the images enhanced using UM [15] and CLAHE [16]; but still it fails to suppress the background effectively. In the case of UM [15], CLAHE [16], QF [17] and AQWF [18] lower values of *PSNR* clearly explains that the noise levels in the transformed images are not reduced. The tabulated results clearly shows that the proposed Volterra filter yields higher values of *CII* and *PSNR* in comparison to other mammogram enhancement techniques.

5 Conclusion

Unlike linear filters, non-linear filters provide better image filtering results not only by suppressing effect of noises but also by preserving edges. In this paper a novel design of Volterra filter is proposed for enhancement of digital mammograms. The proposed filter provides very promising results, not only in terms of improvement in contrast (of the foreground) but also suppressed the background noise. Filter will serve as a pre-processor for segmentation of mammogram. It also preserves the boundary and fine details of the tumor region which will help radiologists for accurate diagnosis of early stages of breast cancer.

References

1. Pisano, E.D., Hendrick, E., et al.: Diagnostic Accuracy of Digital versus Film Mammography: Exploratory Analysis of Selected Population Subgroups in DMIST. Radiology, 376–383 (2008)
2. Tang, H., Zhuang, T., Wu, E.X.: Realization of Fast 2-D/3-D Image Filtering and Enhancement. IEEE Transactions on Medical Imaging 20(2), 132–140 (2001)
3. Jusman, Y., Isa, N.A.M.: A Proposed System for Edge Mammogram Image. In: Proc. of the 9th World Scientific and Engineering Academy and Society (WSEAS) International Conference on Applications of Electrical Engineering, pp. 117–123 (2010)
4. Zheng, J., Fuentes, O., Leung, M.: Super Resolution of Mammograms. In: IEEE Symposium on Computational Intelligence in Bioinformatics and Computational Biology (CIBCB), Montreal, pp. 1–7 (2010)
5. Al-Kindi, S.G., Al-Kindi, G.A.: Breast Sonogram and Mammogram Enhancement Using Hybrid and Repetitive Smoothing-Sharpening Technique. In: Proc. of the 1st Middle East Conference on Biomedical Engineering (MECBME), Sharjah, pp. 446–449 (2011)
6. Mathews, V.J., Sicuranza, G.L.: Volterra and General Volterra Related Filtering. In: IEEE Winter Workshop on Nonlinear Digital Signal Processing, pp. T_2.1–T_2.8 (1993)
7. Salmond, D.J., et al.: Novel Approach to Non-linear/Non-Gaussian Bayesian State Estimation. In: Proc. of the IEEE on Radar and Signal Processing, vol. 140, pp. 107–113 (1993)
8. Pitas, I., Venetsanopoulos, A.N.: Order Statistics in Digital Image Processing. Proc. of the IEEE 80(12), 1893–1921 (1992)
9. Stevenson, R., Arce, G.: Morphological Filters: Statistics and Further Syntactic Properties. IEEE Transactions on Circuits and Systems 34(11), 1292–1305 (1987)
10. Mathews, V.J.: Adaptive Volterra Filter. IEEE Signal Processing Magazine 8(3), 10–26 (1991)
11. Sicuranza, G.L.: Quadratic Filters for Signal Processing. Proc. of the IEEE 80(8) (1992)
12. Sicuranza, G.L.: Volterra Filters for Image and Video Processing. In: Proc. of the First International Workshop on Image and Signal Processing and Analysis, Pula, pp. 15–26 (2000)
13. Morrow, W.M., et al.: Region Based Contrast Enhancement of Mammograms. IEEE Transactions on Medical Imaging 11, 392–406 (1992)
14. Suckling, J., et al.: The Mammographic Image Analysis Society Mammogram Database. In: Proc. of 2nd Int. Workshop Digital Mammography, York, U.K., pp. 375–378 (1994)

15. Rogowska, J., Preston, K., Shasin, D.: Evaluation of Digital Unsharp Masking and Local Contrast Stretching as Applied to Chest Radiology. IEEE Transactions on Information Technology in Biomedical Engineering 35(2), 236–251 (2009)
16. Pisano, E.D., et al.: Contrast Limited Adaptive Histogram Equalization Image Processing to Improve the Detection of Simulated Spiculations in Dense Mammograms. Journal of Digital Imaging 11, 193–200 (1998)
17. Ramponi, G.: Bi-impulse Response Design of Isotropic Quadratic Filters. Proc. of the IEEE 78(4), 665–667 (1990)
18. Zhou, Y., et al.: Mammogram Enhancement Using Alpha Weighted Quadratic Filter. In: Proc. of Annual International Conf. IEEE Engineering in Medicine and Biology Society, Minneapolis, Minnesota, pp. 3681–3684 (2009)

Power Quality Event Classification Using Hilbert Huang Transform

R. Jalaja[1] and B. Biswal[2]

[1] ECE Branch, GMRIT, A.P, India
[2] ECE Dept., GMRIT, A.P, India

Abstract. The objective of this paper is to develop a method based on combination of empirical-mode decomposition (EMD) and Hilbert transform for power quality events classification. Non-stationary power signal disturbance waveform can be considered as superimposition of various undulating modes and EMD is used to separate out these intrinsic modes known as intrinsic mode functions (IMF). Hilbert transform is applied to all the IMF to achieve instantaneous amplitude and frequency. Relevant feature vectors are extracted to do the automatic classification. Time frequency analysis shows clear visual detection, localization and classification of the different power signal disturbances. A balanced neural tree is used to classify the power signal patterns.

Keywords: Non-Stationary power signals, EMD (Empirical Mode Decomposition), Hilbert Transform, Balanced Neural Tree.

1 Introduction

Recently Power Quality (PQ) and related power supply issues have become quite a serious problem both for the end user as well as the utilities. The PQ issues and related phenomena can be attributed to the use of solid-state switching devices, unbalanced and non-linear loads etc. These devices introduce distortions in the phase, frequency and amplitude of the power system signal thereby deteriorating PQ. Hence analysis of PQ related issues are indispensable and this has been the focus of the researchers in the past decade. Time–frequency analysis has been successfully used in dealing with rapidly varying transient signals [1].

The time–frequency transform would provide direct information about the frequency components occurring at any given time. *Fourier Transform* tells us what frequency components are present but do not tell us when it happens and for how long. Although FT is one of the fast technique but its efficiency is limited to stationary signals only. Most PQ events are non-stationary and hence require technique that would not only provide frequency information but also capture the timing of occurrence of the disturbance. *Short Time Fourier Transform* provides frequency as well as time information. The non-stationary nature of the signal is well defined. However, due to the constant window length, some characteristics of the signal are not detected well. Different types of disturbances would require windows of different length. Choosing the best window length could be a problem.

S.C. Satapathy et al. (Eds.): Proc. of Int. Conf. on Front. of Intell. Comput., AISC 199, pp. 153–161.
DOI: 10.1007/978-3-642-35314-7_18

Wavelet Transform [2] provides time and frequency information of the signal by convolving the dilated and translated wavelet with the signal. By allowing variations in time and frequency plane, a multi-resolution analysis can be obtained. The main disadvantage of wavelet transform is its degraded performance under noisy situation. *Stockwel Transform* most commonly known as S-transform is yet another technique which is being widely used by PQ engineers. The S-transform is an extension of wavelet transform and is based on localizing Gaussian window. Here, the modulating sinusoids are fixed with respect to time axis while the Gaussian window scales and moves [3].

Current advances in signal analysis have led to the development of a new method for non-stationary signal analysis called *Hilbert Huang Transform (HHT)*. Together, the EMD and the Hilbert transform are labeled as the Hilbert-Huang transform. In the proposed work, clear visual localization, detection has been investigated thoroughly for each of the power signal disturbances using HHT.

2 Empirical Mode Decomposition

This work presents a new data analysis method based on the Empirical Mode Decomposition (EMD) method, which will generate a collection of intrinsic mode functions (IMF). The decomposition is based on the direct extraction of the energy associated with various intrinsic time scales. Expressed in IMFs, they have well-behaved Hilbert transforms, from which the instantaneous frequencies can be calculated. Thus, we can localize any event on the time as well as the frequency axis.

In general, signals will consist of more than one oscillatory component. The idea of the EMD [4] is to repeatedly apply a process known as sifting to separate out the fastest oscillatory mode, then the next fastest, and so on until the signal has been entirely broken down into simple oscillatory components, which Huang calls intrinsic mode functions (IMFs).

An *Intrinsic Mode Function (IMF)* is a function that satisfies two conditions:

1. For a data set, the number of extrema and the number of zero crossings must be either equal or differ at most by one.
2. At any point, the mean value of the envelope defined by the local maxima and the envelope defined by the local minima is zero.

The detailed description of the sifting process is given step wise:

1) The decomposition method requires use of envelopes defined by the local maxima and minima separately.
2) After identifying the local extrema, cubic spline functions are used for connecting local maximas as the upper envelope and local minimas as the lower envelope.
3) The mean value of envelopes is defined as M_1. The difference between the original data and M_1 is the first component K_1.

$$X(t) - M_1 = K_1. \tag{1}$$

4) If K_1 satisfies the two IMF conditions then K_1 is the first IMF else if K_1 is not an IMF then it is treated as original signal and steps from (1) to (3) are repeated to get component K_{11}.

$$K_1 - M_{11} = K_{11}. \tag{2}$$

5) After repeated sifting i.e. up to n times, K_{1n} becomes an IMF

$$K_{1(n-1)} - M_{1n} = K_{1n}. \tag{3}$$

Then it is designated as
$$C_1 = K_{1n}. \tag{4}$$

6) C_1 is the first IMF component from the original data. Separate C_1 from *X(t)*

$$R_1 = X(t) - C_1. \tag{5}$$

7) Now treating R_1 as the original data and repeating the above processes second IMF can be obtained.
8) The above procedure is repeated *q* times and *q* IMFs of signal *X(t)* are obtained.
9) The decomposition process can be stopped when R_q becomes a monotonic function from which no more IMF can be extracted.

The essence of the method is to identify the intrinsic oscillatory modes by their characteristic time scales in the data empirically, and then decompose the data accordingly.

Simulation Results. In our study we have discussed different types of practical and synthetic power signal problems such as voltage sag, voltage swell, momentary interruption, harmonics, flicker, multiple notches, multiple spikes and transients which are analyzed with MATLAB software. The EMD output shows the plot of the IMF Components in fig-1&2 which are obtained by the decomposition of a given input signal.

3 Hilbert Transform

The Hilbert transform is commonly used to generate a complex time series or analytic signal. The benefit is that instantaneous attributes can be derived from complex traces [5]. However, accurate and meaningful computation of these attributes requires that the input signal's start and end have zero amplitude, and it contains no trend that introduces a nonzero mean. In this regard, perhaps the most significant seismic use for the EMD is to prepare a signal for input to the Hilbert transform. The conventional Hilbert transform of a continuous signal *x(t)* is:

$$y(t) = \int_{-\infty}^{\infty} \frac{x(\tau)}{t-\tau} d\tau. \tag{6}$$

The transfer function of the discrete Hilbert transform is defined as:

$$H(\omega) = \begin{cases} j, 0 < \omega < \pi \\ 0, \omega = 0 \;\&\; \omega = \pi \\ -j, -\pi < \omega < 0 \end{cases} \tag{7}$$

The method for computing the discrete Hilbert transform is based upon its transfer function and utilizing the discrete Fourier transform (DFT) as a tool. Steps for Hilbert transform calculation are:

1) Compute the DFT of the signal $x(n)$ where $n=1, 2 \ldots N$, $X(k) = DFT\,[x(n)]$.
2) X is multiplied by the mask H where H is defined as
   $H = \{0, j, j \ldots 0, -j, -j \ldots\ldots\ldots -j\}$ *if n is even.*
   $H = \{0, j, j \ldots j, -j, -j \ldots\ldots\ldots -j\}$ *if n is odd.*
3) Compute the inverse DFT to obtain $x = IDFT\,[X.*H]$.

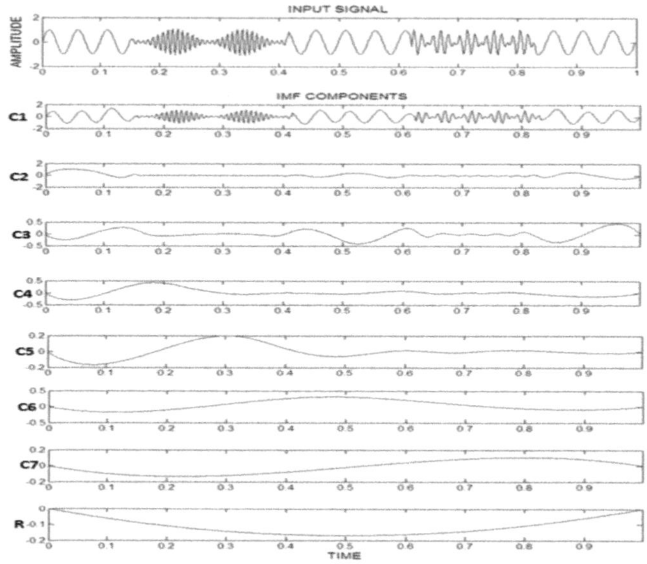

Fig. 1. Decomposition of the Signal with Flicker and Harmonics

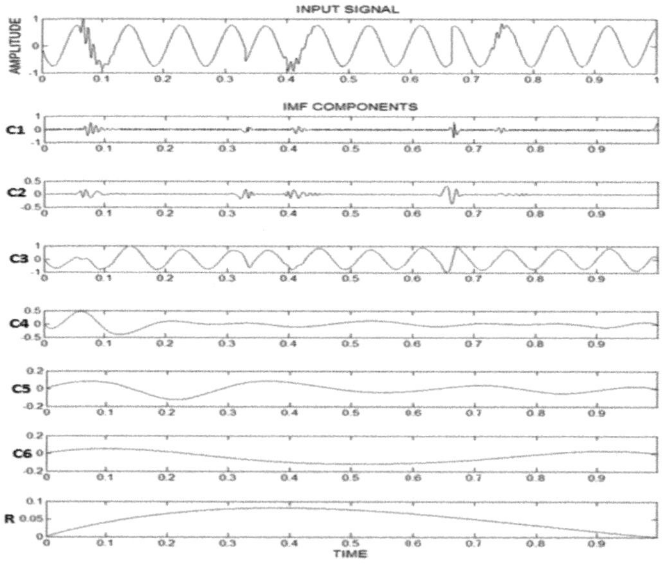

Fig. 2. Decomposition of the signal with multiple disturbances

Non-Stationary signal may not be represented well by sinusoidal components and since frequency is defined well for sinusoidal components it loses its effectiveness for non-stationary signal. This has given rise to notion of *Instantaneous Frequency.* Instantaneous Frequency (IF) has a meaning for mono-component signal, comprising of a single frequency or a narrow band of frequencies. This motivates to decompose a signal into number of mono-component modes for which IF can be defined. The Hilbert Transform of the signal *X(t)* results in an analytical signal *Z(t)* defined as:

$$Z(t) = X(t) + jY(t) = a(t)e^{j\theta(t)}.$$

In which $a(t) = \left[X(t)^2 + Y(t)^2\right]^{1/2}$, $\theta(t) = \arctan\left(\dfrac{Y(t)}{X(t)}\right)$. **(8)**

Where *Y(t)* is Hilbert transformed signal, *a(t)* is instantaneous amplitude and $\theta(t)$ is instantaneous phase. The analytic signal *Z(t)* has a real part *X(t)* which is the original data, and an imaginary part *Y(t)* which contains the Hilbert transform. The imaginary part is a version of the original real sequence with a 90° phase shift. The Hilbert transformed series has the same amplitude and frequency content as the original real data and includes phase information that depends on the phase of the original data. The instantaneous amplitude is the amplitude of the complex Hilbert Transform; the IF is the time rate of change of the instantaneous phase angle. IF is evaluated as:

$$\omega = \frac{d\theta}{dt}. \quad \textbf{(9)}$$

IF given in the above equation is a single valued function of time. At any given time, there is only one frequency value; therefore, it can only represent one component, hence 'mono-component'. This motivates to extract the mono-component signals (IMFs) from the original signal.

4 Hilbert Spectrum

Hilbert spectrum provides an intuitive visualization of what frequencies occurred during the signal duration, and also shows at a glance where most of the signal energy is concentrated in time and frequency plane. It would be ideal for non-stationary data analysis. After performing the Hilbert transform on each IMF component, we can express the data in the following form:

$$X(t) = \sum_{i=1}^{n} a_i(t)e^{j\theta_i(t)} = \sum_{i=1}^{n} a_i(t)\exp\left(j\int \omega_i(t)dt\right). \quad (10)$$

Above equation enables us to represent the amplitude and the IF as functions of time in a three-dimensional plot, in which the amplitude can be contoured on the time-frequency plane. This time-frequency distribution of the amplitude is designated as the Hilbert amplitude spectrum *H(w,t),* or simply Hilbert spectrum. If squared amplitude is more desirable, commonly to represent energy density, then the squared values of amplitude can be substituted to produce the Hilbert energy spectrum.

Simulation Results. This section presents the HHT output which shows the plot of the Energy spectrum, Magnitude response of a given input signal in the time-frequency co-ordinate system. Here the Hilbert Energy Spectrum is plotted in a time-frequency plane where the energy concentration is represented in terms of intensity of the color shown in fig-3&4.

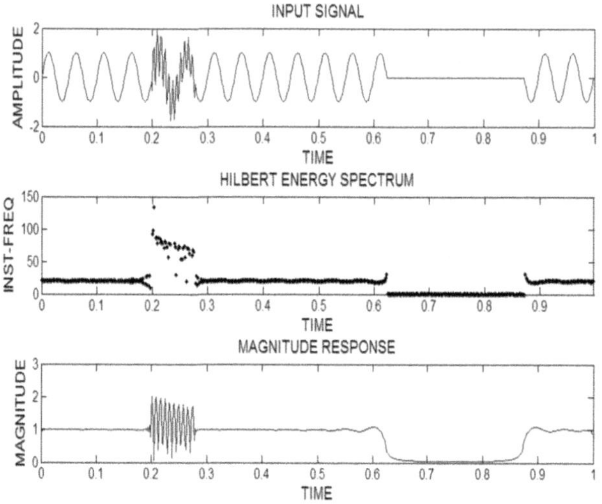

Fig. 3. Detection and visual localization of the Signal with Transient and Mom-Interruption

5 Balanced Neural Tree

In this work a new NT architecture called Balanced NT (BNT) [6] is proposed to reduce the size of the tree (both in depth and in the number of nodes), and to improve the classification of PQ events with respect to a standard NT [7]. To achieve this result, two main improvements are proposed: (a) *Perceptron Substitution* which aims to balance the tree structure by substituting the last trained perceptron with a new perceptron that equally distributes the patterns among the classes, if the current training set is largely misclassified into a reduced number of classes and (b) *Pattern Removal* consists of the introduction of a new criterion for the removal of tough training patterns that cause an over-fitting problem. The proposed novelties aim to define a new training strategy that does not require the definition of complex network topologies [8]. Two main phases can be distinguished, training phase and classification phase.

In the training phase [9], the BNT is constructed by partitioning a training set consisting of feature vectors and their corresponding class labels to generate the tree in a recursive manner. The perceptron [10,11] is trained with the patterns of the TS until the variation of a given error $\bar{e}$ remains in the range [*-toler, +toler*] for more than a given number *wait* of epochs. Let us define the error $\bar{e}$ as the mean error computed on all output neurons and patterns,

$$\bar{e} = \frac{1}{QM} \sum_{q=1}^{Q} \sum_{i=1}^{M} e_i^q \quad (11)$$

Where Q is the total number of patterns at the current node and the error e_i^q is the difference of output o_i^q and the target's class t_i^q which are computed as follows:

$$e_i^q = t_i^q - o_i^q, i = 1,...., M . \quad (12)$$

And

$$t_i^q = \begin{cases} 1 \text{ If } i = i_q \\ 0 \text{ Otherwise} \end{cases}, \quad o_i^q = 1 \Big/ \left[1 + \exp\left(-\sum_{j=1}^{N} w_{ij} x_j^q \right) \right]. \quad (13)$$

Where w_{ij} are the elements of the weight matrix W.

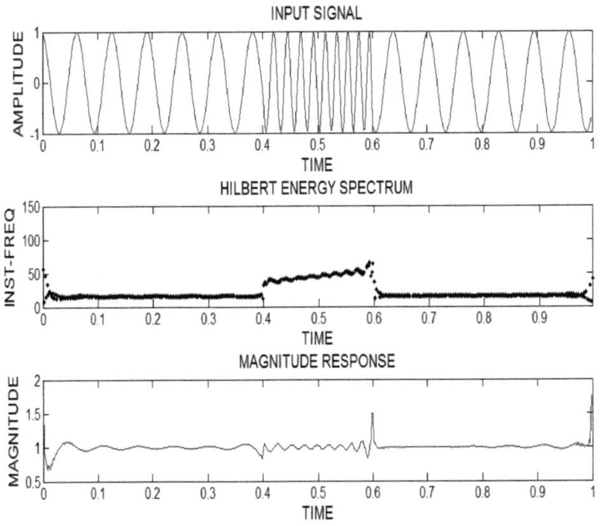

Fig. 4. Detection and visual localization of the Signal with Up Chirp

In the classification phase [12], the unknown patterns are presented to the root node. The class is obtained by moving down the tree. Starting from the root, the activation value of the current node provides the next node to be considered until reaching a leaf node that assigns the class of the input pattern. Each node applies the "winner-takes-all" rule.

Simulation Results. Here the test is characterized by a training set (TS) consisting of 900 patterns of each disturbance, distributed on a 2-D feature space with the features: Entropy Vs Standard Deviation that are extracted from the Hilbert Transformed Signal. The BNT that has been constructed constitutes of a root node, six internal nodes and eight leaf nodes shown in fig-5.

6 Conclusion

EMD is a promising method for non-stationary signal processing. It is used as a tool to extract IF information of each mode, thereby making it an important tool in the assessment of PQ events. The results reported here are believed to provide with new insights on EMD and its use. Apart from assessment, detection capability validates the potential of the algorithm. Finally, the Hilbert Energy Spectrum makes use of the Hilbert Transform which is an essential tool for conversion of signals into analyzable forms and to provide a useful, visual, qualitative understanding of a signal i.e. decomposed by EMD. The features that are extracted are applied to a Balanced Neural Tree for non-stationary power signal disturbance classification.

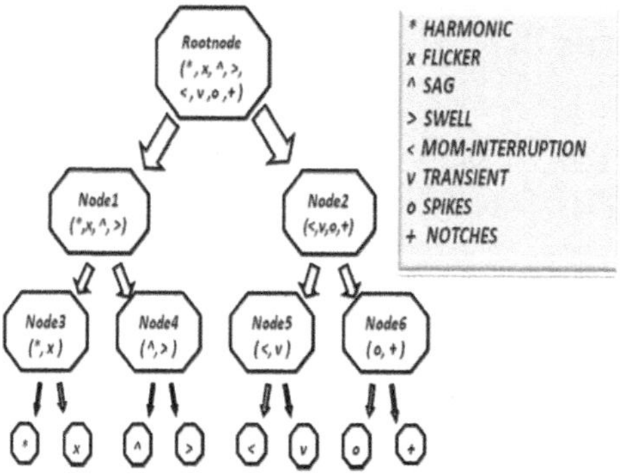

Fig. 5. Balanced Neural Tree.

Table 1. Classification Accuracy Table.

Disturbances	Classification Accuracy
Harmonic	95.75
Flicker	98
Sag	98
Swell	91.45
Momentary Interruption	100
Transient	100
Spikes	100
Notches	100
Overall Accuracy	97.9

References

1. Huang, N.E., Shen, Z.: The Empirical Mode Decomposition and the Hilbert Spectrum for Nonlinear and Non-Stationary Time Series Analysis. The Royal Society of London 454, 903–995 (1998)
2. Santoso, S., Grady, W.M., Powers, E.: Characterization of Distribution Power Quality Events with Fourier and Wavelet Transforms. IEEE Trans. Power Del. 15(1), 247–254 (2000)
3. Mishra, S., Bhende, C.N.: Detection and Classification of Power Quality Disturbances using S-Transform and PNN. IEEE Trans. Power Del. 23(1), 280–287 (2008)
4. Kopsinis, Y., McLaughlin, S.: Development of EMD based Denoising Methods inspired by Wavelet Thresholding. IEEE Transactions on Signal Processing 57, 1351–1362 (2009)
5. Jayasree, T., Devaraj, D., Sukanesh, R.: Power Quality Disturbance Classification using Hilbert Transform and RBF Networks. Neuro-computing 73, 1451–1456 (2010)
6. Micheloni, C., Kumar, S., Foresti, G.L.: A Balanced Neural Tree for Pattern Classification. Neural Networks 27, 81–90 (2012)
7. Foresti, G., Pieroni, G.: Exploiting Neural Trees in Range Image Understanding. Pattern Recognition Letters 19(9), 869–878 (1996)
8. Maji, P.: Efficient Design of Neural Network Tree using a Single Splitting Criterion. Neuro-computing 71, 787–800 (2008)

9. Rasoul, S., Landgrebe, D.: A Survey of Decision Tree Classifier Methodology. IEEE Transactions on Systems, Man, and Cybernetics 21(3), 660–674 (1991)
10. Lau, C., Widrow, B.: Special Issue on Neural Networks. Proceedings of the IEEE 78 (1990)
11. Atlas, L., Cole, R., Muthusamy, Y., Lippman, A., Connor, J., Park, D., et al.: A Performance Comparison of Trained Multilayer Perceptrons and Trained Classification Trees. Proceedings of the IEEE 78(10), 1614–1619 (1990)
12. Foresti, G.L., Micheloni, C.: Generalised Neural Tree for Pattern Classification. IEEE Transactions on Neural Networks 13(6), 1540–1547 (2002)

Colour Image Segmentation with Integrated Left Truncated Bivariate Gaussian Mixture Model and Hierarchical Clustering

G.V.S. Rajkumar [1], K. Srinivasa Rao [2] and P. Srinivasa Rao [3]

[1] Department of Information Technology, GITAM University, Visakhapatnam, Andhra Pradesh, India
gvsrajkumar@gmail.com
[2] Department of Statistics, Andhra University, Visakhapatnam, Andhra Pradesh, India
ksraoau@yahoo.co.in
[3] Department of Computer Science and Systems Engineering, Andhra University, Visakhapatnam, Andhra Pradesh, India
peri.srinivasarao@yahoo.com

Abstract. Image segmentation plays a dominant role in image analysis and image retrievals. Much work has been reported in literature regarding image segmentation based on Gaussian mixture model (GMM). The main drawback of GMM is regarding the assumption that each image region is characterized by Gaussian component, in which the feature vector is mesokurtic and having infinite range. But in colour images the feature vector is represented by Hue and Saturation which are non- negative and may not be symmetrically distributed. Hence the image segmentation can not be accurate unless the non-negative nature of the feature vector is included. In this paper an image segmentation method is developed and analyzed with the assumption that the bivariate feature vector consisting of Hue and Saturation of each pixel follows a left truncated bivariate Gaussian mixture model. In this method the number of components (Image regions) are determined by Hierarchical clustering. The segmentation algorithm is proposed under Bayesian frame with maximum likelihood. The experimentation with six images taken from Berkeley dataset reveals that the proposed image segmentation method outperforms the existing image segmentation method with GMM and finite left truncated bivariate Gaussian mixture model with K-means.

Keywords: Image Segmentation, Bivariate Gaussian Mixture model, Image Quality Metrics, Hierarchical clustering, EM- algorithm.

1 Introduction

Image retrieval and image segmentation become an important aspect in computer vision and machine learning with the evaluation of new information technology more effective and efficient methods are developed for human computer interaction [18]. Image segmentation is an important technology for image processing and

S.C. Satapathy et al. (Eds.): Proc. of Int. Conf. on Front. of Intell. Comput., AISC 199, pp. 163–170.
DOI: 10.1007/978-3-642-35314-7_19

understanding. Over the last three decades, the interest of researchers around the world has been focused on image segmentation. Lot of research activities have been carried out on gray scale image segmentation. Colour features have been found to be effective in image segmentation [5] [6] [7][15] [17]. Choice of a colour space is the main aspect of colour feature extraction.

We can generate colour spaces such as HSI, CIE -Lab, and CIE-Luv by nonlinear transformation of the RGB space. The HSI offers the advantage that separate channels outline certain colour properties, namely Intensity (I), Hue (H), and Saturation (S).Since H and S are functions of I, we consider the feature vector for characterizing the colour image with Hue and Saturation. It is also supported by the arguments given in text book "Digital Image Processing" [12]. Hence, in this paper we consider the feature vector for characterizing the colour image is a bivariate vector consisting of Hue and Saturation.

In colour image segmentation, model based image segmentation methods are more efficient than edge based or threshold or region based methods [8].In model based image segmentation it is customary to consider that the whole image is characterized by a finite Gaussian mixture model. That is, the feature vector of each image region follows a Gaussian distribution [3] [5] [6] [9] [10] [11] [13] [16] [19] [20] [21]. The image segmentation methods based on Gaussian mixture model work well only when the feature vector of the pixels are having infinite range and the distribution of the feature vector is symmetric and meso-kurtic. But in many colour images the feature vector represented by Hue and Saturation will have finite values (say nonnegative) and may not be mesokurtic and symmetric. Hence, to have an accurate image segmentation of these sorts of colour images it is needed to develop and analyze image segmentation methods based on truncated bivariate mixture distributions. With this motivation, in this paper some image segmentation technique based on truncated bivariate Gaussian mixture distribution are developed and analyzed.

Here, it is assumed that the feature vector in different image regions follows a left truncated bivariate Gaussian distribution and the feature vector of the whole image is characterized by a finite left truncated bivariate Gaussian mixture model. This assumption is made since the Hue and Saturation values of the pixel which represents the bivariate feature vector can take non negative values only and hence, the range of the Hue and Saturation values are to be left truncated at zero. The effect of the truncated nature of Hue and Saturation values cannot be ignored, since the leftover probability is significantly higher than zero in the left tail end of the distribution. This left truncated nature of the bivariate feature vector can approximate the pixels of the colour image more close to the reality.

The model parameters are estimated by using Expectation Maximization (EM) algorithm. The initialization of the model parameters for carrying the EM-algorithm is done through feature vector of the pixel intensities of the image regions obtained through Hierarchical clustering and moment method of estimation. An image segmentation algorithm with component likelihood maximization under Bayesian frame work is developed and analyzed.

The performance of the developed segmentation algorithm is compared with finite Gaussian mixture model with K-means and also with finite left truncated bivariate

Gaussian mixture distribution with *K*-means algorithm by obtaining segmentation performance measures. The performance of reconstructed images are studied by computing the image quality metrics [2].

2 Estimation of the Model Parameters by EM- Algorithm

In this section we discuss the estimates of the model parameters through EM-algorithm. Here, it is assumed that the feature vector of each image region follows a left truncated bivariate Gaussian distribution with joint probability density functions of the form

$$g_i(x_s, y_s; \theta) = \frac{\exp\left\{\frac{-1}{2(1-\rho_i^2)}\left[\left(\frac{x_s-\mu_{1i}}{\sigma_{1i}}\right)^2 - 2\rho_i\left(\frac{x_s-\mu_{1i}}{\sigma_{1i}}\right)\left(\frac{y_s-\mu_{2i}}{\sigma_{2i}}\right)+\left(\frac{y_s-\mu_{2i}}{\sigma_{2i}}\right)^2\right]\right\}}{2\pi\sqrt{1-\rho_i^2}\,\sigma_{1i}\sigma_{2i}\int_0^\infty\int_0^\infty f_i(x,y;\theta)\,dx\,dy} \tag{2.1}$$

where,

$$f_k(x_s, y_s; \theta) = \frac{1}{2\pi\sqrt{1-\rho_i^2}\,\sigma_{1i}\sigma_{2i}}\exp\left\{\frac{-1}{2(1-\rho_i^2)}\left[\left(\frac{x_s-\mu_{1i}}{\sigma_{1i}}\right)^2 - 2\rho_i\left(\frac{x_s-\mu_{1i}}{\sigma_{1i}}\right)\left(\frac{y_s-\mu_{2i}}{\sigma_{2i}}\right)+\left(\frac{y_s-\mu_{2i}}{\sigma_{2i}}\right)^2\right]\right\}$$

and $0 < x < \infty$; $0 < y < \infty$

As a result of this, the feature vector of the entire image follow a finite left truncated bivariate Gaussian distribution with probability density function

$$h(x, y; \theta) = \sum_{i=1}^{K} \alpha_i g_i(x, y; \theta) \tag{2.2}$$

where, $g_i(x_s, y_s; \theta)$ is as given in equation (2.1) and $0 < \alpha_i < 1$, $\sum_{i=1}^{K} \alpha_i = 1$.

The parameters α_k, μ_{1k}, μ_{2k}, σ_{1k}^2, σ_{2k}^2, and ρ_k, for $k = 1,2,...,K$ are obtained by using the EM-algorithm . The updated equations of the parameters in each image region are obtained for the EM-algorithm. The parameters α_k, μ_{1k}, μ_{2k}, σ_{1k}^2, σ_{2k}^2, and ρ_k, for $k = 1,2,...,K$ are taken as given in [14].

3 Initialization of the Parameters using Hierarchical Clustering

To utilize the EM-algorithm we have to initialize the parameter α_k and the model parameters μ_{1k}, μ_{2k}, σ_{1k}^2, σ_{2k}^2, and ρ_k which are usually considered as known apriori. The initial values of α_i can be taken as $\alpha_i = \frac{1}{K}$, where, *K* is the number of image regions obtained from the Hierarchical clustering algorithm [4]. After obtaining the final value for the number of regions *K*, we obtain the initial estimates of μ_{1k},

μ_{2k}, σ_{1k}^2, σ_{2k}^2, and ρ_k for the kth region using the segmented region values with the moment method of estimation given by [1] for truncated bivariate normal distribution with initial parameters. After getting these initial estimates for μ_{1k}, μ_{2k}, σ_{1k}^2, σ_{2k}^2, and ρ_k, we obtain the final refined estimates of the parameters through EM-algorithm given in section 2.

4 Segmentation Algorithm

In this section, we present the image segmentation algorithm. After refining the parameters the prime step is image segmentation by allocating the pixels to the segments. This operation is performed by segmentation algorithm. The image segmentation algorithm consists of four steps

Step 1) Obtain the number of image regions using hierarchical clustering algorithm.
Step 2) Obtain the initial estimates of the model parameters using hierarchical clustering and moment estimates for each image region as discussed in section 3.
Step 3) Obtain the refined estimates of the model parameters μ_{1k}, μ_{2k}, σ_{1k}^2, σ_{2k}^2, ρ_k and α_k for k= 1,2,…,K by using the EM-algorithm with the updated equations given by in section 2.
Step 4) Assign each feature vector to the corresponding jth region (segment) according to the maximum likelihood of the jth component L_j.

That is, (x_s, y_s) is assigned to the jth region for which L_j is maximum.

where,

$$L_j = \max_{j\in k}\left\{\frac{\exp\left\{\frac{-1}{2(1-\rho_k^2)}\left[\left(\frac{x_s-\mu_{1k}}{\sigma_{1k}}\right)^2 - 2\rho_k\left(\frac{x_s-\mu_{1k}}{\sigma_{1k}}\right)\left(\frac{y_s-\mu_{2k}}{\sigma_{2k}}\right)+\left(\frac{y_s-\mu_{2k}}{\sigma_{2k}}\right)^2\right]\right\}}{2\pi\sigma_{1k}\sigma_{2k}\sqrt{1-\rho_k^2}\int_0^\infty\int_0^\infty f_k(x,y,\theta)\,dx\,dy}\right\}$$

5 Experimental Results and Performance Evalution

To demonstrate the utility of the image segmentation algorithm developed in this section, an experiment is conducted with six images taken from Berkeley image data set (http://www.eecs.berkeley.edu/Research/Projects/CS/Vision/bsds/BSDS300/html). The images namely, OSTRICH, POT, TOWER, BEARS, DEER and BIRD are considered for image segmentation. The feature vector consisting of Hue and Saturation values of the whole image is assumed that it follows a mixture of left truncated bivariate Gaussian distribution. That is the whole image is a collection of K-components and the feature vectors in each component follows a left truncated bivariate Gaussian distribution. The number of image regions of each image considered for experimentation is determined by hierarchical clustering algorithm. The number of image regions for each image obtained through hierarchical clustering for the images under study are given in Table1.

Table 1. Estimated value of *K* (By Hierarchical Clustering)

IMAGE	OSTRICH	POT	TOWER	BEARS	DEER	BIRD
Estimate of *K*	2	3	4	3	3	2

From Table 1, it is observed that the images, OSTRICH and BIRD have two segments each, the images POT, BEARS and DEER have three segments each and the image TOWER has four segments. The initial values of the model parameters μ_{1i}, $\mu_{2i}, \sigma_{1i}^2, \sigma_{2i}^2, \rho_i$ and α_i , for i =1,2,..*K* for each image region are computed by using the method given in section 3. Using these initial estimates and the updated equations of the EM-algorithm given in section 2, the final estimates of the model parameters for each image are obtained. Using the estimated probability density function and image segmentation algorithm, the image segmentation is done for the six images under consideration. The original and segmented images are shown in Figure 1. The performance of the developed image segmentation method is studied by obtaining the image segmentation performance measures namely, probabilistic rand index (PRI), global consistency error (GCE) and the variation of information (VOI). A comparative study of the developed algorithm based on finite left truncated bivariate Gaussian mixture model with hierarchical clustering (FLTBGMM-H) with the image segmentation algorithms based on finite GMM with *K*-means algorithm and finite left truncated bivariate Gaussian mixture model with *K*-means is carried. The image segmentation performance measures are computed for the three methods and presented in Table 3.

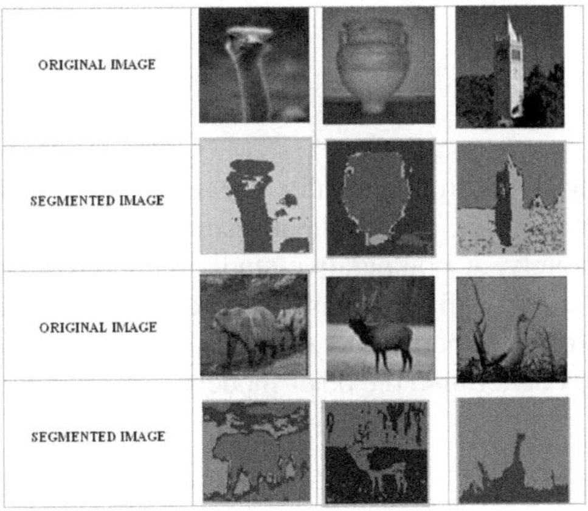

Fig. 1. Original and Segmented images

Table 2. Comparative Study of Image Quality Metrics

IMAGES	METHOD	PERFORMACE MEASURES		
		PRI	GCE	VOI
OSTRICH	GMM-*K*	0.9234	0.4317	2.2761
	FLTBGMM -*K*	0.9782	0.4037	1.7611
	FLTBGMM -H	0.9810	0.3587	0.9481
POT	GMM-*K*	0.9456	0.4281	2.5973
	FLTBGMM -*K*	0.9796	0.4131	1.9263
	FLTBGMM -H	0.9801	0.3895	1.6415
TOWER	GMM-*K*	0.9615	0.4469	3.7121
	FLTBGMM -*K*	0.9816	0.4302	2.8194
	FLTBGMM -H	0.9821	0.3725	1.6554
BEARS	GMM-*K*	0.9121	0.4418	3.2693
	FLTBGMM -*K*	0.9831	0.4337	2.6421
	FLTBGMM -H	0.9834	0.4331	2.6386
DEER	GMM-*K*	0.9774	0.4829	2.2863
	FLTBGMM -*K*	0.9847	0.4030	1.3947
	FLTBGMM -H	0.9849	0.3995	1.2987
BIRD	GMM-*K*	0.9673	0.4671	2.7197
	FLTBGMM -*K*	0.9705	0.4226	2.3244
	FLTBGMM -H	0.9722	0.4170	2.3100

From the Table 3, it is observed that the PRI values of the proposed algorithm for the six images are more than that of the values from the segmentation algorithm based on finite Gaussian mixture model with *K*-means and finite left truncated bivariate Gaussian mixture model with *K*-means and close to 1. Similarly GCE and VOI values of the proposed algorithm for the images are less than that of finite GMM with *K*-means algorithm and finite left truncated bivariate Gaussian mixture model with *K*-means. This reveals that the proposed algorithm performs better than the existing algorithms based on the finite Gaussian mixture model and finite left truncated bivariate Gaussian mixture model.

Using the estimated probability density function of the images under consideration the images are retrieved and are shown in Figure 2. The image quality metrics with respect to the estimated models, the finite Gaussian mixture model with *K*-means and finite left truncated bivariate Gaussian mixture model with *K*-means and are presented in Table 4.

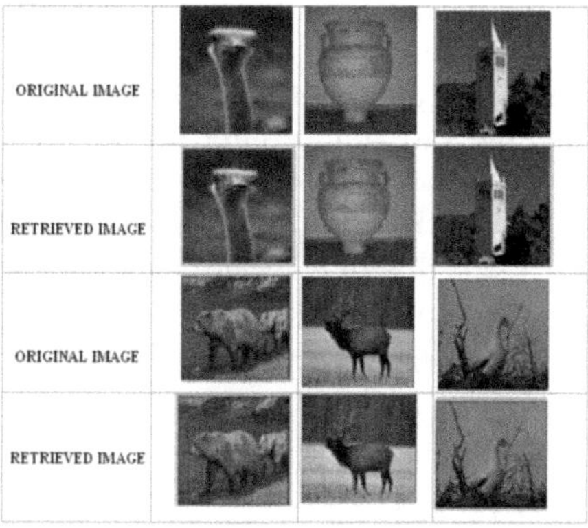

Fig. 2. The Original and Retrieved Images

Table 3. Table 4: Comparative Study of Image Quality Metrics

IMAGE	QUALITY METRICS	GMM-*K*	FLTBGMM-*K*	FLTBGMM-H
OSTRICH	Maximum Distance	0.4013	0.5067	0.5107
	Image Fidelity	0.7910	0.8076	0.9116
	Mean Square Error	0.0932	0.0793	0.0330
	Signal to Noise Ratio	13.3781	13.9959	15.1734
	Image Quality Index	0.8102	0.8492	0.8910
POT	Maximum Distance	0.3290	0.3957	0.3978
	Image Fidelity	0.6729	0.6736	0.6937
	Mean Square Error	0.0738	0.0467	0.0435
	Signal to Noise Ratio	11.7401	13.0240	13.1034
	Image Quality Index	0.6075	0.6174	0.6310
TOWER	Maximum Distance	0.8481	0.8757	0.9583
	Image Fidelity	0.5217	0.5884	0.7635
	Mean Square Error	0.2101	0.1792	0.0676
	Signal to Noise Ratio	8.8488	8.8724	10.9233
	Image Quality Index	0.5173	0.6271	0.7741
BEARS	Maximum Distance	0.5387	0.8765	0.8813
	Image Fidelity	0.4277	0.6536	0.6588
	Mean Square Error	0.0872	0.0484	0.0413
	Signal to Noise Ratio	9.1217	10.7550	10.7573
	Image Quality Index	0.5906	0.5951	0.6067
DEER	Maximum Distance	0.6217	0.6474	0.6592
	Image Fidelity	0.3982	0.4470	0.4640
	Mean Square Error	0.0828	0.0547	0.0510
	Signal to Noise Ratio	10.0629	11.3918	11.9536
	Image Quality Index	0.3763	0.3840	0.4131
BIRD	Maximum Distance	0.8429	0.9129	0.9321
	Image Fidelity	0.1920	0.2349	0.2552
	Mean Square Error	0.2013	0.0900	0.0894
	Signal to Noise Ratio	8.9231	9.3864	9.4108
	Image Quality Index	0.3481	0.4160	0.5479

From the Table 4, it is observed that all the image quality measures for the six images are meeting the standard criteria. This implies that using the proposed algorithm the images are retrieved accurately. A comparative study of the proposed algorithm with that of the algorithms based on finite Gaussian mixture model and finite left truncated bivariate Gaussian mixture model and *K*-means reveals that the proposed model in retrieving the images is better than the other models.

6 Conclusion

In this paper an image segmentation method based on finite left truncated bivariate Gaussian mixture model with hierarchical clustering is developed and analyzed. The model parameters are estimated by EM-algorithm and a segmentation algorithm with component maximum likelihood is developed. The performance of this algorithm is studied by conducting an experiment with six images. The probability density functions of the images are also estimated. The image segmentation performance measures are computed for the six images. From a comparative study it is observed that the results obtained for the colour image segmentation method based on finite left truncated bivariate Gaussian mixture model with hierarchical clustering are better than the results obtained for image segmentation method based on finite left truncated bivariate Gaussian mixture model with *K*-means. It is further observed that the segmented images using the method discussed in this paper are having clear boundaries.

The performance analysis revealed that the hierarchical clustering used for the initial segmentation has significant influence on the performance of the image segmentation and image retrievals. This image segmentation method is much useful for segmentation and retrieval of the images in medical diagnosis, film and video production, remote sensing, robotics, security monitoring, etc., where, the colour image is characterized by Hue and Saturation values. This segmentation technique is also useful for denoising the image by filtering the background noise which is an important aspect of content based image retrieval. The integration of heuristic segmentation methods (*K*-means and Hierarchical) with model based image segmentation reduces the computational time and complexities in colour image segmentation.

References

[1] Muthen, B.: Moments of the censored and truncated bivariate normal distribution. British Journal of Mathematical and Statistical Psychology (43), 131–143 (1990)

[2] Eskicioglu, M.A., Fisher, P.S.: Image Quality Measures and their Performance. IEEE Transactions on Communications 43(12) (1995)

[3] Haralick, Shapiro: Survey: Image segmentation Techniques. In: Proc. of Int. Conf. CVGIP 1985, vol. 29, pp. 100–132 (1985)

[4] Johnson, S.C.: A Tutorial on Clustering Algorithms (1967), `http://home.dei.polimi.it/matteucc/Clustering/tutorial_html/hierarchical.html`

[5] Kato, Z., Pong, T.C.: A markov random field image segmentation model using combined color and texture features. In: Proc. of Int. Conf. on Computer Analysis of Images and Patterns, pp. 547–551 (2001)

[6] Kato, Z., Pong, T.-C., Qiang, S.G.: Unsupervised segmentation of color textured images using a multilayer MRF model. In: Proc. of Intl. Conf. on Image Processing, vol. 1, pp. 961–964 (2003)

[7] Kato, Z., Pong, T.-C.: A Markov random field image segmentation model for color textured images. Image and Computing Vision 24(10), 1103–1114 (2006)

[8] Lucchese, L., Mitra, S.K.: Color image segmentation: A state-of art survey. Proc. of Indian National Science Academy (INSA-A) 67-A, 207–221 (2001)

[9] Paulinas, M., Usinskas, A.: A survey of genenetic algorithms applications for image enhancement and segmentation. Information Technology and Control 36(3), 278–284 (2007)

[10] Sojodishijani, O., Rostami, V., Ramli, A.R.: Real Time Colour Image Segmentation with Non-Symmetric Gaussian Membership Functions. In: Proc. of 5th Int. Conf. on Computer Graphics, Imaging and Visualisation, pp. 165–170 (2008)
[11] Pal, S.K., Pal, N.R.: A Review on Image Segmentation Techniques. Pattern Recognition 26(9), 1277–1294 (1993)
[12] Gonzalez, R.C., Woods, R.E.: Digital Image Processing. A text book from Pearson education, India (2001)
[13] Farnoosh, R., Yari, G., Zarpak, B.: Image Segmentation using Gaussian Mixture Models. IUST International Journal of Engineering Science 19(1), 29–32 (2008)
[14] Rajkumar, G.V.S., Srinivasa Rao, K., Srinivasa Rao, P.: Studies on Colour Image Segmentation method based on finite left truncated bivariate Gaussian mixture model with K-Means. Global Journal of Computer Science and Technology X1(XVIII), 21–30 (2011)
[15] Randen, Husoy, J.: Filtering for texture classification: A comparative study. IEEE Trans. on Pattern Analysis and Machine Intelligence 21(4), 291–310 (1999)
[16] Raut, S., Raghuvanshi, M., Dharaskar, R., Raut, A.: Image Segmentation- A state-of-Art Survey for Prediction. In: Proc. of Int. Conf. on Advanced Computer Control, pp. 420–424 (2009)
[17] Shivani, G., Manika, P., Shukhendu, D.: Unsupervised segmentation of texture images using a combination of gab or and wavelet features. In: Proceedings of the 4th Indian Conference on Computer Vision, Graphics & Image Processing, pp. 370–375 (2004)
[18] Bhattacharyya, S.: A Brief Survey of Color Image Preprocessing and Segmentation Techniques. Journal of Pattern Recognition Research, 120–129 (2011)
[19] Sujaritha, M., Annadurai, S.: Color Image segmentation using Adaptive Spatial Gaussian Mixture Model. International Journal of Signal processing 6(1), 28–32 (2010)
[20] Wu, Y., et al.: Unsupervised Color Image Segmentation Based on Gaussian Mixture Models. In: Proceedings of 2003 Joint Conference At The 4th International Conference on Information, Communication and Signal Processing, vol. 1, pp. 541–544 (2003)
[21] Fei, Z., Guo, J., Wan, P., Yang, W.: Fast automatic image segmentation based on Bayesian decision-making theory. In: Proc. of Int. Conf. on Information and Automation, pp. 184–188 (2009)

EOST-An Access Method for Obfuscating Spatio-Temporal Data in LBS

Ashwini Gavali[1], Suresh Limkar[2], and Dnyanashwar Patil[3]

[1] Department of Computer Engineering, GHRCEM, Pune, India
[2] Department of Computer Engineering, AISSMS's IOIT, Pune, India
[3] Department of Computer Engineering, DYPCOE, Ambi, Pune, India
{dnyane.ash,sureshlimkar,dnyane.ash}@gmail.com

Abstract. Most widely use of mobile communication devices and the technical improvements of location techniques are fostering the development of new applications that use the physical position of users to offer location-based services for business, social, or informational purposes. Since the development of location-based services, privacy-preserving has gained special attention and many algorithms aiming at protecting user's privacy have been created such as obfuscation or k-anonymity. The OST-tree capable of obfuscating the spatio-temporal data of users. Also it is easy for the adversary to infer a user's exact position in the obfuscated area if the probability distribution of user's position is uniformly distributed in case of OST- tree. So this problem can be addressed by proposing EOST-tree, which obfuscates the spatiotemporal data for the probability distribution of user's position belongs to a region is not uniformly distributed. As in real life, the region where a user belongs to depends on many factors related to geography.

Keywords: LBS, obfuscation, privacy-preserving, spatio-temporal indexing.

1 Introduction

Now days due to the rapid development of global positioning system (GPS), there are large numbers of mobile users and users are expected to increase more. Among the various services for mobile phone, the location-based service (LBS) is the most promising one since it supplies users with many value-added services. In order to benefit from these services, users, however, have to reveal their sensitive information such as their current locations. Such novel services pose many challenges because users are not willing to reveal their sensitive information but still want to benefit from these useful services.

As there exist a problem of privacy preservation. To solve this privacy-preserving problem, a variety of algorithms are suggested to hide personal information of users but still allow them to use services with acceptable quality. The general idea of these algorithms is to obfuscate the user's position [1, 2, 3, 4,5], or to anonymize location information [6, 7,8,9]. But it has two limitations. First, all of them deal with only spatial obfuscation, not temporal one. Second, these are separated from the database

S.C. Satapathy et al. (Eds.): Proc. of Int. Conf. on Front. of Intell. Comput., AISC 199, pp. 171–179.
DOI: 10.1007/978-3-642-35314-7_20

level. Due to which they works at two phases: retrieving the exact location of user on the database level first, and then obfuscating this information on the algorithm level [10]. This two-phase process is time-consuming.

The goal of this paper is to define the new index structure which is capable of obfuscating spatio-temporal data as well as to provide an improvement over querying costs and user's privacy protection. Towards this goal, EOST-tree is designed to feature service provider classification by letting users specify authorizations and put the privacy information on the tree nodes. Also try to maintain QOS issue with less revelation of user's private information. Furthermore, because this index structure embeds privacy information on its node, the process of calculating the obfuscated data can be done in only one phase: traversing the index structure to retrieve the appropriately obfuscated data. This one-phase process can reduce the processing time.

The rest of this paper is organized as follows. In section 2, we briefly summarize the background work and related problems within existing system. Next, section 3 presents our proposed approach for preserving privacy in LBS with defined problem statement and scope with an index structure of proposed system along with its experimental setups. And comparative study with our solution is discussed in section 4. Finally, section 5 presents concluding remarks as well as our future work.

2 Background Work

Among the most popular techniques to protect user's location privacy, obfuscation based techniques [10, 11] have gained much interest due to its implementation simplicity. Location obfuscation aims at hiding user's exact location by decreasing the quality of user's location information. In [11] propose obfuscation techniques by enlarging the area containing user's real location. However, these techniques just deal with geometry of the obfuscated region, not concerning about what is included inside (i.e., the geographic feature). Of late, the semantic-aware obfuscation technique introduced in [12, 13, 14] considers sensitive feature types inside an obfuscated region. But, this technique does not concern about how big the area of the obfuscated region is. EOST-Tree proposed approach, classifies service providers in the way that the more reliable the service providers, the smaller area of the obfuscated region they can obtain.

Recently many researchers have focused on indexing the present and future positions of moving objects [15, 16, 17]. With the former, the main idea is that the bounding rectangle is a temporal function, and thus can enclose moving objects. The most popular category is parametric spatial access. The most popular access method in this category, TPR-tree [15, 18, 19], inherits the idea of parametric bounding rectangles in R-tree [20,21,22] to create time-parameterized bounding rectangles (TPBR). However, the TPBR bear two crucial limitations that dramatically affect the performance of TPR-tree: overlapping and high storage cost. Very recently, the OST-tree [23] embeds the user's privacy policy [24, 25, 26] into its nodes and obfuscates spatio-temporal data. But, since OST-tree is based on TPR-tree and concerns only with geometry-based obfuscation, it has high storage cost and quite low privacy protection. That means currently there does not exist any spatio-temporal index structure that can effectively handle geographic-aware obfuscation [26]. Towards this goal, in this paper, there is proposal of the EOST-tree, a structure originally based on TPR (OST)-tree, but with essential modifications to support geographic-aware obfuscation.

3 Overview of Proposed System

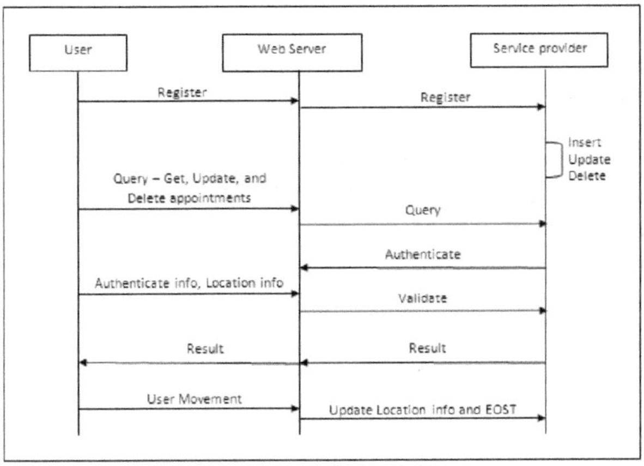

Fig. 1. Block diagram

3.1 Problem Statement

As from previous discussion existing systems have different drawbacks such as querying process is two step process which is time consuming. Another drawback is that no moving object index has yet been reported in the literature that achieves the goal of obfuscating user's position. Now a day there is requirement of suitable spatio-temporal index. Existing index structure can take only access or reject action, they create clusters but not obfuscating customer's location So as to overcome above, there is use of our proposed system, EOST.

3.2 Scope

The proposed EOST system covers the range of topics, many shared with Low querying cost and privacy protection in location based services. Some of the most important are improvement of query process, to optimize process, reduction of query cost as well as privacy protection through complicating user positions and extend probability distribution of users positions.

3.3 Temporal Obfuscation

Many of the research activities are there in the area of spatial obfuscation [1, 5, 11], but, no mature proposals for obfuscating the temporal data of users exist. Similar to spatial obfuscation, temporal obfuscation [11, 23] will degrade the exact value of time t_0 to the vague temporal value $[t^{[}, t^{]}]$, where $t^{[} < t_0 < t^{]}$.

The obfuscated value of timestamp t_0 is the temporal interval $[t^{[},t^{]}]$ which includes the real timestamp t_0 with the probability:

$$P(t_0 \in [t^{[},t^{]}])=1 \tag{1}$$

3.4 Spatio-temporal Obfuscation

The obfuscated value of user's exact position (x_u, y_u) at a timestamp t_0 is a rectangular area (x_c, y_c, w, h) centered on the geographical coordinates (x_c, y_c) with width w, height h, at a temporal interval $[t^{[},t^{]}]$, which includes the user's exact position (x_u, y_u) at a real timestamp t_0 with the probability:

$$P((x_u, y_u) \in Rectangle(x_c, y_c, w, h)\ AND\ t_0 \in [t^{[},t^{]}])=1 \tag{2}$$

3.5 Authorization

An authorization α is a 4-tuple $<id_{sp}, id_{user}, \Delta s, \Delta t>$ where id_{sp} is the identity of service provider, id_{user} is the identity of user, Δs, Δt is the degree of accuracy of user's position (spatial data) and time, respectively. This authorization can be expressed as α1 = <#SP101, #U232, 700m^2, 4m>. If the user's exact position in the next 15 minutes is located at a coordinate $<x_0, y_0>$, the result returned from the next 12 to 16 minutes to the service provider is a rectangle which has the area of 700 square meters and contains the coordinate $<x_0, y_0>$ in case of time and position, respectively.

3.6 Index Structure

The base structure of the EOST-tree is that of the TPR-tree for indexing the spatio-temporal data. However, in order to specify the authorization and the degree of accuracy of user's position and time, the node structure will be modified to attach more information. Specifically, in addition to the tpbr, each node contains a pointer p pointing to the list of entries. Each entry has the form of a 4-tuple $<id_{sp}, id_{user}, \Delta s, \Delta t>$, indicating that a service provider with the identity id_{sp} can access sensitive information of a user with the identity id_{user} at the degree of accuracy of user's position and time specified by the value Δs and Δt, respectively. Figure 2 illustrates the structure of the EOST tree. For the illustration purpose, the values of authorizations α_i (i=1..5) in this figure are α_1 = <#SP101, #U232, 1600m^2, 3m>, α2 = <#SP101, #U134, 600m^2, 3m>, α3 = <#SP102, #U232,500m^2, 3m>, α4 = <#SP101, #U135, 550m2, 4m>, , α5 = <#SP103, #U232, 0m2, 0m> and α6 = <#SP103, #U235, 250m2, 0m>.

Our aim is to develop an index structure that can incorporate the accuracy degree of user's position. Therefore, this accuracy degree parameter must be in the hierarchical form. The EOST-tree achieves this hierarchy well. More specifically, when traversing from the root node to a leaf node in the EOST tree, the degree of accuracy of user's position increases because the area of the bounding rectangle is smaller and vice versa. For example, in the traversal path N1-N2-N6 (see Figure 2), the areas of the returned rectangles reduce from $1600m^2$ to $600m^2$ and $250\ m^2$ corresponding to α1, α2, and α6. This means that the degree of accuracy of user's position increases. Based on this property, if service providers have a higher level of trust from a user, their identities will be placed on the node nearer to the leaf node and vice versa. For instance, the service provider with the identity #SP103 has the highest level of trust from a user with the identity #U235, and so it can obtain the user's exact position (*Δs=0*). This service provider's identity is, therefore, placed on the leaf node.

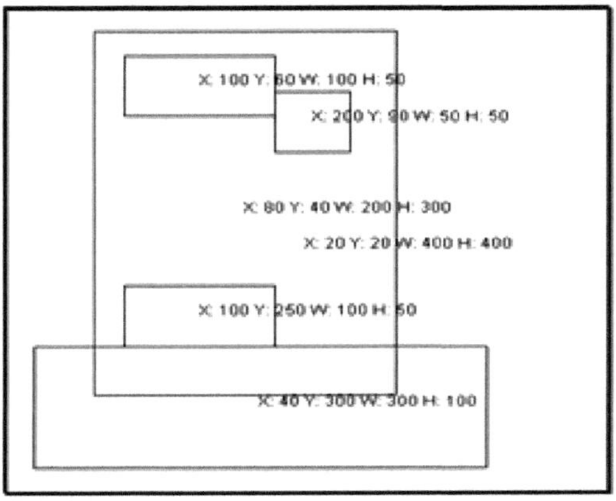

Fig. 2. EOST-tree structure. Here N1(X:20 Y:20 W:400 H:400), N2(X:80 Y:40 W:200 H:300), N3(X:40 Y:300 W:300 H:100), N4(X:100 Y:80 W:100 H:50), N5(X:100 Y:250 W:100 H:50), N6(X:200 Y:90 W:50 H:50).

3.6.1 Privacy Information Overlaying and Insertion

The privacy information overlaying and insertion process happen in parallel. We traverse the EOST-tree from the root node down to the leaf node to place the new object in the suitable leaf node (by applying the insertion algorithm shown in [15, 21]) and, at the same time, recursively compare the degree of accuracy of user's position (Δs) with a spatial extent of each node (N■) in the insertion path to find the appropriate node overlaying privacy information. We have two possible scenarios for this comparison:

• Case 1: If (N is the appropriate sub-tree) and (Δs≥ N■), we overlay α on N and continue the insertion process.
• Case 2: If (Δs< N■), depend on the level of N, we have two scenarios: If N is a non-leaf node, we choose an appropriate sub-tree rooted at N (complying with the algorithm ChooseSubtree of R*-trees [21]) and continue the overlaying process. If N is a leaf node, we overlay and insert the new object into this node. If a moving object has already existed in the index structure and the user wants to add new policies, we find the appropriate node in the insertion path to overlay privacy information.

3.6.2 Privacy Analysis

For obfuscation techniques, the *relevance* [11] is used to measure the location privacy protection. The lower the relevance, the higher the location privacy protection is, and thus the lower the probability an adversary can infer the user's exact location For the approach that separates the algorithm from database level, the relevance is:

$$R_s = \frac{(Ai \cap Af)^2}{Ai * Af} \tag{3}$$

Where A_i is the location measurement [11] and A_f is the obfuscated region created by the privacy-preserving algorithm. To calculate the relevance of proposed approach, we can simply replace A_f by Δs in (3). We extend the relevance concept to use for both spatial and temporal privacy protection as follows:

$$R_{st} = \frac{(A_i \cap \Delta s)^2}{A_i.\Delta s} \cdot \frac{1}{\Delta t} \tag{4}$$

Where Δs and Δt are the degree of accuracy of user's position and time, respectively. From (3) and (4), we can see that $R_s \geq R_{st}$ (since $A_f = \Delta s$ and $\Delta t \geq 1$), meaning that the degree of privacy protection of proposed approach is higher than that of approach separating the algorithm from database level.

3.7 Experiment Setup

EOST-tree is implemented in Java, and all experiments are conducted on a Core 2 Duo PC, running Windows XP Professional with 1GB of RAM, 160GB of HDD, and the disk page size of 4KB. The query cost of EOST-tree is lowest (as illustrated in Figure 3).Figure 4 shows that the update cost in EOST-tree achieves considerable improvement over TPR-tree and OST-tree. Because in EOST-tree, given the key, an update needs to traverse only one path. On the contrary, in TPR tree and OST-tree, an update may traverse multiple paths because of the overlaps among TPBR. As the dataset grows, more overlap happens and thus we have results in a higher update cost.

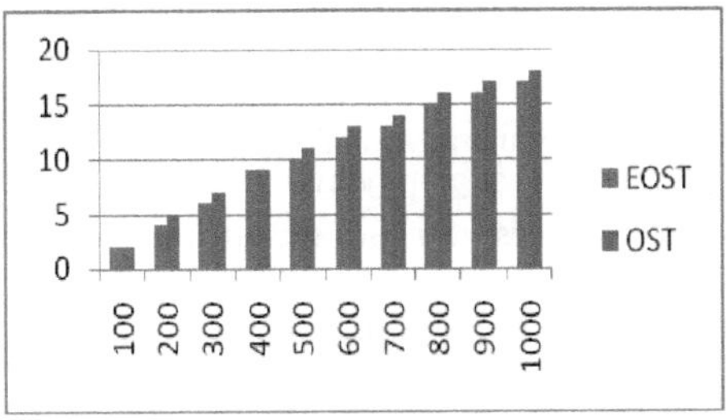

Fig. 3. Query cost

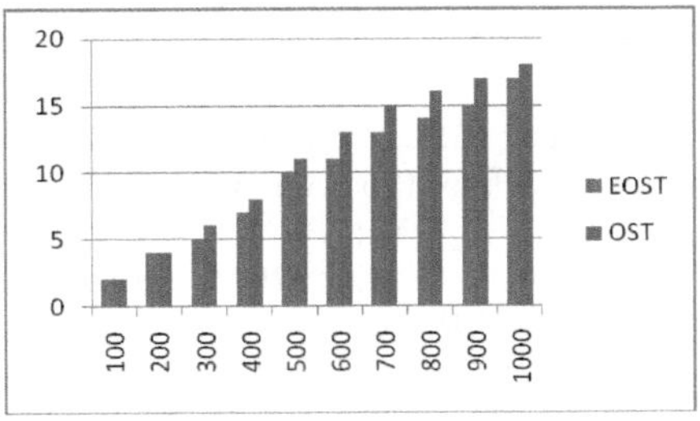

Fig. 4. Update cost

4 Comparative Study

The OST-tree requires more nodes to contain the same number of moving objects pointer. If the pair value <id_{sp}, id_{user}> of the query's authorization is matched with that of some internal node, we will stop at this node and return the result without further traversing on the OST-tree. Hence, the OST-tree requires less disk accesses than that of the TPR-tree but more as compared to EOST. By incorporating the time into the privacy model, the average relevance of proposed approach is smaller than that of the obfuscation algorithm. The insert cost of OST-tree is higher than that of EOST-tree. Given the mobility of users, the update cost of OST-tree is higher than that of EOST tree, because OST-tree has to incur the additional cost of updating the authorization (moving α from current node to another node corresponding to the newly updated position of a user). In general, the query cost of EOST-tree is better than that of TPR-tree and OST tree. Hence, EOST-tree is better than TPR-tree and OST tree in cases users just want to reveal a low degree of accuracy of their locations to service providers.

5 Conclusion and Future Work

In this work, we have introduced the EOST-tree capable of obfuscating the spatio-temporal data of users. The process of calculating obfuscated data can be done in only one phase and one phase reduces processing time. It specifies authorizations and put the privacy information on tree nodes. The EOST-tree requires less storage space and update overhead, it achieves the lower querying cost and higher privacy protection comparing to the OST-tree. In real life, the region where a user belongs it is easy for the adversary to infer a user's exact position in the obfuscated area if the probability distribution of user's position is uniformly distributed. In the future, we will address the quality of LBS problem due to the different shapes of the returned regions wrt. the EOST-tree and other access methods.

References

1. Ardagna, C.A., Cremonini, M., Vimercati, S.D.C., Samarati, P.: An Obfuscation-Based Approach for Protecting Location Privacy. TDSC 8(1), 13–27 (2009)
2. Mohamed, F.M.: Privacy in Location-based Services: State-of-the-art and Research Directions. Tutorial, MDM, Germany (2007)
3. Jafarian, J.H., Ravari, A.N., Amini, M., Jalili, R.: Protecting Location Privacy through a Graph-Based Location Representation and a Robust Obfuscation Technique. In: Lee, P.J., Cheon, J.H. (eds.) ICISC 2008. LNCS, vol. 5461, pp. 116–133. Springer, Heidelberg (2009)
4. Jafarian, J.H., Ravari, A.N., Amini, M., Jalili, R.: Protecting Location Privacy through a Graph-Based Location Representation and a Robust Obfuscation Technique. In: Lee, P.J., Cheon, J.H. (eds.) ICISC 2008. LNCS, vol. 5461, pp. 116–133. Springer, Heidelberg (2009)
5. Ardagna, C.A., Cremonini, M., De Capitani di Vimercati, S., Samarati, P.: An Obfuscation-based Approach for Protecting Location Privacy. In: 7th Framework Programme (FP7/2007-2013) under grant agreement no. 216483 "PrimeLife"
6. Truong, A.T., Truong, Q.C., Dang, T.K.: An Adaptive Grid-Based Approach to Location Privacy Preservation. In: Nguyen, N.T., Katarzyniak, R., Chen, S.-M. (eds.) Advances in Intelligent Information and Database Systems. SCI, vol. 283, pp. 133–144. Springer, Heidelberg (2010)
7. Gruteser, M., Grunwald, D.: Anonymous Usage of Location-Based Services Through Spatial and Temporal Cloaking. In: MOBISYS (2003)
8. Bugra, G., Ling, L.: Protecting Location Privacy with Personalized k- Anonymity: Architecture and Algorithms. IEEETMC 7(1), 1–18 (2008)
9. Truong, A.T., Truong, Q.C., Dang, T.K.: An Adaptive Grid-Based Approach to Location Privacy Preservation. In: Nguyen, N.T., Katarzyniak, R., Chen, S.-M. (eds.) Advances in Intelligent Information and Database Systems. SCI, vol. 283, pp. 133–144. Springer, Heidelberg (2010)
10. Dinh, L.V.N., Aref, W.G., Mokbel, M.F.: Spatio-temporal Access Methods-Part 2. IEEE Data Engineering Bulletin (2010)
11. Ardagna, C.A., Cremonini, M., Vimercati, S.D.C., Samarati, P.: An Obfuscation-Based Approach for Protecting Location Privacy. IEEE Transactions on Dependable and Secure Computing 8(1) (January- February 2011)
12. Damiani, M.N., Bertino, E., Silvestri, C.: Protecting Location Privacy through Semanticsaware Obfuscation Techniques. In: IFIPTM, Norway, pp. 231–245 (2008)
13. Dang, T.K.: Semantic Based Similarity Searches in Database Systems (Multidimensional Access Methods, Similarity Search Algorithms). PhD thesis, FAW-Institute, Johannes Kepler University of Linz, Austria (May 2003)
14. Damiani, M., Bertino, E., Silvestri, C.: PROBE: an Obfuscation System for the Protection of Sensitive Location Information in LBS. TR2001-145, CERIAS (2008)
15. Saltenis, S., Jensen, C.S., Leutenegger, S.T., Lopez, M.A.: Indexing the Positions of Continuously Moving Objects. In: ACM SIGMOD, USA, pp. 331–342 (2000)
16. Kwon, D., Lee, S., Lee, S.: Indexing the Current Positions of Moving Objects Using the Lazy Update R-tree, supported by the Brain Korea 21 Project
17. Dang, T.K., Küng, J., Wagner, R.: The SH-tree: A Super Hybrid Index Structure for Multidimensional Data. Springer, Heidelberg (2001)
18. Tao, Y., Sun, J., Papadias, D.: The TPR *- Tree: An Optimized Spatio- Temporal Access Method for Predective queries. In: VLDB Conference, Berlin, Jerminy (2003)

19. Location-Tracking Applications, Published by the IEEE COMPUTER SOCIETY _ 1540-7993/04/$20.00 ©, IEEE _ IEEE Security & Privacy (2004)
20. Guttman, A.: R-trees: A Dynamic Index Structure for Spatial Searching. In: ACM SIGMOD, USA, pp. 47–57 (1984)
21. Beckmann, N., Kriegel, H.-P., Schneider, R., Seeger, B.: The R*-tree: An Efficient and Robust Access Method for Points and Rectangles. In: ACM SIGMOD, pp. 322–331 (1990)
22. Lee, M.L., Hsu, W., Jensen, C.S., Cui, B., Teo, K.L.: Supporting Frequent Updates in R-Trees: A Bottom-Up Approach. Technical Report (April 2004)
23. To, Q.C., Dang, T.K., Küng, J.: OST-tree: An Access Method for Obfuscating Spatiotemporal Data in Location-based Services. In: NTMS, France (2011)
24. Atluri, V., Adam, N.R., Youssef, M.: Towards a unified index scheme for mobile data and customer profiles in a location-based service environment. In: NG2I (2003)
25. Dang, T.K., To, Q.C.: An Extensible and Pragmatic Hybrid Indexing Scheme for MAC-based LBS Privacy-Preserving in Commercial DBMSs. In: ACOMP, pp. 58–67 (2010)
26. To, Q.C., Dang, T.K., Küng, J.: B^{ob}-Tree: An Efficient B^+-Tree Based Index Structure for Geographic-Aware Obfuscation. In: Nguyen, N.T., Kim, C.-G., Janiak, A. (eds.) ACIIDS 2011, Part I. LNCS, vol. 6591, pp. 109–118. Springer, Heidelberg (2011)
27. Atluri, V., Shin, H.: Efficient Security Policy Enforcement in a Location Based Service Environment. In: DBSEC, USA, pp. 61–76 (2007)
28. Gruteser, M., Grunwald, D.: Anonymous Usage of Location-Based Services through Spatial and Temporal Cloaking. In: MOBISYS (2003)
29. Duckham, M., Kulik, L.: A Formal Model of Obfuscation and Negotiation for Location Privacy. In: Gellersen, H.-W., Want, R., Schmidt, A. (eds.) PERVASIVE 2005. LNCS, vol. 3468, pp. 152–170. Springer, Heidelberg (2005)
30. Ardagna, C.A., Cremonini, M., Damiani, E., De Capitani di Vimercati, S., Samarati, P.: Supporting location-based conditions in access control policies. In: Proc. of ACM ASIACCS 2006, Taipei, Taiwan (March 2006)
31. Bettini, Wang, X.S., Jajodia., S.: Protecting privacy against location-based personal identification. In: Proc. of the 2nd LDB LDB Workshop on Secure Data Management, Trondheim, Norway (2005)
32. Limkar, S., Kadam, N., Jha, R.K.: Access Control Based on Location and Time. In: Das, V.V., Ariwa, E., Rahayu, S.B. (eds.) SPIT 2011. LNICS, vol. 62, pp. 102–107. Springer, Heidelberg (2012)
33. Limkar, S.V., Jha, R.K., et al.: Geo-Encryption: A New Way to Secure Critical National Infrastructure. In: International Conference on Information Technology, New Generations, ITNG (2011)
34. Jha, R., Dalal, U.: WiMAX System Simulation and Performance Analysis under the influence of Jamming. Wireless Engineering and Technology (WET) Journal by Scientific Research 1(1), 20–26 (2010)
35. Jha, R., Dalal, U.D.: A Journey on WiMAX and Its Security Issues. International Journal of Computer Science and Information Technologies 1(4), 256–263 (2010)

Discovering Web Usage Patterns - A Novel Approach

K. Sudheer Reddy[1], Ch.N. Santhosh Kumar[2], V. Sitaramulu[2], and M. Kantha Reddy[3]

[1] Dept. of CSE,
Acharya Nagarjuna University, Guntur, AP, India
sudheercse@gmail.com
[2] Department of Computer Science & Engineering
Swarna Bharathi Institute of Science & Technology, Khammam, AP, India
vsitaramu.1234@gmail.com, santhosh_ph@yahoo.in
[3] IUCEE, India
kanthareddy_m@yahoo.com

Abstract. Pattern mining is one of the most pivotal steps in data mining; pattern mining immediately comes after the preprocessing phase of WUM. Pattern discovery deals with the sorted set of data items presented as part of the sequence. Pattern mining, users can recognize the web paths follow on a web site easily. The aim of this research discovers the patterns which are most relevant and interesting by using a Web usage mining process. The server web logs aids are the input to this process. Our target is to discover users' behavior, who has visited the web sites for less number of times. We have enlightened a method for clustering, based on the pattern summaries. We have conducted intense experiments and the results are shown in this paper.

Keywords: Web usage mining, preprocessing, pattern discovery, sequential patterns, clustering, patterns summary.

1 Introduction

Analyzing the behavior of the web users' is also known as Web Usage Mining (WUM). WUM is an active research area which entails in adapting the mining techniques to the records of access log files. These access web log files collect numerous types of data include IP address of the host, the requested URL. The date and other required information about the user navigation into web. The techniques of WUM provide most interesting knowledge about the numerous web user behaviors in order to excerpt relationships in the recorded data. Amongst the techniques available, the sequential patterns are predominantly well adapted to the web log study. Sequential patterns extraction on a web access log file, is theoretical to provide the thoughtful relationship: "On SRKREC Web Site, 23% of users visited the homepage consecutively, the available resources page, the RSC offers, the RSC missions and finally the past RSC competitive selection". Exhibiting this type of behavior is an assumption, because pattern extraction on a web access log file also infers, managing several problems, as listed below:

S.C. Satapathy et al. (Eds.): Proc. of Int. Conf. on Front. of Intell. Comput., AISC 199, pp. 181–188.
DOI: 10.1007/978-3-642-35314-7_21

- The number of records in the web server log file is lowered due to user's computer cache and the proxies.
- The entries of the log file can be reduced and also reduce the user navigations is possible with the aid of research engines. As a result of this, the user can directly access a definite portion of the web site.
- The number of portions visited on the site is compared to the entire site.
- The user's representativeness who navigates the web through that part is compared to the whole site users.

If the web caching problems are to be solved [5], the representativeness requires a sturdy study. To exemplify our goal, let's consider sequential patterns we are supposed to get. Due to the minor size of the "job offer" part of the web site, users requesting a page on that part represent only 0.3% of users on the entire web site. In the similar way, users navigating on the "research" part of the research assignment represent only 0.003% of all the users. So, the study of WUM on this type of site has to manage this specific representativeness in order to provide sufficient results. Our objective is to showcase that a classical pattern mining technique is unable to provide web users behaviors with such a weak support.

Furthermore, we present a unique method for discovering behavior of all web users of a Web site. We tag our test and experiments and then conclude the paper.

2 Principle

We propose a methodology and describe the outline as mentioned here: discovering the clusters of the web users (web users are typically grouped by the user's behavior) and then analyzing the user navigations by means of the sequential pattern mining process. Therefore, our methodology relies on two steps. The first step targets at splitting the web log into sub-logs, hypothetical to represent several separated actions. The second step targets at analyzing user behavior recorded in each sub-log.

The key principle involved in our method is described as given below:

1. Extracting the patterns on the original log.
2. These extracted patterns can be clustered.
3. Dividing the web log according to the clusters obtained, each sub-log encompasses user sessions from the original web access log, approving at least one of the user behavior of the cluster which permits to create this sub-log. A distinct sub-log is created then to collect the user sessions from the original sub-log which doesn't correspond to the cluster from an earlier step.
4. Apply the whole process recursively, for each sub-log.

The below Figure1 graphically illustrates the proposed method. Initially, sequential patterns are to be obtained and the same are to be clustered, it is shown from C1 to Cn in the figure. Then, the web access log is split into various sub-logs, from SLog1 to SLogn upon these clusters. To finish, a sub-log SLogn+1 is created for the several user sessions which cannot be consistent with a behavior from the web log. The quality of the results produced by our methodology will depend on the sub-log file. In detail, the initial sub-logs comprise the most represented categories of the web users. Hence, they are interesting, but the most interesting patterns discovery will derive from the study of the uncluster sessions of the sub-log SLogn+1. Seeing this sub-log as a new original web log, and recursively repeating the process will allow users to discover behaviors with the minimal support. To acquire reliable results, our method suitable on a "the quality of the split proposed for a log". The split depends on on the clustering done on the discovered patterns in the original log. In the following section, we describe briefly the methodology that we have used in cluster patterns.

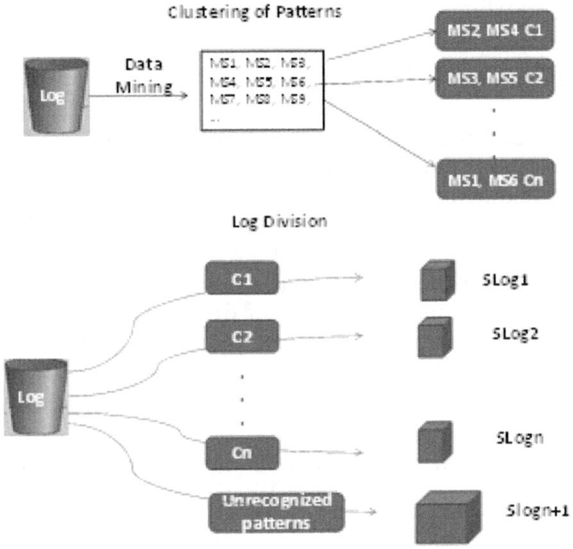

Fig. 1. The Principle of Discovery

3 Clustering Based on Pattern Generalization

We have deliberated several methods of clustering for sequential patterns discovery. We propose and describe here the utmost efficient method for sequential pattern clustering that we have used. The clustering approach which is used in this study is grounded on a method developed by [8] in 2000 for indexing the web sequences in the perception of Web-based recommender systems. The efficiency is based on the neural approach for such method and its effectiveness relies on usage of summarized descriptions for sequential patterns: these descriptions are based on Web access sequences generalization.

3.1 Neural Method

The proposed neural clustering method is based on [8] a framework for supporting the reuse of the various past experiences using the integrated object oriented organization. This approach has been successfully applied on browsing behaviors of thematic repertory, in huge organizations. This is purely based on hybrid model and composed from the connexionist part [5] and unadulterated flat memory compound of patterns' groups.

A threshold TSi is linked to each prototype, which will be altered during the knowledge step. If a user pattern is introduced in the network, drops in the influence region of a prototype, then the prototype will be automatically activated. Such region is determined by set of input vectors acceptable a distance measure which is lower than the threshold. In case, if there is an inactivated prototype, a new prototype is created.

Hence, the structure of a prototype-based network is an evolutionary one in the sagacity that the numerous prototypes at the hidden level is not the priori fixed and might be enlarged during learning step. The prototype is characterized by using its influence region, reference vector, and a set of representing patterns.

4 Experiments and Results

The methods of extraction are written in object oriented language, C++ on a Pentium (3.2 Ghz) PC running a Linux system. The algorithm we have employed is the PSP algorithm [12] for extracting patterns. The neural method and GUI are grasped in Java. For the SRKREC's site, the data was composed over a period of 45 days, while for the other intranets, over a period of 70 days. The narrative of the characteristics (refer Figure 3) is: the number of lines in the web access log is indicated by N, S can be the number of user sessions, the number of filtered URLs is U, the average session length is denote as L, the average number of session URLs is SU. Through our experiments, we are able to bring into interval frequent behaviors, with a comparative representativeness getting feebler and weaker, depending on the sub-log's depth.

	www.srkrec.ac.in	www.srkrec.ac.in/intranet
N	12 57 24	17 167 81
S	287 493	437 648
U	46 218	61 398
L	3.5	2.9
SU	4.6	3.2

Fig. 2. Log file characteristics

C1: The user behavior listed here is related to the higher education prospects offered by the SRKREC. The users visit and read higher education page, and then the web page describing the competitive selection and lastly the web pages describing the education opportunities.

<(trv/higheredu/educon.html) (trv/higheredu/educon/oppot.html) (highedu/inplo/index.html) (highedu/inplo/listings/index.html) > (support:0.28%).

Fig. 3. C1 characteristics

C2: This behavior is based on a search for a security fleabag in the system. Generally, these web attacks are programmed once and further shared and used by different groups.

The discovered user behaviors by employing our method cover more than 75 surfing goals on main SRKREC web site and more 130 goals on the intranet site of SRKREC. We stated three important goals here, from job opportunities requests to activities of hacking. Thus, these discovered behaviors demonstrate the success of our methodology in discovering the behaviors.

<(lscripts/root.exe) (c/winnt/system32/cmd.exe)
(..%255c../..%255c../winnt/system32/cmd.exe)
(..%255c../..%255c/..%c1%1c../..%c1%1c../..%c1%1c../winnt/system32/cmd.exe)
(winnt/system32/cmd.exe) (winnt/system32/cmd.exe)(winnt/system32/cmd.exe)>
(support: 0.04%).

Fig. 4. C2 characteristics

4.1 APRIORI ALGORITHM

Apriori is a classic and most sought after algorithm for learning association rules [6],[11] in data mining area is the Apriori algorithm. Apriori approach is designed to work on databases containing various transactions. Apriori uses breadth-first search (BFS) and a tree structure is applied to efficiently count the candidate item sets. The algorithm generates candidate item sets which are of length k from item sets of length k - 1. Then candidates that have infrequent sub patterns are to be pruned. According to the downward closure principle, the candidate set comprises all k-length frequent item sets. Afterwards, it scans transaction database to fix frequent item sets among the candidates.

The Apriori algorithm is an efficient for finding frequent item sets. A level-wise search being implemented using frequent item sets and can be further optimized.

The efficient Apriori algorithm we have introduced is based on Apriori algorithm but introduced efficiency while generating candidates.

The proposed algorithm has several distinctions from the traditional ones.

Unlike other techniques and algorithms, the proposed algorithm mainly focuses on discovering the frequent patterns. This algorithm has two main advantages when compared with the previous algorithms:

1. The linear time and liner space are being used in the algorithm for building and storing the sequences.
2. The new inclusions are continuously being added to the web logs and the sessions which are having removed files should be removed.

The modified Apriori algorithm is described in step-by-step as given below:

Table 1. Modified apriori algorithm – step-by-step process

Step 1	Split the database D into partitions of size n, these partitions are applied on apriori algorithm generation
Step 2	Use the apriori_generation module as mentioned in the algorithm for applying each partition of size n
Step 3	The candidate generations are being applied on each partition as performed in step1. Scan every partition for generating an itemset count. The output of this phase is finding the itemset count.
Step 4	For pruning the itemset, apply min_support module.
Step 5	This same process can be continued until there are no frequent items located in a partition. This process can be repeated from step 2 till step 4.

The steps given in the above matrix are transformed into a computational algorithm by using several procedures and recursive functions.

The following notations used in the algorithm listed below:

- D indicates the database transactions
- L1 denotes the frequent data item sets found in D
- Assuming K=2
- Ck : candidate itemset of size k
- Lk: frequent itemset of size k

```
Procedure Divide_npartitions ( D, size)
{       if Lk-1 • Ø then
        n_partitions = size / ksize;
           Divide_npartitions (D, n_partitions)
// the database has divided into n no of partitions
        Divide_npartitions (D, size - n_partitions)
        Ck=apriori_generation (Lk-1 , n_partitions, D , size-n_partitions)
                For each partition p  D // scan D for partitions
                        Cp = subset (Ck , p)  // get the sub partition p i.e,
candidates
                        For each candidate c   Cp
                           c.count++;
                          Lk = { c   Cp / c.count • min_sup}
                             k++;
}Return L = UK Lk
Function apriori_generation (Lk-1: frequent (k-1) itemsets , n_partitions, D ,
size-n_partitions)
{       for each itemset l1 Є Lk-1;
        for each itemset l2 Є Lk-1;
        if l1[1] = l2[1] ^ l1[2] = l2[2]......... l1[k-2] = l2[k-2] ^ l1[k-1] = l2[k-1]
then
                        c = apply join on l1 , l2;
        for each partition D
                count the frequent itemsets (k-1) itemsets
                for each (k-1) subset s of c
                        if s does not contain k-1 then
                                delete c
                        else
                                add c to Ck
} Return Ck
End procedure Divide_npartitions
```

Fig. 5. Modified Apriori algorithm based on partitioning technique for improving the effectiveness

In our experiments, the web log files that we have used have been collected from SRKREC's Web server (www.srkrec.ac.in/intranet) during the months of January and February 2012. The size of the two log files is 2.8 MB. We had close to more than two thirds of web requests for the main Web site. We sketch the important characteristics of the initial dataset in the Table 2 given below.

Table 2. Initial Log File description

Characteristic	www.srkrec.ac.in/intranet	Total
Log file size	2812 MB	2812 MB
Number of requests	3046	3046
% of requests	100%	100%

We have applied the data preprocessing methodology on the raw web log files, as discussed and deliberated in chapter 5. Once, the data preprocessing step is successful, the size of the structured web log file, which has user sessions and user visits are reduced to only 532 MB. This presents a total of 32578 visits, from which only 15,246 contained at least two pages. We have selected this visits set with at least two pages, an input for the frequent pattern mining applications.

Table 3. Characteristics of the Structured Web Log File

Characteristic	Value
Structured Log file size	532 MB
Number of sessions	3046
Number of visits	32578
Number of visits (length >-=2 pages)	15246

We have examined methods by extracting the frequent patterns with very low support from the dataset described before. The support value is varied from 0.01% to 0.001% and we have measured algorithm's response time as presented in 7.4. With this, we perceive that the proposed partitioning method turns as a complement for the traditional pattern discovery methods. For 0.02% to 0.06% very low support values, TANASA and WAP-mine are unable to extract any patterns. For the support value below 0.02%, the execution time for partitioning model grows exponentially.

5 Conclusion and Future Work

In this research paper, we offer a sophisticated method for extracting of the all users' behavior of a Web site. Our methodology has the distinguished feature to divide the log file recursively in order to discover the users' behavior and to characterize them as clusters (analogous behaviors are grouped into a cluster). For this perseverance, we have to offer a detailed clustering method, which is devoted to sequential patterns. The key benefit of our proposed method is to study the Web Usage Mining with minimal support as a composite problem that can be solved by succeeding partitions.

The problem therefore, moves from one single open problem to n number of problems. We can resolve and the problem that has to be recursively partitioned.(we can solve a problem, by the application of partitioning method recursively) By furthering in this approach, we could establish that the border between the data quantity and quality of results can occasionally be pushed vertebral by extracting user behaviors with a minimal representativeness.

References

[1] Benedek, A., Trousse, B.: Adaptation of Self-Organizing Maps for CBR case indexing. In: 27th Annual Conference of the Gesellschaft fur Klassifikation, Cottbus, Germany (March 2003)

[2] Fayad, U.M., Piatetsky-Shapiro, G., Smyth, P., Uthurusamy, R. (eds.): Advances in Knowledge Discovery and Data Mining. AAAI Press, Menlo Park (1996)

[3] Giacometti, A.: Modèles hybrides de l'expertise, novembre, PhD Thesis, ENST Paris (1992) (in French)

[4] Cooley, R., Mobasher, B., Srivastava, J.: Data preparation for mining world wide web browsing patterns. Knowledge and Information Systems 1(1), 5–32 (1999)

[5] Jaczynski, M.: Modèle et plate-forme à objets pour l'indexation des cas par situation comportementales: application à l'assistance à la navigation sur le web, décembre, PhD thesis, Université de Nice Sophia-Antipolis (1998) (in French)

[6] Malek, M.: Un modèle hybride de mémoire pour le raisonnementà partir de cas, PhD thesis, Universitẽ Joseph Fourrier (Octobre 1996) (in French)

[7] Masseglia, F., Poncelet, P., Cicchetti, R.: An efficient algorithm for web usage mining. Networking and Information Systems Journal (NIS) (April 2000)

[8] Srikant, R., Agrawal, R.: Mining Sequential Patterns: Generalizations and Performance Improvements. In: Apers, P.M.G., Bouzeghoub, M., Gardarin, G. (eds.) EDBT 1996. LNCS, vol. 1057, pp. 3–17. Springer, Heidelberg (1996)

[9] Tanasa, D., Trousse, B.: Web access pattern discovery and analysis based on page classification and on indexing sessions with a generalised suffix tree. In: Proceedings of the 3rd International Workshop on Symbolic and Numeric Algorithms for Scientific Computing, Timisoara, Romania, pp. 62–72 (October 2001)

[10] W3C. httpd- log files (1995), `http://www.w3.org/Daemon/User/Config/Logging.html`

[11] Masseglia, F., Cathala, F., Poncelet, P.: The PSP Approach for Mining Sequential Patterns. In: Żytkow, J.M. (ed.) PKDD 1998. LNCS, vol. 1510, pp. 176–184. Springer, Heidelberg (1998)

Automatic Clustering Based on Cluster Nearest Neighbor Distance (CNND) Algorithm

Arghya Sur[1], Aritra Chowdhury[1], Jaydeep Ghosh Chowdhury[1], and Swagatam Das[2]

[1] Dept. of Electronics and Telecomunication Engg, Jadavpur University, Kolkata 700032, India
[2] Electronics and Computer Sciences Unit, Indian Statistical Institute, Kolkata, India
{arghyasur1991,jaydeep197}@gmail.com, arit0001@yahoo.co.in, swagatam.das@isical.ac.in

Abstract. This article describes a simple and fast algorithm that can automatically detect any number of well separated clusters, which may be of any shape e.g. convex and/or non-convex. This is in contrast to most of the existing clustering algorithms that assume a value for the number of clusters and/or a particular cluster structure. This algorithm is based on the principle that there is a definite threshold in the intra-cluster distances between nearest neighbors in the same cluster. Promising results on both real and artificial datasets have been included to show the effectiveness of the proposed technique.

Keywords: Cluster nearest Neighbour, Clustering, Automatic Clustering, Various shaped clusters.

1 Introduction

Clustering is the process, by which a set of objects are partitioned into a number of groups, such that the similarity between objects of the same group is maximum and that between objects belonging to different groups are minimum. This measure of similarity is defined mathematically and, the objects are assigned to these groups, known as 'clusters' based on this measure. In the past few decades, cluster analysis has played a central role in diverse domains of science and engineering [1].

An important consideration with clustering algorihms is to determine the appropriate number of clusters from a given dataset, where the number of groups in the dataset is unknown *apriori*. In many clustering algorithms, this number is specified by the user, and accordingly the algorithms partition the dataset. It is a challenge to find the optimal number of clusters automatically.

Clustering algorithms can be hierarchical or partitional [2]. Hierarchical clustering algorithms recursively find nested clusters either in agglomerative mode (starting with each data point in its own cluster and merging the most similar pair of clusters successively to form a cluster hierarchy) or in divisive (topdown) mode (starting with all the data points in one cluster and recursively dividing each cluster into smaller clusters). Partitional clustering algorithms, on the other hand, attempts to partition the data set directly into a set of disjoint clusters by trying to optimize certain criteria (e.g. a squared-error function).

S.C. Satapathy et al. (Eds.): Proc. of Int. Conf. on Front. of Intell. Comput., AISC 199, pp. 189–196.
DOI: 10.1007/978-3-642-35314-7_22

Centroid based clustering algorithms are a class of partitional algorithms , such as K-means and fuzzy c-means. These algorithms attempt to iteratively find cluster centers and assign points to one of the centers. Even though K-means was first proposed over 50 years ago, it is still one of the most widely used algorithms for clustering. They require the number of clusters beforehand and hence are non-automatic. Also, they only detect spherical and similar sized clusters accurately due to the "uniform effect" [3].

In this article, we propose a simple and fast algorithm to automatically detect the appropriate number of clusters from an unlabelled dataset. The algorithm performs equally well towards both hyper-spherical shaped clusters and shell-type or solid clusters of any arbitrary shape. This algorithm is based on the principle that there is a definite threshold in the intra-cluster distances between nearest neighbors in the same cluster. Using this threshold, we can group together objects in the dataset which have their cluster nearest neighbor distance less than the threshold.

2 Proposed Algorithm

In this algorithm, we use the Euclidean Distance as distance measure. The basic idea is that, if a subset of points in the dataset belongs to the same cluster, then, if the nearest neighbor of that subset in the set of remaining un-clustered data points (points which have not been assigned to any cluster as yet) is not in that cluster, the cluster consists only of that subset of points and nothing else. Nearest neighbor of a set S' in a set S is defined as the point in S-S', whose distance from its nearest neighbor in S' is minimum. When S' is the current cluster, and S is the set of remaining un-clustered data points, we call this point as the cluster nearest neighbor (CNN) and the distance as, cluster nearest neighbor distance (cnn_dist). This algorithm runs iteratively over all points in the dataset forming clusters along the way.

First, an initial starting point in a new cluster C is randomly initialized in the set of remaining un-clustered data points. Let this set be denoted by S_REM. Then the nearest neighbor of that point in S_REM is calculated using a standard nearest neighbor search algorithm like k-d tree [4] and kNN search. Here, we have used kNN search algorithm for this purpose. Thus, the current cluster C consists of these two points now. Then, iteratively, we find the CNN of C and determine if the CNN should be in C or not. The cluster completes when either the CNN is found not to belong in C, or when S_REM becomes empty. The condition to determine if CNN should belong to C depends on the number of points in C ($clus\_len$) and a threshold. Let POP be the set of all points in the dataset and min $\_len$ be the minimum number of elements that must always be present in a cluster. Then the following conditions determine if CNN of C belongs to C.

Cond1: If $length(C) <= min\_len$,

CNN should belong to C, if nearest neighbor distance of C in POP-C is equal to cnn_dist. Else, get the cluster C', which contains the nearest neighbor of C in POP-C.

Cond2: Else,

CNN should belong to C, if $cnn\_dist <= threshold * mean(D)$, where D is the set of cnn_dist of all previous points in the current cluster and threshold is a parameter initialized externally. For our experiments, we have initialized $threshold = 1.2$.

The complete algorithm is as follows:

1. Initialize min_len=$ceil(1\%\ of\ length(pop))$ where, $length(S)$ is the number of elements in set S. Ceil function gives the smallest integer greater than or equal to the argument.
2. Initialize $i = 1$.
3. Initialize a starting point for the ith cluster C_i. This can be done randomly from S_REM. Initialize $D = \emptyset$.
4. Remove that point from S_REM.
5. If S_REM not equal to $\emptyset$, Find CNN and cnn_dist of C_i. Else, *stop*.
6. Take $C = C_i$ and determine if CNN of C should belong to C according to conditions Cond1 and Cond2.
7. If $length(C) <= min\_len$ and CNN shouldn't belong to C, merge C' and C and let the merged cluster be C. This is to ensure that no fewer than min_len elements exist in a cluster.
8. If CNN shouldn't belong to C, set $min\_len = ceil(1\%\ of\ length(C))$. Go to step 9. Else, set $C = C \cup \{CNN\}$. Set $C_i = C$. Go to step 5.
9. Set $i = i + 1$. Adapt threshold by the formula $threshold_i = mean(T, cnn\_dist)$, where T is the set of $threshold_j,\ j = 1\ to\ i - 1$. If $length(S\_REM) > 0$, go to step 3. Else, go to step 5.

3 Experimental Results

3.1 Datasets

- **Artificial Datasets:**
  - *i.* *Dataset1:* This dataset consists of 788 points as represented Fig. 1(a). There are a total of 7 groups in the population of varying shapes and size.
  - *ii.* *Dataset2:* This spiral dataset consists of 312 points as represented Fig. 1(b). There are 3 groups in the population.
  - *iii.* *Dataset3:* This is a dataset consisting of 240 points spread over two groups as shown in Fig. 1(c).
  - *iv.* *Dataset4:* This is a dataset with 1000 points with two clusters- an ellipse and a circle of different radii. This is shown in Fig. 1(d).
  - *v.* *Dataset5:* This is a 3-d dataset with 500 points. There are two groups, one, a shell-shaped cluster and another, a vertical ellipsoid. The 3d representation is shown in Fig. 1(e).

- **Real life Datasets:** These 3 real-life datasets are obtained from [5].
  - *i.* *Wine Dataset:* The Wine recognition dataset consists of 178 instances having 13 features obtained from a chemical analysis of wines. The wines were grown in the same region in Italy but were derived from three different cultivars. The analysis determined the quantities of 13 constituents found in each of the three types of wines.
  - *ii.* *Iris Dataset:* This dataset consists of 150 points equally distributed over 3 groups, viz. *Setosa*, *Versicolor* and *Virginica*. It is represented by 4 feature values. *Versicolor* and *Virginica* overlap while *Setosa* can be linearly separated.
  - *iii.* *LiverDisorder:* This dataset consists of 345 instances with each having six features. There are 2 groups in the dataset.

The artificial datasets are shown in figure 1(a) to (e). Datasets 1-4 are 2-d while Dataset5 is a 3-d dataset.

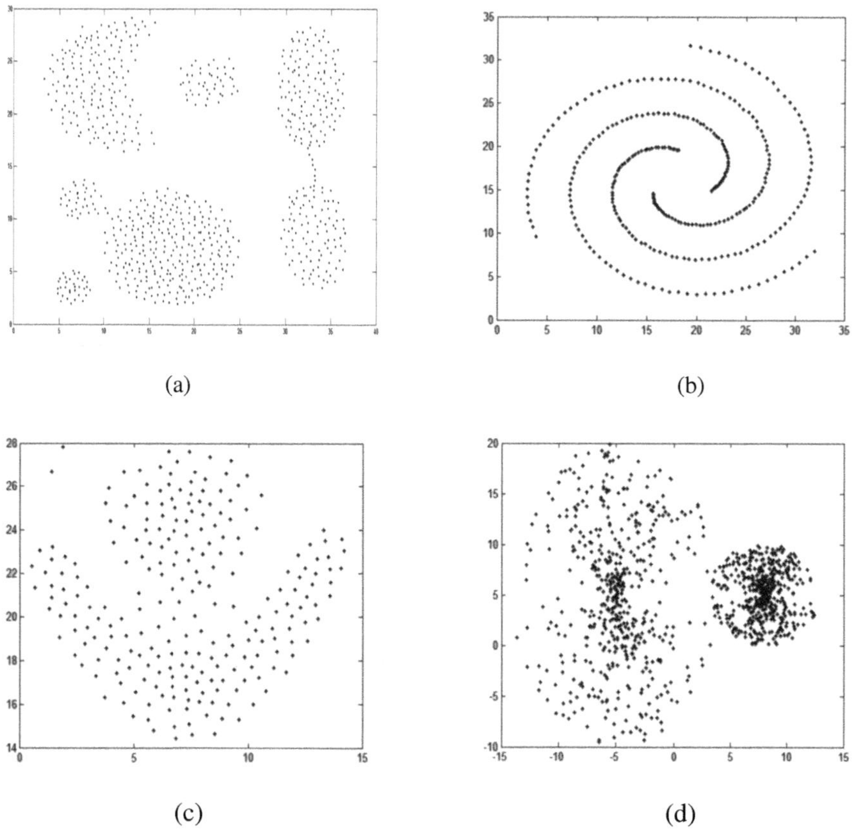

Fig. 1. (a)Dataset1 (b)Dataset2 (c)Dataset3 (d)Dataset4 (e)Dataset5

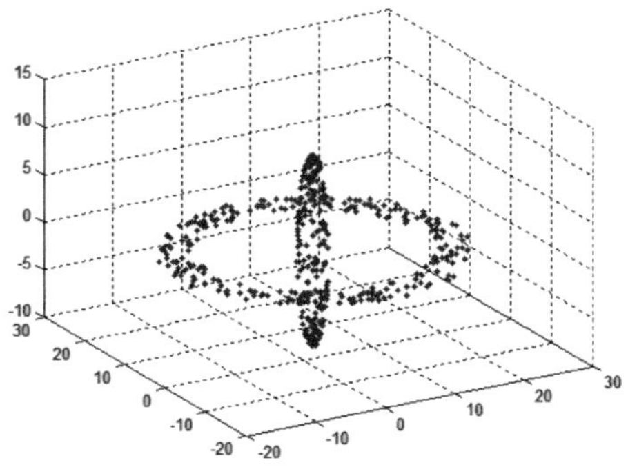

(e)

Fig. 1. *(continued)*

3.2 Results

This section compares the performance of the proposed clustering algorithm with a few standard clustering algorithms like k-means [6], fuzzy C-means [7] and Hierarchical Agglomerative Clustering. The contestant algorithms were mostly non-automatic. K-means and fuzzy c-means require the number of clusters beforehand while hierarchical clustering requires the maximum number of possible clusters. The figures 2(a)-(e) are the clustered representations of the figures 1(a)-(e) obtained by CNND clustering algorithm. These sets of data are used in order to represent all kinds of data adequately.

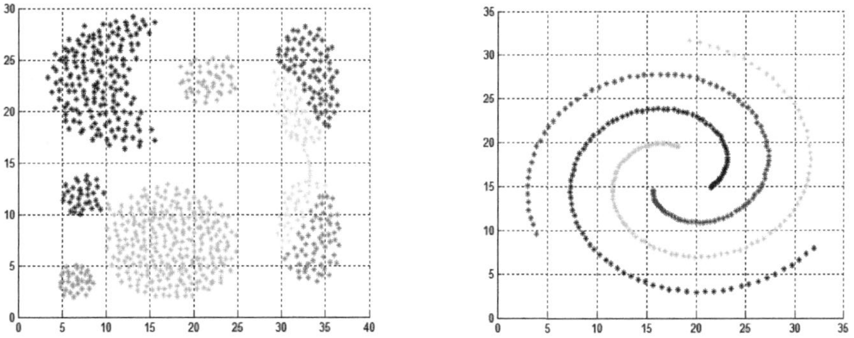

(a) (b)

Fig. 2. (a) Clustered Representation of Dataset1 (b) Clustered Representation of Dataset2 (c) Clustered Representation of Dataset3 (d) Clustered Representation of Dataset4 (e) Clustered Representation of Dataset5

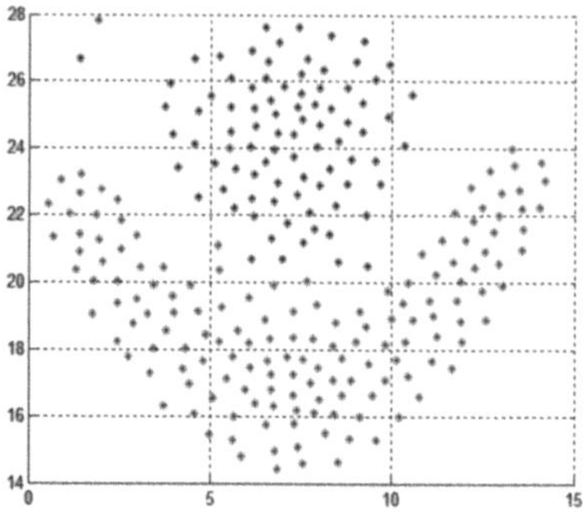

(c)

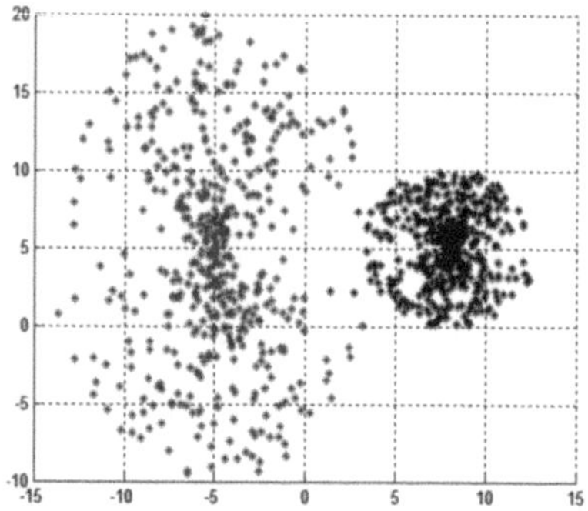

(d)

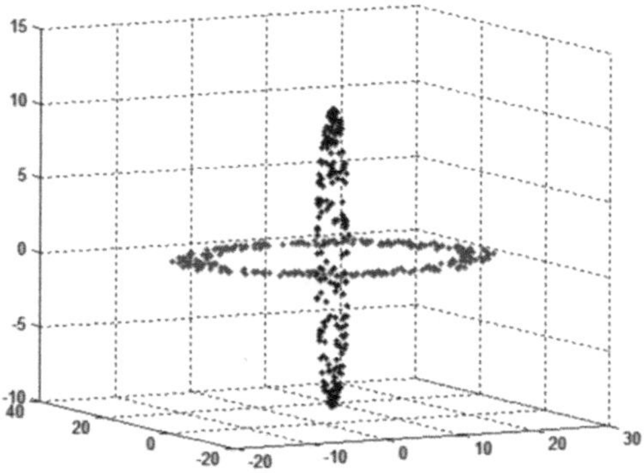

(e)

Fig. 2. (*continued*)

3.3 Minkowski Score

If T is the "true" solution and S is the solution obtained experimentally, then the Minkowski score [8] is defined as,

$$MS(T,S) = \sqrt{\frac{n_{01}+n_{10}}{n_{11}+n_{10}}}, \tag{1}$$

where, n_{01} represents the number of pairs of elements that are in the same cluster only in S, n_{10} represents the number of pairs of elements that are in the same cluster only in T and n_{11} represents the number of pairs of elements that are in the same cluster in both S and T.

The minkowski scores of the datasets obtained for four algorithms are shown in Table 1. The actual no. of clusters (AC) versus the obtained number of clusters (OC) in CNND algorithm is shown in Table 2. All other algorithms, being non-automatic, AC and OC are same trivially.

Table 1. Minkowski Scores of all the experimental datasets

Data Set	CNND clustering	K-means	Fuzzy C-means	Hiererchical
Dataset1	**0.3567**	0.6561	0.6129	0.5859
Dataset2	**0**	1.16	1.16	**0**
Dataset3	**0**	0.7142	0.6913	0.9259
Dataset4	**0**	0.218	0.2512	1
Dataset5	**0**	1	1	**0**
Wine Dataset	**0.5844**	0.9124	0.9255	1.373
Iris Dataset	**0.5309**	0.6047	0.6047	0.8241
Liver Disorder	**0.9786**	0.9846	0.9891	**0.9786**

Table 2. Number of actual clusters(AC) vs Number of obtained Clusters(OC) for CNND Algorithm

Data Set	AC	OC
Dataset1	7	8
Dataset2	3	**3**
Dataset3	2	**2**
Dataset4	2	**2**
Dataset5	2	**2**
Wine Dataset	3	**3**
Iris Dataset	3	4
Liver Disorder	2	**2**

4 Conclusion

From Table 1, we can see that CNND algorithm beats all the other clustering algorithms in terms of minkowski score. Also, Minkowski score of 0 obtained in 4 out of 5 artificial datasets clearly show that this algorithm is robust towards different cluster shapes and sizes, whereas k-means and fuzzy c-means give good results only in case of spherical clusters of nearly equal sizes. Hierarchical clustering is also robust towards different shaped clusters but in closely spaced or semi-overlapping clusters, it fails to give good clustering results. In this area too, however, our CNND algorithm shows quite promising results. Table 2 shows that it also finds the appropriate number of clusters in a dataset with a good accuracy. However in completely overlapped clusters, this algorithm will fail to give accurate results because it depends on finding the boundary between two clusters.

References

[1] Jain, A.K., Dubes, R.C.: Algorithms for Clustering Data. Prentice-Hall, Englewood Cliffs (1988)

[2] Jain, A.K., Murty, M.N., Flynn, P.J.: Data clustering: a review. ACM Computing Surveys 31(3), 264–323 (1999)

[3] Xiong, H., Wu, J.J., Chen, J.: K-means clustering versus validation measures: A data-distribution perspective. IEEE Trans. Systems, Man, and Cybernetics-Part B: Cybernetics 39(2), 318–331 (2009)

[4] Saha, S., Bandyopadhyay, S.: A symmetry based multiobjective clustering technique for automatic evolution of clusters. Pattern Recognition 43, 738–751 (2010)

[5] Frank, A., Asuncion, A.: UCI Machine Learning Repository. University of California, School of Information and Computer Science, Irvine, CA (2010), `http://archive.ics.uci.edu/ml`

[6] Jain, A.K.: Data clustering: 50 years beyond K-means. Pattern Recognition Lett. (2009), doi:10.1016/j.patrec.2009.09.011

[7] Cannon, R.L., Dave, J.V., Bezdek, J.C.: Efficient implementation of the fuzzy c-means clustering algorithms. IEEE Trans. Pattern Analysis and Machine Intelligence PAMI-8, 248–255 (1986)

[8] Ben-Hur, A., Guyon, I.: Detecting Stable Clusters Using Principal Component Analysis in Methods of Molecular Biology. Humana Press (2003)

Data Preprocessing Using Intelligent Agents

Sharon Christa[1], V. Suma[2], and Lakshmi Madhuri[2]

[1] Department of Information Science and Engineering, Dayananda Sagar College of Engineering, Bangalore
[2] Research and Industry Incubation Centre, Dayananda Sagar College of Engineering, Bangalore

Abstract. Knowledge discovery in data base aims to analyze the databases for meaningful patterns and information regarding the same. Data mining is the process in KDD that identifies patterns and knowledge from various sources of database or even from different database. Even though data mining has a very significant role in the real world, the mining of data in the database does not yield correct results due to various real world problems like data inconsistencies. There comes the significance of data preprocessing where the inconsistencies of data is removed. This paper provides the design and development of intelligent software that uses agent based data preprocessing environment to improve the data processing activities, which in turn enhances the quality of data to be mined.

1 Introduction

The advancement in intelligent applications has enabled easy handling of huge complex data, which in turn enables the organizations to have local databases all over the world. With the emergence of WWW, online transactions etc. data accumulates in data source, which is an asset [1]. Unanalysed data is not only valueless but also a liability. Knowledge discovery in data base (KDD) is a domain which aims to analyze the databases for meaningful patterns and information regarding the same [2]. Thus, obtained information is very useful in various fields such as stock market prediction, banking sector etc [3].

Further, authors in [3] recommend data mining technique due to its varied benefits of identification of patterns and knowledge from dynamic data, from various sources of database or even from different database. Nevertheless, the significance of data mining in accurate decision making and in other related database issues, the mining of data in the database does not yield correct results due to various real world problems.

Unlike theoretical explanations KDD has to overcome lot of hurdles. Data inconsistencies are quiet normal in the real world scenario that occurs due to various reasons such as transmission errors, incorrect data, error while entering data to the database, inconsistent formats for input fields, equipment malfunction, and ignoring the modifications in data etc. [4] [5]. Existence of inconsistencies in data affects the mining quality which inturn affect the quality of the knowledge obtained. Data preprocessing is therefore performed to overcome the data inconsistencies. Data preprocessing is the initial set of processes that is performed prior to data mining in KDD [6].

S.C. Satapathy et al. (Eds.): Proc. of Int. Conf. on Front. of Intell. Comput., AISC 199, pp. 197–204.
DOI: 10.1007/978-3-642-35314-7_23

This research hence aims to provide an overview of the data preprocessing activities that are required to be performed in order to deal with the large datasets. The organization of this paper is as follows: Section 2 is literature survey that briefing about the related work carried out in this domain by various researchers. Section 3 explains the design model and architecture for the data preprocessing system. Section 4 provides implementation and results while Section 5 summarizes of the entire work.

2 Literature Survey

KDD is bound to overcome various issues related to the real time data which may be incomplete, inconsistent and noisy and hence, data stored in database often results in errors such as out of range values, missing values, impossible data combinations etc. Furthermore, use of inconsistent, incomplete data in KDD results in inefficient time consumption and unreliable results [5]. Author in [6] therefore recommends the implementation of efficient strategy to overcome the aforementioned issues using data preprocessing techniques. However, selection of data preprocessing task is yet another challenge [7].

Data preprocessing includes data cleaning, data integration, data transformation, and data reduction. Data cleaning activity eliminates the noise present in the data set. Data integration enables merging of data from multiple sources into a coherent data source. Data transformation transforms the data into a suitable form for data mining and data reduction reduces the data size so that KDD yields the same result [4].

Every process in data preprocessing needs to be performed with utmost attention since each process is responsible in providing an efficient result. From the point of integrating various data sources to the point of deciding the right strategy to be selected for data cleaning plays an important role in KDD results.

The current data mining tools like Weka, RapidMiner, TANAGRA, Orange, and KNIME neither support real time modeling nor data preprocessing where the updated and inconsistent data is considered. However, in practice, a data miner carries out remodeling to ensure that the updated knowledge is obtained. The aforementioned limitation leads to a hitch in many applications like in medical, stock exchange and finance since it is critical in having updated knowledge in these domains because it affects the human life and business performances [8].

Successful implementation of the data preprocessing requires an in depth assessment and knowledge of the various tools and algorithms available in the market. However, current era of information age has forced the industries to deal with gigabytes and terabytes of database in lieu of megabytes of database [9]. World is going to witness storage of data in Exabyte in near future. Hence, as data sets increase in size, data preprocessing process become less and less efficient. Data preprocessing may therefore sometimes demand manual processing of data, which is not advisable. Incorporating the most recent knowledge model in the current KDD is further expensive. Additionally, it is necessary to note that currently available data mining tools are not appropriate for novice users [10].

Agent based systems have proven to be an effective approach to overcome the drawbacks cited above. Agents have become popular paradigm in computing because of its unique characteristics such as autonomous, flexible, adaptive and intelligent [10]. Intelligent agents are capable of behaving rationally [11]. According to Maes, Intelligent Agent (IA) is software entity that can be used to perform the operations independently. He further feels that agents can replace user or another program. Since, each agent has specific characteristics; they vary depending on the problem domain [12]. In a multi agent system, agents communicate, co-operate and co-ordinate with the other agents. Each agent in the system acts autonomously, and co-operates with other agents. They work together for the tasks to be performed to achieve the goal of the system [13].

The intelligent agents work with human intelligence in many aspects. The agent which is involved in processing of data preprocessing performs productive task, retrieves useful knowledge with less noise and reduced processing time when compared to the available data mining tools [12]. Authors in [14] introduce the implementation of black box approach in agent technology to hide the complexity of the system and also to make the system scalable. The co-operating agents work together as a team and co-ordinates their activity to manage their task for solving the problem. Thus the whole data preprocessing system designed and implemented, has therefore adopted the black box approach using intelligent agents.

Due to the varied benefits of data preprocessing agents, this research aims to develop an effective preprocessing model for efficient knowledge discovery.

3 Design Model

Since, KDD has a very critical impact in the current industrial environment, efficient data preprocessing has become the elementary strategy. Nevertheless, the existence of several data mining tools, they neither support data preprocessing nor can they efficiently process and manage the complex data. The aim of this research is to optimize the data preprocessing for better results. In order to accomplish the aforementioned objective, secondary data was collected for the analysis and evaluation purpose. Secondary data are published data which are obtained through publications, through technical journals, books, magazines, newspapers, reports published by research scholars etc. Methods for data preprocessing and their responsibilities in the currently available data mining tool is analyzed along with the agent simulation tools. The analysis and evaluation of secondary data indicates that the data preprocessing phase consumes most of the data mining time and effort. Therefore, it is required to optimize the process of data preprocessing in order to give better results. Based on the analysis and evaluation, this research is focuses on data preprocessing using intelligent agents.

Potential features of an agent based data preprocessing tool include propose the processing techniques most suitable to the data, preprocess incoming new data according to user profile, learn from the preprocessing experience, suggesting possible knowledge that can be extracted from the data with the help of the experience shared, suitable for novice user.

Figure 1 depicts the data preprocessing architecture using intelligent agents that optimizes the performance of data preprocessing.

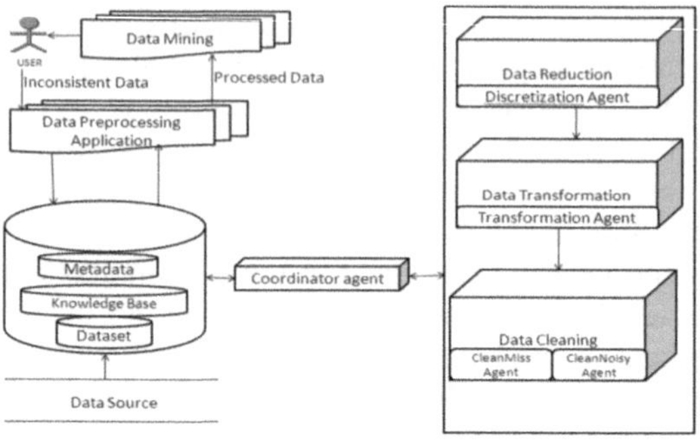

Fig. 1. Architecture of data preprocessing application using intelligent agents

Since, agents are intelligent systems that can perform operations with some level of intelligence and understanding capabilities, incorporating intelligent agents to the data pre-processing system enables to achieve efficient data preprocessing.

Data preprocessing application acts as an interface that process the data to be mined. The dataset having inconsistency is stored in the database with the metadata of it. The whole data is analyzed to find the inconsistencies, duplicate fields, multiple data formats in same attribute, missing data field. Coordinator agent performs of the above stated activities. Each operation will be accordingly performed by the corresponding such as data transformation by transformation agent, data reduction by discretization agent, data cleaning by clean miss and clean noisy agent.

Data integration process combines data from multiple sources like data cubes, multiple databases, and flat files. It performs schema integration and also objects matching. It makes use of the metadata for the integration. It performs correlation analysis and also chi-square test to handle redundancy. Data transformation involves smoothing, aggregation, generalization, normalization and also attributes construction. These methods help to make data more appropriate for mining.

An important part in data preprocessing is data reduction which reduces the data size [15]. Data warehouses can be extremely large, yet obtaining solutions swiftly is important without having to sacrifice the accuracy in the results. In the early stages of data analysis, interactive response time is very critical. Thus, the size of the dataset really matters. Data reduction, thus, becomes a pressing need for the industries to achieve effective data mining [16]. Strategies in data reduction are data cube aggregation, attribute subset selection, dimensionality reduction, numerosity reduction, and discretization and concept hierarchy generation. This is performed to reduce the time taken for complex data analysis and also to reduce mining huge amounts of data.

Data cleaning is a two step process which includes handling missing values and handling noisy data. It is executed with discrepancy detection that makes use of the metadata which give the knowledge about domain and data type. While scanning the dataset, if the tuples have no recorded value then various strategies are used like ignore the tuple, fill each missing values manually, use the most probable value, use the attribute mean to fill in case of numeric data, use a constant like unknown or infinity. Binning, regression and clustering are performed to remove the errors and smoothing the data.

The architecture is designed with five agents: coordinator agent, discretization agent, transformation agent, clean miss agent, clean noisy agent where each agent is responsible for some specified task. Coordinator agent is responsible for coordinating the various tasks that needs to be performed in a cooperative problem solving between the user and other agents. It can determine the required preprocessing task, which can be generated automatically based on meta-knowledge in the coordinator agent. CleanMiss Agent and clean noisy agent handle missing and noisy data by using various types of techniques based on type of missing and noisy cases. Transformation Agent is used to transform the data into appropriate forms for mining. The role of reduction agent is to discretize the data by using discretization techniques selected.

4 Implementation

Each phase of the preprocessing activity is assigned to an agent who responds to the instructions from the coordinator agent. Coordinator agent determines the preprocessing tasks to be performed in the dataset after analyzing it. The database further has the repository of each of the preprocessing activities. The architecture of the preprocessing system is implemented using NetBeans IDE in which each agent is developed.

The application is uploaded and tested with excel information data file with a maximum size of 10 MB. Table 1 depicts the subset of a sample dataset that is used in testing the implemented data preprocessing system. Once the data is uploaded using the user interface, the file has been stored in a specific location. Once the file is uploaded, the inconsistencies are determined by the coordinator agent using association rule and it updates the noisy information. Subsequently, this information will be represented in a file format which will be converted to another format that can be uploaded in the data mining tools available. As an instance, the dataset in excel file format is converted to .arff format suitable for WEKA. All updated data information will consequently be saved in the data repository. Agents will autonomously determine the operations to be performed and executes the same which eventually gets stored in the repository.

Table 1. Sample data subset used in testing the data preprocessing system

Emp_id	Name	Salary	Email	Age
-1009	Manju	10,000	manju@gmail.com	23
1059	$unandha	15,000	sunandha.allur@argmail.com	20
2007	Manjesh	-20,000	manjesh123@mr.com	26
0	Radhika Sharma	18,000	radhika.sharma@arm.com	25
-2985	Seena Roy	20,000	seema.roy@win.com	12
1383	Meenakshi Sing	26,000	m_sing@mming.com	99

The Figures 2 through Figure 6 depict the various activities involved in data preprocessing operations.

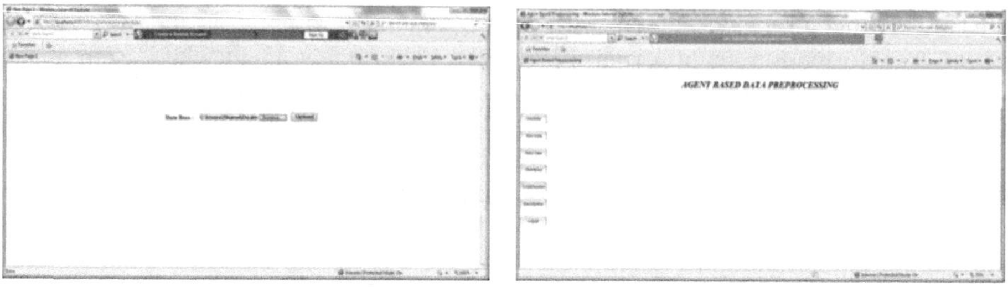

Fig. 2. User Interface and dataset uploading option

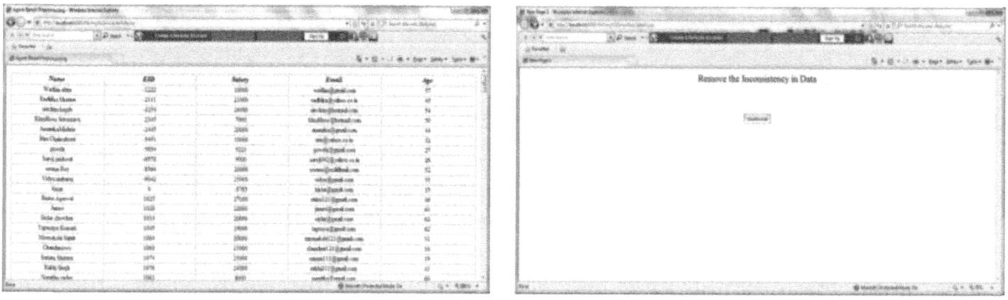

Fig. 3. Data analysis and finding the correlation in data

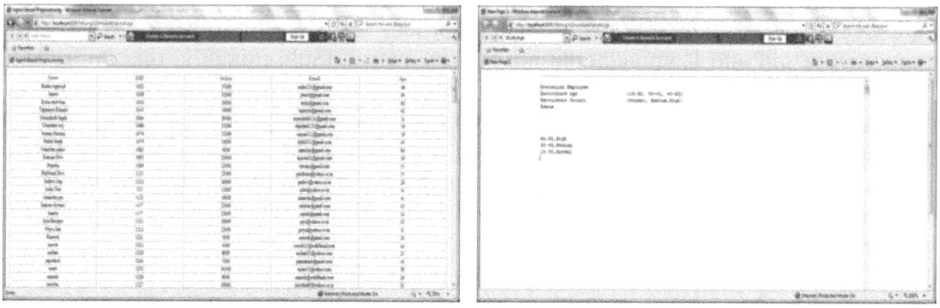

Fig. 4. Data after data cleaning and data transformation

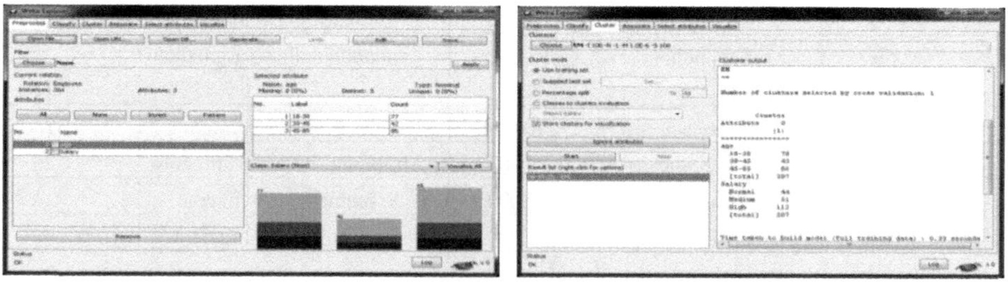

Fig. 5. Analyzing the performance of data preprocessing by uploading in WEKA tool

4.1 Inference

Figure 2 through Figure 6 depicts the implementation of the data preprocessing system. Figure 3 shows the analysis and result. Figure 4 depicts the data after cleaning and data transformation. Figure 5 and Figure 6 shows the analysis of the dataset after data preprocessing in WEKA tool. When compare to the result obtained with the dataset before preprocessing, the result obtained after the data preprocessing is more consistent and has no outliers.

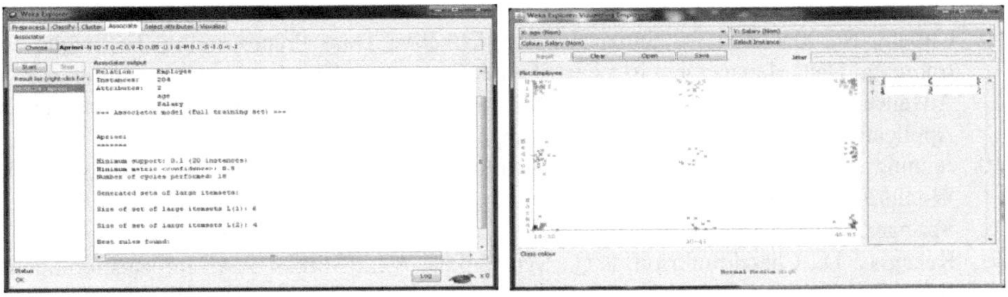

Fig. 6. Analyzing the performance of data preprocessing by uploading in WEKA tool

5 Conclusion

Data mining technique has various benefits like identification of patterns etc. Nevertheless, the significance of data mining in accurate decision making and in other related database issues, the mining of data in the database does not yield correct results due to various real world problems like inconsistent, unclean data. Unanalyzed data is not only valueless but also a liability. Hence, there is a need for efficient strategy to overcome the aforementioned issues using data preprocessing techniques. This paper focuses on development of intelligent agents for effective data preprocessing system. The implementation result indicates that When compare to the result obtained with the dataset before preprocessing, the result obtained after the data preprocessing is more consistent and has no outliers which in turn enhances the quality of data to be mined.

References

1. Mohtar, I.A.: Multiagent Approach to Stock Price Prediction, University Kebangsaan, Malaysia (2006)
2. Bo, Y., Ya-Dong, W., Xiao-Hong, S.: Research and design of distributed training algorithm for neural network. In: Proceedings of the International Conference on Machine Learning and Cybernetics (2005)
3. Kaya, M., Alhajj, R.: Fuzzy OLAP association rules mining-based modular reinforcement learning approach for multi agent systems (2005)
4. Cao, L., Gorodetsky, V., Mitkas, P.A.: Agent Mining: The Synergy of Agents and Data Mining. IEEE Intelligent Systems (2009)

5. Lourenço, A., Gonçalves, J., Belo, O.: Agent-based knowledge extraction services inside enterprise data warehousing systems environments, pp. 887–891. IEEE (2001)
6. Christa, S., Madhuri, L., Suma, V.: Intelligent Data Preprocessing Software. TP/ITC/2012-13/118. Indian Technology Congress, Bangalore (2012)
7. Li, C., Gao, Y.: Agent-based pattern mining of discredited activities in public services. In: Proceedings of the 2006 IEEE/WI C/ACM International Conference on Web Intelligence and Intelligent Agent Technology (2006)
8. Ahmad, A.M., Nordin, N.A., Saaim, E.H.M., Samaon, F.S., Ibrahim, M.D.: An architecture design of the intelligent agent for speech recognition and translation. IEEE (2004)
9. Christa, S., Madhuri, L., Suma, V.: An Effective Data Preprocessing Technique for Improved Data Management in a Distributed Environment. In: International Conference on Advanced Computing and Communication Technologies for High Performance Applications, International Journal of Computer Applications, Cochin (2012)
10. Jennings, N.R., Wooldridge, M.: Application of Intelligent Agents. In: Jennings, N.R., Wooldridge, M.J. (eds.) Agent Technology: Foundations, Applications and Markets. Springer, Berlin (1998)
11. Kehagias, D., Chatzidimitriou, K.C., Symeonidis, A.L., Mitkas, P.A.: Information agents cooperating with heterogeneous data sources for customer-order management. In: ACM Symposium on Applied Computing, pp. 52–57 (2004)
12. Maes, P.: Agents that Reduce the Work and Information Overload. Com. of ACM 36(7), 29–39 (1994)
13. Christa, S., Madhuri, L., Suma, V.: A Comparative Analysis of Data Mining Tools in Agent Based Systems. In: International Conference on Systemics, Cybernetics and Informatics, Hyderabad (2012)
14. Ni, J., Zhang, C.: A human-friendly mass for mining stock data. In: Proceedings of the 2006 IEEE/WIC/ACM International Conference on Web Intelligence and International Agent Technology Workshops, pp. 19–22 (2006)
15. Christa, S., Madhuri, L., Suma, V.: Data Preprocessing Model Using Intelligent Agents. In: International Conference on Information Systems Design and Intelligent Applications, Visakhapatnam (2012)
16. Bordetsky, A.: Agent-based support for collaborative data mining in systems management. In: Proceedings of the 33th Hawaii International Conference on System Sciences (2001)

A Trigram HMM-Based POS Tagger for Indian Languages

Kamal Sarkar* and Vivekananda Gayen

Computer Science & Engineering Department,
Jadavpur University,
Kolkata – 700 032, India
jukamal2001@yahoo.com, Vivek3gayen@gmail.com

Abstract. We present in this paper a trigram HMM-based (Hidden Markov Model) part-of-speech (POS) tagger for Indian languages, which will accept a raw text in an Indian language (typed in corresponding language font) to produce a POS tagged output. We implement the trigram POS Tagger from the scratch based on the second order Hidden Markov Model (HMM). For handling unknown words, we introduce a prefix analysis method and a word-type analysis method which are combined with the well known suffix analysis method for predicting the probable tags. Though our developed systems have been tested on the data for four Indian languages namely Bengali, Hindi, Marathi and Telugu, the developed system can be easily ported to a new language just by replacing the training file with the POS tagged data for the new language. Our developed trigram POS tagger has been compared to the bigram POS tagger defined as a baseline.

Keywords: Part-of-speech tagging, Second order Hidden Markov Model, Deleted interpolation, Indian Languages.

1 Introduction

Part-of-Speech (POS) tagging is the task of assigning grammatical categories (noun, verb, adjective etc.) to words in a natural language sentence [1]. POS tagging can be used in parsing, word sense disambiguation, information extraction, machine translation, question answering, chunking etc.

Since the most previous POS taggers [2] [3] [4] are experimented on datasets which are not publicly available, it poses various difficulties such as comparisons of the present works to the past works. So, our primary motivations are (1) to develop a POS tagger for Indian Languages, which can accept a raw text in one of Indian languages and (2) to report results after testing our developed POS tagger on a publicly available dataset named NLTK (Natural Language Toolkit) dataset to allow the researchers to easily compare their systems with our developed systems. The salient features of our developed POS tagger are as follows:

---

* Corresponding author.

S.C. Satapathy et al. (Eds.): Proc. of Int. Conf. on Front. of Intell. Comput., AISC 199, pp. 205–212.
DOI: 10.1007/978-3-642-35314-7_24

- Our developed POS tagger uses prefix analysis, word-type analysis and suffix analysis methods for predicting the probable tags of an unknown word (the word which is not present in the training dataset).
- Our developed POS tagger can be used for POS tagging for multiple Indian Languages. Since our system has been developed on the Visual Basic platform, it has a good user interface to submit the file to be tagged and to get the tagged output in another file that can be directly redirected to other NLP applications.

A substantial amount of research work has already done in POS tagger developments for Indian languages [5]. The various supervised POS tagging methods for Bengali have been presented in [2][3] [4 [6 [7] [8]. The descriptions of Hindi POS taggers have been presented in [9][10]. A rule based approach to morphological analysis and POS tagging in Tamil Language via Projection and Induction Techniques has been presented in [11]. A SVM Based part-of-Speech tagger for Malayalam has been presented in [12]. The hybrid POS tagger for three Indian languages namely Hindi, Bengali and Telugu presented in [13] combines HMM based approach and rule based approach.

A TnT tagger version of a POS Tagger for three Indian Languages namely Hindi, Bengali and Telugu has been presented in [14]. The work presented in [14] uses only suffix analysis for handling unknown words.

Section 2 presents the background on HMM based POS tagging. The evaluation and results are presented in section 3.

2 HMM Based POS Tagging

A POS tagger based on Hidden Markov Model (HMM) [15] assigns the best sequence of tags to an entire sentence. Generally, the most probable tag sequence is assigned to each sentence following the Viterbi algorithm [16]. The task of Part of Speech (POS) tagging is to find the sequence of POS tags t_1^n that is optimal for a given word sequence, w_1^n. The tagging problem becomes equivalent to searching for $\text{argmax}_{t_1^n} \; P(w_1^n \mid t_1^n)P(t_1^n)$ (by the application of Bayes' law), that is, we need to compute:

$$\hat{t}_1^n = \text{argmax}_{t_1^n} \; P(w_1^n \mid t_1^n)P(t_1^n) \tag{1}$$

Where: where t_1^n is a tag sequence and w_1^n is a word sequence, $P(t_1^n)$ is the prior probability of the tag sequence and $P(w_1^n|t_1^n)$ is the likelihood of the word sequence. Equation (1) is too hard to compute directly. HMM taggers make Markov assumption which states that the probability of a tag is dependent only on a small, fixed number of previous tags. A bigram tagger considers that the probability of a tag depends only on the previous tag. For our proposed trigram model, the probability of a tag depends on two previous tags and $P(t_1^n)$ is computed as:

$$P(t_1^n) \approx \prod_{i=1}^{n} P(t_i \mid t_{i-1}, t_{i-2}) \tag{2}$$

Depending on the assumption that the probability of a word appearing is dependent only on its own part-of-speech tag, $P(w_1^n|t_1^n)$ can be simplified to:

$$P(w_1^n|t_1^n) \approx \prod_{i=1}^{n} P(w_i\,|t_i) \quad (3)$$

Plugging the above mentioned two equations (2) and (3) into (1) results in the following equation by which a bigram tagger estimates the most probable tag sequence:

$$\hat{t}_1^n = \text{argmax}_{t_1^n}\ P(t_1^n\,|w_1^n) \approx \text{argmax}_{t_1^n}\ \prod_{i=1}^{n} P(w_i\,|t_i)P(t_i|t_{i-1}) \quad (4)$$

Where: the tag transition probabilities, $P(t_i|t_{i-1})$, represent the probability of a tag given the previous tag. The word likelihood probabilities, $P(w_i|t_i)$, represent the probability of a word given a tag.

Considering a special tag t_{n+1} to indicate the end sentence boundary and two special tags t_{-1} and t_0 at the starting boundary of the sentence and adding these three special tags to the tag set [1], gives the following equation for part of speech tagging:

$$\hat{t}_1^n = \text{argmax}_{t_1^n} P(t_1^n\,|w_1^n) \approx \text{argmax}_{\hat{t}_1^n}[\prod_{i=1}^{n} P(w_i\,|t_i)P(t_i|t_{i-1},t_{i-2})]P(t_{n+1}\,|t_n) \quad (5)$$

The equation (5) is still computationally expensive because we need to consider all possible tag sequence of length *n*. So, dynamic programming approach is used to compute the equation (5).

At the training phase of HMM based POS tagging, observation probability matrix and tag transition probability matrix are created.

2.1 Computing Tag Transition Probabilities

As we can see from the equation (4) to find the most likely tag sequence for a sentence (considered as an observation sequence), we need to compute two kinds of probabilities: tag transition probabilities and word likelihoods or observation probabilities.

Our developed trigram HMM tagger requires to compute tag trigram probability, $P(t_i|t_{i-1},t_{i-2})$, which is computed by the maximum likelihood estimate from tag trigram counts. To overcome the data sparseness problem, tag trigram probability is smoothed based on the bigram and unigram probabilities using the following equation:

$$P(t_i|t_{i-1},t_{i-2}) = \lambda_1\hat{P}((t_i|t_{i-1},t_{i-2}) + \lambda_2\hat{P}((t_i|t_{i-1}) + \lambda_3\hat{P}(t_i) \quad (6)$$

$\hat{P}((t_i|t_{i-1},t_{i-2})$, $\hat{P}((t_i|t_{i-1})\ and \hat{P}(t_i)$ are the maximum likelihood estimates from counts for tag trigram, tag bigram and tag unigram respectively:

$$\hat{P}((t_i|t_{i-1},t_{i-2}) = \frac{C(t_{i-2},t_{i-1},t_i)}{C(t_{i-2},t_{i-1})}, \qquad \hat{P}((t_i|t_{i-1}) = \frac{C(t_{i-1},t_i)}{C(t_{i-1})}, \qquad \hat{P}(t_i) = \frac{C(t_i)}{N}$$

Where: $C(t_{i-2},t_{i-1},t_i)$ indicates the count of the tag sequence <t_{i-2}, t_{i-1}, t_i> and λ_1, λ_2, λ_3 ($\lambda_1 + \lambda_2 + \lambda_3 = 1$) are the weights for the maximum likelihood estimates of trigram, bigram and unigram tag probabilities respectively computed based on corpus statistics. The values of the parameters: $\lambda_1, \lambda_2, \lambda_3$ are estimated using a smoothing technique called the deleted interpolation proposed in [1].

2.2 Computing Observation Probabilities

The observation probability of a word is computed using the following equation:

$$P(w|t) = \frac{C(w,t)}{C(t)} \quad (7)$$

Using a equation (7), a observation probability matrix is created where each row is labeled with a word and each column is labeled with a tag. Each cell of the matrix contains the probability of a word given a tag. The observation probability of a word *w* given tag *t* may be zero when the pair <*w, t*> is not present in the training corpus although the word may be present in the corpus along with some tag other than *t*. In this case, to avoid sparseness of data, maximum negative value (negative infinity) is assigned to the cell.

2.3 Viterbi Decoding

The task of a decoder is to find the best hidden state sequence given an input HMM and a sequence of observations.

The Viterbi algorithm is the most common decoding algorithm used for HMM based part-of-speech tagging. This is a standard application of the classic dynamic programming algorithm [15]. The Viterbi algorithm that we use, takes as input a single HMM and a set of observed words $O = (o_1 o_2 o_3 \dots o_t)$ and returns the most probable state sequence, $Q = (q_1 q_2 q_3 \dots q_t)$, together with its probability.

Given a tag transition probability matrix and the observation probability matrix, Viterbi decoding (used at the testing phase) accepts an untagged text document in Indian language and finds the most likely tag sequence for each sentence in the input document.

We have used the Viterbi algorithm presented in [16] for finding the most likely tag sequence for a given sentence. One of the important problems to apply Viterbi decoding algorithm is how to handle unknown words in the input test sentence. The unknown words are the words which are not present in the training set and hence their observation probabilities are not known. To handle this problem, we estimate the observation probability of an unknown word by analyzing prefix, word-type and suffix of the unknown word.

2.4 Handling Unknown Words

For unknown words, observation probabilities are not available in the observation probability matrix which is created using the equation (7). We estimate the observation probability of an unknown word in the following ways:

Prefix analysis is done first for estimating the observation probability of an unknown word, if the prefix analysis fails, word-type analysis is done and when both fail, suffix analysis is done.

Prefix Analysis. The observation probability of an unknown word is estimated based on its prefix analysis. This is based on the hypothesis that a slightly modified form of an unknown word may be present in the training data whereas the unknown word itself is absent in the training data, for example, when X is the plural form of a common noun and Y is the singular form of the same common noun, X may be present in the training data, but Y may not be present in the training data. For such cases, we may estimate observation probability of an unknown word by matching carefully the unknown word with the closest known words (whose observation probability is known). To do so, an unknown word U is aligned from left to right with a known word K and if the match between U and K exceeds some threshold value, the observation probability of K is taken as that of U. For predicting the probable tags for an unknown word, some rules are framed based on heuristics that use prefix information constrained by the length of the unknown word. The minimum allowable length L_{min} of U is set to 4(in terms of number of characters), that is, an unknown word whose length L_u is less than L_{min} would not be assigned the observation probability through prefix analysis because the prefix of the word is too short to be confident about it. Where the length of the unknown word is greater than or equal to L_{min}, the allowable prefix mismatch between U and K is varied. When the length of U is relatively shorter, the maximum prefix mismatch between U and K that we allow is set to only one character (relatively tough matching). But, when U is relatively long, the criteria of prefix matching is little bit relaxed. However, we have devised three conditions for setting the observation probability of U based on its length and prefix matching between U and K, that is, the observation probability of U is set to that of K when any one of the following conditions holds:

1. $L_{min} \leq L_u \leq L_{mid1}$ and if U and K differ only in the last character
2. $L_{mid1} < L_u \leq L_{mid2}$ and if U and K differ in the last one or two characters
3. $L_u > L_{mid2}$ and if U and K differ in the last one or two or three characters

For the best results, we set $L_{min} = 4$, $L_{mid1} = 8$, $L_{mid2} = 12$.

Word-Type Analysis. The word-types identified by the surface level information of some unknown words help to narrow down the choices of the probable tags for those words. For example, the 4 digit number, hyphenated word etc. We have identified a list of word-types (instead of individual words) based on surface level features by which we can predict the nature of the word and search the table which contains pre-computed information such as the word type, possible tags along with the observation probabilities(probability of a word-type given a tag). The table contains the probability, P (w-type| t_i) where w-type is a word type and t is a tag. P(w-type| t_i) is used in place of $P(w_i|t_i)$ which is required for the HMM based tagging(as specified in the equation (4)). Data sparseness problem is handled in the similar way presented in section 2 in the context of computing observation probabilities of individual words. We have identified the various types of words: four digit number, number started with a digit and ended with an alphabet, hyphenated word, the word fully numeric but the number of digits is not four.

The word-type analysis method functions in the following way:

For an unknown word, its word type is determined by analyzing its surface level features and if its word type matches with any of the word-type pre-computed and stored in the table, the observation probability retrieved from the table is set to the observation probability of the unknown word.

Suffix Analysis. The observation probabilities of unknown words are decided according to the suffix of a word. We find the observation probabilities of unknown words using suffix analysis of all rare words (frequency <=3) in the corpus since unknown words are infrequent and using suffixes of infrequent words in the lexicon is a better approximation for unknown words [1]. The term suffix as used in this context means "a sequence of characters occurring at the end of a word" which is not necessarily a linguistically meaningful suffix. The maximum length of suffix is set to 10 for which we get the best results on our training corpus. The probability of a tag given a suffix of length i is computed as: P(t |suffix-of-len(i)). These probabilities are smoothed using successively shorter and shorter suffixes [1]. This can be formulated in recursive way as:

$$P(t|\text{ suffix-of-len}(i)) = \frac{\hat{P}(t|\text{ suffix-of-len}(i)) + \theta_i\, P(t|\text{ suffix-of-len}(i-1))}{1+\theta_i} \quad (8)$$

Where: $\hat{p}$ is the maximum likelihood probability based on the count of <tag, suffix> pair in all rare words (frequency <=3) in the corpus.

All θ_i are set to the standard deviation of the unconditioned maximum likelihood probabilities ($\hat{p}(t_i)$) of the tags in the training corpus [1]. P(t| suffix-of-len(i)) gives an estimate of $P(t_i|w_i)$. But for HMM based tagging we need to compute the likelihood $P(w_i|t_i)$ which is computed from $P(t_i|w_i)$ using Bayesian inversion that uses Bayes rule and prior $P(t_i)$.

3 Evaluation and Results

3.1 Evaluation

Accuracy of tagging is computed as the ratio of number of matched tags to the total number of tags with duplicates:

$$\text{Accuracy} = \frac{\textit{number of matched tags}}{\textit{total number of tags in the testing corpus}} \times 100\%$$

3.2 Results

We test our developed POS tagger on NLTK (Natural Language Toolkit) dataset[1] which are freely available for download. The NLTK POS tagged corpora for Indian Languages, downloaded from NLTK website consists of POS tagged text in four different languages namely Bengali, Hindi, Marathi and Telugu. The POS tagged data for Bengali contains a total of 895 Bengali (typed in Unicode) sentences tagged using

[1] `http://nltk.googlecode.com/svn/trunk/nltk_data/index.xml`

26 POS tags. Similarly, the POS tagged data for Hindi contains a total of 539 sentences tagged using 25 tags, the POS tagged data for Marathi contains 1196 sentences tagged using 27 tags and the POS tagged data for Telugu contains 994 sentences tagged using 24 tags.

For system evaluation, 10-fold cross validation using NLTK data set is done and the results obtained after running the system on 10 different folds are averaged to find the final results for the system.

We consider bigram tagger as the baseline tagger to which our trigram tagger is compared. In bigram tagger, when transition probabilities are computed, only one previous tag is considered whereas the trigram tagger considers two previous tags while computing the transition probabilities.

To judge the effectiveness of our introduced new features, prefix analysis and word-type analysis for handling unknown words, two versions of a trigram tagger are developed: version 1 is a trigram tagger that considers only suffix analysis for handling unknown words [17] and version 2 is a trigram tagger that considers prefix, word-type and suffix analysis for handing unknown words. For each language, the performances of these two versions are also compared. Table 1 shows the performances of our developed POS tagger for four Indian Languages namely Bengali, Hindi, Marathi and Telugu.

Table 1. Performance comparisons of the POS tagger for Bengali, Hindi, Marathi and Telugu

Systems	Bengali *Accuracy (%)*	Hindi *Accuracy (%)*	Marathi *Accuracy (%)*	Telugu *Accuracy (%)*
Trigram tagger version 2	79.65	84.80	84.00	80.00
Trigram tagger version 1	78.68	83.79	83.16	79.00
Bigram tagger	74.33	79.59	79.29	74.79

The table shows that the trigram POS tagger performs better than the bigram POS tagger for all four languages. It is also evident from the table that, for each of four languages considered in our experiments, the trigram tagger that uses prefix, word-type and suffix analysis for handing unknown words performs better than the trigram tagger that uses only suffix analysis for handing unknown words.

4 Conclusion

This paper describes a trigram HMM based POS tagger for Indian Languages namely Bengali, Hindi, Marathi and Telugu. The POS tagger has been developed using Visual Basic platform so that a suitable user interface can be designed for accepting the input in Unicode from the novice users. The system has been designed in such a way that only changing the training corpus in a file can make the system portable to a new Indian language.

References

1. Brants, T.: TnT – "A statistical part-of-speech tagger". In: Proc. of the 6th Applied NLP Conference, pp. 224–231 (2000)
2. Dandapat, S., Sarkar, S., Basu, A.: Automatic part-of-speech tagging for bengali: an approach for morphologically rich languages in a poor scenario. In: Proceedings of the Association for Computational Linguistic, pp. 221–224 (2007)
3. Ekbal, A., et al.: Bengali part of speech tagging using conditional random field. In: Proceedings of the 7th International Symposium of Natural Language Processing (SNLP 2007), Pattaya, Thailand, December 13-15, pp. 131–136 (2007)
4. Ekbal, A., Bandyopadhyay, S.: Part of speech tagging in bengali using support vector machine. In: IEEE International Conference on Information Technology, ICIT 2008, pp. 106–111 (2008)
5. Kumar, D., Josan, G.S.: Part of speech taggers for morphologically rich indian languages: a survey. International Journal of Computer Applications (0975-8887) 6(5) (2010)
6. Ali, H.: An unsupervised parts-of-speech tagger for the bangla language, Department of Computer Science, University of British Columbia (2010)
7. Chakrabarti, D.: Layered parts of speech tagging for bangla, Language in Indian. Special Volume: Problems of Parsing in Indian Languages (May 2001), `http://www.languageinindia.com`
8. Antony, P.J., Soman, K.P.: Parts of speech tagging for Indian languages: a literature survey. International Journal of Computer Applications (0975-8887) 34(8) (November 2011)
9. Shrivastava, M., Bhattacharyya, P.: Hindi POS Tagger Using Naive Stemming: Harnessing Morphological Information Without Extensive Linguistic Knowledge, Department of Computer Science and Engineering, Indian Institute of Technology, Bombay. In: Proceeding of the ICON (2008)
10. Ray, P.R., Harish, V., Sarkar, S., Basu, A.: Part of Speech Tagging and Local Word Grouping Techniques for Natural Language Parsing in Hindi, Department of Computer Science & Engineering, Indian Institute of Technology, Kharagpur, INDIA 721302, `http://www.mla.iitkgp.ernet.in/papers/hindipostagging.pdf`
11. Selvam, M., Natarajan, A.M.: Improvement of Rule Based Morphological Analysis and POS Tagging in Tamil Language via Projection and Induction Techniques. International Journal of Computers 3(4) (2009)
12. Antony, P.J., Santhanu, P.M., Soman, K.P.: SVM Based Parts Speech Tagger for Malayalam. In: International Conference on-Recent Trends in Information, Telecommunication and Computing, ITC 2010 (2010)
13. Pattabhi, R.K.R.T., Vijay Sundar Ram, R., Vijayakrishna, R., Sobha, L.: A Text Chunker and Hybrid POS Tagger for Indian Languages, AU-KBC Research Centre. MIT Campus, Anna University, Chromepet, Chennai (2007)
14. Rao, D., Yarowsky, D.: Part of Speech Tagging and Shallow Parsing of Indian Languages, Department of Computer Science, Johns Hopkins University, USA, The Proceedings of the Workshop on Shallow Parsing in South Asian Languages (2007), `http://shiva.iiit.ac.in/SPSAL2007/final/iitmcsa.pdf`
15. Jurafsky, D., Martin, J.H.: Speech and Language Processing An Intoduction to Natural Language Processing, Computational Linguistics and Speech Recognition. Preason Education Series (2002)
16. Viterbi, A.J.: Error bounds for convolutional codes and an asymptotically optimum decoding algorithm. IEEE Transaction on Information Theory IT-13(2), 260–269 (1967)
17. Sarkar, K., Gayen, V.: A Practical Part-of-Speech Tagger for Bengali. In: Third International Conference on Emerging Applications of Information Technology (EAIT 2012) (accepted, 2012)

Locating Candidate Tables in a Spreadsheet Rendered Web Page

V.R. Nanjangud[1] and K.K. Achary[2]

[1] Computer Centre, Mangalore University
Mangalagangotri 574 199, India
[2] Department of Statistics, Mangalore University
Mangalagangotri 574 199, India
{vrn,kka}@mangaloreuniversity.ac.in

Abstract. A method to locate web table(s) is presented in this paper. Web page is captured as a spread sheet grid of textual elements (web sheet) with all visual attributes retained, using a spread sheet software. The leaf tables in that web page are captured in a separate sheet using DOM analysis (DOM sheet). Locating a table in a web sheet consists of two sub tasks namely locating the start point and the end point of the table. Start point is located by text comparison of the table elements from DOM sheet with that of web sheet. End point is located by navigating through the web sheet with located start point. Rows, columns information needed for navigation are used from DOM sheet. This method is tested for arbitrarily selected 60 URLs containing 450 leaf tables and in more than 90% of the cases, tables were located correctly.

Keywords: information extraction, web mining, web table location, spreadsheet grid.

1 Introduction

Internet and WWW have revolutionised the way one looks at information. Search engines such as Google has become integral part of information usage in modern societies. In spite of huge information (trillions of web pages) and availability of powerful search engines, extracting a specific piece of information still needs human intervention. There have been many efforts in automating information extraction from web pages [1, 4 and 6]

Web tables provide one rich source of structured information. It is often desirable to access the web table programmatically, as if it is a relational database table. Web tables in which information can be expressed and extracted using relational database techniques are called as genuine or meaningful tables. There are other tables called decorative table in which information is laid out in tabular fashion without any relational association amongst the table elements. Genuine and decorative tables together constitute candidate tables in a web page.

The steps involved in web table information extraction may be listed as location, detection and interpretation. Locating a table in a web page is to identify the regions which are genuine tables. This task can be further subdivided into identifying

S.C. Satapathy et al. (Eds.): Proc. of Int. Conf. on Front. of Intell. Comput., AISC 199, pp. 213–220.
DOI: 10.1007/978-3-642-35314-7_25

candidate tables and then filtering decorative tables from candidate tables to retain genuine tables. Detection step detect label (attribute) and data (value) cells in located tables. The last step of interpretation is to represent information in the table (like relational schema) to facilitate extraction of queried information.

We propose a spread sheet grid model of web page as input to the table processing tasks. Spread sheet software can retrieve the web page programmatically, analyse the retrieved web page and finally render it to the spreadsheet grid as matrix of textual elements. Tables located in this matrix have following features:

1. Web table elements are laid out in tubular (two dimensional) structure – label and data elements appear in separate cells and are addressable and processable by a program.
2. Elements of table retain all the features of visual appearance such as font size, font type, color etc. which helps to build a very rich feature set for future table processing tasks.
3. Non-table data is also available. This may be useful to locate headings containing keyword such as 'table' outside the table area. It also helps to have access to footnotes pointed out by the elements. It may be noted that footnote pointer such as *, #, numbers usually appearing as superscript to elements can be extracted vide point no.2.
4. Relative position of candidate table is available with respect to the entire web page.

Thus we note this web sheet representation gives access to rich feature set with respect to content, layout and attributes of the elements of a table. However locating candidate tables in this model assume significance as table boundary <table> </ table> tags are not visible to the program.

We construct another sheet- DOM sheet containing all leaf tables for a given web page. This is done by analyzing DOM representation of web page and using getelementsbytagname method on this. This data structure of tables, thus obtained can be analysed to get the rows, columns of a table. Thus the DOM sheet gives list of leaf tables with its content in plain text and number of rows, columns for each table.

We use simple string comparison to locate the start of a table in a web sheet. We try to locate the point in web sheet that matches into the first element of a table as given in DOM sheet. If no match is found for a given table, then that table is dropped and does not figure in the output list of located tables.

Dropped table is considered to be correctly dropped if it is a decorative table; otherwise it is considered to be incorrectly dropped. However at this stage we are interested in keeping the incorrectly dropped tables for minimum. We report this figures for an arbitrarily selected set of 60 URLs containing 490 leaf tables [5].

Table location is well researched problem. The methods rely either on underlying HTML source file or on visual cues of page segmentation in a rendered web page.

[2] use hypertext processing module to analyse the HTML text and extract table tags. These extracted table tags are filtered using heuristics to generate candidate tables. Table with single element (i.e. of size 1x1) and table containing too much hyperlink, forms and figures are filtered. The model proposed contains only textual elements and forms, figures are eliminated. Table location algorithm filters candidate tables of 1x1, treating it as trivial. The model may be used to count row wise hyperlink (or any other attributes such as underline), column wise hyperlink or total cells with hyperlinks. The model facilitates choice of any filtering criteria without adding/modifying any html processing.

[3] mentions the wish list to express the queries 'all items in the second column of table' 'all italicised bold faced strings' and indicates that it may be hard to express in terms of DOM- or token level representation. They have proposed document preprocessing and reasoning at wrapper learning time to accommodate appearance /format based extraction. In the proposed grid models appearance/format features are available by default and no additional processing is needed to answer queries mentioned above.

[8] make use of <table> </ table> to extract candidate tables and classify candidate tables as genuine or otherwise using classifier. The feature list inputted to the classifier is drawn from layout, content and word group. We would like to reiterate this feature list or any other such feature list can be generated using the proposed model.

[7] also make use of <table> tags to extract candidate tables and use decision tree classifier. Their algorithm processes the genuine tables and brings it to a form where it is in a uniform two dimensional structure. The detail of this transformation is not indicated. The table located in web sheet is laid out in tabular form and it is easy to identify spanning cells, as they appear as merged cells.

2 The Proposed Method

The problem considered in this paper is to locate the candidate tables in the web sheet. We explain briefly the two input sheets namely the web sheet and DOM sheet as context information, before we state the problem.

2.1 Spread Sheet Rendered Web Page (Web Sheet)

We make use of the spreadsheet software Microsoft Excel to load the web page to a spreadsheet programmatically. The textual content is rendered to a spreadsheet stripping out images and scripts. Figure1 shows a web sheet for indicated URL.

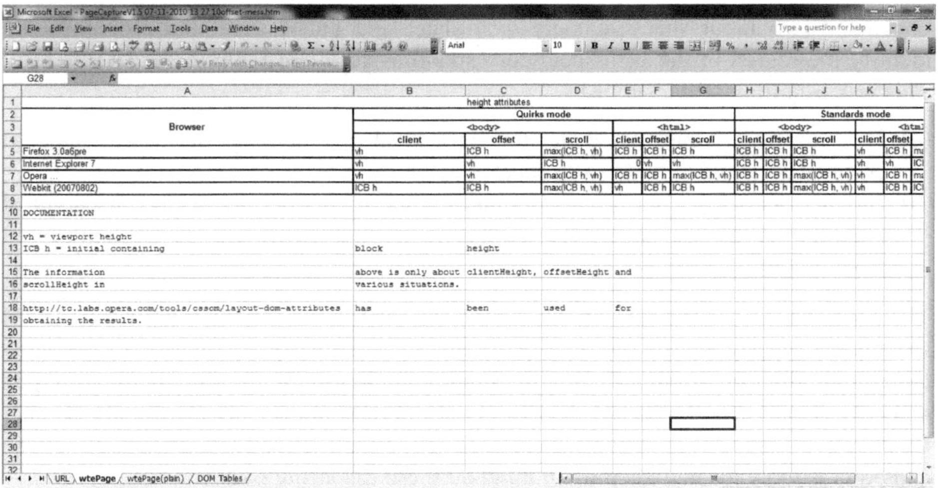

Fig. 1. web sheet for the URL http://lists.w3.org/Archives/Public/www-archive/2007Aug/att-0003/offset-mess.htm

Tables located in the web sheet have following features:

1. Web table elements are laid out in tabular (two dimensional) structure – label and data elements appear in separate cells and are addressable and processable by a program. The attribute COLSPAN (ROWSPAN) appears as merged columns (rows) as seen in Fig1.
2. Elements of table retain all the features of visual appearance such as font size, font type, color etc. which helps to build a very rich feature set for future table processing tasks.
3. Content outside table region is also available in the web sheet. It helps to have access to footnotes pointed out by the elements. The content 'height attribute' and 'DOCUMENTATION' outside table region may be noted in Fig1.
4. Relative position of candidate table is available with respect to the entire web page.

2.2 DOM Sheet

The boundaries of table namely <table> </table> in a web sheet are not visible to the program. We construct another sheet by analyzing the page using DOM model, and capture all leaf tables using getelementsbytagname method on web page document. This collection of tables is analysed and list of leaf tables along with contents (in plain text) is written to another sheet – DOM sheet. The number of rows computed by <tr> tags is recorded as number of rows – DOM Rows. The number of columns may vary from row to row and the highest value of row wise column value is recorded as number of columns – DOM Columns.

2.3 Problem Formulation

Given the two input sheets namely web sheet and DOM sheet the problem is to identify the region in the web sheet corresponding to a leaf table in the DOM sheet. This task is split into two subtasks

(a) For any leaf table listed in the DOM sheet, locating the start point of corresponding table in the web sheet.
(b) Having located the start point of a table in the web sheet, locate the end point of that table in the web sheet. This implies locating last row and last column of that table in a web sheet.

The second task needs some explanation. The number of rows (columns) in DOM sheet are as per number of <tr> tags (<td> or <th>). However when actually rendered in the spread sheet it will take into account attributes such as columnspan, rowspan and span more rows (columns) than reported in DOM sheet. The presence of other tags such as <br> may result in additional rows than reported in DOM sheet. The end point is to be located keeping in mind all these rendering issues.

3 The Proposed Solution

The pseudo code for the main algorithm to solve this problem is outlined below.

```
Begin
For each table in DOM sheet
        Locate the start point for that  table in the web sheet
        If start point of the table found
                  Locate the end point
                  Update the output list
         Else
                  Drop that table.
        End If
Next For
End
```

We make use of the following data items to locate the start point of a table, for a given cell in the web sheet

1. FirstToFind - First element (Row 1, Column 1) of a leaf table obtained from DOM Sheet
2. NextToFind – Second element (Cell (1, 2) or Cell (2, 1)) of a leaf table obtained from DOM sheet.
3. MatchFirst - True if the value in the cell matches FirstToFind.
4. MatchNext – True if the value in next cell matches NextToFind.

We define measure called Match Factor corresponding to the perceived degree of match. It can take values 0 (drop the table), 0.5(partial match) and 1 (for perfect match).

We drop the table if both FirstToFind and NextToFind are blank. Match is attempted if at least one of them is non blank. Table 1 shows truth table for computation of match factor.

Table 1. Truth Table for computing Match Factor

FirstToFind	NextToFind	MatchFirst	MatchNext	Match Factor
Blank	Non-blank	X	True	0.5
Non-blank	Blank	True	X	0.5
Non-blank	Non-blank	True	True	1.0
Blank	Blank	X	X	0.0
Blank	Non-blank	X	F	0.0
Non-blank	Blank	F	X	0.0

For each table, the data items are computed and match factor evaluated. All leaf tables with non-zero match factor are included in the output list. We create two output sheets – one containing candidate tables and lists – other containing tables with many rows and many columns.

We need to compute the end points for a table whose start point has been located. We use the number of rows – DOM Rows and number of columns – DOM columns as available in the DOM sheet. The following two pseudo codes outline the end Column, end Row algorithm.

Begin
Commence with start point on web sheet
For DOMRows times
    Navigate through web sheet column wise for DOMColumns times
    Get the last column reached
    Update the start point by navigating one column below
End For
Set the lowest value of columns got as end column
End

Begin
Commence with start point on web sheet
For DOMColumns times
    Navigate through web sheet row wise for DOMColumns times
    Get the last row reached
    Update the start point by navigating one row across
End For
Set the modal value of rows got as end row
End

4 Experimental Results

An automated tool is developed using Microsoft Excel software to implement the proposed algorithm .The results obtained for an arbitrarily selected list of 60 URLs, some of which are drawn from Millard [5], containing 490 leaf tables is summarised in the Table 2.

Table 2. Experimental results on the sample dataset

URLs	Leaf Tables in DOM Sheet	Dropped Tables		Output		End row/ column errors		Genuine Tables
		Correctly	Incorrectly	Tables and Lists	Tables	Row	column	
60	450	122	20	310	221	4	0	232

The error in locating tables is around 5% for the sample data set. It may be noted that there is no error in locating end column of tables, However there are four tables where error in detecting the last row has been found. This is due to the presence of tags such as <br> and <p> which introduce additional rows when rendered by spreadsheet software. It may be noted that table location has filtered 122 decorative tables correctly and 20 incorrectly. We also note that only 9 tables out of 221 are decorative to be classified/filtered by subsequent tasks.

The following are the reasons noted for dropping tables incorrectly:

1. Presence of two or more successive blanks in first row. Some tables had multiple blank rows, perhaps to improve the readability which also resulted in dropping of genuine table (URL: http://nucleardata.nuclear.lu.se/NuclearData/toi/listnuc.asp?sql=&A1=100&A2=102).
2. Presence of <br> tag in first two elements. This results in splitting of single cell in DOM sheet into multiple cells in web sheet. Hence comparison fails and results in dropping of genuine tables (URL: http://www.sbilife.co.in/sbilife/content/9_4144).
3. Date format mismatch. The cell content such as 'Dec 2005' was rendered as 'Dec 2005' in web sheet while it was 'Dec-05' in DOM sheet resulting in mismatch(URL: http://tradeself.com/).
4. Tables in frame based pages may be dropped incorrectly as scanning is from left to right and top to bottom for the entire page as against the scanning frame wise(URL: http://tradeself.com/). This happens when end point for a table in a frame is below the start point of a table in adjoining frame.

5 Conclusion

We have presented a method to locate candidate tables in a spreadsheet rendered Web page whose output could be used for further table processing tasks. The web sheet captures tabular layout of contents along with its visual features. It also gives access to the neighborhood data around the candidate tables. We also capture information pertaining to leaf tables in DOM sheet.

An algorithm to locate leaf tables in web sheet is presented. Start point of a table is located by a simple text comparison of textual elements from DOM sheet. End point is located by navigating through the web sheet by rows/columns available in the DOM sheet for that table.

Results indicate success rate of more than 90% for the location task for the arbitrarily selected data set. There are no errors reported in locating end column, however an error of less than 2% is found in locating end rows of table.

References

1. Chang, C.-H., Kayed, M., Girgis, M.R., Shaalan, K.: A Survey of Web Information Extraction Systems. IEEE Transactions on Knowledge and Data Engineering, 1411–1428 (2006)
2. Chen, H.-H., Tsai, S.-C., Tsai, J.-H.: Mining tables from large scale HTML texts. In: Proc. 18th COLING, pp. 166–172. Morgan Kaufmann (2000)
3. Cohen, W.W., Hurst, M., Jensen, L.S.: A flexible learning system for wrapping tables and lists in HTML documents. In: Proc. 11th WWW, pp. 232–241. ACM (2002)

4. Laender, A.H.F., Ribeiro-Neto, B.A., da Silva, A.S., Teixeir, J.S.: A Brief Survey of Web Data Extraction Tools. SIGMOD Record 31(2), 84–93 (2002)
5. Millard, B.T.: Collections of Interesting Data Tables (2007), `http://projectcerbera.com/web/study/2007/tables/` (accessed August 2, 2009)
6. Muslea, I.: Extraction Patterns for Information Extraction Tasks: A Survey. In: Proc. AAAI 1999 Workshop Machine Learning for Information Extraction, pp. 1–6 (1999)
7. Tengli, A., Yang, Y., Ma, N.L.: Learning table extraction from examples. In: Proc. 20th COLING, pp. 987–993 (2004)
8. Wang, Y., Hu, J.: A Machine Learning Based Approach for Table Detection on the web. In: Proceedings of the 11th International Conference on World Wide Web, pp. 242–250 (2002)

e -Library Content Generation Using WorldNet Tf-Idf Semantics

Mukesh Kumar and Renu Vig

University Institute of Engineering and Technology,
Panjab University, Chandigarh, India
mukesh_rai9@yahoo.com, renuvig@hotmail.com

Abstract. Electronic library is the collection of digital information related to an individual domain and in turn to all domains. A focused crawler traverses the Web looking for the pages most relevant to a domain and at the same time discarding the irrelevant pages and hence is helpful for generating the-e contents for digital library related to a particular domain. In this paper a focused crawling technique to generate online contents for e-library based upon WorldNet semantics is proposed. The applicability of the proposed approach is shown by retrieving the documents which are highly related to a single domain. The quality of the pages included into the library is derived from the relevancy measure of the page with the content of domain related pages.

Keywords: Focused Web crawler, information retrieval, Tf-Idf, semantics, search engine, indexing.

1 Introduction

Digital Libraries are being created today for diverse communities and in different fields e.g. education, science, culture, development, health, governance, sports and so on. With the availability of several free digital Library software packages at the recent time, the creation and sharing of information through the digital library collections has become an attractive and feasible proposition for library and information professionals around the world. Currently World Wide Web contains billions of publicly available documents. Besides its huge size the Web is characterized by its huge growth and change rates. It grows rapidly in terms of new servers, sites and documents. The addresses of documents and their contents are changed, and documents are removed from the web. As more information becomes available on the Web it is more difficult to find relevant information from it. "WordNet® [16,18] is a large lexical database of English. Nouns, verbs, adjectives and adverbs are grouped into sets of cognitive synonyms (synsets), each expressing a distinct concept. Synsets are interlinked by means of conceptual-semantic and lexical relations. The resulting network of meaningfully related words and concepts can be navigated with the browser. WordNet [17] is also freely and publicly available for download. WordNet's structure makes it a useful tool for computational linguistics and natural language processing" [18]. Web search engines such as Goggle, AtlaVista provides

S.C. Satapathy et al. (Eds.): Proc. of Int. Conf. on Front. of Intell. Comput., AISC 199, pp. 221–227.
DOI: 10.1007/978-3-642-35314-7_26

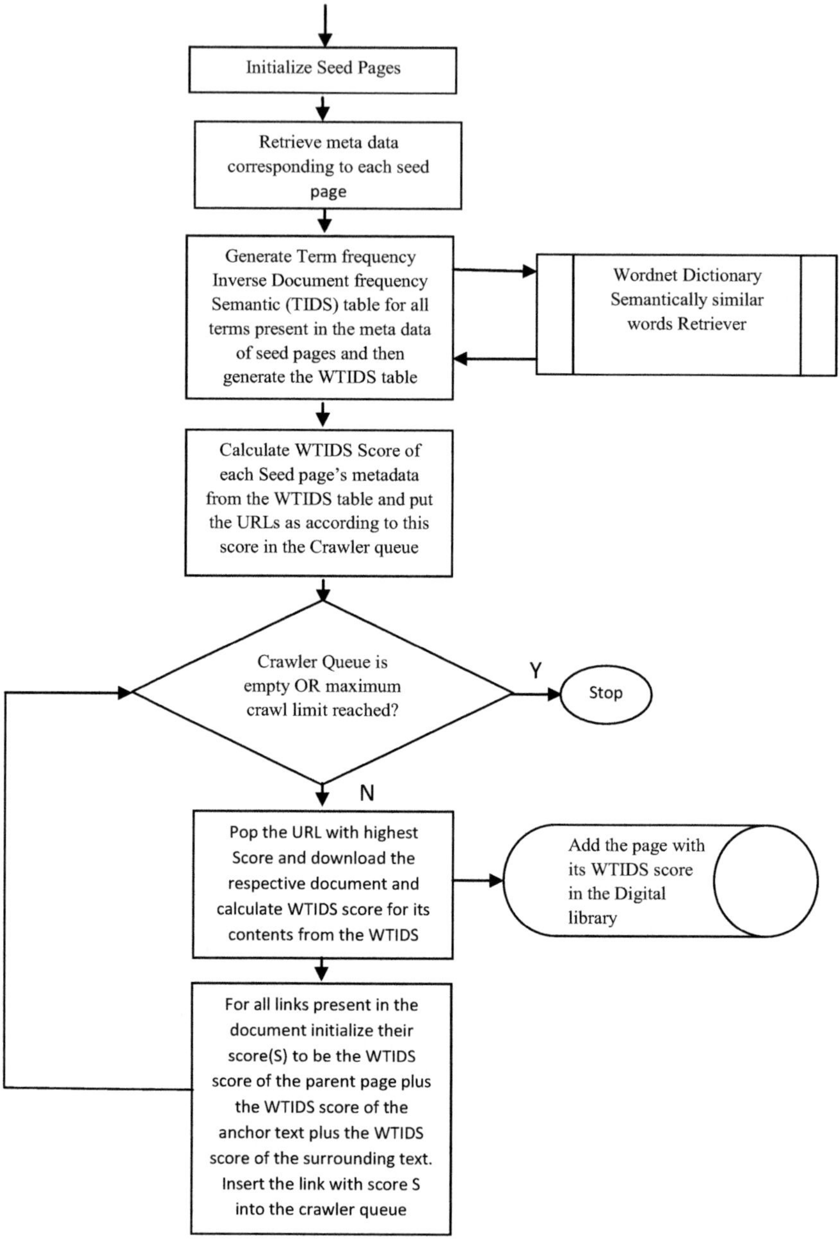

Fig. 1. e-Library online content generation process

access to the Web documents. A search engine's crawler collects Web documents and periodically revisits the pages to update the index of the search engine. Due to the Web's immense size and dynamic nature no crawler is able to cover the entire Web and to keep up all the changes. Focused crawlers are designed to download Web documents that are relevant to a predefined domain, and to avoid irrelevant areas of the Web. The benefit of the focused crawling [14] approach is that it is able to find a large proportion of relevant documents on that particular domain and is able to effectively discard irrelevant documents and hence leading to significant savings in both computation and communication resources, and high quality retrieval results. In some early works on the subject of focused collection of data from the Web, Web crawling was simulated by a group of fish migrating on the Web, Bra and.Post [12]. In the so called fish search, each URL corresponds to a fish whose survivability is dependent on visited page relevance and remote server speed. Page relevance is estimated using a binary classification by using a simple keyword or regular expression match. Only when fish traverse a specified amount of irrelevant pages they die off. The fish consequently migrate in the general direction of relevant pages which are then presented as results. Cho, Molina and Page [5] proposed calculating the PageRank given by Page et al. [6] score on the graph induced by pages downloaded so far and then using this score as a priority of URLs extracted from a page. They show some improvement over the standard breadth-first algorithm. The improvement however is not large. This may be due to the fact that the PageRank score is calculated on a very small, non-random subset of the web and also that the PageRank algorithm is too general for use in topic-driven tasks. Ehrig and Meadche [2] considered an ontology-based algorithm for page relevance computation. After pre-processing, entities (words occurring in the ontology) are extracted from the page and counted. Relevance of the page with regard to user selected entities of interest is then computed by using several measures on ontology graph (e.g. direct match, taxonomic and more complex relationships). Most of the existing focused crawlers [1, 3, 4, 8, 11, 13, 7] are based on simple keyword matching or some very complex machine learning techniques for guiding the future crawls. The work proposed in this paper is application oriented extension of work done in [8, 9, 10 and 19].

2 Proposed Work

Tf-Idf (Term frequency–Inverse document frequency) weight is a statistical measure used to evaluate how important a word is to a document in a collection or corpus. The importance increases proportionally to the number of times a word appears in the document but is offset by the frequency of the word in the corpus or in turn to the domain. If we are having a corpus of documents which are all highly related with a specific domain then the Tf-Idf score of a term in a document gives the importance of that term for that document with respect to the whole corpus. Now if we add Tf-Idf score obtained by a term for all documents in the corpus, then the resulting score can be seen as a meaningful, semantic, score for that term with respect to the whole corpus. Based upon this thought a TIDS (Term frequency–Inverse document frequency Definition Semantic) Score Table is constructed. The semantically similar words corresponding to each term in TIDS table are retrieved from the Wordnet and assigned the same score as that of the TIDS score to enforce the term semantic measures and put in the WTIDS (Wordnet TIDS) table whose entries are supposed to help the crawler for deciding the future crawls. The TIDS Score Table generation algorithm is given in Algorithm 1 and WTIDS table generation is given in Algorithm 2. The initial collection of Web pages related to the Sports domain (Seed pages) is generated from the hierarchical categories of ODP (Open Directory Project) from http://dmoz.org. ODP provides the categorical collection of URLs that are manually edited and not biased by any commercial user. From here we can find individual categories link. These URLs are put in the Relevant_Page_Set.

Algorithm 1:TIDS Score Table Generation

1. Initialize Relevant_Page_Set.
2. Generate Tf-Idf Score Inverted Index Table for all the meta data terms for the Relevant_Page_Set.
3. For each term t in the Tf-Idf Score Inverted Index Table Do
   - 3.1. Calculate sum of the Tf-Idf score obtained by t in all documents from Tf-Idf Score Inverted Index Table, let it be TIDS_Score.
   - 3.2. Insert entry <t, TIDS_Score> into TIDS Score Table.
   - 3.3. Normalize the TIDS_Score values in TIDS Score Table.

Algorithm 2: WTIDS Score Table Generation

1. For each term t present in the TIDS table do Step 2.
2. Retrieve all the semantically similar words for t and assign them the score as that of t and put in the WTIDS along with its score.

According to the TIDS Score Table Generation Algorithm Tf-Idf score of the collection is calculated. The term frequency $\text{tf}_{t,d}$ of term t in document d is defined as the number of times that t occurs in d, df_t is the document frequency of t, means the number of documents that contain t. The df_t is an inverse measure of the informativeness of t also $\text{df}_t \leq N$ where N is the total number of documents in the Relevant Page Set. Then the idf (inverse document frequency) of t is given by

$$\text{idf}_t = \log \ (N/\text{df}_t) \tag{1}$$

The Tf-Idf weight of a term t in the document d ($\text{w}_{t,d}$) is the product of its tf weight and its idf weight and will be given by

$$\text{w}_{t,d} = \log(1 + \text{tf}_{t,d}) \times \log \ (N / \text{df}_t) \tag{2}$$

The TIDS_Score of a term t is given by

$$\text{TIDS\_Score}(t) = \sum_{d \in Relevant\_Page\_Set} \text{tf.idf}_{t,d} \tag{3}$$

The proposed system flows like the Fig.1. The WTIDS score table is created for all terms present in the Seed's metadata. All the seeds are pushed into the priority queue as according to the WTIDS score. The crawler picks the URL with highest score and downloads the corresponding document from the Web and put that into the library.

All the links present in the downloaded document are the possible candidates for the next crawl and hence are put into the crawl queue as according to the WTIDS score calculated for it anchor ,parent and the surrounding text. The crawl proceeds for yhe time till the crawl queue is not empty or the maximum crawl limit is not reached.

3 Experimental Results

The proposed system for generating e-library contents is tested for generating contents for Mathematics domain. The initial collection of Web pages (Seed pages) related to sports domain is generated from the hierarchical categories of ODP (Open Directory Project) from http://dmoz.org as suggested by Rungsawang, N.Angkawattanawit [15]. ODP provides the categorical collection of URLs that are manually edited and not biased by any commercial user. From here we can find individual categories link. The categories ending with "Maths", and "Mathematics", were retrieved from the ODP. 115 such links were retrieved, which further acted as the seed pages for the proposed system. Precision, the percentage of the pages relevant to the domain out of total pages retrieved by the system, is calculated by placing different values for relevancy value i.e. WTIDS score. On the other hand a keyword based crawler is implemented for the same set of seed pages. It contains 200 keywords related to the mathematics domain, any page that contains 25% of the keywords is supposed to be relevant to the domain, and is to be stored in to the digital library. The two crawlers were run for collecting 2000 pages from the Web using same set of the seed pages, and WTIDS score for each page for each crawler is found and is plotted as graph in Fig 2, which is a precision versus relevancy limit graph. The graph shows that WTIDS approach outperform keywords based approach for each relevancy limit.

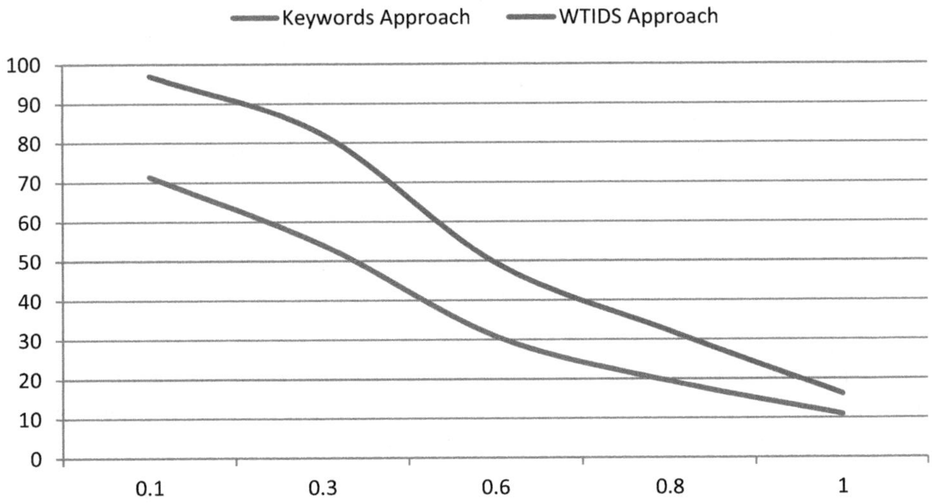

Fig. 2. Precision graph for different values of relevancy for the two approaches

If we suppose all the documents having a WTIDS score of .5 or more are related to the mathematics domain. Then the average precision value for WTIDS approach comes out to be comes out to be 45% of the total number of the pages downloaded while it is 26 % for the keyword approach.

4 Conclusion

Online e-library content generation system using WordNet and Tf-Idf semantics is proposed. The results for the proposed system are plotted against the keyword based approach which shows the proposed approach outperforms the keyword approach for every relevancy limit and hence justifying its use.

References

[1] Brin, S., Page, L.: The anatomy of a large scale hypertextual web search engine. Computer Networks and ISDN Systems 30, 107–117 (1998)
[2] Ehrig, M., Maedche, A.: Ontology-Focused Crawling of Web Documents. In: Proceedings of the Symposium on Applied Computing 2003, Melbourne, FL, USA (2003)
[3] Cho, J., Hector, G.-M.: Parallel Crawlers. In: Proceedings of the World Wide Web conference (WWW), Honolulu, Hawaii (2002)
[4] Cho, J., Garcia-Molina, H.: The evolution of the web and implications for an incremental crawler. In: Proceeding of 26th International Conference on Very Large Database, Cairo, Egypt, pp. 200–209 (2000)
[5] Cho, J., Garcia-Molina, H., Page, L.: Efficient Crawling Through URL Ordering. In: Proceedings of the Seventh International World Wide Web Conference, Brisbane, Australia, pp. 379–388 (1998)
[6] Page, L., Brin, S., Motwani, R., Winograd, T.: The PageRank Citation Ranking: Bringing Order to the Web. Technical report, Stanford Digital Library Technologies Project, pp. 1–17 (1998)
[7] Ester, M., Groß, M., Kriegel, H.-P.: Focused Web Crawling: A Generic Framework for Specifying the User Interest and for Adaptive Crawling Strategies. In: Proceedings of the 27th International Conference on Very Large Database, VLDB 2001, Roma, Italy, pp. 633–637 (2001)
[8] Kumar, M., Vig, R.: Design of CORE: context ontology rule enhanced focused web crawler. In: Proceedings of the International Conference on Advances in Computing, Communication and Control, pp. 494–497. ACM, New York (2009) ISBN: 978-1-60558-351-8, doi:10.1145/1523103.1523201
[9] Kumar, M., Vig, R.: Term-Frequency Inverse-Document Frequency Definition Semantic (TIDS) Based Focused Web Crawler. In: Krishna, P.V., Babu, M.R., Ariwa, E. (eds.) ObCom 2011, Part II. CCIS, vol. 270, pp. 31–36. Springer, Heidelberg (2012)
[10] Goyal, N., Kumar, M., Vig, R.: Consistency Enforcement Using Ontology on Web. Journal of Computers 5(10), 1520–1526 (2010), ISSN 1796-203X, doi:10.4304/jcp.5.10.1520-1526
[11] Boldi, P., Codenotti, B., Santini, M., Vigna, S.: Ubicrawler: a scalable fully distributed web crawler. Software Practice & Experience 34(8), 711–726 (2004)

[12] De Bra, P.M.E., Post, R.D.J.: Information retrieval in the World-Wide Web: Making client-based searching feasible. Computer Networks and ISDN Systems 27(2), 183–192 (1994)

[13] Chakrabarti, S., van den Berg, M., Domc, B.: Focused crawling: a new approach to topic-specific Web resource discovery. In: Proceedings of the 8th International World Wild Web Conference, Toronto, Canada, pp. 1623–1640 (1999)

[14] Pirkola, A.: Focused Crawling: A Means to Acquire Biological Data from the Web. In: VLDB 2007, Vienna, Austria (2007)

[15] Rungsawang, A., Angkawattanawit, N.: Learnable topic-specific web crawler. Journal of Networks and Computer Applications, 97–114 (2005)

[16] Miller, G.A.: WordNet: A Lexical Database for English. Communications of the ACM 38(11), 39–41 (1995)

[17] Fellbaum, C.: WordNet: An Electronic Lexical Database. MIT Press, Cambridge (1998)

[18] `http://wordnet.princeton.edu/`

[19] Singh, J., Kumar, M.: A Meta Search Approach to Find Similarity between Web Pages Using Different Similarity Measures. In: Unnikrishnan, S., Surve, S., Bhoir, D. (eds.) ICAC3 2011. CCIS, vol. 125, pp. 150–160. Springer, Heidelberg (2011)

An Automatic Legal Document Summarization and Search Using Hybrid System

Selvani Deepthi Kavila[1], Vijayasanthi Puli[1],
G.S.V. Prasada Raju[2], and Rajesh Bandaru[3]

[1] Department of Computer Science and Engineering,
Anil Neerukonda Institute of Technology and Sciences,
Sangivalasa, Visakhapatnam, AP, India
{selvanideepthi14,vijjusanny83}@gmail.com
[2] Department of Computer Science, SDE,
Andhra Univeristy, Visakhapatnam, India
gsvprajudr9@yahoo.co.in
[3] Department of Computer Science and Engineering
VITAM Engineering College, Visakhapatnam, AP, India
b.rajesh68@yahoo.com

Abstract. In this paper we propose a hybrid system for automatic text summarization and automatic search task related to legal documents in the legal domain. Manual summarization requires much human effort and time. For this reason automatic text summarization is introduced which saves the legal expert time. The summarization task involves the identification of rhetorical roles presenting the sentences of a legal judgement document. The search task involves the identification of related past cases as per the given legal query. For these two tasks we have introduced hybrid system which is the combination of different techniques. The techniques involved in our hybrid system are keyword or key phrase matching technique and case based technique. We have implemented and tested and required results are produced.

Keywords: Automatic Text Summarization, Automatic Search, Legal domain, Keyword/phrase matching, Case based technique.

1 Introduction

Automatic text summarization is a technique, where a computer summarizes a text. Summarization provides the possibility of finding the main points of text so that the user will spend less time on reading the whole document. Automatic summarization involves reducing a text document into a short set of words or paragraph that conveys the main meaning of the text [7]. It consists of different type of approaches or methods for summarization process [8] such as Abstract text summarization and Extract text summarization

Abstract summarization is a technique, which rewrites the original text into a shorter version by replacing the wordy concept with shorter ones. It reduces the

S.C. Satapathy et al. (Eds.): Proc. of Int. Conf. on Front. of Intell. Comput., AISC 199, pp. 229–236.
DOI: 10.1007/978-3-642-35314-7_27

sentences from the original text without changing their meaning and Extract summarization is a technique, which reuses the most important sentences form the original text for text summarization. By considering existing words, sentences etc. from the original text to form summary. In this paper we used extract text summarization technique on legal document.

Case based system, which automatically searches the existed cases by using Key phrase and provides the required result. It consists of Case retrieval measures and Case result representation.

a) Case retrieval measures

As per the legal expert request the case based system searches the documents of past cases based on the key phrase matching.

b) Case result representation

Depending upon the new case request Case based system returns a list of past cases. A summary of a judgement helps in organizing a large volume of documents and finding the relevant judgements for their cases. For this reason, the information frequently summarized by legal experts. But due to manual summarization by legal experts it requires much human time and expertise to provide manual summaries for legal documents. In automatic legal document summarization, by using extraction technique it extracts the main points from the legal document and provides the summary. So that it saves the time to provide summaries of legal documents and a lawyer can spends less time on reading the whole document.

In this paper we explained hybrid system in section 2, for the summarization and search of legal document. Section 3 discusses our proposed frame work. Section 4 shows the Procedure for Summarization of legal documents, Section 5 describes Case based system, Section 6 explains the rhetorical roles of legal document. Section 7 shows the experimental Results and Section 8 consists of conclusion and future work.

2 Hybrid Systems

Hybrid system is the combination of methods and techniques from artificial intelligence. In our paper we used the techniques like

- Keyword/ Key phrase matching technique
- Case based technique

By using the combination of these techniques we implemented automatic summarization for a legal document. All these techniques are explained clearly in respective places.

3 Frame Work

Fig1.frame work of hybrid system. The input of system is a request by the legal expert and a number of past cases which are nothing but judgement documents. First of all, the past cases are loaded, then the roles will be decided and by pre-processing it, the structured text will be maintained. When the user searches for a particular case

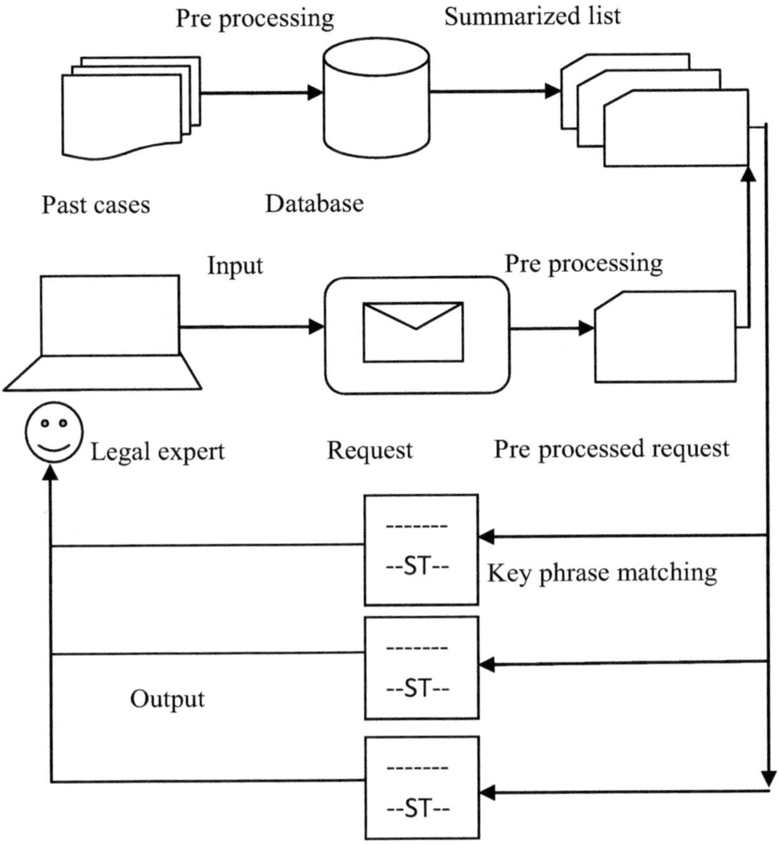

Fig. 1. Frame work of hybrid system

details depending upon the key phrase match, the list of summarized documents will be displayed. Complete explanations of the summarization approaches and case based system techniques are presented in the next sections.

4 Procedure for Legal Document Summarization

a) Pre-processing

Pre-processing is a primary step to load the text into the proposed system and make some process that transfer the text into the html format and it carries out the sentences in a paragraph format that improves the accuracy of the system to distinguish similar words.

b) Segmentation

Segmentation is the process of dividing text into meaningful units, Such as words, sentences, or topics. Word segmentation is the problem of dividing a string into its component words. Sentence segmentation is the problem of dividing a string into its component sentences.

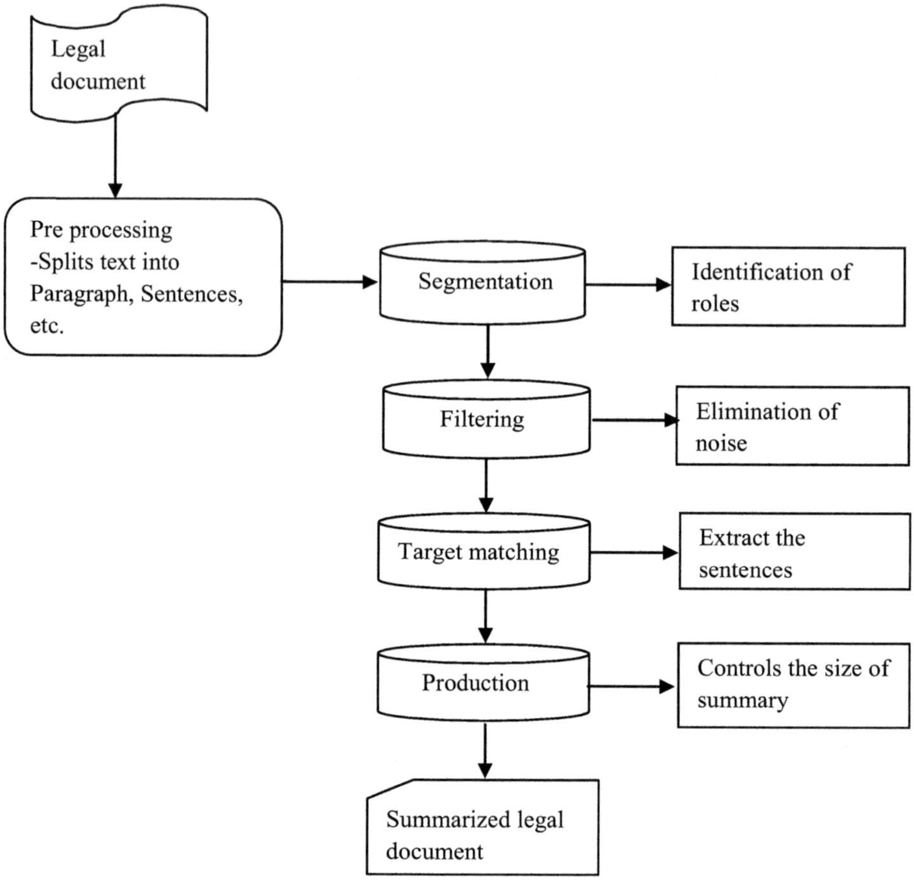

Fig. 2. Procedure for generating summarization for legal documents

c) Filtering

Filtering identifies parts of the text which can be eliminated without loosing relevant information for summary [1]. It reduces the noise from the text like eliminating repeated sentences, etc...

d) Target Matching

Determines whether the sentence contains target entity from the document. By validating the words the line count will be calculated it is nothing but sentence length.

$$L.\ Scorelen\ (Si) = Li$$

e) Production

This is the final step, it eliminates all unimportant elements and concentrates on size of the text. Finally it provides the summarized text.

5 Case Based System

Case based system is the one of the emerging paradigm for designing intelligent systems. Retrieval of similar cases is a primary step in case based system and the similarity measure plays a very important role in case retrieval. Many methods have been developed to find the relevance of past cases to request in database in order to search [4]. Some case based systems have been developed to search the solution by suggesting the most informative questions [2][5][6]. The similarity measurements used in these systems usually are based on keyword matching.

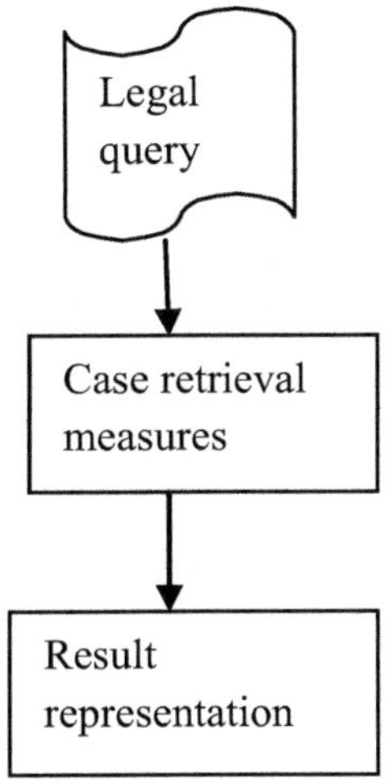

Fig. 3. Case Based System

a) Case Retrieval Measures

When a new request regarding a new case given by legal expert, most case based systems judges similarity of input and the past documents which are in database. It matches on the basis of agreements of words, word pairs, and lines. A text search system stores texts and complimentary term lists prepared there from in respective database. Legal queries are inputted in the form which sets of keywords are extracted. After searching the texts that will be stored in the database with respect to the keywords extracted from each input query of keywords are determined.

b) Result Representation

When the given input, new case key phrase matches with the past cases which are in database then the matched case list will be returned. When the legal expert clicks on a particular case then the summarized legal text will be displayed.

6 Rhetorical Roles

In our paper, we focused on developing a fully automatic summarization system for a legal domain. The fundamental need of a legal judgement is to legitimize a decision from authoritative source of law. In our concept different rhetorical roles presented. In our paper, our classifications have been enhanced into thirteen labelled elements for a more structured presentation. The labels showed in Table 1. To identify the labels, we need to maintain a collection of features, which includes all important features like structure identification, abbreviated words ,length of word, etc,. Many of the judgements do not even follow the general structure of the legal document. To overcome this we implemented hybrid system to provide structured text. We have introduced a new label Positive & Negative, which illustrates the points which have been taken into consideration for the given judgement are considered as positive, and the points which have not been taken into consideration represented as negative points. To extract positive and negative sentences we have maintained a positive points list like facts, witness, etc, and negative points like failure, but ,etc,. Like this the roles will get identify.

Table 1. Proposed system rhetorical roles for legal judgements

Label	Description
Appeal no :	Provides the case number.
Year :	Provides year
Case :	Provides the case details
Judges :	Provides the judges name who handled this case
Petitioner :	Provides the petitioner name
Respondent :	Provides the Respondent name
Counsel for the appellant :	Provides the name of the counsel for the appellant
Counsel for the respondent :	Provides the name of the respondent name
Judgement by :	Provides the name of the judge who gave judgement
Sections :	Provides the sections details which are provided by judgement document
Facts :	Provides the facts of the case which are provided in judgement document.
Positive & Negative :	The points which have been taken into consideration for the given judgement are considered as positive and the points which have not been taken into consideration are considered as negative points.
Judgement:	Final decision given by judge

7 Experimentations and Results

Our corpus presently consists of 100 legal documents related to criminal and civil, collected from AP lawyer's archive, Out of which 35 were annotated. Each document in corpus contains an average of 30 to 35 words in a sentence. In our paper, judgement document divided into criminal and civil sections. Our legal document contains the unstructured information related to petitioner, respondent, year , judge, judgement, section details etc . In our experiment we are able to provide structured information by using key phrase matching and case based technique and also succeeded in providing related documents list as per the given search item.

Here, we experimented on unstructured legal judgement document to provide the automatic summary and also showed the how the case has been searched shown in below screen shots.

1. Screen1 indicates the unstructured judgement document given as input.
2. Screen2 indicates the summarized document which is nothing but result.
3. Screen3 indicates the legal query which have been given for searching of past cases
4. Screen 4 indicates the results which have been provided after searching, the result is nothing but the list of relevant cases as per the legal query.
5. Screen 5 indicates the summarized structure of required document.

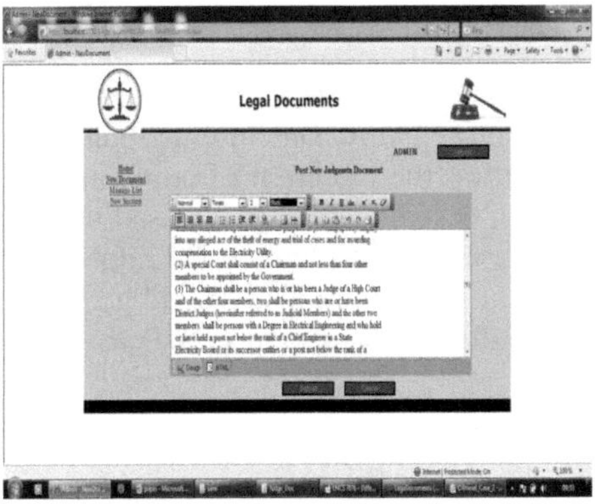

Screen 1

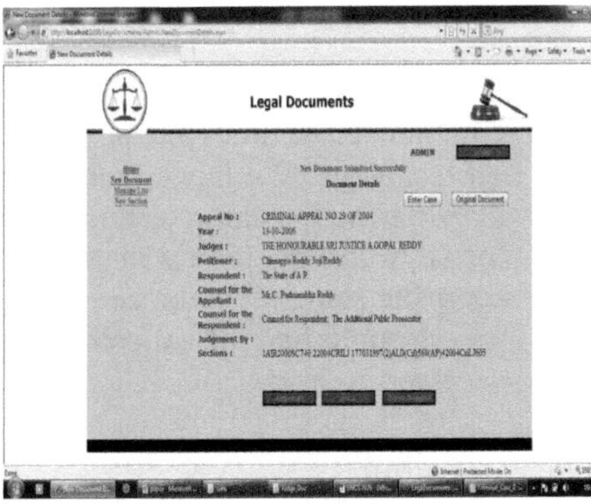

Screen 2

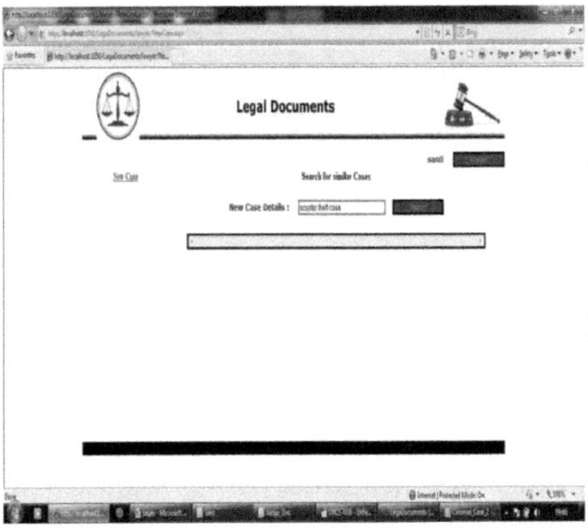

Screen 3

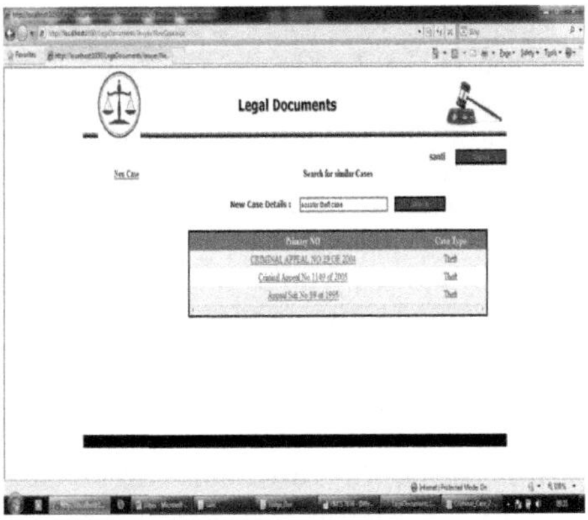

Screen 4

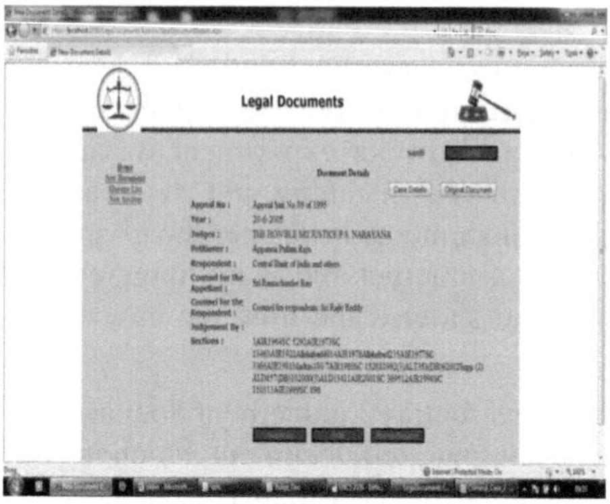

Screen 5

In our evaluation, we obtained more than 80% of the result in structured summarization of the document and 90% in automatic search.

8 Conclusion and Future Work

In this paper we have introduced hybrid system for summarization of legal document. We have implemented this system on the basis of keyword matching technique and case based technique. We have done our research in the direction of combination of keyword/key phrase matching technique and Case based technique, we used the combination of keyword/key phrase matching and Case based technique for implementation of automatic summarization and we succeed. Future work also should go through this hybrid system to find the expected judgement for a new case from the past cases.

References

1. Farzindar, A., Lapalme, G.: Legal Text Summarization by Exploration of the Thematic Structures and Argumentative Roles (2004)
2. Bridge, D., Goker, M.H., Mcginty, L., Smyth, B.: Case-based recommender systems. Knowl. Eng. Rev. 20(3), 315–320 (2005)
3. Wang, D., Li, T., Zhu, S., Gong, Y.: iHelp: An Intelligent Online Helpdesk System. IEEE Transactions on Systems, Man, and Cybernetics-Part B: Cybernetics (April 11, 2010)
4. Jagadish, H.V., Ooi, B.C., Tan, K.-L., Yu, C., Zhang, R.: idistance: An adaptive b+-tree based indexing method for nearest neighbour search. ACM Trans. Database Syst. 30(2), 364–397 (2005)
5. Doyle, M., Cunningham, P.: A Dynamic Approach to Reducing Dialog in On-Line Decision Guides. In: Blanzieri, E., Portinale, L. (eds.) EWCBR 2000. LNCS (LNAI), vol. 1898, pp. 49–60. Springer, Heidelberg (2000)
6. Mirzadeh, N., Ricci, F., Bansal, M.: Feature selection methods for conversational recommender systems. In: Proc. IEEE Int. Conf. e-Technol., e-Commerce e-Service, pp. 772–777 (2005)
7. Al-Hashemi, R.: Text Summarization Extraction System (TSES) Using Extracted Keywords. International Arab Journal of e-Technology 1(4) (June 2010)
8. Gholamrezazadeh, S., Salehi, M.A., Gholamzadeh, B.: A Comprehensive Survey on Text Summarization Systems. 978-1-4244-4946-0/09/$25.00 ©2009 IEEE

Minimum Spanning Tree Based Clustering Using Partitional Approach

Damodar Reddy Edla* and Prasanta K. Jana

Department of Computer Science & Engineering
Indian School of Mines, Dhanbad
Jharkhand-826 004, India
dr.reddy.cse@gmail.com, prasantajana@yahoo.com

Abstract. Graph-based clustering techniques have widely been researched in the literature. MST-based clustering is the well known graph-based model in producing the clusters of arbitrary shapes. However, the MST-based clustering methods suffer from high computational complexity (i.e. quadratic). In this paper, we propose a partitional approach not only to speed up the MST-based clustering, but also to identify the outlier points. Initially, a squared error clustering algorithm is used as a pre-processing stage for MST-based clustering. Then the MST-based approach is applied on the representative points (centroids) of the sub-clusters produced by the squared error clustering method. The local outlier factor is used to deal with the outliers. We have performed wide-ranging experiments on several synthetic and real world data sets. The results of the multi-dimensional data are evaluated using the computation time of the algorithms.

Keywords: Clustering, minimum spanning tree, squared error method, local outlier factor.

1 Introduction

Clustering [1] refers to the process of partitioning the homogeneous objects/points into meaningful groups called clusters such that the objects within a cluster are similar to each other whereas the objects of different clusters are dissimilar to each other. The applications of clustering are involved in a wide variety of domains such as image processing [2], geology [3], chemistry [4], economic science [5] and many more. Based on the characteristics of the data and measures of similarity, the clustering algorithms are broadly divided into various types, such as partitonal, hierarchical, density-based and grid-based. Among the above, hierarchical clustering algorithms were extensively researched despite their quadratic computational complexity and a large number of algorithms [6], [7] have developed in this direction. Minimum spanning tree (MST) based clustering [8] is the well known model of this kind that has researched over the decades and various algorithms [9], [10], [11], [12], [13] of

* Corresponding author.

S.C. Satapathy et al. (Eds.): Proc. of Int. Conf. on Front. of Intell. Comput., AISC 199, pp. 237–244.
DOI: 10.1007/978-3-642-35314-7_28

this kind were proposed in the literature. Although, most of these algorithms are efficient in producing the complex clusters, they are vulnerable from the view of the computational complexity. A brief survey of the existing minimum spanning tree based clustering algorithms is as follows.

MST-based clustering has been initiated by Zahn [8] by constructing an MST over the given dataset and then removes the inconsistent (longer) edges to create the connected components. Repeated application of this approach eventually leads to form different clusters that are represented by the connected components. However, the main problem of eliminating the inconsistent edges from the MST remains unsolved. Wang et al. [9] proposed a divide-and-conquer approach for MST-based clustering by using an efficient implementation of the cut and the cycle property of the MST. This algorithm performs better than $O(n^2)$ time. Zhong et al. [10] developed a two-round MST-based clustering technique which is robust to the varied cluster sizes, shapes and densities. It also discovers the number of clusters. However, it is not robust for the outliers. It also runs in $O(n^2)$ time. Chowdhury et al. [11] developed a clustering algorithm based on the minimum spanning tree and Bayes classifier [12]. This method extracts the clusters by finding the valley regions in the feature space. The performance of this scheme is similar to that of a Bayes classifier as the number of objects goes to infinity under a smooth assumption. Here, the proposed method begins with no information about the object classes or clusters of the given data. The proposed method also presents a way of finding the "valley regions" in multivariate histogram. However, the difficulty of reducing the large number of computations to form the clusters in higher dimensions has not been focused in this algorithm. Zhong et al. [13] designed an efficient hierarchical clustering algorithm based on the MST and MST-based graph which have been used for the split-and-merge process.

Motivated with the above methods, we propose here, a novel clustering algorithm to enhance the MST-based clustering and locate the outlier points using the local outlier factor (*LOF*) proposed by Breunig et al. [14].

The rest of the paper is structured as follows. Section 2 provides the necessary terminologies, namely, minimum spanning tree and the local outlier factor (*LOF*). Section 3 presents the proposed MST-based algorithm. Then, the experimental results of synthetic and real world data are illustrated in section 4. Finally, the paper is concluded in section 5 followed by the useful references.

2 Preliminaries

In this section, we provide the concise definitions of the supportive preliminaries, namely, minimum spanning tree and local outlier factor (*LOF*).

2.1 Minimum Spanning Tree

Given $G = (V, E)$ a connected, undirected, weighted graph and a function $w: E \rightarrow R$, that assigns a weight $w(e)$ to each edge e (Assume all the edge weights are real numbers). Then the minimum spanning tree [15] of the graph G is the spanning tree T that minimizing the function w.

$$i.e. \quad w(T) = \sum_{e \in T} w(e) \tag{1}$$

In other words, the minimum spanning tree of a weighted graph G is the minimum-weight spanning tree of that graph where a spanning tree is an acyclic sub-graph of the graph G that contains all the vertices from G. The key feature of MST is that the elimination of inconsistent (usually longer) edges using a threshold value result into several connected components which are known as the clusters. The computational cost of constructing the minimum spanning tree is quadratic. The cluster formation using the MST is illustrated in the Fig. 1(a-b).

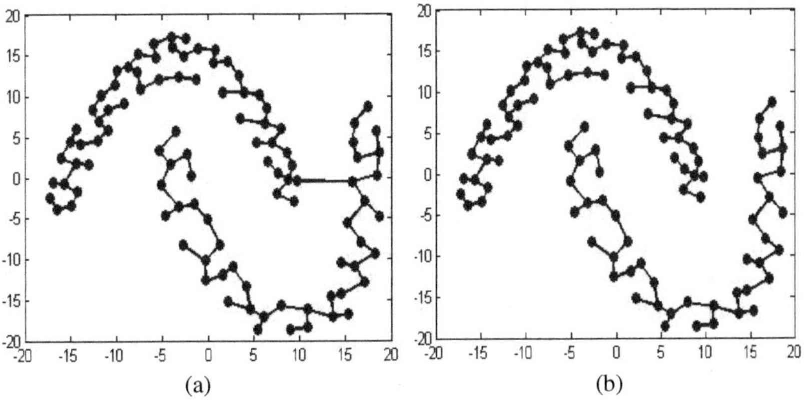

Fig. 1. MST-based clustering. (a) minimum spanning tree of the given data; (b) two clusters after the elimination of a longest edge.

2.2 Local Outlier Factor

There are several outlier identification schemes found in the literature. In this paper, we use the local outlier factor (*LOF*) proposed by Breunig et al. [14]. The *LOF* computes the degree to which a point is an outlier. It is formally defined as follows.

Initially, the *local neighborhood* of a point $x \in S$ with respect to the minimum points threshold mp is defined as follows:

$$N(x, mp) = \{y \in S / d(x, y) \leq d(x, x_{mp})$$

where x_{mp} is the mpth nearest neighbor of x. Thus $N(x, mp)$ contains at least mp points. The density of a point $x \in S$ is defined as follows.

$$density(x, mp) = \left(\frac{|N(x, mp)|}{\sum_{y \in N(x, mp)} d(x, y)} \right) \tag{2}$$

If the distances between x and its neighboring points are small, then the density of x is high. Then the *average relative density* (*ard*) of x is calculated as below.

$$ard\,(x, mp) = \frac{density(x, mp)}{\left(\dfrac{\sum_{y \in N(x,mp)} density(x, y)}{|N(x, mp)|}\right)} \tag{3}$$

Now, the local outlier factor (*LOF*) of x is defined as the inverse of the *average relative density* of x.

$$\text{i.e., } LOF\,(x, mp) = \frac{1}{ard\,(x, mp)} \tag{4}$$

3 Proposed Algorithm

The main idea behind the proposed method is as follows. The algorithm is comprises of two phases. In the phase-I, the squared error clustering algorithm is applied to produce the sub-clusters. Then the outlier points are located from these sub-clusters. In the phase-II, the MST is constructed using the centroids of the sub-clusters which do not hold any outliers. Then, the clusters are formed using the traditional MST-based clustering approach. The above two phases are briefly described as follows.

Initially, the squared error clustering algorithm (for instance, K-means) is applied on the given data of n points to produce K' (>K, the number of clusters) sub-clusters $SC_1, SC_2, \ldots, SC_{K'}$. As K value is unknown, K' value is to be chosen sufficiently larger than K. Here, all the K' sub-clusters are sorted in the non-decreasing order of their cardinalities. At this stage, the outlier points, if any, can be found in the smaller size sub-clusters.

Hence, the sorted sub-clusters are verified for the possibility of holding the outliers using the local outlier factor (*LOF*). The *LOF* value is computed for all the points in the first sub-cluster of the sorted list (i.e. the sub-cluster with lesser number of points). If the sub-cluster holds an outlier, then all the other points of it are also the outliers. Then we proceed to the next sub-cluster of the sorted list and repeat the process. If any sub-cluster in the sorted list does not contain any outlier points, then we stop computing the *LOF*. It is obvious to note that we need not to compute the *LOF* for the points of all the sub-clusters. The sub-clusters which represent the outliers are removed and decrease K' accordingly. Next, the K' centroids $sc_1, sc_2, \ldots, sc_{K'}$ of the sub-clusters $SC_1, SC_2, \ldots, SC_{K'}$ are computed. Now, the minimum spanning tree is constructed for the K' (which is significantly lesser than n) points $sc_1, sc_2, \ldots, sc_{K'}$. The clusters $C_1, C_2, \ldots, C_K$ are produced from the connected components in usual MST-based clustering approach. At the end, the sub-clusters that represent the outliers are also shown. The pseudo code of the algorithm is as follows.

Algorithm MST-Cluster (S, K', μ)

Input: A set S of the given points, number of sub-clusters K', a threshold value μ.
Output: A set of clusters $C_1, C_2, \ldots, C_K$ and outliers $o_1, o_2, \ldots, o_q$.

Functions and variables used:

***SEC* (S, K'):** A function to find the sub-clusters $SC_1, SC_2, \ldots, SC_{K'}$ using any squared error clustering method.
Sort ($SC_1, SC_2, \ldots, SC_{K'}$): A function to sort the sub-clusters $SC_1, SC_2, \ldots, SC_{K'}$ in the non-increasing order of the cardinalities and store them in the array *sorted*.
Centroid (SC): A function to find the centroid of the sub-cluster SC.
S': A set of centroids $sc_1, sc_2, \ldots, sc_{K'}$ of the sub-clusters $SC_1, SC_2, \ldots, SC_{K'}$.
***MST* (S'):** A function, when called, constructs the minimum spanning tree of the set S' of the centroid points of the sub-clusters produced by the function *SEC*.
***LOF* (p):** A function to compute the local outlier factor of the point p.
O: A set of outliers. i, j, l, m: Temporary variables.

Step 1: Call *SEC* (S, K') to produce the sub-clusters $SC_1, SC_2, \ldots, SC_{K'}$ using the squared error clustering algorithm..
Step 2: Call *Sort* ($SC_1, SC_2, \ldots, SC_{K'}$) to sort the sub-clusters $SC_1, SC_2, \ldots, SC_{K'}$ and store them in the array *sorted*.
Step 3: $i \leftarrow 1$;
Step 4: Call *LOF* (p_j) for some $j = 1, \ldots, m$. where $p_j \in sorted[i]$.
Step 5: **If** *LOF* (p_j) ≈ 1 for any point of the sub-cluster $sorted[i]$ **then**

{
Add all the points of $sorted[i]$ to the outlier set O.
i.e. $O \leftarrow O \cup p_j$; $\forall\, j = 1, \ldots, m$.
Remove the sub-cluster $sorted[i]$.
$K' \leftarrow K' - 1$;
$i \leftarrow i + 1$;
Go to step 4 to repeat the same process with the next sub-cluster sorted[i]of the sorted lsit.
}

Step 6: Call *Centroid* (SC_l), $\forall\, l = 1, 2, \ldots, K'$ to compute $sc_1, sc_2, \ldots, sc_{K'}$, the centroids of the K' sub-clusters $SC_1, SC_2, \ldots, SC_{K'}$ and store them in S'.
i.e. $S' = \{ sc_1, sc_2, \ldots, sc_{K'} \}$;
Step 7: Call *MST* (S') to construct the minimum spanning tree of the set of centroids S'.
Step 8: Remove the inconsistent edges whose weights are greater than or equal to μ from the minimum spanning tree to produce the clusters $C_1, C_2, \ldots, C_K$.
Step 9: Output the outlier points of set O along with the clusters $C_1, C_2, \ldots, C_K$.
Step 10: Stop.

Time Complexity: Initially, the squared error clustering algorithm is applied on the given data of n points. This requires linear time as the required number of sub-clusters K' is supplied a priori. The sub-clusters are sorted in non-increasing order of their cardinalities. This is achieved in $O(K' \log K')$ time. Then, the local outlier factor (*LOF*) is computed only for the sub-clusters of outliers which require linear time. The centroids of all the K' sub clusters are computed in $O(K')$ time. Finally, the minimum spanning tree of the K' centroids is constructed in $O(K'^2)$ time where K' is significantly smaller compare to n. Therefore, the overall time complexity of the algorithm is maximum of $\{O(K'^2), O(n)\}$.

4 Experimental Analysis

We have performed a large number of experiments with the proposed algorithm on various synthetic and real world data. The experiments are performed in MATLAB on an Intel Core 2 Duo Processor machine with T9400 chipset, 2.53 GHz CPU and 2 GB RAM running on the platform Microsoft Windows Vista. For the visualization purpose, first we show the results of the proposed algorithm on four synthetic data sets. Then the algorithm is experimented on various multi-dimensional real world data sets taken from UCI machine learning repository [16]. We compare the runtime of the proposed algorithm with few existing MST-based techniques in case of four multi-dimensional real world data sets. The experimental results are as follows.

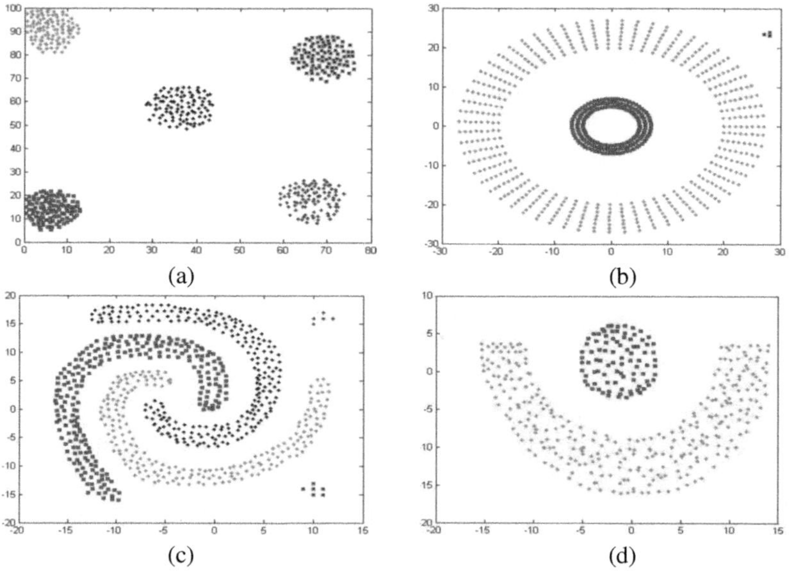

Fig. 2. Results of the proposed algorithm. (a) 5-group data of size 505; (b) cluster-inside-cluster data of size 696; (c) 3-spiral data of size 611; (d) moon data of size 395.

Initially, we have considered four synthetic data sets, namely, 5-group, cluster-inside-cluster, 3-spiral and moon. In case of the 5-group data, the proposed method produced five well separated clusters as shown in the above Fig. 2(a). Then, the proposed method is applied on cluster-inside-data. It has successfully produced two clusters where one cluster lies inside the other one as depicted in the above Fig. 2(b). Similarly, the proposed method results into the desired clusters in case of the other two synthetic data sets as shown in the Figs. 2(c-d). It can also be noted from the Figs. 2(b-c) that the proposed algorithm is able to detect the outlier points.

Next, the proposed algorithm is experimented on four real world data sets [16], namely, iris, pima-India-diabetes, blood transfusion and yeast. The proposed algorithm is compared with the existing graph-based techniques, namely, SFMST [17], MSDR [18], MinClue [19] and SC [20] using the runtime of the algorithms. The Figs. 3(a-d) illustrates that the proposed scheme is consistently faster than the existing graph based clustering algorithms in case of all the four real world data.

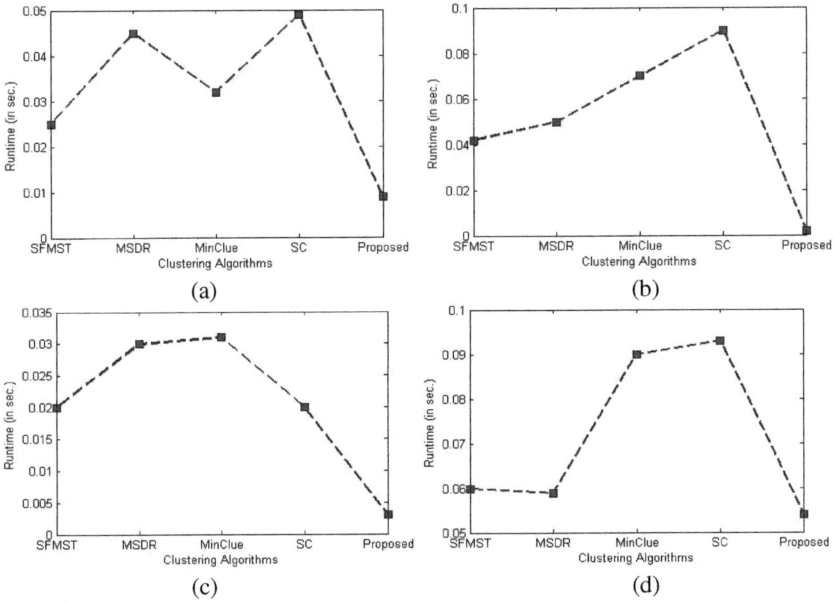

Fig. 3. Runtime comparison of the proposed algorithm with SFMST, MSDR, MinClue and SC for real world data (a) iris; (b) p.i. diabetes; (c) blood transfusion; (d) yeast.

5 Conclusion and Future Work

We have proposed a novel approach to improve the computational cost of the traditional minimum spanning tree based clustering algorithm using the squared error clustering and local outlier factor (*LOF*). The proposed method is able to deal with the outlier points. The experiments carried out on various artificial and real world data shows the efficiency of the proposed scheme over the existing graph-based techniques. Although the proposed algorithm is faster, it obviously has the common threshold problem. In future, we attempt to automate the threshold value.

Acknowledgments. We sincerely thank the Council of Scientific & Industrial Research (CSIR), New Delhi, India, for supporting this work under the grant (No. 25(0177)/ 09/EMR-II).

References

1. Jain, A.K., Dubes, R.C.: Algorithms for Clustering. Prentice Hall, New Jersey (1988)
2. Juang, L., Wu, M.N.: Psoriasis Image Identification using *K*-means Clustering with Morphological Processing. Measurement 44, 895–905 (2011)
3. Parks, J.M.: Cluster Analysis Applied to Multivariate Geologic Problems. The Journal of Geology 74(5), 703–715 (1966)
4. Reyes, S., Nino, A., Munoz-Caro, C.: Customizing Clustering Computing for A Computational Chemistry Environment - The Case of the DBO-83 Nicotinic Analgesic. Molecular Structure: THEOCHEM 727(1-3), 41–48 (2005)
5. Garibaldi, U., Costantini, D., Donadio, S., Viarengo, P.: Herding and Clustering in Economics: The Yule-Zipf-Simon Model. Computational Economics 27, 115–134 (2006)
6. Wang, B., Rahal, I., Dong, A.: Parallel Hierarchical Clustering using Weighted Confidence Affinity. International Journal of Data Mining, Modelling and Management 3(2), 110–129 (2011)
7. Jiang, D., Pei, J., Zhang, A.: DHC: A Density-based Hierarchical Clustering Method for Time Series Gene Expression Data. In: 3rd IEEE Symposium on Bioinformatics and Bioengineering, USA, pp. 1–8 (2003)
8. Zahn, C.T.: Graph-Theoretical Methods for Detecting and Describing Gestalt Clusters. IEEE Transactions on Computers 20(1), 68–86 (1971)
9. Wang, X., Wang, X., Wilkes, D.M.: A Divide-and-Conquer Approach for Minimum Spanning Tree-based Clustering. IEEE Transactions on Knowledge and Data Engineering 21, 945–958 (2009)
10. Zhong, C., Miao, D., Wang, R.: A Graph-Theoretical Clustering Method based on Two Rounds of Minimum Spanning Trees. Pattern Recognition 43, 752–766 (2010)
11. Chowdhury, N., Murthy, C.A.: Minimal Spanning Tree based Clustering Technique: Relationship with Bayes Classifier. Pattern Recognition 30(11), 1919–1929 (1997)
12. Langley, P., Iba, W., Thompson, K.: An Analysis of Bayesian Classifiers. In: 10th National Conference on Artificial Intelligence, California, pp. 223–228 (1992)
13. Zhong, C., Miao, D., Franti, P.: Minimum Spanning Tree based Split-and-Merge: A Hierarchical Clustering Method. Information Sciences 181(16), 3397–3410 (2011)
14. Breunig, M.M., Kriegel, H., Ng, R.T., Sander, J.: LOF: Identifying Density-based Local Outliers. In: ACM SIGMOD International Conference on Management of Data, Dallas, TX, pp. 93–104 (2000)
15. Cormen, T.H., Leiserson, C.E., Rivest, R.L., Stein, C.: Introduction to Algorithms. MIT Press, USA (2001)
16. UCI Machine Learning Repository, `http://archive.ics.uci.edu/ml/dataset`
17. Paivinen, N.: Clustering with A Minimum Spanning Tree of Scale-Free-Like Structure. Pattern Recognition Letters 26, 921–930 (2005)
18. Grygorash, O., Zhou, Y., Jorgensen, Z.: Minimum Spanning Tree-based Clustering Algorithms. In: IEEE International Conference on Tools with Artificial Intelligence, pp. 73–81. IEEE Computer Society, USA (2006)
19. He, Y., Chen, L.: MinClue: A MST-based Clustering Method with Auto-Threshold Detection. In: IEEE International Conference Cybernetics and Intelligent Systems, Singapore, pp. 229–233 (2004)
20. Ng, A.Y., Jordan, M.I., Weiss, Y.: On Spectral Clustering: Analysis and an Algorithm. In: International Conference on Advances in Neural Information Processing Systems, pp. 849–856. MIT Press, USA (2001)

Evaluating the Genuinity of Opinions Using Decision Tree

Heeralal Evanthjain[1], Shanmugasundaram Hariharan[1], Rajasekar Robertsekar[2], and Thirunavukkarasu Ramkumar[2]

[1] TRP Engineering College (SRM Group), Tamilnadu, India
[2] A.V.C College of Engineering, Tamilnadu, India
{h.evanthjain,robertrajsekar,mailtos.hariharan}@gmail.com
ramooad@yahoo.com

Abstract. Due to the rapid growth of internet technologies and communication networks, product vendors and companies are identifying web as a platform for launching and promoting their products globally. The user communities of web utilize these forums as a tool for posting and delivering their opinion about products. In such circumstances, techniques based on opinion mining play significant roles which include, understanding the opinions posted by the users, analyzing user reviews, extracting useful patterns, assisting decision making process etc. This paper analyses the genuinity of opinions expressed by web users.

Keywords: Opinion, product reviews, genuinity, customers, decision tree.

1 Introduction

Opinion mining, which is also called sentiment analysis, involves building a system to collect and examine opinions about the product made in blog posts, comments, reviews or tweets [2]. The importance of opinion mining can be revealed in various perspectives[5,6]. In a marketing manager perspective, the opinion mining assists in concluding the success of new product launch by identifying which versions of a product get popularity earlier and even identify whether the product features are liked or disliked based on the demographical location. For example, a review might be broadly positive about a digital camera, but be specifically negative about how heavy it is. Being able to identify this kind of information in a systematic way gives the vendor a much clearer picture about public opinion than surveys, because the data is created by the customer. In buyer's perspective, opinion mining enables the new user in purchasing a product based on the opinion expressed by various users in social web forums [1].

Identifying the genuineness of the opinions and forming candidate reviews are important stages in mining process [7,8]. Instead of considering these opinions posted

S.C. Satapathy et al. (Eds.): Proc. of Int. Conf. on Front. of Intell. Comput., AISC 199, pp. 245–252.
DOI: 10.1007/978-3-642-35314-7_29

by all the users, a model for identifying the trustiness of opinion posted by the potential user would be highly desirable and an effective one. This paper adopts a classification model based on decision tree for classifying the effective opinions based on the user characteristics. Hence the problem of considering irrelevant opinions is eliminated and while calculating the product recommendation, trusted opinions alone are accounted.

The paper is organized as follows. Section 1 presented some basic information on opinion mining. Work pertaining to the opinion mining is presented in section 2. Proposed work is discussed in section 3 and section 4 concludes the work.

2 Previous Work

Internet has brought a major drift in user community. Customers online sell products through web forums. Such environments are not supported by human sales experts, sometimes. In such situations recommender applications help to identify the products and/or services that fits the user needs. In order to successfully apply recommendation technologies we have to develop an in-depth understanding of decision strategies of users. These decision strategies are explained in different models of human decision making [3].

Research on such social networking has advanced significantly in recent years which have been highly influenced by the online social websites. A product review by the user is a more accurate representation of its real-world performance and web-forums are generally used to post such reviews. Though commercial review websites allow users to express their opinions in the way they feel, the number of reviews that a product receives could be very high[9]. Opinion mining techniques can be used to analyze the user-reviews, classify the content as positive or negative, and thereby find out how the product fares. There exist solutions to provide recommendation to products available on the web by analyzing the context to score the sentences for each review by identifying the opinion and feature words [4].

Comparing consumer opinion concerning one's own products and those of the competitors to find their strengths and weaknesses is a crucial activity for marketing specialists in the production industry to overcome the requirements of marketing intelligence and product benchmarking. The proposed architecture [5] includes a wide variety of state-of-the-art text mining and natural language processing techniques. Furthermore, the key elements of applications for mining large volumes of textual data for marketing intelligence are reasoned: a suite of powerful mining, visualization technologies and an interactive analysis environment that allows for rapid generation and testing of hypothesis. The concluding results show that recent technologies look promising, but are still far from a semantically correct textual understanding.

3 Proposed System

The architecture of the proposed system is shown in the following figure. In the proposed system, users are categorized as administrator and web users. The following roles are played by the web users.(i) able to view the product information such as model, image, price and version if any.(ii) able to view the review posted by customer who bought the product.(iii) able to view the recommendation chart for the purchasing of product.(iv) able to post opinion along with the required information such as age, gender, location, qualification and income for the purpose of classification. The administrator can able to perform the following roles.(i) designing and updating the product catalogue.(ii) identifying the splitting attribute, decision tree can be constructed on the basis of the chosen attribute and novel rules are extracted for forming opinion genuineness (iii) pre processing of genuineness opinion, mining and product recommendation are also performed as roles by the administrator.

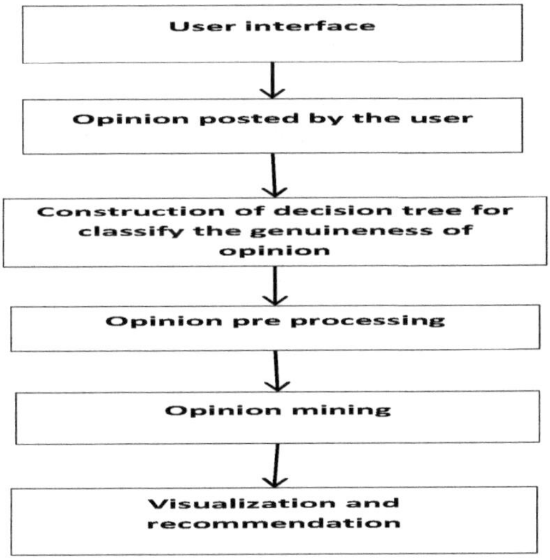

Fig. 1. Proposed System Architecture

The following example shows the attributes of customers who are posted opinion for a digital camera through a social web forum. Information pertaining to attributes age, income, qualification and location of the users are extracted for classifying the genuineness of opinion. Attribute which is having higher information gain or entropy measure has been designated as splitting attribute. The splitting attribute minimizes the information needed to classify the samples in the resulting partitions and reflects the least randomness or impurity among partition. Attributes such as product id, customer id are excluded in the classification task since they are used only for the purpose of product and customer identification. The attribute gender is also neglected because the consumer may belong to both genders. The classification label genuineness is having two values. (i).Genuineness = "Yes", represents the opinions of potential customers whose opinions are accounted for the mining task. (ii).The class label Genuineness ="No" consolidates the opinions not a worthy one.

The expected information needed to classify a given opinion posted by the user is calculated using expression 1.

$$info(D) = -\sum_{i=1}^{m} p_i \, log(p_i) \quad (1)$$

where p_i is the probability that an arbitrary opinion in dataset D belongs to class Genuineness ="Yes" or Genuineness ="No". A log function is used, because the information is encoded in bits. Info(D) is just the average amount of information needed to identify the class label of a opinion in dataset D.

Table 1. Data For Example

AGE	INCOME	QUALIFY	LOCATION	Genuineness
Old	High	DEGREE	METRO	Yes
Old	High	HSE/DIPLAMO	URBAN	Yes
Middle age	High	HSE/DIPLAMO	METRO	Yes
Youth	High	SSLC	METRO	Yes
Old	Low	DEGREE	METRO	Yes
Youth	High	DEGREE	METRO	No
Middle age	Medium	SSLC	URBAN	Yes
Youth	Medium	HSE/DIPLAMO	METRO	No
Youth	Medium	SSLC	METRO	No
Old	Medium	DEGREE	METRO	Yes
Old	Medium	DEGREE	RURAL	Yes
Old	High	SSLC	RURAL	No
Youth	Low	HSE/DIPLAMO	METRO	Yes
Youth	Medium	DEGREE	RURAL	No
Middle age	Low	SSLC	METRO	Yes
Middle age	Medium	DEGREE	RURAL	Yes
Middle age	High	HSE/DIPLAMO	METRO	Yes
Middle age	High	SSLC	URBAN	No
Middle age	High	DEGREE	URBAN	Yes
Middle age	Low	HSE/DIPLAMO	RURAL	No

Each attribute has *v* distinct values, for an example, Attribute "age" can have three distinct values such as {"age = Old" , "age = Middle age" and "age = Youth"} observed from data set D. Such distinct values of attributes are used to split Dataset D into *v* partitions or subsets{ $D_1, D_2, \ldots, D_v$} (see expression 2).

$$info_A(D) = \sum_{j=1}^{m} \frac{|Dj|}{|D|} \times info(Dj) \quad (2)$$

The term $\frac{|Dj|}{|D|}$ act as the weight of the jth partition. $Info_A(D)$ is the expected information required to classify a opinion from Dataset 'D' based on the partitioning by Attribute A. Info(Dj) is the is just the average amount of information needed to identify the class label of a opinion in dataset D based on the distinct values of that attribute A.

Information gain is defined as the difference between the original information requirement (i.e., based on just the proportion of classes) and the new requirement (i.e., obtained after portioning on A) using expression 3.

$$Gain(A) = info(D) - info_A(D) \tag{3}$$

Using the above three expressions, entropy value of information needed to identify the class label of a opinion for entire dataset 'D', entropy values for the attributes based on their distinct values and Information Gain are calculated as follows:

Info(D) = - (13/20) log (13/20) – (7/20) log(7/20) = 0.2811

Info(age) = ((6/20)(-(5/6)log(5/6) – (1/6) log(1/6)) +
(8/20)(-(6/8)log(6/8) –(2/8)log(2/8)) +
(6/20)(-(2/6)log(2/6) –(4/6)log(4/6)))
= 0.2392

Info(location) = ((11/20)(-(8/11)log(8/11) – (3/11) log(3/11)) +
(4/20)(-(3/4)log(3/4) – (1/4) log(1/4)) +
(5/20)(-(2/5)log(2/5) – (3/5) log(3/5)))
= 0.2618

Info(qualify) = ((8/20)(-(6/8)log(6/8) – (2/8) log(2/8)) +
(6/20)(-(4/6)log(4/6) –(2/6)log(2/6)) +
(6/20)(-(3/6)log(3/6) –(3/6)log(3/6)))
= 0.2708

Info(income) = ((9/20)(-(6/9)log(6/9) – (3/9) log(3/9)) +
(7/20)(-(4/7)log(4/7) –(3/7)log(3/7)) +
(4/20)(-(3/4)log(3/4) –(1/4)log(1/4)))
= 0.2769

Gain(age) = 0.2811 – 0.2392 = 0.0419
Gain(location) = 0.2811 - 0.2618 = 0.0193
Gain(qualify) = 0.2811-0.2708 = 0.0103
Gain(income) = 0.2811-0.2769= 0.0042

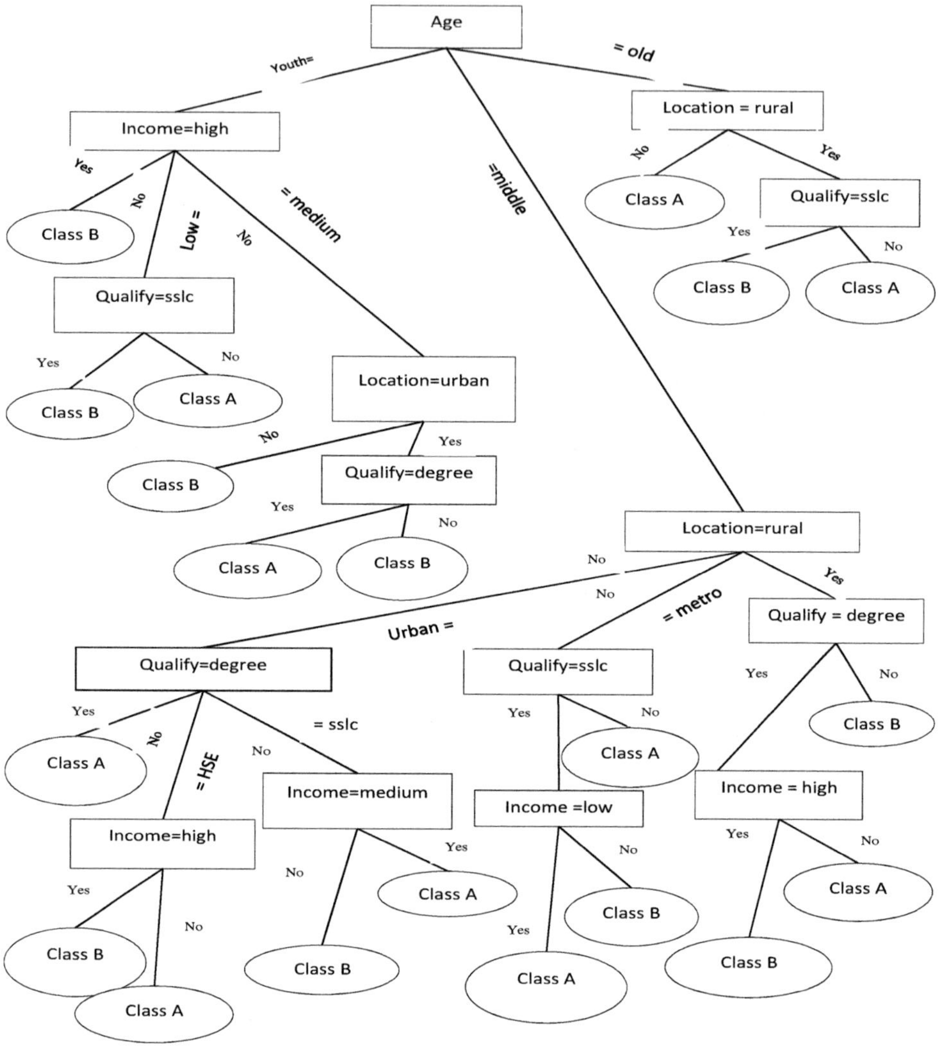

Fig. 2. Decision tree on the basis of splitting attribute 'age'

IF Age = Old AND Location != Rural THEN Opinion_genuineness = Yes

IF Age = Old AND Location = Rural AND Qualify != sslc THEN Opinion_genuineness = Yes

IF Age = Old AND Location = Rural AND Qualify= sslc THEN Opinion_genuineness = No

IF Age = Youth AND Income = High THEN Opinion_ genuineness = No

IF Age = Youth AND Income = Low AND Qualify = sslc THEN Opinion_genuineness = No

IF Age = Youth AND Income = Low AND Qualify ! = sslc THEN Opinion_genuineness = Yes

IF Age = Youth AND Income = Medium AND Location != urban THEN Opinion_genuineness = No

IF Age = Youth AND Income = Medium AND Location = urban AND Qualify = Degree THEN Opinion_genuineness = Yes

IF Age = Youth AND Income = Medium AND Location = urban AND Qualify != Degree THEN Opinion_genuineness = No

If Age = MiddleAged AND Location = Rural AND Qualify != Degree THEN Opinion_genuineness = No

If Age = MiddleAged AND Location = Rural AND Qualify = Degree AND Income = High THEN Opinion_genuineness = No

If Age = MiddleAged AND Location = Rural AND Qualify = Degree AND Income != High THEN Opinion_genuineness = Yes

If Age = MiddleAged AND Location = Metro AND Qualify != sslc THEN Opinion_genuineness = Yes

If Age = MiddleAged AND Location = Metro AND Qualify = sslc AND Income = Low THEN Opinion_genuineness = Yes

If Age = MiddleAged AND Location = Metro AND Qualify = sslc AND Income != Low THEN Opinion_genuineness = No

If Age = MiddleAged AND Location = Urban AND Qualify = Degree THEN Opinion_genuineness = yes

If Age = MiddleAged AND Location = Urban AND Qualify = Hsc AND Income != High THEN Opinion_genuineness = Yes

If Age = MiddleAged AND Location = Urban AND Qualify = Hsc AND Income = High THEN Opinion_genuineness = No

If Age = MiddleAged AND Location = Urban AND Qualify = sslc AND Income = Medium THEN Opinion_genuineness = Yes

If Age = MiddleAged AND Location = Urban AND Qualify = sslc AND Income != Medium THEN Opinion_genuineness = No

Fig. 3. Extracting rules from decision tree

From the above calculation, age has the highest gain value. Therefore age is taken as a splitting attribute (i.e.) root node for the decision tree (refer Fig 2). The knowledge represented in decision trees can be extracted and represented in the form of classification IF-THEN rules. One rule is created for each path from the root to a leaf node. Each attribute-value pair along a given path forms a conjunction in the rule antecedent ("IF" part). The leaf node holds the class prediction, forming the rule consequent ("THEN" part). The IF-THEN rules may be easier for humans to understand, particularly if the given tree is very large.

4 Conclusion

This paper adopts a classification model based on decision tree for classifying the effective opinions. Through the proposed approach, irrelevant opinions could be eliminated. The proposed architecture intends to calculate the product recommendation by taking into consideration only the trusted opinions.

Acknowledgments. The authors would thank the anonymous reviewers for providing valuable suggestions. The authors would like to thank the competent authorities of both the institutions for providing us valuable support to carry out the work successfully.

References

1. Prabowo, R., Thelwall, M.: Sentiment Analysis: A Combined Approach. J. Informetrics 3(2), 143–157 (2009)
2. Melville, P., Gryc, W., Richard, D., Lawrence, S.: Sentiment analysis of blogs by combining lexical knowledge with text classification. In: Proc. 15th ACM SIGKDD International Conference on Knowledge Discovery and Data Mining (2009)
3. Mandl, M., Felfernig, A., Teppan, E., Schubert, M.: Consumer decision making in knowledge-based recommendation. J. Intell. Inf. Syst. 37, 1–22 (2011)
4. Hariharan, S., Ramkumar, T.: Mining Product Reviews in Web Forums. J. Information Retrieval and Research 1(2), 1–17 (2012)
5. Auinger, A., Fischer, M.: Mining consumers' opinions on the web, http://research.fh-ooe.at/files/publications/884_08_Opinion_Mining.pdf
6. Hu, M., Liu, B.: Mining and summarizing customer reviews. In: Proc. Tenth ACM SIGKDD International Conference on Knowledge Discovery and Data Mining (2004)
7. Sun, J., Long, C., Zhu, X., Huang, M.: Mining Reviews for Product Comparison and Recommendation. J. on Computer Science and Computer Engineering with Applications (39), 33–40 (2009)
8. Zhan, J., Loh, H.T., Liu, Y.: Gather customer concerns from online product reviews - A text summarization approach. Expert Systems with Applications: An International Journal 36(2), 2107–2115 (2009)
9. http://www.cs.cornell.edu/People/pabo/movie-review-data/

Comparative Study of Artificial Emotional Intelligence Structuring Social Agent Behavior Based on Graph Coloring Problem

A. Kavitha and S. Sandhya Snehasree

Anil Neerukonda Institute of Technology and Sciences,
Sangivalasa, Bheemunipatnam Mandal, Visakhapatnam Dist.

Abstract. Building Software Models based on social agent behavior is very non-static and highly complicated. There have been various architectures proposed to model computer software to collaborate these multi-agent models. The commutation frameworks are efficiently managed by the most emerging technique reputation models. This paper is being proposed to formulate a social agent behavior based on three factors - ethical constraints, stress avoidance, and emotional interference. The framed scenario is applied to solve the Graph Coloring Problem. The agents make decisions based on the information obtained from environment. An agent applies reactive rules to move from Artificial Emotional Intelligence under Ethical Constraints in Formulating Social Agent Behavior - one color lattice to another and to satisfy as many violated constraints as possible and finally reach a zero position where it has no violated constraints at all. There four different reactive behaviors were specified.

Keywords: Multi-agent based systems, social impulses, and cultural algorithm, Graph Coloring Problem.

1 Introduction

The graph coloring problem is a class of Constraint Satisfaction Problems (CSP) which could be applied to practical applications such as scheduling, time tabling, frequency allocation, map coloring, pattern matching, satellite range scheduling, and analysis of networks. The Graph Coloring Problem is an NP-hard problem. For coordinating the problem solving process the Multi-agent systems can be leveraged. We study this in the concept of Graph Coloring Problem. The Multi-agent Graph Coloring Problem model is based on the research work of [1] which presents a formalization of GCP as a Multi-Agent problem. In Artificial Intelligence research, agent-based system technology has become a center of attraction as a new paradigm for conceptualizing, designing, and implementing software systems. Agents are said to be sophisticated computer programs, that act in open and distributed environments, in order to solve a number of complex problems by self-governing on behalf of their users. Increasingly however, applications require multiple agents that can work

S.C. Satapathy et al. (Eds.): Proc. of Int. Conf. on Front. of Intell. Comput., AISC 199, pp. 253–263.
DOI: 10.1007/978-3-642-35314-7_30

together. A Multi-Agent System (MAS) provides interaction between the agents to solve the problems which are beyond the capacity of an individual knowledge, when individual problem solvers are considered. The Multi-agent system is completely loosely coupled network.

The main goal of Multi-agent systems is to provide methods that allow us to construct complex systems composed of self-governing agents who, while operating on local knowledge composing of only limited abilities, are nonetheless capable of passing a law for the desired global behaviors. Multi-agent system, as typified by ant colonies, the Economy and the Immune system, aims at reverse engineering emergent phenomena. In [1] the authors have introduced an ERA framework (Environment Reactive Rules and Agents) for solving GCP problems. In GCP, the edges are divided into two different constraints, namely local constraints and neighbor constraints or global constraints. The unsatisfied neighbor constraints are to be minimized to solve the GCP problem. When the problem of local minimum occurs then random move is preferred, the agent can also compromise to overcome from the local minimum. Agent compromises when two neighbors have the same violated neighbor constraints. In this case, the corresponding two agents will act as a single agent for making their decisions.

The main aim of a graph coloring problem is to take or assume a limited set of predefined colors, and assign color to the vertices such that no two adjacent vertexes have the same color. The GCP serves as a theoretical basis for many Constraint Satisfaction Problems (CSP) including but not limited to many social modeling problems.

The remaining paper is so organized that section II spares light on the prior research proposals and existing view of points. Section III explains the formulation of the considered problem and approach of the experiment and its setup paving way to understand the results. This is supported by the next section IV exhibiting the conducted experimental results and then we conclude.

2 Literature Survey

There are many resolutions for this problem. Some of them are greedy constructive approaches, local search algorithms, such as Tabu Search or Simulated Annealing. Evolutionary methods are also used such as Genetic Algorithms, Scatter Search, Ant Colony Optimization systems and hybrid Strategies. The authors [1] have developed an ERA framework (Environment, Reactive rules and Agents). Based on the information obtained from environment the agents are able to make decisions. An agent applies some reactive rules to move from one color lattice to another and to minimize the number of violated constraints as possible and finally reach a zero position where it has no violated constraints at all. The favorite tools of logicians and mathematicians, such as first order logic, are based on abstract principles of Ethical reasoning which are not readily applicable. Throughout intellectual history, philosophers have proposed many theoretical frameworks, such as Aristotelian virtue theory [23], the ethics of respect for persons [24], act utilitarianism [25], utilitarianism [26], and prima facie duties [27] and no universal agreement exists on which ethical theory or approach is the best.

In multi-agent based evolutionary models the agents 'are capable to make better decisions which play an important role in the convergence of the whole model. Whatever the decision may be, whether better or worse, is mostly domain specific and many Artificial Intelligence (AI) algorithms has been put together or been invented over the last few decades. Case Based Reasoning (CBR), Reinforced Learning (RL), Artificial Neural Networks (ANN) are few to name. Evolutionary Algorithms (EA) like Genetic Algorithms (GA), Ant-Colony Optimization (ACO), Particle Swarm Optimization (PSO) , Cultural Algorithm (CA) [20] all occupy the attention of their own fitness and objective functions for measuring the quality of their corresponding population. Several Evolutionary Algorithms have been incorporated CBR, RL, ANN and hybrid mechanisms as their choice engine. In CBR methodology the agents or the population make use of the old patterns encountered in the environment for decision making according to need, required for future generations.

CBR methods in ACO, in which the agents sustain a case base of their own and again reuses the same case in future which may be helpful in the process of decision making. It is considered to be an interesting example of hybridization of evolutionary algorithm where ACO and PSO were engaged in tandem to derive the topology and weight distribution of ANN. The list of combinations and hybridization of AI algorithms into Evolutionary models is enormous.

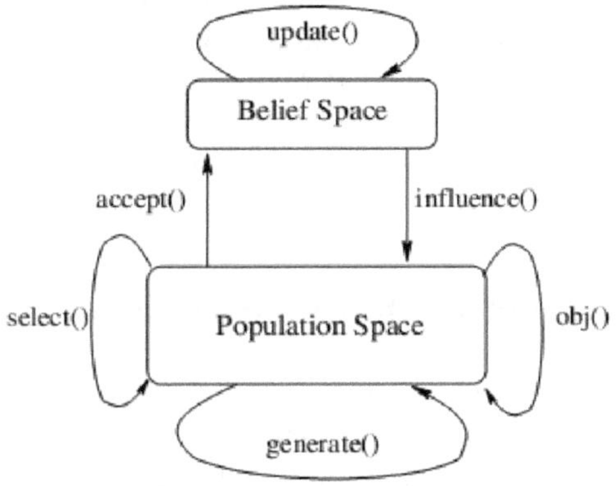

Fig. 1. Basic framework of cultural algorithm

The figure 1 illustrates the basic framework of a cultural algorithm. Cultural Algorithms proposed by Reynolds in 1994 are a class of models derived from the cultural evolution process. Population Space and Belief space are said to be the two spaces in Cultural Algorithms. Firstly, the individuals in the population space are been evaluated by using an action obj(). Then the Acceptance functions accept () will determines which individuals has to impact the Belief space. The update () is then used to update the experiences of those chosen elites of the Belief space. This represents the development of beliefs. Next, these beliefs are used to affect the evolution of the population. New individuals are generated under the control of the beliefs.

The two feedback paths of information, one through the accept ( ) and influence ( ) functions, and the other through individual experience and the obj( ) function creates a system of twofold inheritance of both population and belief. The population component and the belief space communicate with each other and hold each other, in a manner analogous to the evolution of human culture. Cultural Algorithms operate at two levels. (1) a micro-evolutionary level, that has a genetic material that an offspring inherits from its parents, and (2) a macro-evolutionary level, which deals with the knowledge acquired by the individuals through generations. This knowledge is helpful to guide the behavior of the individuals that belong to a particular or a certain population. Cultural Algorithm is basically a global optimization technique which consists of an evolutionary population space whose experiences are integrated into a Belief Space which influences the search process to converge the problem in a direct way. Cultural algorithm was applied to guide the social models of [21] and [22]. Multi-population multi objective cultural algorithms, information are exchanged among sub-populations by individuals. A cultural algorithm (CA) utilizes a set of knowledge sources, each related to knowledge observed in various animal species. These knowledge sources are then united to express the decisions of the individual agents in solving optimization problems. The use of CA frameworks allows for sharing the knowledge of each individual agent community with the overall population and also permits the global knowledge to determine the fruition of that population. The three social impulses namely Stress, Emotion and Ethics are been derived from the cultural algorithm. The Multi agent simulation is been embedded in a cultural algorithm for the purpose to implant social interaction and cultural learning into the system and see how various socio-economic factors change the social agent, behavior.

3 The Graph Problem Model Formulation

Formally we present our model as a tuple. G(V,E) is the graph where V is the set of vertexes and $E \subset V \times V$ is the set of Edges. The problem is to find a Mapping M: $V \rightarrow S$ such that M (u) $\neq$ M(v) if (u ,v) $\in$ E. In this context a set of violated constraints B can be defined so that B = {(u, v) | (u, v) $\in$ E and M (u) =M (v)}. S= {1, 2…k}$\subset$N is a set of available colors, where N is the set of all natural numbers. With respect to B the problem can be now redefined as to find the Mapping M that reduces B to Ø, an empty set. The goal of a graph coloring problem is to allocate color to the vertexes of a graph from a restricted set of predefined colors, such that no two flanking vertexes have the similar color.

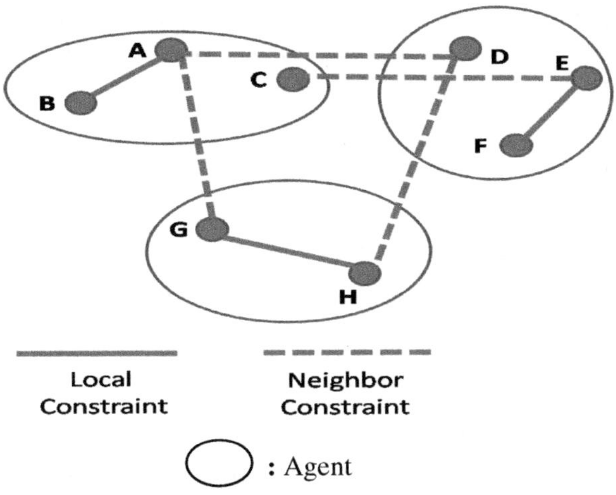

Fig. 2. Model Architecture

Agent: The vertex set in the graph is divided into partitions with 3 vertices, each of the partition is said to be an agent. **Neighboring Constraint:** If (edge is represented as e) $e_n = (x, w) \in B$ and $x \in a_1$ and $w \in a_2$ where a_1 and a_2 are two different agents, e_n is defined as the neighbor constraints. **Local Constraints:** For an agent a={x, y, z} if $e_1 = (x, y) \in B$ then e_1 is said to be the local constraint for the agent a.

Agents: The vertex set (V) of the graph is divided into partitions of 3 (or less) vertexes each. Each of this partition forms an Agent. Therefore a graph with 191 vertexes will have 63 agents with 3 vertex child components and 1 agent with only 2 components. (as shown in the figure)

Constraints: Since Graph Coloring Problem is a CSP problem the constraints must be mentioned and represented in the agents' environment. In this context the Constraints are defined in terms of the edges linking the vertexes. If the edge e connects the vertexes v1 and v2 then it is given as e = (v1, v2) is a constraint between v1 and v2 and the satisfaction of e depends on the color assignment of v1 and v2.

Least Move: At any given state (i.e. color lattice) an agent may have certain number of violated local constraint and neighbor constraint. At this point if the agent decides to make a least move then it moves to another lattice where it has the minimum possible violated neighbor constraints and no violated local constraint.

Better Move: In this reactive move the agent tries to move to a lattice where the number of violated local constraint is zero and the number of violated neighbor constraint is lower than what it had before the move.

Random Move: Without considering any possibility of violations the agent has to make a move randomly. This move is said to be a trivial move where there may be or may not be a chance to minimize the number of violations.

Compromise Move: In this move two agents are grouped together and create a new agent called Compromise Agent. It is similar to that of a regular agent but the number of vertices is more in Compromise move. The number of moves made is less in order to reduce the violations.

By keeping in mind all these moves it is essential to find out an effective ratio for all these moves to optimize the problem. In [1] the authors had made a ratio to yield optimal efficiency. The ratio is

Random: Compromise: Least: Better=5:100:950:950

Reference: 978-1-4244-8126-2/10 IEEE.

The concepts of social impulses have been applied to the multi-agent GCP problem by introducing the notion of society. The social impulses that the agents possess may affect the agent's judgement either negatively or positively. The three impulses considered are Stress, Ethics and Emotion. These words have some special significance in this context.

Stress: The number of violated constraints that agents have is the measure of stress.

Ethics: The willingness of an agent to abide by standard rules or social trends. Agents stress condition may have an impact on its ethics level.

Emotion: The agent's willingness to make a move that may or may not be socially beneficial. This particular social impulse is dependent on its peers.

4 The Experimental Results

The objective of this research is to account for the impact of agent's social behavior on the convergence of Social Models. The particular tables are chosen from [4]

Table 1. Graphs used from Stanford graph base

Graph	Vertex Count	Edge Count	Colour Number
Anna	138	493	11
Jean	80	54	10
miles250	128	387	8
queen6_6	36	290	7
myciel6	95	755	7
myciel7	191	2360	7

The communication model for various agents is represented in different graph topologies with different internetworking. For evaluating how miscellaneous agent orientation affects the convergence of this social impulse based model we have devised a test framework.

The conducted experiments are being depicted with the below representations. We have achieved a good result for the following datasets when two of the social impulses Emotion, Ethics are combined, rather when compared to Ethics individually.

Table 2. Combination of Emotion and Ethics

Dataset	**Anna**	**Queen6-6**	**Myciel7**
Vertex count	138	36	191
Initial iterations	56	63	377
Later iterations	16	25	249

By the above data, it is observed that the combination of Emotion and Ethics has performed better in which the violations have been reduced so far. The variation is been represented in graph1. Here there is a high variation between vertices and edges and a minimum number of colors are considered [table1]. For such a combinations Emotion and Ethics are doing better.

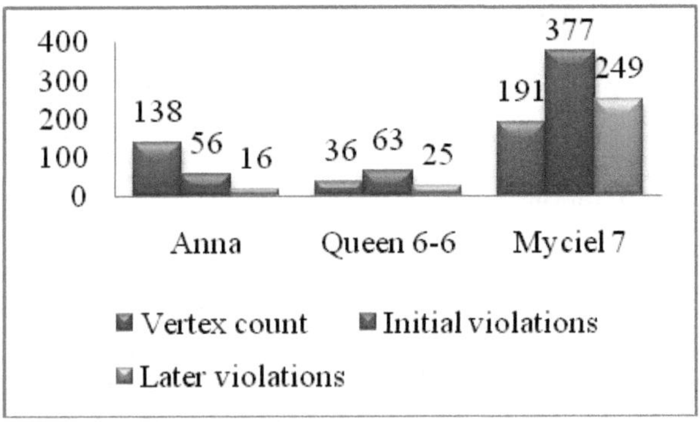

Graph 1. Representing the combination of Emotion and Ethics

In table3 we have represented the data only when Ethics is considered. For the datasets Miles250 and Myciel6 the result was good considering only Ethics.

Table 3. Only when Ethics is considered

Dataset	**Miles250**	**Myciel6**
Edge count	387	755
Initial iterations	64	121
Later iterations	22	46

For the above data a graph was been drawn which concludes that there are more reduced violations for the datasets Miles250 and Myciel6 when only Ethics is considered. When this is data is taken into account we can observe that the vertices are consistent, edges are more and limited set of colors are chosen to obtain the optimal solution [table1]. In such cases Ethics is performing better. The variations are shown in the graph below:

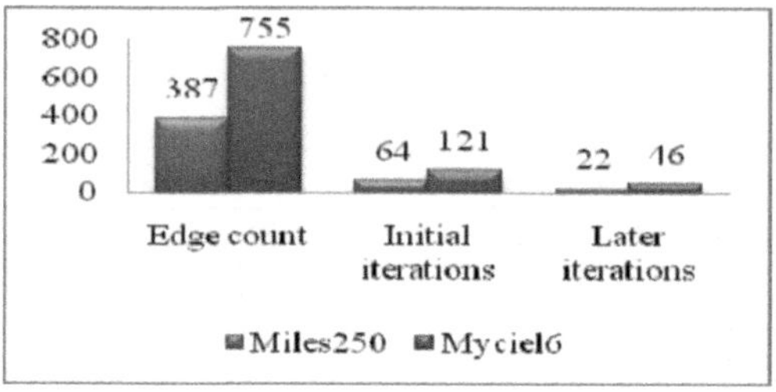

Graph 2. Representation of Ethics

In table4 the combination of Stress and Emotion gives better result when compared to any other combination of the three social impulses for the dataset Jean. The number violations have been reduced and it is represented in the below table.

Table 4. Combination of Stress and Emotion

Dataset	Jean
Color	10
Initial iterations	04
Later iterations	00

Based on the above data a graph is plotted in which we can easily identify the reduced violations. It is observed that after some iteration there are no violations at all. In this case [table1] it is observed that the vertices are less, edges are minimum and the number of colors are more. When compared with the number of vertices and edges, the colors assigned for them are more. It is shown in the graph below:

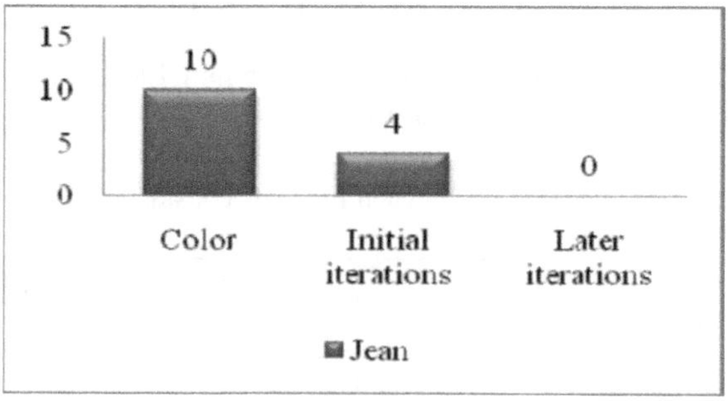

Graph 3. Representing the combination of Stress and Emotion

In table4 the combination of Stress and Emotion gives better result when compared to any other combination of the three social impulses

5 The Future Work and Conclusion

With varying edge count, ethics give a better result whereas with varying vertex count combination of ethics and emotion gives a better result. In this paper we introduced the notion of using social impulses in social models. We formalized and defined three such impulse factors, stress, emotion and ethics and observed their combined effect on a multi-agent based graph coloring algorithm.

In this paper we have done a comparative study on the three social impulses namely Ethics, Emotion and Stress. When the three factors are considered individually, Ethics is the only factor which is achieving a better result when compared to Stress and Emotion. Next, we have focused on the different combinations of these factors such as Stress and Ethics, Stress and Emotion, Emotion and Ethics and also the combination of all the three factors Stress, Ethics and Emotion. It has been found that the combination of these social impulses produces much better results when compared to the social impulse Ethics.

Though the conclusion implies ethics should be capitalized for better convergence, it is important to realize this is not always the case in social modeling. For example, for the dataset Queen6-6 Ethics has achieved a very poor result whereas the same problem is been minimized by using the combination of Emotion and Ethics. Many other graph instances with interesting network topology should also be investigated.

The combination of Emotion and Ethics as well, gives an efficient result for the queen6-6 problem. Intelligent social agents are not always driven by ethics, but also they are influenced by their desire, social peer pressure and other objectives, which we collectively termed as emotion, that may be not relevant or yet worse hazardous to their real objective. The social impulse that triggers the agent's emotions is termed as stress. An agent not emotionally sound may be in a stressed condition and makes wrong move irrespective of its current situation. An emotionally sound agent on the other-hand is expected to make moves and thus converge quickly. The intention of a social model can thus be characterized as to minimize the overall stress level of its agents, which in turn would allow them to reach their objectives quickly by making more ethical moves. In case, if still there is a chance to minimize we then go for other social impulses Emotion, Stress or the combination of all these factors.. At this point we only used the 3 social impulses and defined them from a very basic and conventional manner. Other social impulses like peer-pressure and better formalization and characterization of the social terminologies are necessary to better understand the human social behavioral factors in social modeling.

We further investigated how the social network topology may impact the convergence of the problem. We had done a comparative study on the three social factors. Combinations of these different social impulses produce better results instead when each of them is considered individually.

References

[1] Tang, Y., Liu, J., Jin, X.: Agent Compromises in Distributed Problem Solving. In: Liu, J., Cheung, Y.-M., Yin, H. (eds.) IDEAL 2003. LNCS, vol. 2690, pp. 35–42. Springer, Heidelberg (2003)

[2] Kobti, Z., Rahaman, S., Snowdon, A.W., Kent, R.D.: A Reputation Model Framework for Artificial Societies: A Case Study in Child Vehicle Safety Simulation. In: Bergler, S. (ed.) Canadian AI. LNCS (LNAI), vol. 5032, pp. 185–190. Springer, Heidelberg (2008)

[3] Kobti, Z., Rahaman, S., Snowdon, A.W., Kent, R.D.: A cultural algorithm to guide driver learning in applying child vehicle safety restraint. In: IEEE World Congress on Computational Intelligence, Vancouver, BC, Canada, July 16-21, pp. 1111–1118 (2006)

[4] `http://mat.gsia.cmu.edu/COLOR/instances.html`

[5] Brelaz, D.: New méthods to color vertices of a graph. Communications of ACM 22, 251–256 (1979)

[6] Hertz, A., De Werra, D.: Using Tabu search techniques for graph coloring. Computing 39, 345–351 (1987)

[7] Chams, M., Hertz, A., De Werra, D.: Some experiments with simulated annealing for coloring graphs. EJOR 32, 260–266 (1987)

[8] Fleurent, C., Ferland, J.A.: Genetic and hybrid algorithms for graph coloring. In: Laporte, D.G., Osman, I.H. (eds.) Metaheuristics in Combinatorial Optimization, Annals of Operations Research, vol. 63, pp. 437–441 (1996)

[9] Hamiez, J.-P., Hao, J.-K.: Scatter Search for Graph Coloring. In: Collet, P., Fonlupt, C., Hao, J.-K., Lutton, E., Schoenauer, M. (eds.) EA 2001. LNCS, vol. 2310, pp. 168–179. Springer, Heidelberg (2002)

[10] Dorigo, M., Maniezzo, V., Colorani, A.: The ant system: An Autocatalytic optimizing process. Technical Report 91-016 Revised

[11] Galinier, P.: Hybrid Evolutionary Algorithms for graph coloring. J. Combin. Optim. 3(4), 379–397 (1999)

[12] Dorne, R., Hao, J.K.: A New Genetic Local Search Algorithm for Graph Coloring. In: Eiben, A.E., Bäck, T., Schoenauer, M., Schwefel, H.-P. (eds.) PPSN 1998. LNCS, vol. 1498, pp. 745–754. Springer, Heidelberg (1998)

[13] Aamodt, A., Plaza, E.: Case-Based Reasoning: Foundational Issues, Methodological Variations, and System Approaches. Artificial Intelligence Communications 7(1), 39–52 (1994)

[14] Aamodt, A.: Towards robust expert systems that learn from experience - an architectural framework. In: The 3rd Workshop on European Knowledge Acquisition for Knowledge-Based Systems, Paris, pp. 311–326 (July 1989)

[15] Sadeghi, Z., Teshnehlab, M.: Ant colony clustering by expert ants. In: Proceedings of International Workshop on Data Mining and Artificial Intelligence (DMAI 2008), Khulna, Bangladesh, December 24-27, pp. 94–100 (2008)

[16] Conforth, M., Meng, Y.: Reinforcement Learning for Neural Networks using Swarm. In: 2008 IEEE Swarm Intelligence Symposium, St. Louis, MO, USA, September 21-23, pp. 1–7 (2008)

[17] Haykin, S.: Neural Networks: A Comprehensive Foundation. Prentice Hall (1999) ISBN: 9780780334946

[18] Kennedy, J., Eberhart, R.C.: Particle swarm optimization. In: Proceedings of IEEE International Conference on Neural Networks, Piscataway, NJ, pp. 1942–1948 (1995)

[19] Kobti, Z., Reynolds, R.G., Kohler, T.: A multi-agent simulation using cultural algorithms: the effect of culture on the resilience of social systems. In: Proceedings of the IEEE Conference on Evolutionary Computation, CEC 2003, December 8-12, vol. 3, pp. 1988–1995 (2003)

[20] Kobti, Z., Rahaman, S., Snowdon, A.W., Kent, R.D.: A Reputation Model Framework for Artificial Societies: A Case Study in Child Vehicle Safety Simulation. In: Bergler, S. (ed.) Canadian AI. LNCS (LNAI), vol. 5032, pp. 185–190. Springer, Heidelberg (2008)

[21] Kobti, Z., Rahaman, S., Snowdon, A.W., Dunlop, T., Kent, R.D.: A cultural algorithm to guide driver learning in applying child vehicle safety restraint. In: IEEE World Congress on Computational Intelligence, Vancouver, BC, Canada, July 16-21, pp. 1111–1118 (2006)

[22] Ross, W.D. (ed.): Aristotle, Nicomachean Ethics. Oxford University Press (1924)

[23] Kant, I.: Groundwork of the Metaphysic of Morals. In: Gregor, M.J. (trans.) Practical Philosophy. Cambridge University Press (1996)

[24] Bentham, J.: Introduction to he Principles of Morals and Legistlation. In: Harrison, W. (ed.). Hafner Press (1948)

[25] Mill, J.S.: Utilitarianism. In: Sher, G. (ed.) Hackett (1979)

[26] Ross, W.D.: The Right and the Good. Oxford University Press (1930)

[27] Andrews, C.M.: Complexity-based Ethics

[28] Di Paolo, E.A.: Artificial Life and Historical Processes. In: Kelemen, J., Sosík, P. (eds.) ECAL 2001. LNCS (LNAI), vol. 2159, pp. 649–658. Springer, Heidelberg (2001)

[29] Andres, C., Lewis, E.J.E.: ibid (2004)

Frequent Itemset Extraction over Data Streams Using Chernoff Bound

K. Jothimani and S. Antony Selvadoss Thanamani

Research Department of Computer Science,
NGM College, 90, Palghat Road,
Pollachi - 642 001
Coimbatore District,
Tamilnadu, India
jothi1083@yahoo.co.in
selvdoss@gmail.com

Abstract. Mining data streams poses many new challenges amongst which are the one-scan nature, the unbounded memory requirement and the high arrival rate of data streams.In this paper we revise a Chernoff Bound based Sliding-window approach called CBSW+ which is capable of mining frequent itemsets over high speed data streams. The new method keeps the advantages of the previous CBSW also resolve the drawbacks and produce the runtime memory consumption. In the proposed method we design a synopsis data structure to keep track of the boundary between maximum and minimum window size prediction for itemsets. Conceptual drifts in a data stream are reflected by boundary movements in the data structure.

Keywords: Chernoff Bound, Data Streams, Mining Frequent Itemsets.

1 Introduction

Unlike mining static databases, mining data streams poses many new challenges. First, it is unrealistic to keep the entire stream in the main memory or even in a secondary storage area, since a data stream comes continuously and the amount of data is unbounded. Second, traditional methods of mining on stored datasets by multiple scans are infeasible, since the streaming data is passed only once. Third, mining streams requires fast, real-time processing in order to keep up with the high data arrival rate and mining results are expected to be available within short response times. In addition, the combinatorial explosion1 of itemsets exacerbates mining frequent itemsets over streams in terms of both memory consumption and processing efficiency. Due to these constraints, research studies have been conducted on approximating mining results, along with some reasonable guarantees on the quality of the approximation [6].

Frequent itemset mining is a traditional and important problem in data mining. An itemset is frequent if its support is not less than a threshold specified by users.

S.C. Satapathy et al. (Eds.): Proc. of Int. Conf. on Front. of Intell. Comput., AISC 199, pp. 265–272.
DOI: 10.1007/978-3-642-35314-7_31

Traditional frequent itemset mining approaches have mainly considered the problem of mining static transaction databases [1]. Many applications generate large amount of data streams in real time, such as sensor data generated from sensor networks, online transaction flows in retail chains, Web record and click-streams in Web applications, call records in telecommunications, and performance measurement in network monitoring and traffic management. Data streams are continuous, unbounded, usually come with high speed and have a data distribution that often changes with time. Hence, it is also called streaming data [5].

Nowadays, data streams are gaining more attention as they are one of the most used ways of managing data such as sensor data that cannot be fully system supervision (*e.g.* web logs), require novel approaches for analysis. Some methods have been defined to analyse this data, mainly based on sampling, for extracting relevant patterns [5, 10]. They have to tackle the problem of handling the high data rate, and the fact that data cannot be stored and has thus to be treated in a *one pass* manner [1].

With the rapid emergence of these new application domains, it has become increasingly difficult to conduct advanced analysis and data mining over fast-arriving and large data streams in order to capture interesting trends, patterns and exceptions. From the last decade, data mining, meaning *extracting useful information or knowledge from large amounts of data*, has become the key technique to analyse and understand data. Typical data mining tasks include association mining, classification, and clustering. These techniques help find interesting patterns, regularities, and anomalies in the data. However, traditional data mining techniques cannot directly apply to data streams. This is because mining algorithms developed in the past target disk resident or in-core datasets, and usually make several passes of the data.

For the window-based approach, we regenerate frequent itemsets from the entire window whenever a new transaction comes into or an old transaction leaves the window and also store every itemset, frequent or not, in a traditional data structure such as the prefix tree, and update its support whenever a new transaction comes into or an old transaction leaves the window.[20] In fact, as long as the window size is reasonable, and the conceptual drifts in the stream is not too dramatic, most itemsets do not change their status (from frequent to non-frequent or from non-frequent to frequent) often. Thus, instead of regenerating all frequent itemsets every time from the entire window, we shall adopt an *incremental* approach.

2 Related Work

Most strategies in the literature use variations of the sliding window idea: a window is maintained that keeps the most recently read examples, and from which older examples are dropped according to some set of rules. The contents of the window can be used for the three tasks: 1) to detect change (e.g., by using some statistical test on different sub windows), 2) obviously, to obtain updated statistics from the recent examples, and 3) to have data to rebuild or revise the model(s) after data has changed. The simplest rule is to keep a window of some fixed size, usually determined a priori by the user. This can work well if information on the timescale of change is available, but this is rarely the case. Normally, the user is caught in a tradeoff without solution: choosing a small size (so that the window reflects accurately the current distribution) and choosing a large size (so that many examples are available to work on, increasing accuracy in periods of stability). A different strategy uses a decay function to weight the importance of examples according to their age (see e.g. [3]): the relative contribution of each data item is scaled down by a factor that depends on elapsed time. In this case, the tradeoff shows up in the choice of a decay constant that should match the unknown rate of change.

In a sliding window model, knowledge discovery is performed over a fixed number of recently generated data elements which is the target of data mining. Two types of sliding widow, i.e., transaction-sensitive sliding window and time-sensitive sliding window, are used in mining data streams. The basic processing unit of window sliding of transaction-sensitive sliding window is an expired transaction while the basic unit of window sliding of time-sensitive sliding window is a time unit, such as a minute or 1 h. The sliding window can be very powerful and helpful when expressing some complicated mining tasks with a combination of simple queries [17].

For instance, the itemsets with a large frequency change can be expressed by comparing the current windows or the last window with the entire time span. For example, the itemsets have frequencies higher than 0.01 in the current window but are lower than 0.001 for the entire time span. However, to apply this type of mining, mining process needs different mining algorithms for different constraints and combinations.[15,18] The flexibility and power of sliding window model can make the mining process and mining algorithms complicated and complex. To tackle this problem, we propose to use system supports to ease the mining process, and we are focusing on query languages, system frameworks, and query optimizations for frequent itemset mining on data streams.

Continuous sliding-window queries over data streams have been introduced to limit the focus of a continuous query to a specific part of the incoming stream transactions. The window-of-interest in the sliding-window query model includes the most-recent input transactions. In a sliding window query over n input streams, S_1 to S_n, a window of size w_i is defined over the input stream S_i. The sliding window w_i can be defined over any ordered attribute in the stream tuple. As the window slides, the query answer is updated to reflect both the new transactions entering the sliding-window and the old transactions expiring from the sliding-window. Transactions enter and expire from the sliding-window in a First-In-First- Expire (FIFE) fashion

3 Preliminaries

In this section we describe our algorithms for dynamically adjusting the length of a data window, make a formal claim about its performance, and derive an efficient variation. We will use Chernoff's bound in order to obtain formal guarantees, and a streaming algorithm. However, other tests computing differences between window distributions may be used. The inputs to the algorithms are a confidence value δ ^{2} (0;

1) and a (possibly infinite) sequence of real values $x_1, x_2, x_3, \ldots, x_t, \ldots$ The value of x_t is available only at time t. Each x_t is generated according to some distribution D_t, independently for every t. We denote with μ_t and δ^{2t} the expected value and the variance of x_t when it is drawn according to D_t,. We assume that xt is always in [0; 1]; by an easy rescaling, we can handle any case in which we know an interval [a; b] such that $a \leq x_t \leq b$ with probability 1. Nothing else is known about the sequence of distributions D_t,; in particular, μ_t and δ^{2t} are unknown for all t.

Let I = {x1, x2, ..., xz} be a set of items (or attributes). An itemset (or a pattern) X is a subset of I and written as X =xi xj…xm. The length (i.e., number of items) of an itemset X is denoted by |X|. A transaction, T, is an itemset and T supports an itemset, X, if X⊆T. A transactional data stream is a sequence of continuously incoming transactions. A segment, S, is a sequence of fixed number of transactions, and the size of S is indicated by s. A window, W, in the stream is a set of successive w transactions, where w ≥ s. A sliding window in the stream is a window of a fixed number of most recent w transactions which slides forward for every transaction or every segment of transactions. We adopt the notation Il to denote all the itemsets of length l together with their respective counts in a set of transactions (e.g., over W or S). In addition, we use Tn and Sn to denote the latest transaction and segment in the current window, respectively. Thus, the current window is either W = < Tn-w+1, ..., Tn > or W = < Sn-m+1, ..., Sn >, where w and m denote the size of W and the number of segments in W, respectively.

3.1 Window Initialization Using Binomial Sampling

CBSW is able to adapt its window size to cope with a more efficient transition detection mechanism. It viewed as an independent Bernoulli trial (*i.e.*, a sample draw for *tag i*) with success probability $p_{i,t}$ using Equation (1) [16]. This implies that the number of successful observations of items *i* in the window W_i with epochs (*i.e.*, $W_i = (t - w_i, t)$) is a random variable with a binomial distribution $B(w_i, p_{i,t})$. In the general case, assume that *item i* is seen only in subset of all epochs in the window *Wi*. Assuming that, the item probabilities within an approximately sized window calculated using Chernoff, are relatively homogeneous, taking their average will give a valid estimate of the actual probability of *tag i* during window W_i [16].

The derived binomial sampling model is then used to set the window size to ensure that there are enough epochs in the window W_i such that *tag i* is read if it does exist in the reader's range. Setting the number of epochs within the smoothing window according to Equation (3) ensures that *tag i* is observed within the window W_i with probability $>1 - \delta$ [16]

$$W_i \geq [\ (\ 1/p_i^{avg}\ )\ ln\ (1/\delta)] \tag{1}$$

3.2 Window Size Adjustment

In order to balance between guaranteeing completeness and capturing tag dynamics the CBSW algorithm uses simple rules, together with statistical analysis of the underlying data stream, to adaptively adjust the cleaning window size.

Assume Wi = (t - wi, t) is tag i current window, and let W1i′ = (t - wi, t - wi/2) denote the first half of window Wi and W2i′ = (t - wi/2, t) denote the second half of the window Wi. Let |S1i| and|S2i| denote the binomial sample size during W1i′ and W2i′ respectively. Note that the mid value in inclusive on both range as shown in Figure 1.

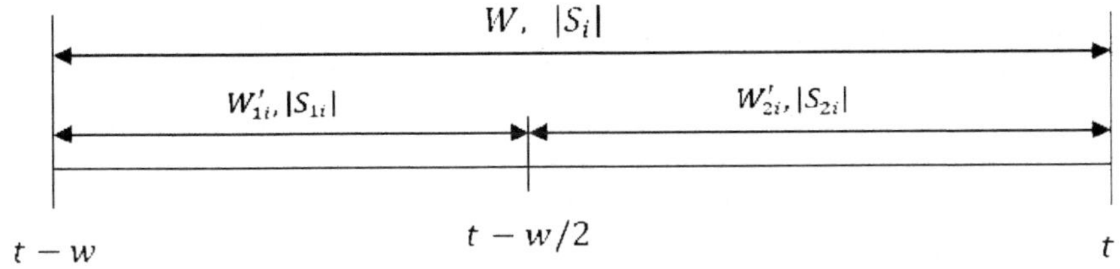

Fig. 1. Illustration of the transaction itemsets in the smoothing window

The window size is increased if the computed window size using Equation (1) is greater than the current window size and the expected number of observation samples($|S_i| >$ $w_i\, p_i^{avg}$)is less than the actual number of observed samples. Low expected observation samples indicates that the probability of detection is p_i^{avg} low, in this case we need to grow the window size to give more opportunity for the poor performing tag to be detected. Otherwise, if the expected observation sample is equal or greater than the actual sample size it means that, the p_i^{avg} is good enough and we do not have to increase the window size. This rule ensures that the window size is increased only when the read rate is poor.

The following code describes the CBSW algorithm. It mainly deals with the window size prediction based on the Chernoff bound method. Initially all new transactions are stored into windows and whenever high speed data stream arrives the window size will be automatically regenerated.

The CBSW Algorithm

Input: T = set of all observed transaction IDs

δ = required data streams

Output: t = set of all present frequent itemset IDs

Initialise: $\forall i \in T, w_i \leftarrow 1$

while(*getNextTransaction*) **do**

  for (*i* in *T*)

    processWindow(W_i) $\rightarrow, p_{i,t}$'s, p_i^{avg}, $|S_i|$

    if (itemExist($|S_i|$)

      output i

    end if

    $w_i^* \leftarrow$ requiredWindowSize(p_i^{avg} , δ)

    if (itemexists ^ $|S_{2i}| = 0$)

    $w_i \leftarrow \max(\min\{ w_i/2, w_i^* \}, 3)$

    else if (detectTransaction($|S_i|$, w_i , p_i^{avg}))

      $w_i \leftarrow \max\{(w_i - 2), 3\}$

else if ($w_i^* > w_i \wedge |S_i| < w_i p_i^{avg}$)
$w_i \leftarrow \min\{(w_i + 2), w_i^*\}$
end if
end for
end while

Also we observed the variation within the window could also be caused by missing itemsets and it is not necessarily happened only due to transition. Hence, to reduce the number of false positive due to transition and the number of false negative readings, which will be further introduced in case of wrong transition detection, the window size is reduced additively by reducing the window size. Setting the minimum window size can be balanced between maintaining the smoothing effect of the algorithm and reducing the false positive errors. Similar to FIDS, CBSW also slides its window per single transaction and produces output readings corresponding to the midpoint of the window after the entire window has been read.

4 Chernoff Bound Based Sliding Window(CBSW+) Algorithm

In this section we present our CBSW+ algorithm. CBSW+ uses the simplified Chernoff bound concepts to calculate the appropriate window size for mining frequent itemsets. It then uses the comparison of the two window sub-range observations and itemset counts when a transition occurs within the window and then adjusts the window size appropriately.

The CBSW+ Algorithm

Input: T = set of all observed transaction IDs
δ = required datastreams
Output: t = set of all present frequent itemset IDs
Initialise: W as an empty list of segments
Initialize WIDTH, VARIANCE and TOTAL
for each t > 0
do SETINPUT(x_t;W)
output i as TOTAL/WIDTH and ChangeAlarm
SetInput(item e, List W)
InsertElement(e;W)
repeat DeleteElement(W)
for every split of W into W = W0 ·W1
InsertElement(item e, List W)
create a new segment s with content e and capacity 1
W ← W ∪ f{s} (i.e., add e to the head of W)
update WIDTH, VARIANCE and TOTAL
output t
CompressSegments(W)

Delete Element(List W)
remove a segment from tail of ListW
update WIDTH, VARIANCE and TOTAL
ChangeAlarm true

Our first version of CBSW is computationally expensive, because it checks exhaustively all "large enough" sub windows of the current window for possible cuts. Furthermore, the contents of the window is kept explicitly, with the corresponding memory cost as the window grows. To reduce these costs we present a new version CBSW+ using ideas developed in data stream algorithms [15][18][20] to find a good cut point quickly. We next provide a sketch of how this algorithm and these data structures work. Our data structure is a variation of exponential histograms [20], a data structure that maintains an approximation of the number of 1's in a sliding window of length W with logarithmic memory and update time. We adapt this data structure in a way that can provide this approximation simultaneously for about O(logW) subwindows whose lengths follow a geometric law, with no memory overhead with respect to keeping the count for a single window. That is, our data structure will be able to give the number of 1s among the most recently $t - 1$, $t - b_{cc}$, $t - b_{c2c}$,. . . , $t - b_{cic}$, . . . read bits, with the same amount of memory required to keep an approximation for the whole W. Note that keeping exact counts for a fixed window size is provably impossible in sub linear memory. We go around this problem by shrinking or enlarging the window strategically so that what would otherwise be an approximate count happens to be exact.

5 Conclusion

In this paper, we study the problem of mining frequent itemsets over the sliding window of a transactional data stream. Based on the improvement theory CBSW, we revise and propose an algorithm called CBSW+ for finding frequent itemsets through an approximating approach. CBSW+ conceptually divides the sliding window into segments and handles the sliding of window in a segment-based manner with the help of Chernoff bound method. We also introduce the concept of simplified Chernoff Bound, which makes CBSW+ capable of approximating itemsets dynamically by choosing different parameter-values for different itemsets to be approximated .Compared with other sliding window based mining techniques, we save memory and improve speed by dynamically maintaining all transactions in high speed streams. A segmented based data structure was designed to dynamically maintain the up to date contents of an online data stream by scanning it only once, and revised method CBSW+ was proposed to mine the frequent itemsets in sliding window.

References

1. Raissi, C., Poncelet, P., Teisseire: Towards a new approach for mining frequent itemsets on data stream. J. Intell. Inf. Syst. 28, 23–36 (2007)
2. Xu Yu, J., Chong, Z., Lu, H., Zhang, Z., Zhou, A.: A false negative approach to mining frequent itemsets from high speed transactional data streams. Information Sciences 176, 1986–2015 (2006)
3. Giannella, G., Han, J., Pei, J., Yan, X., Yu, P.: Mining frequent patterns in data streams at multiple time granularities. In: Next Generation Data Mining. MIT, New York (2003)
4. Li, H.F., Lee, S.Y., Shan, M.: An efficient algorithm for mining frequent itemsets over the entire history of data streams. In: Proceedings of the 1st International Workshop on Knowledge Discovery in Data Streams (2004)
5. Jin, R., Agrawal, G.: An Algorithm for In-Core Frequent Itemset Mining on Streaming Data
6. Domingos, P., Hulten, G.: Mining high-speed data streams. In: Proceedings of the ACM Conference on Knowledge and Data Discovery, SIGKDD (2000)
7. Cheng, Ke, Y., Ng, W.: A survey on algorithms for mining frequent itemsets over data streams. Knowl. Inf. Syst. (2007)
8. Charikar, M., Chen, K., Farach, M.: Finding frequent items in data streams. Theory Comput. Sci. 312, 3–15 (2004)
9. Agrawal, R., Imielinski, T., Swami, A.: Mining association rules between sets of items in large databases. In: Proceedings of the ACM SIGMOD International Conference on Management of Data, Washington, DC, pp. 207–216 (1993)
10. Chernoff, H.: A measure of asymptotic efficiency for tests of a hypothesis based on the sum of observations. The Annals of Mathematical Statistics 23(4), 493–507 (1953)
11. Charikar, M., Chen, K., Farach-Colton, M.: Finding Frequent Items in Data Streams. In: Widmayer, P., Triguero, F., Morales, R., Hennessy, M., Eidenbenz, S., Conejo, R. (eds.) ICALP 2002. LNCS, vol. 2380, pp. 693–703. Springer, Heidelberg (2002)
12. Calders, T., Dexters, N., Goethals, B.: Mining Frequent Items in a Stream Using Flexible Windows
13. Sun Maria, X., Orlowska, E., Li, X.: Finding Frequent Itemsets in High-Speed Data Streams
14. Han Dong, X., Ng, W., Wong, K., Lee, V.: Discovering Frequent Sets from Data Streams with CPU Constraint. This paper appeared at the AusDM 2007, Gold Coast, Australia. Conferences in Research and Practice in Information Technology (CRPIT), vol. 70 (2007)
15. Agrawal, R., Srikant, R.: Fast algorithms for mining association rules. In: Proc. Int. Conf. Very Large Data Bases (VLDB 1994), pp. 487–499 (1994)
16. Zaki, J.M., Hsiao, C.: CHARM: An efficient algorithm for closed itemset mining. In: Proc. SIAM Int. Conf. Data Mining, pp. 457–473 (2002)
17. Agrawal, R., Srikant, R.: Mining sequential patterns. In: Proc. Int. Conf. Data Engineering (ICDE 1995), pp. 3–14 (1995)
18. Kuramochi, M., Karypis, G.: Frequent subgraph discovery. In: Proc.Int. Conf. Data Mining (ICDM 2001), pp. 313–320 (2001)
19. Li, H.F., Lee, S.Y., Shan, M.K.: Online Mining (Recently) Maximal Frequent Itemsets over Data Streams. In: Proceedings of the 15th IEEE International Workshop on Research Issues on Data Engineering, RIDE (2005)
20. Indyk, P., Woodruff, D.: Optimal approximations of the frequency moments of data streams. In: Proceedings of the Thirty-seventh Annual ACM Symposium on Theory of Computing, pp. 202–208 (2005)

Isolated Word Recognition Using Enhanced MFCC and IIFs

S.D. Umarani[1], R.S.D. Wahidabanu[1], and P. Raviram[2]

[1] Government College of Engineering, Salem - 636011, India
umaraviram@gmail.com, rsdwb@yahoo.com
[2] Department of CSE
Mahendra Engineering College, Tiruchengode- 637503, India
drpraviram@gmail.com

Abstract. The main objective of this paper is to design a noise-resilient and speaker independent speech recognition system for isolated word recognition. Mel-frequency Cepstral Coefficients (MFCCs) has been used for feature extraction. Noise robust performance of MFCC under mismatched training and testing conditions is enhanced by the application of wavelet based denoising algorithm and also to make MFCCs as robust to variation in vocal track length (VTL) an invariant-integration method is applied. The resultant features are called as enhanced MFCC Invariant-Integration Features (EMFCCIIFs). To accomplish the objective of this paper, classifier called feature-finding neural network (FFNN) is used for the recognition of isolated words. Results are compared with the results obtained by the traditional MFCC features. Through experiments it is observed that under mismatched conditions, the EMFCCIIFs features remains high recognition rate under low Signal-to-noise ratios (SNRs) and their performance are more effective under high SNRs too.

Keywords: Isolated word, Denoising, Invariant-integration, MFCC, IIF, FFNN, SNR.

1 Introduction

Speech processing and Automatic Speech Recognition by machines is one of the most attractive areas of research over the past five decades [1]. In recent years performance of the Automatic Speech Recognition (ASR) systems has extremely improved. Speaker-independent voice recognition systems have a very strong probability of becoming a necessity in the workplace in the future [2]. Although many technological improvements have been done in ASR, recognition accuracy is still far from human levels. The performance of ASR degrades when the training and testing environments are differing. These environments are speaker variation, channel distortion, reverberation, noise etc [3, 4]. As voice based systems are being transferred to real applications, robustness of this system is necessary so that the performance should not get affected, when the quality of the input speech is corrupted or when the training and testing environment is varying.

S.C. Satapathy et al. (Eds.): Proc. of Int. Conf. on Front. of Intell. Comput., AISC 199, pp. 273–283.
DOI: 10.1007/978-3-642-35314-7_32

MFCC is a standard and popular feature coefficient used in the speech recognition and it can obtain more accurate results under a clean testing environment. But, accompany with deterioration of the environmental noisy condition as well as the effects of inter-speaker variability originating from different vocal tract lengths (VTLs), the performance of MFCC in speech recognition capabilities decrease dramatically [5]. This shows the unsuitability of the usage of MFCC in the strong noisy conditions and speaker-independent voice recognition systems. In traditional MFCC algorithm, FFT is applied to compute the signal frequency spectrum and it is applicable for any stationary signal. However, the method is not suitable in processing the non stationary speech signal [6]. But as the wavelet transform composes diverse time-frequency spaces with appropriate time-frequency resolution under the principle of uncertainty, here it is used overcome the shortages of FFT [7].

The effects of inter-speaker variability originating from different vocal tract lengths (VTLs), reflects as translations in the subband- index space of TF representations of a speech signal [5]. This translation is measured and based on that features are extracted with invariance to that translation and thus increases the robustness against VTL changes. Techniques called translation invariant transformations [8] and generalized cyclic transformations (GCT) [9] have been used for feature extraction that is robust against vocal tract length changes in speaker independent speech recognition. A method "invariant integration", integrates regular nonlinear functions of the features over the transformation group for which an invariance should be achieved [10] is applied in this paper to make the speech recognition as robust to VTL changes.

In this paper to achieve a noise robust and speaker independent isolated word recognition, existing traditional MFCC features is modified by using "Bark wavelet" and are enhanced by the denoising algorithm called adaptive wavelet thresholding. Then the method of invariant integration is applied for enhanced MFCCs to make the features as robust to VTL changes also. These features are called as Enhanced MFCC IIFs feature (EMFCCIIFs). This method of feature-extraction makes the speech recognition as robust to both variation in VTL and noise. Because as the resultant feature dimension of this method is large, an appropriate feature selection method is described. Through experiments it is proved that the resulting feature sets lead to better recognition results than the standard mel-frequency cepstral coefficients (MFCC) under matching training and testing conditions.

2 MFCC with Adaptive Wavelet Thresholding and Invariant Integration

The mel-Cepstrum, introduced by Davis and Mermelstein [11], are coefficients of representation of the short-term power spectrum of a sound, based on a linear cosine transform of a log power spectrum on a nonlinear Mel scale of frequency. Mel-Frequency Cepstral Coefficients (MFCCs) are based on the human peripheral auditory system and are proved more efficient. The overall process of the MFCC is shown in Fig.1a [12].

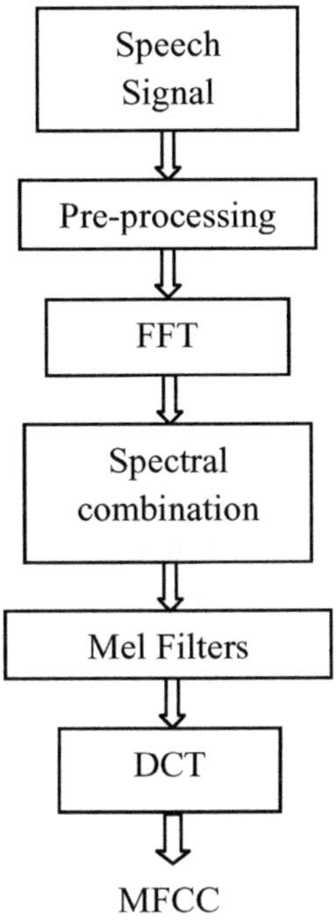

Fig. 1a. Existing method of computation of (MFCC)

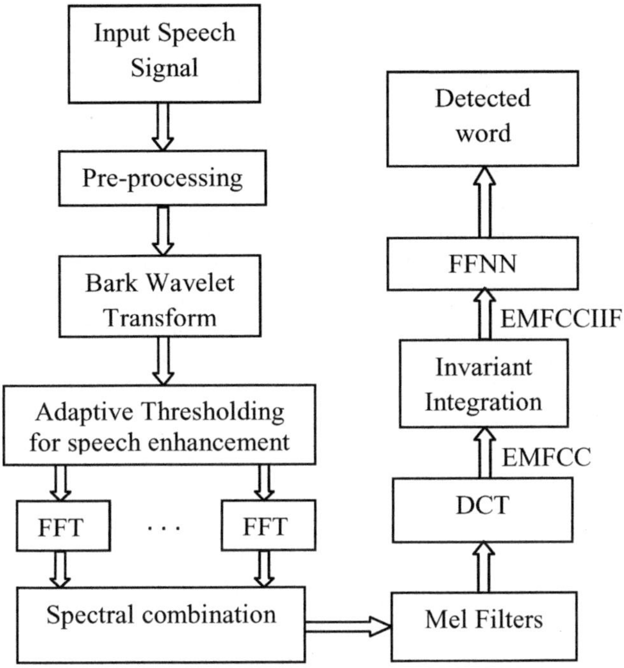

Fig. 1b. Proposed method of computation of EMFCCIIFs

The robustness of MFCC against noise and variation in VTL is improved by the application of enhancement method called adaptive wavelet thresholding and Invariant integration. From very many of these features, the most important are selected by fast algorithm, called substitution algorithm and finally, the activities of the feature cells are classified in a linear manner in order to recognize the word by the FFNN. Fig. 1b shows the block diagram of steps involved in the computation of EMFCCIIFs.

The calculation process is as follows.

Step 1: Preprocessing

As preprocessing increases the recognition rates considerably, initially preprocessing is done, which is a process of sampling followed by Exerting pre-emphasis, framing and window adding processing to the original speech signal. Let, $X(t), 0 \leq t \leq T-1$ be the continuous time voice signal to be sampled to obtain discrete signal for the upcoming stages. Here the sampling is done at a sampling frequency of $f_s = \frac{1}{\Delta t} Hz$, where Δt is called sampling interval in seconds. Thus obtained discrete signal $x(n) = X(t = n.\Delta t), n = 0,1,2,3, \ldots. N$.

Pre-emphasis: The purpose of pre-emphasis filter is to eliminate the effect of dc offsets and increase the amplitude of the speech signal at frequencies where SNR is low. This step processes the passing of signal through a filter which emphasizes higher frequencies. This process will increase the energy of signal at higher frequency.

$$y(n) = x(n) - a(x(n-1)) . \quad (1)$$

Let us consider a = 0.95, which make 95% of any one sample is presumed to originate from previous sample.

Framing: Framing is applied to decompose the speech signal into a series of overlapping Frames. Framing involves separating the sample data into specific sizes of frames of N samples, with adjacent frames being separated by M (M < N).

Windowing: Since speech is non-stationary and as the parameters are estimated in short-term such as the Fourier spectrum, requires that a speech segment is chosen for analysis. Also to minimize the signal discontinuities at the beginning and end of each frame the effective cross-multiplication of signal by a window function is done. If the signal in a frame is denoted by $y(n)$, n = 0,…N-1, then the signal after windowing is $y(n) * w(n)$, where $w(n)$ is the window used. Here Hamming window is applied, equation is given as:

$$\mathrm{w(n)} = \left[0.54 - 0.46\cos\left(\frac{2\pi \mathrm{n}}{\mathrm{N}-1}\right)\right] 0 \le \mathrm{n} \le \mathrm{N}-1 . \quad (2)$$

where N = number of samples in each frame and w (n) = Hamming window.

Frame shifting: Overlapping of windows has been done to improve the continuity in time domain after transformation. This overcomes the problems raised by window artifacts and non stationary channel noise. An overlap of half the window size (or less) is typical. The overlap reduces the troubles that may occur due to signal data discontinuity.

Step 2: Translation by Bark wavelet

Before going to adaptive thresholding, for denoising the preprocessed signal, it is converted to wavelet format. In this work the input signal is processed by Bark wavelet transform. The relation of the linear frequency and Bark frequency is [7],

$$\mathrm{b} = 13\arctan(0.76\mathrm{f}) + 3.5\arctan\left(\frac{\mathrm{f}}{7.5}\right)^2 . \quad (3)$$

where b is Bark frequency and f is the linear frequency in Hz.

In the linear frequency the Bark wavelet function is described as,

$$\mathrm{W_k(f)} = \mathrm{c_2}.\, 2^{-4\left(13\arctan(0.76\mathrm{f})+3.5\arctan\left(\frac{\mathrm{f}}{7.5}\right)^2-(\mathrm{b_l}+\mathrm{k}\Delta\mathrm{b})\right)^2} . \quad (4)$$

where c2 is the normalization factor.

Now with input signal s(t), the signal of the k-th frame, which had been transformed by Bark wavelet is given by,

$$\mathrm{s_k(t)} = \int_{-\infty}^{\infty} \mathrm{S(f)}.\, \mathrm{W_k(f)}.\, \mathrm{e}^{\mathrm{j}2\pi \mathrm{f}}\mathrm{df} \quad (5)$$

Where, S(f) is the spectrum of speech signal s(t).

Step 3: Denoising by thresholding

In order to enhance the ability to resist the noises of different environments, ZHANG Jie et al. [13] has introduced an adaptive wavelet thresholding approach. The major noises that may occur are white noise and colored noise. When the background noise is white noise, including color noise with flatting spectrum amplitude, then the threshold function is given by,

$$\lambda = \delta\sqrt{2\log_{10}^{N}}\ g(i) \tag{6}$$

Where, $g(i)$ is the inverse proportion function of variable i. Now based on the type of scale and the range of noise variance, the threshold function will vary. So it is called as adaptive threshold algorithm associated to the scale. By selecting appropriate g(i) function, the performance of denoising can be enhanced for the white noise. With lots of tests in the laboratory, Zhang Jie et al. [13] has introduced the inverse proportion function as,

$$g(i) = \frac{1}{\left[2^{\frac{i-1}{2}}\ \ln(i+1)\right]} \tag{7}$$

When the background noise is colored noise, the spectrum amplitude is not flat. For this the threshold function is given by,

$$\lambda' = \delta\sqrt{2\log_{10}^{N}}\ \varphi(\gamma) \tag{8}$$

Where, γ denote the flatness and $\varphi(\gamma)$ is the correction function. As different colored noise has different flatness of the spectrum amplitude, via repetitious tests, $\varphi(\gamma)$ is introduced by Zhang Jie et al. [13] as follows,

$$\varphi(\gamma) = \frac{2}{\left[\log_{10}^{0.05\gamma}\right]} \tag{9}$$

These threshold functions adjust adaptively so that it can adapt all kinds of noisy environment and the noise present in the transformed signal get removed.

Step 4: Fast Fourier Transform

The next processing step is the Fast Fourier Transform, which converts each frame of N samples from the time domain into the frequency domain and is given by,

$$S_K(n) = \sum_{l=0}^{N-1} W_K(l)\, S(l) e^{j2\pi nl/N}\ . \tag{10}$$

In the equation above, N is the FFT's length of filling zeros. $W_K(l)$ is the discrete form of $W_k(f)$ in equation (4), $S(l)$ is the speech signal frequency spectrum, $S_K(n)$ is the Kth sub-band's speech spectrum.

Step 5: Spectral combination

The frequency synthesis is done to obtain s(n) which is the spectrum of frequency synthesis as follows,

$$s(n) = \sum_{k=0}^{K-1} s_k(n) \tag{11}$$

After this the scale of frequency is converted from linear to Mel scale.

Step 6: Mel Filter Bank Processing

This non-linear frequency resolution of FFT can be approximated by using the mel-scale [14], which can be modeled by filters whose magnitude frequency response is triangular in shape and equal to unity at the Centre frequency and decrease linearly to zero at centre frequency of two adjacent filters. Then, each filter output is the sum of its filtered spectral components. After that the following equation is used to compute the Mel for given frequency f in HZ,

$$F(Mel) = \left[2595 * log_{10}\left[1 + \frac{f}{700}\right]\right] . \tag{12}$$

After Mel scaling Log energy of the signal is computed as

$$d(m) = lg(\textstyle\sum_{n=0}^{N-1} |s(n)|^2 H_m(n)) \ (0 \le m \le L) . \tag{14}$$

where N is the sample number of $s(n)$. $H_m(n)$ is Mel frequency triangle band pass filters. L is its number.

Step 7: Discrete Cosine Transform:

In final step, the speech features are extracted by converting back log mel spectrum to time domain by applying the inverse DFT on the compressed spectrum which computes only the cosine components due to the fact that power spectral values are real and even. It is given by,

$$\mathrm{y(n)} = \textstyle\sum_{\mathrm{n=0}}^{\mathrm{N-1}} \mathrm{d(m)} \frac{\cos 2\pi \mathrm{fn}}{\mathrm{N}}, (\, 0 \le m \le L - 1\,) . \tag{15}$$

The result is called the Enhanced Mel Frequency Cepstrum Coefficients (EMFCC). As a next step EMFCCs are combined with IIFs.

Step 8: Invariant Integration Feature (IIF)

Invariant integration is a productive approach for calculating separable features that are invariant to a selected group of transformations. The theory of the IIFs is their invariance to translations along the subband axis [10]. Let $y_k(n)$ denote the magnitudes of the obtained subband coefficients of EMFCC at the final frame rate, where n is the time and k the subband index, with $1 \le n \le N$ and $1 \le k \le K$. Given the indices vector $k = (k_1, k_2, \dots, k_M)$, $K \in \mathbb{N}_0^M$, an integer exponent vector $= (l_1, l_2, \dots, l_M)$, $l \in \mathbb{N}_0^M$, and a temporal offset vector $m \in \mathbb{N}^M$, a (contextual) monomial m with M components is defined as

$$m(n;w,k,l,m) := \left[\prod_{i=1}^{M} y_{k_i+w}^{l_i}(n+m_i)\right]^{1/\gamma(m)} . \quad (16)$$

where $w \in \mathbb{N}_0$ is a spectral offset parameter that is used for the ease of notation in the following, $\gamma(m)$ is the order of a monomial m defined as

$$\gamma(m) := \sum_{i=1}^{M} l_i \ . \quad (17)$$

The term (17), which occurs in the exponent in (16), operates as a normalizing term with respect to the order of the monomials. Now, an Enhanced MFCC Invariant Integration Features (EMFCCIIFs) $A_m(n)$ is defined as

$$\mathrm{A_m} := \frac{1}{2\mathrm{W}+1}\sum_{\mathrm{w}=-\mathrm{W}}^{\mathrm{W}} \mathrm{m}(\mathrm{n};\mathrm{w},\mathbf{k},\mathbf{l},\mathbf{m}) . \quad (18)$$

Apparently, the parameter space of the IIFs is quite large and choosing an appropriate set of features is non-trivial. In this work, the Feature-finding Neural Network (FFNN) [15] approach is used for feature-selection.

3 Classifications and Optimization of the Type of Secondary Features

As FFNN has high recognition rate than the classical HMM and **DTW** recognizers and yields similar recognition rates, for classification and optimization of the enhanced features the Feature-Finding Neural Network (FFNN) [16] is used. Its architecture is depicted in Fig. 2. It consists of a linear single-layer perceptron in conjunction with secondary feature extraction and an optimization rule for the feature set. The enhanced features are the inputs to FFNN. From very many of these features, the most important are selected by fast algorithms called substitution rule. Finally, the activities of the feature cells are classified in a linear manner in order to recognize the word.

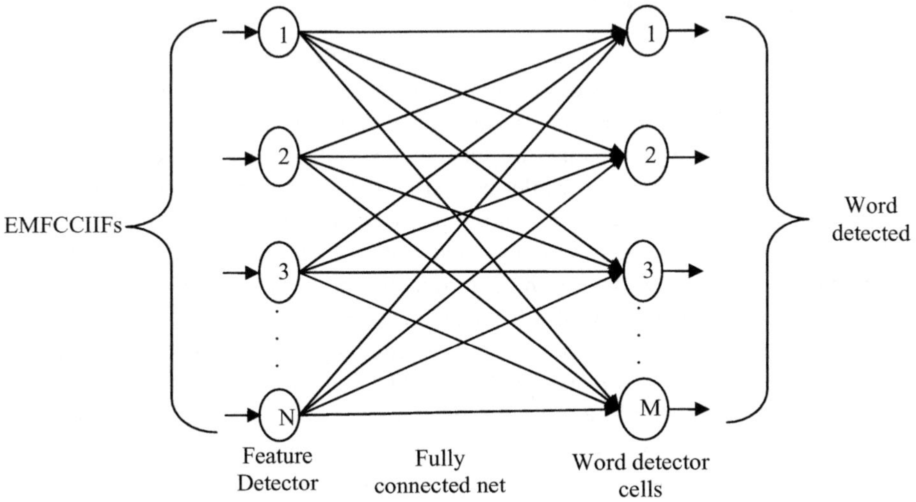

Fig. 2. System structure of the Feature Finding Neural

Gramss [15] (1991) proposed training algorithms for the FFNN system namely growth and substitution rule. As the accuracy and speed of recognition by using the substitution rule is higher than the growth rule, it is used in this research work. This rule starts with full-size net and randomly chosen features. During one iteration cycle, the feature cell with the lowest relevance is exchanged for an additional randomly chosen feature cell. The steps involved in optimization of features using substitution rule is discussed below.

1. Choose M secondary features arbitrarily.
2. Find the optimal weight matrix W using all M features and the M weight matrices that are obtained by using only $M - 1$ features, thereby leaving out every feature once.
3. Measure the relevance R of each feature i by

$$R_i = \text{E(without feature i)} - \text{E(with all features)} \tag{19}$$

4. Discard the least relevant feature $\text{j} = \text{argmin}(R_i)$ from the subset and randomly select a new candidate.
5. Repeat from point 2 until the maximum number of iterations is reached.
6. Recall the set of secondary features that performed best on the training / validation set and return it as result of the substitution process.

The resulting activity vectors of the feature detector layer are classified in a linear, optimal (least mean square) manner. This implies that the substitution algorithms yield the pseudoinverse solution [15]. Although the classification is performed by a linear neural network, the whole classification process is highly nonlinear due to the second order characteristics of the extracted features. In the experiments, a set of 110 secondary features are optimized over 2000 iterations.

4 Experimental Evaluation

This section presents the experimental evaluation of the enhanced MFCCIIFs features for speaker -independent speech recognition using a Feature-Finding Neural Network (FFNN). The EMFCCIIFs/FFNN system was evaluated in a task where the acoustic conditions of training and testing are mismatched, i.e., the training data set was recorded under a clean condition while the testing data sets were mixed with white and color noise at various noise levels. In this work the Speech Separation Challenge database [17] has been used for training and testing. The training set has totally five different sets 600 speech data each. The training can be done with all these files but testing was done only with 20 files from each group.

During experiment, the speech sampling frequency is 11.025 kHz, frame length is 256 points and the frame shift is 128. First, MFCC method is implemented along with adaptive wavelet thresholding. The filterbank used for the computation of the MFCCs had 24 filters. Since the speech data file's lengths are different, the extracted coefficients are made the time-normalization process in order to conveniently processing the case. That is, single pronunciation of each word will be normalized into 12 dimensions MFCC speech feature vector series. Then the enhanced MFCC features are made invariant to translation effect by applying the Invariant Integration Method.

All experiments use 30 IIFs of order one with 110 subbands [10]. This has yielded 110 enhanced IIFs and is given as input to FFNN. In FFNN the substitution rule has been used. In this case, one starts with 110 randomly taken features from the IIFs. The number of features remains constant throughout the experiment. Low-relevance features are replaced with features with higher relevance. This substitution algorithms yield the pseudo inverse solution for the linear classifier of FFNN [18]. After about 1000 substitutions, features are optimized and faultless recognition is achieved.

Experiments were conducted to compare the performance of enhanced IIFs with the enhanced MFCC features for three different word counts (10, 20 and 30) of the speech data files in different SNR environments (including -15dB, -20dB, -25dB, -30dB and clean).

5 Results and Discussions

The EMFCCIIFs is implemented. Performance of the algorithm for three different word counts and for five different noise conditions including clean speech are tabulated in Table 1. The combination of IIFs-based ASR system with MFCC and adaptive wavelet thresholding increased the accuracy during the matching scenario by about 1 to 3 percentage points and it has yielded at least 1 to 2 percentage points of improvement when the enhancement is introduced for maximum noise disturbance of 30 dB compared with MFCC with adaptive wavelet thresholding. The improvement under low noise disturbance, i.e. at 15dB, is also same 1 to 2 percentage points. It can be observed that, in the both matching and mismatching scenario, all enhanced IIFs accuracies are higher than the corresponding MFCC with enhancement accuracies. This is plotted in Fig. 3 to Fig. 5 for different word length.

Table 1. The performance % of proposed method for various noise conditions and for different word counts

No. of Words	Method	SNR				
		-15dB	-20dB	-25dB	-30dB	Clean
10	EMFCC	92.72	93.20	95.14	96.01	97.91
	EMFCCIIFS	94.23	95.00	96.56	97.45	98.11
20	EMFCC	91.78	92.17	93.26	94.35	95.01
	EMFCCIIFS	93.31	94.09	95.20	96.19	97.84
30	EMFCC	90.33	91.52	92.19	93.47	94.80
	EMFCCIIFS	91.56	92.42	93.60	94.48	96.01

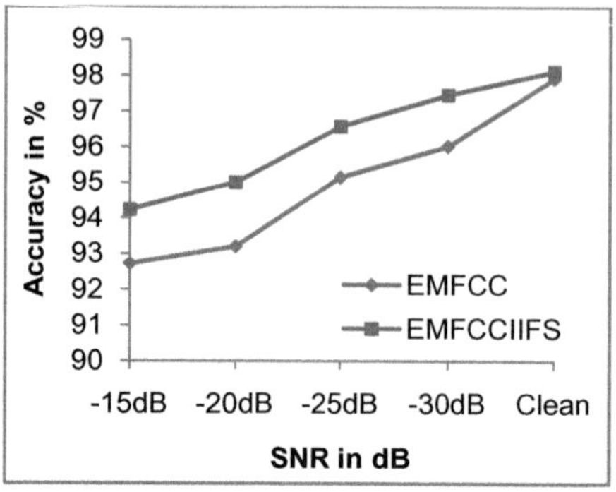

Fig. 3. Comparison of Accuracy for 10 Words

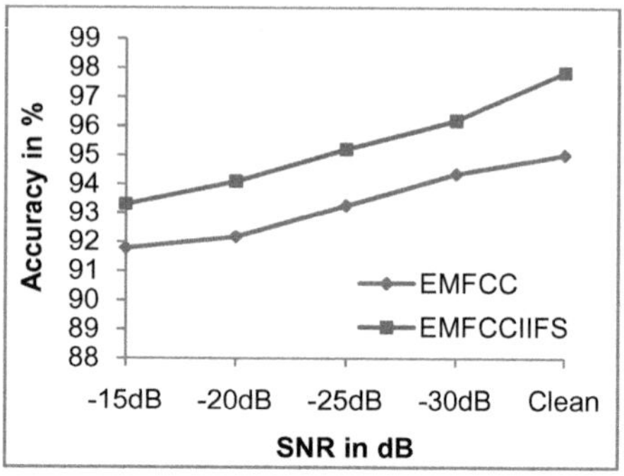

Fig. 4. Comparison of Accuracy for 20 Words

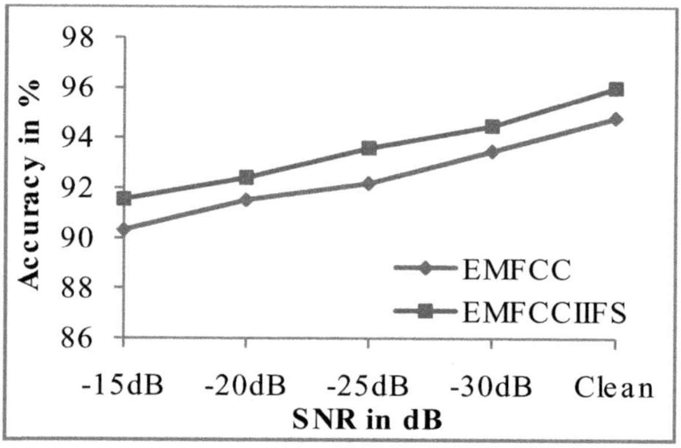

Fig. 5. Comparison of Accuracy for 30 Words

6 Conclusions

A noise-resilient and speaker independent speech recognition system for isolated word recognition has been designed and implemented. Noise robust performance of MFCC is enhanced by the application of adaptive wavelet thresholding and resultant features are made as robust to variation in VTL by the invariant-integration. Classifier called feature-finding neural network (FFNN) is used for the recognition of isolated words. Results are compared with the results obtained by the traditional MFCC and EMFCC features. Through experiments it is observed that under mismatched conditions, the combined EMFCC and IIFs features remains high recognition rate under low Signal-to-noise ratios (SNRs) and their performance are more effective under high SNRs too.

References

1. Juang, B.H., Rabiner, L.R.: Automatic Speech Recognition—A Brief History of the Technology, 2nd edn. Elsevier Encyclopedia of Language and Linguistics (2005)
2. Hermansky, H., Morgan, N.: RASTA processing of speech. IEEE Trans. Speech, Audio Processing 2(4), 578–589 (1994)
3. Acero, A., Stern, R.M.: Environmental robustness in automatic speech recognition. In: International Conference on Acoustics, Speech, and Signal Processing (ICASSP 1990), vol. 2, pp. 849–852 (1990)
4. Aldibbiat, N.M.: Optical wireless communication systems employing Dual Header Pulse Interval Modulation (DH-PIM). Sheffield Hallam University (2001)
5. Acero, A.: Acoustical and Environmental Robustness in Automatic Speech Recognition. Ph.D Thesis, Carnegie Mellon University, Pittsburgh, Pennsylvania (1990)
6. Muller, F., Mertins, A.: Contextual invariant-integration features for improved speaker-independent speech recognition. Speech Communication 53(6), 830–841 (2011)
7. Li, Q.: Solution for pervasive speaker recognition. SBIR Phase I Proposal, Submitted to NSF IT.F4, Li Creative Technologies, Inc., NJ (2003)
8. Zhang, X., Meng, W.: The Research of Noise-Robust Speech Recognition Based on Frequency Warping Wavelet. In: Grimm, M., Kroschel, K. (eds.) Source: Robust Speech Recognition and Understanding, p. 460. I-Tech, Vienna (2007) ISBN 987-3-90213-08-0
9. Burkhardt, H., Muller, X.: On invariant sets of a certain class of fast translation invariant transforms. IEEE Transactions on Acoustic, Speech, and Signal Processing 28(5), 517–523 (1980)
10. Muller, F., Belilovsky, E., Mertins, A.: Generalized cyclic transformations in speaker independent speech recognition. In: IEEE Automatic Speech Recognition and Understanding Workshop, Merano, Italy, pp. 211–215 (2009)
11. Muller, F., Mertins, A.: Invariant-integration method for robust feature extraction in speaker-independent speech recognition. In: Int. Conf. Spoken Language Processing (Interspeech 2009-ICSLP), Brighton (2009)
12. Davis, S.B., Mermelstein, P.: Comparison of parametric representations for monosyllabic word recognition in continuously spoken sentences. IEEE Trans. Acoustics, Speech and Audio Processing 26, 357–366 (1980)
13. Muda, L., Begam, M., Elamvazuthi, I.: Voice Recognition Algorithms using Mel Frequency Cepstral Coefficient (MFCC) and Dynamic Time Warping (DTW) techniques. Journal of Computing 2(3), 138–143 (2010)
14. Zhang, J., Li, G.-L., Zheng, Y.-Z., Liu, X.-Y.: A Novel Noise-robust Speech Recognition System Based on Adaptively Enhanced Bark Wavelet MFCC. In: Sixth International Conference on Fuzzy Systems and Knowledge Discovery (FSKD 2009), pp. 443–447 (2009)
15. Mammone, R.J., Zhang, X., Ramachandran, R.P.: Robust speaker recognition: A feature-based approach. IEEE Signal Processing Magazine 13, 58–70 (1996)
16. Gramss, T., Strube, H.W.: Recognition of isolated words based on psychoacoustics and neurobiology. Speech Commun. 9, 35–40 (1990)
17. Gramss, T.: Word recognition with the Feature Finding Neural Network (FFNN). In: IEEE-SP Workshop Neural Networks for Signal Processing, Princeton, New Jersey (1991)
18. Cooke, M., Lee, T.-W.: Speech separation challenge, `http://www.interspeech2006.org`
19. Kohonen, T.: Self-Organization and Associative Memory, 2nd edn. Springer Series in Information Sciences, vol. 8, ch. 5, 7 (1988)

Fault Detection in Synchronous Generators Based on Independent Component Analysis

Narri Yadaiah[1] and Nagireddy Ravi[2]

[1] Dept. of Electrical and Electronics Engineering
Jawaharlal Nehru Technological University
Hyderabad-500 085, A.P, India
svpnarri@yahoo.com.
[2] Pulichintala Hydro Electric Scheme
Andhra Pradesh Power Generation Corporation Limited
Nalgonda – 503 246, A.P, India
nagireddyravi@rediffmail.com

Abstract. A novel technique for fault detection and classification in the synchronous generator is proposed. In this paper, a new statistical method based on independent component analysis is presented. The proposed fault detection scheme identifies external and internal faults of synchronous generator. This characterization of fault transients will aid in the development of a protection relay for synchronous generator and the proposed method is effective in detecting faults and has great potential in power engineering applications.

Keywords: Synchronous Generator, Internal fault, External Fault, Independent Component Analysis.

1 Introduction

Synchronous Generator is an important element of power systems and their protection is essential. The trend toward a deregulated global electricity market has put the electric utilities under severe stress to keep the machines continuously in service to give uninterrupted power supply to the customers. Hence the synchronous generator is the most critical equipment of the power industry and fault detection is important to prevent outages and black outs.

The faults in the synchronous generator are classified as an internal faults and external faults. The internal faults are phase to phase and phase to earth faults in stator winding, where as external faults are the faults those occur outside the generator which are due to short circuit, over loading and unbalanced loading. Traditionally, differential protection using electro mechanical, solid state and numerical relays are the most common methods are in use for synchronous generator protection [1]. The conventional schemes consists of differential protection scheme and stator earth fault detection schemes are found to be slow in clearing the faults unless the fault develops and current reaches operating value. The faults which are close to neutral cannot be detected by earth fault relays as there will not be sufficient voltage to drive the fault current. To overcome these limitations several techniques have been developed and proposed.

S.C. Satapathy et al. (Eds.): Proc. of Int. Conf. on Front. of Intell. Comput., AISC 199, pp. 285–292.
DOI: 10.1007/978-3-642-35314-7_33

In [2], a new protection algorithm of a synchronous generator using a fuzzy logic controlled neural network was presented. This technique utilized the fault generated transients for the identification of internal and external faults.

In recent years, artificial neural network technique has been proposed on fault detection of synchronous generator. These schemes are provided in detection of various internal faults, external faults and ground faults which are close to neutral [3].

In [4], a protection scheme for synchronous generator is developed based on advance signal processing technique such as wavelet transforms.

The objective of this paper is to implement a fault detection scheme based on Independent component analysis (ICA) which is a most emerging application area of great potential in fault transient detection and identification. ICA considers change in scale and extracts features of the transient signal. The I^2 statistics of ICA is used for the detection of threshold for the classification of internal and external faults. Hence the proposed technique will aid in automatic detection and classification of faults in synchronous generators.

2 Modeling of Synchronous Generator

A power system network consisting of synchronous generator which is fed to the load through a step down transformer and a transmission line is shown in Fig. 1 is considered for simulation. The typical parameters of synchronous generator are shown in Table 1.

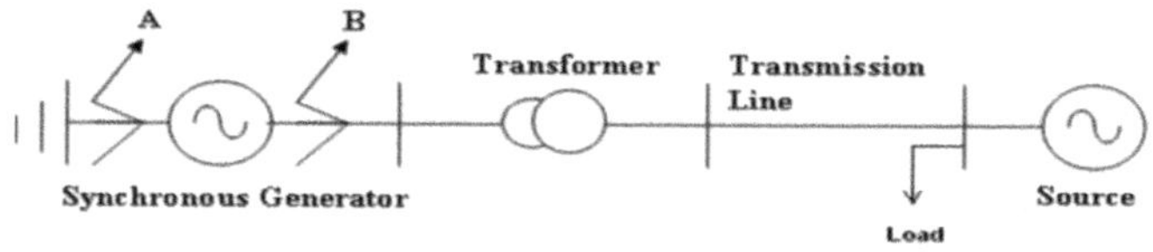

Fig. 1. A synchronous generator connected to power system network.

The internal faults and external faults are created at position A and B respectively. In order to illustrate the internal faults such as stator winding faults in the synchronous generator the equivalent circuit [5] shown in Fig. 2 is implemented.

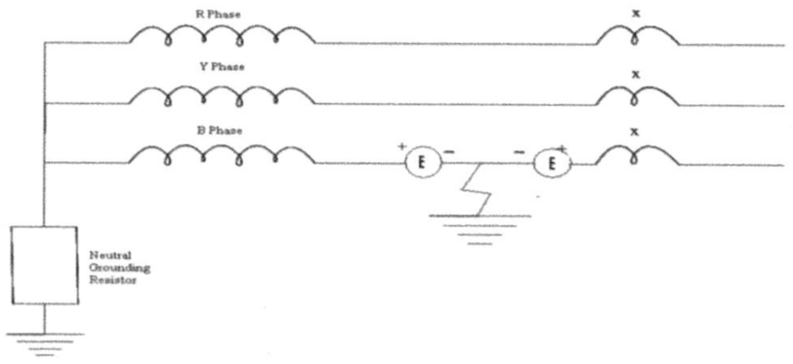

Fig. 2. Equivalent circuit for internal fault

The stator winding faults are created by adding two series voltage sources with reverse polarities to each other in the faulty phase at the external terminals. The magnitude of these voltage sources is equal to the electro motive force of the healthy portion of the windings. The sub-transient, transient and synchronous reactances of synchronous generator are reduced by a value of x which is equal to the sub transient reactance of the healthy portion of the windings.

Table 1. Synchronous Generator Parameters

S. No	Parameter	Rated Value
1	Rated Power (P)	200 MVA
2	Rated Voltage (V)	13.8 KV
3	Rated Frequency (f)	50 Hz
4	Stator Resistance	0.0028544 pu
5	d- axis synchronous reactance (X_d)	1.305 pu
6	q- axis synchronous reactance (X_q)	0.474 pu
7	d- axis sub transient reactance (X_d'')	0.252 pu
8	q-axis sub transient reactance (X_q'')	0.243 pu

3 Theory of ICA

Independent component Analysis is a statistical method for transforming an observed multidimensional vector into components that are mutually as independent [6]. The applications of ICA can be found in many different areas such as audio processing, biomedical signal processing, image processing, telecommunications, econometrics and another emerging application area of great potential is fault detection for power system components.

3.1 Definition of ICA

The basic ICA model can be expressed as in equation 1.

$$X = AS \tag{1}$$

Where $X = (x_1, x_{2,\dots} x_m)$ is the data matrix of measured variables and $S = (s_1, s_{2,\dots} s_n)$ is independent component matrix and A is the unknown matrix. ICA model is a statistical model where the independent components are latent variables which are not directly observed. The observed random vector is used to estimate A and S. The goal of ICA is to find the un-mixing matrix W that will give S. The independent components can be obtained using equation 2.

$$S = WX \tag{2}$$

where the separate matrix $W = A^{-1}$.

3.2 Fixed- Point Algorithm for ICA

For performing ICA, there are several ways to measure independence using different algorithms such as minimization of mutual information and maximization of non-gaussianity. In this paper Fast ICA algorithm is used, which is based on fixed point iteration used for estimation of independent components. Fast ICA algorithm is an efficient algorithm in dimensionality reduction [7]. The maximization of non-gaussianity is obtained by fixed point algorithm using kurtosis. Kurtosis is a classical measure of non gaussianity and it is defined in equation. 3.

$$K(X) = E\{x^4\} - 3(E\{x^2\})^2 \tag{3}$$

The obtained data is centralized and whitened for simpler and stable. The centralization is that data set is made whose mean is zero, the whitening is to avoid correlation.

The new events are detected by calculating the I^2 of the mixed matrix using equation 4. I^2 statistics describes the features of current signal.

$$I^2(k) = \sum_{i=1}^{n} (S_i(k) - \hat{S}_i(k))^2 \tag{4}$$

where $\hat{S}(k) = WX_{new}(k)$. The fault is considered at the instant of k if I^2 (k) > α where α is the threshold of I^2 statistic.

3.3 Algorithm of ICA

The generalized algorithm for fault detection in any power system component using Independent Component Analysis is as follows:

Step 1: Acquire the current signals from the power system component under normal operating condition.

Step 2: The data is centralized such that the mean is zero and whitened to avoid cross correlation.

Step 3: Apply ICA to the data obtain in step 2 and determine the matrix W and calculate independent components

Step 4: Determine the threshold limits for all independent components.

Step 5: The new signal is monitored and transformed to independent components using matrix W and If any of the independent component is outside the specified threshold then the fault is detected and classified.

4 Proposed fault Detection Scheme

The proposed fault detection scheme acquires three phase current I_R, I_Y, I_B signals at the external terminals of synchronous generator from the secondary side of current transformers. The fault signals are analyzed using independent components of the acquired current signals. The faults are detected by the threshold value of I and it is used to capture the signal variability. The algorithm for the proposed ICA based fault detection scheme is summarized in the following algorithm.

4.1 Fault Detection Algorithm of Synchronous Generator Using ICA

Step 1: Obtain the current signals from the R, Y, B terminals of synchronous generator during normal operation.

Step 2: Construct the data matrix $X \in R^{m \times n}$ under normal operating conditions, the data set is to be centralized to zero mean and whitened.

Step 3: Establish ICA model under normal operating conditions.

Step 4: Obtain Independent components based on fixed point iteration of FastICA algorithm.

Step 5: Fault detection is done by calculating the I^2 statistics by certain threshold. if the monitored signal is greater than the specified threshold then the fault is identified.

5 Results and Discussions

A power system network with synchronous generator, three phase step up transformer connected to load and source through transmission line shown in the Fig. 1 is considered for simulation studies. This power system network is developed in MATLAB7® using SIMULINK software [8].

The simulation is carried out for 0.2 sec for the power system network considered and the data is captured for 10 cycles. The internal and external faults are set to occur at instant of 0.06 sec and cleared at 0.07 sec. The three phase current signals I_R, Iy, I_B are obtained from the stator terminals of synchronous generator.

5.1 Internal Faults in Synchronous Generator

The internal faults such as single line to ground fault (L-G), line to line fault (L-L) and three phase fault (L-L-L) of synchronous generator are considered. The synchronous generators are subjected to most common type of fault which is stator ground fault. The internal faults are simulated at 0.1% of the stator windings from generator neutral. The L-G fault near the generator neutral is very difficult to detect as the L-G fault near neutral is high impedance fault. It is difficult to observe the change in signal at the time of fault in time domain analysis whereas the magnitude change in the output of ICA is clearly visible. The distinguishing of various types of internal faults is clearly observed from the magnitude of I^2 statistics. The time response signal of the stator currents and I^2 statistics of the ICA for the internal faults of L-G, L-L and L-L-L are shown in Fig. 3, 4 and 5.

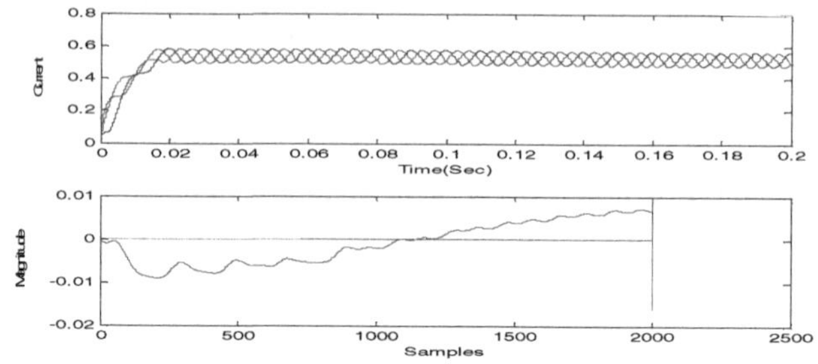

Fig. 3. Stator Currents and I^2 statistics for internal L-G fault

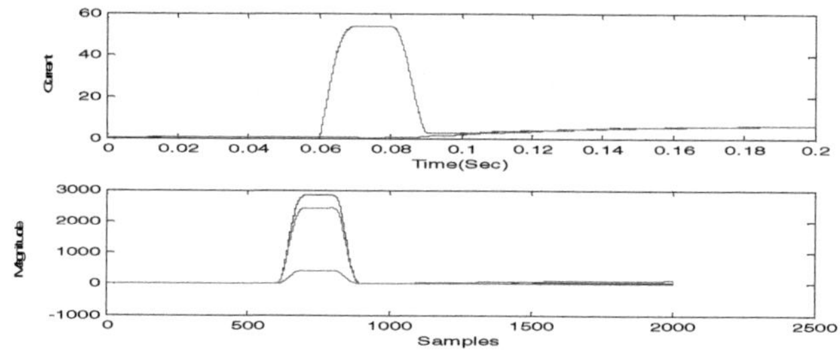

Fig. 4. Stator Currents and I^2 statistics for internal L-L fault

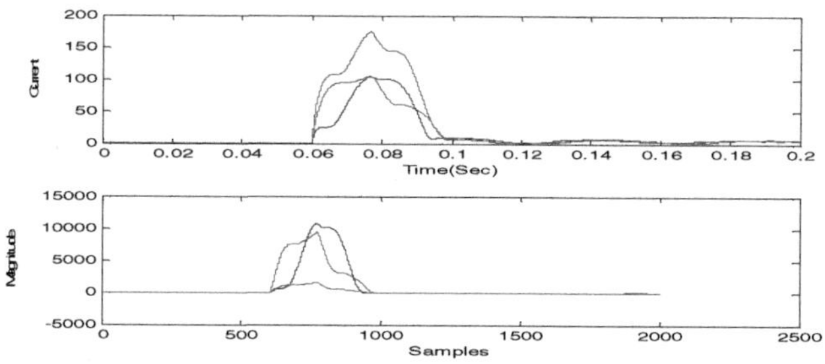

Fig. 5. Stator Currents and I^2 statistics for internal L-L-L fault

5.2 External Faults in Synchronous Generator

The external faults such as L-G, L-L and L-L-L are created outside the stator winding of synchronous generator. External faults are the faults that occur on any one of the equipment such as transformers, transmission lines, bus, load etc in the power system network. The ICA technique is applied to the current signals obtained at the terminals of synchronous generator. The distinguishing of external fault and internal faults is clearly observed from the values of I^2 statistics. The time response signal of the stator currents and of I^2 statistics of the ICA for the external faults of L-G, L-L and L-L-L are shown in Fig 6, 7 and 8. The value of I^2 statistics provides the information of the signals to detect the various types of faults which determine the condition of the system.

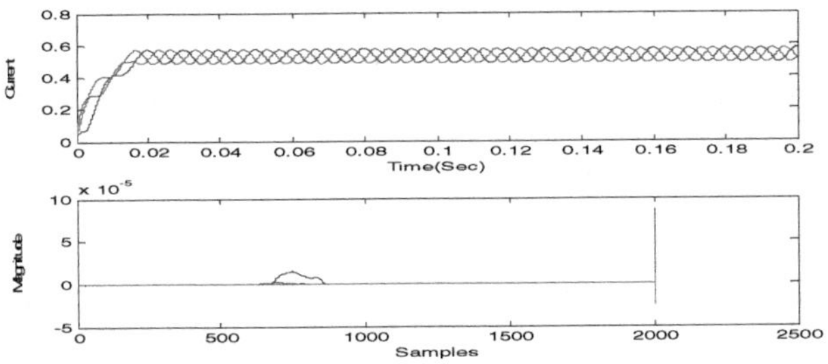

Fig. 6. Stator Currents and I^2 statistics for external L-G fault

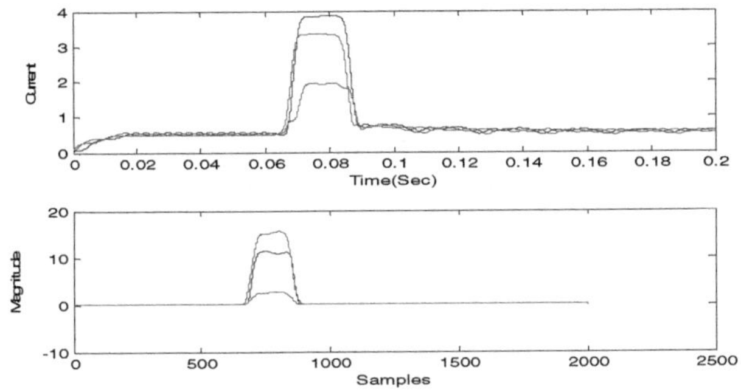

Fig. 7. Stator Currents and I^2 statistics for external L-L fault

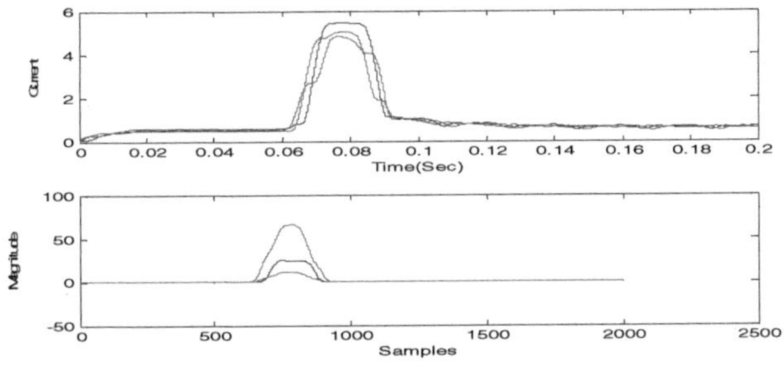

Fig. 8. Stator Currents and I^2 statistics for external L-L-L fault

6 Conclusions

This paper proposes an approach to detect faults using ICA which is one of the statistical techniques. The proposed methodology detects all the faults and discriminates the internal and external faults. The I^2 statistic of ICA model is used for detecting the faults. The results obtained make it possible to validate the method which is applied for detecting and localizing the faults.

References

1. IEEE Power Engineering Education Committee: IEEE Tutorial on the Protection of Synchronous generators (1995)
2. Bo, Z.Q., Wang, G.S., Wang, P.Y., Weller, G.: Non-Differential Protection of Geneartor using Fuzzy Neural Networks. In: International Conference on Power System Technology, pp. 1072–1076 (1998)
3. Megahed, A.I., Mallik, O.P.: An Artificial Neural Network Based Digital Differential Protection Scheme for Synchronous Generator Stator Winding Protection. IEEE Transaction on Power Delivery 14(1), 86–91 (1999)
4. Gaffor, S.A., Ramana Rao, P.V.: A New Wavelet Based Ground Fault Protection Scheme for Turbo Generators. In: Second International Conference on Industrial and Information Systems, Sri Lanka, pp. 353–355 (2007)
5. Taalab, A.I., Dawarish, H.A., Kawady, T.A.: ANN – Based Novel Fault Detector for Generator Windings Protection. IEEE Transaction on Power Delivery 14(3), 824–830 (1999)
6. Bingham, E., Hyvarinen, A.: A fast Fixed-Point Algorithm For Independent Component Analysis of Complex valued Signals. International Journal of Neural Systems 10(1), 1–8 (2000)
7. Shi, Z., Tang, H., Yiyuan: A fixed – point algorithm for complexity pursuit. International Journal of Neural Computing, 529–536 (2005)
8. MATLAB Documentation – Wavelet Tool Box, Ver. 7, The Mathworks Inc., Natick, MA (2008)

Modified Harmony Search for Global Optimization

Suresh Chandra Satapathy[1] and Anima Naik[2]

[1] ANITS, Vishakapatnam
sureshsatapathy@ieee.org
[2] MITS, Rayagada, India
animanaik@gmail.com

Abstract. Harmony search (HS) is a meta-heuristic optimization method imitating the music improvisation process where musicians improvise their instruments' pitches searching for a perfect state of harmony. HS is a reliable, accurate and robust optimization technique scheme for global optimization over continuous spaces. This paper presents an, improved variants of HS algorithm, called the Modified Harmony search (MHS). Performance comparisons of the proposed methods are provided against the original HS and two improved variant of HS such as Improved Harmony search (IHS) and global-best Harmony search (GHS). The Modified Harmony search algorithm on several benchmark optimization problems shows a marked improvement in performance over the traditional HS, HIS and GHS.

Keywords: Meta-heuristic, IHS, GHS.

1 Introduction

Harmony search(HS)[1-5], one of the Swarm Optimization algorithms, was proposed by Z. W. Geem et al inspired by the process of music.

As HS's staggering convergence rate but low accurateness, a host of improving algorithms are proposed. A great improvement is done by Mahdavi et al. [6] that is named under Improved Harmony Search (IHS). The proposed algorithm includes dynamic adaptation for both pitch adjustment rate and bandwidth values.

Another important improvement was done by Omran et al.[7] which named as Global-best harmony search (GHS) inspired by Particle Swarm Optimization (PSO) concepts. The authors proposed GHS to overcome the expected limitation of HIS. The limitation is the difficulty of determining the lower and upper bound of automatic bandwidth (bw). Therefore, they incorporate the PSO concept by replacing the bw parameter altogether and adding a randomly selected decision variables from the best harmony vector in HM. However, the research just ignored the behavior of instruments themselves, which could find the rules behind their behaviors.

In this paper, for the sake of HS's low accurateness, we propose a modified HS and its performance has been shown by evaluating number of benchmark functions.

The remaining of the paper is organized as follows: in Section 2, we give a brief description of Harmony search and two improve variants of HS, IHS and GHS. In

S.C. Satapathy et al. (Eds.): Proc. of Int. Conf. on Front. of Intell. Comput., AISC 199, pp. 293–301.
DOI: 10.1007/978-3-642-35314-7_34

Section 3, we describe the proposed Modified Harmony search (MHS). In Section 4, experimental settings and numerical results are given. The paper concludes with section 5.

2 Harmony Search Algorithm

Harmony search (HS) [9] is a new meta-heuristic optimization method imitating the music improvisation process where musicians improvise their instruments' pitches searching for a perfect state of harmony. The HS works as follows:

Step 1: Initialize the problem and HS parameters:

The optimization problem is defined as Minimize (or maximize) $f(x)$ such that $LB_i \leq x_i \leq UB_i$ .Where $f(x)$ is the objective function, x is a candidate solution consisting of N decision variables (x_i) , and LB_i and UB_i are the lower and upper bounds for each decision variable, respectively. In addition, the parameters of the HS are specified in this step. These parameters are the harmony memory size (HMS), harmony memory considering rate (HMCR), pitch adjusting rate (PAR) and the number of improvisations (NI).

Step 2: Initialize the harmony memory:

The initial harmony memory is generated from a uniform distribution in the ranges $[LB_i, UB_i]$ where $j \leq i \leq N$. This is done as follows:

$$x_i^j = LB_i + r * (UB_i - LB_i)\text{,j=1,2,...........,HMS where } r \sim U(0,1). \quad (1)$$

Step 3: Improvise a new harmony

Generating a new harmony is called improvisation. The new harmony vector $x' = (x_1', x_2', \ldots\ldots, x_N')$, is generated using the following rules: memory consideration, pitch adjustment and random selection. The procedure works as follows:

for each $i \in [1, N]$ do
  if $U(0,1) \leq HMCR$ then
   begin
    $x_i' = x_i^j$ where $j \sim U(1, \ldots \ldots \ldots, HMS)$
    if $U(0,1) \leq PAR$ then
     begin
    $x_i' = x_i' \pm r * bw, where\ r \sim U(0,1)$ and bw is an arbitrary distance bandwidth
      endif
    else
     $x_i' = LB_i + r * (UB_i - LB_i)$
    endif
 done

Step 4: Update harmony memory

The generated harmony vector, $x' = (x_1', x_2', \ldots\ldots, x_N')$, replaces the worst harmony in the HM, only if its fitness (measured in terms of the objective function) is better than that of the worst harmony.

Step 5: Check the stopping criterion

Terminate when the maximum number of improvisations is reached. The HMCR and PAR parameters of the HS help the method in searching for globally and locally improved solutions, respectively. PAR and bw have a profound effect on the performance of the HS. Thus, fine tuning these two parameters is very important. From these two parameters, bw is more difficult to tune because it can take any value from $(0,\infty)$.

For the Improved Harmony search and global-best Harmony search the paper [7].

Table 1. Benchmark functions used in experiment

Name of function	Mathematical representation	Dim	Range of search	Value
Sphere	$f(x)=\sum_{i=1}^{D} x_i^2$	30	[-100,100]	$f_{min}=0$
Step	$f(x)=\sum_{i=1}^{D}(\lfloor x_i+0.5\rfloor)^2$	30	[-100,100]	$f_{min}=0$
Griewank	$f(x)=\frac{1}{4000}\sum_{i=1}^{D} x_i^2-\prod_{i=1}^{D}\cos(\frac{x_i}{\sqrt{i}})+1$	30	[-600,600]	$f_{min}=0$
Ackley	$f(x)=-20\exp\left(-0.2\sqrt{\frac{1}{D}\sum_{i=1}^{D} x_i^2}\right)-\exp\left(\frac{1}{n}\sum_{i=1}^{D}\cos(2*pi*x_i)\right)+20+e$	30	[-32,32]	$f_{min}=0$
Rastrigin	$f(x)=\sum_{i=1}^{D}[x_i^2-10\cos(2\pi x_i)+10]$	30	[-5.12,5.12]	$f_{min}=0$
Schwefel 1.2	$f(x)=\sum_{i=1}^{D}(\sum_{j=1}^{i} x_j)^2$	30	[-100,100]	$f_{min}=0$
Six Hump Camel Back	$f(x)=4x_1^2-2.1x_1^4+\frac{1}{3}x_1^6+x_1x_2-4x_2^2+4x_2^4$	2	[-5,5]	$f_{min}=-1.03163$
Schwefel 2.22	$f(x)=\sum_{i=1}^{D}\lvert x_i\rvert+\prod_{i=1}^{D}\lvert x_i\rvert$	30	[-10,10]	$f_{min}=0$
Sumsquares	$f(x)=\sum_{i=1}^{D} ix_i^2$	30	[-10,10]	$f_{min}=0$
Schwefel 2.21	$f(x)=\max_i\{\lvert x_i\rvert, 1\le i\le D\}$	30	[-100,100]	$f_{min}=0$
Bohachevsky	$f(x)=x_1^2+2x_2^2-0.3\cos(3\pi x_1)-0.4\cos(4\pi x_2)+0.7$	2	[-100,100]	$f_{min}=0$

Table 1. *(continued)*

Bohachevsky2	$f(x) = x_1^2 + 2x_2^2 - 0.3\cos(3\pi x_1) * \cos(4\pi x_2) + 0.3$	2	[-100,100]	$f_{min} = 0$
Bohachevsky3	$f(x) = x_1^2 + 2x_2^2 - 0.3\cos((3\pi x_1) + (4\pi x_2)) + 0.3$	2	[-100,100]	$f_{min} = 0$
Multimod	$f(x) = \sum_{i=1}^{D} \lvert x_i \rvert \prod_{i=1}^{D} \lvert x_i \rvert$	30	[-10, 10]	$f_{min} = 0$
Noncontinuous Rastrigin	$f(x) = \sum_{i=1}^{D} [y_i^2 - 10\cos(2\pi y_i) + 10]$ Where $y_i = \begin{cases} x_i & \lvert x_i \rvert < 0.5 \\ \frac{round(2x_i)}{2} & \lvert x_i \rvert \geq 0.5 \end{cases}$	30	[-5.12,5.12]	$f_{min} = 0$
Zakharov	$f(x) = \sum_{i=1}^{D} x_i^2 + (\sum_{i=1}^{D} 0.5ix_i)^2 + (\sum_{i=1}^{D} 0.5ix_i)^4$	30	[-5, 10]	$f_{min} = 0$
Price	$f(x) = (2x_1^3 x_2 - x_2^3)^2 + (6x_1 - x_2^2 + x_2)^2$	2	[-10,10]	$f_{min} = 0$
Matyas	$f(x) = 0.26(x_1^2 + x_2^2) - 0.48x_1x_2$	2	[-10,10]	$f_{min} = 0$
Martin and Maddy	$f(x) = (x_1 - x_2)^2 + [{}^{(x_1 + x_2 - 10)}/_{3}]^2$	2	[0,10]	$f_{min} = 0$
Shubert	$f(x) = (\sum_{i=1}^{5} icos((i+1)x_1 + i))(\sum_{i=1}^{5} icos((i+1)x_2 + i))$	2	[-10 10]	$f_{min} = -186.73$

3 The Proposed Modified Harmony Search

Inspired by the concept of Differential Evolution with Random Scale Factor (DERSF) [10], a new variation of HS is proposed in this paper. In the GHS the x_k^{best} is chosen always. This term is scaled in this new approach by scale factor in a random manner in the range (0.5, 1) by using

$$0.5 * (1 + rand(0,1)$$

where rand (0, 1) is a uniformly distributed random number within the range [0, 1]. The mean value of the scale factor is 0.75. This allows for stochastic variations in the amplification of the difference vector and thus helps retain population diversity as the search progresses. Even when the tips of most of the population vectors point to locations clustered near a local optimum due to the randomly scaled difference vector, a new trial vector has fair chances of pointing at an even better location on the multimodal functional surface. Therefore the fitness of the best vector in a population is much less likely to get stagnant until a truly global optimum is reached.

The new approach, called Modified -best harmony search (MHS), modifies the pitch adjustment step of the HS such that the new harmony can mimic the best harmony in the HM. The MHS has exactly the same steps as the GHS with the exception that

for each $i \in [1, N]$ do
  if $U(0,1) \leq HMCR$ then
    begin
      $x_i' = x_i^j$ where $j \sim U(1, \dots \dots \dots, HMS)$
      if $U(0,1) \leq PAR(t)$ then
      begin
      $x_i' = 0.5 * (1 + rand(0,1) * x_k^{best}$, where best is the index of the best harmony in the HM and $k \in U(1, N)$
      endif
  else
    $x_i' = LB_i + r * (UB_i - LB_i)$
  endif
done

4 Experimental Settings and Numerical Result

Here we have compares the performance of the modified Harmony search(MHS) with that of Harmony search (HS), global-best harmony search (GHS) and the improved harmony search (IHS) algorithms. As IHS algorithm does not show better performance than GHS [8] in maximum cases, we have considered only HS, GHS and MHS for comparison of their performance. We have taken HMCR = 0.9,PAR_{min} = 0.01 and PAR_{max} =0.99. For all algorithms, HMS = 20. All functions were implemented in 30 dimensions except for the two-dimensional functions. Unless otherwise specified, these values were used as defaults for all experiments which use static control parameters. The initial harmony memory was generated from a uniform distribution in the ranges specified. The benchmark functions given in table 1 have been used to compare the performance of the different methods

Table 2. Performance comparison of HS, GSH and MSH in terms of mean and standard deviation

Name of function	HS	GSH	MSH	Value
Sphere	115.7260±24.8535	0.0537±0.0971	$7.8470e^{-037}$ ±1.124 e^{-36}	0
Step	85.4500±25.8385	0±0	0±0	0
Griewank	1.6498±02381	0.6712±0.0671	0±0	0
Ackley	3.5391±0.3769	0.0357±0.0575	$2.6645e^{-15}$ ±$2.5122e^{-15}$	0
Rastrigin	14.3941±1.1408	0.0752±0.0253	0±0	0
Schwefel 1.2	$4.7722e^{+003}$ ± 118.9107	$1.6851e^{+003}$ ±625.5 763	$1.2464e^{-23}$ ±$1.5711e^{-23}$	0

Table 2. (*continued*)

Six Hump Camel Back	-1.0316±0	-1.0316±0	-1.0316±0	-1.0316
Schwefel 2.22	1.5500±0.2597	0.1431±0.1165	$5.2063e^{-37}$ $\pm 8.7336e^{-36}$	0
Sumsquares	0.6449±0.1423	0.0060±0.0030	$7.2206e^{-20}$ $\pm 9.6697e^{-20}$	0
Schwefel 2.21	7.3217±0.6745	3.5931±1.0873	$4.5880e^{-36}$ $\pm 4.5380e^{-36}$	0
Bohachevsky	0.3192±0.1222	0.0412±0.0068	0±0	0
Boha-chevsky2	0.0284±0.0136	0.0619±0.0056	0±0	0
Boha-chevsky3	0.0019±0.0012	0.0127±0.0076	0±0	0
Multimod	$.661e^{-16}$ $\pm 3.68e^{-16}$	$7e^{-16}$ $\pm 4.3526e^{-16}$	$6.1456e^{-24}$ $\pm 2.1256e^{-24}$	0
Noncontinuous Rastrigin	6.2563±2.5931	$1.5771e^{-7}$ $\pm 2.51e^{-7}$	0±0	0
Zakharov	7.3521±3.1569	0.0665±0.0153	$3.7325e^{-31}$ $\pm 2.1523e^{-31}$	0
Price	0.0323±0.0095	$2.25e^{-5}$ $\pm 6.48e^{-6}$	$7.0585e^{-13}$ $\pm 1.7441e^{-13}$	0
Matyas	0.0522±0.0304	0.0052±0.0027	1.6670e-251±0	0
Martin and Maddy	$2.75e^{-4}$ $\pm 1.008e^{-4}$	$1.72e^{-4}$ $\pm 1.002e^{-4}$	$1.5663e^{-4}$ $\pm 1.1234e^{-4}$	
Shubert	-186.7202±0.0039	-186.72±0.0067	-186.7284±0.0033	

The result reported here by averages and standard deviations over 20 simulations. Each simulation was allowed to run for maximum 20,000 function evaluations of the objective function. Table 2 summarizes the results obtained by applying the three approaches to the benchmark functions in terms of average and standard deviation.

Results

Here we use *t*-distribution test which is quite popular among researchers. Researchers in evolutionary computing [11] to compare the means of the results produced by the best and the second best algorithms (with respect to their final accuracies). In Table 3 we report the statistical significance level of the difference of the means of two best

Table 3. Results of the Unpaired *t*-test Between the best and the Second Best Performing of table 2

Name of function	Standard Error	t	95% confidence internal	Two tailed P	Significance
Sphere	0.0217	2.4733	-0.0976 to -0.0098	<0.05	Significant
Step					NA
Griewank	0.0150	44.7347	-0.7016 to -1.6408	<0.0005	significant
Ackley	0.0129	2.7765	-0.0617 to -0.0097	<0.05	Significant
Rastrigin	0.0057	13.2927	-0.0867 to -0.0637	<0.0005	significant

Table 3. (*continued*)

Schwefel 1.2	139.8836	0.2420	-316.9705to 249.2782		Not significant
Six Hump Camel Back					NA
Schwefel 2.22	0.0261	5.4932	-0.1958 to -0.0904	<0.0005	significant
Sumsquares	6.7082e-4	8.9442	-0.0074 to -0.0046	<0.0005	significant
Schwefel 2.21	0.2431	14.7787	-4.0852 to -3.1010	<0.0005	significant
Bohachevsky	0.0015	27.0959	-0.0443 to -0.0381	<0.0005	significant
Bohachevsky2	0.0030	9.3389	-0.0346 to -0.0222	<0.0005	significant
Bohachevsky3	2.6833e-4	7.0809	-0.0024 to -0.0014	<0.0005	significant
Multimod	1.0953e-7	7.7827	-1.0741e-6 to -6.307e-7	<0.0005	significant
Noncontinuous Rastrigin	5.1227e-4	2.8074	-0.0025 to -4.013 e-4	<0.05	Significant
Zakharov	0.0034	19.4377	-0.0734 to -0.0596	<0.0005	significant
Price	0.0036	4.2181	-0.0224 to -0.0079	<0.0005	significant
Matyas	6.0374e-4	8.6130	-0.0064 to -0.0040	<0.0005	significant
Martin and Maddy	0.0062	0.4610	--0.0153 to -0.0016		Not significant
Shubert	0.0017	3.7125	-0.0096 to -0.0028	<0.0005	significant

algorithms. Here we have done Unpaired t test at a 0.05 level of significance by two-tailed test. From table it is clear that MHS is significant in most of the cases. Therefore, it is evident that the MHS algorithm has a good performance then HS and GHS .

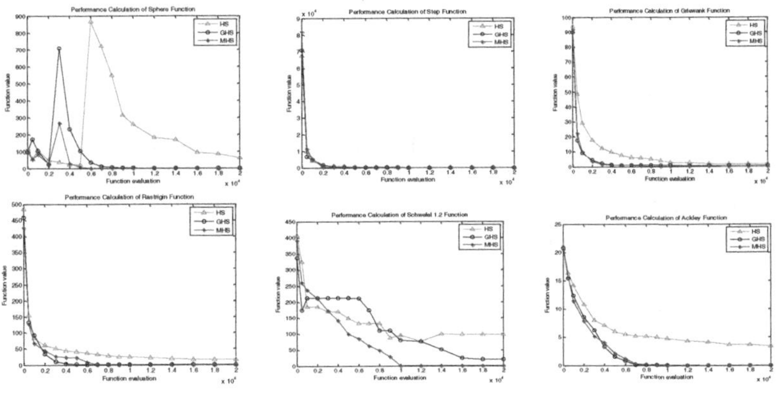

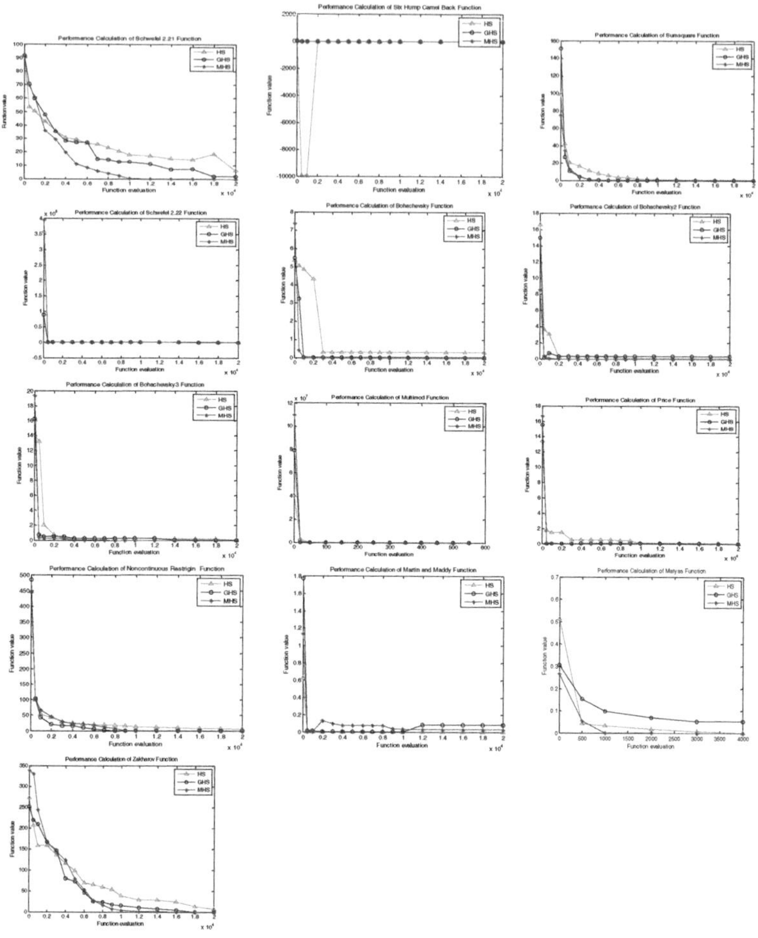

Performance Comparisons

5 Conclusion

This paper proposed a new version of harmony search. The approach modifies the pitch-adjustment step of the HS such that a new harmony is affected by the best harmony in the harmony memory. In addition, the new modification allows the MHS to work efficiently on both continuous and discrete problems. The approach was tested on twenty benchmark functions where it generally outperformed the other approaches

References

1. Geem, Z., Kim, J., Loganathan, G.: A New Heuristic Optimization Algorithm: Harmony Search. Simulation 76(2), 60–68 (2001)
2. Geem, Z.W., et al.: Optimal Design of Water Distribution Networks Using Parameter-Setting-Free Harmony Search for Two Major Parameters. Journal of Water Resources Planning and Management 137(4), 377–380 (2011)

3. Taleizadeh, A.A., et al.: Multiple-buyer multiple-vendor multi-product multi-constraint supply chain problem with stochastic demand and variable lead-time: A harmony search algorithm. Applied Mathematics and Computation 217(22), 9234–9253 (2011)
4. Gil-Lopez, S., Landa-Torres, I., Del Ser, J., Salcedo-Sanz, S., Manjarres, D., Portilla-Figueras, J.A.: A Novel Grouping Heuristic Algorithm for the Switch Location Problem Based on a Hybrid Dual Harmony Search Technique. In: Cabestany, J., Rojas, I., Joya, G. (eds.) IWANN 2011, Part I. LNCS, vol. 6691, pp. 17–24. Springer, Heidelberg (2011)
5. Sui, J., et al.: Mine ventilation optimization analysis and airflow control based on harmony annealing search 6(6), 1270–1277 (2011)
6. Mahdavi, M., Fesanghary, M., Damangir, E.: An improved harmony search algorithm for solving optimization problems. Applied Mathematics and Computation 188(2), 1567–1579 (2007)
7. Omran, M.G.H., Mahdavi, M.: Global-best harmony search. Applied Mathematics and Computation 198(2), 643–656 (2008)
8. Omran, M.G.H., Mahdavi, M.: Global-best harmony search. International Journal of Applied Mathematics and Computation (2007), doi:10.1016/j.amc.2007.09.004
9. Lee, K., Geem, Z.: A new meta-heuristic algorithm for continuous engineering ptimization: harmony search theory and practice. Computer Methods in Applied Mechanics and Engineering 194, 3902–3933 (2005)
10. Das, S., Konar, A., Chakraborty, U.K.: Two Improved Differential Evolution Schemes for Faster Global Search. In: GECCO 2005, Washington, DC, USA, June 25-29 (2005), ACM 1-59593-010-8/05/0006...$5.00
11. Das, S., Abraham, A., Chakraborty, U.K., Konar, A.: Differential evolution using a neighborhood-based mutation operator. IEEE Trans. Evol. Comput. 13, 526–553 (2009)

In Time Access of Biomedical Data through Ant Colony Optimization

A. Haritha[1], L. Pavan Krishna[2], Y. Suresh[1],
K. Pavan Kumar[1], and P.V.S. Lakshmi[1]

[1] Department of Information Technology,
PVP Siddhartha Institute of Technology, Vijayawada -7
[2] Senior Backup and Storage Consultant, Fujitsu, Sydney, Australia
haritha_akkineni@yahoo.com

Abstract. Health care is becoming one of the country's top priorities. This means that there is an increasing demand for quality medical devices, disease management plans, equipment and procedures, as well a need to improve cost-effectiveness. To cope up with this demand there is a need to manage the biomedical data in an effective manner. This progression involves the use of many distributed resources, such as high performance computational resources to analyze the biomedical data, mass storage systems to store them ,the medical instruments ,and advanced visualization and rendering tools. Grids offer the computational power, security and availability needed by such novel applications. The real time biomedical data acquisition plays a prominent role in times of emergency in health care environment. Grid is a distributed environment where data has to be accessed from different computational resources which may limit the in time accessibility of data. In this paper we are focusing on the need for storing the biomedical data on the grids and proposing Ant Colony Optimization algorithm which could be one of the possible solutions to make the crucial data on grids arrive in time before the treatment of the emergency patient is started.

Keywords: Grid, Ant colony optimization, Electronic health records, pheromone.

1 Introduction

Over the last decade, there has been a tremendous progress in information technology which has brought opportunities in improving the state of art of medical services which has been successfully prolonging human lives.

One scenario is that patients digital health record can easily be shared among Hospitals and medical centers via internet, enabling the examination performed in one location while clinical diagnosis by physicians in another location. Electronic processing of health data is important to assist health care. The group of technologies like, health care information systems, electronic medical records and mobile devices for health care plays a major role. Telemedicine is the use

S.C. Satapathy et al. (Eds.): Proc. of Int. Conf. on Front. of Intell. Comput., AISC 199, pp. 303–310.
DOI: 10.1007/978-3-642-35314-7_35

of devices and communications systems, where physiological measurements in patients are transferred using communication technologies like Internet [1].

Health care information systems include computers resources to manage patients information, medical decision supporting data, standards of medical data, images and signals. A system known as M-Health can be used to capture and transfer the physiological measurements in patients which helps high throughput. It uses high throughput computing, mobile devices and physiological sensors[7]. GRID computing provides a powerful alternative to handle large amounts of data and processing. It has potential to be used in medical research, As it includes common clinical trials involving large patients databases, and physiological and genomics modeling [2]. This needs intensive computer processing.

A computer system that is supposed to be used only by those authorized must attempt to detect and exclude the unauthorized. Its access is, therefore, usually controlled by insisting on the authentication procedure to establish, with some degree of confidence, the identity of the user, hence granting those privileges as may be authorized to that identity[2].

In order to provide emergency medical services the patient related information is crucial. This information contains four characteristics for decision making like

(a) Relevance.
(b) Completeness
(c) Structured and automatic ranking
(d) In time arrival of data

The first three characteristics can be achieved using different ranking and mining algorithms.

Our focus is on the fourth characteristics "In time arrival of data". To satisfy that we are proposing a methodology that uses Ant Colony Optimization through which an optimal path can be achieved in grids.

The emergency medical condition can be treated as a hard time situation which requires in time decision. The scheduling and load balancing problems faced in the grid environments are defined as the Nondetermistic Polynomial complete problems which means there is no exact algorithm to solve them in a polynomial time (Blum & Roli, 2003). The only way to solve these problems is to use approximate (heuristic) algorithms such as Genetic Algorithm (GA), SimulatedAnnealing (SA), Tabu Search and recently Ant Colony Optimization (ACO) . We are applying ACO which could be a better facilitator in providing optimal path to reach the destination in time on the health grids.

2 Role of ACO in Grid Environment

2.1 The Gust of Grid Computing

The gust of grid computing has been blowing everywhere. Grid computing is an innovative approach that leverages existing IT infrastructure to optimize computing resources and manage data and computing workloads. According to

Gartner, "a grid is a collection of resources owned by multiple organizations that is coordinated to allow them to solve a common problem".

As grid-connected computers often numbering in the thousands are usually geographically dispersed and heterogeneous, with different processing speeds and operating systems, different grids and networks typically bear little likeness to one another. That can make finding and accessing grid and network resources a costly and time-consuming challenge for the researchers, research institutes and businesses that need them.

Traditionally, the Grid is perceived as an integration platform for data and computing resources as determined by its definite characteristics: "*coordination of resources that are not subject to centralized control*", "*use of standard open and general-purpose protocols and interfaces*", and "*delivery of non-trivial qualities of service*" [4].

2.2 Ant Colony Optimization

Ant Colony Optimization [6] has been inspired by the foraging behavior of real ants. Ants randomly explore the surroundings of the anthill; when they find food, they return to the nest depositing a pheromone trail, a trace of a chemical substance that can be smelled by other ants. Ants can follow various paths to the food source and back, but it has been observed that, thanks to the reinforcement of the pheromone trail by successive passages, only the shortest path remains in use, since ants prefer to follow stronger pheromone concentrations. Pheromone reinforcement is autocatalytic, since the shortest the path, the least time will be taken to travel back and forth, and therefore, while ants on longer paths are still in transit, the ants on the shortest path can restart the route again, reinforcing the pheromone trail on the shortest path. Over time, the majority of the ants will travel on that path, while a minority will still choose alternative paths. The behavior of this minority is important, since it allows to explore the environment to find even better solutions, which initially were not considered. The choice of the path is therefore probabilistic and, while it is strongly influenced by the pheromone intensity, it still allows for random deviations from the current best solution.

The ACO algorithm replicates this behavior, adding some features to make it more efficient in the computer implementation. We implemented the algorithm using ants as a set of concurrent and asynchronous agents. They construct a solution visiting a series of nodes in the grid. They select the move along an edge to the next node to visit according to two parameters: trails and attractiveness. As real ants, also artificial ants will prefer in most cases a deterministic choice of the path, based on the selection of the path with the strongest pheromone and on the highest attractiveness. Yet, in a fraction of cases, the choice will be made probabilistically, though guided by attractiveness and trails[3].

2.3 Algorithm

The given algorithm would be advantageous in finding the optimal path. We can transfer our required information through optimal path to make information reach in time [8].

I. **Algorithm initialization:** Initialize the pheromone values and set parameters.
II. **Initialize the ants:** A group of M artificial ants are used in the algorithm. In each iteration, each ant randomly selects a constructive direction and builds a sequence of tasks.
III. **Constructing the solution:** M ants set out to build M solutions to the problem based on amount of pheromone or trial present on possible paths from starting node and heuristic values using the selection rule of the ACS algorithm.
IV. **Trial Update(Local pheromone update):** Soon after an ant maps a service instance s_{ji} to task T_i , the corresponding pheromone value is updated by a local pheromone updating rule.
V. **Trial Update (Global pheromone update):** After all ants have completed their solutions at the end of each iteration, the tour is analyzed for optimality pheromone values corresponding to the best-so-far solution are updated by a global pheromone updating rule.
VI. **Terminating condition:**If the test is passed, the algorithm will be ended. Otherwise, It repeats from Step II.

As given in the algorithm we can find out the best possible path in the grid environment through which information can be transmitted [5]. Each edge is associated with a *static value* based on the edge-cost

$$\eta(i,j) = \frac{1}{d_{i,j}}$$

Each edge of the graph is augmented with a trace $\tau(i,j)$ deposited by ants. Initially, 0. Trace is dynamic and it is learned at run-time. Each ant tries to produce a complete tour, using the probability depending on $\eta(i,j)$ and $\tau(i,j)$ to choose the next city. Alpha and beta are parameters that control the relative importance of pheromone and visibility.

The transition Probability of the ant k to go from city i to j building its route is given by:

$$p_{ij}^k(t) = \begin{cases} \frac{[\tau_{ij}(t)]^\alpha [\eta_{ij}]^\beta}{\sum\limits_{k \in allowed_k} [\tau_{ij}(t)]^\alpha [\eta_{ij}]^\beta}, & if\ j \in allowed_k. \\ 0, & otherwise. \end{cases}$$

After the ants in the algorithm ended their tours the pheromone trial $\tau(i,j)$ values on every edge are updated according to the following formula

$$\tau_{ij}(t+1) = \rho\tau_{ij}(t) + \Delta\tau_{ij}$$

Where ρ local pheromone decay parameter and it ϵ to 0,1. Where Δ is quantity of per unit length of pheromone trial laid on edge(i,j) by k^{th} ant between time t and (t+n) in a popular ant cycle is given by:

$$\Delta\tau_{i,j}^{k} = \begin{cases} \frac{Q}{L_k}, & if\ (i,j) \in bestTour \\ 0, & otherwise. \end{cases}$$

Where Q is constant. L_k is the tour length of kth ant.

3 Results and Discussion

The algorithm for Ant colony consisting of the strategies described are implemented using MATLAB. The results obtained for the implementation of ACO show that they provide good solutions for small size problems tested, with optimal solutions for the tested problems consisting of 16 computational resources on the health grids. The figures shown are done for 16 computational resources set on the grids.

A number of iterations are conducted based on the probability selection of the ants using the algorithm and it shows that a final optimal solution is attained. The samples of the paths which were obtained from various iteration times are shown in the above figures. A conclusion can be drawn that path shown in Fig. 4 is the optimal path for fetching the data of the patient on the health grid framework. Further improvements can be obtained by applying time greedy heuristics which biases the ants to select services instances with shorter execution time.

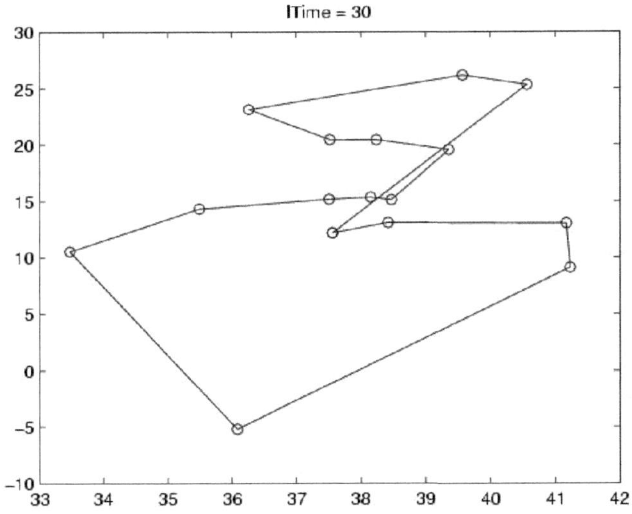

Fig. 1. At Iteration time: 30ms

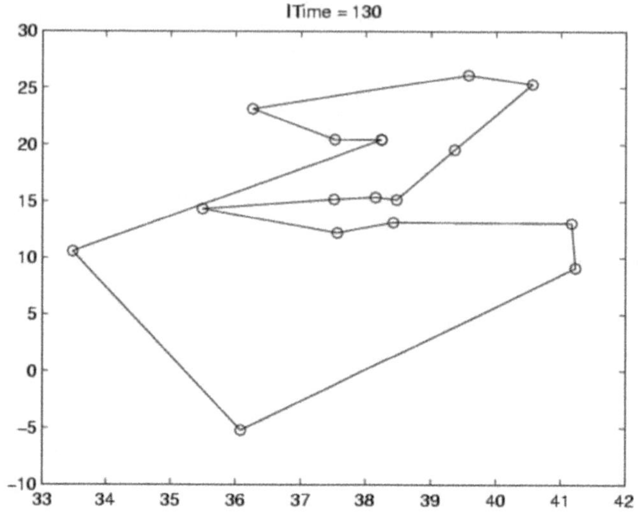

Fig. 2. At Iteration time: 130ms

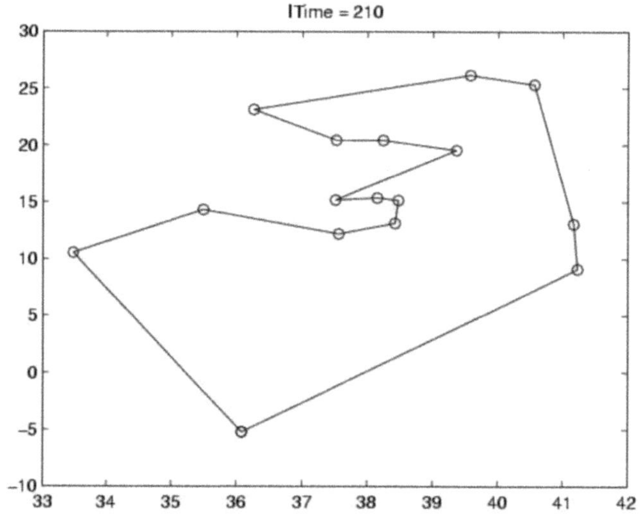

Fig. 3. At Iteration time: 210ms

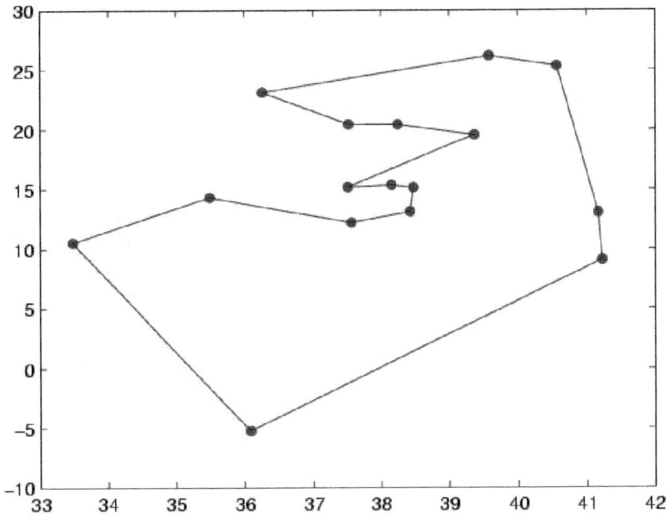

Fig. 4. Optimal path for the 16 computational resources on a health grid

4 Conclusion

Biomedical data requires enormous computing power which can be effectively solved by the Grid computing. Medical data needs to be accessed effectively from the grids which can be achieved using Ant Colony optimization. It has been observed that out of several iterations conducted probabilistically on 16 computational resources in the health grids the path which resulted in the maximum iteration time can be considered as the optimal path for movement of the health data. Further improvements of the approach can be made by considering the heuristic as greedy by time where the issue of in time access of health data can be improved.

References

1. Moorman, P.W., Branger, P.J., Van der Kan, W.J., Van der Lei, J.: Electronic Messaging between Primary and Secondary Care: A Four-year case report. J. Ahmed Inform. Assoc. 8(4), 372–378 (2001)
2. Risk, M., Castrilloy, F.P., Francisco, J., Eijo, G., Ortegay, C.S., Fernandezy, M.B., Diazy, A.P., del Solary, M.R., Pollany, R.R.: CardioGRID: a framework for the analysis of cardiological signals in GRID computing. In: Network Operations and Managent Symposium, LANMOS 2009, pp. 1–4 (2009)
3. Rizzoli, A.E., Montemanni, R., Oliverio, F., Gambardella, L.M.: Ant Colony Optimisation for real-world vehicle routing problems: from theory to applications. Swarm Intelligence 1(2), 135–151 (2007)
4. Foster, I.: What is the Grid? A 3-point Check List. Grid Today 1(6) (June 2002)

5. Duan, H., Ma, G., Liu, S.: Experimental study of adjustable parameters in basic Ant colony Optimization Algorithm. In: 2007 IEEE Congress on Evolutionary Computation (CEC 2007), pp. 147–156 (2007)
6. Dorigo, M., Di Caro, G., Gambardella, L.M.: Ant algorithms for discrete optimization. Artificial Life 5, 137–172 (1999)
7. Mageean, R.J.: Study of "Discharge Communication" from Hospital. Br. Med. J. (CLIN. RES. ED.) 293(6557), 1283–1284 (1986)
8. Chen, W.-N., Zhang, J.: An Ant Colony optimization Approach to a GRID Workflow Scheduling Problem with various QoS Requirement. IEEE Transactions on Systems, Man and Cybernetics - Part C: Appilcations & Reviews 39(1), 29–43 (2009)

Simulation of Frequency Dependent Transmission Line for Identification of Faults and Switching over Voltages

Rajashree Dhua[1] and Chiranjib Koley[2]

[1] Electrical Engineering Department,
Dr. Sudhir Chandra Sur Degree Engineering College affiliated to WBUT
and approved by AICTE, Dumdum, Kolkata, India
rajashreedhua@gmail.com
[2] Electrical Engineering Department, National Institute of Technology, Durgapur, India
chiranjib@ieee.org

Abstract. Though transmission lines are designed to ensure a reliable supply of energy with the highest possible continuity, but about 85-87% of faults in power system occur in transmission lines. Faults can occur due to external causes or internal failures. Identification of type and location of fault is extremely necessary to reduce the outage time and maintenance works. Switching phenomenon occurring in transmission lines often produces similar types of transients as that of faults, making the identification task even more difficult. The present paper discusses about the simulation of symmetrical, unsymmetrical faults, arc faults and a special case of switching transient on a frequency dependent line model. Fault identification has been performed with the help of Short Time Fourier Transform (STFT). STFT has been used to characterize the transient waveforms occurring due to disturbances in time and frequency domain. The study reveals that the amplitude and frequency of the first predominant peak present in the STFT coefficients after the disturbance occurs can help to identify the type of disturbance.

Keywords: Electro Magnetic Transients Programming (EMTP), switching over voltages, transmission line fault identification, Short Time Fourier Transform (STFT).

1 Introduction

Transmission lines are the backbone of the power system which help in the proper and timely delivery of power. Therefore, reliability and continuity of the power supply is one of the major challenges faced, because transmission lines are most prone to faults. Whenever a fault occurs it is necessary to timely locate and identify the fault. Switching is also one of the causes of transients originating in transmission lines and it is also reviewed in the simulation procedure.

Faults occur in transmission lines mostly because these are exposed to severe environmental conditions like falling of a tree branch, storm and lightning strokes. Insulation failure and broken insulators also cause faults in the transmission lines [1], [2].

S.C. Satapathy et al. (Eds.): Proc. of Int. Conf. on Front. of Intell. Comput., AISC 199, pp. 311–320.
DOI: 10.1007/978-3-642-35314-7_36

In an electric power system, a fault is any abnormal flow of electric current. For example, a short circuit is a fault where current flows bypassing the normal load. In three-phase systems, a fault may involve one or more phases and ground, or may occur only between phases. In a "ground fault" or "earth fault", current flows into the earth. If the fault affects all the three phases of the transmission line then a three phase fault occurs which is a symmetrical fault, however a symmetrical fault rarely occurs in practice as majority of faults occurring in transmission lines are unsymmetrical in nature. It must be noted that, although symmetrical faults are less prevalent in transmission lines, but these are the most severe of all faults.

The first real-time digital simulator of transmission line was developed in the year of 1988, which was a constant parameter, balanced, lossless and distortion less line [3]. In the work [3] the lossy line was approximated by treating the line as lossless and adding lumped resistances at both ends and in the middle. However, the distributed natures of all the system parameters (R, L, G and C) as well as their frequency dependence [4] make the solution of the transmission line in time domain very difficult. Much effort has been devoted over a long time to the development of frequency dependant line models for digital computer transient simulation [5-9].

Multiconductor transmission lines usually run distances long enough to make their lumped parameter modeling inaccurate. Approximate models that can fake the distributed nature of the line parameters can be obtained by using several cascaded lumped parameter 'pi' section models. A more accurate model, which is referred to as the constant parameter (CP-model) line model, can be obtained by lumping the resistance and modeling the remaining loss-less part, by using the method of Bergeron [10]. This model incorporates traveling wave delays via a simple equivalent circuit containing a current source and a constant resistance (line's characteristic impedance, Z_0) at each end of the line. The current sources depend upon the voltage and current values from the remote end of the line, with a certain time delay that is determined by the traveling wave velocity and the line length.

Variations of line parameters, such as R, L and C as a function of frequency, are simply ignored when building the CP-model of the line [3]. In order to address this deficiency, a frequency dependent line model (FD-model) is developed by J. Marti [9]. FD-model essentially uses the same equivalent circuit as the CP-model, except for the fact that the characteristic impedance Z_0, appearing at each end of the line, is replaced by a properly chosen network equivalent having approximately the same frequency spectrum as that of Z_0. In addition, the current source values are no longer simple time delayed functions of remote line end variables, but involve more complicated convolutions [9]. Provided that the required accuracy of the fitting functions that approximate the frequency response of Z_0 and the propagation function are attained [9],[11], FD-model of the line has been used for transient simulation studies [9].

As lumped parameter model being inaccurate and not suitable for the simulation of different kinds of transients the distributed parameter frequency dependent model [9] has been used. Therefore, J. Marti's line model [9] is adopted in the real time digital simulator for electromagnetic transient simulation. This new digital line model operating in real time offers the user a more sophisticated and accurate model instead of being restricted to a simple constant parameter model.

For identification of transmission line faults several algorithms using Wavelet Transform (WT) and Neural Network (NN) have been reported [12]. In the work [12], [13], [14] fault classification algorithms have been reported with the help of NN. Due to non-stationary nature of the transient voltage/current signal WT has been successfully used for fault classification [15]. Furthermore, combined techniques such as ANN and fuzzy logic [16], [17], [18]; ANN and WT [12] have already been used. In the proposed work a common test system earlier used by various researchers [13],[18] for the simulation of faults in transmission lines is considered, which is a three phase single circuit 100 km transmission line fed from one end. As the recorded voltage waveforms for different fault conditions are non-stationary in nature, i.e. the harmonic content change with time, the Short Term Fourier Transform (STFT) has been performed in order to study the variation of the transient behavior closely in time-frequency domain. In the present work, STFT has been applied in order to monitor the variation of the spectral component continuously, where, the variation of this spectral component has been found to contain information about the type and location of fault.

Section 2 gives a brief theory of the distributed parameter model of the proposed transmission line. Section 3 describes the test system. Section 4 discusses about the simulation of faults, Section 5 gives the Short Time Fourier Transform (STFT) analysis, Section 6 describes the analysis of STFT coefficients for identification of power system disturbances and Section 7 provides the conclusions of the proposed work.

2 Distributed Parameter Model of a Transmission Line for Simulation of Faults

A distributed parameter based long transmission line model is shown in Fig. 1 (a). In the model, voltage and current along the line are functions of the time (t) and distance (x). The voltage is represented by v(x, t) and the current by i(x, t) and these quantities can be related to the parameters of the line by the so-called Telegrapher's Equations [14]: Now with the help of the Telegrapher's Equations, voltage (v (x, t)) and current (i(x, t)) can be expressed as,

$$\frac{\partial v}{\partial x} + L\frac{\partial i}{\partial t} = -Ri \tag{1}$$

$$C\frac{\partial v}{\partial t} + \frac{\partial i}{\partial x} = 0 \tag{2}$$

Partial differentials (1) and (2) can be solved using the method proposed by Collatz [19]. Here modal transformation technique has been applied on distributed parameters as in [9], [14], [20]. Taking help of the above equations and taking the three phase distributed parameter line under consideration, the modeling of the distributed parameter line for simulation has been carried out. The parameters which are not considered during simulation studies are [20], Variation of temperature, Effect of sag and Effect of corona. These parameters are neglected because these do not significantly affect the different types of transients considered in the proposed work.

3 Test System

The test system is a 100 km long three phase transmission line fed by a 400 kV source at one end and terminated by load impedance as shown in Fig. 1 (b).

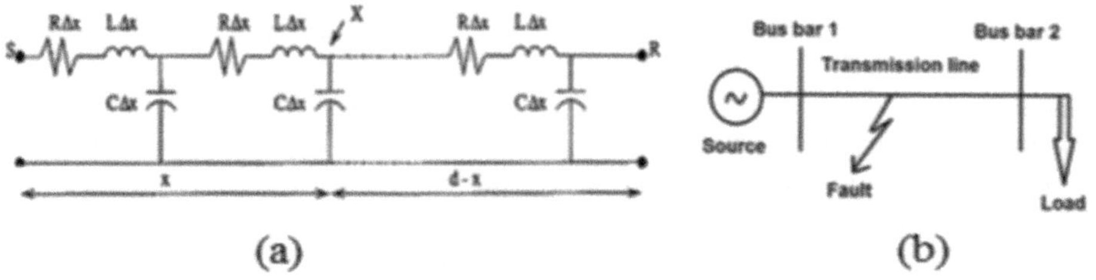

Fig. 1. (a) Distributed parameter model of long transmission line and (b) transmission line fed from one end

The line parameters considered for simulation study are as follows: line length =100 km, source voltage (V_s) = 400 kV, source impedance (Z_S) = (0.2+j4.5) Ω per phase, positive-sequence line parameters: R=2.34Ω, L=95.10mH, C=1.24μF, zero-sequence line parameters: R=38.85Ω, L=325.08mH, C=0.845μF, load impedance (Z_L) = 300Ω-1200Ω per phase with 0.7–0.9 power factor (p.f.) lagging [13],[18].

4 Simulation of Faults

The different types of faults simulated along with their brief description are presented in the following section.

4.1 Symmetrical and Unsymmetrical Faults

In the work symmetrical faults such as three phase fault (L-L-L) and three phase to ground fault (L-L-L-G), are considered. In L-L-L all the three phases are short circuited with some resistances and in the case of L-L-L-G; the three phases are short-circuited and then grounded with some resistances. For simulating unsymmetrical faults such as single-line-to-ground fault (L-G), Line-to-line fault (L-L), and double line to ground (L-L-G) the lines are short circuited in a similar fashion with the relevant nodes.

As in the practical scenario, faults can occur at any location, any time (any inception angle) and the short circuit resistance can vary from as low as few ohms to few hundreds of ohms, simulations are performed by considering the variation of these parameters in the following ranges. Location: 20%, 40%, 60%, 80%, and 100%, Fault Inception Angle (FIA): 20°, 60°, 90°, 135°, and 225°, and Fault resistance: 1 Ω, 10Ω, 50Ω, 100Ω, and 200Ω.

4.2 Arc Faults

Apart from the symmetrical faults, arc faults also occur in the power system, which may be a static arc fault or a dynamic arc fault. An arc fault mainly consists of very high frequency transients or spikes which can cause severe damage if it gets transmitted to the consumer end because every power device is sensitive to high voltage spikes and it can cause permanent damage to the device.
In the work, arc fault having dynamic arc characteristics as represented by equation (3) [21]-[25] has been considered.

$$\frac{dg}{dt} = \frac{1}{\tau}(G - g) \tag{3}$$

where, g time varying arc conductance, G stationary arc conductance and τ time constant. The value of the parameters for simulation of this fault are as follows; fault initiation at 1 ms, Tdynamic (time constant) =3ms, Kdynamic=30.3, Lstatic=0.1m, Gstatic=1e-008 S, Line length=50 km (midpoint of the line).

4.3 Switching over Voltage Simulation

The transients occurring due to switching phenomenon or switching over voltages are often similar to that of other transmission line fault transients; hence a separate study is necessary.

In the work, load rejection over voltages, which originate when a loaded system is suddenly unloaded has been simulated. The simulation network for estimating switching over voltages due to load rejection has been performed with the same line parameters [13], [18]. The 350 MW (0.8 p.f.) load [20] has been suddenly switched off at the crest of the voltage in phase A at 10 ms by opening a switch.

The work takes into account all the above mentioned disturbances that occur in the power system and develops a test system of a frequency dependent transmission line model which incorporates these disturbances with the help of the (Electromagnetic Transients Programming) EMTP software [26].

All the simulations have been done for a duration of 30 ms (1.5 cycles) with a sampling frequency of 100 kHz. The obtained transient voltage of phase A at the source end for symmetrical and unsymmetrical fault simulation is shown in Fig. 2.

From the simulation graphs in Fig. 2, it can be concluded that transients in the transmission line decrease in magnitude with increase in the fault resistance. The initial magnitude of transients decrease with increase in the fault location, however the transients persist over a larger duration of time. The peak magnitude of transients increases with increase in the inception angle of the fault. Though the time domain transients are different for different types of faults, but the transient behavior needs to be quantified by some parameters so that the fault can be identified automatically from the recorded voltage waveforms without the help of any human expert.

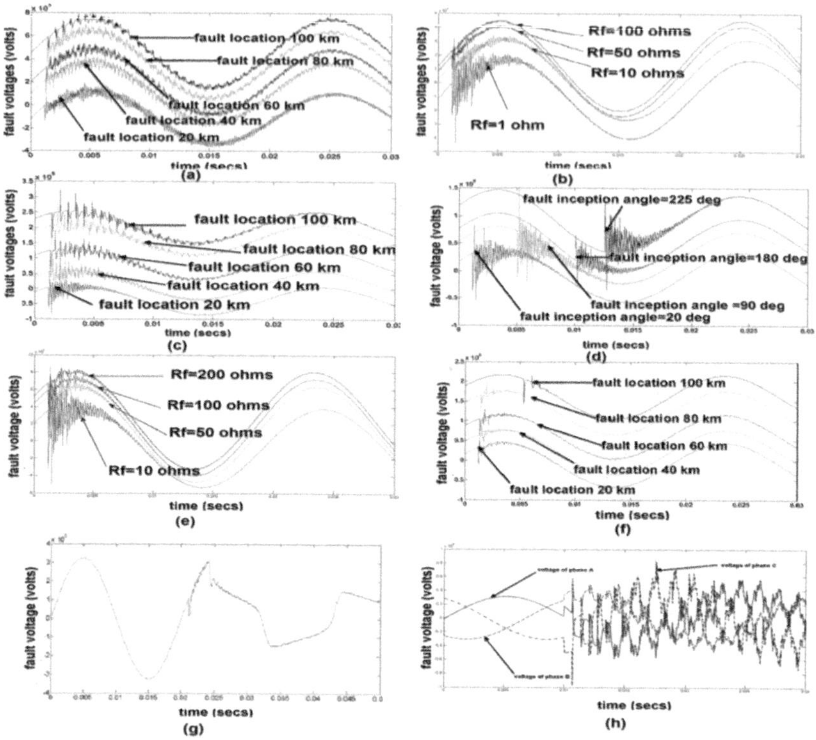

Fig. 2. Time domain plots of different transient voltage waveforms at different conditions (a) line to ground fault (AG fault) with variation of location, (b) line to ground fault (AG fault) with variation of fault resistances, (c) line to line fault (AB fault) with variation of location, (d) line to line fault (AB fault) with variation of fault inception angle, (e) three phase fault (ABC fault) with variation in fault resistance (f) three phase fault (ABC fault) with variation in fault location, (g) dynamic arc fault characteristics for a line to ground fault (AG fault) at 50 km (Tdynamic=3 ms, Lstatic= 0.1m) (h) Switching over voltages of phase A, B and C due to sudden load rejection.

5 Short Time Fourier Transform (STFT)

The Short time Fourier transform (STFT), is a Fourier-related transform used to determine the sinusoidal frequency and phase content of local sections of a signal as it changes over time. This method is also called windowing, because it works by choosing a time function or window, which is essentially non-zero only on a finite interval [27]. The STFT represents a sort of compromise between time-based and frequency-based views of a signal. It provides some information about both when and at what frequencies a signal event occurs. In the discrete time case, the data to be transformed could be broken up into chunks or frames (which usually overlap each other). Each chunk is Fourier transformed, and the complex result is added to a matrix, which records the magnitude and phase for each point in time and frequency.

Through STFT, instead of considering the entire signal, sections of the signals are analyzed and frequency and amplitude of the first predominant peak are taken as features for identification of location and type of fault. The number of windows is 11, with a sampling frequency of 100 kHz.

From Fig. 3 (a), and (b it is observed that the frequency of the predominant harmonic varies with the location of fault. The inception angle can be evaluated through Discrete Wavelet Transform (DWT).

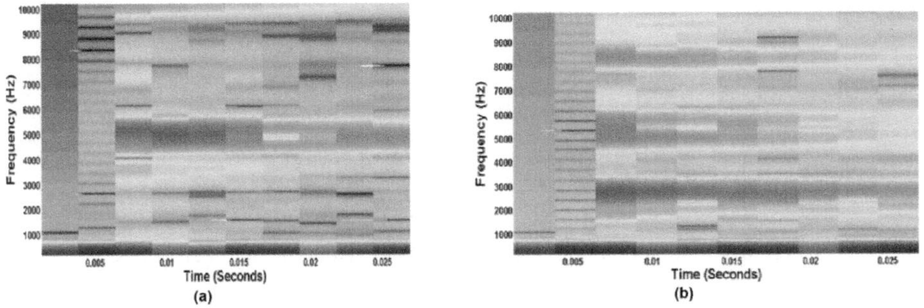

Fig. 3. Spectrograms for ground fault (AG) with variation in fault location, fault location and fault inception angle (a) at 25 km, Rf =1 ohm and inception angle=90°, (b) at 50 km, Rf =1 ohm and inception angle=90° (c) at 75 km, Rf=1 ohm and inception angle=90°, (d) at 50 km, Rf =10 ohms and inception angle=90° (e) at 50 km, Rf =50 ohms and inception angle=90°, (f) at 50 km, Rf =1 ohm and inception angle=20°.

6 Identification of Faults

The features that are extracted from the STFT of transient waveforms are the amplitude and frequency values of the first predominant peak. The variation of the above features with fault location, fault inception angle and fault resistance for symmetrical and unsymmetrical faults, are analyzed. Fig. 4, shows the variation of the amplitude of the first predominant peak w.r.t frequency at different disturbance conditions occurring in the power system.

Amplitude versus frequency plots for different faults establish the fact that arc faults and switching transients can be easily separated because of their distinctly different frequencies and amplitudes. The variations in the amplitude and frequency in a particular type of disturbance are due to variations of the disturbance parameters, like distance, fault/switching inception angle, fault resistance etc. It can be further observed that, when a fault does not involve ground i.e. in the case of AB and ABC type of faults, the amplitude and frequency of the first peak shows a higher value in most of the cases i.e. when fault parameters like fault resistance, fault inception angle and fault location varies.

On the other hand, when a fault involves ground, the amplitue and frequency are of lower magnitudes as compared to those of faults not involving ground. Further investigation during separation of particular fault type like AB from ABC or AG from ABG and ABCG reveals that it is difficult to separate the faults, when variation of the all fault parameters are considered. Fig. 5 shows the scatter plot of faults involving ground i.e. AG, ABG and ABCG.

From the scatter plot of Fig. 5 it can be observed that even though AG fault can be some how separated with some resonable accuracy from ABG and ABCG fault, but separation of ABG fault from ABCG is difficult. Similar observations can be made from Fig. 5, for the faults, which do not involve ground i.e. the AB and ABC faults.

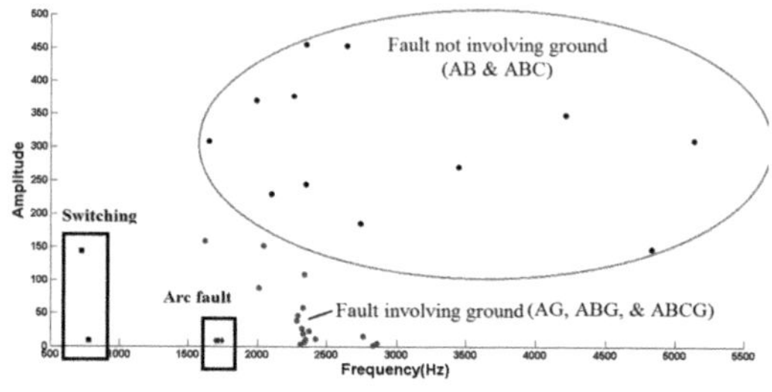

Fig. 4. Variation of amplitude and frequency with type of disturbances

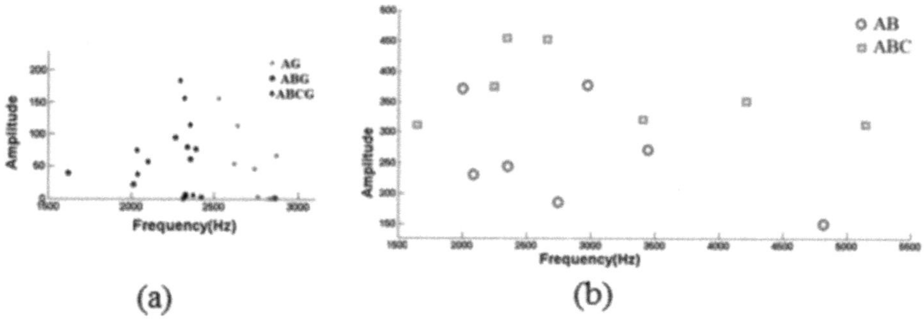

Fig. 5. Variation of amplitude and frequency with type of faults involving (a) ground and (b) not involving ground

7 Conclusions

In the work frequency dependent transmission line model has been used to simulate most of disturbances occurring in transmission lines. During simulation, variation of fault parameters are considered, the simulation study is also further extended to dynamic arc fault. The transient study reveals that the transient behavior of voltage waveforms measured at the source end are different for different fault conditions, if the fault parameters like location, fault resistance and fault inception angle remain same. It has also been observed that, these fault parameters influences the transient characteristics in a similar manner irrespective of the type of fault.

Due to the non-stationary nature of the transient waveforms originating whenever disturbance occurs, STFT has been applied on the time domain signals, which provides idea about the frequency components and their respective magnitudes.

The detailed analysis with the help of STFT reveals that the amplitude and frequency of the first predominant peak (present after the disturbance occurs) can be used to identify the type of disturbances i.e. whether it is an arc fault or switching transients or faults involving ground and not involving ground. But these two parameters fail to identify the exact type of fault i.e. AB or ABC with reasonable accuracy.

Further investigation on several features extracted from STFT may help to identify the exact type of fault. Use of classifiers may also improve the detection accuracy.

References

1. Gaunt, C.T.: Causes of faults on a transmission line in Mozambique- Case study
2. Ross, I.K.P.: "Voltage Sags: An Explanation-Causes, Effects and Correction - Part I. Electricity Today Magazine, 37–39 (November/December 2007)
3. Mathur, R.M., Wang, X.: Real-Time Digital Simulator of the Electromagnetic Transients of Power Transmission lines. IEEE/PES 1988, Summer Meeting, Portland, Oregon, July 24-29, SM-584-5 (1988)
4. Hedman, D.E.: Propagation on Overhead Transmission lines. Pt.2–Earth conduction effects and Practical Results. IEEE Trans. PAS-84, 205–211 (1965)
5. Budner, A.: Introduction of Frequency-Dependent Line Parameters into an Electromagnetic Transients Program. IEEE Trans. PAS-89, 88–97 (1970)
6. Snelson, J.K.: Propagation of Travelling Waves on Transmission lines–Frequency Dependent Parameters. IEEE Trans. PAS-91, 85–91 (1972)
7. Meyer, W.S., Dommel, H.W.: Numerical Modelling of Frequency-Dependent Transmission Line Parameters in an Electromagnetic Transients Program. IEEE Trans. PAS-93, 1401–1409 (1974)
8. Semlyen, A., Dabuleanu, A.: Fast and Accurate Switching Transient Calculations on Transmission Lines with Ground Return Using Recursive convolutions. IEEE Trans. PAS-94, 561–571 (1975)
9. Marti, J.R.: Accurate Modelling of Frequency-Dependent Transmission Lines in Electromagnetic Transients Simulation. IEEE Trans. PAS-101, 147–157 (1982)
10. Bergeron, L.: Du Coup de Belier en Hydraulique au Coup de Foudre en Electricite Dunod, France (1949)
11. Marti, L.: Low-order Approximation of Transmission Line Parameters for Frequency-Dependent Models. IEEE Trans. PAS-102, 3582–3589 (1983)
12. Kale, V.S., Bhide, S.R., Bedekar, R.P., Mohan, G.V.K.: Detection and classification of faults on parallel transmission lines using wavelet transform and neural network. International Journal of Electric and Electronics Engineering 1, 364–368 (2008)
13. Mahanty, R.N., Dutta Gupta, P.B.: Application of RBF neural network to fault classification and location in transmission lines. IEE Proc.-Gener. Transm. Distrib. 151(2), 201–212 (2004)
14. Aggarwal, R.K., Xuan, Q.Y., Johns, A.T., Dunn, R.W., Bennett, A.: A novel fault classification technique for double circuit lines based on a combined unsupervised/supervised neural network. IEEE Transactions on Power Delivery 14(4), 1250–1256 (1999)
15. Abur, A., Ozgun, O., Magnago, F.H.: A Wavelet Transform-Based Method for Improved Modeling of Transmission Lines. IEEE Transactions on Power Systems 18(4), 1432–1438 (2003)
16. Das, B., Reddy, J.V.: Fuzzy-logic-based fault classification scheme for digital distance protection. IEEE Transactions on Power Delivery 20, 609–616 (2005)

17. Wang, H., Keerthipala, W.W.L.: Fuzzy-neuro approach to fault classification for transmission line protection. IEEE Transactions on Power Delivery 13(4), 1093–1104 (1998)
18. Jaya Bharata Reddy, M., Mohanta, D.K.: Performance Evaluation of an Adaptive-Network-Based Fuzzy Inference System Approach for Location of Faults on Transmission Lines Using Monte Carlo Simulation. IEEE Transactions on Fuzzy Systems 16(4), 909–919 (2008)
19. Collatz, L.: The Numerical Treatment of Differential Equations. Springer (1966)
20. Das, J.C.: Switching Transients and Temporary Over voltages. In: Transients in Electrical Systems: Analysis, Recognition and Mitigation, ch. 4, 7, pp. 65–66, 81–84, 155–157, 168–170
21. Maezono, P.K., Altman, E., Brito, K., dos Santos Mello Maria, V.A., Magrin, F.: Very High-Resistance Fault on a 525 kV Transmission Line – Case Study
22. Elkalashy, N.I., Lehtonen, M., Darwish, H.A., Izzularab, M.A., Taalabl, A.-M.I.: Modeling and Experimental Verification of High Impedance Arcing Fault in Medium Voltage Networks. IEEE Transactions on Dielectrics and Electrical Insulation 14(2), 375–383 (2007)
23. Goda, Y., Iwata, M., Ikeda, K., Tanaka, S.-I.: Arc Voltage Characteristics of High Current Fault Arcs in Long Gaps. IEEE Transactions on Power Delivery 15(2), 791–795 (2000)
24. Kizilcay, M., La Seta, P.: Digital simulation of fault arcs in medium-voltage Distribution networks. In: 15th PSCC, Liege, Session 36, Paper 3, August 22-26, pp. 1–7 (2005)
25. Johns, A.T., Aggarwal, R.K., Song, Y.H.: Improved techniques for modeling fault arcs on faulted EHV transmission systems. IEE Proc. -Gener. Transm., Distrib. 141(2), 148–154 (1994)
26. Alternative Transient Program, User Manual and Rule Book, EMTP Center, Leuven, Belgium (1987)
27. Maher, R.C.: FFT based filtering and the Short time Fourier Transform (STFT). ECEN4002/5002 DSP Laboratory (Spring 2003)

A New Improved Particle Swarm Optimization Technique for Reactive Power Compensation of Distribution Feeders

Kamal K. Mandal[1,*], D. Jana[2], B. Tudu[1], and B. Bhattachary[3]

[1] Dept. of Power Engineering, Jadavpur University, Salt lake Campus, Kolkata-700098
kkm567@yahoo.co.in
[2] Dept. of Electrical Engineering, Camellia Institute of Engineering, Madhyamgram Campus, Kolkata-700129, India
djana_143@yahoo.co.in
[3] Dept. of Electrical Engineering, Techno India, Kolkata-700091
bidishna_inf@yahoo.co.in

Abstract. Optimal reactive power compensation is one of the fundamental issues in the operation of power systems. This paper presents a new improved particle swarm optimization technique called black-hole particle swarm optimization (BHPSO) for optimal reactive power compensation of distribution feeders to avoid premature convergence. The performance of the proposed algorithm is demonstrated on a sample test system. The results obtained by the proposed methods are compared with other methods. The results show that the proposed technique is capable of producing comparable results.

Keywords: Reactive Power Compensation, Black-hole Particle Swarm Optimization (BHPSO), Power Loss, Voltage Profile.

1 Introduction

Optimal reactive power compensation plays a very important role in the economic operation of distribution systems. Capacitors have been very commonly used in distribution systems to provide reactive power compensation. They are used to reduce power losses, to improve power factor and to maintain voltage profile within acceptable limits. The voltage profiles throughout the electric power system network have to be kept at acceptable levels to ensure network reliability. The benefits of compensation are highly governed by the optimal location, optimal size of the capacitors and associated cost. Thus, the objective of optimal reactive power planning is to minimize system losses while satisfying various operating constraints under a certain load pattern.

The optimal reactive power compensation is a complex combinatorial optimization problem and several optimization techniques and algorithms have been applied over

---

* Corresponding author.

S.C. Satapathy et al. (Eds.): Proc. of Int. Conf. on Front. of Intell. Comput., AISC 199, pp. 321–328.
DOI: 10.1007/978-3-642-35314-7_37

the years to solve it Some of them are dynamic programming [1], heuristic numerical algorithm [2], mixed integer programming technique[3], fuzzy-reasoning method [4], evolutionary programming [5], An algorithm based on particle swarm optimization technique was proposed by Yu et al [6] for capacitor placement considering harmonic distortion. A new method based on plant growth algorithm was proposed by Wang et al [7]. Chiou et al [8] used variable scale differential evolution technique to find optimal solution for capacitor placement problem in large distribution systems.

Particle swarm optimization (PSO) is one of the comparatively new combinatorial metaheuristic techniques and is based on the social metaphor of bird flocking or fish schooling [9]. A new algorithm based on particle swarm optimization technique called black-hole particle swarm optimization (BHPSO) is proposed in this paper for optimal reactive power compensation. The proposed algorithm was applied on a simple test system to determine its effectiveness. The results have been compared with other evolutionary methods and it is found that it can produce comparable results.

2 Problem Formulation

The primary objective of optimal reactive power compensation is to minimize the total annual cost of the system while satisfying some operating constraints under a certain load pattern. The mathematical model of optimal reactive power compensation can expressed as follows:

$$\min F = \min(COST) \tag{1}$$

where COST includes cost of power loss and capacitor placement. The voltage magnitude at each bus must be maintained within its limits and is expressed as

$$V_{\min} \le |V_i| \le V_{\max} \tag{2}$$

where $|V_i|$ is the voltage magnitude of i th bus i, $V_{\min}$ and $V_{\max}$ are the minimum and maximum bus voltage limits respectively.

A set of simplified feeder-line flow formulation is assumed for simplicity. Considering the one-line diagram shown depicted in Fig.1, the following set of equations may be used for power flow calculation [4].

$$P_{i+1} = P_i - P_{Li+1} - R_{i,i+1}\left[\frac{\left(P_i^2 + Q_i^2\right)}{|V_i|^2}\right] \tag{3}$$

$$Q_{i+1} = Q_i - Q_{Li+1} - X_{i,i+1}\cdot\left[\frac{\left(P_i^2 + Q_i^2\right)}{|V_i|^2}\right] \tag{4}$$

$$|V_{i+1}|^2 = |V_i|^2 - 2(R_{i,i+1}.P_i + X_{i,i+1}.Q_i) + (R_{i,i+1}^2 + X_{i,i+1}^2)\frac{(P_i^2 + Q_i^2)}{|V_i|^2} \tag{5}$$

where P_i and Q_i are the real and reactive powers flowing out of i th bus respectively. P_{Li} , Q_{Li} are the real and reactive load powers at the i th bus respectively. The resistance and reactance of the line section between buses i and i+1 are denoted by $R_{i,i+1}$ and $X_{i,i+1}$ respectively.

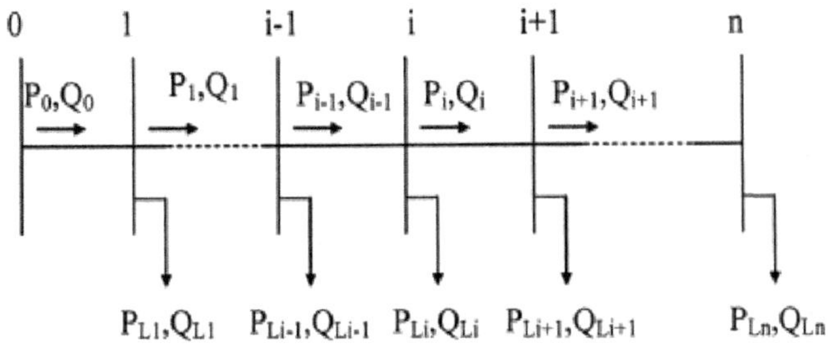

Fig. 1. Convergence characteristics for optimal fuel cost

The power loss of the line section connecting buses i and $i+1$ can be calculated as

$$P_{Loss}(i, i+1) = R_{i,i+1} \cdot \frac{P_i^2 + Q_i^2}{|V_i|^2} \tag{6}$$

The total power loss of the feeder $P_{T,Loss}$ may then be determined by summing up the losses of all line sections of the feeder. The loss is given by

$$P_{T,Loss} = \sum_{i=0}^{n-1} P_{Loss}(i, i+1) \tag{7}$$

The objective of placing compensating capacitor along distribution feeders is to lower the total power loss and keep the bus voltages within their specified limits while minimizing the total cost. Considering the practical capacitors, there exists a finite number of standard sizes which are integer multiples of the smallest size Q_0^c . In general, capacitors of larger size have lower unit prices. The available capacitor size is usually limited to

$$Q_c^{\max} = LQ_0^c \tag{8}$$

where L is an integer. Therefore, for each installation location, there are L capacitor sizes $\{Q_0^c, 2Q_0^c, \ldots\ldots\ldots, LQ_0^c\}$ available. Let $\{K_1^c, K_2^c, \ldots\ldots\ldots\ldots, K_L^c\}$ be their corresponding equivalent annual cost per kVAr. Therefore, the total annual cost function due to capacitor placement and power loss may be found as

$$COST = K_P P_{T,Loss} + \sum_{i=1}^{n} K_i^c Q_i^c \tag{9}$$

where K_P is the equivalent annual cost per unit of power loss in \$/ (kW – year) and i = 1, 2, . . . , n are the indices of buses selected for compensation. The bus reactive compensation power is limited to

$$Q_i^c \le \sum_{i=1}^{n} Q_{Li} \tag{10}$$

3 Particle Swarm Optimization (PSO)

In this section, we briefly describe classical PSO technique and the proposed black-hole particle swarm optimization (BHPSO) technique.

3.1 Classical PSO

Particle Swarm Optimization (PSO) is one of the recent developments in the category of heuristic optimization technique. Kennedy and Eberhart [9] originally developed the PSO concept based on the behaviour of individuals (i.e. particles or agents) of a swarm or group. In a PSO algorithm, the particles fly around the multidimensional search space in order to find the optimum solution. Each particle adjusts its position according to its own experience and the experience of neighbouring particle.

Let in a physical d-dimensional search space, the position and velocity of the i-th particle (i.e. i th individual in the population of particles) be represented as the vectors $X_i = (x_{i1}, x_{i2}, ..., x_{id})$ and $V_i = (v_{i1}, v_{i2}, ..., v_{id})$ respectively. The previous best position of the i th particle is recorded and represented as $pbest_i = (pbest_{i1}, pbest_{i2}, ..., pbest_{id})$. The index of the best particle among all the particles in the group is represented by the $gbest_d$. The modified velocity and position of each particle can be calculated using the current velocity and the distance from $pbest_{id}$ to $gbest_d$ as shown in the following formula:

$$V_{id}^{k+1} = w * V_{id}^k + c_1 * rand(\ ) * (pbest_{id} - X_{id}^k) + c_2 * rand(\ ) * (gbest_d - X_{id}^k) \tag{11}$$
$$i = 1, 2, ..., N_p, \qquad d = 1, 2, ..., N_g$$

where N_p is the number of particles in a swarm or group, N_g is the number of members or elements in a particle, V_{id}^k is the velocity of individual i at iteration k, w is the weight parameter or swarm inertia, c_1 , c_2 are the acceleration constant and $rand(\ )$ is uniform random number in the range [0 1].

The updated velocity can be used to change the position of each particle in the swarm as depicted in (6) as:

$$X_{id}^{k+1} = X_{id}^{k} + V_{id}^{k+1} \tag{12}$$

In general, the inertia weight w the inertia weight w is set according to the following equation:

$$w = w_{\max} - \frac{w_{\max} - w_{\min}}{iter_{\max}} \times iter \tag{13}$$

where $iter_{\max}$ is the maximum iteration number and $iter$ is the current iteration number.

3.2 Black-Hole PSO (BHPSO)

A new black-hole theory was proposed by Stephen Hawking in 2004 which is different than the theory which he had proposed in 1975. According to this new theory, black-holes do not capture all the matter nearby. On the contrary, after a period of time, they release some inhaled matter and energy [10].

This concept is used in the proposed technique based on particle swarm optimization called the black-hole PSO (BHPSO). Here, a black-hole is generated randomly for each particle near the current best particle. It provides another direction to the particle to converge. In this paper, the black-hole is assumed to be a circular region with the position of current best particle as the center and the radius is proportional to the distance between the positions of the current particle and the global best particle. The new position can be found out as follows:

$$X_{id}^{k+1} = X_{id}^{k} + V_{id}^{k+1} \qquad for\ h_{id}^{k} \geq p \tag{14}$$

$$X_{id}^{k+1} = gbest_d + 2R_{id}^{k}(r_1 - 0.5) \quad for\ h_{id}^{k} \langle\ p \tag{15}$$

$$\text{with } R_{id}^{k} = \mu \left| gbest_d - x_{id}^{k} \right| \tag{16}$$

where h_{id}^{k} is the probability of the of the particle i at the iteration drawn from the uniform distribution in the range of [0 1], R_{id}^{k} is the radius of the black-hole, p is the probability threshold and μ is the specified portion over the range [0 1]. Here p and μ are the two user defined parameter.

4 Results and Discussions

The proposed algorithm was implemented using in house Matlab code on 3.0 MHz, 2.0 GB RAM PC. To demonstrate the effectiveness and feasibility of the proposed algorithm, it was applied on a sample test system. The test system [4], [11] under consideration consists of a 23 kV, 9 section feeder.

The equivalent unit cost per unit of power loss considered for the present problem is \$168/(kW-year) [4], [11]. Feeder impedance, three-phase load, available capacitor sizes and other data are from [4],[11] and not reproduced here. The limits on bus voltages are as follows: The limits on bus voltages are as follows:

$$V_{min} = 0.90\ p.u.$$
$$V_{max} = 1.10\ p.u.$$

The following parameters for PSO were selected by trail and error method after several runs. The optimization is done with a randomly initialized population of 40 swarms. The maximum iteration was set at 300. Values of c_1, c_2 are set a 2. The values of p and μ are set at 0.004 and 0.002 following several runs.

It is considered that all the buses were available for compensation. The annual costs, system power loss both before and after compensation, capacitor addition at the desired location are shown in Table 1.

Table 1. Results including voltage profile, annual cost, and capacitor and power loss

Bus No.	Uncompensated Voltage (p.u)	Placed (Qc) (kVar)	Compensated Voltage (p.u)
0	1	0	1.0000
1	0.9929	0	0.9998
2	0.9874	3600	1.0044
3	0.9634	1350	0.9932
4	0.9619	1500	0.9830
5	0.9480	600	0.9617
6	0.9072	600	0.9550
7	0.8890	150	0.9405
8	0.8587	450	0.9177
9	0.8375	450	0.9010
Total cap. size Size (Mvar)		8.700	
Total Loss (MW)	0.7837	0.6729	
Annual cost in (\$/year)	131,675	114,808	
CPU time (sec)		90.12	

It is seen from Table 3 that voltage profile for all the buses are with the system limits. The annual cost is \$114,808 while the system power loss is 0.6729 MW in comparison with uncompensated cases where the annual cost is \$131,675 and power loss is 0.7836 MW. The computation time is found to be 90.12 sec.

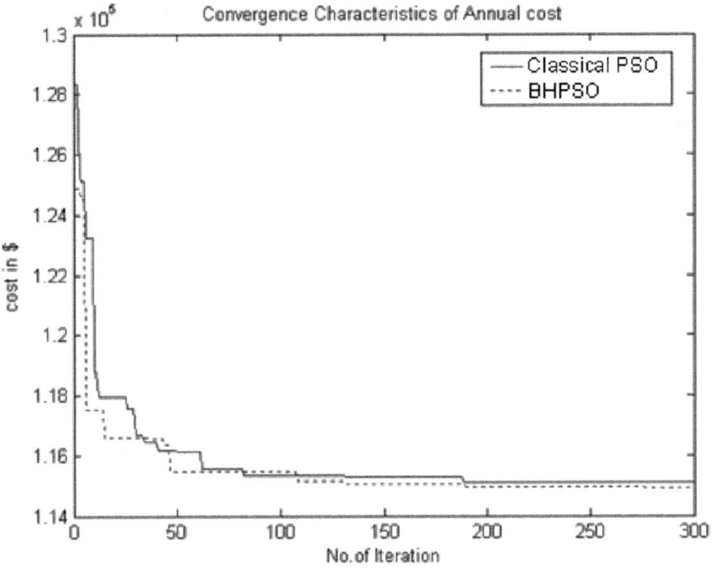

Fig. 2. Convergence characteristics for annual cost

Fig.2 shows the convergence characteristics for optimal annual cost. It also compares the convergence characteristics for classical PSO and the proposed BHPSO.

Table 2. Comparison of results with different methods

	Fuzzy Reasoning	DE	GA	ACSA	Proposed Method
Total Loss (MW)	0.7048	0.6763	0.676 6	0.6753	0.6729
Annual cost in ($/year)	119,420	115,47 1	115,5 72	115,395	114,808

The result is also compared with other methods like fuzzy reasoning [4], Differential Evolution (DE), Genetic Algorithm (GA), Ant Colony Search Algorithm (ACSA) [11] and is shown in Table 2. From Table 2, it is clearly seen that proposed method can produce better results.

5 Conclusion

Optimal reactive power compensation is one of the important tasks in the operation of distribution systems. The basic objective is to reduce power losses as well as to improve voltage profile. In this paper, an algorithm based on a novel improved particle swarm optimization technique has been successfully applied to avoid the premature convergence. To evaluate the performance of the proposed algorithm, it has been applied on a sample test system. The results obtained by the proposed method have been compared with other population based algorithms like fuzzy reasoning, DE, GA, ACSA. The results show that the proposed algorithm is indeed capable of obtaining good quality solution.

Acknowledgments. We would like to acknowledge and thank Jadavpur University, Kolkata, India for providing all the necessary help to carry out this work. This work is supported by the program University with Potential for Excellence (UPE) –Phase II funded by UGC, Govt. of India.

References

1. Duran, H.: Optimum number, location, and size of shunt capacitors in radial distribution feeder: A dynamic programming approach. IEEE Trans. on Power Apparatus and Systems 87(9), 1769–1774 (1983)
2. Baghzouz, Y., Ertem, S.: Shunt capacitor sizing for radial distribution feeders with distorted substation voltages. IEEE Trans. on Power Delivery 5, 650–657 (1990)
3. Baran, M.E., Wu, F.F.: Optimal Sizing of Capacitors Placed on a Radial Distribution System. IEEE Trans. Power Delivery (1), 1105–1117 (1989)
4. Su, C.T., Tasi, C.C.: A new fuzzy reasoning approach to optimum capacitor allocation for primary distribution systems. In: Proc. 1996 IEEE on Industrial Technology Conference, pp. 237–241 (1996)
5. Lai, L.L., Ma, J.T.: Application of evolutionary programming to receive power planning-comparsion with nonlinear programming approach. IEEE Trans. on Power Systems 12, 198–204 (1997)
6. Yu, X., Xiong, X., Wu, Y.: A PSO based approach to optimal capacitor placement with harmonic distortion consideration. Electric Power System Research 71, 27–33 (2004)
7. Wang, C., Cheng, H.Z.: Reactive power optimization by plant growth simulation algorithm. IEEE Trans. on Power Systems 23(1), 119–126 (2008)
8. Chiou, J.-P., Chang, C.-F., Su, C.-T.: Capacitor placement in large scale distribution system using variable scaling hybrid differential evolution. Electric Power and Energy Systems 28, 739–745 (2006)
9. Kennedy, J., Eberhart, R.: Particle swarm optimization. In: Proc. IEEE Conf. Neural Networks (ICNN 1995), Perth, Australia, vol. IV, pp. 1942–1948 (1995)
10. Hogan, J.: Hawking cracks blach hole paradox. New Scientists (2004), `http://www.newscientists.com`
11. Su, C.T., Chan, C.F., Chiou, J.P.: Capacitor placement in distribution system employing Ant Colony Search Algorithm. Electric Components and Systems 33, 931–946 (2005)

A New Optimal Solution to Environmentally Constrained Economic Dispatch Using Modified Real Coded Genetic Algorithm

Debashis Jana[1,*] and Kamal K. Mandal[2]

[1] Dept. of Electrical Engineering, Camellia Institute of Engineering, Kolkata-700129
djana_143@yahoo.co.in
[2] Dept. of Power Engineering, Jadavpur University, Salt lake Campus
Kolkata-700098
kkm567@yahoo.co.in

Abstract. This paper presents a novel optimization algorithm for environmentally constrained economic dispatch (ECED) problem using modified real coded genetic algorithm (MRCGA). The ECED problem is formulated as a non-linear constrained multi-objective optimization dilemma satisfying both equality and inequality constraints. The regenerating population procedure is added to the conventional RCGA in order to improve escaping the local minimum solution by a new combination of crossover and mutation technique. To solve ECED problem the predictable RCGA is customized specially by the concept of self adaptation of mutation distribution followed by polynomial mutation approach with arithmetic crossover. To test performance compatibility between them, a six units system is being considered and the better simulation results produce improved solution compare to different methods.

Keywords: Environmentally Constrained Economic Dispatch (ECED), Modified Real Coded Genetic Algorithm (MRCGA), Improved Crossover and Mutation Combination, Multivariate q-Gaussian Distribution, Self Adaptation.

1 Introduction

Environmentally constrained economic dispatch (ECED) deals with generation dispatch of available power generating units satisfying power demand under economical consideration of operating fuel cost and clean environment. The conscience to reduce the electric utility bill or falling of coal reserve level or emission which is accountable for global warming explain the primary subject that energy usage must be economised with lesser emission. Hence to tackle with the problem one key thing above all is ECED which is the best way to allocate generations among the accessible generating units in a power plant such that the cost of generation and emission become optimum satisfying equality and inequality constraints.

---

[*] Corresponding author.

S.C. Satapathy et al. (Eds.): Proc. of Int. Conf. on Front. of Intell. Comput., AISC 199, pp. 329–337.
DOI: 10.1007/978-3-642-35314-7_38

There are numerous efforts employed to solve the ECED problem for better quality solution with good computational effectiveness. Some of these methods include linear and non-linear goal programming techniques [1] evolutionary programming [2], Fuzzy logic control genetic algorithm [3]. Kumar et al. [4] proposed an efficient optimization based on real-coded genetic algorithm where simulated binary crossover and polynomial mutation are used. M. A. Abido [5] presented superior results of ECED problem using a Non Dominated Sorting Genetic Algorithm (NSGA) and multi-objective evolutionary programming to establish the pareto-optimal set. The algorithm was upgraded by Dev et al. [6] to its improved version of NSGA-II. Rughooputh et al. [7] obtained promising result for solving environmental/economic dispatch problem using NSGA-II. Wu et al. [8] discussed ECED problem using multi-objective differential evolution.

Realizing the versatility of RCGA, an attempt has been made in this paper to solve ECED problem based on MRCGA. MRCGA basically enhances the searching of most fitted chromosome for improved solution which will not be attracted by local minima. The proposed metaheuristic technique was applied on a sample test system consisting of six generators to demonstrate its effectiveness.

2 Problem Formulation of ECED

The following objectives and constraints has been taken into account in the problem formulation for ECED.

2.1 Economic Load Dispatch

The fuel cost function of generating unit is usually defined by the following equation:

$$F_1 = \sum_{i=1}^{NG} F_i\left(P_i\right) = \sum_{i=1}^{NG}\left(a_i P_i^2 + b_i P_i + c_i\right) \quad (\$/h) \tag{1}$$

where $F_i(P_i)$ is the ith unit fuel cost function for P_i output; a_i, b_i, c_i are fuel cost coefficients of ith unit; NG is number of dedicated units. The entire power generation must equal to total power demand and real power losses by transmission lines.

$$\sum_{i=1}^{NG} P_i - P_D - P_L = 0 \tag{2}$$

where P_D is the entire load power demand, P_i is the output power of the ith unit and P_L represents total transmission network losses which is expressed using B-coefficients and unit power output as below.

$$P_L = \sum_{i=1}^{NG}\sum_{j=1}^{NG} P_i B_{ij} P_j + \sum_{i=1}^{NG} B_{oi} P_i + B_{oo} \tag{3}$$

For stable operation of the system, real power output of each generating unit should work between the restricted zone of upper and lower limit as follows

$$P_i^{\min} \le P_i \le P_i^{\max} \tag{4}$$

where P_i^{min} and P_i^{max} are lower and upper hurdle of generation of the ith generating unit . Assuming power loadings of first *(NG – 1)* unit as specified, the power level of NG^{th} unit (i.e. Slack Generator) is introduced by the following equation.

$$P_N = P_D + P_L - \sum_{i=1}^{NG-1} P_i \tag{5}$$

The P_L is function of all the generating units together with that of the dependent unit.

$$P_L = \sum_{i=1}^{NG-1}\sum_{j=1}^{NG-1} P_i B_{ij} P_j + 2P_N\left(\sum_{i=1}^{NG-1} B_{Ni} P_i\right) + B_{NN} P_N^2 + \sum_{i=1}^{NG-1} B_{oi} P_i + B_{ON} P_N + B_{oo} \tag{6}$$

Now rearranging and expanding (5) by (6)

$$B_{NN} P_N^2 + \left(2\sum_{i=1}^{NG-1} B_{Ni} P_i + B_{ON} - 1\right)P_N + \left(P_D + \sum_{i=1}^{NG-1}\sum_{j=1}^{NG-1} P_i B_{ij} P_j + \sum_{i=1}^{NG-1} B_{oi} P_i + \sum_{i=1}^{NG-1} P_i + B_{oo}\right) = 0 \tag{7}$$

The loading of the dependent unit (i.e. NG^{th} unit) can be found by solving (7).

2.2 Economic Emission Dispatch

The economic emission dispatch is modelled for NOx gas using 2nd order polynomial function also right as economic load dispatch problem.

$$F_2 = \sum_{i=1}^{NG} F_{Xi}(P_i) = \alpha_i P_i^2 + \beta_i P_i + \gamma_i \quad (kg/h)\, or\, (t/h) \tag{8}$$

where F_2 is the total amount of emission released from the system in (kg/h) or (ton/h), $Fx_i(P_i)$ is the emission function of the ith generator for P_i output.

2.3 Environmentally Constrained Economic Dispatch

ECED seeks equilibrium between fuel cost and emission i.e.

$$Minimize \qquad C\,(F_1, F_n) \tag{9}$$

where 'n' depends on number of objective function. This equation is optimized subject to above constraints as given in (2) and (4). The above multi-objective optimization is done introducing a price penalty factor (PPF) that transforms multi-objective into a single scalar objective function as follows, where no. of function is two and balance between the two objectives is controlled by weighting vector W.

$$Minimize \quad C = w_1 * (F_1) + w_2 * (PPF) * (F_2)\,(\$/h) \tag{10}$$

The value of PPF has been find out following a general procedure as discussed step wise by Kulkarni et al. [9] using (11)

$$PPF\,[i] = \frac{\sum_{i=1}^{NG}\left(a_i P_i^{\max 2} + b_i P_i^{\max} + d_i\right)}{\sum_{i=1}^{NG}\left(\alpha_i P_i^{\max 2} + \beta_i P_i^{\max} + \lambda_i\right)} . \tag{11}$$

3 RCGA with Continuous Variables Framework

Binary coded genetic Algorithm (BCGA) use a coding of variables and they work with a discrete search space. But GA has also been demonstrated to work directly with continuous variables instead of discrete variables. After creating population of random sets of points, a reproduction operator i.e. roulette-wheel selection operator may be used to pick good strings in the population. In order to create new strings, the crossover and mutation operators used in binary-coded version cannot be used efficiently. In this paper to optimize objective function we worked with RCGA using the new improved combination of arithmetic crossover and polynomial mutation which will be cultivated elaborately afterwards.

3.1 New Improved Combination of Arithmetic Crossover and Polynomial Mutation

The use of real valued representation in the GA suggests a number of benefits in numerical function optimization over binary encoding. Besides efficient floating-point internal computer representations or lossless in precision by discretisation to binary or other values, there is greater freedom to use of different genetic operators efficiently also. For the genetic search to be successful, every promising point in the search space must be reachable from the initial population through crossover only. Hence the crossover operator has noteworthy impact on genetic algorithm performance. It is well known that production of new offspring from the mating of selected pair of parents or mating pairs for crossover. Here we used an arithmetic crossover operator that characterizes a linear combination of two chromosome vectors to produce two new children according to the following equations.

$$\begin{aligned} Child\ 1 &= u * father + (1-u) * mother \\ Child\ 2 &= (1-u) * father + u * mother \end{aligned} \tag{12}$$

where 'u' is a random weighting factor in the range of [0,1], selected by user. In this particular solution approach two chromosome vectors selected randomly for crossover, $C_i^{new\_genx}$ and $C_j^{new\_genx}$ will generate two offspring, C_i and C_j which is a linear combination of their parents.

Thus the equation (12) can be rewritten as follows

$$\begin{aligned} C_i &= u * C_i^{new\_genx} + (1-u) * C_j^{new\_genx} \\ C_j &= (1-u) * C_i^{new\_genx} + u * C_j^{new\_genx} \end{aligned} \tag{13}$$

where $u = (1 + 2 * \alpha) * rand - \alpha$, here α is chosen as 0.5.

The mutation operator is employed to inject new genetic material into the population. Even so in order to prevent the permanent loss of any particular gene value, it is applied to each new structure individually. Though, mutation is characteristically a secondary operator and cannot be trusted upon for reaching the complete search space. But proposed method is being aimed to add effective value for a complete search space. Currently in the real coded GA, the polynomial mutation [10] has been used satisfactorily. A solution is generally produced in the m-dimensional real-valued search space by the mutation operator from the individual $\vec{x}_i$, where i = 1, ..., μ, as follows:

$$\vec{x}_i = \vec{x}_i + c\,\vec{z}$$

where c is the matrix which show the mutation strength in each coordinate j = 1, . , m and $\vec{z}$ is an m-dimensional random vector created from a provided multivariate distribution. Here, an m-dimensional random vector is produced from the multivariate Gaussian distribution [11]. We put forward to breed the random mutation vector $\vec{z}$ from an isotropic q-Gaussian distribution as $\vec{z} \sim r\,\vec{u}_n$, where $\vec{r}$ is a uniform random vector acquired by sampling a random vector with Gaussian distribution. This random vector also can be denoted by $\vec{z} \sim \chi_q{}^m$, while an m-dimensional random vector generated by sampling *m* independent *q*-Gaussian random variables $\chi_q(0,1)$. In q-Gaussian distribution self-adaption [11] is made of by the parameter q which illuminates the shape of the distribution. Then the muted child is computed by this concept as follows

$$P_{gi} = P_{gi} + \delta\left(P_{gi}^{\max} - P_{gi}^{\min}\right) \tag{14}$$

A random number u_n is generated between 0 and 1 and a calculation is done for the variable δ

$$\delta = \begin{cases} \left[2u_n + (1-2u_n)(1-\phi)^{\eta m+1}\right]^{1/(\eta m+1)} - 1 & \text{if} \quad u_n \le 0.5 \\ 1 - \left[2(1-u_n) + 2(u_n - 0.5)(1-\phi)^{(\eta m+1)}\right]^{1/(\eta m+1)}, & \text{otherwise} \end{cases} \tag{15}$$

where computation is made for ϕ as below

$$\phi = \min\left[\left(P_{gi} - P_{gi}^{\min}\right), \left(P_{gi}^{\max} - P_{gi}\right)\right] / \left(P_{gi}^{\max} - P_{gi}^{\min}\right)$$

and the parameter ηm is the distribution index for mutation and acquires any non-negative value. In the above mentioned solutions, the perturbance may be adjusted by changing ηm and p_m with generations as given below.

$$p_m = 1/N + \left(gen / gen_{\max}\right)(1 - 1/N)$$
$$\eta m = \eta m_{\min} + gen\,,$$

where p_m is known as the probability of mutation, ηm_{min} is the user defined minimum value for ηm and $N = NG\text{-}1$ is the maximum number of decision variables in the solution vector. The proposed solution approach illuminate that no solution would be produced outside the range of Pg_i^{min} and Pg_i^{max}.

4 Simulation Results

The proposed algorithm based on MRCGA has been applied to ECED problem with simple test case of six-unit system for power demand of 900 MW. In the test case we have used smooth fuel and emission level function to demonstrate the performance of MRCGA which has been implemented in Matlab 7.8.0 (R2009a) on a PC (Core 2 duo, RAM 2 GB, 2.10 GHz). In the computation population size and maximum number of generation have been selected as 100 and 200 respectively with crossover probability of 0.8 and the value of q in the interval of $1 < q < 3$. The six generating units with different operating capacity shown in Table 1, fuel cost and emission level co-efficient which is given in Table 2 for a particular B co-efficient matrix in [7].

Table 1. Generating Capacity Limit

Unit	P_1	P_2	P_3	P_4	P_5	P_6
P_{min} (MW)	10	10	35	35	130	125
P_{max} (MW)	125	150	225	210	325	315

Table 2. Generator Characteristics

Unit	Fuel Cost Coefficient			Emission Coefficient		
	a_i	b_i	c_i	α_i	β_i	γ_i
1	0.15247	38.53973	756.79886	0.00419	0.32767	13.85932
2	0.10587	46.15916	451.32513	0.00419	0.32767	13.85932
3	0.02803	40.39655	1049.99770	0.00683	-0.54551	40.26690
4	0.03546	38.30553	1243.53110	0.00683	-0.54551	40.26690
5	0.02111	36.32782	1658.56960	0.00461	-0.51116	42.89553
6	0.01799	38.27041	1356.65920	0.00461	-0.51116	42.89553

At first fuel cost and then emission level is minimized individually and found subsequent emission and fuel cost respectively which are depicted in Table 3 and Table 4 respectively. Table 3-5 shows actual generation schedule, power loss and fuel cost for demand of 900 MW. During minimum fuel cost estimation for $w_1=1$ and $w_2=0$, we find fuel cost of ***$49298.0611*** per hour by MRCGA and corresponding emission was *850.66193 Kg* per hour. The convergence characteristic is shown in Fig. 1. During minimum emission estimation for $w_1=0$ and $w_2=1$, we find emission quantity of ***749.48503 Kg*** per hour and corresponding fuel cost was *$51008.36538* per hour. Hence a comparison is given as depicted in Table 3 and Table 4. It is noted that fuel cost or emission reduction is followed by significant reduction of power loss.

Table 3. Minimum Fuel Cost Estimation

Unit Output (MW)	MRCGA	DE/BBO [12]	NSGA-II [7]	FCGA [3]
P_1	102.99296	101.58489	102.963	101.11
P_2	70.16013	72.59232	74.235	67.64
P_3	60.17861	62.07912	66.003	50.39
P_4	140.34827	144.09377	140.316	158.80
P_5	324.99548	325.00000	324.888	324.08
P_6	252.10839	252.13712	248.416	256.56
Total Output (MW)	950.78385	957.48724	956.822	958.58
P Loss (MW)	50.78385	57.48724	56.822	58.58
Fuel Cost ($/h)	**49298.0611**	**49,615.05371**	**49,620.824**	**49655.40**
NOx (Kg/h)	850.66193	857.092187	849.326	877.61

Table 4. Minimum Emission Level Estimation

Unit Output (MW)	MRCGA	DE/BBO [12]	NSGA-II [7]	FCGA [3]
P_1	124.99977	125.00000	124.998	133.31
P_2	111.30621	113.05141	109.893	110.00
P_3	110.75468	111.27399	111.081	100.38
P_4	141.59720	143.19041	141.961	119.27
P_5	249.97226	253.61250	254.36	250.79
P_6	224.30559	223.56209	226.578	251.25
Total Output (MW)	962.93571	969.69042	968.87	965.00
P Loss (MW)	62.93571	69.69042	68.87	65.00
Fuel Cost ($/h)	51008.36538	51368.2467	51254.195	53299.64
NOx (Kg/h)	**749.48503**	**759.8670138**	**760.052**	**785.64**

So the two results show that when we go for single objective optimization other function will give increased value with respect to another. To find out best compromise value we have executed more than 30 trials to get the optimum point by choosing different values of weighting factor, w. At first on search for best compromise solution the value of w_1 is increased from 0 to 1 and w_2 is decreased simultaneously from 1 to 0. After that in similar fashion the results have been tabulated by vice-versa process and examine the nature of gross cost value of

Table 5. Best Compromise Solution of Fuel Cost and Emission Level

Unit Output(MW)	MRCGA	DE/BBO[12]	NSGA-II[7]	FCGA [3]	NR [13]
P_1	124.99887	125.0000	120.0587	111.40	122.004
P_2	94.63170	96.0320	85.202	69.33	86.523
P_3	99.67527	100.4221	89.565	59.43	59.947
P_4	139.87153	141.5235	140.278	143.26	140.959
P_5	266.21355	270.6546	288.614	319.40	325.000
P_6	229.31098	227.7011	233.687	252.11	220.063
Total Output (MW)	954.70189	961.3335	957.405	954.92	954.498
P Loss (MW)	54.70189	61.3335	57.405	54.92	54.498
Fuel Cost ($/h)	50278.2115	50622.181947	50,126.059	49674.28	50807.24
NOx (Kg/h)	755.63074	766.2497	784.696	850.29	864.060
PPF_{NOx} ($/h)	47.82205	47.82224	47.82224	47.82224	47.82224
NOx ($/h)	36135.8150	36643.7811	37525.9204	40662.7724	41321.2846
Total Cost ($/h)	**86414.0266**	**87265.9630**	**87651.9794**	**90337.0524**	**92128.5246**

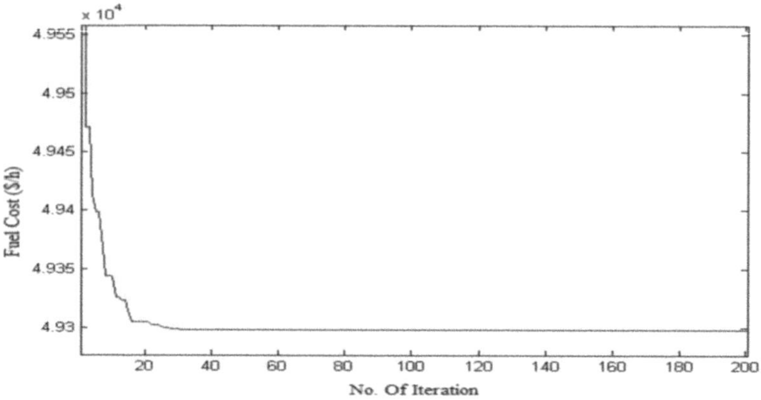

Fig. 1. Convergence characteristic for minimum fuel cost estimation

succeeding and preceding trails. The point vector w_1=0.02 and w_2= 0.98 where one trial reached to the best solution which was find out during the entire trail process and it is seen that Table 5 produce better and acceptable result compared to other methods.

5 Conclusions

The proposed technique presented in this paper is applied to ECED problem as formulated multi-objective optimization with objectives of operating fuel cost and emission level but it was converted into single objective by linear combination of two objectives as weighted sum by using weighting factor giving the bi-objective presentation. As it is earlier said that the technique which has been modified with new combination of crossover and mutation operations, has performed satisfactorily and efficiently in six units test systems. We obtained the best solution by varying the weighting factor for several trial solutions. Hence except this limitation to get best compromise solution by single simulation run, the numerical results shows improved solution with key effectiveness.

Acknowledgement. We would like to acknowledge and thank Jadavpur University, Kolkata, India for providing all the necessary help to carry out this work.

References

1. Nanda, J., Kothari, D.P., Lingamurthy, K.S.: Economic-emission load dispatch throughgoal programming techniques. IEEE Transaction on Energy Conversion 3(1), 26–32 (1988)
2. Sinha, N., Chakrabarti, R., Chattopadhyay, P.K.: Evolutionary programming techniques for economic load dispatch. IEEE Trans. Evol. Comput. 7(1), 83–94 (2003)
3. Song, Y.H., Wang, G.S., Wang, P.Y., Johns, A.T.: Environmental/economic dispatch using fuzzy logic controlled genetic algorithms. IEE Proceedings Generation, Transmission and Distribution 144(4), 377–382 (1997)

4. Sushil, K., Naresh, R.: Non Convex Economic Load Dispatch using an Efficient Real Coded Genetic Algorithm. Applied Soft Computing 9, 321–329 (2009)
5. Abido, M.A.: A Niched Pareto genetic algorithm for multi-objective environmental/economic dispatch. International Journal of Electrical Power and Energy System 25(2), 79–105 (2003)
6. Deb, K., Agrawal, S., Pratap, A., Meyarivan, T.: A Fast Elitist Non-dominated Sorting Genetic Algorithm for Multiobjective Optimization: NSGA-II. In: Deb, K., Rudolph, G., Lutton, E., Merelo, J.J., Schoenauer, M., Schwefel, H.-P., Yao, X. (eds.) PPSN 2000. LNCS, vol. 1917, pp. 849–858. Springer, Heidelberg (2000)
7. Rughooputh, H.C.S., King, R.T.F.A.: Environmental/economic dispatch of thermal units using an Elitist Multiobjective Evolutionary Algorithm. In: IEEE Conf. on Industrial Technology, Maribor, Slovania, December 10-12, vol. 1, pp. 48–53 (2003)
8. Wu, L.H., Wang, Y.N., Yuan, X.F., Zhou, S.W.: Environmental/economic power dispatch problem using multi-objective differential evolution algorithm. Electric Power Systems Research 80(9), 1171–1181 (2010)
9. Kulkarni, P.S., Kothari, A.G., Kothari, D.P.: Combined Economic and Emission Dispatch Using Improved Backpropagation Neural Network. Electrical Machines and Power Systems 28, 31–44 (2000)
10. Deb, K., Goyal, M.: A combined genetic adaptive search (GeneAS) for engineering design. Comput. Sci. Inform. 26(4), 30–35 (1996)
11. Beyer, H.G., Schwefel, H.S.: Evolution strategies: a comprehensive introduction. Natural Computing 1(1), 3–52 (2002)
12. Bhattacharya, A., Chattopadhyay, P.K.: Solving Economic emission Load Dispatch problems using hybrid differential evolution. Applied Soft Computing 11, 2526–2537 (2011)
13. Dhillon, J.S., Parti, S.C., Khotari, D.P.: Stochastic economic load dispatch. Electric Power Systems Research 26(3), 179–186 (1993)

A New Substitution Block Cipher Using Genetic Algorithm

Srinivasan Nagaraj[1], D.S.V.P. Raju[2], and Kishore Bhamidipati[3]

[1] Dept. of CSE, GMRIT, Rajam
[2] CSE, Andhra University, Visakhapatnam
[3] Dept. of CSE, MIT, Manipal
{sri.mtech04,kishore.gmr}@gmail.com

Abstract. In cryptography, a **substitution block cipher** is a method of encryption by which units of plain text are replaced with cipher text according to a regular system. The receiver deciphers the text by performing an inverse substitution. If the cipher operates on single blocks, it is termed as **simple substitution block cipher**. We proposed an algorithm which considers a random matrix key which on execution of a sequence of steps generates a sequence. Based on the equality of values, this sequence is being divided into basins. Each basin represents one block of data on which the genetic algorithm operations like crossover and mutation are performed. Each block of plain text is replaced by summation of ASCII value of plain text and the sequence is generated to form the cipher text. Thus, the cipher text obtained is very difficult to be broken without knowing the key, which provides high security.

Keywords: Genetic Algorithm (GA), random matrix key, basins, sequence generation.

1 Introduction

1.1 Cryptography

There are many aspects to security and many applications, ranging from secure commerce and payments to private communications and protecting passwords. One essential aspect for secure communications is that of cryptography. But it is important to note that while cryptography is necessary for secure communications, it is not sufficient by itself. There are some specific security requirements, including:

- ***Authentication.*** The process of proving one's identity
- ***Privacy/confidentiality.*** Ensuring that no one can read the message except the intended receiver.
- ***Integrity.*** Assuring the receiver that the received message has not been altered in any way from the original.
- ***Non-repudiation.*** A mechanism to prove that the sender really sent this message.

S.C. Satapathy et al. (Eds.): Proc. of Int. Conf. on Front. of Intell. Comput., AISC 199, pp. 339–347.
DOI: 10.1007/978-3-642-35314-7_39

1.2 Background

Several solutions have been proposed in this area. In 1993, for the first time, the paper by Spillman [8] presented a genetic algorithm based approach for the cryptanalysis of substitution cipher. The paper has explored the possibility of random type search to discover the key (or key space) for a simple substitution.

1.2.1 Block Ciphers

In cryptography, a **block cipher** is a symmetric key cipher operating on fixed-length groups of bits, called *blocks*, with an unvarying transformation. A block cipher encryption algorithm might take (for example) a 128-bit block of plaintext as input, and output a corresponding 128-bit block of cipher text. The exact transformation is controlled using a second input the secret key. Decryption is similar: the decryption algorithm takes, in this example, a 128-bit block of cipher text together with the secret key, and yields the original 128-bit block of plain text. A message longer than the block size (128 bits in the above example) can still be encrypted with a block cipher by breaking the message into blocks and encrypting each block individually. To overcome this issue, modes of operation are used to make encryption probabilistic.

2 Genetic Algorithms

A GA is general method of solving problems to which no satisfactory, obvious, solution exists. The Genetic Algorithms (GAs) are adaptive heuristic search algorithm based on the evolutionary ideas of natural selection and genetics. Although randomized, GAs are by no means random, instead they exploit historical information to direct the search into the region of better performance within the search space. It is better than conventional AI in that it is more robust. Unlike older AI systems, they do not break easily even if the inputs changed slightly, or in the presence of reasonable noise.

The algorithm evolves through these operators: **Selection** (which equates to survival of the fittest), **Reproduction** (which is usually the first operator applied on a population selects good chromosomes in a population to form the mating pool), **Crossover** (which represents mating between individuals) and **Mutation** (which introduces random modifications).

2.1 GA's Operations in Our Model

2.1.1 Crossover Operator

- Prime distinguished factor of GA from other optimization techniques.
- Two individuals are chosen from the population using the selection operator
- The values of the two strings are exchanged up to this point
- The two new offspring created from this mating are put into the next generation of the population
- By recombining portions of good individuals, this process is likely to create even better individuals

Parent 1: 8 9 1 3|6 5 4 2 7
Parent 2: 1 5 4 9|7 8 2 3 6
Child 1: 8 9 1 3|5 4 7 2 6
Child 2: 1 5 4 9|8 3 6 2 7

The first part of first child is the first part of first parent and second part of first child is the remaining digits as the order of second parent, and the first part of second child is the first part of second parent and second part is the remaining digits as the order of first parent.

2.1.2 Mutation Operator

- With some low probability, a portion of the new individuals will have some of their bits flipped.
- Its purpose is to maintain diversity within the population and inhibit premature convergence.
- Mutation and selection (without crossover) create a parallel, noise-tolerant, hill-climbing algorithms

Child: 1 5 4 9 8 3 6 2 7
Mutated child: 1 5 3 9 8 4 6 2

2.2 Tradeoffs between Cryptography and GA

Genetic algorithms (GA's) are a class of optimization algorithms attempt to solve problems through modeling a simplified version of genetic processes.

Table 1.

Parameter	GA	GA in Cryptanalysis
Gene	A Single bit in Chromosome	A Single bit in key
Chromosome	Any possible solution	Any possible key
Population	Group of Chromosomes	Group of keys
Cost value	A function to evaluate the performance	Letter frequency analysis
Generations	Number of generation	Number of Iterations

3 Existing System

Several well-known algorithms such as substitution techniques, transposition techniques, RSA, Deffi-Hellmen, DES, Triple DES Etc. algorithms can be used to perform this encryption and decryption. But there is no guarantee that Cipher text generated by them is safe from the intruders. In this situation, we propose a new technique that gives support for the cipher text which is created by conventional encryption algorithm. Using this technique, one extra feature is added to the conventional encryption technique. In this, we generated the random keys used to generate randomized cipher text and original cipher text.

4 Proposed System

In cryptography, a **substitution block cipher** is a method of encryption by which units of plain text are replaced with cipher text according to a regular system. The receiver deciphers the text by performing an inverse substitution.
One of the main problems with simple substitution ciphers is that they are so vulnerable to frequency analysis. Therefore, to make ciphers more secure, cryptographers have long been interested in developing enciphering techniques that are immune to frequency analysis.

In this paper the Cryptanalysis of substitution block cipher by using Genetic algorithm is presented. This algorithm considers a random matrix key which on execution of a sequence of steps generates a sequence which is divided into basins. Each basin represents one block of data on which the genetic algorithm operations like crossover and mutation are performed. Each block of plain text is undergoes operations which further forms the cipher text. Thus, the cipher text obtained is very difficult to be broken without knowing the key, which provides high security.

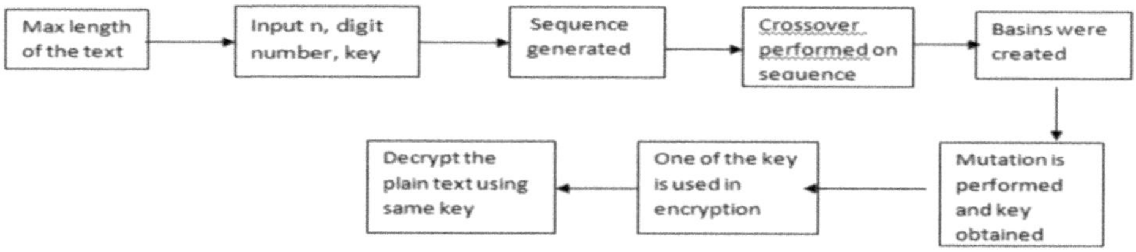

Fig. 1. Encryption and Decryption process

4.1 Proposed Algorithm

1) Consider the sequence for n= 0 to 26 values.
2) Converting n: 0-26 to ternary vector. Let it be X.
3) Representing the values in matrix form.
4) Calculate R-1 and store it in R.
5) Multiply R with key considered.
6) Divide R with 2 and store in R.
7) Add R and a and store in R.
8) Apply mod function on R and store it in R.
9) Remove negative integers and replace with positive numbers.
10) Converting this output to integer from. Let this be S.
11) Cross over is applied and sequence is generated S.
12) Generating different basins by placing equality of values of the sequence in one basin. Represent this by array of B.
13) Mutation is performed on B and key is produced.
14) Converting the plain text to ASCII values.

15) Considering the first character of the plain text and applying a mod function of the order of number of basins formed.
16) Depending on the output of mod function, the corresponding basins is used as key.
17) Adding the key to ASCII values of plain text.
18) Converting the value to binary format.
19) Transposition takes place in each character after all process is over i.e. change one bit LSB.
20) The binary value is converted to decimal value.
21) If the decimal values exceed the printable range then we have to store the respective indexes and subtract 127 from the value.
22) Converting the outputs to characters of the alphabets to get cipher text.

4.2 Example

Step1. Consider the sequence for n= 0 to 26 values.
Step2. Convert the sequence to ternary form of a 3 digit number

i.e. 0 ------- 000
1 ------- 001
2 ------- 002
3 ------- 010
4 ------- 011
5 ------- 012
6 ------- 020
7 ------- 021
8 ------- 022
.
.
26 ------ 222

Step3. Represent above ternary form in 27x3 matrixes which is denoted with ‘R’.
Step4. Subtract 1 from each element of the above matrix and the resulting matrix R is
Step5. Consider random matrix of (digit no) * (digit no) which is denoted with A, and multiply with ‘R’. Step3, Step4 and Step5 are shown in left to right order below

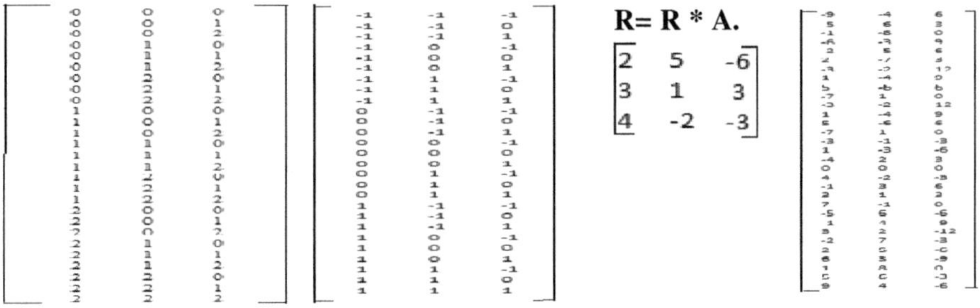

Step6. R= R DIV 2, Step7. R = R+A, Step8. R MOD 3, Step9. Remove negatives

$$
\begin{bmatrix}
-4 & -2 & 3 \\
-2 & -3 & 1 \\
0 & -4 & 0 \\
-3 & -1 & 4 \\
-1 & -2 & 3 \\
1 & -3 & 1 \\
-1 & -1 & 6 \\
0 & -2 & 4 \\
2 & -3 & 3 \\
-3 & 0 & 0 \\
-1 & 0 & -1 \\
0 & -1 & -3 \\
-2 & 1 & 1 \\
0 & 0 & 0 \\
2 & -1 & -1 \\
0 & 1 & 3 \\
1 & 0 & 1 \\
3 & 0 & 0 \\
-2 & 3 & -3 \\
0 & 2 & -4 \\
1 & 1 & -6 \\
-1 & 3 & -1 \\
1 & 2 & -3 \\
3 & 1 & -4 \\
0 & 4 & 0 \\
2 & 3 & -1 \\
4 & 2 & -3
\end{bmatrix}
\quad
\begin{bmatrix}
6 & 1 & -9 \\
5 & 0 & -10 \\
5 & 0 & -10 \\
9 & 4 & -6 \\
9 & 4 & -6 \\
8 & 3 & -7 \\
13 & 8 & -2 \\
11 & 6 & -4 \\
11 & 6 & -4 \\
6 & 1 & -9 \\
7 & 2 & -8 \\
5 & 0 & -10 \\
9 & 4 & -6 \\
9 & 4 & -6 \\
9 & 4 & -6 \\
13 & 8 & -2 \\
11 & 6 & -4 \\
12 & 7 & -3 \\
7 & 2 & -8 \\
7 & 2 & -8 \\
5 & 0 & -10 \\
10 & 5 & -5 \\
9 & 4 & -6 \\
9 & 4 & -6 \\
13 & 8 & -2 \\
13 & 8 & -2 \\
12 & 7 & -3
\end{bmatrix}
\quad
\begin{bmatrix}
0 & 1 & 0 \\
2 & 0 & -1 \\
2 & 0 & -1 \\
0 & 1 & 0 \\
0 & 1 & 0 \\
2 & 0 & -1 \\
1 & 2 & -2 \\
2 & 0 & -1 \\
2 & 0 & -1 \\
0 & 1 & 0 \\
1 & 2 & -2 \\
2 & 0 & -1 \\
0 & 1 & 0 \\
0 & 1 & 0 \\
0 & 1 & 0 \\
1 & 2 & -2 \\
2 & 0 & -1 \\
0 & 1 & 0 \\
1 & 2 & -2 \\
1 & 2 & -2 \\
2 & 0 & -1 \\
1 & 2 & -2 \\
0 & 1 & 0 \\
0 & 1 & 0 \\
1 & 2 & -2 \\
1 & 2 & -2 \\
0 & 1 & 0
\end{bmatrix}
\quad
\begin{bmatrix}
0 & 1 & 0 \\
2 & 0 & 1 \\
2 & 0 & 1 \\
0 & 1 & 0 \\
0 & 1 & 0 \\
2 & 0 & 1 \\
1 & 2 & 2 \\
2 & 0 & 1 \\
2 & 0 & 1 \\
0 & 1 & 0 \\
1 & 2 & 2 \\
2 & 0 & 1 \\
0 & 1 & 0 \\
0 & 1 & 0 \\
0 & 1 & 0 \\
1 & 2 & 2 \\
2 & 0 & 1 \\
0 & 1 & 0 \\
1 & 2 & 2 \\
1 & 2 & 2 \\
2 & 0 & 1 \\
1 & 2 & 2 \\
0 & 1 & 0 \\
0 & 1 & 0 \\
1 & 2 & 2 \\
1 & 2 & 2 \\
0 & 1 & 0
\end{bmatrix}
$$

Step10. Converting into sequence:

3,19,19,3,3,19,17,19,19,3,17,19,3,3,3,17,19,3,17,17,19,17,3,3,17,17,3.

Step11. Dividing into Basins:

B[0] = (0,3,4,9,12,13,14,17,22,23,26)
B[1] = (1,2,5,7,8,11,16,20)
B[2] = (6,10,15,18,19,21,24,25)

Step12. Crossover:

B[0] = (0,3,4,9,12,8,11,16,20)
B[1] = (1,2,5,7,13,14,17,22,23,26)
B[2] = (6,10,15,18,19,21,24,25)

Crossover:

B[0] = (0,3,4,9,12,8,11,16,20)
B[1] = (1,2,5,7,13,19,21,24,25)
B[2] = (6,10,15,18,14,17,22,23,26)

Step13. Mutation:

B[0] = (20,3,4,9,12,8,11,16,0)
B[1] = (1,24,5,7,13,19,21,2,25)
B[2] = (6,10,22,18,14,15,23,26)

Representing steps from 15 to 21 in a Tabular format

Table 2. Encryption Process

Plain text	W	E	L	C	o	m	e
ASCII	119	101	108	99	111	109	101
Key	25	5	8	9	12	14	16
Add	144	106	116	108	123	123	117
Binary	0100100000	0001101010	0001110100	0001101100	0001111011	0001111011	0001110101
LSB	0010010001	0001101011	0001110101	0001101101	0001111001	0001111010	0001110100
Decimal	145	107	117	109	122	122	116
Cipher text	2	K	U	M	z	z	t

Table 3. Decryption Process

Cipher text	2	K	U	M	Z	z	t
ASCII	145	107	117	109	122	122	116
Binary	0010010001	0001101011	0001110101	0001101101	0001111000	0001111010	0001110100
LSB	0100100000	0001101010	0001110100	0001101100	0001111011	0001111011	0001110101
Decimal	144	106	116	108	123	123	117
Key	25	5	8	9	12	14	16
Sub	119	109	108	99	111	109	101
Plain text	w	E	L	C	O	m	e

4.3 Computing Power

1) Converting n: 0-26 to ternary vector. Let it be X.
2) Calculate R-1 and store it in R.
3) Multiply R with key considered.
4) Divide R with 2 and store in R.
5) Add R and store in R.
6) Apply mod function on R and store it in R.
7) Remove negative integers and replace with positive numbers.
8) Converting this output to integer from. Let this be S.
9) Cross over is applied and sequence is generated S.
10) Generating different basins by placing equality of values of the sequence in one basin. Represent this by array of B.
11) Mutation is performed on B and key is produced.
12) Converting the plain text to ASCII values.
13) Considering the first character of the plain text and applying a mod function of the order of number of basins formed.
14) Depending on the output of mod function, the corresponding basins are used as key.
15) Adding the key to ASCII values of plain text.
16) Converting the value to binary format.
17) Transposition takes place in each character after all process is over i.e. changing one bit LSB.
18) The binary value is converted to decimal value.
19) If the decimal values exceed the printable range then we have to store the respective indexes and subtract 127 from the value.
20) Converting the outputs to characters of the alphabets to get cipher text.

5 Results

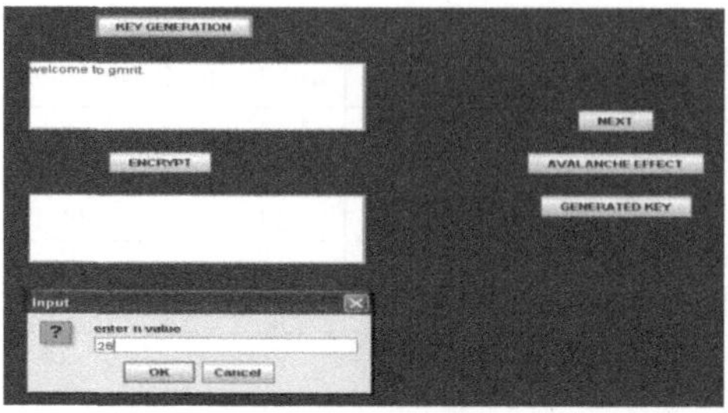

Fig. 2. Enter the n value

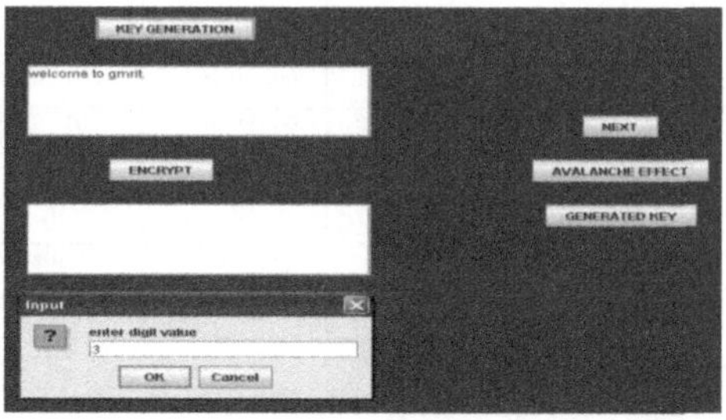

Fig. 3. Enter digit no

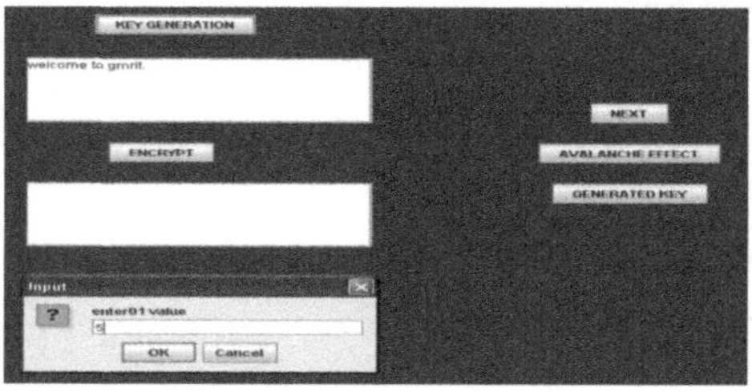

Fig. 4. Enter the values of random matrix

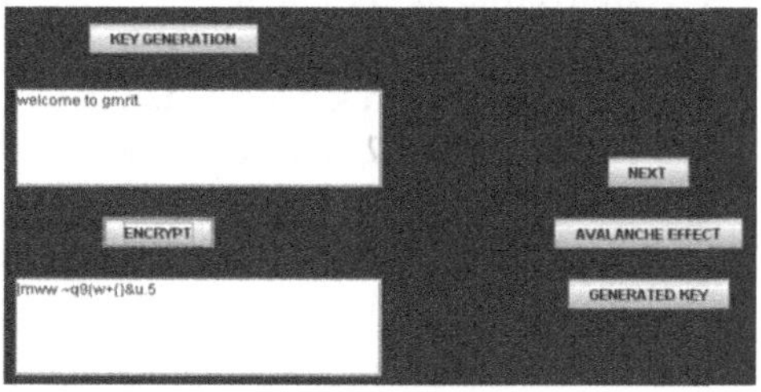

Fig. 5. Encryption

Fig. 6. Decryption

6 Conclusion

The primary goals of this work were to produce a performance evaluation of traditional cryptanalysis methods and genetic algorithm (GA) based methods, and to determine the validity of typical GA-based methods in the field of cryptanalysis. We implemented the technique of producing the cipher text by including Genetic Algorithm operations such as (Mutation and Crossover). We proposed an algorithm which considers a random matrix key which on execution of a sequence of steps generates a sequence. Based on the equality of values, this sequence is being divided into basins. Each basin represents one block of data on which the genetic algorithm operations like crossover and mutation are performed. Each block of plain text is replaced by summation of ASCII value of plain text and the sequence is generated to form the cipher text. We therefore assessed the generation of cipher text in more secured manner by overcoming the faults of identifying the key by 3rd party users. In the future work, there is a planning to design a sophisticated software based on this technique which will targeted to use in highly secure multimedia data transmission applications.

References

1. Gorodilov, A., Morozenko, V.: Genetic Algorithms for finding the key's length and crypto analysis of the permutation cipher. International Journal Information Theories and Applications 15 (2008)
2. Delman, B.: Genetic Algorithms in Cryptography, published in web (July 2004)
3. Whitley, D.: A Genetic Algorithm Tutorial, Computer Science Department, Colorado State University, Fort Collins, CO 80523
4. Spillman, R., Janssen, M., Nelson, B., Kepner, M.: Use of a Genetic Algorithm in the Cryptanalysts of Simple Substitution Ciphers. Cryptologia 16(1), 31–34 (1993)
5. Stallings, W.: Cryptography and Network Security: Principles and Practices, 3rd edn. Pearson Education (2004)
6. Bose, R.: Introduction to Cryptography – – Tata Mc-Grew–hill Publisher ltd. (2001)
7. Koblitz, N.: A course in number theory and Cryptography. Springer-Verlag, New York, Inc. (1994)
8. Nalani, N., Raghavendra Rao, G.: Cryptanalysis of Simplified Data Encryption Standard via Optimisation Heuristics. IJCSNS 6(1B) (January 2006)
9. Simmons, S.: Algebric Cryptoanalysis of Simplified AES. Proquest Science Journals 33(4), 305 (2009)
10. Ravi, S., Knight, K.: Attacking Letter Substitution Ciphers with Integer Programming. Proquest Science Journals 33(4), 321 (2009)
11. Kumar, A., Kumar, A.: Development of New Cryptographic Construct using Palmprint Based Fuzzy voult. EURASIP Journal on Adv. in Signal Processing 21, 234–238 (2009)
12. Wang, B., Wu, Q., Hu, Y.: A Knapsack Based Probabilistic Encryption Scheme (March 2007), `http://www.citeseer.ist.psu.edu`
13. Bluekrypt 2009: Cryptographic Key length Recommendations
14. Blum, L., Blum, M., Shub, M.: A simple unpredictable pseudo random number generator. SIAM J. Compute 15(2), 364–383 (1986)
15. Canetti, R., Krawczyk, H.: Universally Composable Notions of Key Exchange and Secure Channels. In: Knudsen, L.R. (ed.) EUROCRYPT 2002. LNCS, vol. 2332, pp. 337–351. Springer, Heidelberg (2002)

Parametric Performance Evaluation of Different Types of Particle Swarm Optimization Techniques Applied in Distributed Generation System

S. Kumar, S. Sau, D. Pal, B. Tudu , K.K. Mandal, and N. Chakraborty

Power Engineering Department,
Jadavpur University, Kolkata: 700098, India
{sajjan.pradhan48,susmita.sau,dptndpal0}@gmail.com,
{bhimsen_ju,kkm567}@yahoo.co.in,
chakraborty_niladri@hotmail.com

Abstract. This paper presents performance comparative study of various particle swarm optimization (PSO) techniques for the placement of generator units in the distributed generation (DG) system. For the installation of generator units in the distributed generation system, it is very important to know the generator sizing and its placement in the network system for reducing the line losses and hence the cost. Various PSO techniques such as Canonical PSO, Hierarchical PSO (HPSO), Time varying acceleration coefficient (TVAC) PSO, Self-organizing hierarchical particle swarm optimizer with time-varying acceleration coefficients (HPSO-TVAC), Stochastic inertia weight (Sto-IW) PSO and Time varying inertia weight (TVIW) PSO have been used for comparative study. Here the main objective function (OF) is to minimize the system cost. These techniques have been tested on the standard IEEE-14 bus, IEEE-30 bus and IEEE-57 bus network system by the use of MATLAB software.

Keywords: Distributed Generation (DG), Particle Swarm Optimization (PSO), Sizing, Location, IEEE-14, IEEE-30, IEEE-57.

1 Introduction

As the demand of electricity consumption increases, conventional power system leads to several disadvantages like considerable amount of transmission loss, transmission line congestion, increasing environmental impact etc. These problems can be solved by the introduction of Distributed Generation (DG). Distributed Generation (DG) system has many small generators of size 2-50 MW which are installed on various strategic points throughout the system, so that each generator provides power to small number of consumers nearby. These generators may be renewable or/and nonrenewable sources of energy like wind generators, photo voltaic (PV) cell, mini/micro hydel power plant, gas turbines, fuel cell, combined cycle plants etc. depending on the system structure, resources availability and system reliability. For

S.C. Satapathy et al. (Eds.): Proc. of Int. Conf. on Front. of Intell. Comput., AISC 199, pp. 349–356.
DOI: 10.1007/978-3-642-35314-7_40

maximum reliability, technical and economical advantages and benefits, proper sizing or capacity of distributed generators, number of such units, its proper allocation in the power systems, types of network connection, types of generating unit and technology to be used etc. are very important. Among these factors, the problem of installation of DG units at proper location and sizing is of great importance.

Various optimization techniques such as evolutionary programming (EP), genetic algorithm (GA) and others have been used for these types of problems by different researchers. Ghosh *et al.* discussed the sizing and allocation of DG in three separate steps using conventional N-R load flow analysis method [1]. The authors first found the weighting factor (WF) then DG location in the network system and then the size of DG in three separate steps. T. K. A. Rahman *et al.* discussed the sizing and allocation of DG in two separate steps using the evolutionary programming (EP) techniques [2]. The authors first solved for the optimal location and then for DG size. The problem of sizing and allocation of DG in a single step using the combination of load flow and particle swarm optimization algorithm was addressed by M. F. AlHajri *et al.* [3].

In this paper, different types of improved versions of particle swarm optimization (PSO) techniques such as Canonical PSO, Hierarchical PSO (HPSO), Time varying acceleration coefficient (TVAC) PSO, Self-organizing hierarchical particle swarm optimizer with time-varying acceleration coefficients (HPSO-TVAC), Stochastic inertia weight (Sto-IW) PSO and Time varying inertia weight (TVIW) PSO have been used for determining the best location and the optimal size of DG simultaneously to reduce the overall system cost as well as loss. A comparative study of these techniques has been also done. All these techniques are applied to IEEE-14 bus, IEEE-30 bus and IEEE-57 bus system.

2 Particle Swarm Optimization

Particle Swarm Optimization (PSO) is a meta-heuristic technique which can easily optimize non-linear, complex and real world optimization problems. This technique was first proposed by Kennedy and *Eberhart* in 1995 [4]. This methodology is based on very simple concept like bird flocking and fish schooling.

In technical term, each bird or fish is called "Particle" and its flock is called "Particle Population". All particles move around the wide area of search space according to objective function (OF). Movement of each particle is based on its personal experience as well as neighbors' experiences.

In the PSO, first of all we randomly initialize the particles position according to problem constraints. The set of all particles positions is called initial population or initial swarm. After that we generate random velocities for each particle. According to the objective function, objective value is evaluated. In the initial condition, position corresponds to optimum value is called personal best or "pbest" (pb) as well as global best (gb) or "gbest" (only for initial condition). Then the particles' velocities (v_i) and positions (x_i) are updated according to personal influence and social influence. In the mathematical form, velocity is updated according to the following expression:

$$\underbrace{v_i^{t+1} = v_i^t}_{Inertia} + \underbrace{c_1 U_1^t\,(pb_i^t - x_i^t)}_{Personal\ Influence} + \underbrace{c_2 U_2^t\,(gb^t - x_i^t)}_{Social\ Influence} \tag{1}$$

And position is updated according to following expression:

$$x_i^{t+1} = x_i^t + v_i^{t+1} \tag{2}$$

Where, c_1 is called *cognitive parameter*
c_2 is called *social parameter* } Both are called *acceleration coefficients.*
U_1^t & U_2^t are two *random numbers* varies between 0 to 1.

In the next iteration, updated velocities and positions are used as the present velocities and positions. Now these particle positions are used for the calculation of new value. In that condition, position of particle corresponds to optimum value is called new "gbest" and position of particle corresponds to optimum value evaluated by itself, is called new "pbest". And these above processes are repeated until stopping criteria (limitation of maximum iteration) is satisfied [5-8].

2.1 Different Types of PSOs

Basic PSO may not be suitable for all type of problems. For the application of PSO in the different types of problems, many variants of the original algorithm have been proposed. Depending on the variations in the constants or solution techniques, different types of PSOs have been proposed.

Canonical Particle Swarm Optimization: This technique was introduced by *Clerc and Kennedy* in 2002 [9]. They introduced a "*Constriction Factor (X)*" into the basic PSO to control the convergence properties of the particles. After introducing the Constriction Factor (*X*), the velocities update formula becomes

$$v_i^{t+1} = X\,(v_i^t + c_1 U_1^t\,(pb_i^t - x_i^t) + c_2 U_2^t\,(gb^t - x_i^t)) \tag{3}$$

Where,

$$X = 2k \Big/ (|2 - \varphi - \sqrt{(\varphi^2 - 4\varphi)}\,|) \tag{4}$$

Here, *k* is a random number which varies from 0 to 1, i.e. $k \in [0, 1]$, $\boldsymbol{\varphi} = \boldsymbol{c_1} + \boldsymbol{c_2}$ and φ should be greater than 4. Generally, *k* is set to 1 and both $\boldsymbol{c_1}$ & $\boldsymbol{c_2}$ are set to 2.05, giving as a result *X* equal to 0.729 [10].

Self-organizing Hierarchical Particle Swarm Optimization (HPSO): This technique was introduced by Kennedy and Eberhart [4]. In this technique, they proposed the velocity update formula without addition of previous velocity i.e. they exclude the inertia term. So, the new velocity update formula becomes

$$v_i^{t+1} = c_1 U_1^t \, (pb_i^t - x_i^t) + c_2 U_2^t \, (gb^t - x_i^t) \tag{5}$$

Hence, they observed that in the absence of the previous velocity term, particles rapidly rush to a local optimum solution and stagnate due to the lack of momentum. So, during the velocity calculation, if a particle's new velocity becomes zero (in any dimension), then in that condition, that velocity is reinitialized to some value according to the maximum allowable velocityV_{max}. Finally, the re-initialization of velocity is also linearly decreased from V_{max}at the beginning of the run to (0.1*V_{max}) at the end.

Time-Varying Acceleration Coefficients Particle Swarm Optimization (TVAC-PSO): In the PSO technique, particles are guided by personal influence as well as social influence which depend on cognitive parameter (c_1) and social parameter (c_2) respectively. These two parameters are called acceleration coefficients. Therefore, proper control of these two components is very important to find the optimum solution accurately and efficiently.

In this technique, the acceleration coefficients are varied (cognitive and social parameter) with respect to time or iteration. At the beginning, large value of cognitive component and small value social component is considered. And in the latter stage, small cognitive component and large social component is selected. Due to this type of selection, particles are allowed to move around the search space, instead of moving toward the population best at the beginning and on the other hand, particles try to converge to the global optima in the latter part of the optimization.
Mathematically these variations can be represented as

$$c_1 = c_{1i} + \frac{c_{1f} - c_{1i}}{it_{max}} . it \tag{6}$$

$$c_2 = c_{2i} + \frac{c_{2f} - c_{2i}}{it_{max}} . it \tag{7}$$

It is observed that when c_1 varies from 2.5 to 0.5 and c_2 varies from 0.5 to 2.5 gives better result and we have consider c_{1i} = 2.5, c_{1f} = 0.5, c_{2i} = 0.5 and c_{2f}= 2.5 for better results.

Self-organizing Hierarchical Particle Swarm Optimization with Time-Varying Acceleration Coefficients (HPSO-TVAC): In this technique, the combined effect of both HPSO and TVAC PSO is considered simultaneously for better control at the beginning as well as in the latter stage [11]. The velocity update, cognitive and social parameters are governed by (5), (6) & (7) respectively. In this technique, if a particle's new velocity becomes zero in any dimension, then the velocity is reinitialized as per the rule described in the HPSO technique.

Stochastic Inertia Weight Particle Swarm Optimization (Sto-IW PSO): This technique was first introduced by *Eberhart and Shi* [12]. In this technique the *inertia weight* is randomly selected according to a uniform distribution in the range [0.5, 1.0]. This range was inspired by *Clerc and Kennedy's* constriction factor. In this version, both acceleration coefficients c_1 and c_2 are set to 1.494 [10].

Time Varying Inertia Weight Particle Swarm Optimization (TVIW-PSO): This technique was introduced by *Shi and Eberhart* in 1998 [13]. They introduced a *"time varying inertia weight, w(t)"* into the basic PSO to control the diversification-intensification behavior of the original PSO. The velocity update rule becomes

$$v_i^{t+1} = w(t).v_i^t + c_1U_1^t\,(pb_i^t - x_i^t) + c_2U_2^t\,(gb^t - x_i^t) \tag{8}$$

Generally *time varying inertia weight, w(t)* varies linearly from an initial value to final value. In most of the cases, both c_1 & c_2 are set to 2 [10].

Time varying inertia weight means the value of inertia weight varies from iteration to iteration. If its value continuously decreases then this is called "*Dec-IW*" and it was proposed by *Shi and Eberhart* [13]. If the value of *w(t)* continuously increases then this is called "*Inc-IW*" and it was proposed by *Zheng et al.* [14]. Normally, the starting value of the inertia weight is set to 0.9 and the final to 0.4 for *Dec-IW* and from 0.4 to 0.9 for *Inc-IW* [6], [10].

Mathematically it can be written as:

For *Dec-IW*,

$$w(it) = w_{max} - \frac{w_{max} - w_{min}}{it_{max}}.it \tag{9}$$

For *Inc-IW*,

$$w(it) = w_{min} + \frac{w_{max} - w_{min}}{it_{max}}.it \tag{10}$$

Where,

$w(it) \Rightarrow$ Inertia weight varies with respect to iteration number
$w_{max} \Rightarrow$ Maximum value of inertia weight i.e. 0.9
$w_{max} \Rightarrow$ Minimum value of inertia weight i.e. 0.4
$it \Rightarrow$ Iteration number
$it_{max} \Rightarrow$ Maximum number of iteration

3 Problem Formulation

The main objective of this work is to minimize the cost as well as loss by the installation of generator units of optimal sizing at optimal location in the network system. Here, the objective function is overall system cost and the cost due the energy wasted in the transmission network. The loss in the network system is calculated by N-R load flow analysis method. And after that, this loss is converted into the cost by the introduction of a term *Weighting Factor*. This method of sizing and placement of DG is tested on the standard IEEE-14, IEEE-30 and IEEE-57 network system.

4 Results and Discussion

Here a single DG is installed on the standard IEEE network system. The problem of sizing and placement of DG in the network system is addressed simultaneously. The algorithms are implemented for obtaining location and size of the DG at the same time. Initially with the help of Newton-Raphson (N-R) technique, the basic power flow equations are solved for the respective network system. Then the optimization is done by the different types of PSOs considering different constraints. The MATLAB software of version 7.8.0 (R2009a) is used for implementing all these algorithms. A computer of Windows 7 Professional as an operating system, Dual Core @ 1.73 GHz Processor and 3 GB of RAM is used to run these MATLAB Programs.

Initially these methods are implemented with different values of population and number of iteration and it is found that for the present study, population size of 50 and maximum number of iterations of 200 is giving the desired results. The results obtained by all these techniques for different network systems are given in table 1.

Table 1. Results obtained by different types of PSOs

Network System	Techniques	DG size (MW)	Location in the system	Value of OF	No. of iterations to converge (approx.)	Time taken in 200 iterations (Seconds)
IEEE-14 bus	Canonical PSO	35.5426	3	2288.3	57	34.508619
	HPSO	35.5426	3	2288.3	45	33.939571
	HPSO TVAC	35.5426	3	2288.3	75	33.160691
	Sto-IW PSO	35.5426	3	2288.3	55	34.734024
	TVAC PSO	35.5336	3	2288.3	182	33.971738
	TVIW PSO	35.5426	3	2288.3	140	32.452982
IEEE-30 bus	Canonical PSO	47.3421	5	2847.8	167	122.029345
	HPSO	47.3421	5	2847.8	85	121.668147
	HPSO TVAC	47.3421	5	2847.8	85	122.705320
	Sto-IW PSO	47.3421	5	2847.8	140	121.404472
	TVAC PSO	47.5684	5	2847.9	200	120.521975
	TVIW PSO	47.3421	5	2847.8	200	120.625879
IEEE-57 bus	Canonical PSO	13.0876	34	4115.8	105	439.897621
	HPSO	13.0876	34	4115.8	36	435.596834
	HPSO TVAC	13.0876	34	4115.8	58	445.774441
	Sto-IW PSO	13.0876	34	4115.8	124	442.126232
	TVAC PSO	13.6344	34	4115.9	200	443.153731
	TVIW PSO	13.0876	34	4115.8	164	440.746614

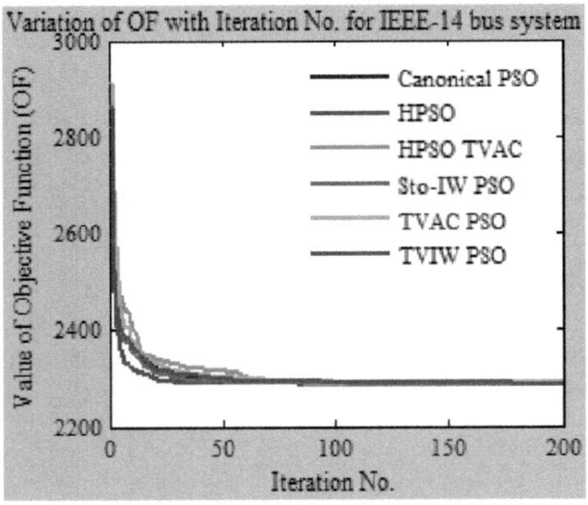

Fig. 1. Convergence characteristics of different types of PSOs for IEEE-14 bus network system

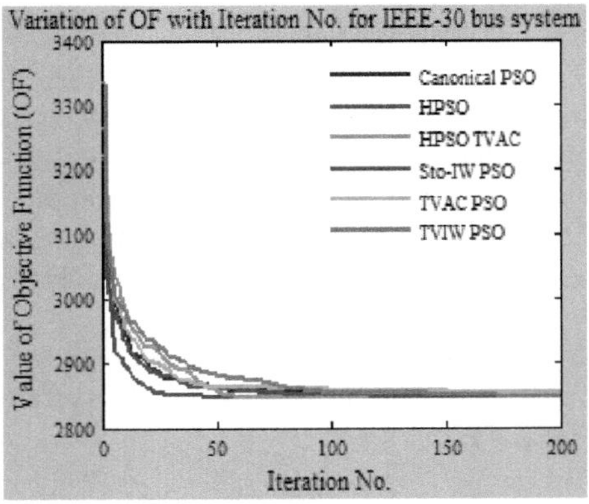

Fig. 2. Convergence characteristics of different types of PSOs for IEEE-30 bus network system

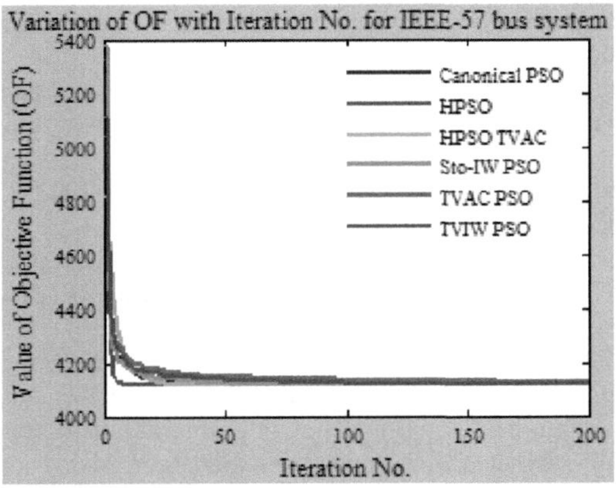

Fig. 3. Convergence characteristics of different types of PSOs for IEEE-57 bus network system

From the above results, it is seen that 35.5426 MW, 47.3421 MW and 13.0876 MW are the optimum values of generator size and bus number 3, 5 and 34 seems to be the optimum location of generator unit for the IEEE-14 bus, IEEE-30 bus and IEEE-57 bus network system respectively.

It is also seen that though all the techniques are capable of giving global solution quite effectively but compared to other methods HPSO techniques converges rapidly in less time as well as in less number of iterations. Since the time consumed to perform the optimization of this type of complex problem is one of the major constraints to be considered, HPSO technique can be applied with quite effectively and with great confidence in this type of problem.

5 Conclusion

In this work, the optimum value of DG size and its location in the network systems are obtained simultaneously. Here, different types of improved version of PSO techniques are used to solve the problem. All these techniques are applied to IEEE-14 bus, IEEE-30 bus and IEEE-57 bus network system. Among these techniques, it is seen that HPSO techniques performs better for all systems.

Acknowledgment. This work is supported by "University Grants Commission (UGC)" under the scheme of "University with Potential for Excellence, phase – II (UPE – II)".

References

1. Ghosh, S., Ghoshal, S.P., Ghosh, S.: Optimal sizing and placement of distributed generation in a network system. Electrical Power and Energy Systems 32, 849–856 (2010)
2. Rahman, T.K.A., Rahim, S.R.A., Musirin, I.: Optimal allocation and sizing of embedded generators. In: Proceedings of the National Power and Energy Conference, PECon 2004, pp. 288–294 (2004)
3. Krueasuk, W., Ongsakul, W.: Optimal Placement of Distributed Generation Using Particle Swarm Optimization. In: Australian Universities Power Engineering Conference, December 10-13 (2006)
4. Kennedy, J., Eberhart, R.: Particle Swarm Optimization, pp. 1942–1948. IEEE (1995)
5. Swarm Intelligence Tutorial, `http://www.swarmintelligence.org/tutorials.php`
6. de Oca, M.A.M.: Particle Swarm Optimization Introduction. IRIDIA-CoDE (May 7, 2007)
7. Wikipedia, `http://en.wikipedia.org/wiki/Particle_swarm_optimization`
8. Wong, L.Y., Rafidah, S., Rahim, A., et al.: Distributed Generation Installation Using Particle Swarm Optimization. In: The 4th International Power Engineering and Optimization Conf (PEOCO 2010), Shah Alam, Selangor, Malaysia, June 23-24 (2010)
9. Clerc, M., Kennedy, J.: The particle swarm-explosion, stability, and convergence in a multidimensional complex space. IEEE Transactions on Evolutionary Computation 6(1), 58–73 (2002)
10. de Oca, M.A.M., Stützle, T., Birattari, M., Dorigo, M.: A Comparison of Particle Swarm Optimization Algorithms Based on Run-Length Distributions. In: Dorigo, M., Gambardella, L.M., Birattari, M., Martinoli, A., Poli, R., Stützle, T. (eds.) ANTS 2006. LNCS, vol. 4150, pp. 1–12. Springer, Heidelberg (2006)
11. Ratnaweera, A., Halgamuge, S.K., Watson, H.C.: Self-Organizing Hierarchical Particle Swarm Optimizer with Time-Varying Acceleration Coefficients. IEEE Transactions on Evolutionary Computation 8(3) (June 2004)
12. Eberhart, R., Shi, Y.: Tracking and optimizing dynamic systems with particle swarms. In: Proceedings of the 2001 IEEE Congress on Evolutionary Computation, pp. 94–100. IEEE Press, Piscataway (2001)
13. Shi, Y., Eberhart, R.: A modified particle swarm optimizer. In: Proceedings of the 1998 IEEE World Congress on Computational Intelligence, pp. 69–73. IEEE Press, Piscataway (1998)
14. Zheng, Y.L., Ma, L.H., Zhang, L.Y., Qian, J.X.: Empirical study of particle swarm optimizer with an increasing inertia weight. In: Proceedings of the 2003 IEEE Congress on Evolutionary Computation, pp. 221–226. IEEE Press, Piscataway (2003)

Implementation of Adaptive MMSE Rake Receiver

A.M. Borkar, U.S. Ghate, and N.S. Panchbudhe

DBACER, Nagpur, Maharashtra, India
{borkar.atul,nilesh.panchbudhe}@gmail.com,
ujwalghate@rediffmail.com

Abstract. Several types of Rake Receivers like A-Rake, S-Rake, P-Rake, Adaptive frequency Rake, Time frequency Rake, Conventional MMSE Rake and Adaptive MMSE rake are used for WCDMA. In this paper we observed that the BER performance of the Adoptive MMSE Rake receiver gives better result in WCDMA. The comparative analysis proved that the Adaptive MMSE Rake Receiver is much better than Conventional Rake Receiver. Both Genetic Algorithm (GA) and as well as Conventional algorithm are derived for WCDMA environment.

Keywords: Adaptive MMSE Rake Receiver, Conventional Rake Receiver, Genetic Algorithm, Conventional Algorithm.

1 Introduction

We provide a complete optimization theoretical framework for the finger selection problem for MMSE S-Rake receivers. First, we formulate the optimal MMSE S-Rake as a no convex, integer-constrained optimization, in which the aim is to choose the finger locations of the receiver so as to maximize the overall Signal-Plus- Interference-Noise-Ratio (SINR). While computing the optimal finger selection is NP- hard, we present several relaxation methods to turn the (approximate) problem into convex optimization problems that can be very efficiently solved by interior-point methods, which are polynomial time in the worst case, and are very fast in practice.

These optimal finger selection relaxations produce significantly higher average SINR than the conventional one that ignores the correlations, and represent a numerically efficient way to strike a balance between SINR optimality and computational tractability. Moreover, we propose a genetic algorithm (GA) based scheme, which performs finger selection by iteratively evaluating the overall SINR expression. Using this technique, near-optimal solutions can be obtained in many cases with a degree of complexity that is much lower than that of optimal search.

2 MMSE Rake Receiver with Conventional Algorithm

Instead of the solving the problem in [1], the "conventional" finger selection algorithm chooses the M paths with largest individual SINRs, where the SINR for the lth path can be expressed as

S.C. Satapathy et al. (Eds.): Proc. of Int. Conf. on Front. of Intell. Comput., AISC 199, pp. 357–363.
DOI: 10.1007/978-3-642-35314-7_41

$$\mathrm{SINR_l} = \frac{E\mathrm{l}(\alpha \mathrm{l}(\mathrm{l}))2}{(8\mathrm{l}(\mathrm{MAI}))T\mathrm{A}28\mathrm{l}(\mathrm{MAI}) + \sigma_n^2)} \quad (1)$$

for l = 1, . . . , L.

This algorithm is not optimal because it ignores the correlation of the noise components of different paths. Therefore, it does not always maximize the overall SINR of the system given in [2]. For example, the contribution of two highly correlated strong paths to the overall SINR might be worse than the contribution of one strong and one relatively weaker, but uncorrelated, path. The correlation between the multipath components is the result of the MAI from the interfering users in the system.

3 MMSE Rake Receiver with Genetic Algorithm

The GA is an iterative technique for searching for the global optimum of a cost function [3]. The name comes from the fact that the algorithm models the natural selection and survival of the fittest [4]. We propose a GA (Genetic Algorithm) based approach to solve the finger selection problem, which directly uses the exact SINR expression and does not employ any relaxation technique in MMSE receiver. The GA is an iterative technique for searching for the global optimum of a cost function. The name comes from the fact that the algorithm models the natural selection and survival of the fittest. The GA has been applied to a variety of problems in different areas. Also, it has recently been employed in the multi-user detection problem. The main characteristics of the GA algorithm are that it can get close to the optimal solution with low complexity, if the steps of the algorithm are designed appropriately.

In order to be able to employ the GA for the finger selection problem we need to consider how to represent the chromosomes, and how to implement the steps of the iterative optimization scheme in MMSE. By choosing the fitness function, the fittest chromosomes of the population correspond to the assignment vectors with the largest SINR values. Now, we can summarize our GA-based finger selection scheme as follows: Generate Ni pop different assignments randomly and select N pop of them with the largest SINR values.

1. Pairing: Pair N good of the finger assignments according to the weighted random scheme.

2. Mating: Generate two new assignments from each pair.

3. Mutation: Change the finger locations of some assignments randomly except for the best assignment. The GA has been applied to a variety of problems in different areas [3] [5]. Also, it has recently been employed in the multi-user detection problem [6][7]. The main characteristics of the GA algorithm are that it can get close to the optimal solution with low complexity, if the steps of the algorithm are designed appropriately.

4 Simulation Result of MMSE Rake Receiver with Genetic Algorithm

We plot the SINR of the proposed suboptimal and conventional techniques for different numbers of fingers, where there are 50 multipath components and Eb/N0 = 20. The number of chips per frame, Nc, is set to 75, and all other parameters are kept the same as before. In this case, the optimal algorithm takes a very long time to simulate since it needs to perform exhaustive search over many different finger combinations and therefore it was not implemented. The improvement using convex relaxations of optimal finger selection over the conventional technique decreases as M (M x L selection matrix X follows: M of columns of X is the unit vectors e1….eM) increases since the channel is exponentially decaying and most of the significant multipath components are already combined by all the algorithms. M finger of MMSE Rake Receiver, xi = 1 *i*th path is selected, and xi = 0 otherwise;

$$\sum_{i=1}^{L} x_i = M \tag{2}$$

Also, the GA based scheme performs very close to the suboptimal schemes using convex relaxations after 10 iterations with Nipop = 128, Npop = 64, Ngood = 32, and 32 mutations. Finally, we consider an MAI-limited scenario, in which there are 10 users with E1 = 1 and Ek = 10 k = 1, and all the parameters are as in the previous case. Then, as shown in Figure, the improvement by using the suboptimal finger selection algorithms increase significantly. The main reason for this is that the suboptimal algorithms consider (approximately) the correlation caused by MAI whereas the conventional scheme simply ignores it.

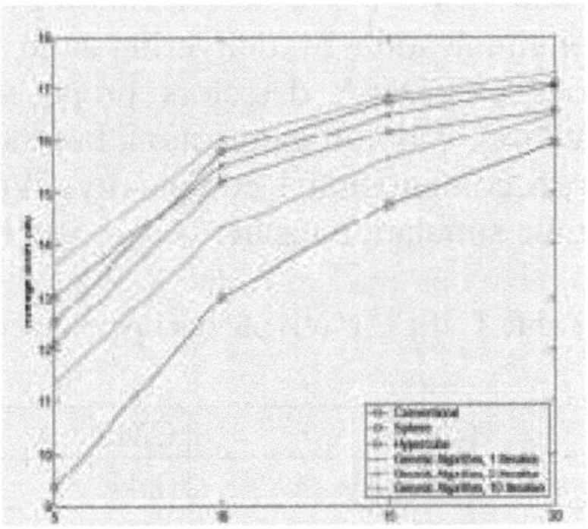

Fig. 1. Number of Fingers

Fig 1: Average SINR versus number of fingers M . There are 10 users with each interferer having 10dB more power than the desired user.,where Eb is the bit energy. The channel has L = 15 multipath components and the taps are exponentially decay ing. The IR-UWB sy stem has Nc = 20 chips per frame and Nf = 1 frame per symbol.

There are 5 equal energy users in the system and random TH and polarity codes are used. Optimal and suboptimal finger selection algorithms for MMSE-SRake receivers in an IR-UWB system have been considered. Since UWB systems have large numbers of multipath components, only a subset of those components can be used due to complexity constraints. Therefore, the selection of the optimal subset of multipath components is important for the performance of the receiver. We have shown that the optimal solution to this finger selection problem requires exhaustive search which becomes prohibitive for UWB systems.

Moreover, we have proposed a GA based iterative finger selection scheme, which depends on the direct evaluation of the objective function. A feasible implementation of multipath diversity combining can be obtained by a selective-Rake (SRake) receiver, which combines the M best, out of L, multipath components [8] . Those M best components are determined by a finger selection algorithm. For a maximal ratio combining (MRC) Rake receiver, the paths with highest signal-to-noise ratios (SNRs) are selected, which is an optimal scheme in the absence of interfering users and intersymbol interference (ISI) [9][10]. For a minimum mean square error (MMSE) Rake receiver, the "conventional" finger selection algorithm can be defined as choosing the paths with highest signal- to-interference plus- noise ratios (SINRs). This conventional scheme is not necessarily optimal since it ignores the correlation of the noise terms at different multipath components. The finger selection problem is also studied in the context of WCDMA downlink equalization.

5 Comparison of Adaptive (Proposed) MMSE Rake Receiver with Conventional Rake

In this session we are comparing the Adaptive MMSE Rake receiver with Conventional Rake. Ultra wideband (UWB) is a new technology that has the potential to revolutionize wireless communication by delivering high data rates with very low power densities. Multiuser DSCDMA detectors proposed in [11] [12] [13], for DSCDMA can be extended to UWB communication, but the major drawback of these techniques is the very high computational complexity. We choose the IEEE UWB channel parameters to get the simulation result.

Table 1. IEEE UWB channel parameters

Parameter	CM1	CM2	CM3
Cluster arrival rate, ^ (1/ns)	0.0243	0.41	0.06672
Ray arrival rate, ʎ (1/ns)	2.52	0.51	2.12
Cluster decay factor, γ	7.15	5.52	1.41
Ray decay factor, y	4.31	6.72	7.91
Std. dev. of cluster, ϭζs (dB)	3.39412	3.39415	3.39413
Std. dev. of ray, s ϭξ (dB)	3.39412	3.39413	3.39412
Std. dev. of total MP, ϭg (dB)	3	3	3

6 Simulation Result of Adaptive (Proposed) MMSE Rake Receiver with Conventional Rake

Simulations were carried out to evaluate and compare the bit error probability performance of the proposed adaptive MMSE Rake receiver in multipath channels with AWGN. The system for simulations considered in this paper is, synchronous WCDMA UWB with the following specifications. All users have equal power with Gold sequence of spreading gain 31 as spreading code. Binary phase shift keying with sampling frequency of 50 GHz, chip time of 0.5 nsec and second derivative of Gaussian pulse of width 0.5 nsec used. Random binary data is generated for each user; the data is spread with the respective spreading code followed by modulation with second derivative of the Gaussian pulse. Each user undergoes a different UWB channel. Channel models CM1, CM2 and CM3 from IEEE P802.15 are used.

Channel model parameters are listed in table. The number of multipath is selected in such a way that 90 percent of the transmitted energy is captured. Proposed adaptive MMSE Rake receiver and conventional adaptive MMSE Rake (C-Rake) receiver use training signals of 500 bits followed by decision directed operation. Proposed MMSE Rake receiver does not require spreading code of any user, whereas, it is assumed that C –Rake receiver knows spreading code of the user of interest. Bit error probability is averaged over 500 realizations for each user with 2000 bits/channel. Initial value of w = [0, 0, 0....0] T and r = [0, 0, 0....0] T. μ = 0.01, and 0.001 gives best performance for C-Rake and proposed adaptive MMSE Rake receiver respectively. To verify and investigate receiver performance bit error probability vs. Eb /N0 for K = 5, L = 10, 15 and 20 is considered. Simulation results for CM1, CM2 and CM3 respectively. It shows that the proposed detectors BER performance is better than that of C-Rake receiver in all three channel models.

It is observed that proposed detector gives better BER performance even for small number of Rake fingers (L = 10), where as for C-Rake receiver even for L = 20 BER performance is still inferior to proposed receiver. It is also observed that, for higher SNR (> 6 *d*B) proposed detector BER performance is much better than C-Rake receiver indicating that proposed detector has better MAI and multipath effect cancellation capability. Proposed detector gives an improvement of 2 *d*B at 10-2 BER, and substantial improvement for BER < 10-3 Fig shows simulation results for bit error probability vs. number of users with E*b* /N*o* = 20 *d*B for CM1, CM2 and CM3 respectively. It is observed that the proposed detector performs much better than C-RAKE even for large number of users. This improved performance is once again attributed to the better MAI cancellation capability in multipath environment. The number of users Supported by the above discussed detectors is summarized in Table.

Table 2. Number of users supported for Eb /N0 = 20 dB

BER		10-2	10-3	10-4
CM1	Proposed	16	10	8
	C-Rake	8	5	2
CM2	Proposed	16	10	8
	C-Rake	7	4	2
CM3	Proposed	13	10	9
	C-Rake	7	4	2

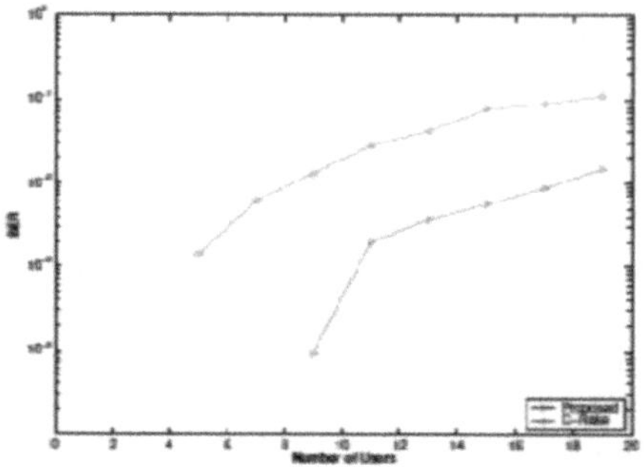

Fig. 2. BER vs. No. of users for CM1 with E*b* /N*o* = 2*d*B

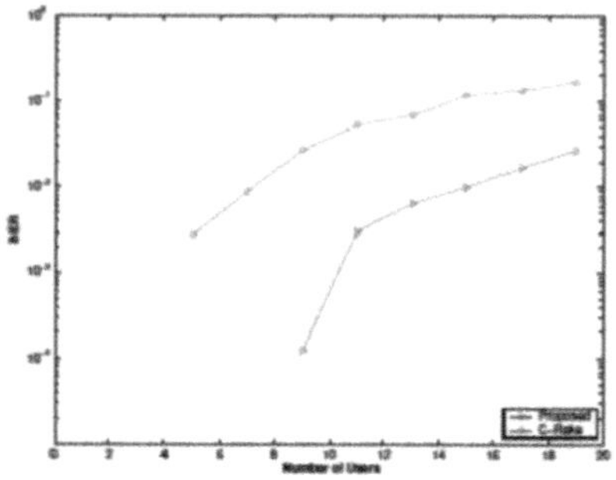

Fig. 3. BER vs. No. of users for CM2 with E*b* /N*o* = 20*d*B

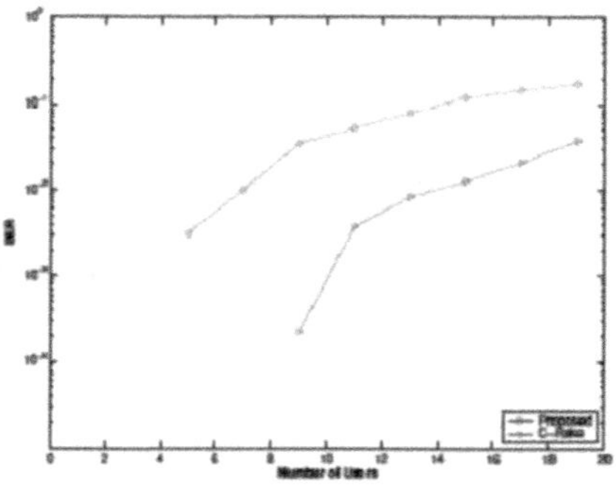

Fig. 4. BER vs. No. of users for CM3 with E*b* /N*o* = 20*d*B

7 Conclusions

We have derived that the Adaptive MMSE Rake receiver for WCDMA UWB multipath channels and studied its BER performance in multiuser environment with AWGN. It is observed that the BER performance of the Adoptive MMSE Rake receiver is much better in comparison with conventional MMSE Rake receiver. Proposed receiver given as improvement of 2 *d*B at BER of 10-2 and substantial improvement for BER < 10-3 in all three channel models (CM1-CM3). Further, it offers significant improvement in MAI cancellation in multipath channels. We have shown by simulation results that the number of users supported by the proposed receiver at BER of 10-3 with E*b* /N*o* = 20 *d*B is two times that of the conventional Rake receiver with the same computational complexity.

References

[1] Fishler, E., Poor, H.V.: On the tradeoff between two types of processing gain. IEEE Transactions on Communications 53(10), 1744–1753 (2005)

[2] Zhiwei, L., Premkumar, A.B., Madhukumar, A.S.: Matching pursuit-based tap selection technique for UWB channel equalization. IEEE Communications Letters 9, 835–837 (2005)

[3] Haupt, R.L., Haupt, S.E.: Practical Genetic Algorithms. John Wiley & Sons Inc., New York (1998)

[4] Goldberg, D.E.: Genetic Algorithms in Search, Optimization, and Machine Learning. Addison- Wesley, Reading (1989)

[5] Mitchell, M.: An Introduction to Genetic Algorithms. MIT Press, Cambridge (1996)

[6] Juntti, M.J., Schlosser, T., Lilleberg, J.O.: Genetic algorithms for multiuser detection in synchronous CDMA. In: Proc. IEEE International Symposium on Information Theory, p. 492 (1997)

[7] Yen, K., Hanzo, L.: Genetic-algorithm-assisted multi-user detection in asynchronous CDMA communications. IEEE Transactions on Vehicular Technology 53(5), 1413–1422 (2004)

[8] Cassioli, D., Win, M.Z., Molisch, A.F.: The ultra-wide bandwidth indoor channel: From statistical model to simulations. IEEE Journal on Selected Areas in Communications 20, 1247–1257 (2002)

[9] Win, M.Z., Winters, J.H.: Analysis of hybrid selection/maximal-ratio combining of diversity branches with unequal S NR in Rayleigh fading. In: Proc. IEEE 49th Vehicular Technology Conference (VTC 1999), Houston, TX, vol. 1, pp. 215–220 (Spring 1999)

[10] Yue, L.: Analysis of generalized selection combining techniques. In: Proc. IEEE 51st Vehicular Technology Conference (VTC 2000), Tokyo, Japan, vol. 2, pp. 1191–1195 (Spring 2000)

[11] Xu, Z., Tsatsanis, M.K.: Blind adaptive algorithms for minimum variance cdma receivers. IEEE Transactions on Aerospace and Electronic Systems 26(2), 423–427 (1990)

[12] Honig, M., Madhow, U., Verdu, S.: Blind Adaptive multiuser detection. IEEE Transactions on Information Theory 41, 944–960 (1950)

[13] Li, Q., Rusch, L.A.: Multiuser detection for ds-cdma uwb in the home environment. IEEE Transactions on Communications 49(1), 180–193 (2001)

[14] Molisch, A.F., Foerster, J.R., Pendergrass, M.: Channel models for ultra wideband personal area networks. IEEE Wireless Communications 10(1), 14–21 (2003)

[15] Int. J. of Advanced Networking and Applications 2(2), 621–625 (2010); Adaptive MMSE Rake Receiver for WCDMA, J. Jeya A Celin Professor of Information Technology, Noolul Islam University, Tamilnadu, India

Efficient and Automatic Reconfiguration and Service Restoration in Radial Distribution System Using Differential Evolution

D. Pal, S. Kumar, B. Tudu, K.K. Mandal, and N. Chakraborty

Power Engineering Department,
Jadavpur University, Kolkata: 700098, India
{dptndpal0,sajjan.pradhan48}@gmail.com,
{bhimsen_ju,kkm567}@yahoo.co.in,
chakraborty_niladri@hotmail.com

Abstract. This paper addresses two complex optimization problems in the form of radial distribution system reconfiguration and service restoration using a novel optimization technique called differential evolution. For distribution feeder reconfiguration (DFR) problem, the close and open statuses of sectionalizing and tie switches are changed to find minimum loss configuration. During any sudden outage of any section of the distribution system, the quickness of the restoration is checked with the help of basic optimization technique while feeding all the load points. A standard IEEE 3 feeder, 16 bus distribution system is chosen to simulate the dual problem of optimization. The feasibility and novelty of the optimization is also checked in a comparatively more complex IEEE 33 bus distribution system. Differential Evolution is chosen to find alternative topologies for feeder system and simplified forward Dist-Flow Equation is implemented to do power flow study and it is seen that differential evolution is quite capable of solving this type of complex, non-linear optimization problem with less time which is a basic requirement for the service restoration (SR) of the network system.

Keywords: Radial distribution system, Network reconfiguration, Service restoration, Power loss reduction, Differential Evolution, Dist flow equation.

1 Introduction

In distribution system automation, the topological structure of distribution system is changed from a distant control center. The topological structure of any radial system can be changed by altering the statuses of (normally close) sectionalizing switches and (normally open) tie-switches for the purpose of reconfiguration and service restoration of the network system. The reconfiguration procedure is done at the time of service maintenance and service testing to find minimum loss configuration or optimum alternative configuration by exchanging the heavily loaded section with lightly loaded portion. Service restoration is a process of restoring power flow immediately after any kind of disturbance in the power system. In both cases, any kind of islanding

S.C. Satapathy et al. (Eds.): Proc. of Int. Conf. on Front. of Intell. Comput., AISC 199, pp. 365–372.
DOI: 10.1007/978-3-642-35314-7_42

of region and looping in between the sub-tree structures have to be avoided. To check the feasibility of the procedure, customer reliability factors are also considered over different topological structures. Because of many candidates switching, the reconfiguration and service restoration are complex, combinatorial, non-differentiable constrained optimization problem. The first ever approach on distribution feeder configuration is done by Civanlar *et al.* [1]. Ji-Pyng Chiou *et al.* have come up with this loss minimization for network configuration using modified Differential Evolution [2]. Ahmed A. Hossam-Eldin *et al.* [3] have worked on a twofold objective of feeder reconfiguration and service restoration using simulated annealing method. To observe any kind of voltage instability due to branch exchange in reconfiguration, Marcos A. N. Guimaraes *et al.* have introduced a voltage stability calculation in reconfiguration procedure [4].

2 Radial Distribution System and Problem Statement

The term "radial" in radial distribution system comes from the process of nuclear fission of radioactive heavy material. In that process the radioactive material splits into lighter nucleuses and radiates huge energy. The same radiation of electric energy happens from one end of the radial distribution system. Here the power system is handled in tree topology and looping of any topological structure is strictly avoided.

For paper work, the test systems IEEE 3 feeder, 16 bus system (13 sectionalizing switches, 3 tie switches) and IEEE 33 bus distribution system (32 sectionalizing switches, 5 tie switches) are operated in radial structure to execute dual problem of reconfiguration and service restoration.

2.1 Problem Statement

To calculate active power, reactive power, voltage and power loss the simplified forward dist-flow equations are chosen [2]. This work aims to minimize the power loss, subject to operating constraints under certain load pattern. The mathematical term is expressed as follow:

$$f_1 = \min\left(P_{T,loss}\right) \tag{1}$$

Where $P_{T,loss}$ is the total loss in all connected branches to the system. Here, node voltage value should be maintained as per the upper (V_{max}) and lower (V_{min}) limits of node voltage and ($I_{i,j}$) current in each branch should be under the maximum current capacity of that branch.

Equality Constraint:

$$V_{min} \le V_i \le V_{max} \tag{2}$$

Inequality Constraint:

$$\left|I_{i,j}\right| \le I_{max} \tag{3}$$

Power Flow Equations

Power flow in Radial distribution networks are expressed with the help of Dist-Flow equations. A set of simplified feeder-line flow formulas are employed.
The equations are as follow:

$$P_{i+1} = P_i - P_{L,\ i+1} - R_{i,i+1} \times [(P_i^2 + Q_i^2)/|V_i|^2] \tag{4}$$

$$Q_{i+1} = Q_i - Q_{L,\ i+1} - X_{i,i+1} \times [(P_i^2 + Q_i^2)/|V_i|^2] \tag{5}$$

$$|V_{i+1}|^2 = |V_i|^2 - 2 \times (R_{i,i+1} \times P_i + X_{i,i+1} \times Q_i) + (R_{i,i+1}^2 + X_{i+1}^2) \times \left[\frac{(P_i^2 + Q_i^2)}{|V_i|^2}\right] \tag{6}$$

P_i and Q_i are real and reactive power flowing out of ith bus. P_{Li} and Q_{Li} are active and reactive power loads at i^{th} bus. $R_{i,\ i+1}$ and $X_{i,\ i+1}$ are resistance and reactance in between i^{th} and $i+1^{th}$ bus.
Power loss in the section connecting i^{th} and $i+1^{th}$ bus is computed as:

$$P_{Loss}(i, i+1) = R_{i,i+1} \times [(P_i^2 + Q_i^2)/|V_i|^2] \tag{7}$$

The total loss ($P_{T,loss}$) is found out by summing all branch losses in the feeder section.

$$P_{T,loss} = \sum_{i=0}^{n-1} P_{Loss}(i, i+1) \qquad i = 0,1,2, \ldots\ldots n \tag{8}$$

3 Differential Evolution

The formulation of simple differential evolution is proposed by Storn and Price [7]. Differential Evolution is a population (P) and generation-based (G) optimization method, and here population is formulated on the control variables of switching statuses. During the optimization process, target vectors (s_i^k) , donor vectors (w_i^k) and trial vector $\left(u_i^{k+1}\right)$ are created to find the best suitable variable value for fitness function [7].

Step1) Initialization

$$s_i^k = s_{min} + \sigma_i \times (s_{max} - s_{min}) \tag{9}$$

$i = 1,2 \ldots N;$ $\qquad 0 \le \sigma i \le 1;$ $\qquad$ k=1, 2......G

step2) Mutation

$$w_i^k = s_j^k + F \times s_m^k - s_l^k \tag{10}$$

$i = 1, 2, 3, \ldots\ldots..N;$ $\qquad F\in [.4, .5]$ $\qquad k = 1,2, \ldots\ldots. G$

Step 3) Cross-over ($\sigma_{1i} \in [0,1]$)

$$u_i^{k+1} = \begin{cases} w_i^k & \text{if } (\sigma_{1i} \le CR)\,;\ CR \in [.8\,,.9\,] \\ s_i^k & \text{if } (\sigma_{1i} > CR) \end{cases} \tag{11}$$

Step 4) Evolution and Selection

$$s_i^{k+1} = \begin{cases} u_i^{k+1} \text{ if } \left(f\left(u_i^{k+1}\right) \le f\left(s_i^k\right)\right) \\ s_i^k \text{ if } \left(f\left(u_i^{k+1}\right) > f\left(s_i^k\right)\right) \end{cases} \tag{12}$$

Here *(f)* is the objective function or fitness value. Among target and trail vectors new target vectors (s_i^{k+1}) are created for the next generation or for the next iteration.

4 Optimization Procedure

To simulate the procedure, the validity of the switching combination and reliability factors are to be maintained.

4.1 Checking Validity

Every switching combination can be represented into branch vs. node matrix form [8]. Here the radial distribution system works as a directed tree. Outgoing branch from any node is represented using '-1'. Incoming branch to any node is represented by '+1'. If a node is not connected with any branches, then the connection is represented by '0'. In a matrix, if any node contains all zeros in its column then it is presumed as islanding of that node. Hence the switching combination is invalid. If certain matrix representation is not suffered of this above explained problem, then the respective switching combination is valid one.

4.2 Penalty Function

A penalty function is added with fitness value if some invalid switching combination gives infeasible result [9].

$$\text{Penalty Function} = \lambda_1 \times \sum_{i=1}^{node} (V_{max} - V_i)^2 + \lambda_2 \times \sum_{i=1}^{branch} (I_{max} - I_i)^2 \tag{13}$$

Fitness function is formulated on penalty value in case of infeasible solution.

λ_1 and λ_2 are user defined penalty factors.

4.3 Voltage Stability Index for Distribution System

Branch exchange phenomena in reconfiguration and service restoration, may cause serious voltage stress on some nodes in the system. To see the node voltage stability, different authors have proposed different methods. A power flow method based stability index calculation is adapted in this work [10].

It is a calculation of stability index of mth node on the basis of voltage of kth node and active and reactive power of mth node. If stability index of node gives higher value than zero then it is secure node in the voltage profile point of view.

$$SI_m = |V_k|^2 - 4 \times (P_m r_n + Q_m x_n)|V_k|^2 - 4 \times (P_m r_n - Q_m x_n)^2 \tag{14}$$

4.4 Reliability Factors for Customer Satisfaction

Either process of reconfiguration or the strategy of service restoration, main purpose of these two is to satisfy the customers. Customer's reliability on power system may be hampered due to capacity reduction of feeder and distribution transformer, voltage deviation and switching surge [11].

Feeder Capacity Margin (FCM)

$$\text{Min } f_3 = 1 - \min_i \left\{ \frac{I_{i\,Rated} - I_{iLoad}}{I_{iRated}} \right\} \qquad i = 1,2, \ldots \ldots \ldots Nbr \tag{15}$$

Transformer Capacity Margin (TCM)

$$\text{Min } f_4 = 1 - \min_i \left\{ \frac{S_{i\,Rated} - S_{iLoad}}{S_{iRated}} \right\} \qquad i = 1,2, \ldots \ldots \ldots N_t \tag{16}$$

Maximum Voltage Deviation (MVD)

$$f_5 = \max|V_i - V_{rated}| \qquad i = 1,2, \ldots \ldots \ldots \ldots n \tag{17}$$

Minimum Switching

Reconfiguration and service restoration should be done using minimum number of switching operation so that there would not be any kind of switching surge in the system.

5 Simulated Results and Comparative Study

Simulated results of reconfiguration and service restoration of both the systems are discussed below.

5.1 System 1: IEEE 3 Feeder, 16 Bus Distribution System

For system 1, IEEE 3 feeder, 16 bus distribution system, there are 13 sectionalizing switches and3 tie switches. The programming is done on the active power inputs of 10, 16, 6 MW and reactive power inputs of 6, 10, 4 MVAr [12]. 100MVA is base MVA and 20kV is base kV.

Table 1. Comparative Result of system 1 (3 feeder system) in DFR

Comparative Result	Original Configuration	Minimum Loss Configuration
Switches open	15,21,26	17,18,23
Loss(kW)	579.9	421.37
Feeder Capacity Margin (p.u)	0.6075	0.7475
Transformer Capacity Margin (p.u)	1	1
Maximum Voltage Deviation (p.u)	0.059015	0.0448307
% Loss minimization	-	27.33

Table 2. Result of system 1 (3 feeder system) in SR

Conditions	Cut off switch	Out of service nodes	Sectionalizing off	Tie off	Total Loss(kW)	Reliability Factors	Run Time (sec)
G=100 P=10 F=.5 CR=.9	18	9,11,12	14,17,18	-	431.46	FCM=.7475 TCM=1 M.V.D=.039	13.55
	14	7	17,19,14	-	443.69	FCM=.6075 TCM=1 M.V.D=.056	11.21
	17	10	17,19,23	-	433.6	FCM=.6075 TCM=1 M.V.D=.056	9.15

From the comparative study on the loss reduction for system 1, it is found out that in optimized configuration the loss is reduced about 27.33 % from its original configuration. In case of service restoration, switches 18, 14 and 17 are cut in different cases to see the restoration of the system in minimum time.

5.2 System 2: IEEE 33 Bus Distribution System

For system 2, IEEE 33 bus distribution system, there are 32 sectionalizing switches and 5 tie switches. The programming is done on the active power input of 2520 kW

Table 3. Result of system 2 (33 bus distribution system) on DFR

Conditions	Sectionalizing Off	Tie Off	Loss (kW)	Critical Nodes	Reliability Factor	Time(s)
	Original Configuration	-	384.75	3,6,20	FCM=1 TCM=1 MDV=.05	-
G=100,P=40 F=.5, CR=.9	8,15,18	33,36	163.34	6,20		66.81

Table 4. Result of system 2 (33 bus distribution system) on SR

Outage Branch	Sectionalizing Off	Tie off	Loss (kW)	Critical Nodes	Reliability Factors	Time (s)
3	3,12,17,18,21	-	194.95	3	FCM=1 TCM=1 MDV=.05	60.62
8	8,11,18,22	37	155.94	3,6		66.55
17	2,8,9,17	37	201.73	20,21,22		69.73
28	11,18,28	35,37	205.05	3,6		45.58

and reactive power input of 1073 kVAr. 10MVA is base MVA and 12.66 kV is base kV. The current limits for the branches 1 to 9 and 10 to 37 are taken 400A and 200A respectively [3].

In case of reconfiguration in system 2, for constant input the optimize configuration gives 163.34 kW of loss and that is near about 60% lower than its loss from original configuration. In service restoration, 3, 18, 17, 28 are cut off in different cases to see the same restoration of system in minimum time.

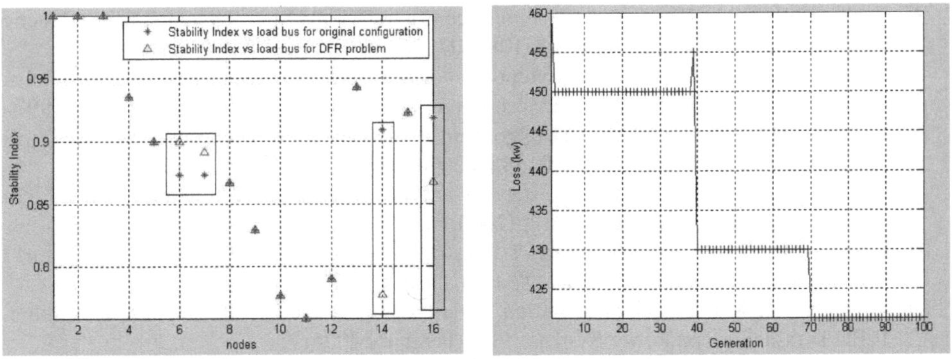

Fig. 1. Stability Index vs. Node and Optimization Curve Loss vs. Iteration of System 1

From Stability Index vs. nodes graph node number 11 is more insecure in term of voltage stability. The system is optimized that can be observed from optimization curve (Figure 1 for system 1, 3 feeders, 16 bus system).

6 Conclusion

It can be concluded that the loss reduction and quick responsive feature of the network system can be achieved using DE. For both the feeder systems (3 feeder system and 33 bus distribution system) restoration is achieved in more or less than one minute. This kind of intellectual programming can be put into the controlling strategy of distribution feeder system quite efficiently.

Acknowledgments. The work described in this article has been supported by "University Grants Commission (UGC)" under the scheme of "University with Potential for Excellence, Phase-II (UPE-II)".

References

1. Civanlar, S., Grainger, J.J., Yin, H., Lee, S.S.H.: Distribution feeder reconfiguration for loss reduction. IEEE Transactions on Power Delivery 3(3) (July 1988)
2. Chiou, J.-P., Chang, C.-F., Su, C.-T.: Variable scaling hybrid differential evolution for solving network reconfiguration of distribution system. IEEE Transaction on Power Systems 20(2) (2005)
3. Hossam-Eldin, A.A., Abdelaziz, A.R., Abu Fard, A.-E.I.: A Simualted Annealing–Based Automation of Distribution Systmes. In: UPEC 2010, vol. 31 (2010)
4. Guimaraes, M.A.N., Lorenzeti, J.E.C., Castro, C.A.: Reconfiguration of distribution systems for voltage stability margin-enhancement using tabu search. In: 2004 Intematlonal Conference on Power Syslem Technology - POWERCON 2004, Singapore (2004)
5. Baran, M.E., Wu, F.F.: Network Reconfiguration in Distribution Systems for loss Reduction and Load Balancing. IEEE Transactions on Power Delivery 4(2) (April 1989)
6. Subburaj, P., et al.: Distribution System Reconfiguration for Loss Reduction using Genetic Algorithm. J. Electrical Systems 2-4, 198–207 (2006)
7. Storn, R., Price, K.: Differential Evolution – A Simple and Efficient Heuristic for Global Optimization over Continuous Spaces. Journal of Global Optimization 11, 341–359 (1997)
8. Farahani, H.F.: Presentation of an algorithm to find paths between buses and feeder in radial distribution networks. JBASR; ISSN 2090-4304
9. Chakravorty, M., Das, D.: Voltage stability analysis of radial distribution networks. Electrical Power and Energy Systems 4(2) (2000)
10. Lin, W.-M., Cheng, F.-S., Tsay, M.-T.: Distribution feeder reconf iguration with refined genetic algorithm. IEE Proc. Gener. Transm. Distrib. 147(6) (November 2000)
11. Hsiao, Y.-T.: Multiobjective evolution programming method for feeder reconfiguration. IEEE Transactions on Power Systens 19(1) (February 2004)
12. Akduman, B.: Service restoration in distribution systems using an evolutionary algorithm. In: 7th Mediterranean Conference and Exhibition on Power Generation, Transmission, Distribution and Energy Conversion, Agia Napa, Cyprus, November 7-10 (2010) (Paper No. MED10/177)

Unsupervised Non-redundant Feature Selection: A Graph-Theoretic Approach

Monalisa Mandal and Anirban Mukhopadhyay

Department of Computer Science and Engineering
University of Kalyani
Kalyani-741235, West Bengal, India
{monalisa,anirban}@klyuniv.ac.in

Abstract. In this article a graph-theoretic approach for non-redundant unsupervised feature selection has been presented. The input data matrix is first converted into a weighted undirected complete feature-graph where the nodes represent the features and the edges are weighted according to the dissimilarity of features. Then the densest subgraph having maximum average weight is identified from the original feature graph. The features contained in the reduced subgraph are the final selected features for which average correlation is very less. The proposed method is compared with other dimensionality reduction techniques such as SFS and SBS in terms of entropy, classification accuracy, class separability, average correlation and execution time on several real life data sets.

Keywords: Filter, Unsupervised, Entropy, Class Separability.

1 Introduction

Feature selection has great impact in improving the quality of classification and clustering in machine learning and data mining. But high dimensional data poses a big challenge to feature selection algorithms. Moreover in the absence of class label feature subset selection is a challenging problem.

In a supervised scenario, the correct class of all samples are additionally known and the feature evaluation criteria to generate selected feature set is based on a classifier result. The real life data sets frequently contain attributes that are redundant or have a low information content which attributes introduces noise and may slow down the classification process gradually and also introduces high cross-validation errors. Wrapper methods for supervised classification [1] directly use a specific classifier. As filter methods are independent of the classifier applied subsequently, they have good generalization properties, but may be less effective at decreasing the dimensionality of the feature space and boosting classification accuracy. Generally, they are computationally cheaper than the wrapper approaches.

In unsupervised scenario, the utility of a wrapper-based approach is usually measured in terms of the performance of a clustering method. Unsupervised feature selection has been addressed in several ways such as clustering based [2], [3],

S.C. Satapathy et al. (Eds.): Proc. of Int. Conf. on Front. of Intell. Comput., AISC 199, pp. 373–380.
DOI: 10.1007/978-3-642-35314-7_43

content based [4], for ensemble classifier [5], graph based [6], [7] and feature similarity based [8]. Performance in unsupervised classification is typically measured as the ability of a clustering to reveal groupings (clusters) in a given data set. Basically the resulting clustering solution is evaluated using some cluster validation techniques like entropy (E), class separability (S) and fuzzy feature evaluation index (FFEI) etc [8]. Besides these, there exist many other such validation techniques in the clustering literature, each of which has its own biases, strengths and limitations. In contrast to wrapper approaches, the most common filter strategies are based on feature ranking [9]. In this context, two opposite strategies have been proposed in the literature: those that aim at the removal of redundant features [8] and those that focus on the removal of irrelevant features [10].

The objective of feature selection should be to select the features that are most relevant to classification while minimizing redundancy. Practically, most of the existing methods may have various types of shortcomings: 1) The performance on high-dimensional data sets is not enough satisfactory 2) redundant features are not removed completely 3) a few irrelevant features are eliminated and 4) expensive computation cost for high-dimensional data or noisy data. In our proposed method, a weighted undirected complete graph is formed from the dissimilarity measure (maximal information compression index) between features. Then the densest subgraph regarding to edge weight has been extracted from the original feature set. The attributes contained in the extracted subgraph are the final selected features.

2 Problem Formulation

In this article the goal is to find non-redundant features from a data matrix that means the resultant features are non-correlated. So the problem should be defined in such a way that the correlated features are eliminated gradually. In our proposed technique the problem of unsupervised feature selection is formulated as a problem of densest subgraph finding problem from a weighted undirected graph. The structure of the data matrix can be viewed as a two-dimensional matrix; the rows imply instances and columns imply attributes or features; one extra column is used for representing the corresponding class labels of the instances. A range of some similarity/dissimilarity measures includes correlation coefficient [11], least square regression error [11] and maximal information compression index [8] etc. Using one of theses dissimilarity (negative similarity) measures the symmetric matrix is generated which is termed as a dissimilarity matrix. Let the data set has n features, $\{F = f_1, f_2, f_3, ..., f_n\}$. Calculating pairwise similarity between features of the feature set F generates the $(n \times n)$ symmetric dissimilarity matrix Sm where both n rows and n columns correspond to n features. Therefore from this dissimilarity matrix Sm a weighted complete graph G can be formed. Each node represents a feature so the vertex set of the graph G is $\{V = f_1, f_2, f_3, ..., f_n\}$, i.e. the graph contains total n number of nodes. The value present in intersection of row i and column j in the dissimilarity matrix Sm represents the weight of the edge between node f_i and f_j.

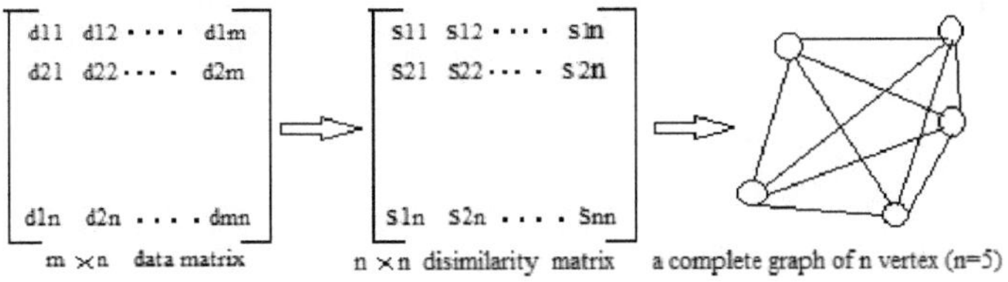

Fig. 1. Data matrix to Dissimilarity matrix to Graph formulation

As each feature has some dissimilarity value with every other feature (present in dissimilarity symmetric matrix Sm), hence the graph G is a complete graph. Fig. 1 demonstrates the process of conversion from data matrix to feature graph.

First the dissimilarity matrix is calculated for the data matrix using maximal Information Compression Index (λ_2) [8] which can be defined by covariance matrix. Let $\sum$ be the covariance matrix of random variables x and y. Therefore, λ_2 can be defined as the smallest eigen value of $\sum$:

$$2\lambda_2(x,y) = (var(x) + var(y) - \sqrt{(var(x) + var(y))^2 - 4var(x)var(y)(1-\rho)^2}. \quad (1)$$

Subsequently, a graph G is formulated from the dissimilarity matrix. For the graph G larger the edge weight means that the features connected by that edge are more dissimilar. Thus finding the most dense subgraph g from graph G is equal to finding the most non-redundant feature set because the features (nodes) contained by the subgraph g will have maximum average edge weight (dissimilarity). Therefore the problem can be defined as most dense subgraph g finding from a complete weighted graph G and the features present in the reduced subgraph g are the required output of our proposed technique.

3 Proposed Method

The conversion of the dissimilarity matrix to a weighted complete graph has already been discussed. In the same way, the graph $G = [V, E, W]$ has been formed from the dissimilarity the matrix Sm where V implies vertex set, E implies edge set and W implies set of corresponding edge weights of the graph G. Algorithm 1 describes the flow of the proposed approach. Both the graph and the dissimilarity matrix are the input to the algorithm and the output is the reduced subgraph whose vertex set is the reduced feature set. The average density of a graph is calculated as the sum of edge weight divided by the number of possible edges in a complete graph of $|V|$ number of vertices. The function $adj(v)$ of a vertex v returns the adjacent edges (direct link) of the vertex v and therefore, the Edge-weighted-degree has been calculated as the sum of the weights of adjacent edges divided by the number of adjacent edges of that vertex. Basically in each iteration a subgraph is generated by removing the vertices

Algorithm 1. Algorithm: Densest subgraph from a weighted complete graph

Input: Graph $G[V, E, W]$ designed from dissimilarity matrix Sm.
Output: Reduced subgraph $g = [V1, E1, W1]$

1: Avg-Density= $\frac{\sum W_{i,j \in V}}{\frac{|V|.(|V|-1)}{2}}$
2: **for** $i = 1 : |V|$ **do**
3: **if** i =isolated-vertex **then**
4: Edge-weighted-degree $(i) = 0$;
5: **else**
6: Edge-weighted-degree$(i) = \frac{\sum W(e)_{e \in E} |\forall e \in adj(i)|}{|adj(i)|}$
7: **end if**
8: **end for**
9: Mean-degree= $\frac{\sum \text{Edge-weighted-degree}}{|V|}$
10: **for** $i = 1 : |V|$ **do**
11: **if** Edge-weighted-degree (i) $\geq$Mean-degree **then**
12: $s_v \leftarrow s_v \cup i$
13: **end if**
14: **end for**
15: Subgraph $g[V1, E1, W1] = Sm(s_v, s_v)$
16: Avg-Density-new $= \frac{\sum W1_{i,j \in V1}}{\frac{|V1|.(|V1|-1)}{2}}$
17: **if** Avg-Density-new>Avg-Density **then**
18: Avg-Density=Avg-Density-new;
19: $G[V, E, W] = g[V1, E1, W1]$
20: *GoTo* Step-2.
21: **else**
22: $g[V1, E1, W1] = G[V, E, W]$;
23: exit with g ;
24: **end if**

having Edge-weighted-degree less than the average Edge-weighted-degree and average density is calculated for the resultant subgraph. If the average density increases then steps are repeated otherwise end with the resultant subgraph $g = [V1, E1, W1]$. At the end the vertex set $|V1|$ of the graph g is the reduced feature set.

4 Datasets and Results

Five real life data sets are used for the comparative study. The data sets are collected from the website : `http://archive.ics.uci.edu/ml/datasets/`. Only the real valued attributes of data sets are taken to evaluate of the algorithms.

Ionosphere: There are 351 number of instances and 34 number of attributes plus the class attribute which includes "Good" or "bad" .

Ecolai: There exist 336 number of samples across 8 number of attributes for which 7 are predictive continuous valued features and one name for class label. Every sample has seven classes.

Parkinson: From voice recording data of the individuals ("name" column) is collected to discriminate healthy people from those with Parkinson Disease. It contains 197 × 23 real valued data having 2 classes.

Wisconsin Prognostic Breast Cancer (WPBC): This is a binary classification task which is required for the data set which contains 198 number of instances and 34 number of attributes.

Connectionist Bench(Sonar, Mines vs. Rocks): The data set contains 208 instances and 60 attributes with two classes.

The above described five real-life data sets are used to evaluate our proposed method first. Subsequently, the size of the resultant feature set has been used as the input of the other two algorithms. The evaluation criteria on which the other two algorithms iterate is feature entropy Eqn. 2. Now each of the three reduced sets are evaluated for checking the effectiveness in terms of entropy, representation entropy, K-nn classification accuracy, Naive Bayes accuracy and execution time. Entropy decides whether data set has well-formed clusters or not and representation entropy [8] is used to determine the amount of information compression possible by dimensionality reduction i.e. non-redundancy. In addition, using 10-fold cross-validation, classification accuracy is computed. Therefore 10 different runs of the validation method give mean accuracy with standard deviation for each of the classifiers. All of the evaluation techniques have been run in a machine with following specifications: Core i3-370M Processor(2.40 GHZ), 3GB RAM, 64-bit Windows 7 Professional.

4.1 Performance Metrics

Entropy [12]: Let M be the number of features and distance between two data points p and q is defined as follows:

$$D_{pq} = \left[\sum_{j=1}^{M} \left(\frac{x_{pj} - x_{qj}}{max_j - min_j} \right) \right]^{\frac{1}{2}}, \tag{2}$$

where x_{pj} is the jth data point of pth attribute, max_j and min_j is the maximum and minimum of jth data point. Now similarity between p and q is computed as $Sim(p,q) = e^{-\alpha D_{pq}}$ where α is positive constant possibly $\frac{-ln(0.5)}{\bar{D}}$. $\bar{D}$ is the average distance computed over all data points. If l be the number of sample points, then the entropy E can be defined as follows:

$$E = -\sum_{p=1}^{l}\sum_{q=1}^{l}(sim(p,q) \times log(sim(p,q)) + (1 - sim(p,q)) \times log(1 - sim(p,q))). \tag{3}$$

Representation Entropy [13]: Let d be the size of feature set, $d \times d$ is the covariance matrix and λ_j is the corresponding eigen values of d features, then $\bar{\lambda_j} = \frac{\lambda_j}{\sum_{j=1}^{d} \lambda_j}$; Therefore the representation entropy is defined as follows:

$$H_R = -\sum_{j=1}^{d} \bar{\lambda_j} log(\bar{\lambda_j}). \quad (4)$$

Class Separability [13]: Class separability is defined by $S = trace(S_w^{-1} S_b)$, where S_w is within scatter matrix calculated as $S_w = \sum_{j=1}^{c} \pi_j \Sigma_j$ and S_b is the between scatter matrix which is calculated as where π_j is the priori probability that a pattern belongs to a particular class, μ_j is the sample mean vector of that class, M_0 in Eqn. 5, is the sample mean vector of the entire data points, Σ_j is the sample covariance matrix of the class.

$$S_b = \sum_{j=1}^{c} (\mu_j - M_0)(\mu_j - M_0)^T, \text{ where } M_0 = \sum_{j=1}^{c} \pi_j \mu_j. \quad (5)$$

It is evident from Table 1 that the entropy produced by our proposed method is less than other algorithms in all data sets, i.e the proposed method generates a feature set which represents the fact that the data set has well formed clusters. Higher is the representation entropy, higher is the expectation of non-redundancy. Excluding Ecoli and Breast Cancer data, in all other cases representation entropy produced by our method is maximum imply maximum non-redundancy than the other two methods. It has already been told that our objective is to select non-redundant feature set which implies non-correlated features. For this reason we have presented the average correlation which is calculated for the reduced set produced by the proposed technique, SFS and SBS. From the table it is clear that excluding Connectionist Bench data, the average correlation of the feature set identified by the proposed method is less than the average correlation of the original dataset for all remaining four datasets. Again the average correlation of the reduced feature set of the proposed method is less than the other two method excluding the Breast Cancer data. In terms of computation time our approach gives the most impressive output than SFS and SBS.

For Supervised Evaluation table 2 shows that with respect to K-nn classifier, our method generate the best accuracy 67.73 (0.014) and 75.58 (0.0113) for Breast Cancer and Connectionist bench data datasets respectively. In all other cases K-nn accuracy more or less same as other methods. When considering Bayes classifier and Ecoli, Breast Cancer, Ionosphere and Connectionist Bench datasets, it is seen that the proposed method gives the best result than the other two algorithms. For parkinson data the Bayes accuracy of the method is 86.24 (0.0099) which is better than SBS and slightly less than SFS. For parkinson, Breast Cancer and Connectionist bench dataset K-nn accuracy of our method outperforms the other two methods and in other datasets the results differ slightly.

Table 1. Comparative score analysis between SFS,SBS and proposed method using various datasets. D= number of features in original data set, d=number of features in reduced data set.

Data Set	Methods	Entropy	Separability	Represen- -tion entropy	Avg Corr	time(sec)
Ecoli	sfs	0.5526	360.8	1.5651	0.1634	56.7879
D=7	sbs	0.5593	374.2	1.6839	0.1533	20.729
d=6	proposed	0.535	372.2	1.6095	0.1228	0.0545
Parkinson	sfs	0.5436	3.115	1.3466	0.4130	171.6298
D=22	sbs	0.5247	3.054	1.4754	0.4373	49.718
d=20	proposed	0.5127	3.116	1.4782	0.3027	0.0933
Breast	sfs	0.5606	0.4216	1.576	0.1245	132.1334
Cancer	sbs	0.5518	0.3867	1.6	0.1739	468.2644
D=31; d=8	proposed	0.5502	0.3305	1.585	0.193	0.1333
Ionosphere	sfs	0.5594	2.2225	2.3575	0.1317	1.706×10^3
D=33	sbs	0.5578	2.346	2.5097	0.0913	1.757×10^3
d=21	proposed	0.5502	1.8283	2.5108	0.1062	0.2043
Connectionist	sfs	0.5605	0.378	0.2862	0.343	484.601
Bench	sbs	0.5543	0.6931	1.6093	0.2448	2.956×10^3
D=60; d=11	proposed	0.5507	0.339	1.6393	0.1113	0.33

Table 2. Supervised Evaluation of SFS,SBS and proposed method

Data Set	Methods	**Supervised evaluation**	
		Knn (%)	Bayes (%)
Ecoli	sfs	79.91 (0.0042)	38.07 (0.0385)
D=7	sbs	75.62 (0.0086)	35.12 (0.0514)
d=6	proposed	76.73 (0.0111)	39.88 (0.0551)
Parkinson	sfs	87.33 (0.0097)	86.67 (0.0084)
D=22	sbs	82.87 (0.0073)	85.13 (0.0128)
d=20	proposed	82.54 (0.008)	86.24 (0.0099)
Breast Cancer	sfs	65.91 (0.0167)	73.94 (0.012)
D=31	sbs	65.05 (0.0161)	72.02 (0.012)
d=8	proposed	67.73 (0.014)	75 (0.0104)
Ionosphere	sfs	85.98 (0.005)	89.32 (0.0024)
D=33	sbs	87.35 (0.0031)	88.72 (0.0051)
d=21	proposed	86.01 (0.0029)	89.89 (0.0028)
Connectionist	sfs	58.56 (0.0117)	56.30 (0.0125)
Bench	sbs	70.29 (0.0147)	61.54 (0.0159)
D=60; d=11	proposed	75.58 (0.0113)	71.73 (0.0091)

5 Conclusion

The objective of our proposed technique is to find non-redundant feature set for which average correlation will be less. In this proposed method, the problem of dimensionality reduction is modeled as densest subgraph finding from a weighted graph. In an unsupervised way, i.e., without using the class label information, our approach identifies a subset of non-redundant features. The results show that in most of the cases our algorithm performs better than the other methods, and most interestingly, it takes negligible time in comparison with other methods.

References

1. Kohavi, R., John, G.: Wrapper for feature subset selection. Artificial Intelligence 97, 273–324 (1997)
2. Jiang, S., Wang, L.: An unsupervised feature selection framework based on clustering. School of Informatics, Guangdong University of Foreign Studies, Guangzhou (2008)
3. Cai, D., Zhang, C., He, X.: Unsupervised feature selection for multi-cluster data. In: KDD 2010, Washington, DC, USA (2010)
4. Dy, J.G., Brodley, C.E., Kak, A., Broderick, L.S., Aisen, A.M.: Unsupervised feature selection applied to content-based retrieval of lung images. IEEE Transaction on Pattern Analysis and Machine Intellegence 25(3), 373–378 (2003)
5. Morita, M., Oliveira, L.S., Sabourin, R.: Unsupervised feature selection for ensemble of classifiers. Frontiers in Handwriting Recognition (2004)
6. Zhang, Z., Hancock, E.R.: A graph-based approach to feature selection. Springer (2011)
7. Bahmani, B., Kumar, R., Vassilvitskii, S.: Densest subgraph in streaming and mapreduce. VLDB Endowment 5(5), 454–465 (2012)
8. Mitra, P., Murthy, C.A., Pal, S.K.: Unsupervised feature selection using feature similarity. IEEE Transaction on Pattern Analysis and Machine Intellegence 24(3), 301–312 (2002)
9. Li, Y., Lu, B., Wu, Z.: A hybrid method of unsupervised feature selection based on ranking. IEEE Computer Society, Washington, DC (2006)
10. Sondberg-Madsen, N., Thomsen, C., Pena, J.M.: Unsupervised feature subset selection. In: Proceedings of the Workshop on Probabilistic Graphical Models for Classification (2003)
11. Chatterjee, S., Hadi, A.S.: Regression Analysis by Example(4e). John Wiley & Sons, Inc.
12. Dash, M., Liu, H.: Unsupervised feature selection. In: Proc. Pacific Asia Conf. Knowledge Discovery and Data Mining (2000)
13. Devijver, P.A., Kittler, J.: Pattern recognition: A statistical approach. Prentice-Hall, Englewood Cliffs (1982)

Soft Set and Genetic Algorithms for Association Rule Mining: A Road Map and Direction for Hybridization

Satya Ranjan Dash[1] and Satchidananda Dehuri[2]

[1] School of Computer Application,
KIIT University, Patia
Bhubaneswar, Odisha, India-751006
sdashfca@kiit.ac.in

[2] Department of Information and Communication Technology,
Fakir Mohan University,
Vyasa Vihar, Balasore, Odisha, India – 756019
satch_d@yahoo.co.in

Abstract. Association rules have relied on user-specified threshold of support and confidence. With no prior/little domain knowledge, if the user is specifying threshold for the mining task; then there is a direct impact on quality of association rules. In this paper, we have discussed some of the early attempts of choosing automatically the user specified threshold (i.e., no user intervention to specify threshold) by soft set and genetic algorithms for association rule mining.

The reason of being restricted with soft set and genetic algorithms is that: association rule using soft set is free from inadequacy of the parameterization tools, which can also deals with uncertainty. Alongside, genetic algorithms can help to user for finding out optimal threshold for generating a number of interesting and novel association rules. Furthermore, we discuss the possibility of hybridization and their future usage in association rule mining.

Keywords: Soft set, Genetic Algorithm, Rule mining.

1 Introduction

Association rule mining(ARM) is a fundamental tasks in many areas like data mining [36], computational biology [39], finance [40], agriculture [41], social science [42] etc. An association rule is considered interesting if it satisfies certain constraint such as predefined minimum support and confidence threshold. The association rules mining method was developed particularly for the analysis of transactional data analysis, whose attribute possess Boolean value. The occurrence of an item can be viewed as Boolean variable and either it occur (denoted as '1'), or not occur (denoted by '0') Association rules are used to represent and identify dependencies between items in a database. These are an expression of the type $X \rightarrow Y$, when $X \rightarrow Y$ are items & $X \cap Y = \phi$. It means that if all the items in X exists in a transaction then all the item in Y are also in the transaction with high probability, but X & Y should not have common item.

S.C. Satapathy et al. (Eds.): Proc. of Int. Conf. on Front. of Intell. Comput., AISC 199, pp. 381–393.
DOI: 10.1007/978-3-642-35314-7_44

Cheung & Fu [14] developed a technique to identify frequent itemsets without the threshold of the support. Zhang et al [15] advocate the fuzzy-logic-based method to acquire user threshold of minimum support for mining association rules. However, most of these approach attempt to avoid specifying the minimum support and some of them even confidence driven method.

Soft set theory [16] proposed by Molodtsov, is a new general method for dealing with uncertain data. Soft sets are called neighborhood systems. Soft set may be redefined as the classification of objects in two distinct classes, which said that it can deal with Boolean-valued information systems. Molodtsov [16] pointed out that one of the main advantages of soft set theory is that it is free from the inadequacy of the parameterization tools. Soft set is also used for representing transactional data.

In other hand genetic algorithms (GAs) can identify association rules without minimum support. GA is efficient for global search work, especially when the search space is too large,(i.e. it is less likely to take the help of a deterministic search method than heuristic method such as GA). GA based approach for ARM does not require users to specify minimum support threshold. Instead of generating an unknown number of interesting rules in traditional mining models, only the most interesting rules are returned by GAs according to the interestingness measure defined by the fitness function.

2 Preliminaries

In this section, we state the preliminaries such as association rule, GAs & soft set theory. Association rule technique is particularly useful for discovery of hidden relations which might be interesting when it comes to large databases [26, 38]. Genetic algorithms are important when discovering association rules because they work with global search to discover the set of items frequency and their relationship [36]. Soft set theory is a relatively new approach to discuss vagueness and the membership is decided by adequate parameters [37].

2.1 Association Rule

Let $I = \{i_1, i_2 \ldots i_N\}$ be a set of N distinct literals called items. Let D be a set of database transactions where each transaction T is a set of items such that $T \subseteq I$. Each transaction is associated with a unique identifier, called TID. Let A, B be a set of items, an association rule is an implication of the form $A \Rightarrow B$, where $A \subset I$, $B \subset I$, and $A \cap B = \phi$. A is called *antecedent* of the rule, and B is called the *consequent* of the rule. The rule $A \Rightarrow B$ holds in the transaction set D with *support* s, where s is the percentage of transactions in D that contain both A and B. In other words, the support of the rule is the probability P $(A \cup B)$. The rule $A \Rightarrow B$ also has another measure called *confidence* c where c is the percentage of transactions in D containing A that also contain B. In other words, the confidence of the rule is the conditional probability P $(B|A)$. The problem of discovering all association rules from transactional database D consists of generating the rules that have a support and confidence greater than predefined thresholds. Such rules are called valid (or strong) rules, and the framework is known as the *support-confidence framework.*

Let $I = \{i_1,..., i_k\}$ be a set of k elements, called *items*. Let $B = \{b_1, ..., b_n\}$ be a set of n subsets of I. We call each $b_i \subseteq I$ a *basket* of items. For example, in the market basket application, the set I consist of the items stocked by a retail outlet and each basket is the set of purchases from one register transaction. Similarly, in the "document basket" application, the set I contains all dictionary words and proper nouns, while each basket is a single document in the corpus. Note that the concept of a basket does not take into account the ordering or frequency of items that might be present. An association rule is intended to capture a certain type of dependence among items represented in the database B. Specifically, we say that $i_1 \rightarrow i_2$ if the following two hold

1. i_1 and i_2 occur together in at least $s\%$ of the n baskets (the *support*).
2. Of all the baskets containing i_1, atleast $c\%$ also contains i_2 (the *confidence*).

This definition is also extended to $I \rightarrow J$, where I and J are disjoint sets of items instead of single items. Let us consider an example of a document basket application. The baskets in this case are many short stories that are available at our disposal, while the items within each basket are the words. A reader might observe that stories which contain the word "sword" also frequently contain the word "blood". This information can be represented in the form of a rule as:

$$\textit{sword} \rightarrow \textit{blood}$$
$$[\textit{support} = 5\%,\ \textit{confidence} = 55\%]$$

Rule support and confidence are the two measures of rule interestingness. They respectively reflect the usefulness and certainty of discovered rules. A support of 5% for an association rule means that 5% of stories under analysis show that "blood" and "sword" occur together. A confidence of 55% means that 55% of the stories that contain the word "sword" also contain the word "blood".

Typically, association rules are considered interesting if they satisfy both minimum support threshold and a minimum confidence threshold. Such threshold can be set by users or domain experts; the confidence measure is merely an estimate of the conditional probability of i_2 given i_1.

2.2 Soft Set Theory

Theory of soft sets is introduced by Molodtsov [16]. This theory is a relatively new approach to discuss vagueness. It is getting popularity among the researchers and a good number of papers are being published every year. In [37] Maji discussed theoretical aspect of soft sets and they introduced several operations for soft sets. In soft set theory membership is decided by adequate parameters, rough set theory employs equivalence classes, whereas fuzzy set theory depends upon grade of membership.

Let U be an initial universe set and let E be a set of parameters. A pair (F,E) is called a soft set (over U) if and only if F is a mapping of E into the set of all subsets of the set U(i.e. F: E→P (U)).

In other words, the soft set is a parameterized family of subsets of the set *U*. Every set $F(\varepsilon), \varepsilon \in E$, from this family may be considered as the set of ε-elements of the soft set (*F,E*), or as the set of ε-approximate elements of the soft set.

A soft set (F; E) describes the attractiveness of the houses which Mr. X is going to buy.

U - is the set of houses under consideration.
E - is the set of parameters. Each parameter is a word or a sentence.
E = {expensive; beautiful; wooden; cheap; in the green surroundings; modern; in good repair; in bad repair}

In this case, to define a soft set means to point out expensive houses, beautiful houses, and so on. It is worth noting that the sets may be arbitrary. Some of them may be empty, some may have nonempty intersection.

Assume that we have a binary operation, denoted by $*$, for subsets of the set *U*. Let (*F, A*) and (*G, B*) be soft sets over *U*. Then, the operation $*$ for soft sets is defined in the following way:

$$(F, A) * (G, B) = (H, A \times B),$$

Where $H(\alpha, \beta) = F(\alpha) * G(\beta)$, $\alpha \in A$ and $\beta \in B$.

This definition takes into account the individual nature of any soft set. If we produce a lot of operations with soft sets, the result will be a soft set with a very wide set of parameters. Sometimes such expansion of the set of parameters may be useful. The resulting soft set points out the houses which are expensive and beautiful, modern and cheap, and so on.

Let U={ h1,h2,h3,h4,h5} be a set of houses under consideration where E={e1,e2,e3,e4,e5} and A={e1,e2,e3,e4} be a subset of parameter for selection of the house. Let

e1 stands for expensive houses,
e2 stands for wooden houses,
e3 stands for houses located in green surroundings,
e4 stands for houses located in the urban area,
e5 stands for the low cost houses,

Let (F, A) be the soft set to categorize the houses with respect to parameters given by set A, such that F(e1)={h1,h3}, F(e2)={h1,h3,h6}, F(e3)={h1,h3,h4,h5}, F(e4)={h1,h2,h3}.

For computer applications it is more appropriate to represent a soft set in tabular form

	h1	h2	h3	h4	h5	h6
e1	1	0	1	0	0	0
e2	1	0	1	0	0	1
e3	1	0	1	1	1	0
e4	1	1	1	0	0	0

2.3 Genetic Algorithm

GA is started with a *set of solutions* (represented by *chromosomes*) called *population*. Solutions from one population are taken and used to form a new population. This is motivated by a hope, that the new population will be better than the old one. Solutions which are selected to form new solutions are selected according to their fitness - the more suitable they are the more chances they have to reproduce.

/*Algorithm GA */

1. [*Start*] Generate random population of *n* chromosomes (suitable solutions for the problem)
2. [*Fitness*] Evaluate the fitness *f(x)* of each chromosome *x* in the population
3. [*New population*] Create a new population by repeating following steps until the new population is complete

    i. [*Selection*] Select two parent chromosomes from a population according to their fitness (the better fitness, the bigger chance to be selected)
    ii. [*Crossover*] With a crossover probability cross over the parents to form a new offspring (children). If no crossover was performed, offspring is an exact copy of parents.
    iii. [*Mutation*] With a mutation probability mutate new offspring at each locus (position in chromosome).
    iv. [*Accepting*] Place new offspring in a new population

4. [*Replace*] Use new generated population for a further run of algorithm
5. [*Test*] If the end condition is satisfied, *stop*, and return the best solution in current population
6. [*Loop*] Go to step 2

Algorithm is started with a set of *solutions* (represented by *chromosomes*) called *population*. Solutions from one population are taken and used to form a new population. This is motivated by a hope, that the new population will be better than the old one. Solutions which are selected to form new solutions (*offspring*) are selected according to their fitness - the more suitable they are the more chances they have to reproduce. This is repeated until some condition (for example number of populations or improvement of the best solution) is satisfied.

3 Association Rules Using Soft Set Theory

The association rules are said to be strong if it meets the minimum confidence threshold. However, while association rules provide means to discover many interesting associations, they fail to discover other, no less interesting associations, which also hidden in the data. Maximal association rules introduced by Feldman[17] is a variant of association rules which is designed to handle the above problem. It allows the discovery of associations pertaining to items that most often do not appear alone, but rather together with closely related items, and hence associations relevant only to these items tend to obtain low confidence. These rules are very important in discovering maximal association, particularly from documents text collection. The idea is inspired from the fact that many interesting rules in databases cannot captured by regular rules. Feldman et al. noted that maximal association rules are not designed to replace regular association rules, but rather to complement them. Every maximal association rule is also regular association, with perhaps different support and confidence [18]. While association rules are based on the notion of frequent itemsets which appears in many records, maximal association rules are based on frequent maximal itemsets which appears maximally in many records [19]. Using only maximal association rules, many interesting regular associations may and will be lost. of items. In maximal association rule, $X \overset{max}{\Longrightarrow} Y$ we are interested in capturing the notion that whenever X appears alone then Y also appears, with some confidence.

Let (F, E) be a soft set over the universe U and $X \subseteq E$. A set of attributes X is said to be supported by a transaction $u \in U$ if $X \subseteq$ co-occurrence (U).

Let (F, E) be a soft set over the universe U and two maximal itemsets $X, Y \subseteq E_i$, where $X \cap Y = \phi$. A maximal association rule between X and Y is an implication of the form $X \overset{max}{\Longrightarrow} Y$. The itemsets X and Y are called maximal antecedent and maximal consequent, respectively.

The applicability of soft set theory for association rules and maximal association rules mining. Pre-requisite of using soft set approach for maximal association rules mining is the transactional dataset need to be transformed into a soft set, where each item is regarded as a parameter (attribute).In the proposed approach, we use the notion of co-occurrence of parameters for association rules mining as used in [25].

$$\begin{matrix} i_1 \rightarrow a_1 \\ i_2 \rightarrow a_2 \\ \ldots \Leftrightarrow I = \{i_1, i_2, .., i_{|A|}\} \rightarrow A = \{a_1, a_2, \ldots, a_{|U|}\} \\ i_{|A|} \rightarrow a_{|A|} \end{matrix}$$

and

$$\begin{matrix} t_1 \rightarrow u_1 \\ t_2 \rightarrow u_2 \\ \ldots \Leftrightarrow D = \{t_1, t_2 \ldots t_{|U|}\} \rightarrow U = \{u_1, u_2, .., u_{|U|}\} \\ t_{|U|} \rightarrow u_{|U|} \end{matrix}$$

Let (F, E) be a soft set over the universe U and u $\in$ U. An items co-occurrence set in a transaction u can be defined as

$$\text{Coo}(u) = \{e \in E : f(u, e) = 1\}$$

Obviously,

$$\text{Coo}(u) = \{e \in E : F(e) = 1\}$$

Let (F, E) be a soft set over the universe U and $X \subseteq E$. The support of a soft set parameters X, denoted by sup(X) is defined by the number of transaction supporting X,

$$I=\{i_1,i_2,\ldots,i_{|A|}\} \Rightarrow A=\{a_1,a_2,\ldots,a_{|A|}\} \Rightarrow E=\{e_1,e_2,\ldots,e_{|E|}\}$$
$$D=\{t_1, t_2 \ldots t_{|U|}\} \Rightarrow U=\{u_1,u_2,\ldots,u_{|U|}\}$$

$$a \in b\, f_i : U \rightarrow V_i \text{ and } \quad f_i(x) \begin{cases} 1 & x \in F(e_i) \\ 0 & x \notin F(e_i) \end{cases}$$

Thus $D = \{t_1, t_2 \ldots t_{|U|}\} \Rightarrow (F, E)$

Here is one data transaction

TID	Items
1	Canada, Iran, USA, crude, ship
2	Canada, Iran, USA, crude, ship
3	USA, earn
4	USA, jobs, cpi
5	USA, jobs, cpi
6	USA, earn, cpi
7	Canada, sugar, tea
8	Canada, USA, trade, acq
9	Canada, USA, trade, acq
10	Canada, USA, earn

Let (F, E) be a soft set over the universe U and X, $Y \subseteq E$, where $X \cap Y = \phi$. An association rule between X and Y is an implication of the form $X \Rightarrow Y$. The itemsets X and Y are called antecedent and consequent, respectively. The support of a association rule $X \Rightarrow Y$, denoted by sup $(X \Rightarrow Y)$ is defined by

$$\sup(X \Rightarrow Y) = \sup(X \cup Y) = |\{u : X \cup Y \subseteq \text{Coo}(u)\}|$$

and the confidence of a association rule $X \Rightarrow Y$, denoted by conf $(X \Rightarrow Y)$.

$$\text{conf}(X \Rightarrow Y) = \frac{\sup(X \cup Y)}{\sup(X)} = \frac{|\{u : X \cup Y \subseteq Coo(u)\}|}{|\{u : X \subseteq Coo(u)\}|}$$

The soft set representation of the above table is

$$(F, E)= \left\{ \begin{array}{l} \text{Canada}=\{1,2,7,8,9,10\}, \text{Iran}=\{1,2\}, \\ \text{USA}=\{1,2,3,4,5,6,8,9,10\}, \text{Crude}=\{1,2\}, \\ \text{Ship}=\{1,2\}, \text{earn}=\{3,10\}, \text{jobs}=\{4\}, \text{cpi}=\{3,10\}, \\ \text{Tea}=\{7\}, \text{Sugar}=\{7\}, \text{trade}=\{8,9\}, \text{acq}=\{8,9\} \end{array} \right\}$$

4 Association Rules Using Genetic Algorithm

Association Rules Mining through Genetic Algorithms: EARMGA, GAR and GENAR. We have used three GAs in the literature to achieve the association rules mining task:

- EARMGA [43]: It is based on the discovery of quantitative association rules.
- GAR [44]: It searches for frequent itemsets by dealing with numerical domains.
- GENAR [45]: It directly mines association rules by handling numerical domains.

A chromosome in EARMGA encodes a generalized k-rule, where k indicates the desired length. Since we may handle association rules with more than one item in the consequent, the first gene stores an index representing the end of the antecedent part. In order to uniquely encode a rule into a chromosome, both antecedent attributes and consequent attributes are sorted two-segmentally in an ascending order.

Then the remaining k genes encode items. Each item is represented by a pair of values, where the first value is an attribute's index ranged from 1 to the maximum number of attributes in the database, whereas the second stands for a gapped interval. The authors have defined a gapped interval as the union of a finite number of base intervals obtained once a uniform discretization process has been accomplished over all attributes in the database. Notice that we do not need to partition the domains of categorical attributes because here the lower and the upper bounds basically coincide. Nevertheless, a base interval is always represented by an integer number apart from the kind of attributes we deal with. As a consequence, a gapped interval is a set of these integers. Now we will give some details of the genetic operators applied to each chromosome:

- Selection: it is achieved by computing the fitness value along with a random number, so that the chromosome will be selected only if this product is less than a given probability of selection (ps).
- Crossover: all the selected chromosomes have the chance to reproduce offspring at a probability of crossover (pc). This operation simply consists of exchanging a segment of genes between the first chromosome and the second one and vice-versa, depending on two crossover-points randomly generated.
- Mutation: by considering both a probability of mutation (pm) and the fitness value, a chromosome is altered in the way that the boundary between antecedent attributes and consequent attributes could be changed within the same rule. In addition, the operator randomly chooses a gene and modifies the attribute's index

along with the gapped interval associated with it. Notice that the new gapped interval is always a union of base intervals which now form a sub-domain of the new attribute.

Finally, the association rules mining problem has been restated by the authors of this work because the algorithm searches for k-association rules only by considering fitness values given by a measure of interest known as positive confidence [40].

By contrast, GAR follows different strategies. First, a chromosome is composed of a variable number of genes, between 2 and n, where n is the maximum number of attributes. However, as we find frequent itemsets with this method, it is afterwards necessary to run another procedure in order to generate association rules. Moreover, it is unnecessary to discrete a priori the domain of the attributes since each gene is represented by an upper and a lower bound along with an identifier for that attribute. To briefly recall the genetic operators for this method:

- Selection: it simply selects a percentage (ps) of the chromosomes in the current population which have the best fitness. These ones will be the first individuals of the new-made population.
- Crossover: the new-made population is completed by reproducing offspring until reaching a desired size. To do that, the parents are randomly chosen at a probability of pc. Then, we only obtain two different offspring when their parents have genes containing the same attribute. In that case, their intervals could simply be exchanged considering all the possible combinations between them, but, in the end, two chromosomes should always be generated. Finally, only the best one will be added to the population.
- Mutation: as usual, at the probability of pm, it alters one gene in such a way that each limit could randomly decrease or increase its value. Its fitness function tends to reward frequent itemsets that have a high support as well as a high number of attributes. In addition, it punishes frequent itemsets which have already covered a record in the database and whose intervals are too large.
- GENAR was the first attempt by the same authors of GAR to handle continuous domains. Here a chromosome is encoded as an association rule which contains intervals as in the case of GAR. Nevertheless, the length of the rules is always fixed to the number of attributes and only the last attribute forms the consequent. Similar considerations can be taken into account regarding the definition of genetic operators, except for cross-over which employs a one-point strategy to reproduce offspring chromosomes [28].By contrast, its fitness function only considers the support count for the rules and punishes those which have already covered the same records in the database.
- The Genetic Association Rules algorithm is based in the theory of evolutionary algorithms and it is an extension of the GENAR algorithm presented in [35], that search directly for the association rules, so it is necessary to prepare the data to indicate to the tool which attributes form part of the antecedent and which one is the consequent. Nevertheless, this process is not necessary in Genetic Association Rules, because the algorithm finds the frequent itemsets and the rules are built departing from them.

Algorithm: Genetic Association rules

```
1. nItemset = 0
2. while (nItemset < N) do
3.          nGen = 0
4.          generate first population P(nGen)
5. while (nGen < NGENERATIONS) do
6.          process P(nGen)
7.          P(nGen+1) = select individuals of P(nGen)
8.          complete P(nGen+1) by crossover
9.          make mutations in P(nGen+1)
10.          nGen++
11.         end_while
12.                   I[nItemset] = choose the best of P(nGen)
13.          penalize records covered by I[nItemset]
14.                   nItemset++
15.         end_while
End
```

The process is repeated until we obtain the desired number of frequent itemsets N. The first step consists in generating the initial population. The evolutionary algorithm takes charge of calculating the fitness of each individual and carries out the processes of selection, crossover and mutation to complete the following generation. At the end of the process, in step 12, the individual with the best fitness is chosen and it will correspond with one of the frequent itemsets that the algorithm returns. The operation made in step 13 is very important. In it, records covered by the obtained itemset in the previous step are penalized. Since this factor affects negatively to the fitness function we achieve that in the following evolutionary process the search space tends to not be repeated.

An individual in Genetic Association Rules is a k-itemset where each gene represents the maximum and minimum values of the intervals of each attribute that belongs to such k-itemset.

5 Summary and Future Research

When more data is collected and accumulated, extensive data analysis would not be easier without effective and efficient data mining methods. The association rule algorithm is adopted to obtain useful clues based on which the soft set and GA is able to proceed its searching tasks in a more efficient way. Since the 'standard" soft set (F,E) over the universe U can be represented by a Boolean-valued information system, thus a soft set can be used for representing a transactional dataset. Therefore, one of the applications of soft set theory for data mining is for mining association rules. . In addition an association rule algorithm is employed to acquire the insights for those input variables most associated with the outcome variable before executing the evolutionary process. These derived insights are converted into GA's seeding chromosomes.

The association rules obtained by genetic association rule extraction methods maintain a high confidence and a good coverage of the database, providing the user with high quality rules. Genetic association rule extraction methods help us to obtain a reduced set of association rules, although the number of rules is restricted by the population size. Moreover, these rules consider few attributes in the antecedent, giving the advantage of easier understanding from a user's perspective. The runtime of the genetic association rule extraction methods scales quite linearly when we increase the size of the problem. Our future work includes: i) hybridization of soft set and genetic algorithms for association rule mining and validation with a real dataset, ii) comparative study with other benchmarks algorithms developed so far, and iii) developing a new associative based classifier by using the best attributes of soft set, genetic algorithms, and association rule mining.

References

[1] Borgelt, C.: Association Rule Induction (2005), `http://fuzzy.cs.uni-magdeburg.de/~borgelt`

[2] Kuok, C., Fu, A., Wong, M.: Mining fuzzy association rules in databases. SIGMOD Record 27(1), 41–46 (1998)

[3] Agrawal, R., Imielinski, T., Swami, A.: Mining association rules between sets of items in large databases. In: Proc. 1993 ACMSIGMOD Int. Conf. Management Data, Washington, DC, pp. 207–216 (1993)

[4] Maeda, A., Ashida, H., Taniguchi, Y., Takahashi, Y.: Data mining system using fuzzy rule induction. In: Proc. IEEE Int. Conf. Fuzzy Syst. FUZZ IEEE 1995, pp. 45–46 (March 1995)

[5] Wei, Q., Chen, G.: Mining generalized association rules with fuzzy taxonomic structures. In: Proc. NAFIPS 1999, New York, pp. 477–481 (June 1999)

[6] Au, W.H., Chan, K.C.C.: An effective algorithm for discovering fuzzy rules in relational databases. In: Proc. IEEE Int. Conf. Fuzzy Syst. FUZZ IEEE 1998, pp. 1314–1319 (May 1998)

[7] Flockhart, I.W., Radcliffe, N.J.: A genetic algorithm-based approach to data mining. In: Proc. 2nd Int. Conf. Knowledge Discovery Data Mining (KDD 1996), Portland, OR, August 2-4, p. 299 (1996)

[8] Raymer, M.L., Punch, W.F., Goodman, E.D., Kuhn, L.A.: Genetic programming for improved data mining: An application to the biochemistry of protein interactions. In: Proc. 1st Annu. Conf. Genetic Programming 1996, Stanford Univ., CA, July 28-31, pp. 375–380 (1996)

[9] Ryu, T., Eick, C.F.: MASSON: Discovering commonalties in collection of objects using genetic programming. In: Proc. 1st Annu. Conf. Genetic Programming 1996, Stanford Univ., CA, July 28-31, pp. 200–208 (1996)

[10] Teller, A., Veloso, M.: Program evolution for data mining. Int. J. Expert Syst. 8, 216–236 (1995)

[11] Fayyad, U.M., Piatetsky-Shapiro, G., Smyth, P., Uthurusamy, R. (eds.): Advances in Knowledge Discovery and Data Mining. AAAI/MIT Press, Menlo Park, CA (1996)

[12] Xu, K., Wang, Z., Leung, K.S.: Using a new type of nonlinear integral for multiregression: An application of evolutionary algorithms in data mining. In: Proc. IEEE Int. Conf. Syst., Man, Cybern., pp. 2326–2331 (October 1998)

[13] Noda, E., Freitas, A.A., Lopes, H.S.: Discovering interesting prediction rules with a genetic algorithm. In: Proc. IEEE Congr. Evolutionary Comput., CEC 1999, pp. 1322–1329 (July 1999)
[14] Cheung, Y., Fu, A.: Mining frequent itemsets without support threshold: With and without item constraints. IEEE Transactions on Knowledge and Data Engineering (2004)
[15] Zhang, S., Lu, J., Zhang, C.: A fuzzy-logic-based method to acquire user threshold of minimum-support for mining association rules. Information Sciences (2004)
[16] Molodtsov, D.: Soft set theory-first results. Computers and Mathematics with Applications 37, 19–31 (1999)
[17] Feldman, R., Aumann, Y., Amir, A., Zilberstein, A., Klosgen, W.: Maximal association rules: a new tool for mining for keywords co-occurrences in document collections. In: The Proceedings of the KDD, pp. 167–170 (1999)
[18] Amir, A., Aumann, Y., Feldman, R., Fresco, M.: Maximal association rules: a tool for mining associations in text. Journal of Intelligent Information Systems 25(3), 333–345 (2005)
[19] Guan, J.W., Bell, D.A., Liu, D.Y.: The rough set approach to association rule mining. In: The Proceedings of the Third IEEE International Conference on Data Mining (ICDM 2003), pp. 529–532 (2005)
[20] Freitas, A.: A genetic algorithm for generalized rule induction. In: Engineering Design and Manufacturing. AISC, pp. 340–353. Springer, Berlin (1999)
[21] Han, J., Pei, J., Yin, Y.: Mining frequent patterns without candidate generation. In: Proceedings of the ACM SIGMOD International Conference on Management of Data, Dallas, TX, USA, pp. 1–12 (2000)
[22] Park, J., Chen, M., Yu, P.: Using a hash–based method with transaction trimming for mining association rules. IEEE Trans., Knowledge and Data Eng. 9(5), 813–824 (1997)
[23] Aggarawal, C., Yu, P.: A new framework for itemset generation. In: Proceedings of the PODS Conference, Seattle, WA, USA, pp. 18–24 (June 1998)
[24] Dubois, D., Prade, H.: Rough fuzzy sets and fuzzy rough sets. International Journal of General Systems 17, 191–208 (1990)
[25] Bi, Y., Anderson, T., McClean, S.: A rough set model with ontologies for discovering maximal association rules in document collections. Knowledge-Based Systems 16, 243–251 (2003)
[26] Aggarawal, C., Yu, P.: A new framework for itemset generation. In: Proceedings of the PODS Conference, Seattle, WA, USA, pp. 18–24 (June 1998)
[27] Agrawal, R., Imielinski, T., Swami, A.: Mining association rules between sets of items in large databases. In: Proceedings of the 1993 ACM SIGMOD International Conference on Management of Data, Washington, D.C., pp. 207–216 (May 1993)
[28] Au, W., Chan, C.: An evolutionary approach for discovering changing patterns in historical data. In: Proceedings of SPIE, pp. 398–409 (2002)
[29] Silverstein, C., Brin, S., Motwani, R.: Beyond market baskets: Generalizing association rules to dependence rules. In: Data Mining and Knowledge Discovery, pp. 39–68 (1998)
[30] Toivonen, H.: Sampling large databases for association rules. In: Proceedings of the 22nd VLDB Conference, pp. 134–145 (1996)
[31] Webb, G.: Efficient search for association rules. In: Proceedings of ACM SIGKDD, New York, pp. 99–107 (2000)
[32] Park, J., Chen, M., Yu, P.: Using a hash–based method with transaction trimming for mining association rules. IEEE Trans. Knowledge and Data Eng. 9(5), 813–824 (1997)
[33] Zhang, C., Zhang, S., Webb, G.: Identifying approximate itemsets of interest in large databases. Applied Intelligence 18, 91–104 (2003)

[34] Piatetsky-Shapiro, G.: Discovery, analysis and presentation of strong rules. In: Piatetsky-Shapiro, G., Frawley, W. (eds.) Knowledge Discovery in Databases, pp. 229–248. AAAI Press/MIT Press, Cambridge, MA (1991)

[35] Mata, J., Alvarez, J.L., Riquelme, J.C.: Mining Numeric Association Rules with Genetic Algorithms. In: 5th International Conference on Artificial Neural Networks and Genetic Algorithms, praga ICANNGA, pp. 264–267 (2001)

[36] Ghosh, S., Biswas, S., Sarkar, D., Sarkar, P.P.: Mining Frequent Itemsets Using Genetic Algorithm. International Journal of Artificial Intelligence & Applications (IJAIA) 1(4), 133–143 (2010)

[37] Maji, P.K., Biswas, R., Roy, A.R.: Soft set theory. Computers and Mathematics with Applications 45, 555–562 (2003)

[38] Agrawal, R., Imielinski, T., Swami, A.: Mining association rules between sets of items in large databases. In: Proceedings of the ACM SIGMOD International Conference on Management of Data, pp. 207–216 (1993)

[39] Handl, J., Kell, D.B.: Multi-objective Optimization in Bioinformatics and Computational Biology. Transactions on Computational Biology, 283 (2007)

[40] Ibrahim, C.: Consumption universes based supermarket layout through association rule mining and multi dimensional scaling. Expert Systems with Applications (February 2012)

[41] Delgado, G., Aranda, V., Calero, J., Sánchez-Marañón, M., Serrano, J.M., Sánchez, D., Vila, M.A.: Using fuzzy data mining to evaluate survey data from olive grove cultivation. Computers and Electronics in Agriculture 65(1), 99–113 (2009)

[42] Abdullah, Z., Herawan, T., Ahmad, N., Deris, M.M.: Mining significant association rules from educational data using critical relative support approach. Procedia - Social and Behavioral Sciences 28, 97–101 (2011)

[43] Yan, X., Zhang, C., Zhang, S.: Genetic algorithm-based strategy for identifying association rules without specifying actual minimum support. Expert Systems with Applications 36(2), 3066–3076 (2009)

[44] Mata, J., Alvarez, J., Riquelme, J.: An Evolutionary Algorithm to Discover Numeric Association Rules. ACM Symposium on Applied Computing (March 2002)

[45] Mata, J., Alvarez, J., Riquelme, J.: Mining Numeric Association Rules with Genetic Algorithms. In: 5th International Conference on Artificial Neural Networks and Genetic Algorithms, Taipei, Taiwan (April 2001)

Solving Composite Test Functions Using Teaching-Learning-Based Optimization Algorithm

R.V. Rao and G.G. Waghmare

Department of Mechanical Engineering,
S.V. National Institute of Technology, Surat-395007, India
ravipudirao@gmail.com

Abstract. Multimodal function optimization has attracted a growing interest especially in the evolutionary computation research community. Multimodal optimization deals with optimization tasks that involve finding all or most of the multiple solutions (as opposed to a single best solution). The challenge is to identify as many optima as possible to provide a choice of good solutions to the designers. A composite function is a combination of the two or more functions. The Teaching-Learning-Based Optimization (TLBO) algorithm is a teaching-learning process inspired algorithm based on the effect of influence of a teacher on the output of learners in a class. In this paper, the TLBO algorithm has been tested on six composite test functions for numerical global optimization. The TLBO algorithm has outperformed the other six algorithms for the composite test problems considered.

Keywords: Teaching-Learning-Based Optimization, Large-Scale optimization, Composite benchmark functions.

1 Introduction

Knowledge of multiple solutions to an optimization task is especially helpful in engineering, when due to physical (and/or cost) constraints; the best results may not always be realizable. In such a scenario, if multiple solutions (local and global) are known, the implementation can be quickly switched to another solution and still obtain an optimal system performance. Multiple solutions could also be analyzed to discover hidden properties (or relationships), which makes them high-performing. In addition, the algorithms for multimodal optimization usually not only locate multiple optima in a single run, but also preserve their population diversity, resulting in their global optimization ability on multimodal functions. Moreover, the techniques for multimodal optimization are usually borrowed as diversity maintenance techniques to other problems [14].

In the past four decades, different kinds of optimization algorithms have been designed and applied to solve real parameter functions optimization problems. Some of the popular approaches are real-parameter EAs, evolution strategies (ES), differential evolution (DE), Particle swarm optimization (PSO), evolutionary programming (EP), classical methods such as quasi-Newton method (QN), hybrid

S.C. Satapathy et al. (Eds.): Proc. of Int. Conf. on Front. of Intell. Comput., AISC 199, pp. 395–403.
DOI: 10.1007/978-3-642-35314-7_45

evolutionary-classical methods, other non-evolutionary methods such as simulated annealing (SA), tabu search (TS) and others [11]. There are several different composite test functions for multimodal optimization available in the literature. Liang et al. [5] presented a novel composite test functions for numerical global optimization. Suganthan et al. [11] introduced a problem definition and evaluation criteria for special session on real-parameter optimization. Rao et al. [9, 10] and Rao and Patel [7] developed a new optimization method, Teaching-Learning-Based Optimization (TLBO), as an innovative optimization algorithm inspiring the natural phenomena, which mimics teaching-learning process in a class between the teacher and the students (learners). Most real world problems have no clear structure and it is necessary that further research on evolutionary computation is required [5]. Hence in this paper TLBO has been used for testing of the complex composite test problems and results have been compared with six other evolutionary algorithms.

The remainder of this paper is structured as following: Section 2 describes the TLBO algorithm whereas problem formulation and description is given in Section 3. Section 4 provides results and discussion. In section 5, the paper has been concluded.

2 Teaching-Learning-Based- Optimization (TLBO) Algorithm

TLBO is a teaching-learning process inspired algorithm proposed by Rao et al. [9, 10], and Rao and Patel [7] based on the effect of influence of a teacher on the output of learners in a class. In this optimization algorithm a group of learners is considered as population and different subjects offered to the learners are considered as different design variables of the optimization problem and a learner's result is analogous to the 'fitness' value of the optimization problem. The algorithm describes two basic modes of the learning: (i) through teacher (known as teacher phase) and (ii) interacting with the other learners (known as learner phase). Working of both these phases is explained below [7-10].

2.1 Teacher Phase

It is the first part of the algorithm where learners learn through the teacher. This phase produces a random ordered state of points called learners within the search space. At any iteration i, assume that there are 'm' number of subjects (i.e. design variables), 'n' number of learners (i.e. population size, k=1,2,…,n) and $M_{j,i}$ be the mean result of the learners in a particular subject 'j' (j=1,2,…,m) The best overall result $X_{total\text{-}kbest,i}$ considering all the subjects together obtained in the entire population of learners can be considered as the result of best learner *kbest*. However, as the teacher is usually considered as a highly learned person who trains learners so that they can have better results, the best learner identified is considered by the algorithm as the teacher. The difference between the existing mean result of each subject and the corresponding result of the teacher for each subject is given by,

$$Difference\_Mean_{j,k,i} = r_i\ (X_{j,kbest,i} - T_F M_{j,i}) \quad (1)$$

Where, $X_{j,kbest,i}$ is the result of the best learner (i.e. teacher) in subject j. T_F is the teaching factor which decides the value of mean to be changed, and r_i is the random number in the range [0, 1]. Value of T_F can be either 1 or 2. The value of T_F is decided randomly with equal probability as,

$$T_F = round\ [1+rand(0,1)\{2\text{-}1\}] \quad (2)$$

Based on the $Difference\_Mean_{j,k,i}$, the existing solution is updated in the teacher phase according to the following expression.

$$X'_{j,k,i} = X_{j,k,i} + Difference\_Mean_{j,k,i} \quad (3)$$

Where $X'_{j,k,i}$ is the updated value of $X_{j,k,i}$. Accept $X'_{j,k,i}$ if it gives better function value. All the accepted function values at the end of the teacher phase are maintained and these values become the input to the learner phase. The learner phase depends upon the teacher phase.

2.2 Learner Phase

It is the second part of the algorithm where learners increase their knowledge by interaction among themselves. A learner interacts randomly with other learners for enhancing his or her knowledge. A learner learns new things if the other learner has more knowledge than him or her. Considering a population size of 'n', the learning phenomenon of this phase is expressed below.

Randomly select two learners P and Q such that $X'_{total\text{-}P,i} \neq X'_{total\text{-}Q,i}$ (where, $X'_{total\text{-}P,i}$ and $X'_{total\text{-}Q,i}$ are the updated values of $X_{total\text{-}P,i}$ and $X_{total\text{-}Q,i}$ respectively at the end of teacher phase)

$$X''_{j,P,i} = X'_{j,P,i} + r_i\ (X'_{j,P,i} - X'_{j,Q,i}) \quad (4)$$
$$\text{If } X'_{total\text{-}P,i} < X'_{total\text{-}Q,i}$$

$$X''_{j,P,i} = X'_{j,P,i} + r_i\ (X'_{j,Q,i} - X'_{j,P,i}) \quad (5)$$
$$\text{If } X'_{total\text{-}Q,I} < X'_{total\text{-}P,i}$$

Accept $X''_{j,P,i}$ if it gives a better function value.

The next section presents the details of complex composite benchmark functions attempted by TLBO algorithm.

3 Problem Formulation and Description

Liang et al. [5] proposed a general framework to construct novel and challenging composite test functions possessing many desirable properties. The idea behind this is to compose the standard benchmark functions to construct a more challenging function with a randomly located global optimum and several randomly located deep local optima. Gaussian functions are used to combine these benchmark functions and blur the individual function's structure. The details are described below [5].

$F(x)$: new composite function.
$f_i(x)$: i^{th} basic function used to construct the composite function.
n: number of basic functions. The bigger n is the more complex F(x) is.
D: dimension.
$[X\min, X\max]^D$: $F(x)$'s search range
$[x\min, x\max]^D$: $f\,x$'s search range
M_i: orthogonal rotation matrix for each $f_i(x)$
o_i: new shifted optimum position for each $f_i(x)$
o_{iold} : old optimum position for each $f_i(x)$

$$F(x) = \sum_{i=1}^{n}\{w_i \times [f_i'((x - o_i + o_{iold}/\lambda_i M_i))]\} + f\_bias \tag{6}$$

w_i: weight value for each $f_i(x)$, calculated as below:

$$w_i = exp\left(\frac{\sum_{k=1}^{D}(x_k - o_{ik} + o_{ikold})^2}{2D\sigma_i^2}\right)$$
$$\begin{cases} w_i & if\ w_i = \max(w_i) \\ w_i(1 - \max(w_i)^{10}) & if\ w_i \neq \max(w_i) \end{cases} \tag{7}$$

then normalized the weight

$$w_i = w_i / \sum_{i=1}^{n} w_i \tag{8}$$

σ_i : used to control each $f_i(x)$'s coverage range, a small σ_i gives a narrow range for $f_i(x)$.

λ_i : used to stretch or compress the function, $\lambda_i > 1$ means stretch, $\lambda_i < 1$ means compress. Since different basic function has different search range, in order to make full use of the basic function.

$$\lambda_i = \sigma_i.(X_{max}-X_{min}/x_{maxi}-x_{mini}) \tag{9}$$

o_i define the global and local optima's position, bias define which optimum is global optimum. The smallest bias i corresponds to the global optimum. Using o_i, $bias_i$, a global optimum can be placed anywhere.

If $f_i(x)$ are different functions, different functions have different properties and height, in order to get a better mixture, estimate the biggest function value f_{maxi} for 10 functions $f_i(x)$, then normalize each basic functions to similar height as below.

$$|f_{maxi}| = C * f_i(x) / |f_{maxi}| \tag{10}$$

C is a predefined constant.

$$|f_{maxi}| \text{ is estimated using } |f_{maxi}| = f_i((z/\lambda_i) * M_i) \tag{11}$$

$$z = X_{max} \tag{12}$$

3.1 Basic Functions

The basic functions used to construct the composite functions are given below.

Sphere function

$$f(x) = \sum_{i=}^{D} x_i^2 , \; x \in [-100{,}100]^D$$

Rastrigin function

$$f(x) = \sum_{i=1}^{D} (x_i^2 - 10\cos(2\pi x_i) + 10)$$

Weierstrass function

$$f(x) = \sum_{i=1}^{D} \left(\sum_{k=0}^{k_{max}} [a^k \cos(2\pi b^k (x_i + 0.5))] \right) - D \sum_{k=0}^{k_{max}} [a^k \cos(2\pi b^k \, . \, 0.5)]$$

A = 0.5, b = 3, $k_{\max} = 20$, $x \in [\,-0.5, 0.5]^D$

Griewank's Function

$$f(x) = \sum_{i=1}^{D} \frac{x_i^2}{4000} - \prod_{i=1}^{D} \cos\left(\frac{x_i}{\sqrt{i}}\right) + 1 \;, x \in [-0.5, 0.5]^D$$

Ackley's function

$$f(x) = -20\exp\left(-0.2 \sqrt{\frac{1}{D} \sum_{i=1}^{D} x_i^2} \right) - \exp\left(\frac{1}{D} \sum_{i=1}^{D} \cos(2\pi x_i) \right) + 20 + e, x \in [-32{,}32]^D$$

3.2 Composite Test Functions

f_i, σ_i, λ_i settings for the Composite Functions are given below [5]:

3.2.1. Composite Function 1 (CF1)

$f_1, f_2, \ldots f_{10}$: Sphere Function
$[\sigma_1, \sigma_2, \ldots, \sigma_{10}] = [1, 1, \ldots, 1]$, $[\lambda_1, \lambda_2, \ldots, \lambda 10] = [5/100, 5/100, \ldots, 5/100]$

3.2.2. Composite Function 2 (CF2)

$f_1, f_2, \ldots f_{10}$: Griewank's Function
$[\sigma_1, \sigma_2, \ldots, \sigma_{10}] = [1, 1, \ldots, 1]$, $[\lambda_1, \lambda_2, \ldots, \lambda_{10}] = [5/100, 5/100, \ldots, 5/100]$

3.2.3. Composite Function 3 (CF3)

$f_1, f_2, \ldots f_{10}$: Griewank's Function
$[\sigma_1, \sigma_2, \ldots, \sigma_{10}] = [1, 1, \ldots, 1]$, $[\lambda_1, \lambda_2, \ldots, \lambda_{10}] = [1, 1, \ldots, 1]$

3.2.4. Composite Function 4 (CF4)

$f_{1\text{-}2}(x)$: Ackley's Function, $f_{3\text{-}4}(x)$: Rastrigin's Function, $f_{7\text{-}8}(x)$: Griewank's Function, $f_{9\text{-}10}(x)$: Sphere's Function, $[\sigma_1, \sigma_2, \ldots, \sigma_{10}] = [1, 1, \ldots, 1]$
$[\lambda_1, \lambda_2, \ldots, \lambda_{10}] = [5/32, 5/32, 1, 1, 5/0.5, 5/0.5, 5/100, 5/100, 5/100, 5/100]$

3.2.5. Composite Function 5 (CF5)

$f_{1\text{-}2}(x)$: Rastrigin's Function, $f_{3\text{-}4}(x)$: Weierstrass Function, $f_{5\text{-}6}(x)$: Griewank's Function, $f_{7\text{-}8}(x)$: Ackley's Function, $f_{9\text{-}10}(x)$: Sphere's Function, $[\sigma_1, \sigma_2, \ldots, \sigma_{10}] = [1, 1, \ldots, 1]$, $[\lambda_1, \lambda_2, \ldots, \lambda_{10}] = [1/5, 1/5, 5/0.5, 5/0.5, 5/100, 5/100, 5/32, 5/32, 5/100, 5/100]$

3.2.6 Composite Function 6 (CF6)

$f_{1\text{-}2}(x)$: Rastrigin's Function, $f_{3\text{-}4}(x)$: Weierstrass Function, $f_{5\text{-}6}(x)$: Griewank's Function, $f_{7\text{-}8}(x)$: Ackley's Function, $f_{9\text{-}10}(x)$: Sphere's Function

$[\sigma_1, \sigma_2, \ldots, \sigma_{10}] = [0.1, 0.2, 0.3, 0.4, 0.5, 0.6, 0.7, 0.8, 0.9, 1]$

$[\lambda_1, \lambda_2, \ldots, \lambda_{10}] = [0.1*1/5, 0.2*1/5, 0.3*5/0.5, 0.4*5/0.5, 0.5*5/100, 0.6*5/100, 0.7*5/32, 0.8*5/32, 0.9*5/100, 1*5/100]$

CF5 and CF6 use the same optima's position o and the same orthogonal matrixes $M_1, M_2, \ldots, M_n$. The difference between CF5 and CF6 is the values of σ and λ, which makes CF6, has a narrower coverage area for the basic function with the global optimum and a flatter coverage area for the basic function with the local optima. In this way, the complexity of the function is increased.

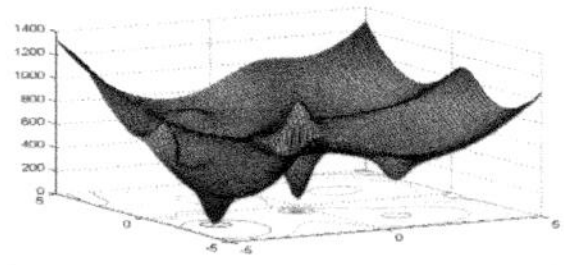

Fig. 1. Composite Function1

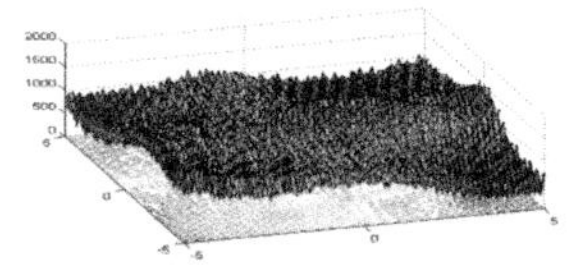

Fig. 2. Composite Function 2

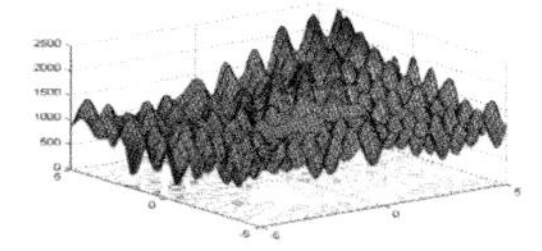

Fig. 3. Composite Function 3

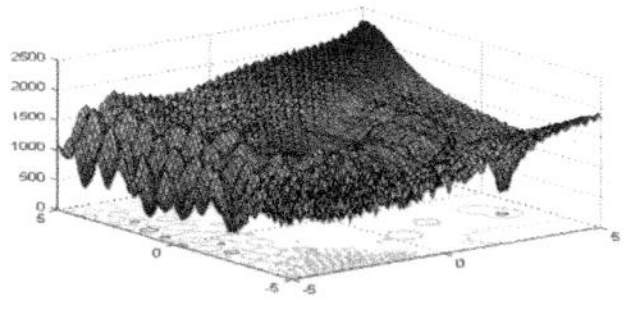

Fig. 4. Composite Function 4

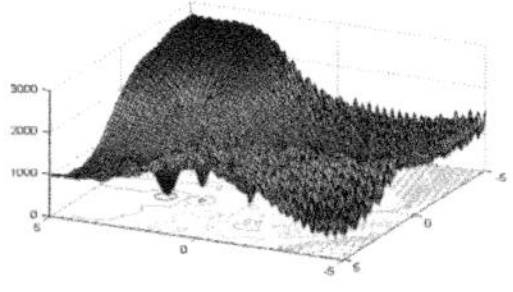

Fig. 5. Composite Function 5

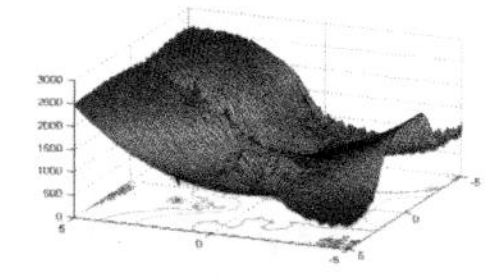

Fig. 6. Composite Function 6

3.3 Parameters Settings for the Composite Test Function

By controlling f_i, σ_i, λ_i, $bias_i$, o_i and M_i, can obtain different composite functions with different desired properties. Parameter settings for the composite functions are as follows [5].

Basic function number n =10, Dimensions D = 10, C=2000, Search range: $[-5, 5]^D$ f_bias = 0, bias = [0, 100, 200, 300, 400, 500, 600, 700, 800, 900].

Hence the first function $f_i(x)$ is always the function with the global optimum, as its bias is zero always. o_1 , o_2, ..., o_9 are all generated randomly in the search range, except o_{10} is set [0, 0, ...,0] for trapping algorithms which have a potential to converge to the center of the search range. $M_1, M_2, \ldots, M_n$ are $D*D$ orthogonal rotation matrixes obtained by using Salemon's (1996) method [12].

Table 1. Results obtained by using the seven algorithms on six Composite Functions

Composite functions		**PSO**	**CPSO**	**CLPSO**	**CMA-ES**	**G3-PCX**	**DE**	**TLBO**
CF1	Mean	1.0000 e+002	1.5626 e+002	**5.7348 e-008**	1.0000 e+002	6.0000 e+001	6.7459 e-002	3.1186 e-001
	Std Dev.	8.1650 e+002	1.3427 e+002	1.9157 e-007	1.885 e+002	6.9921 e+001	1.1057 e-001	3.0421 e-001
CF2	Mean	1.5591 e+002	2.4229 e+002	1.9157 e+001	1.6199 e+002	9.2699 e+001	2.8759 e+001	**1.702 e+001**
	Std Dev.	1.3176 e+002	1.4895 e+002	1.4748 e+001	1.5100 e+002	9.9067 e+001	8.6277 e+001	7.2188 e+000
CF3	Mean	1.7203 e+002	3.6264 e+002	1.328 e+002	2.1406 e+002	3.1980 e+002	1.4441 e+002	**1.2381 e+002**
	Std Dev.	3.2869 e+001	1.9631 e+002	2.0027 e+001	7.4181 e+001	1.2519 e+002	1.9401 e+001	6.0392 e+001
CF4	Mean	3.1430 e+002	5.2237 e+002	3.2232 e+002	6.1640 e+002	4.9296 e+002	3.2486 e+002	**2.9439 e+002**
	Std Dev.	2.0066 e+001	1.2209 e+002	2.7461 e+001	6.7192 e+002	1.4249 e+002	1.4784 e+001	3.1580 e+001
CF5	Mean	8.3450 e+001	2.5556 e+002	5.370 e+000	3.5853 e+002	2.6021 e+001	1.0789 e+001	**5.1815 e+000**
	Std Dev.	1.0011 e+002	1.7563 e+002	2.6056 e+000	1.6826 e+002	4.1579 e+001	2.6040 e+000	1.6165 e+000
CF6	Mean	8.6142 e+002	8.5314 e+002	5.0116 e+002	9.0026 e+002	7.7208 e+002	4.9094 e+002	**2.3018 e+002**
	Std Dev.	1.2581 e+002	1.2798 e+002	7.7800 e-001	8.3186 e-002	1.8939 e+002	3.9461 e+002	4.8365 e+001

PSO: Particle Swarm Optimization, CPSO: Cooperative PSO, CLPSO: Comprehensive Learning PSO, CMA-ES: Evolution Strategy with Covariance Matrix Adaptation, G3-PCX: G3 model with PCX crossover, DE: Differential Evolution, TLBO: Teaching-Learning-Based Optimization

4 Results and Disscussion

Table 1 shows the results obtained by using the seven algorithms on six composite functions. For each test function each algorithm is run 20 times and the maximum fitness evaluations are 50000 for all algorithms. The mean values of the results are recorded in Table 1. As can be seen from the results, the TLBO algorithm outperforms the other six algorithms on all benchmark problems except for test problem 1. In that problem, CLPSO and DE are better than TLBO. However, the performance of the TLBO is better than the others (except CLPSO and DE) for test problem 1.

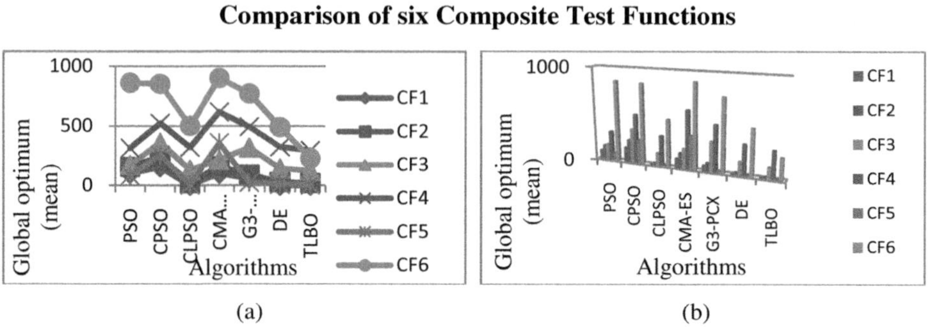

Fig. 7. Comparison of Six Composite Test Functions (a) Line chart (b) Column chart

Fig. 7 shows the comparison of six composite test functions. It can be seen from the Fig.7 that TLBO finds the global optimum solution with better mean for all the composite test functions except for test function 1. The TLBO algorithm has outperformed the other algorithms when optimizing the composite test functions and obtained first rank between the six composite test functions except CF1 for which TLBO obtained third rank among the six composite functions. TLBO algorithm has reduced the global optimum mean 490.94 to 230.18 for most complex composite test function 6 thereby giving improvement over 53%.

5 Conclusion

Real world optimization problems often contain multiple global or local optima. Multimodal optimiztion aims to locate all of the global optima of a multimodal function. The challenge in multimodal optimization is to identify as many optima as possible to provide a choice of good solutions to the designers. The performance of the TLBO algorithm has been checked with the well known optimization algorithms such as PSO, CPSO, CLPSO, CMA-ES, G3-PCX and DE by experimenting with different composite test functions with different characteristics. As can be seen from the results, the TLBO algorithm has outperformed all the algorithms for the composite functions CF2, CF3, CF4, CF5 and CF6. However TLBO has shown inferior results for the composite function 1(CF1) compared to that given by DE and CLPSO, but still the result of TLBO is better than those given by PSO, CPSO, CMA-ES, and G3-PCX. Therefore, it can be stated that the TLBO algorithm is accurate, effective and efficient and has great potential for solving multimodal problems. The TLBO is going to be tried on more complex composite functions in the near future.

References

1. Ahrari, A., Atai, A.A.: Grenade explosion method-A novel tool for optimization of multimodal functions. Applied Soft Computing 10, 1132–1140 (2010)
2. Akay, D., Karaboga, A.: Modified Artificial Bee Colony algorithm for real-parameter optimization. Information Sciences 192, 120–142 (2012)

3. Bergh, F., Engelbrecht, A.P.: A cooperative approach to particle swarm optimization. IEEE Transactions on Evolutionary Computation 8(3), 225–239 (2004)
4. Coello, C.A.C., Pulido, G.T., Lechuga, M.S.: Handling multiple objectives with particle swarm optimization. IEEE Transactions on Evolutionary Computation 8(3), 256–279 (2004)
5. Liang, J.J., Suganthan, P.N., Deb, K.: Novel Composition Test Functions for Numerical Global Optimization. IEEE Trans. on Evolutionary Computation 5(1), 1141–1153 (2005)
6. Leung, Y.W., Wang, Y.P.: An orthogonal genetic algorithm with quantization for global numerical optimization. IEEE Trans. on Evolutionary Computation 5(1), 41–53 (2001)
7. Rao, R.V., Patel, V.: An elitist teaching-learning-based optimization algorithm for solving complex constrained optimization problems. International Journal of Industrial Engineering Computations 3(4), 535–560 (2012)
8. Rao, R.V., Patel, V.: Multi-objective optimization of two stage thermoelectric cooler using a modified teaching–learning-based optimization algorithm. Engineering Applications of Artificial Intelligence (2012), doi:10.1016/j.engappai.2012.02.016
9. Rao, R.V., Savsani, V.J., Vakharia, D.P.: Teaching-learning-based optimization: A novel method for constrained mechanical design optimization problems. Computer-Aided Design 43, 303–315 (2011)
10. Rao, R.V., Savsani, V.J., Balic, J.: Teaching-learning-based optimization algorithm for unconstrained and constrained real parameter optimization problems. Engineering Optimization, doi:10.1080/0305215X.2011.652103
11. Suganthan, P.N., Hansen, N., Liang, J.J.: Problem definition and evaluation criteria for the CEC 2005 special session on real-parameter optimization, Technical Report, Nanyang Technological University, Singapore (2005)
12. Soloman, R.: Reevaluating genetic algorithm performance under coordinate rotation of benchmark functions. BioSystems 39, 263–278 (1996)
13. Akbari, R., Hedayatzadeh, R., Ziarati, K., Hassanizadeh, B.: A muti-objective artificial bee colony algorithm. Swarm and Evolutionary Computation 2, 39–52 (2012)
14. Wong, K., Wu, K., Peng, C., Zhang, Z.: Evolutionary multimodal optimization using the principle of locality. Information Sciences 194, 138–170 (2012)

A Review on Application of Particle Swarm Optimization in Association Rule Mining

Singhai Ankita[1], Agrawal Shikha[2], Agrawal Jitendra[1], and Sharma Sanjeev[1]

[1] SOIT, Rajiv Gandhi Proudyogiki Vishwavidyalaya, Bhopal (M.P)
ankita_singhai29@yahoo.com, {jitendra,sanjeev}@rgtu.net
[2] UIT, Rajiv Gandhi Proudyogiki Vishwavidyalaya, Bhopal (M.P)
shikha@rgtu.net

Abstract. Data mining, the extraction of hidden predictive large amounts of data and picking out the relevant information from large databases, is a powerful new technology with great potential to help companies focus on the most important information in their data warehouses. Data mining tools predict future trends and behaviors, allowing businesses to make proactive, knowledge-driven decisions. Association Rule Mining has become one of the core data mining tasks that used to show the relationship between data items. These relationships are not based on inherent properties of the data themselves like functional dependencies, but based on co-occurrence of the data items. Association rules are frequently used in telecommunication network, market and risk management, advertising and inventory control. Recently many advance techniques are researched for making association rule mining more efficient to proposing a new perspective development in the field of data mining. One of the latest topics in this area is mining the hidden pattern from existing collection of databases by implementing particle swarm optimization (PSO) approach for increasing mining efficiency, extending the notion of association rules, enhancing the parameter such as support and confidence. In this article, the various advancements in association rule mining using particle swarm optimization is discussed.

Keywords: Association Rule Mining, Particle Swarm Optimization, Quantum Swarm Evolutionary, Support Vector Machine, Rough Particle Swarm Optimization, Comprehensive Learning Particle Swarm Optimization, ACO, Genetic Algorithm.

1 Introduction

1.1 Association Rule Mining

Association rule mining is one of the most important and well researched techniques which come under descriptive category of data mining. Association rules aims in extracting important correlation, frequent pattern, association or casuals structures among the set of items in the transactional databases, relational databases or other information repositories. The first announce of association rule mining is introduced

S.C. Satapathy et al. (Eds.): Proc. of Int. Conf. on Front. of Intell. Comput., AISC 199, pp. 405–414.
DOI: 10.1007/978-3-642-35314-7_46

by R. Agrawal et.al. [1]. Shichao Zhang et al [20] have given association mining methods and the importance of rule interestingness measures. Association rule, basically extracts the patterns from the database based on the two measures such as minimum support and minimum confidence. The support and confidence measures are described as stated in [10] for mining frequent itemset mining and association rule generation.

Formally, association rules are defined as follows: Let $I = \{i_1, i_2, \ldots i_n\}$ be a set of items. Let D be a set of transactions, where each transaction T is a set of items such that $T \subseteq I$. Each transaction is associated with a unique identifier TID. A transaction T is said to contain X, a set of items in I, if $X \subseteq T$. An association rule is an implication of the form "$X \rightarrow Y$", where $X \subseteq I$; $Y \subseteq I$, and $X \cap Y = \Phi$. The rule $X \rightarrow Y$ has support s in the transaction set D if s% of the transactions in D contains X U Y. In other words, the support of the rule is the probability that X and Y hold together among all the possible presented cases. It is said that the rule $X \rightarrow Y$ holds in the transaction set D with confidence c if c% of transactions in D that contain X also contain Y. In other words, the confidence of the rule is the conditional probability that the consequent Y is true under the condition of the antecedent X.

Support: It is the probability of item or item sets in the given transactional data base:

$$\text{Support}(X) = n(X) / n \qquad (1)$$

Where n is the total number of transactions in the database and n(X) is the number of transactions that contains the item set X

Confidence: It is conditional probability, for an association rule X=>Y and defined as

$$\text{Confidence}(X \Rightarrow Y) = \text{support}(X \text{ and } Y) / \text{support}(X) \qquad (2)$$

1.2 Particle Swarm Optimization

Particle Swarm Optimization is an artificial intelligence technique, capable of optimizing a non-linear and multidimensional problem which usually reaches good solutions efficiently while requiring minimal parameterization. Particle Swarm Optimization (PSO) concept and algorithm were introduced by James Kennedy and Russel Eberhart in [11]. It belongs to the class of swarm intelligence algorithms, which are inspired from the social dynamics and emergent behavior that arise in socially organized colonies like social behavior of bird flocking or fish schooling. The main strength of PSO is its fast convergence, which compares favorably with many global optimization algorithms [2].The basic concept of the algorithm is to create a swarm of particles which move in the problem space around them searching for their goal. Each particle are tried to achieve best result by updating its position and speed according to its own past one as well as information of current particle which is best among all particle in swarm.

Like neural networks, computation in the PSO paradigm is based on a collection (called a swarm) of fairly-primitive processing elements (called particles). PSO is initializes by random particles and each particle moves with an adaptable velocity within the search space, and retains a memory of the best position it ever encountered and also calculates its value on the basis of some fitness function. Each particle has its

P_{best} value, which is best among all the previous and current value of that particle. The system stores G_{best} value, which is the best among all P_{best} value of all particles. Then every particle is updating its position and velocity on the basis of its P_{best} value and system G_{best} value.

$$V_{(t+1)} = V_{(t)} + C1 \,.\, R1.\, (P_{best} - X_{(t)}) + C2 \,.\, R2 \,.\, (G_{best} - X_{(t)}) \tag{3}$$

$$X_{(t+1)} = X_{(t)} + V_{(t+1)} \tag{4}$$

Where
$V_{(t)}$: velocity of the tth particle
$X_{(t)}$: position of the tth particle
R_1, R_2: random number in (0,1)
C_1 : acceleration constant for the cognitive component in (0,2)
C_2 : acceleration constant for the social component in (0,2)

The velocities of particles in each dimension are clamped to a maximum velocity V_{max}. If the sum of accelerations would cause the velocity of that dimension to exceed V_{max}, which is a parameter specified by the user, then the velocity of that dimension is limited to V_{max}. This method is called "V_{max} method" [18].

2 Particle Swarm Optimization in Association Rule Mining

Association rule mining is to find out association rules that satisfy the predefined minimum support and confidence from a given database. Methodology for mining association rules is usually decomposed into two sub-problems:

1. Generating those itemsets whose occurrence is greater or equal to user specified minimal support, are called frequent or large itemsets. That further divided into two sub-problems: candidate large itemsets generation process and frequent itemsets generation process.
2. Generating rules from those frequent itemsets that have minimum confidence.

There are many different basic algorithms for association rule mining. They use different strategies and data structure although their resulting sets of the rules are all the same but their computational efficiencies and memory requirements are different such as Apriori series approach acquires downward closure property which finds frequent itemsets by generating confined candidate, Frequent Pattern Growth tree structured approach which adopts a divide and conquer strategy. However there are some bottlenecks of the Apriori algorithm are as follows:

- The complex candidate generation process that uses most of the time, space and memory.
- The multiple scan of the database.
- The set of threshold values of support and confidence
- The quality of rules

For improving Apriori algorithm, many new techniques were implemented with some modifications or improvements in it such as hash based technique, transaction reduction, partitioning, sampling, dynamic itemset counting.

Particle swam optimization is a new heuristic optimization method based on swarm intelligence. To extract the knowledge, a database may be considered as a large search space and a mining algorithm as a search strategy. PSO makes use of particles moving in an n-dimensional space to search the solutions for an n-variable function optimization problem. The datasets are the sample space to search and each attribute is a dimension for the PSO-miner. Particle Swarm Optimization is mostly implemented to Classification technique and Clustering technique of data mining in different ways for improving algorithm's performance in respect to different parameters, described by David Martens et.al. [7], but rarely implemented in Association Rule mining technique of data mining. Particle Swarm Optimization is implemented to Association Rule Mining in two ways:

1. Generate rules by implementing PSO in the traditional algorithm of association rule mining
2. Optimization of Association Rule generated by traditional algorithm using PSO

First approach is discussed in Section 2.1 and second approach is discussed in Section 2.2.

2.1 Generate Rules by Implementing PSO in the Traditional Algorithm of Association Rule Mining

In this approach, association rules are generated by hybrid the concept of PSO algorithm and traditional association rule mining algorithm like Apriori, FP growth etc. Beauty in this approach is that it does not depends on traditional rule mining algorithms fully so does not inherits all the drawbacks of traditional rule mining algorithm.

In 2008, Osama M Badawy et al [17] introduces Swarm Intelligence based algorithm for mining quantitative association rules a hybrid of PSO/ACO algorithm. This algorithm designs to discover optimized intervals in numeric attributes, which deals directly with both continuous and nominal attribute values, because most of the databases in real life contain both types of attributes. The algorithm uses best support for generating frequent itemset instead of using user specified minimum support and confidence, can interact with the dataset without the need of any preprocessing for the data. The algorithm has two main steps; first it discovers association rules containing nominal attributes only. Then, it adds attributes with continuous values to the discovered rules. The results shows that the algorithm is very competitive and accurate compared to other quantitative association rule algorithms such as GAR [4] which deals only with numeric value and also require minimum support or confidence, QUANTMINER [13] which require specifying rule template to work. Most of the association rule algorithms uses scheme to discretise all numeric attributes that leads to loss of knowledge, which is overcome by proposed algorithm. This algorithm has longer execution time than ordinary algorithm of association rules such as Apriori algorithm. In 2008, Bilal Alatas and Erhan Akin [3] proposed a novel particle swarm optimization algorithm for numeric association rule mining, named as Rough Particle Swarm Optimization Algorithm (RPSOA). RPSOA uses both rough as well as conventional decision variables and particles that are based on the notion of rough pattern. Decision variables represent the items and intervals that will be considered as a one value, namely rough value that represent an interval or set of values for an attribute by considering only lower and upper bounds that are relevant in computation. RPSOA is database independent, does not depend on support or confidence thresholds which are hard to choose for each database. It mines the accurate and comprehensible rules that have high support and confidence values according to the synthetically created sets, with small number of attributes without redundancy. Association rule mining is limited to the manual discretization handle by the user in traditional algorithm, which needs to create intervals of numeric attributes before mining process that is hard problem. By considering that problem, RPSOA efficiently utilized in automatic mining of numeric association rules by using various operators and evaluation measures such as fitness function, mutation, refinement of bound intervals.

Population based evolutionary algorithm PSO is initialized with a population of candidate solution and the activities of the population are guided by some behavior rules. In the weighted variant of PSO, concept of inertia weight uses whose suitable value provides a balance between the global and local exploration ability of the swarm. A Nonlinear inertia weight adaptation strategy was proposed in [5] which is based on the concept of decrease strategy [6, 21]. In this strategy lower value of inertia weight chooses during the early iterations and maintains its higher value than linear model. Experimentally Nonlinear strategy enables particles to search the solution space more aggressively to look for better areas, thus avoids local optimum effectively. In 2010, Guo-Rong Cai et al [8] presents a fuzzy association rule mining algorithm by using above mentioned Nonlinear Particle Swarm Optimization (NPSO) to determine appropriate fuzzy membership functions that cover the domain of quantitative attributes. The proposed algorithm construct membership function according to the best particle found by NPSO, extract all frequent itemsets and generate fuzzy association rules. NPSO based approach produces meaningful and more interesting fuzzy association rules and large 1-itemsets than former methods and has reasonable efficiency.

In 2011, Mourad Ykhlef [15] presents a new algorithm to extract the best rules in a reasonable time of execution but without assuring always the optimal solutions, called as Quantum Swarm Evolutionary approach based QEA-RM. It is based on Quantum Swarm Evolutionary Algorithm (QSE) [25] for mining association rules. QSE is a hybridization of Quantum Evolutionary Algorithm (QEA) [9] and particle swarm optimization (PSO). QEA approach based on quantum computing and is better than classical evolutionary algorithm like Genetic Algorithm (GA) that uses Q-bit as a probabilistic representation, instead of using binary, numeric or symbolic representation. Q-bit defined as a small unit of information, which represents a linear superposition of states in search space probabilistically. Q-bit representation has a better characteristic of population diversity than chromosome representation used in GA. QSE is an mechanism of quantum bit expression called quantum angle, which adopts the improved PSO to update Q-bit of QEA automatically that is better than QEA. QSE-RM generated rules with better fitness by considering the concept of string of Q-bits called multiple Q-bit than the fitness of rules given by non parallel version of Genetic Algorithms (GA-PVMINER) [12], where both belong to class of evolutionary algorithm that gives good solution and may be non optimal ones but in reasonable time. Some more hybridization will be added in QSE-RM for implementing parallelism.

In 2011, R.J. Kuo, C.M. Chao et al [19] introduced a novel algorithm for association rule mining in order to improve computational efficiency as well as to automatically determine suitable threshold values. The proposed algorithm comprises two parts, preprocessing and mining. The preprocessing part provides procedures related to calculating the fitness values of the particle swarm. Thus, the data are transformed and stored in a binary format. Then, the search range of the particle swarm is set using the IR (itemset range) value. In the mining part of the algorithm, the PSO algorithm is employed to mine the association rules. First, the algorithm proceeds with particle swarm encoding, that is similar to chromosome encoding of genetic algorithms. The next step is to generate a population of particle swarms according to the calculated fitness value. Finally, the PSO searching procedure proceeds until the stop condition is reached, which means the best particle is found. PSO algorithm determines support and confodence quickly and objectively. This algorithm is better than the traditional Apriori algorithm since it does not need to subjectively set up the threshold values for minimal support and confidence that affect the quality of association rule mining. This save computation time and enhance performance.

Mining quantitative Association Rule which included numeric and discrete attributes is a hard optimization problem and some researchers tried to mine AR's using global optimization algorithms called as QAR mining algorithm. The framework of this algorithms is to find the best itemset (rule) with best support (and confidence) in the first run, then penalizes recovered records and finds the next rule during new run. This process continues till it finds all N best rules. This sequential manner require multiple run to find all rules that decreases the chance of parallelism in distributed and parallel environments. In 2012, Zahra Karimi-Dehkordi et al [26] address the above mentioned issues of that QAR mining algorithm and propose parallel version of stochastic numeric PSO variant association rule mining by implementing multi agent architecture to find optimized rules simultaneously using a dynamic priority approach. Each agent is responsible for finding different optimized rule. Every run consists of several synchronization points at which the agents start to diverge by marking recovered records. The best agent marks its recovered records and other agents re-evaluate their rules to penalize marked records. Then the second best rule is selected and marking and re-evaluating is done. This prioritizing process continues till all except one agent completes recovering. As priorities changes at each synchronization point, it is also known as dynamic priority schema. The key point of this algorithm is re-initializing particles for escaping recovered rules proportional with recovered percent and roulette wheel selection to improve re-initialization, random flying mode to increases PSO diversity.

In 2012, Nandhini M et al [16] introduced a technique to reduce the quantity of the rules by combining mining and post-mining techniques without compromising the usefulness factor and thus improves the computational efficiency of rule mining. Particle swarm optimization is used in the mining process to compute feasible threshold values of support and confidence parameters, for obtaining strong rules which improves efficiency. In the post-mining process, domain ontology is designed to map the database, which helps in providing a formal, explicit specification of a shared conceptualization. The proposed methodology involves two steps. First, the association rules are generated from the database using particle swarm optimization. This step provides a mechanism to effectively choose the threshold support and confidence values by taking support and confidence of best particle as an optimal value for minimal support and confidence. In the second step, generated rules are reduced using post-mining techniques. Use of the operators such as pruning, conforming, unexpected antecedent and unexpected consequent over rule schemas improves the effectiveness of the post-mining process. Combining PSO and domain ontology interactively, reduced excessive amount of rules without compromising the usefulness of rules. Based on the user knowledge and the domain ontology, most interesting rules are discovered.

2.2 Optimization of Association Rule Generated by Traditional Algorithm Using PSO

In this approach, generated association rules by traditional association rule mining algorithm like Apriori, FP growth then that generated association rules are optimized by PSO algorithm. This approach totally based on traditional algorithm, so inherits some drawbacks of traditional algorithm.

In 2011, Shruti Mishra et al [22] presents the algorithm that uses the Particle Swarm Optimization approach for optimizing frequent pattern generated by fuzzy logic based various frequent pattern mining technique rather than other evolutionary algorithm over fuzzy dataset. Evolutionary algorithm requires a particular representation and specific method for cross-over, mutation and selection, while PSO only involves a single operator for updating solutions which is easier and effective to implement and able to produce good solution at a very low computational cost. In this model, the original dataset is fuzzified and categorized those data into High and Low. Frequent pattern mining algorithm is applied over it to generate all possible frequent patterns, rather to use any input threshold value as selection criteria, it calculate mean squared residue score for the frequent pattern. Then PSO is applied on frequent pattern for optimizing it by considering mean squared residue score as the fitness function. By analyzing the result obtained by various fuzzified frequent pattern mining algorithms such as Fuzzy Apriori algorithm, Fuzzy Vertical data format, Fuzzy FP-growth and the result obtained by PSO based Fuzzy FP-growth algorithm, yields better result and the number of frequent patterns generated is more and the runtime of this algorithm is much better than other fuzzy logic based traditional algorithm. In 2012, Shruti Mishra et al [23] proposes an algorithm which generate frequent patterns using Frequent Pattern (FP) growth from fuzzy datasets and then optimized by CLPSO algorithm for generating best individual frequent patterns out of entire sets of patterns. Comprehensive Learning Particle Swarm Optimization (CLPSO) uses all particles best information to update a Particle's velocity and also preserves the diversity of swarm to be preserved to discourage premature convergence. CLPSO have some selection measure called mean squared residue (MSR) score. In the proposed model, the gene expression data matrix is fuzzified using the fuzzy framework. Then various frequent pattern mining algorithms like Apriori algorithm, vertical data format, FP growth etc are used to generate frequent patterns and then using the mean squared residue score as the selection measure the

CLPSO algorithm is applied which optimize the generated rules. The results of the CLPSO based Fuzzy FP growth algorithm is much better as compared to the PSO based fuzzy FP growth algorithm [22] as well as the traditional algorithm such as Apriori algorithm, vertical data format, FP growth etc. The accuracy of finding the best patterns was much more in CLPSO than in traditional PSO. This algorithm outperformed the traditional PSO algorithm in terms of the generation of the number of frequent pattern, runtime and accuracy of generating best individual patterns with a comparatively lower MSR value. In the existing versions of the PSO algorithm with different neighborhood structures and the multi swarm PSOs, the swarms are predefined or dynamically adjusted according to the distance, which limit the freedom of sub-swarms. But Dynamic Multi Swarm Particle Swarm Optimization (DMS-PSO) resolved this problem of local PSO by adopting its neighborhood structure dynamic and randomized. In 2012, Shruti Mishra et al [24], proposed DMS-PSO based fuzzy frequent pattern mining. The frequent patterns obtained were considered as the set of initial population or particles. Then DMS-PSO is applied on that frequent pattern by considering the selection criteria as a mean squared residue (MSR) score rather using the threshold value, as the lower the mean squared residue the large the volume of frequent patterns are always preserved. DMS-PSO based fuzzy FP growth technique finds the best individual frequent patterns, the runtime of the proposed algorithm was much better as compared to the traditional PSO based fuzzy FP growth [22] as well as Genetic Algorithm

In 2012, Mohammad Javad Abdi [14] developed a diagnosis model based on association rules (ARs), particle swarm optimization (PSO) and support vector machines (SVMs) to diagnose erythemato-squamous diseases. The classification of erythemato-squamous diseases involves the simultaneous discrimination of numerous diseases. In order to tackle this problem, a number of multiclass classification strategies can be adopted. "One against all" is one of the most popular which is included in that proposed algorithm. Each classifier tries to solve a binary classification problem by differentiating one disease from all others. Then in the classification stage, the winner takes all rules is utilized to decide which disease is assigned to each patient. This represents that the winning disease is the one that corresponds to the SVM classifier of the ensemble that shows the highest output. SVM classifier is a supervised learning algorithm based on statistical learning theory, whose aim is to determine a hyper plane that optimally separates two classes using training dataset. In this proposed model, AR is used to select the optimal feature subset from the original feature set then a PSO based approach is developed to find the best parameters of kernel function for parameter determination of SVM. This system is designed to optimize the SVM classifier accuracy by automatically: Reducing the number of features with AR, Estimating the best values for confidence and support parameters of SVM by PSO.

3 Conclusions and Future Work

Association Rule Mining is one of the core data mining tasks that used to show the relationship between data items. These relationships are not based on inherent properties of the data themselves like functional dependencies, but based on co-occurrence of the data items. This paper gives a brief survey of implementation of particle swarm optimization in various ways to association rule mining technique, the modifications made to the association rules according to the applications they were used and its effective results. Thus particle swarm optimization is to be the most effective technique for making association rule algorithm more efficient.

Association Rule Mining attracts many researchers interest, some more new methods and variant of Particle Swarm Optimization will be introduces to enhance it by focusing many issue of association rule mining technique such as efficiency ,computing time, Predictive accuracy, comprehensibility, interestingness, interpretation, privacy-preserving, repeated disk access overhead, inconsistencies, negative occurrence of the attribute, dynamic changes in the user's requirements.

References

[1] Agrawal, R., Imielinski, T., Swami, A.: Mining association rules between sets of items in large databases. In: Proceeding of ACM SIGMOD International Conference Management of Date, Washington, DC, pp. 207–216 (1993)

[2] Abraham, A., Guo, H., Liu, H.: Swarm Intelligence: Foundations, Perspectives and Applications. In: Nedjah, N., de Macedo Mourelle, L. (eds.) Swarm Intelligent Systems. SCI, vol. 26, pp. 3–25. Springer, Heidelberg (2006)

[3] Ansaf, S.A., Christl, V., Cyril, N.: QuantMiner; A Genetic Algorithm for Mining Quantitative Association Rules. In: Proceeding of the 20th International Conference on Artificial Intelligence, IJCAI, Hyberadad, India (2007)

[4] Alatas, B., Akin, E.: Rough Particle Swarm Optimization and its application in data mining. In: Proceeding of Soft Computing, pp. 1205–1218. Springer (2008)

[5] Cai, G.-R., Chen, S.-L., et al.: Study on the Nonlinear Strategy of Inertia Weight in Particle Swarm Optimization Algorithm. In: International Conference on Natural Computation, pp. 683–687. IEEE (2008)

[6] Chatterjeea, A., Siarry, P.: Nonlinear Inertia Weight Variation for Dynamic Adaptation in Particle Swarm Optimization. Computers & Operations Research, 859–871 (2006)

[7] Martens, D., Baesens, B., Fawcett, T.: Editorial survey: swarm intelligence for data mining. Springer (2011)

[8] Cai, G.-R., Li, S.-Z., Chen, S.-L.: Mining Fuzzy Association Rules by Using Nonlinear Particle Swarm Optimization. In: Cao, B.-Y., Wang, G.-J., Chen, S.-L., Guo, S.-Z. (eds.) Quantitative Logic and Soft Computing 2010. AISC, vol. 82, pp. 621–630. Springer, Heidelberg (2010)

[9] Han, K.H., Kim, J.H.: Quantum-inspired Evolutionary Algorithm for a class of combinatorial optimization. IEEE Transaction on Evolutionary Computation 6(6), 580–593 (2002)

[10] Han, J., Kamber, M.: Data Mining: Concepts and Techniques, 2nd edn. Elsevier (2006)

[11] Kennedy, J., Eberhart, R.C., et al.: Particle swarm optimization. In: Proceedings of International Conference on Neural Networks, vol. 4, pp. 1942–1948. IEEE, Perth (1995)

[12] Lopes, H.S., Araujo, D.L.A., Freitas, A.A.: A parallel genetic algorithm for rule discovery in large databases. In: IEEE Systems, Man and Cybernetics Conference, pp. 940–945

[13] Mata, J., Alvarez, J.L., Riquelme, J.C.: An Evolutionary algorithm to discover numeric association rules. In: Proceeding of the ACM Symposium on Applied Computing, SAC. ACM (2002)

[14] Abdi, M.J., Giveki, D.: Automatic detection of erythemato-squamous diseases using PSO–SVM based on association rules. In: Proceeding of Engineering Application of Artificial Intelligence. Elsevier (2012)
[15] Ykhlef, M.: A Quantum Swarm Evolutionary Algorithm for mining association rules in large databases. Elsevier (2011)
[16] Nandhini, M., Janani, M., Sivanandham, S.N.: Association rule mining using swarm intelligence and domain ontology. IEEE (2012)
[17] Badawy, O.M., Sallam, A.-E.A., Habib, M.I.: Quantitative Association Rule Mining Using a Hybrid PSO/ACO Algorithm, PSO/ACO-AR (2008)
[18] Eberhart, R.C., Kennedy, J.: A new optimizer using particle swarm theory. In: Proceedings of the 6th International Symposium on Micro Machine and Human Science, Nagoya, Japan, pp. 39–43 (1995)
[19] Kuo, R.J., Chao, C.M., Chiu, Y.T.: Application of Particle Swarm Optimization to association rule mining. In: Proceeding of Applied Soft Computing, pp. 326–336. Elsevier (2011)
[20] Zhang, S., Wu, X.: Fundamentals of association rules in data mining and knowledge discovery. In: WIREs Data Mining Knowledge Discovery, vol. 1, John Wiley & Sons, Inc., Wiley Online Library (March/April 2011)
[21] Shi, Y., et al.: A Modified Particle Swarm Optimizer. In: Proceeding ICES, pp. 69–73. IEEE, Los Alamitos (1998)
[22] Mishra, S., Mishra, D., Sarapathy, S.K.: Particle Swarm Optimization based Fuzzy Frequent Pattern Mining from Gene Expression Data. In: International Conference on Computer and Communication Technology, pp. 15–20. IEEE (2011)
[23] Mishra, S., Sarapathy, S.K., Mishra, D.: CLPSO- Fuzzy Frequent Pattern Mining from Gene Expression Data, pp. 807–811. Elsevier (2012)
[24] Mishra, S., Mishra, D., Satapathy, S.K.: Fuzzy Frequent Pattern Mining from Gene Expression Data using Dynamic Multi-Swarm Particle Swarm Optimization, pp. 797–801. Elsevier (2012)
[25] Wang, Y., Feng, X.Y., Huang, Y.X., Zhou, W.G., et al.: A Novel Quantum Swarm Evolutionary Algorithm for Solving 0-1 Knapsack Problem. In: Proceeding of Advances of Natural Computation. Springer (2006)
[26] Karimi-Dehkordi, Z., Nematbakhsh, M., Baraani-Dastjerdi, A., Ghassem-Aghaee, N.: Stochastic Mining of Quantitative Association Rules Using Multi Agent Systems. Proceeding of ARPN Journal of System and Software, AJSS Journals 2(2) (2012)

Evolutionary Algorithm Approach to Pupils' Pedantic Accomplishment

Devasenathipathi N. Mudaliar[1,2] and Nilesh K. Modi[3]

[1] MCA Department, SVIT Vasad, India
[2] R & D Centre, Bharathiar University, Coimbatore
[3] MCA Department, SVICS, Kadi, India

Abstract. Group learning helps pupils in boosting their learning power by creating interactions among them. However, creating groups among pupils with appropriate coupling and cohesion is still a challenge. Pupils' groups are formed with some constraints and group formation performed by a single individual is customarily prejudiced in one way or other. In this paper, an approach has been proposed using evolutionary algorithm to increase the pupils' pedantic accomplishment. This approach helps in optimal pupil group formation on the basis on of their previous examination scores. To justify the proposal, a study was carried out among a class of pupils pursuing post graduation. The semester examination results of pupils before and after group learning were compared. More than 66.07% of pupils scored better than their previous semester examination which positively proved the proposed approach.

Keywords: Group Formation, Genetic Algorithm, Group Learning, Optimization, Performance Prediction, Academic Improvement.

1 Introduction

Group learning is that approach of learning where pupils are collaborated into different groups and each group dawns in the direction of learning. In case of individual learning pupils learn on an individual basis having least interactions among their peers. However, many academicians feel that when pupils (in groups) interact and discuss about problems and their solutions of any given topic they grasp more understanding of it than learning individually (K. Shin-ike and H. Iima, 2009) [5].

Unfortunately, it is noted that not all groups perform better in group learning environment proving that there are some issues yet to be discovered in the creating of pupil groups. In addition to this very few literatures has focused on the approaches applied for forming groups among pupils. Group learning may not succeed because pupils may be randomly grouped together without any application of group formation technique.

The authors suggest a broad group formation technique among pupils studying in a class in a regular college. A formula was devised to measure the increase in academic score of a given pupil among a group of pupils using their previous (latest) score. The main objective was to increase the sum of difference between the previous score and the predicted (calculated) score. Many approaches were available to solve this

S.C. Satapathy et al. (Eds.): Proc. of Int. Conf. on Front. of Intell. Comput., AISC 199, pp. 415–423.
DOI: 10.1007/978-3-642-35314-7_47

problem including exhaustive search. But exhaustive search option did not seem possible because of the dynamic group size and the size of the class. This created the need for using heuristic search techniques. Although these techniques do not assure of searching the best possible answer, they search an acceptable answer by requiring less computing power. Examples of heuristic search techniques include Ant colonization algorithms, particle swarm optimization, genetic algorithm, etc. Among all heuristic techniques, genetic algorithm was chosen for its simplistic and yet optimizing nature. Experimenting using other heuristic techniques mentioned above to create pupils groups was considered as part of future work.

The remaining of paper is divided as follows: The second section presents an overview of the literatures published in connection to the pupils' group formation using heuristics. Section three explains algorithm formation of creating pupils' groups using genetic algorithm, while section four elucidates on the application of proposed algorithm in the specific case. The fifth section presents the attained results after the experimental work. Lastly sixth section concludes the paper in addition to providing a glimpse of the future possible experimental work in connection to the current work done.

2 Background Work

There are many approaches followed by academic researchers for group formation among pupils. The parameters for group formation include pupil grasping level, characteristics of assignment, interaction level among pupils, etc (Yen-Ting Lin, Yueh-Min Huang, Shu-Chen Cheng, 2010) [12]. However, in context to group formation by pupils grasping level, the total number of pupils in the class and balanced grasping level among pupils of group can be considered as base. The below text gives a brief overview of the research work published in connection to group formation among pupils.

The finding presented by (Julian, Demetrio & Rosa, 2012) [4] considers three characteristics for group formation among pupils viz. pupil knowledge levels, pupils' communicative skills and pupil leadership skill. They proposed a generic method based on genetic algorithms for getting inter-homogenous and intra-homogenous groups. They measured all pupils' outcomes and compared them with randomly formed groups and organized groups of pupils. To find if there was a notable difference in the learning processes among groups of pupils formed by the two methods, an analogous hypothetic test was performed. They found that there was a 0.11 difference in favor of the proposed method compared with the random method, while 0.15 differences compared with the self-organized method. Their research work proved that their proposed method succeeded in achieving inter-homogenous and intra-homogenous groups in addition to the characteristics envisaged affecting positively, the growth of actions among the group members.

In the research article published by (Pedro P., Alvaro O. and Pilar R., 2010) [9], the authors have believed that learning styles provide proper ways to group pupils. Also they found that heterogeneous groups performed better than groups of pupils with similar characteristics. They defined pupils' attributes to compare the groups' performance. Their method based on Far-so-close algorithm found different groups based on four factors viz. the collection of pupils, the number of pupils in each group, the pair threshold and group threshold. They developed a tool called TOGETHER (a visualization tool) which chose every time a different starting pupil and used a randomly chosen pupil ordering every time. The tool TOGETHER had pupil centered use and teacher centered use. In the pupil centered use of the proposed tool, 44% of pupils answered all 10 questions correctly and none of them failed in more than 6 questions. Similarly in teacher centered use, teachers entered the learning style of pupils previously collected. They found that most of the teachers found useful to use this tool.

The authors (K. Shin-ike and H. Iima., 2009) [5] in their research paper proposed a method to improve the learning effect of collaborative learning. The authors used Synthetic Personality Inventory (SPI) test to measure the academic backgrounds and personalities of pupils. The personality test result was categorized into emotional aspect, an active aspect, a volitional aspect and a characteristic aspect. Linguistic ability and non-linguistic ability questions formed measuring factors of academic backgrounds. A neural network model was applied to predict the learning results of pairs of pupils in group learning using correctness rates. Further, genetic algorithm (as stochastic local search method) was applied with the predicted results to determine the optimal pairs of pupils. Thirty pupils (4 female and 26 male) participated in this experiment and the authors confirmed that their proposed method was effective.

The authors (Hwang, G.J., Yin, P.Y., Hwang C.W., & Tsai C.C., 2008) [2] proposed an enhanced genetic algorithm in order to organize cooperative learning methods using multiple grouping factors. They expressed that teachers/instructors use various sets of grouping criteria according to situations/backgrounds in a group learning context. The authors formulated a Multi-Criteria Group Composition (MCGC) problem to meet their research objective. The evaluation of proposed algorithm was done by comparing the results of series of conducted experiments with already employed methods. They found 3 constraints in the process of group formation among pupils viz. each pupil should be assigned to exactly one group, difference of number of pupils among groups should not be more than one and there is at least one pupil in the group who has understood the concept well. The authors used Roulette-Wheel selection method for selecting chromosomes for crossover and mutation. In addition to this, a web-based interface was implemented and integrated with their distance learning system, for helping teachers. The limitations of their algorithm included no guarantee of confirming the solutions as best possible solution and the proposed method was slower than the greedy method.

The authors (Isotani et al, 2009) [3] have tried to present an ontology (Reality based system) which would work as a framework for group formation in collaborative learning environment. The authors hypothesized that knowing pupils' needs beforehand increases the benefits of collaboration learning and help in gaining individually and as a group. They used ontology to represent learning theories, in order to obtain the needs of the pupils. Using this approach, the authors were able to investigate the individual and group interactions to ascertain if pupils gained expected benefits or not, which in turn helped in better group formation. The authors focused on learning scenario designing concepts, which had higher impact on changes in pupils' learning. Learning goal, role play and instructional-learning formed the main concepts of group formation. There were 2 pairs of quality instructors and 20 participants (from 7 different countries). The experiment consisted of two phases. First phase was planning the collaborative learning session and second phase was actual implementation of it. Instructors dealt with the group problem and used their own methods in the first phase. Then they merged selected collaborative sessions. Instructors were asked to give content learned by participants, choice of forming groups, individual and group goals and creation of sequence activities. Then the same tasks were performed using the authors proposed ontology methods. After the experiment, instructors agreed that use of ontology helped very much in the planning phase. In addition to this, instructors were able to create and share collaborative learning sessions. In the second phase, the experimental group performed better than control groups in various dimensions, thereby positively proving the authors hypothesis. Thus, the proposed ontology helped in decision making when, how and why to use the learning theories in collaborative learning.

3 Genetic Algorithm and Pupils' Group Formation

A method which tries to copy the evolution process of nature and is a search algorithm and an optimization method is Genetic Algorithm (GA). A large number of real world problems have been solved using GA (P.J. Bentley, 1997) [8]. Every possible solution for a given problem in GA is formed as a chromosome, which in turn is again composed of genes. Operators like crossover and mutation are applied onto these chromosomes to produce better chromosomes (solutions which can solve the problem). Figure 1 represents typical functioning of a genetic algorithm.

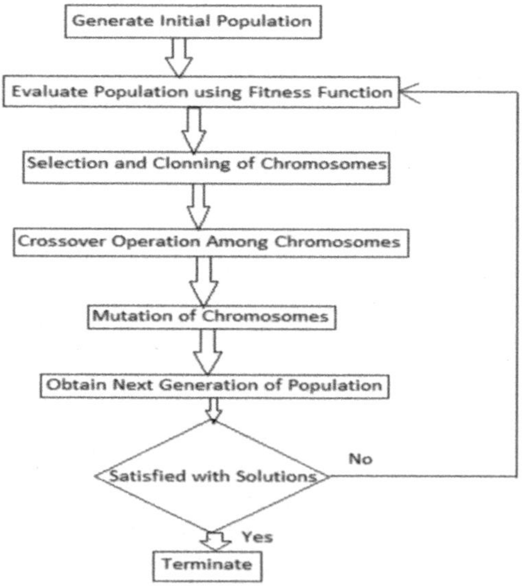

Fig. 1. Representation of *(Genetic Algorithm)* Process

From the figure it is clear that an initial set of chromosome called initial population is formed as the first step of GA process. Each chromosome is a valid solution for the given problem. Next these initial set of chromosomes is evaluated for goodness of fit (called as fitness value) using some fitness function and they (chromosomes) can be arranged according to the fitness value. Some chromosomes from the initial population are selected using techniques like Roulette Wheel selection, Tournament selection, Rank selection, etc. The selected chromosomes are then cloned and sent for crossover and mutation and the next generation of population is evolved, which replaces the initial population. The process iterates till some fixed number of generations evolutions or till some fitter chromosomes are not obtained.

GA can be helpful in solving the Traveling Salesman Problem (TSP). In TSP, a salesman needs to travel to every city and only once to every city with least traveling cost. In terms of GA, all valid paths are considered as chromosomes and using crossover operators (for TSP like Partially Mapped Crossover, Order Crossover, Edge Recombination Crossover, etc.) and mutation operators (like displacement mutation, insertion mutation, exchange mutation, etc.) onto the initial selected chromosomes, through various generations yield better paths with respect to the fitness function. In TSP, normally the fitness value is the total distance/cost of the given path (chromosome), which is obtained through a fitness function. The lesser the total distance of a path, the more fit is the chromosome. In TSP, the ordering of cities (genes) is important with respect to the fitness value.

Pupils' group formation problem is similar to Traveling Salesman Problem. It has been proved that group members affect the performance of individual pupils (in that group) in a group learning environment (Yen-Ting Lin, Yueh-Min Huang, Shu-Chen Cheng, 2010) [10]. A formula was formulated to compute the increase in a given pupil's academic performance (in terms of percentage) among a given group of pupils (Devasenathipathi N., Nilesh K. Modi, 2011) [1]. To achieve this task, the previous performance (in terms of Cumulative Performance Index) of all the pupils was taken into consideration. Chromosomes represented the order of pupils' future (calculated) performance (in terms of percentage). Each pupils future score acted as genes of chromosomes. The summation of the positive difference between the previous scores and the future scores was the fitness value. Higher this value more fit was the chromosome. It seemed difficult to use traditional crossover operators as a pupil cannot be put in more than one group. As this problem resembles ordering problem (like TSP where the order of cities is important), applying only crossover operators suited for it would be feasible. Similar was the case with mutation operator. Only mutation operators suited for ordering problems (like TSP) would be beneficial.

4 Methodology

Any pupils' performance working in group depends on two factors (Wilkinson, I. A. G., & Fung, I. Y. Y., 2001, Kyparisia Papanikolaou, Evangelia Gouli, 2010, Kuisma, R., 1998) [11] [7] [6], the grasping level of the individual pupil and the contribution level of other pupils of the group. Using these two factors, the authors have formulated a formula that could help in predicting the examination score of any given pupil among other group members. It was constrained that each group should consist of exactly 5 members. However, there were 56 pupils in that particular class. So 10 groups with 5 pupils in each group and one group (11th group) with six pupils was planned. Figure 2 shows the chromosomal representation of Travelling Salesman Problem and chromosomal representation of Pupils' group formation problem (for 56 pupils).

Typical Chromosome of a Travelling Salesman Problem (For 6 Cities)

Mumbai
Pune
Surat
Ahmedabad
Udaipur
Nagpur

Typical Chromosome of Pupils' Group Formation Problem

Pupil 1	Score till last Semester of Pupil 1	Predicted Score of Pupil 1
Pupil 2	Score till last Semester of Pupil 2	Predicted Score of Pupil 2
Pupil 3	Score till last Semester of Pupil 3	Predicted Score of Pupil 3
Pupil 4	Score till last Semester of Pupil 4	Predicted Score of Pupil 4
Pupil 5	Score till last Semester of Pupil 5	Predicted Score of Pupil 5
...	...	...
...	...	...
Pupil 56	Score till last Semester of Pupil 56	Predicted Score of Pupil 56

Fig. 2. Representing Chromosomes of (*Pupils Group Formation Problem*)

The formula for calculating the future score among other group members is as follows (for any given group):

$$((((\sum x_i - x_i)/4) * x_i)/100)/5 + x_i \tag{1}$$

where x_i is the previously obtained score of the pupils among that particular group of pupils.

The fitness function evaluated the total difference between the previous score and the predicted score of all the pupils. Higher positive difference represented more fit chromosomes. We used Partially Mapped Crossover technique for crossing over among chromosomes and Displacement Mutation technique for mutation. The functions coded (in C language) for achieving these tasks are as follows:

1. Initialization of Population (1000 random chromosomes)
2. Fitness value calculation through fitness function
3. Selection of top 500 chromosomes
4. Crossover Performing (PMX)
5. Mutation Performing (Displacement Mutation)

5 Results

The results of the experiment presented in this section can be viewed from three different angles. First angle is the number of pupils matching the predicted score, while second is the number of pupils scoring higher than their previous scores and final angle is the measure representing similarity among the predicted score and the actual score of the pupils. The graph in Figure 3 represents closeness between the actual score of pupils after the experiment and the predicted score of pupils before the experiment.

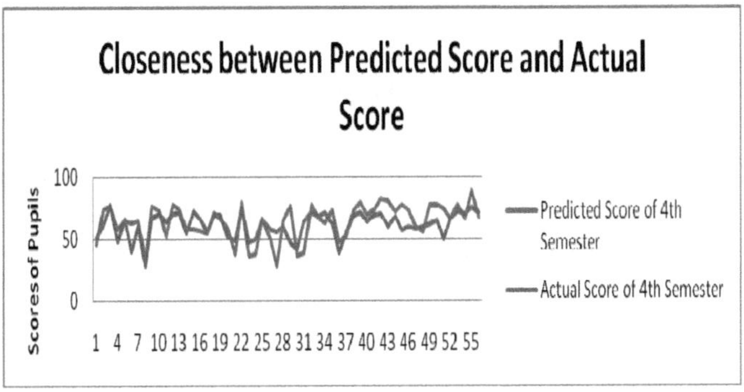

Fig. 3. (Representing closeness between Predicted Score and Actual Score)

Number of Pupils Matching the Prediction:

The total number of pupils matching the prediction is 25 out of 56 pupils (with an error level of 5%). The below table shows the error level (in percentage) and the number of pupils matching the prediction.

Table 1. (Error level in Predicting Scores of Students)

Error Level (in Percentage)	Number of Pupils
0 – 1	3
1 – 2	3
2 – 3	5
3 – 4	3
4 – 5	11

Number of Pupils Scoring Higher than Previous Examinations:

37 pupils out of 56 pupils (66.07%) scored higher than their previous examinations. The below graph in Figure 4 shows the increase in score of pupils after the experiment compared to the previous examinations result.

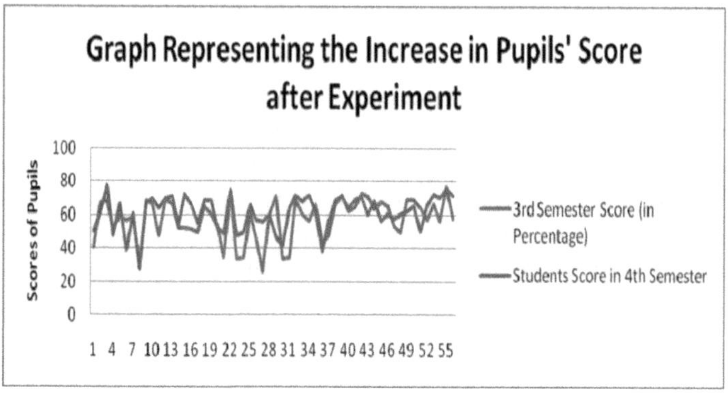

Fig. 4. (Representing Increase in Pupils' Scores after the Experiment)

Similarity Measure among Predicted Score and Actual Score of Pupils:

Karl Pearson's Linear Correlation method was used to calculate similarity measure among pupils' predicted score and the actual score in 4th semester. The result was 0.6371 which means the method adopted is 63.71% reliable about in the pupils score using the given methodology

6 Conclusion and Future Work

We have tried to present a work in relation to the pupils' group formation method using an evolutionary algorithmic approach (more specifically genetic algorithm). Since a formula to predict the score of a pupil among other group members was already proved, it was easy to form the fitness function. The array of pupils' scores (predicted future score) formed a chromosome. Crossover and Mutation operators were applied onto these formed chromosomes and a better chromosome was chosen, which formed the solution. In this paper we formed groups only on the basis of previous scores of pupils. Although only the score of 25 pupils (out of 56 pupils) matched the prediction (with 5% error level), it may be noted that 37 pupils (out of 56 pupils) scored better than previous semester, which positively proved our stated hypothesis.

As part of future work, we would like to form groups of pupils based on many other relevant factors responsible for group formation as well try to use other heuristics based approach. In addition to this, the same kind of experiment can also be performed among people of other industries which may result in better services and production.

References

1. Devasenathipathi, N., Modi, N.K.: Measuring Individual Knowledge Gain Statistically in a Group Learning Environment. National Journal of Computer Science and Technology 3(2), 25–29 (2011)
2. Hwang, G.J., Yin, P.Y., Hwang, C.W., Tsai, C.C.: An Enhanced Genetic Approach to Composing Cooperative Learning Groups for Multiple Grouping Criteria. Educational Technology & Society 11(1), 148–167 (2008)
3. Isotani, et al.: An Ontology Engineering Approach to the Realization of Theory- Driven Group Formation. International Journal of Computer Supported Collaborative Learning 4(4) (2009)
4. Moreno, J., Ovalle, D.A., Vicari, R.M.: A Genetic Algorithm Approach for Group Formation in Collaborative Learning Considering Multiple Student Characteristics. Computers & Education 58(1), 560–569 (2012)
5. Shin-ike, K., Iima, H.: A method for Development of collaborative learning by using a neural network and a genetic algorithm. In: Proceedings of ISADS, pp. 417–422 (2009)
6. Kuisma, R.: Assessing Individual Contribution to a Group Project. In: Watkins, D., Tang, C., Biggs, J., Kuisma, R. (eds.) Assessment of University Students in Hong Kong: How and Why, Assessment Portfolio, Students' Grading - Evaluation of the Student Experience Project, vol. 2, pp. 79–106. City University of Hong Kong, Centre for the Enhancement of Learning and Teaching (1998)
7. Papanikolaou, K., Gouli, E.: Collaboration as an Opportunity for Individual Development. In: International Conference on Intelligent Networking and Collaborative Systems, pp. 54–61 (2010)
8. Bentley, P.J.: The Revolution of Evolution for Real-World Applications. In: Emerging Technologies 1997: Theory and Application of Evolutionary Computation, University College London (1997)
9. Paredes, P., Ortigosa, A., Rodriguez, P.: A Method for Supporting Heterogenous-Group Formation through Heuristics and Visualization. Journal of Universal Computer Science 16(19), 2882–2901 (2010)
10. Bello, T.O.: Effect of Group Instructional Strategy on Students' Performance in Selected Physics Concepts. African Educational Research Network 11(1), 71–79 (2011)
11. Wilkinson, I.A.G., Fung, I.Y.Y.: Small-group composition and peer effects. International Journal of Educational Research (Special issue) 37, 425–447 (2002)
12. Lin, Y.-T., Huang, Y.-M., Cheng, S.-C.: An Automatic Group Composition System for Composing Collaborative Learning Groups using Enhanced Particle Swarm Optimization. Computers & Education 55(4), 1483–1493 (2010)

Identification of System with Non-stationary Signal Using Modified Wilcoxon Approach

Sidhartha Dash and Mihir Narayan Mohanty

ITER, Siksha 'O' Anusandhan University, Jagamara, Bhubaneswar, Odisha
{Sidharthadashiter,mihir.n.mohanty}@gmail.com

Abstract. Non-stationary random signals exhibit time-dependent characteristics and require proper models and corresponding identification methods. The focus is on identification method. We study system identification of the non-stationary parameters in this task. In this paper, the problem of non-causal identification of non-stationary, linear stochastic systems has been considered. A robust system identification approach adapted to chirp signals is proposed. An asymptotically unbiased estimate for the system's transfer function is analyzed. We show that compared to a competing non-stationarity based method, a significantly smaller error variance is achieved and generally shorter observation intervals are required. The adaptive method used here, is Wilcoxon approach based. Also the comparison have been done with Sign WLMS and Sign sign WLMS methods as the modified technique. In case of a time-varying system, faster convergence and higher reliability of the system identification are obtained. The results confirm the advantages of proposed approach. The resulting parallel estimation scheme automatically adjusts its smoothing parameters to the unknown, and possibly time-varying, rate of non-stationarity of the identified system.

Keywords: System Identification, non-stationary signal, adaptive signal processing, chirp impulse response, Wilcoxon norm, Sign Wilcoxon Technique.

1 Introduction

Most physical systems can be represented by models which account for the transformations. Depending on these transformations, linear time-varying (LTV) systems have been represented using narrowband, wideband; or dispersive non-stationary models [1–3].

Identification of the unknown system is analytically tractable only for stationary parameters. In many applications, however, a complete model of parameter variations is not known. The problem is then undertaken for full analytical solution and leading to many heuristic techniques [4]. Also, identification of non-stationary dynamic systems can be carried out using different frameworks, such as the local estimation approach, the basis function approach, or the approach based on Kalman filtering [5]. In spite of methodological differences, the corresponding identification algorithms share one common feature, they all have finite estimation memory. Appropriate

S.C. Satapathy et al. (Eds.): Proc. of Int. Conf. on Front. of Intell. Comput., AISC 199, pp. 425–430.
DOI: 10.1007/978-3-642-35314-7_48

choice of estimation is one of the key issues in identification of non-stationary systems. The best results can be obtained if the estimation of the identification algorithm is selected so as to match the rate of non-stationarity of the analyzed system. If the rate of parameter changes varies with time, then the estimation method should be adjusted accordingly [6].

In machine learning, incomplete data is a major problem. There are many possibilities that can cause the training data to be incomplete, such as mislabeling, biases, non-sufficiency, imbalance, noise, outliers, etc. The Least Mean Square (LMS) algorithm is also used as a learning tool for optimization technique. Still the desire result could not meet this challenge. But the resulting model obtained by this approach is not effective against outliers. LMS uses a gradient-based method of steepest decent and it uses the estimates of the gradient vector from the available data. It incorporates an iterative procedure that makes successive corrections to the weight vector in the direction of the negative of the gradient vector which eventually leads to the minimum mean square error [7-8].

2 Model for System Identification

System identification is one of the important aspect for designing a system. The identification task is to determine a suitable estimate of finite dimensional parameters, which completely characterize the plant. The selection of the estimate is based on comparison between the actual output sample and a predicted value on the basis of input data up to that instant [9].

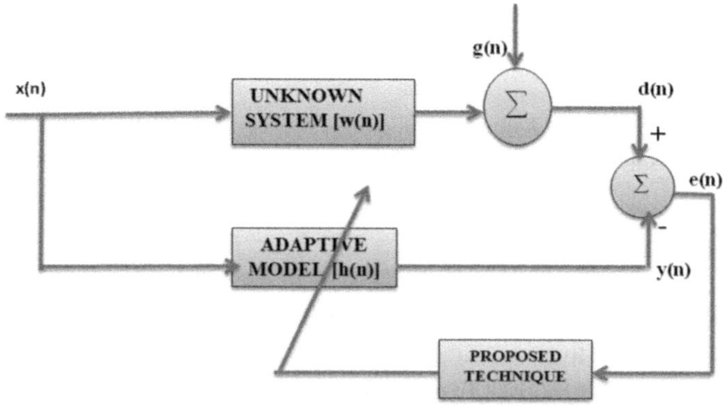

Fig. 1. Block Diagram of Chirp System Identification

The block diagram for system identification is shown in Fig.1. In this case, the input is applied to the unknown system as well as to the adaptive model. The impulse response of the unknown time-varing harmonic system is represented as a chirp signal. A chirp signal is a sinusoid with a frequency that changes continuously over a certain band as well as over a certain time period. In the chirp signal the instantaneous frequency increases from the lower bound of the frequency band to the higher. When applying the signal to a system it gives good control over the excited frequency band, and is therefore often used for system identification. The model is used as the impulse function of the unknown system. The chirp signal *w(n)* is represented as

$$w = A\sin(2\pi(a_0 + a_1 t)) + B\sin(2\pi(b_0 + b_1 t + b_2 t^2)) + \sqrt{2}C \tag{1}$$

where the chirp coefficients A, B and C are taken as A=1, B=2 and C is a random value. The values of chirp coefficients taken in this model are a_0=random number, a_1=0.25 and b_0=random number, b_1=0.5, b_2=0.7.

A white Gaussian noise *g(n)* is added with time-varing harmonic system output which accounts for measurement noise. The desired output *d(n)* is compared with the estimated output *y(n)* of the system identifier to generate the error *e(n)*, which is used by the adaptive algorithm for updating the weights of the model *h(n)*.The desired output of the unknown system is defined as

$$d(n) = x(n) * w(n) + g(n) \tag{2}$$

The estimated output of the adaptive filter is found to be

$$y(n) = h^T(n).x(n) \tag{3}$$

The error signal is expressed mathematically as

$$e(n) = d(n) - y(n) \tag{4}$$

where *h(n)* is the parameter correspond to the impulse response value of the filter at time *n* known as the adaptive weights of the model.

The objective is to minimize the error by using an optimum set of filter coefficient. When the adaptive filter weights are optimized, at that time the model provides the best performance. In this paper, the Wilcoxon technique is used as the adaptive algorithm for the chirp system identification.

3 Proposed Method for Identification

(i) Wilcoxon Technique

Wilcoxon Learning Algorithm is one of the effective method of robust identification as it is insensitive to mislabeling, biases, non-sufficiency, imbalance, noise, outliers in the sysytem. Most of the application areas such as in Communication, machine learning, data mining, adaptive control etc., it has used successfully. The cost function taken in the proposed model is a robust norm called Wilcoxon norm. The weights of the models are updated using conventional LMS,which progressively reduces the norm. The Wilcoxon norm minimization technique is mostly preferred to achieve improved and dynamic performance of a system [9].

The Wilcoxon Norm of a vector is expressed in terms of an increasing score function having zero mean and unit variance. The score function is defined as

$$\Phi(u) : [0,1] \rightarrow R \tag{5}$$

Let the error vector of ith term at j^{th} interval due to application of N input samples to the model be represented as $[e_{1,i}(j), e_{2,i}(j), \ldots\ldots. e_{N,i}(j)]^T$. The error vector is obtained by subtracting estimated output from the desired output. The rank of each error term is obtained by sorting the error terms in increasing order [8]. The score function associated with each error term is expressed as

$$\begin{aligned} s(k) &= \Phi(u) = \sqrt{12}(u) \\ &= \sqrt{12}\left(\frac{k}{N+1} - 0.5\right) \end{aligned} \tag{6}$$

Here k denotes the rank associated with each error term ($1 \leq k \leq N$). At j^{th} interval of each ith term, the wilcoxon norm is evaluated as

$$C_i(j) = \sum_{k=1}^{N} s(k) e_{k,i}(j) \tag{7}$$

In this gradient based technique $C_i(j)$ is defined as the Wilcoxon Norm of the error vector which is also termed as cost function. The weight update equation for the Wilcoxon Learning Algorithm is

$$h(n+1) = h(n) + 2\mu x(n)\sqrt{12}(u) \tag{8}$$

Where μ is defined as the step size and $x(n)$ are the random input samples.

(ii) Modified Wilcoxon Technique

It is known that sign-LMS and sign-sign LMS techniques are faster in terms of convergence compared to conventional LMS methods [10]. This motivates to use signum function in wilcoxon adaptive algorithm. In Sign Wilcoxon approach error vector is represented in terms of its signum value.

A Signum function can be represented as

$$y = sign(t) = t / abs(t) \tag{9}$$

The weight update equation for Sign Wilcoxon Algorithm is defined as

$$h(n+1) = h(n) + 2\mu x(n) sign(e(n))\sqrt{12}u \tag{10}$$

And in Sign-Sign Wilcoxon approach both error vector and score matrix is represented in terms of its signum value. The modified update expression for Sign-Sign Wilcoxon Algorithm is represented as

$$h(n+1) = h(n) + 2\mu x(n) sign(e(n))\sqrt{12} sign(u) \tag{11}$$

Equation (10-11) describes the weight update equation using a fixed step-size parameter μ.

Here this proposed technique is used for system identification problem having harmonic impulse response for different SNR values. The simulation results are compared to conventional Wilcoxon techniques.

4 Result and Discussion

The task is performed in MATLAB-7 environment. It has been simulated by taking Step-Size parameter μ as 0.02. There has been 10000 input data taken which is random value between [-0.5 0.5]. The proposed weight update equation is used for system identification. The impulse response for the unknown system used in this proposed model is taken from equation (1). The simulation results are compared with Wilcoxon Technique for different SNR values. The simulation result is shown between number of iterations and normalized MSD (Mean Square Deviation) in db.

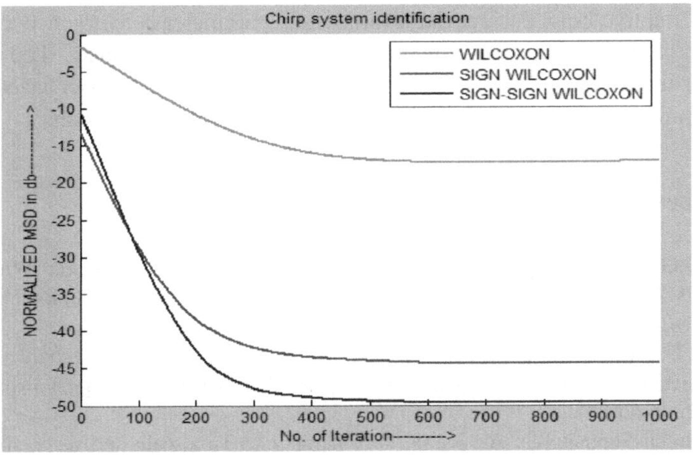

Fig. 2.MSD curve for SNR=10dB

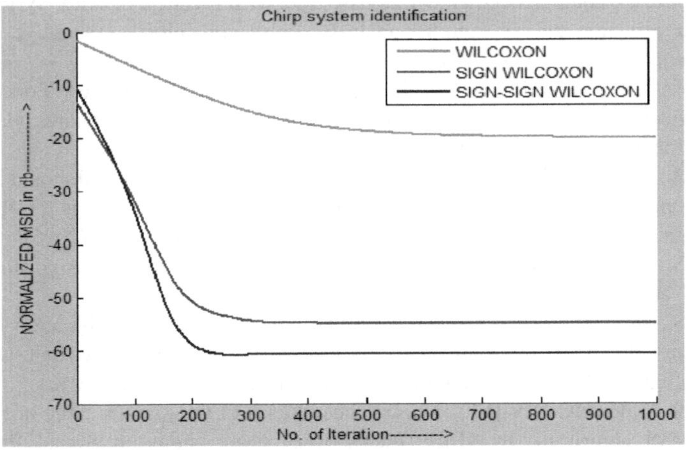

Fig. 3. MSD curve for SNR=20dB

Fig.4-5 is MSD curves for chirp system identification. From the figures, it is found that the Sign-Sign Wilcoxon technique performs better for different SNR values and convergence speed is faster than the Wilcoxon norm and sign Wilcoxon norm. So the proposed algorithm can be a suggested technique in the lower SNR environment as well as for harmonic impulse response systems.

5 Conclusion

The problem of system identification of linear stochastic systems was considered. The resulting parallel estimation scheme automatically adjusts its smoothing bandwidth to the unknown, and possibly time-varying, rate of non-stationarity of the identified system. It can also account for the distribution of measurement noise. We have presented a self-contained asymptotic analysis for identification of stable systems. In the context of adaptive input design, the results imply parameter convergence as well as that the optimal covariance matrix of the parameter estimates is obtained. The performance of adaptive input design is thus the same as for optimal input design based on the true system parameters. The results have been verified on a simulation basis.

References

1. Iem, B.G., Papandreou-Suppappola, A., Faye Boudreaux-Bartels, G.: Wideband Weyl symbols for dispersive timevarying processing of systems and random signals. IEEE Transactions on Signal Processing 50(5), 1077–1090 (2002)
2. Jiang, Y., Papandreou-Suppappola, A.: Discrete timefrequency characterizations of dispersive linear time-varying systems. IEEE Transactions on Signal Processing 55(5), 2066–2076 (2007)
3. Papandreou-Suppappola, A., Ioana, C., Zhang, J.J.: Timescale and dispersive processing for wideband time-varying channels. In: Hlawatsch, F., Matz, G. (eds.) Wireless Communications over Rapidly Time-Varying Channels. Academic Press (2011)
4. Gerencsér, L., Hjalmarsson, H., Mårtensson, J.: Identification of ARX systems with non-stationary inputs and asymptotic analysis with application to adaptive input design. Automatica 45, 623–633 (2009)
5. Niedźwiecki, M.: Identification of time-varying processes. Wiley, New York (2000)
6. Niedźwiecki, M., Gackowski, S.: On noncausal weighted least squares identification of nonstationary stochastic systems. Automatica 47, 2239–2244 (2011)
7. Chen, B., Hu, J., Li, H., Zengqi: A Joint Stochastic Gradient Algorithm and its application to system identification with RBF Networks. In: Sun Proceedings of the 6th World Congress on Intelligent Control and Automation, China, June 21-23 (2006)
8. Dash, S., Mohanty, M.N.: A Comparative Analysis for WLMS Algorithm in System Identification. In: IEEE Conf. ICECT, Kanyakumari, April 6-7 (2012)
9. Dash, S., Mohanty, M.N.: Analysis of Outliers in System identification using WLMS Algorithm. In: IEEE International Conference on Computing, Electronics and Electrical Technologies, ICCEET, Kanyakumari, March 21-22, pp. 802–806 (2012)
10. Shaik, R.A., Reddy, D.V.R.K.: Noise cancellation in ECG signals using normalized Sign-Sign LMS algorithm. In: IEEE International Symposium on Signal Processing and Information Technology, ISSPIT, December 14-17 (2009)

Recent Trends and Developments in Graph Coloring

Malti Baghel, Shikha Agrawal, and Sanjay Silakari

UIT, RGPV, Bhopal M.P., India
{maltibaghel,ssilakari}@yahoo.com, shikha@rgtu.net

Abstract. This paper is intended to give review of various heuristics and metaheuristics methods to graph coloring problem. The graph coloring problem is one of the combinatorial optimization problems used widely. It is a fundamental and significant problem in scientific computation and engineering design. The graph coloring problem is an NP-hard problem and can be explained as given an undirected graph, one has to find the least number of colors for coloring the vertices of the graph such that the two adjacent vertices must have different color. The minimum number of colors needed to color a graph is called its chromatic number. In this paper, a brief survey of various methods is given to solve graph coloring problem. Basically we have categorized it into three parts namely heuristic method, metaheuristic methods and hybrid methods. This paper surveys and analyzes various methods with an emphasis on recent developments.

Keywords: Ant colony optimization, Genetic algorithm, particle swarm optimization, simulated annealing, Tabu search.

1 Introduction

The optimization problem is a challenging field that searches for the best solution among all possible solutions. Optimization problems can be divided into two categories depending on whether the variables are continuous or discrete. Combinatorial optimization problems are the problems where solutions are encoded with discrete values. The graph coloring problem is one of the combinatorial optimization problems. Given an undirected graph $G = \{V, E\}$ where $V = \{v_i,\ i= 1,2......... N\}$ is the set of vertices and $E=\{e_{ij}\}$ is the set of edges, a k-coloring of G is to partition a V into a minimum number (called as chromatic number)of color class $C_{1,}\ C_2.....C_k$ Such that an edge $e_{ij} \in$ E, v_i and v_j are not in the same color class. Such a color class is called as an independent set. Let C (v_i) be the color assigned to the nodev_i, a proper coloring must satisfy follow condition:

$$\forall e_{ij} \in E, c(v_i) \neq c(v_j) \tag{1}$$

Graph coloring is an NP hard problem [19]. Graph coloring has many applications such as map coloring, scheduling, radio frequency assignment, register allocation, pattern matching, sudoku, timetabling and many more. An example of the proper graph coloring problem is shown in fig.1

S.C. Satapathy et al. (Eds.): Proc. of Int. Conf. on Front. of Intell. Comput., AISC 199, pp. 431–439.
DOI: 10.1007/978-3-642-35314-7_49

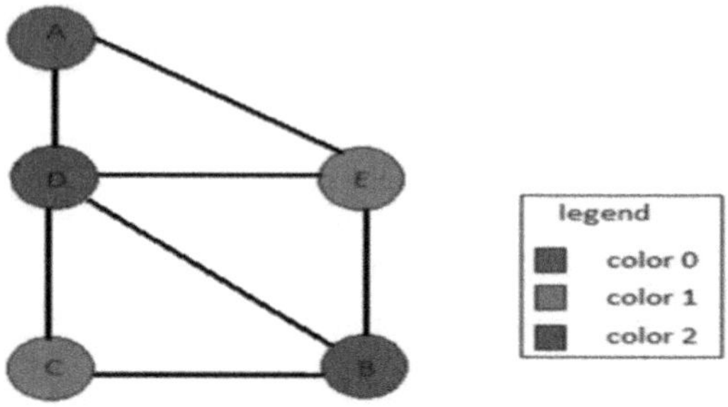

Fig. 1. An instance of a Graph Coloring Problem solved

In the following sections we present different coloring methods grouped in three categories. Heuristic methods in Sect. 2, Metaheuristic methods in Sect. 3. Metaheuristics is divided into two parts. Sect. 3.1 describes Local search methods which again are subdivided into three subsections named as Sect. 3.1.1 as simulated annealing, Sect.3.1.2 as Tabu search and Sect. 3.1.3 Describes other method. Sect. 3.2 is population based methods which again is subdivided into three subsections named as 3.2.1 as Genetic algorithm, 3.2.2 as Ant colony optimization, 3.2.3 as particle swarm optimization. Hybrid methods in Sect.4. The final conclusion is reported in Sect. 5.

2 Heuristic Method

Due to the high computational complexity of the graph coloring problem there is a need of the heuristic methods to determine the suboptimal solutions in polynomial time. The relevant characteristics of heuristic methods include not only computational complexity but also accuracy and level of complexity of graphs for which the method leads to suboptimal solutions. There are different types of heuristic methods. Among them sequential coloring approaches are the simplest heuristic methods. Welsh and Powell [41] proposed Largest-Fit (LF) method in which vertices of a high degree are colored in first priority. LF method can be implemented to run in O (m+n) time by allowing a more flexible choice of color for the vertices of low degree. Despite of its simplicity this method is very effective. Matula et al. [33] proposed a heuristic method alike the LF method named as Smallest Last (SL) method. It is also based on the idea of color vertices with high degree first but it does not have certain faults of LF algorithm such as SL optimally color trees, cycles, unicyclic graph, wheels, complete bipartite graph. The SL method can also be implemented to run in O (m+n) time. Although both the methods are easy to implement and fast by nature but they are quite inefficient in optimal coloring. To improve the coloring of these sequential algorithms, interchanges are executed. By performing interchanges, a previously colored vertex is switched to another class, by allowing the current vertex to be colored without adding new color. Hence Largest Fit with Interchange (LFI) and Smallest Fit with Interchange (SLI) are the final method [33]. By performing interchanges, performance was improved but they are more time consuming. Johnson [28] proposed a heuristic method known as greedy independent sets (GIS) method which is an implementation of the maximum independent set algorithm. In this method, the vertices of graph are analyzed in a certain order and the vertex is assigned a color if the vertex is not adjacent to any vertex with the same color. The GIS method can be implemented to run in O (mn) time. Another technique introduced by Brlaz [4] was based on the idea of reordering the nodes at each stage known as Degree of Saturation (DSATUR) or Saturation LF (SLF). In this, saturation degree is the term by which vertex to color next is chosen. A vertex with maximum saturation degree is given the priority to be placed in the first legal color class. This method can be implemented to run in O ((m+n) log n)) time. All the methods discussed above are based on the idea of choosing a vertex first and then assigning an appropriate color. However there is a more successful method proposed by Leighton [18] known as Recursive Largest Fit (RFL) which is based on the idea where each color is completed before introducing a new one. In this method, vertices of one class got selected at a time. Randomization is also carried to improve the performance of simple heuristics. This concept is reflected in the work proposed by Johnson [10] who introduced the XRLF method. In this method, for each color many candidate classes are created and one is selected with the least degree in the remaining graph.

3 Metaheuristic Methods

Metaheuristic is another way to solve graph coloring problem. Metaheuristic can be defined as high level strategies for exploring search space by using different methods. In the context of graph coloring we can have two types of metaheuristic one is local search method and another is population based method. The local search method includes simulated annealing, tabu search and population based method includes genetic algorithm, ant colony optimization, particle swarm optimization.

3.1 Local Search Methods

Local search is a metaheuristic method for solving computationally hard optimization problems. Local search methods start with a complete assignment of a value to each variable and try to iteratively improve this assignment by improving steps, by taking random steps, or by restarting with another complete assignment. Local search methods can be of different types such as simulated annealing [29], tabu search [17], variable neighborhood search [34], variable search space [21], iterated local search [7] and large scale neighborhood search [40].

3.1.1 Simulated Annealing

In 1987, Simulated annealing (SA) was first applied to graph coloring problem by Chams, Hertz and Werra [31]. In this, initially a neighboring solution is selected. If the neighboring solution is better than the current solution it will be accepted as the starting solution. If it is not better it will be accepted with a certain probability that gradually decreases with a global parameter called Temperature. Initially the temperature is high and it accepts all the solution but gradually temperature decreases which results in accepting only best solution. This process allows the algorithm to avoid local maximum. In 1991, Johnson et al. [10] intensively tested simulated annealing on random graphs. In order to evaluate and compare several graphs coloring heuristics, three simulated annealing heuristic based on three different strategies such as K-fixed penalty strategy, the proper strategy and the penalty strategy. Experiment shows that none of these strategies dominated clearly the others. In 2007, Szymon et al. [38] proposed parallel annealing (PSA). In this, the synchronous master slave model with periodic solution update is being used. Final results are compared with the Parallel genetic algorithm (PGA). Experiment shows that PSA achieved a similar performance level as the PGA. In 2009, Huberto et al. [23] proposed a way to transform the graph coloring problem into satisfiability (SAT) problem. Then a new approach that uses the threshold accepting algorithm (a variant of SA) and the Davis and Putnam algorithm was proposed. This is done because SA is not a complete algorithm and it's not always gets the optimal solutions. The resulting algorithm is a complete and obtains better results than the well known SA algorithm.

3.1.2 Tabu Search

In 1987, Hertz and Werra [1] were the first who applied tabu search (TS) known as Tabucol for finding a solution to the graph coloring problem using k-fixed penalty strategy. In this, instead of making a whole configuration tabu, the only single move is made. The algorithm is compared with simulated annealing and experiment shows that tabu search outperforms simulated annealing on different instances of random dense graph performance. In 1998, tabu search's peak performance is obtained by Dorne and Hao [14]. In their paper a generic tabu search is introduced for three coloring problems such as graph coloring, T coloring and set T coloring. In this the main concept is to integrate features such as greedy initialization, solution regeneration, dynamic tabu tenure, incremental evaluation of solution and constraint handling. Author compared the proposed algorithm with the well known DSATUR and found that the proposed algorithm is more efficient and robust. In 2008, Zufferey et al. [24] proposed a variant of Tabucol known as Partialcol which has a reactive component for adjusting the length of the time certain moves are excluded based on the fluctuations of the objective function and applied it with k coloring approach. The results are competitive when compared with other algorithms. In 2009, Porumbel et al. [11] proposed a new local search algorithm known as position guided tabu search (PGTS) heuristic which besides avoiding local optima also avoids revisiting candidate solution in previously visited regions. A learning process, based on a metric of the search space guides the tabu search towards yet uncensored regions. When compared with other better known local search algorithm PGTS proved to be effective.

3.1.3 Other Methods

Other local search methods include variable neighborhood search and variable search space. In 2003, Avanthay et al. [5] proposed a variable neighborhood search (VNS) algorithm which uses Tabucol as a local search operator and applies different types of perturbation operators. For analyzing the performance of the proposed VNS algorithm they compared the result obtained from Tabucol and the genetic hybrid algorithm (GH) proposed by Galinier and Hao which combine a tabu search with a genetic algorithm. The results achieved are VNS algorithm produces better results than Tabucol but VNS algorithm is not competitive with GH. In 2008, Hertz et al. [21] proposed variable search space (VSS-COL) as a new local search coloring heuristic. The VSS-Col heuristic alternates the use of three different local search heuristics that adopt different strategies: TABUCOL, PARTIALCOL, and a third tabu algorithm [20], which is not competitive by itself but complements adequately the two others. Experiment shows that VSS obtained better results when compared with local search heuristic.

3.2 Population Based Method

Population based methods are a kind of metaheuristic which deals with the set of populations. Population-based algorithms provide a natural, intrinsic way for the exploration of the search space. Population based methods are of different types such as genetic algorithm [25], ant colony optimization [13], particle swarm optimization [27].

3.2.1 Genetic Algorithm

Evolutionary algorithms have been adapted in the context of the graph coloring problem. The first algorithm based on genetic and evolutionary principles was developed in 1991 by Davis [30]. In this paper, genetic algorithm (GA) is used to encode solution as permutations of the vertices known as order based encoding. Then a greedy algorithm is applied to evaluate a solution. Experiment shows that this algorithm is competitive with greedy algorithm but when compared to other algorithm its performance is not competitive. In 1995, Costa et al. [12] were the first who published results of an experiment with genetic local search (GLS). Author tested this evolutionary descent method on random graph and found that it performed better than Tabucol. In 1996, Fleurent and Ferland [6] also experimented with GLS but they used tabu search as a local search operator, instead of a descent method used by Costa et al. Experiment shows that GLS outperforms Tabucol. In 2010, a new evolutionary algorithm known as memetic algorithm (MACOL) was proposed by Zhipeng et al. [42] which uses an adaptive multiparent cross operator (AMPaX) and a distance and quality base replacement criterion for pool updating. Experiment shows that MACOL obtains very competitive results on many of the benchmark graphs.

3.2.2 Ant Colony Optimization

In 1997, Costa and Hertz [9] were the first to apply an ant colony optimization (ACO) to the graph coloring problem. In their work they introduced ANTCOL which embed two graphs coloring constructive heuristics RLF and DSATUR called as ANTRLF and ANTDSATUR respectively. The experiment showed that the result obtained of ANTCOL based on RLF outperforms those based on DSATUR. In 2008, Salari and Eshghi [15] proposed a modification of ANTCOL, a Max-Min ant system algorithm for graph coloring (MMGC) to improve the performance of ANTCOL. The experiment showed that result achieved by MMGC is quite better than by ANTCOL. In 2010, Plumettaz et al. [32] proposed ant local search coloring method (ALS-COL). In most ACO based algorithms, the role of each ant is to create a solution in a constructive way while in this paper the authors proposed the concept of a local search. In the proposed scheme, each ant performs local search and at each step updates the current solution by the use of greedy force. Experiment results show that ALS-COL outperforms PARTIALCOL and other ant colony optimization heuristics for graph coloring problem.

3.2.3 Particle Swarm Optimization

In 2008, Cui et al. [8] applied particle swarm optimization (PSO) to solve the graph coloring problem. In this a modified particle swarm optimization was added to the disturbance factor to improve the performance of the algorithm. The experiment showed that the performance of the modified PSO is better than that of classical PSO. In 2010, Mostafa et al. [16] proposed hybrid algorithm which uses a recombination operator. Author proposed a modified PSO with fuzzy logic to obtain a high performance algorithm for solving planer graph coloring problem. Experiment shows better result and less complexity of time and storage. In 2011, Hsu et al. [22] added a modified turbulence to previous PSO, to solve the graph coloring problem. The proposed model consists of walking one strategy, assignment strategy and turbulent strategy. It solves the planer graph coloring problem using four colors more effectively and accurately.

4 Hybrid Methods

In 1999, Hao and Galinier [35] proposed a hybrid coloring algorithm (HCA) that uses tabu search and genetic algorithm. HCA uses the greedy partition crossover (GPX) operator which combines color classes instead of specific color assignments. Hybrid approach proves to be very powerful. In 2004, Lim and Wang [2] applied various metaheuristic to solve robust graph coloring problem (RGCP). Metaheuristic algorithms used are genetic algorithm, simulated annealing and tabu search. Experimental results on various sizes of input graph provide the performance of these meta-heuristics in terms of accuracy and run time. In 2005, Sivanandam et al. [37] proposed a new permutation based representation of the graph coloring problem. In this a migration model of parallelism for genetic algorithm (PGA) is used with Message Passing Interface (MPI). In addition, three crossover operators are used to name as a Greedy partition crossover (GPX), Uniform independent set crossover (UISX) and Permutation based crossover (PX) are used .The experiment showed that GPX works well in context of convergence and PX in context of execution time. In 2010 Ray et al. [3] proposed a combination of evolutionary algorithm known as genetic algorithm with multi point Guided Mutation for the graph coloring problem. In this, a new operator called double point guided mutation operator is used to increase the performance level of the simple genetic algorithm dramatically. In 2011, David [36] proposed two new metaheuristic algorithms for graph coloring algorithm.

One is population based multiagent evolutionary algorithm (MEA) using a multiagent system where an agent represents a tabu search procedure. The second is a pseudo reactive tabu search (PRTS) introducing a new online learning strategy. Both algorithms empirically outperform basic tabu search algorithm Tabucol on the well established DIMACS instances. In 2011, Qin et al. [26] proposed a hybrid discrete particle swarm algorithm (HPSO) to solve the graph coloring problem. In this, initially a general discrete PSO algorithm is proposed. Then a hybrid discrete PSO algorithm is proposed by combining a local search known as Tabucol. Experiment with a set of eight DIMACS benchmarks was conducted and the computational results show that HPSO is feasible and competitive with other well-known algorithms. In 2011, Titiloye et al. [39] proposed an effective quantum annealing algorithm to solve the k-coloring. The work has been inspired from the quantum mechanics. Compared with simulated annealing, it includes an additional parameter Γ (with the classical temperature parameter). While simulated annealing evolves in a neighborhood of constant radius, Γ is used here to modify the radius of the neighborhood (to control diversity) and to reinforce the evaluation function with a "kinetic" energy relying on interactions between replicas (roughly speaking, it quantifies the similarity of replicas). This kinetic energy aims to help escaping local optima. Experiments show remarkable results on some tested DIMACS graphs.

5 Conclusion

In this paper, a survey of the graph coloring problem is shown. We have divided the survey into three parts heuristic methods, population based methods and hybrid methods. Heuristic are very fast by nature but there results are not satisfactory in terms of quality. Best known method in this category includes DSATUR and RLF. Metaheuristic are very efficient in producing the best results for a large class of graph instances. Among metaheuristic, local search is a key ingredient for coloring graph. Local search methods are applicable for solving small or medium size instances graph. When the graph size becomes large the results are far away from optimal so the population based method are best applicable for solving such problems. Population based methods provide a natural way for the exploration of the search as it deals with the set of solutions.. Hybrid methods are efficient in finding good solutions that cannot be obtained by any complete methods within a feasible time.

References

1. Hertz, A., de Werra, D.: Using tabu search techniques for graph coloring. Comput. 39(4), 345–351 (1987)
2. Lim, A., Wang, F.: Meta-heuristics for robust graph coloring problem. In: Proceedings of 16th IEEE International Conference on Tools with Artificial Intelligence, Florida, pp. 514–518 (2004)
3. Ray, B., Pal, A.J., Bhattacharyya, D., Kim, T.H.: An Efficient GA with Multipoint Guided Mutation for Graph Coloring Problems. Int. J. Signal Process. Image Process. and Pattern Recognit. 3(2), 51–58 (2010)

4. Brelaz, D.: New methods to color the vertices of a graph. Commun. ACM. 22, 251–256 (1979)
5. Avanthay, C., Hertz, A., Zufferey, N.: A variable neighborhood search for Graph coloring. Eur. J. Oper. Res. 151(2), 379–388 (2003)
6. Fleurent, C., Ferland, J.A.: Genetic and hybrid algorithms for graph coloring. Ann. Oper. Res. 63(3), 437–461 (1996)
7. Chiarandini, M., Stutzle, T.: An application of iterated local search to graph coloring. In: Johnson, D.S., Mehrotra, A., Trick, M. (eds.) Proc. of the Computational Symposium on Graph Coloring and its Generalizations, Ithaca, New York, USA, pp. 112–125 (2002)
8. Cui, G., Qin, L., Liu, S., Wang, Y., Zhang, X., Cao, X.: Modified PSO algorithm for solving planar graph coloring problem. Progress Nat. Sci. 18, 353–357 (2008)
9. Costa, D., Hertz, A.: Ants Can Color Graphs. J. Oper. Res. Soc. 48, 295–305 (1997)
10. Johnson, D.S., Aragon, C.R., McGeoch, L.A., Schevon, C.: Optimization by simulated annealing:an experimental evaluation; part II, graph coloring and number partitioning. Oper. Res. 39(3), 378–406 (1991)
11. Porumbel, D.C., Hao, J.-K., Kuntz, P.: Position-Guided Tabu Search Algorithm for the Graph Coloring Problem. In: Stützle, T. (ed.) LION 3. LNCS, vol. 5851, pp. 148–162. Springer, Heidelberg (2009)
12. Costa, D., Hertz, A., Dubuis, C.: Embedding a sequential procedure within an evolutionary algorithm for coloring problems in graphs. J. Heuristics 1, 105–128 (1995)
13. Dorigo, M., Maniezzo, V., Colorni, A.: Positive feedback as a search strategy. Technical Report 91-016, Politecnico di Milano, Italy (1991)
14. Dorne, R., Hao, J.K.: Tabu Search for graph coloring, T-coloring and Set T-colorings. In: Osman, I.H., et al. (eds.) Metaheuristics 1998: Theory and Applications. ch. 3. Kluver Academic Publishers (1998)
15. Salari, E., Eshghi, K.: An ACO Algorithm for the Graph Coloring Problem. Int. J. Contemp. Math. Sci. 3, 293–304 (2008)
16. Erfani, M.: A modified PSO with fuzzy inference system for solving the planar graph coloring problem. Masters thesis, Universiti Teknologi Malaysia, Faculty of Computer Science and Information System (2010)
17. Glover, F.: Future paths for integer programming and links to artificial intelligence. Comput. Oper. Res. 13, 533–549 (1986)
18. Leighton, F.T.: A graph coloring algorithm for large scheduling problems. J. Res. Natl. Bur. Stand. 84(6), 489–505 (1979)
19. Garey, R., Johnson, D.S.: A guide to the theory of NP–completeness. Computers and intractability. W. H. Freeman, New York (1979)
20. Gendron, B., Hertz, A., St-Louis, P.: On edge orienting methods for graph coloring. J. of Comb. Optim. 13(2), 163–178 (2007)
21. Hertz, A., Plumettaz, M., Zufferey, N.: Variable space search for graph coloring. Discret. Appl. Math. 156(13), 2551–2560 (2008)
22. Hsu, L., Horng, S., Fan, P.: Mtpso algorithm for solving planar graph coloring problem. Expert Syst. Appl. 38, 5525–5531 (2011)
23. Ayanegui, H., Chavez-Aragon, A.: A complete algorithm to solve the graph-coloring problem. In: Fifth Latin American Workshop on Non-Monotonic Reasoning, LANMR, pp. 107–117 (2009)
24. Blochliger, I., Zufferey, N.: A graph coloring heuristic using partial solutions and a reactive tabu scheme. Comput. Oper. Res. 35(3), 960–975 (2008)
25. Holland, J.H.: Adaption in natural and artificial systems. The University of Michigan Press, Ann Harbor (1975)

26. Qin, J., Yin, Y.-X., Ban, X.-J.: Hybrid discrete particle swarm optimization for graph coloring problem. J. Comput. 6, 1175–1182 (2011)
27. Kennedy, J., Eberhart, R.C.: Particle swarm optimization. In: Proc. of IEEE Int. Conf. Neural Netw., Piscataway, NJ, USA, pp. 1942–1948 (1995)
28. Johnson, D.S.: Approximation algorithms for combinatorial problems. J. Comp. Syst. Sci. 9, 256–278 (1974)
29. Kirkpatrick, S., Gelatt, C.D., Vecchi, M.P.: Optimization by simulated annealing. Sci. 220, 671–680 (1983)
30. Davis, L.: Order-based genetic algorithms and the graph coloring problem. In: Handbook of Genetic Algorithms, pp. 72–90 (1991)
31. Chams, M., Hertz, A., Werra, D.: Some experiments with simulated annealing for coloring graphs. Eur. J. of Oper. Res. 32(2), 260–266 (1987)
32. Plumettaz, M., Schindl, D., Zufferey, N.: Ant Local Search and its effcient adaptation to graph colouring. Journal of Operational Research Society 61(5), 819–826 (2010)
33. Matula, D.W., Marble, G., Isaacson, D.: Graph coloring algorithms. In: Graph Theory and Computing, pp. 109–122. Academic Press, New York (1972)
34. Mladenovic, N., Hansen, P.: Variable Neighborhood Search. Comput. Oper. Res. 24, 1097–1100 (1997)
35. Galinier, P., Hao, J.K.: Hybrid evolutionary algorithms for graph coloring. J. Comb Optim. 3(4), 379–397 (1999)
36. Chalupa, D.: Population-based and learning-based metaheuristic algorithms for the graph coloring problem. In: Krasnogor, N., Lanzi, P.L. (eds.) GECCO, pp. 465–472. ACM (2011)
37. Sivanandam, S.N., Sumathi, S., Hamsapriya, T.: A hybrid parallel genetic algorithm approach for graph coloring. Int. J. Knowl. Based Intel. Eng. Syst. 9, 249–259 (2005)
38. Lukasik, S., Kokosinski, Z., Swieton, G.: Parallel Simulated Annealing Algorithm for Graph Coloring Problem. Parallel Process. Appl. Math., 229–238 (2007)
39. Titiloye, O., Crispin, A.: Quantum annealing of the graph coloring problem. Discret. Optim. 8(2), 376–384 (2011)
40. Trick, M.A., Yildiz, H.: A Large Neighborhood Search Heuristic for Graph Coloring. In: Van Hentenryck, P., Wolsey, L.A. (eds.) CPAIOR 2007. LNCS, vol. 4510, pp. 346–360. Springer, Heidelberg (2007)
41. Welsh, D.J., Powell, M.B.: An upper bound for the chromatic number of a graph and its application to timetabling problem. Comp. J. 10, 85–86 (1967)
42. Lu, Z., Hao, J.-K.: A memetic algorithm for graph coloring. Eur. J. Oper. Res. 203(1), 241–250 (2010)

A Survey on Anomaly Detection in Network Intrusion Detection System Using Particle Swarm Optimization Based Machine Learning Techniques

Khushboo Satpute[1], Shikha Agrawal[2], Jitendra Agrawal[1], and Sanjeev Sharma[1]

[1] School of Information Technology, Rajiv Gandhi Proudyogiki Vishwavidyalaya, Bhopal
[2] University Institute of Technology, Rajiv Gandhi Proudyogiki Vishwavidyalaya, Bhopal
Khushboosatpute88@gmail.com, {shikha,jitendra,sanjeev}@rgtu.net

Abstract. The progress in the field of Computer Networks & Internet is increasing with tremendous volume in recent years. This raises important issues with regards to security. Several solutions emerged in the past which provide security at the host or network level. These traditional solutions like antivirus, firewall, spyware & authentication mechanism provide security to some extends but they still face the challenges of inherent system flaws & social engineering attacks. Some interesting solution emerged like Intrusion Detection & Prevention Systems but these too have some problems like detecting & responding in real time & discovering novel attacks. Several Machine Learning techniques like Neural Network, Support Vector Machine, Rough Set etc. Were proposed for making an efficient and Intelligent Network Intrusion Detection System. Also Particle Swarm Optimization is currently attracting considerable interest from the research community, being able to satisfy the growing demand of reliable & intelligent Intrusion Detection System (IDS). Recent development in the field of IDS shows that securing the network with a single technique proves to be insufficient to cater ever increasing threats, as it is very difficult to cope with all vulnerabilities of today's network. So there is a need to combine all security technologies under a complete secure system that combines the strength of these technologies under a complete secure system that combines the strength of these technologies & thus eventually provide a solid multifaceted well against intrusion attempts. This paper gives an insight into how Particle Swarm Optimization and its variants can be combined with various Machine Learning techniques used for Anomaly Detection in Network Intrusion Detection System by researchers so as to enhance the performance of Intrusion Detection System.

Keywords: Particle Swarm Optimization, Anomaly Detection, Machine Learning, Supervised Learning, Intrusion Detection.

1 Introduction

In recent year, tremendous increase in the use of internet added an exponential development of interest of people that brings complicated problems and pressure of computer security. Traditional security policies or firewall has difficulty in preventing attacks because of hidden vulnerabilities contained in software application. Some

S.C. Satapathy et al. (Eds.): Proc. of Int. Conf. on Front. of Intell. Comput., AISC 199, pp. 441–452.
DOI: 10.1007/978-3-642-35314-7_50

reliable solution must be available to protect our computer from cyber-attacks & criminal activities, therefore Intrusion Detection System despite the prevention techniques has set a perfect platform to defend the confidentiality, integrity and security aspects of cyber world. IDS analyzes information about users' behavior from various sources such as system table and network usage data .Since the first Intrusion Detection System [1] was proposed, effort had been made to boost IDS efficiency. It deals with huge amounts of data causing slow training and testing process & low detection rate. Thus construction of efficient intrusion detection model is a challenging task. While constructing IDS one needs to consider many issues such as data collection, data preprocessing, intrusion recognition, reporting and response. Artificial Intelligence and Machine Learning technique were used to discover the underlying models from a set of training data & detection model is applied in the execution of some critical procedure such as differentiating between normal and abnormal behavior, but all these fails to achieve high detection accuracy and fast processing and has their own pros and cons, so there is still a need of an efficient IDS. This paper gives a brief overview of the work done in the field of Anomaly Detection in Network Intrusion Detection System (NIDS) using Particle Swarm Optimization based Machine Learning techniques.

Next section covers a brief about IDS, their types and various learning techniques used for classification in Intrusion Detection System. Section 3 introduces PSO & in section 4 work related to Anomaly Detection in Network Intrusion Detection System using Particle Swarm Optimization based Machine Learning techniques is discussed in brief.

2 Intrusion Detection System

Intrusion Detection System (IDS) has quickly established as the most important element of security infrastructure. Intrusion is an attempted act of using computer system resources without privileges, causing incidental damage. Intrusion Detection is the process of monitoring the events occurring in a computer system or network and analyzing them to sign in possible incidents. An ID monitors network traffic, monitoring the events occurring in a computer system or network and analyzing them for sign in possible incidents. If it detects any threat then alerts the system or network administrator. There are two performance evaluation variables criteria, Detection Rate (DR) which is defined as the ratio of number of correctly detected attacks to the total number of attacks & False alarm rate(FAR) which is ratio of the number of normal connection that are misclassified as attacks to total number of normal connections. Intrusion Detection must be able to identify intrusion with high accuracy and it must not confuse normal action with the occurrence of a system with intrusive ones. Construction of efficient Intrusion Detection is a challenging task so it must have a high attack Detection Rate (DR) with low False alarm rate (FAR) at the same time.

2.1 Type of Intrusion Detection System

Intrusion Detection System is broadly classified on the basis of following two criteria

(i) Based on Data Collection mechanism
(ii) Based on Detection Techniques

On the basis of data collection mechanisms IDS is categorized into Host-Based Intrusion Detection System (HIDS) and Network-Based Intrusion Detection System (NIDS). Host-based IDS is dependent for support on capturing local network traffic to the specific host. This local host analyzes and process data which is used to secure the activities of this host and informs about the attacks in the network. HIDS analysis events mainly related to OS information. Network-based intrusion detection system (NIDS) works on network and observes the network traffic. NIDS analyses network related traffic volumes, IP address service port etc. which are able to detect attack from outside, examine packet header and entire packet.

On the basis of detection techniques IDS is classified as Misuse Detection and Anomaly Detection. Misuse Detection it involves searching network traffic for a series of malicious activity within the analyzed data. The main advantage of this technique is that it provides very good detection results for specified, well known attacks & is very easy to develop and understand. However they are not capable of detecting novel attacks. Anomaly intrusion detection system (AIDS) uses normal usage behavior patterns to identify the intrusion. The normal usage patterns are constructed from the statistical measures of the system features. While the anomaly behavior detecting system generates a standard traffic sketch & employs it to detect any abnormal traffic pattern and attempts of intrusion. The three main vital factor's that impact the quality anomaly detection is Feature Selection, Data value normalization and Classification technique. According to the type of processing related to the "Behavioral" model of the target system [2], Anomaly Detection Techniques can be classified into three main categories [3]: Statistical based, Knowledge-based, and Machine Learning based. In the Statistical-based, the behavior of the system is represented by a random view point. On the other hand, Knowledge-based Anomaly network intrusion detection techniques try to capture the claimed behavior from available system data (protocol specifications, network traffic instances, etc.). Finally Machine learning techniques are based on establishing an explicit or implicit model that enables the patterns analyzed to be categorized. The comparison of all the three AIDS as shown in Table 1. As our research is on Machine learning based Anomaly detection system so in the next section a short introduction of Machine Learning Techniques used in Anomaly Intrusion Detection System is described.

Table 1. Comparison of All the Three AIDS

Technique	*Advantages*	*Disadvantage*
Statistical-based:- stochastic behaviour	*Future knowledge about normal activity is not required. Exact and accurate notification about intruder's activities.*	*Parameters and metrics are very difficult to set. Easily influenced could be trained by attackers.*
Knowledge-based:-	*Robustness. Flexibility and scalability.*	*Difficult and time-consuming availability for high-quality knowledge/data.*
Machine learning-based:-	*Flexibility and adaptability. Capture of interdependencies*	*High dependency on the assumption about the behaviour accepted for the system. High resource consuming.*

2.2 Learning Techniques

Learning or training is a process by means of which a network adapts itself to a stimulus by making proper parameter adjustments resulting in production of desired responses. The learning technique can generally classified into two categories as Unsupervised learning and Supervised learning.

Unsupervised algorithm seeks out similarities between pieces of data in order to determine whether they can be characterized as forming a group. These groups are termed clusters, and there are whole families of clustering machine learning techniques. In unsupervised classification, often known as 'cluster analysis' the machine is not told how the texts are grouped. Example of unsupervised learning is the self-organizing map (SOM) and Adaptive Resonance Theory (ART). Supervised Machine Learning is the search for algorithms that reason from externally supplied instances to produce general hypotheses, which then make predictions about future instances. In other words, the goal of supervised learning is to build a concise model of the distribution of class labels in terms of predictor features. The resulting classifier is then used to assign class labels to the testing instances where the values of the predictor features are known, but the value of the class label is unknown. Examples of supervised learning are Rough Set, Support Vector Machine and Neural Network.

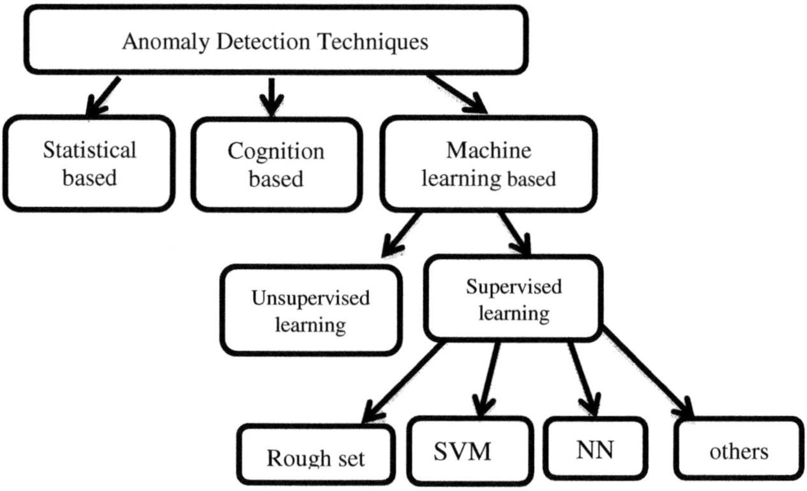

Fig. 1. Anomaly Intrusion Detection Technique

3 Particle Swarm Optimization

Computational Intelligence is the study of the design of intelligent agents. It encompasses Artificial Neural Networks, Fuzzy sets, Evolutionary computation methods, Artificial immune systems, Swarm intelligence and Soft computing. Computational intelligence is known for their ability to adapt and to exhibit fault tolerance, high computational speed and resistance against noisy information. [4] PSO is involved from Swarm intelligence which is an Computational intelligence technique involving the study of collective behavior in decentralized system. In the PSO algorithm, a point in the search space (i.e., a possible solution) is called a particle. The collection of particles in a given iteration is referred to as the swarm. The terms "particle" and "swarm" are analogous to "individual" and "population" used in Evolutionary algorithms such as GAs. At each iteration, each particle in the swarm moves to a new position in the search space. The velocity and position updating equations are:-

$$Vid = wVid + C1Rand(\,)(Pid\text{-}Xid) + C2Rand(\,)(Pgd\text{-}Xid) \quad (1)$$

$$Xid = Xid + Vid \quad (2)$$

Where C1 and C2 are positive constants called learning rates. These represent the weighting of the stochastic acceleration terms that pull each particle towards its pbest and gbest positions. Low values allow particles to roam far from target regions before being tugged back, while high values result in abrupt movement toward, or past target regions. Rand ( ) and Rand ( ) are two random functions in the range [0,1] and w is the inertia weight. Suitable selection of the inertia weight provides a balance between global and local exploration, and results in less iteration on average to find a sufficiently optimal solution. Xi = (xi1, xi2, … , xid) represents the ith particle and Pi = (pi1, pi2, … , pid) represents the best previous position of the ith particle. Vi = (vi1, vi2, … , vid) represents the rate of the position change (velocity) for particle i. PSO is effective in nonlinear optimization problems, it is easy to implement and only a few input parameters needed to be adjusted. Because the update process in PSO is based on simple equations, it can be efficiently used on large data sets .Due to these advantages PSO has been successfully applied to many areas such as function optimization, artificial neural network training, fuzzy system control and all other areas where GA can be applied. Next section describes hybridization of PSO with some of these supervised machine learning classification techniques such as Neural Network (NN), Support Vector Machine (SVM) & Rough Set.

4 Particle Swarm Optimization Based Machine Learning Oriented Network Anomaly Detection System

4.1 Anomaly Detection Using Rough Set & Particle Swarm Optimization Based Approach

Rough Set is a mathematical tool for approximate reasoning for decision support and is particularly well suited for classification of objects. It is an extension of the conventional set theory that supports approximations in decision making & is also being used for feature selection and feature extraction. Most existing IDS use all features in network packet to evaluate, which is a lengthy detection which may degrade the performance of IDS. The effectiveness of rough set theory in intrusion detection is studied by Zainal et. al [5] Feature selection is done prior to training and is applied to classify the data to evaluate the performance. There are feature that is really significant in classifying the data & it also has been proven that there is no single generic classifier that can best classify all the attack types. So to enhance the performance, Rough Set in intrusion detection is combined with different supervised learning such as ANN, SVM & PSO by several researchers. In this section work related to hybridization of rough set with PSO in IDS is discussed.

Zainal et.al [6] used wrapper approach where integration of Rough Set and Particle Swarm Optimization is used to form a 2-tier architecture of feature selection process. At the first stage Rough Set is applied to eliminate redundant and irrelevant features thus reduce number of iterations that Discrete Particle Swarm Optimization (DPSO) has to perform in the next stage to find the optimum feature subset and Support Vector Machine (SVM) classifier is then used to classify the data & the fitness function. Based on the datasets used for the experiment, the results indicate that the feature subset proposed by this hybridization is superior in terms of accuracy and robustness.

Another method proposed by Tian.W et al. [7] proposed Intrusion Detection Method based on Neural Network & PSO Algorithm along with Rough Set. Rough Set is used as a preprocessor of Artificial Neural network (ANN) to select a subset of input attributes and PSO is employed to optimize the parameters of ANN and thus improve the ANN performance in intrusion detection. Experiment shows that the proposed method has higher stability, higher detecting and recognition accuracy.

Similarly Liu.H et.al[8] proposed an intelligent Intrusion Detection method based on Rough Set Theory (RST) and Improved Binary Particle Swarm Optimization with Support Vector Machine (IBPSO-SVM), which combined attribute reduction with parameters optimization. In this first Rough Set Theory is applied to subtract redundant and noisy attributes & to reduce the attribute space of training & test datasets. Then improved BPSO-SVM is applied to optimize parameters in SVM so as to improve the accuracy of SVM classifier. The main purpose of IBPSO-SVM is parameter optimization, so the reduced training dataset is input to search the optimal penalty parameter and kernel parameter respectively. The experimental result on KDD CUP'99 dataset shows that the proposed method is an effective way for intrusion detection, by not only accelerating the training time, but also improving the accuracy of test.

In Wang.H et.al [9] proposed an Intrusion detection reduction model based on Particle Swarm Optimization, in which QPSO is applied to Rough Set attribute reduction algorithm along with Monte Carlo method to simulate the particle position on the measure of quantum uncertainty. The algorithm is faster than GA and has a high rate of network intrusion detection.

4.2 Anomaly Detection Using Neural Network and Particle Swarm Optimization Based Approach

Artificial Neural Network (ANN), coming from the inspiration of biological neural systems, has been successfully applied to a large diversity of application. Unfortunately, these ANN has some inherent defects, such as low learning speed, the existence of local minima, and difficulty in choosing the proper size of the network to suit a given problem. To solve these defects, different variants of neural network were proposed such as Wavelet Neural Network(WNN)& Radial Base Function(RBF).There are lots of training algorithms for training of Neural Network, but all of these algorithms have their disadvantages. Evolutionary algorithm has strong ability of global convergence and strong robustness, and need not be with the feature information, such problems as the gradient derivative. Therefore, its application in the Neural Network learning algorithm, not only can play Neural Network's generalization ability, and the mapping can improve the convergence rate of the neural network and learning ability. PSO is recently applied to train the Evolutionary Algorithm .So many researchers applied neural network in the section 4.2.1 and section 4.2.2 work related to the training of WNN & RBF using Particle Swarm Optimization (PSO)& its variants applied to Network Intrusion Detection System is discussed in detail.

4.2.1 Anomaly Detection Using Wavelet Neural Network and Particle Swarm Optimization Based Approach

Wavelet neural network (WNN) is a combination of wavelet theory with Neural Network. WNN is established as a three-layer structure with input layer, hidden layer & output layer. The wavelet neural network uses nonlinear wavelet bases instead of usually neuron nonlinear motivation function.

Liu. L et.al [10] introduced a novel approach for Anomaly Detection in Network Intrusion Detection System based on Wavelet Neural Network (WNN) using Modified Quantum-Behaved Particle Swarm Optimization (MQPSO) algorithm. The algorithm is trained using Morlet Wavelet. A multidimensional vector composed of WNN parameters was regarded as a particle in learning algorithm. The parameter vector, which has a best adaptation value, was searched globally. The experiment result reveals that the algorithm proposed better training performance, faster convergence as well as better detection rate.

Yuan. L et.al [11] proposed a novel hybrid to optimize Wavelet Neural Network for Network Intrusion Detection System. This new Evolutionary algorithm, which is based on a hybrid of Quantum-Behaved Particle Swarm Optimization (QPSO)and Conjugate Gradient algorithm (CG), is employed to train WNN. In the beginning of run, QPSO has more possibilities to explore a large space and therefore the particles are free to move and sit on various valleys, but as the search progresses it is difficult for QPSO to find a global optimum so the Fletcher-Reeves Conjugate Gradient algorithm is employed. The experiment result of the hybrid algorithm trained WNN on network anomaly detection with the dataset of KDD CUP99 shows that the hybrid algorithm has a better training performance, faster convergence, as well as a better detecting ability for new unknown type attacks.

[12] Liu.Y et al. proposed another anomaly detection method in which Modified QPSO is used to train Wavelet Fuzzy Neural Network(WFNN).Wavelet transform is applied to extract fault characteristics from the anomaly state. In this novel evolutionary technique, a modified QPSO is employed to train WFNN, a decision vector that represents a group of network parameter is initialized, then WFNN is trained on training set and evaluate the fitness value of each particle and update pbest and gbest across population accordingly. Experimental result shows MQPSO-WNN model exhibits superior performance with higher attack detection rate and lower false positive rate.

4.2.2 Anomaly Detection Using Radial Base Function-Neural Network (RBF-NN) & Particle Swarm Optimization Based Approach

Radial Basis Function (RBF) Neural Network is a kind of feed forward neural network. In RBF neural network, the center of radial basis function, the variance of radial basis of function and the weight have to be chosen. If they are not appropriately chosen, the RBF neural network may degrade validity and accuracy of modeling. So PSO is used to optimize the RBF neural network parameters.

In Chen. Z al.[13] PSO is used to optimize RBF-NN Parameters for NIDS by evaluating fitness function, updating particle velocity and position and judging termination criteria. PSO has proved to be competitive with genetic algorithm in parameter optimization. Compared with the results of the conventional RBF neural network model, the experimental results show that the proposed model is superior to the conventional RBF neural network.

A Novel hybrid algorithm [14] based on Radial Basis Function (RBF) neural network is proposed by Yuan. Liu* et.al, for Network Anomaly Detection in which QPSO and Gradient Decent is employed to train RBF neural network. Comparison of RBFNN method using QPSO, GD and QPSO-GD is shown in which QPSO-GD out performs both QPSO, GD in global search ability.

Xu. Ruzhi et al. [15] introduces the hybrid classifier composed by Kernel Principal Component Analysis (KPCA), RBFNN and PSO. KPCA is used to reduce the dimensions of the original sample data. RBF module is core classifier that classifies the data and PSO module is used to optimize the parameter. The training dataset, which had been reduced dimensions by KPCA, was then inputted to RBFNN to get the classification model. The best parameters of classification model had been found by PSO iterations. In the process of intrusion detection experiments, It was reported that total classification accuracy is 98.95% and algorithm also founds global optimum parameter of RBFNN in parameter space.

4.3 Anomaly Detection Using Support Vector Machine and Particle Swarm Optimization Based Approach

The Support Vector Machine (SVM) is a supervised learning method from the field of machine learning applied to both classification and regression based on statistical learning theory. It can find a solution by making a nonlinear transformation of the original input space into a high dimensional feature space where an optimal separating hyper plane can be found, which means that a maximal margin classifier in relation to the training data set can be obtained. Support Vector Machine is effective in reducing the number of alerts, false positive, false negative better, parameter optimization in SVM is very important for its efficiency. A number of methods, such as grid search &evolutionary algorithms have been utilized to optimize the model parameters of SVM. So this section discusses the use of PSO for feature selection & parameter optimization in Network anomaly detection system for SVM.

In Tu. Chung et.al [16] Particle Swarm Optimization (PSO) is used to implement a feature selection, and then fitness values are evaluated with a Support Vector Machines (SVMs) which was combined with one-versus rest method for five classification problem. The Binary Particle swarm Optimization (BPSO) is used to serve as feature selection for classification problem. It helps to improve the performance owing to its smaller number of simple parameter settings. Kernel Adatron (KA) SVM is used to evaluate the fitness values of the PSO, which can be obtained by comparing the characteristic of the general test data .Experimental results show that proposed method simplified feature selection and the total number of parameters needed effectively, thereby obtaining a higher classification accuracy compared to other feature selection methods.

In Ma.Jing et al.[17] a New method of hybrid Intrusion Detection based on hybridization of Binary Particle Swarm Optimization (BPSO) and Support Vector Machine (SVM).Method is proposed for simultaneous feature selection and parameter optimization. In this combinatorial technique, parameters of SVM and dataset features are represented by every particle position (i.e., a binary series). The modified BPSO is used to obtain the best particle position quickly throughout the search space, which cooperates with SVM for evaluating the fitness of the corresponding particle. Consequently, the optimum features and parameters are chosen at the same time. The main purpose is to find out better parameters for SVM and a feature subset involving key features of network intrusion attacks based on the improved BPSO-SVM. Experimental results shows that technique will be useful to reduce the data quantity of large scale dataset and improve the classification ability of the classifier in IDS.

In Zhang et al.[18] presents a Hybrid Quantum Binary Particle Swarm Optimization (QBPSO)-SVM based network intrusion wrapper algorithm.. In QPSO each bit of particle is represented by quabit, which has two basic state'0' and '1'.The quantum superposition characteristic can make a single particle represent several states, thus potentially increases population diversity. The probability representation makes particle mutate according a certain probability to avoid local optimal. When experimented with the classical intrusion feature selection, it was found that there exist correlation relationship among network intrusion features, so Modified QBPSO based wrapper feature selection is superior to those classical intrusion feature selection methods. The paper reported that, the proposed method is an effective and efficient way for feature selection and detection when tested on the data sets of KDD cup 99.

New design of IDS was proposed in Zhou .J et al.[19]which presents optimal selection approach of the SVM parameters based on Particle Swarm Optimization algorithm. PSO parameters selection method not only to ensure that SVM learning ability but also to some extent, improved the generalization ability of SVM and performance of support vector machine classifier. The experimental result shows Particle Swarm Optimization and Support Vector Machine are effective in reducing the number of alerts, false positive, false negative better.

In Wang J et al.[20] Simple Particle Swarm Optimization (SPSO) is used to optimize the SVM model parameters and feature selection for IDS. Support vector machine (SVM) has been employed to provide potential solutions for the IDS problem. Firstly feature selection algorithm select important features, and then built intrusion detection systems using these selected features. The training data set is then separated into attack data sets and normal datasets, which are then subsequently, fed into the hybrid PSO-SVM algorithms. Experiment results show that proposed method is not only able to achieve the process of selecting important features but also to yield high detection rates for IDS.

4.4 Anomaly Detection Using other Machine Learning and Particle Swarm Optimization Approach

Some researches had also applied Machine Learning techniques other than SVM, Rough Set and Neural Network with variance of PSO in Network Anomaly Intrusion Detection System. Some of these methods are discussed below.

In Chen. Yet al.[21] proposed a novel method, in which enhanced Flexible Neural Tree(FNT) based on predefined intrusion operator sets, a Flexible Neural Tree model can be created and evolved. The framework allows input variable selection over layer connections and different activation functions for various nodes involved. The FNT structure is developed using Evolutionary Algorithm and parameters are optimized using Particle Swarm Optimization.

Similarily Chen.Y et al.[22]evaluates the performance of Estimates of Distribution Algorithm (EDA) to train a feed forward Neural Network classifier and Decision Tree, where EDA is a new class of EA's in which search is mainly based on global information about search space. Here Neural Network is trained using PSO. EDA-NN classification accuracy is greater than 95% as achieved good accuracy in true positives and false positive rates.

Another method described in Michailidis et.al [23] implemented and evaluated an Evolutionary Neural Network(ENN) in order to recognize known as well as new and unknown attacks. The analysis engine of the IDS is modeled by the ENN and its ability to predict attacks in a network environment is evaluated. The ENN is trained by a Particle Swarm Optimization (PSO) algorithm using labeled data from the KDD cup `99 competition. The results from the experiments are compared to the results by the same competition and give positive results in the recognition of DoS and Probe attacks.

In Gong. S et al.[24] proposed a novel approach to feature selection based on Genetic Quantum Particle Swarm Optimization(GQPSO) attribute reduction in Network Intrusion Detection. Selection and variation of genetic algorithm with QPSO algorithm a recombined to form GQPSO algorithm; normalized mutual information between attributes defined as GQPSO algorithm fitness function to guide its reduction of attributes to realize the optimal selection of network data feature subset. Experimental result shows that the approach is more effective than QPSO and PSO algorithms in discarding independent and redundancy attributes.

5 Conclusion

Intrusion Detection based upon Particle Swarm Optimization is currently attracting considerable interest from the research community, being able to satisfy the growing demand of reliable and intelligent Intrusion Detection Systems. The main advantage of PSO is that it is easy to implement & only a few input parameters are needed to be adjusted & is effective in nonlinear optimization problem. Also updation of velocity and position in Particle Swarm Optimization is based on simple equations so it can be efficiently used on large data sets. From the survey done in this paper it is revealed that there are several factors that affects the performance IDS. First is selection & extraction of relevant features. If all features are evaluated then it degrade the IDS performance, so to enhance the performance researchers uses several Supervised Machine Learning techniques each of which has its own pros and cons. Also it has been proven that there is no single generic classifier available that can classify all the attack types effectively so hybridization of different Supervised Machine Learning techniques is done by several researchers. Since the single article cannot be a complete review of the research done in the mentioned area , so only hybridization of PSO with Rough-Set, ANN and SVM & some of the other Machine Learning techniques is discussed here. In this paper, the contributions of research work done in recent years, in each method were summarized and existing research challenges are also defined. It is hoped that this survey can serve as a useful guide for the researchers interested in Particle Swarm Optimization Based Machine learning Oriented Anomaly Network Intrusion Detection System.

References

[1] Denning, D.: An intrusion detection model. IEEE Transactions of Software Engineering 13(2), 222–232 (1987)

[2] Lazarevic, A., Kumar, V., Srivastava, J.: Intrusion detection: a survey. In: Managing Cyber Threats: Issues, Approaches, and Challenges, p. 330. Springer (2005)

[3] Garcia-Teodoroa, P., Diaz-Verdejoa, J., Macia-Fernandez, G., Vazquez, E.: Anomaly-based network intrusion detection; technique, systems and challenges. Compuers and Security 28, 18–28 (2009)

[4] Kennedy, J., Eberhart, R.C.: Particle Swarm Optimization. In: Proceedings of the IEEE International Joint Conference on Neural Networks, pp. 1942–1948 (1995)

[5] Zainal, A., Maarof, M.A., Shamsuddin, S.M.: Feature Selection Using Rough Set in Intrusion Detection. In: IEEE TENCON 2006, Hongkong, November 14-17 (2006)

[6] Zainal, A., Maarof, M.A., Shamsuddin, S.M.: Feature Selection Using Rough-DPSO in Anomaly Intrusion Detection. In: Gervasi, O., Gavrilova, M.L. (eds.) ICCSA 2007, Part I. LNCS, vol. 4705, pp. 512–524. Springer, Heidelberg (2007)

[7] Tian, W., Liu, J.: Network Intrusion Detection Analysis with Neural Network and Particle Swarm Optimization Algorithm. In: 2010 Chinese IEEE Control and Decision Conference, CCDC, pp. 1749–1752 (2010)

[8] Liu, H., Jian, Y., Liu, S.: A New Intelligent Intrusion Detection Method Based on Attribute Reduction and Parameters Optimization of SVM. In: Proceedings of the Second International Workshop on Education Technology and Computer Science (ETCS), pp. 202–205 (2010)

[9] Wang, H.-B., Fu, D.-S.: An Intrusion Detection System Model Based on Particle Swarm Reduction. In: Proceedings of 4th the IEEE International Conference on Genetic and Evolutionary Computing, pp. 383–385 (2010)

[10] Liu, L.-L., Liu, Y.: MQPSO based on wavelet neural network for network anomaly detection. In: Proceedings of the 5th International Conference on Wireless Communications, Networking and Mobile Computing (WiCom 2009), pp. 1–5 (2009)

[11] Liu, Y., Ruhui, M.A.: Wavelet Neural Networks Optimized by QPSO for Network Anomaly Detection. Journal of Computational Information Systems 7(7), 2452–2460 (2011)

[12] Liu, Y.: Wavelet fuzzy neural network based on modified QPSO for network anomaly detection. Applied Mechanics and Materials 20-23, 1378–1384 (2010)

[13] Chen, Z., Qian, P., Chen, Z.: Application of PSO-RBF neural network in network intrusion detection. In: Proceedings of the 3rd International Symposium on Intelligent Information Technology Application, pp. 362–364 (2009)
[14] Liu, Y.: QPSO-optimized RBF Neural Network for Network Anomaly Detection. Journal of Information & Computational Science 8(9), 1479–1485 (2011)
[15] Xu, R., Rui, A., Xiao, F.: Research Intrusion Detection Based PSO-RBF Classifier. In: Proceeding of IEEE 2nd International Conference on Software Engineering and Service Science (ICSESS), pp. 104–107 (2011)
[16] Tu, C.-J., Li-Yeh, C., Jun, Y., Cheng, H.: Feature Selection using PSO-SVM. IAENG International Journal of Computer Science 33(1), IJCS_33_1_18 (2007)
[17] Ma, J., Liu, X., Liu, S.: A New Intrusion Detection Method Based on BPSO-SVM. In: Proceedings of the International Symposium on Computational Intelligence and Design, pp. 473–477 (2008a)
[18] Zhang, H., Gao, H.-H., Wang, X.Y.: Quantum Particle swarm optimization based network Intrusion feature selection and Detection. In: Proceedings of the 17th World Congress The International Federation of Automatic Control, Seoul, Korea (2008)
[19] Zhou, T., Li, Y., Li, J.: Research on intrusion detection of SVM based on PSO. In: Proceedings of the International Conference on Machine Learning and Cybernetics, pp. 1205–1209 (2009)
[20] Wang, J., Hong, X., Ren, R.-R., Li, T.-H.: A Real-time Intrusion Detection System based on PSO-SVM. In: Proceedings of the International Workshop on Information Security and Application (IWISA 2009), pp. 319–321 (2009)
[21] Chen, Y., Abraham, A., Yang, J.: Feature Selection and Classification Using Hybrid Flexible Neural Tree. Journal of Neuro Computing 7, 305–313 (2006)
[22] Chen, Y., Zhang, L.: Evolutionary Flexible Neural Networks for Intrusion Detection System. In: Proceedings of the 5th WSEAS International Conference on Applied Computer Science, Hangzhou, China, pp. 428–433 (2006)
[23] Michailidis, E.: Proceedings of the 2008 Panhellenic Conference on Informatics, PCI 2008, pp. 8–12. IEEE Computer Society, Washington, DC (2008)
[24] Gong, S.F., Gong, X., Bi, X.: Feature Selection Method for Network Intrusion Based on GQPSO Attribute Reduction. In: 2011 International Conference on Multimedia Technology (ICMT), pp. 6365–6368 (2011)

Optimal Scheduling of Short Term Hydrothermal Coordination for an Indian Utility System Using Genetic Algorithm

S. Padmini[1], C. Christober Asir Rajan[2], Subhronil Chaudhuri[1], Arkita Chakraborty[1]

[1] Department of Electrical and Electronics Engineering, SRM University, Chennai, India
{padminisp81,subhronil11,arkitavicky}@gmail.com
[2] Department of Electrical and Electronics Engineering Pondichery University, Chennai, India
asir_70@hotmail.com

Abstract. This paper addresses short-term scheduling of two test hydrothermal systems by using Genetic algorithm. Short-term hydrothermal coordination consists of determining the optimal usage of available hydro and thermal resources during a scheduling period of time.Genetic algorithm is applied to determine the optimal hourly schedule of power generation in a hydrothermal power system. The developed algorithm is illustrated for a test system an Indian Utility System which consists of 7 hydro and 4 thermal systems respectively. The effectiveness and stochastic nature of proposed algorithm has been tested with standard test case and the results have been proved to be better than conventional method and results obtained by the proposed method are superior in terms of fuel cost.

Keywords: Short-term Hydrothermal Scheduling, Genetic algorithm, discharge rate.

1 Introduction

Power Systems are large complex networks that deal with the generation, transmission and distribution of power. The power system is mainly expected to supply the ever changing load demand of consumers at an economical rate without wastage of the generation fuel. Thus the short term hydrothermal generation scheduling is one of the major concerns. A new model to deal with this problem and has been proposed using Genetic Algorithm. This model considers a scheduling horizon period of 24 hours.For both the hydro and thermal units, the hourly generation schedules are obtained separately. The main objective of hydrothermal operation is to minimize the total system operating cost, represented by the fuel cost for the systems thermal generation subject to the operating constraints of hydro and thermal plant over the optimization interval. In short range problem the water inflows is considered fully known and is constrained by the amount of water available for draw down in the interval. The short term hydro thermal scheduling problems have been solved by various methods. These methods which have been reported in the literature includes classical methods such as Langrage Multiplier Gradient Search [1] , Evolutionary Programming (EP) [2] and Dynamic programming [3] and stochastic search algorithm such as simulated annealing

S.C. Satapathy et al. (Eds.): Proc. of Int. Conf. on Front. of Intell. Comput., AISC 199, pp. 453–459.
DOI: 10.1007/978-3-642-35314-7_51

(SA) [6], and Particle Swarm Optimization (PSO) [5,10]. These technique can generate high-quality solution within shorter calculation time and more stable convergence characteristic than other stochastic methods. In this paper an Indian Utility System is used to validate the presented algorithm. The results prove that they are better than the conventional method.

2 Problem Formulation

2.1 Objective Function

The main objective of Hydro Thermal Scheduling is minimizing the thermal generation cost by satisfying the hydro and thermal constraints.

The problem formulation is same as that of real power dispatch problem, but emission coefficients in place of fuel coefficients are used and dispatching is done by allocation of power generation across various generation units. Hydro thermal scheduling is the optimization of a problem with non-linear objective function, the objective function to be minimized can be written as:

$$\min F_T = \sum_{t=1}^{T} \sum_{j=1}^{N} F_j(P_s(j,t)) \tag{1}$$

Where

F_T is the total production cost function
$P_s(j, t)$ is the power generation of thermal unit j at time interval t;
$F_j(Ps(j, t))$ isthe production cost for Ps(j, t);
N is the number of thermal units
T is the number of time intervals.

Power balance constraint $$\left(P_{dj} + P_{Loss_j}\right) - \left(PH_j + PT_j\right) = 0 \tag{2}$$

The transmission loss is given by $$P_{Loss} = k\left(PH_j\right) \tag{3}$$

The hydro generation is consider to be a function of discharge rate only

$$q_j = g\left(PH_j\right) \tag{4}$$

Discharge rate limits $$q_{\max} > q_j > q_{\min} \tag{5}$$

Thermal generation limits $$PT_{\max} > PT_j > PT_{\min} \tag{6}$$

Hydro generation limits $$PH_{\max} > PH_j > PH_{\min} \tag{7}$$

Reservoir volume

$$Volume_{j+1} = Volume_j + n_j\left(r_j - q_j - s_j\right) \tag{8}$$

Where

nj is the number of hours in jth interval
rj is the water inflow rate in jth interval
qj is the water discharge rate in jth interval
sj is the water spillage rate in jth interval

Reservoir storage limits

$$Volume_{max} > Volume_j > Volume_{min} \quad (9)$$

Hydro plant power generation equation

$$P_h = C_1 * V^2 + C_2 * q^2 + C_3 * V * q + C_4 * V + C_5 * q + C_6 \quad (10)$$

3 Genetic Algorithm

In the field of artificial intelligence, a genetic algorithm (GA) is a search method that mimics the process of natural evolution. This heuristic is routinely used to generate useful solutions to optimization and search problems. Genetic algorithms belong to the larger class of evolutionary algorithm (EA), which generates solutions to optimization problems using techniques inspired by natural evolution, such as mutation, selection, and crossover.

Genetic algorithms find application in bioinformatics, computational science, engineering, economics, chemistry, manufacturing, mathematics, physics and other fields.

4 Proposed Genetic Algorithm for Hydrothermal Scheduling

The proposed GA based HTS algorithm for short term hydrothermal scheduling rese is presented here.

Step 1: Input parameters of the system are identified and the upper and lower boundaries of each variable are specified.
Step 2: Let the GA process generate a set of discharge values for each plant over the scheduling period
Step 3: Let qj be the dishcarge rate denoting the particles of population to be evolved.It is the discharges of turbines of reservoirs at various intervals.Then knowing the hydro discharges, storage volumes of reservoirs V_j are calculated using Eq.(8).Then P_H *and* P_T is calculated for all the intervals.
Step 4:Evaluate the fitness of each individual in that population.
Step 5:Repeat on this generation until termination (time limit, sufficient fitness achieved, etc.).
Step 6: Select the best-fit individuals for reproduction.
Step 7: Breed new individuals through crossover and mutation operations to give birth to offspring.
Step 8: Evaluate the individual fitness of new individuals.
Step 9: Replace least-fit population with new individuals.

5 Simulation Results

In this study two test cases have been considered. test case system consisting of four hydro unit and seven thermal unit and second test case system consisting of for 7 hydro and 4 thermal Indian utility system unit over a period of 24 hours, For 4 hydro 7 thermal over a period of 24 hourS respectively. Table 1 and 2 presents the optimal hydrothermal generation scheduling using conventional and genetic algorithm respectively. The datas for the case study have been referred from paper[7].

Table 1. Hydrothermal generation(MW) using Conventional Method

Hr	P_{s1}	P_{s2}	P_{s3}	P_{s4}	P_{s5}	P_{s6}	P_{s7}	P_{h1}	P_{h2}	P_{h3}	P_{h4}
1	28.62	20.0	30.00	120	95.32	96.65	75.00	132.3	27.4	149.6	304.6
2	15.00	25.6	64.11	120	95.15	65.17	75.00	164.1	33.5	150.3	292.5
3	33.44	23.4	45.32	25	70.36	50.00	75.00	138.6	30.6	187.1	278.1
4	15.00	57.1	30.00	120	50.00	76.62	99.65	74.2	1.36	200.6	261.3
5	31.92	45.0	30.00	120	65.61	50.00	75.00	74.6	3.5	125.5	243.1
6	15.00	45.6	30.00	120	69.06	89.14	119.1	74.1	38.5	133.4	224.8
7	37.83	20.0	56.00	25	92.41	50.14	75.00	83.6	1.4	118.2	204.2
8	37.56	35.1	38.63	120	50.64	75.68	94.65	75.1	2.9	113.6	183.6
9	23.65	46.6	64.69	25	56.14	50.00	75.00	75.6	5.8	108.3	161.1
10	23.65	20.0	42.45	25	65.64	68.64	93.68	76.10	8.5	152.9	138.6
11	15.00	49.6	37.04	120	50.00	50.00	75.00	76.1	12.4	92.8	114.3
12	39.65	20.0	30.00	25	80.68	50.00	113.6	168.6	14.6	128.5	89.2
13	39.78	34.6	38.15	25	85.45	62.45	75.00	75.1	16.3	78.4	63.4
14	15.00	20.0	30.00	120	50.00	90.69	118.6	77.1	19.2	73.5	36.5
15	16.65	20.0	54.65	120	50.00	50.00	75.00	73.6	23.1	67.2	8.6
16	15.00	20.0	30.00	120	50.00	50.00	75.00	72.9	57.4	125.1	20.5
17	24.15	20.0	52.36	25	68.36	50.00	75.00	100.5	30.85	64.6	50.1
18	34.65	29.8	30.00	120	50.00	50.00	75.00	95.4	35.5	40.4	82.65
19	36.45	31.6	30.00	120	87.14	55.69	77.65	13.6	37.4	46.5	114.4
20	15.00	50.0	30.00	25	50.00	50.00	75.00	88.5	23.98	17.9	147.1
21	40.71	58.6	30.00	25	50.00	50.00	75.00	71.6	57.6	19.8	181.3.
22	15.00	20.0	30.00	120	50.00	50.00	75.00	70.41	27.4	14.7	216.6
23	21.98	34.1	30.00	120	96.65	54.01	75.00	131.1	29.5	8.5	252.5
24	17.79	22.3	69.63	25	88.65	91.03	99.96	68.8	31.3	28.6	290.2

Table 2. Hydrothermal generation(MW) using Genetic Algorithm

Hr	P_{s1}	P_{s2}	P_{s3}	P_{s4}	P_{s5}	P_{s6}	P_{s7}	P_{h1}	P_{h2}	P_{h3}	P_{h4} (MW)
1	38.32	35.64	30.00	120.0	72.71	66.48	75.00	109.6	0.25	140.5	304.1
2	22.11	20.00	30.00	25.0	50.00	50.00	113.6	168.4	2.62	139.5	292.1
3	15.00	20.00	39.03	120.0	50.00	50.00	120.6	74.6	6.24	136.1	278.4
4	15.00	20.00	67.71	120.0	92.14	50.00	75.00	74.1	9.36	209.9	261.7
5	26.65	20.00	57.15	120.0	52.32	50.00	75.00	75.5	11.21	174.6	243.8
6	15.26	28.85	32.68	120.0	50.00	50.00	75.00	134.6	13.54	161.1	224.6
7	38.15	20.00	30.00	1200.	57.58	63.48	110.8	75.1	47.68	152.9	204.3
8	15.00	21.15	59.69	120.0	50.00	80.65	122.5	75.6	11.12	139.6	183.1
9	15.00	26.65	30.00	120.0	89.95	50.00	75.00	75.5	13.10	161.5	161.2
10	23.15	38.64	30.00	120.0	50.00	73.49	75.00	125.3	16.48	152.3	138.6
11	21.36	20.00	38.15	120.0	50.00	50.00	75.00	100.3	19.62	74.02	114.4
12	15.22	20.00	64.65	120.0	75.35	50.00	110.5	75.1	22.32	79.46	89.5
13	23.21	38.68	37.45	120.0	50.00	50.00	75.00	176.6	24.01	61.85	63.5
14	18.69	20.00	38.14	120.0	50.00	64.18	75.00	75.6	27.03	56.56	36.6
15	15.00	35.59	30.65	120.0	91.36	50.00	96.69	74.9	30.59	50.77	8.2
16	15.00	20.00	43.78	120.0	93.13	50.00	75.00	174.8	56.84	65.14	20.4
17	27.64	20.00	37.41	120.0	75.15	50.00	75.00	142.1	31.62	51.85	50.9
18	38.95	34.48	66.58	120.0	90.45	50.00	75.00	73.5	42.31	25.05	82.3
19	43.36	47.56	69.69	120.0	94.65	73.45	75.00	72.6	31.44	13.15	114.1
20	37.15	20.00	30.00	120.0	50.00	50.00	89.65	72.2	33.69	29.22	147.3
21	35.33	20.00	30.00	120.0	50.00	50.00	75.00	110.2	36.74	12.98	181.4
22	15.00	31.36	43.45	120.0	87.96	68.69	75.00	137.5	39.44	20.45	216.5
23	15.00	20.00	30.00	120.0	50.00	50.00	75.00	71.4	54.32	26.25	252.1
24	15.00	20.00	48.36	25.0	68.03	50.00	75.00	70.6	41.12	29.36	292.6

Table 3. Comparison of costs for Indian Utility System

Parameter	Conventional Method (Rs.)	Genetic Algorithm (Rs.)	Difference (Rs.)
Total Operating Cost	1017400.00	1013700.5868	3699.4132

6 Conclusion

In this paper, GA based algorithm has been proposed for solving Short term hydrothermal scheduling problem by taking into account for an Indian Utility System.The results shows that best optimal solutions can be obtained by existing technique. Thus the proposed algorithm justifies its need in saving the fuel cost in power systems.

References

1. Wood, A.J., Wollenberg, B.F.: Power Generation, Operation and Control. John Wiley and Sons, New York (1984)
2. Sinha, N., Chakrabarti, R.: Fast Evolutionary Programming Techniques For Short-Term Hydrothermal Scheduling. IEEE Trans. PWRS 18(1), 214–219 (2003)
3. Ferrero, R.W., Rivera, J.F., Shahidehpour, S.M.: A dynamicprogramming two-stage algorithm for long-termhydrothermal scheduling of multireservoir systems. IEEE Transactions on Power Systems 13(4), 1534–1540 (1998)
4. Gil, E., Rudnick, H.: Short-term hydrothermal generation scheduling model using a genetic algorithm. IEEE Transactions on Power Systems 18(4), 1256–1264 (2003)
5. Mandal, K.K., Basu, M., Chakraborty, N.: Particle swarm optimization technique based short-term hydrothermal scheduling. Applied Soft Computing 8(4), 1392–1399 (2007)
6. Wong, K.P., Wong, Y.W.: Short-term hydrothermal scheduling, part-I: Simulated. Annealing approach. IEE Proc., Part- C 141(5), 497–501 (1994)
7. Christober Asir Rajan, C.: Hydrothermal unit commitment problem using simulated annealing embedded evolutionary programming approach. Electrical Power and Energy Systems 33, 939–946 (2011)
8. Sinha, N., Chakrabarti, R.: Fast Evolutionary Programming Techniques For Short-Term Hydrothermal Scheduling. IEEE Trans. PWRS 18(1), 214–219 (2003)
9. Suman, D.S., Nallasivan Joseph Henry, C., Ravichandran, S.: A Novel Approach for Short-Term Hydrothermal Scheduling Using Hybrid Technique. In: IEEE Power India Conference, April 10-12 (2006)

10. Padmini, S., Rajan, C.C.A., Murthy, P.: Application of Improved PSO Technique for Short Term Hydrothermal Generation Scheduling of Power System. In: Panigrahi, B.K., Suganthan, P.N., Das, S., Satapathy, S.C. (eds.) SEMCCO 2011, Part I. LNCS, vol. 7076, pp. 176–182. Springer, Heidelberg (2011)
11. Sinha, N., Loi-Lei Lai, N.: Meta Heuristic Search Algorithms for Short-Term Hydrothermal Scheduling. In: International Conference on Machine Learning and Cybernetics, Dalian (2006)
12. Hotaa, P.K., Barisala, A.K., Chakrabarti, R.: An improved PSO technique for short-term optimal hydrothermal scheduling. Electric Power Systems Research 79(7), 1047–1053 (2009)
13. Liang, J.J., Qin, A.K., Suganthan, P.N., Baskar, S.: Comprehensive Learning Particle Swarm Optimizer for Global Optimization of Multimodal Functions. IEEE T. on Evolutionary Computation 10(3), 281–295 (2006)
14. Dhillon, J.S., Parti, S.C., Kothari, D.P.: Fuzzy decision–making in stochastic multiobjective short-term hydrothermal Scheduling. IEEE Trans. Proc. - Generation,Transmission and Distribution 149(2), 191–200 (2002)

Automated Design and Optimization of Combinational Circuits Using Genetic Algorithms

K. Sagar[1] and S. Vathsal[2]

[1] Department of Computer Science and Engineering,
Chitanya Bharathi Institute of Technology, Hyderabad, India
kadapasagar@yahoo.com
[2] HOD (EEE) & Dean R&D, JBIET, Moinabad
Hyderabad
svathsal@gmail.com

Abstract. We introduce a method, based on genetic algorithm to automate and optimize the design of combinational circuits. Usually, logic circuits are designed by human beings who have a specific repertoire of conventional design techniques. These techniques limit the solutions that may be considered during the design process in both form and quality. The application of genetic algorithms has allowed the creation of circuits which are substantially superior to the best known human designs. We describe the important issues to consider when solving this circuit design problem : the representation scheme, the encoding scheme, the fitness function. We compare the solutions by our approach against those generated by a human designer. We also show that our approach produces better performances both in terms of quality of solution and in terms of speed of convergence.

Keywords: Circuit design, optimization, genetic algorithms, convergence.

1 Introduction

Central to modern computing is the ability to perform logic. Indeed, logic is the framework on which the very concept of modern computation is built. This logic can be simple or complex, but logic is always present at the heart of whatever computation is taking place. In its most fundamental form, the logic in computers is facilitated by digital logic circuits. Moreover, the basic components of these circuits are known as logic gates.

However, they are most often combined and interconnected in various ways to create more complex circuits. Digital design is one example of a discrete combinatorial system. The characteristics of such a system are that it has a finite collection of discrete elements, which are combined to create new distinct objects. In the case of logic circuit design, gates are the discrete elements, and they are combined to create new circuits which function differently than any of the individual gates. By automating design, the goal is to remove human effort, and human limitations, from the design process. This can be done by taking advantage of what computers do very well, quickly examine a huge number of possible solutions.

S.C. Satapathy et al. (Eds.): Proc. of Int. Conf. on Front. of Intell. Comput., AISC 199, pp. 461–468.
DOI: 10.1007/978-3-642-35314-7_52

In automatically designing logic circuits of this type, techniques from artificial intelligence have been extremely useful. Researchers explain that logic circuit design could be formulated as a puzzle, and puzzle solving is an area of great success within artificial intelligence. Specifically, genetic algorithms have been highly researched as a candidate for automating circuit design. Moreover, there has been a good amount of success with using these algorithms. Genetic algorithms have been able to produce better results than human designers, and in a shorter period of time.

2 Previous Work

The design process for combinational logic circuits has evolved from its first notions [1] to a standard element of undergraduate computing curricula [6]. Standard graphical design aids such as Karnaugh Maps [5] are widely used and tool suitable for computer implementation have evolved from the QuineMcCluskey Method [4].

Louis [7] is one of few sources found in the literature to address the use of GAs for the combinational logic design problem. Louis combines knowledge- based systems with the genetic algorithm, making use of a genetic operator called masked crossover that adapts to the encoding being able to exploit information unused by classical crossover operators [8]. His results, although very encouraging for certain examples, but do not seem to have solved the combinational circuit design problem completely. However his idea of incorporating knowledge about the domain in the genetic operator constitutes a big step toward increasing the power of the GA as a design tool. Unfortunately, the incorporation of knowledge in to the GA decreases its usefulness as a general search tool. Louis overcomes this problem by defining an operator that he claims to be domain independent, but whose efficiency turns out to depend on the representation used.

Koza [6] has used genetic programming to design combinational circuits. He has designed, for example, a two-bit adder, using a small set of gates (AND, OR, NOT), but his emphasis has been on generating functional circuits rather than on optimizing them. In fact, this is also the case in Louis' research, where the main focus was to provide an easier way to generate functional designs using the GA rather than in optimizing a functional design according to certain metrics. In more recent work, has focused more towards the design of an a log circuits in which the goal is to produce their appropriate topology and size so that they are functional given a certain set of components. So far, genetic programming has been considered a more powerful tool in such tasks, because the representation it uses is more powerful for structural design in general.

Miller et al [9] developed (independently) an approach similar to ours, but using a more compact representation that instead of considering the inputs and gates as completely separate elements in the chromosomic string, use a single gene to encode a complete Boolean expression. Miller's notation does not decrease the total length of the chromosome, but it increases the cardinality of the alphabet needed, having as its main drawback the lack of flexibility of the representation to handle a larger number of inputs.

3 Genetic Algorithms for Logic Circuit Design

Up to the present, most research has focused on using local search algorithms for the design of logic circuits. More specifically, genetic algorithms have been the most common choice.

In order to use GAs for this purpose, though, there must be some additional formulation of the problem. As we have seen, GAs use strings as their basic elements, in the same way that biological systems use DNA strands. Therefore, if we are to use GAs for circuit design, all of the information about gates and connections must be encoded in a string. In accordance with the terminology from biology, this string is known as the "genotype".

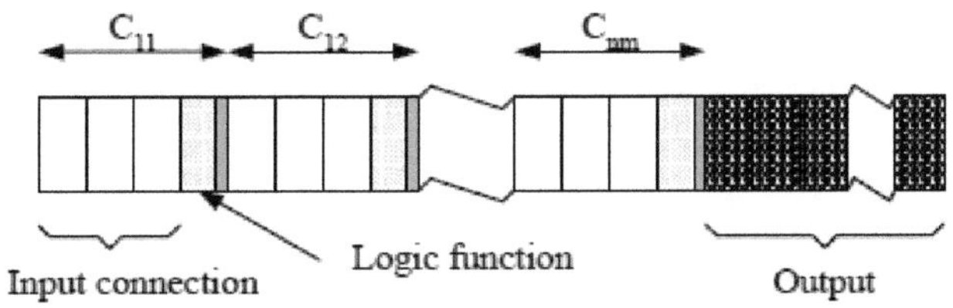

Fig. 1. Representation of genotype

Genotype is structured from phenotype shows in fig. 1. Cells in phenotype are lined from C11 to Cnm and finish by outputs for set to genotype.

The genotype is an encoding of all the relevant information about the circuit. The relevant information, which is encoded, is known as the "phenotype".

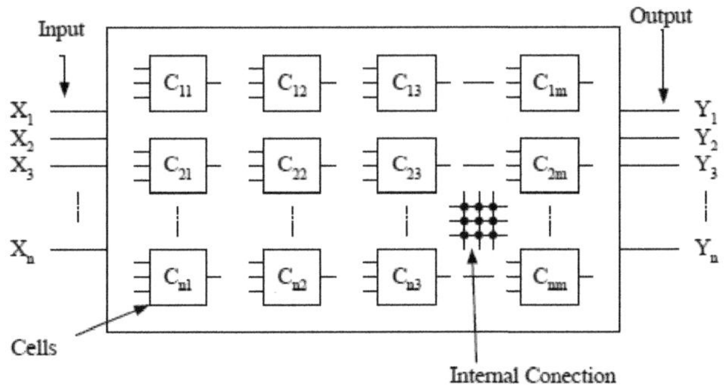

Fig. 2. Representation of phenotype

Phenotype consists of inputs, cells, internal connections, and outputs shown in Fig. 2. Inputs are input signals of a digital logic circuit. Each cell is a logic gate which is connected thru internal connection.

The phenotype includes the gates used in the circuit, the connections between gates and other essential properties. The phenotype can be derived from the genotype, and in turn, the operation of the circuit can be derived from the phenotype.

Attempts were made to explore other circuit formulations , but the one which has gained the most favor is the array formulation. In this formulation, a circuit is conceptualized as an array of logic gates and connections between them.

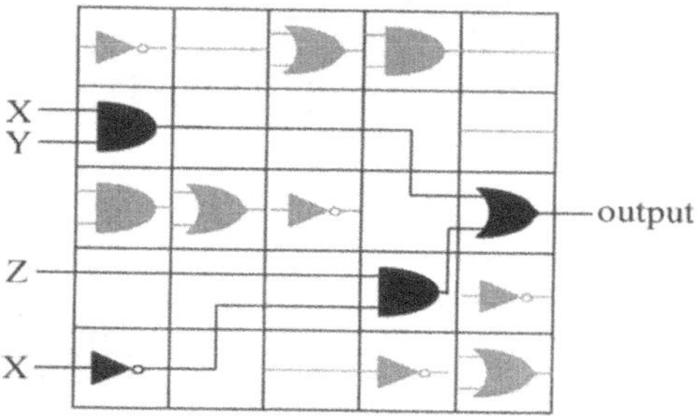

Fig. 3. Array formulation

This is conceptually similar to the way an FPGA is structured. At one end of the array are presented the inputs to the circuit, and at the other end of the array are the outputs. Each gate at a particular location is a member of an array column, and it can get its inputs from any gates in the previous column. The gates in the left-most column get their inputs from any of the circuit inputs, rather than any gates. The benefits of applying GAs to logic circuit design have been as good as expected. By automating the entire process, GAs have been able to quickly develop circuits which are fully functional. Moreover, some circuits which have been developed are superior to those designed by humans.

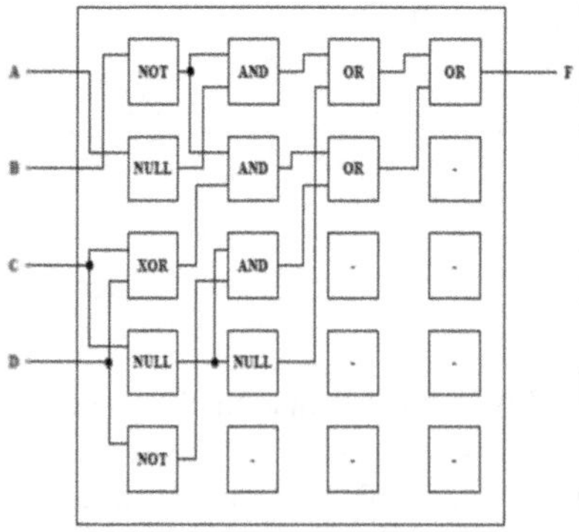

Fig. 4. Example for Array Formulation of Logic Gates

The Fig. 4 describes an example of array formulation for two dimensional template, Gates gets its input from one of the gates from the previous columns, From the above figure , Second column and first of the AND gate is getting the input from an not gate and the other input is directly connected from the input variable 'A' and similarly the other gates in the array gets inputs from the gates in the previous columns and the output is produced by the last gate from the last column of the array.

The genetic algorithm approach is mainly based on its genetic operators like iteration, the number of times the loop is being repeated the probability of getting the appropriate solution is high. But genetic algorithm is a stochastic process where the chance of getting the exact solution is 50/50. The generational process is repeated until termination condition is reached, the common terminating conditions are:

-Fixed number of generation reached.
-An individual is found which satisfies all the minimum condition-Highest ranking individual's fitness is reached or has a plateau such that further iterations does not produce better results
-Combinations of above.

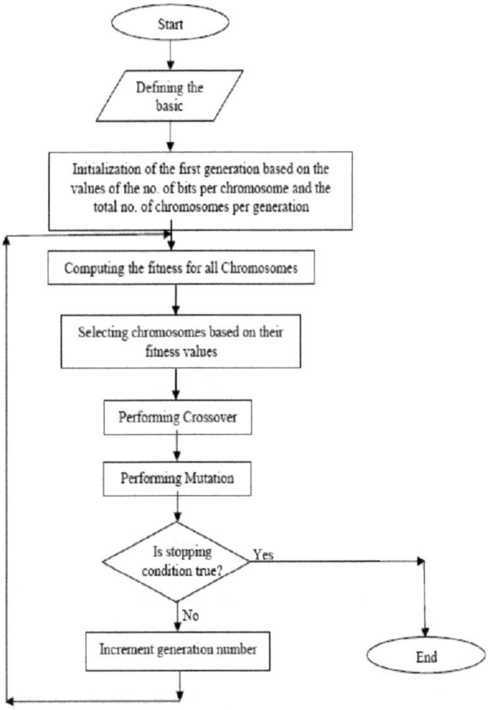

Fig. 5. Design of the Genetic algorithm for Circuit designing

The general concept connected to the design of combinational digital circuits is related to pattern of gates. The size of the gate pattern is determined at the start of algorithm. From Fig. 5, for each pattern t.t there exist places for potential circuit gates. The pattern also consists of k inputs, and f outputs, which represent circuit inputs and circuit outputs. The main aspect of the algorithm is determination such set of gates in the pattern, and such set of connections between gates, that the designed circuit fulfills assumed truth table. Also, depending on assumed optimization criterion, it is required to design the circuit consisting of minimal number of gates, or circuit having minimal time of signal propagation from its input to its output.

4 Experimental Results

The output screen shots which are produced by the algorithm these outputs are produced depending on the truth tables.

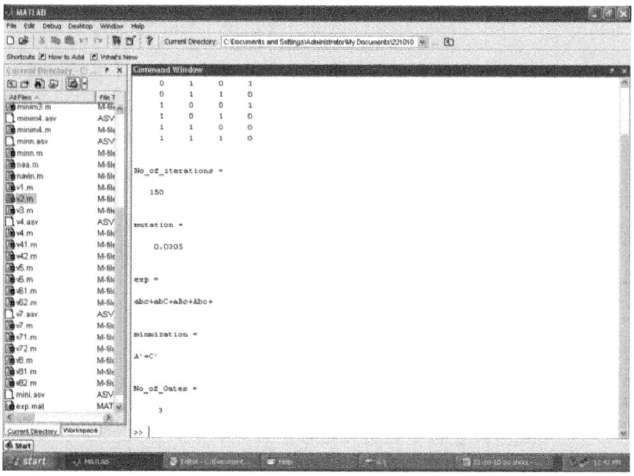

Fig. 6. Output screen of genetic algorithm for variable =3

The above Fig. 6 is the output of the genetic algorithm which consists of truth table , number of iterations, mutation value the minimized expression and number of gates in the expression

The issue is regarding the crossover and mutation rates. Two-point crossover with a probability of 50% and uniform bit mutation with a probability that each string would have a 50% chance of being mutated in at least one position across its length is used. Since mutation was applied on a single-gene basis, we used as our probability of mutation the result of dividing this 50% by the length of the string.

Binary tournament selection with full generational replacement was used in all cases and the termination criterion adopted was a maximum number of generations defined by the user. Termination criterion is not based on the lack of improvement of a certain solution after a number of generations, because it is difficult to define a threshold that could ensure no further improvement of a solution. In some circuits, there would be no improvement of a solution after a relatively large number of generations and then the GA would be able to jump abruptly to a much better solution.

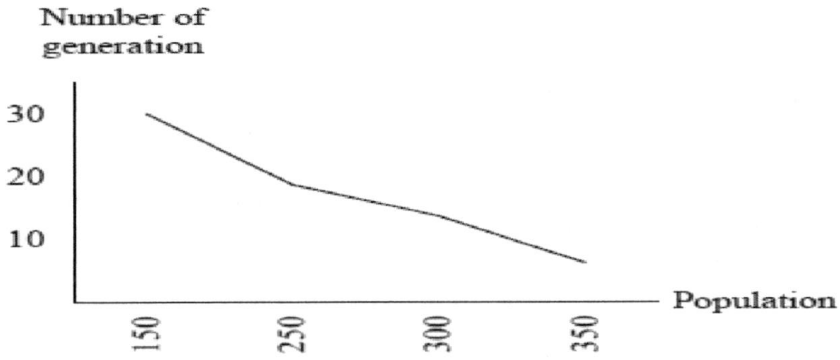

Fig. 7. Number of Generations Vs. Population

Fig. 7 is a comparison between number of generations to number of populations which depicts that with the large number of population the number of generations also increases. The maximum number of generations was arbitrarily set to a reasonably large number, and the population size was chosen based on a number of independent runs. In each case, the value shown for the population size is the one that produced the best results for that particular circuit

The algorithm is compared with one of the existing tool which also used to minimize the given expression, minimization in practical is done by human designers who use different patterns and methods, and each method has its own set of rule and limitations

To avoid the limitation of human designers such as Karnaugh and Tabular method we are proposing a method using genetic algorithm which reduces the number of gates required to design the logical circuits.

Below table represents the comparison of each variable separately with minimization tool and our genetic algorithm.

Table 1. Comparison of genetic algorithm with other approaches w.r.t variables=4

Method	**No. of Variables**	**Expression**	**Number of Gates**
Genetic algorithm	**4**	A'(B^C)+AB'C'+B'D'	12
Expression Minimization Tool	**4**	A'B'C+A'BC'+AB'C'+B'D '	18

The above Table 1 represents the comparison of genetic algorithm for variable=4, with one of the minimization tool it is clearly observed that the expression produced by Genetic algorithm for variable 4 has 12 gates whereas the minimization tool is producing 18 gates.

5 Conclusion

This paper presented how genetic algorithm can be used to design combinational logic circuits. Systematic and Local search techniques of artificial intelligence are studied and have been applied to the problem of genetic based logic circuit design.

We have implemented genetic algorithm using all genetic operators on an input for circuit designing, these genetic operators include selection, fitness function, crossover and mutation. A computer program has been developed which can reduce the number of gates on a particular input .We compared the results produced by our genetic algorithm approach against those generated by Minimization tool.

References

[1] Al-saiari, U.S.: Digital Circuit Design Through Simulated evolution. King Fahd University of petroleum and minerals, Dhahran, Saudi Arabia (November 2003)

[2] Russell, S., Norvig: Artificial Intelligence: A Modern Approach. Prentice Hall, New Jersey (2003)

[3] CoelloCoello, C.A., Christiansen, A.D., Aguirre, A.H.: Design of Combination Logic Circuits through an Evolutionary Multi-objective Optimization Approach. Department of Electrical Engineeringand Computer Science, Tulane University, New Orleans, LA, USA (2000)

[4] McCluskey, E.J.: Minimization of Boolean functions. Bell System Technical Journal (1996)

[5] Karnaugh, M.: A map method for synthesis of combinational logic circuit. Transactions of the AIEE, Communications and Electronics (1993)

[6] Koza, J.R.: Genetic Programming on the programming of computers by means of natural selection. The MIT press, Cambridge (1992)

[7] Louis, Rawlins, G.: Designer Genetic algorithms: "Genetic algorithms in structure design". In: Belew, R.K., Booker, L.B. (eds.) Proceedings of the Fourth International Conference on Genetic Algorithms, San Mateo, California. Morgan Kaufmann Publishers (1991)

[8] Tutorials for Genetic Algorithm, http://www.obitko.com/tutorials/genetic-algorithms

[9] Miller, J.F., Thompson, P., Fogarty: Designing Electronic Circuits Using Evolutionary Algorithms. Arithmetic Circuits: A Case Study. Genetic Algorithm and Evolution Strategy in Eng. and Comp. Sci., 105T–131T (1997)

A Fitness-Based Adaptive Differential Evolution Approach to Data Clustering

G.R. Patra[1], T. Singha[1], S.S. Choudhury[1], and S. Das[2,*]

[1] Department of Electronics and Telecommunication Engineering,
Jadavpur University, Kolkata-700032, India
[2] Engineering and Communication Sciences Unit
Indian Statistical Institute, Kolkata-700108, India
gyana.patra@gmail.com, jcettanmoy@rediff.com,
sschaudhuri@etce.jdvu.ac.in, swagatamdas19@yahoo.co.in

Abstract. Fuzzy clustering helps to find natural vague boundaries in data. The fuzzy c-means (FCM) is one of the most popular clustering methods based on minimization of a criterion function as it works fast in most scenarios. However, it is sensitive to initialization and is easily trapped in local optima. In this work, a fuzzy clustering (FC) algorithm based on Differential Evolution (DE) is proposed. Here we use a DE with Fitness Based Adaptive Technique (FBADE) for the adaptation of DE parameters. 3 well-known data sets viz. Iris, Wine, Motorcycle and 2 synthetic datasets are used to demonstrate the effectiveness of the algorithm. The resulting algorithm is compared with conventional Fuzzy C-Means (FCM) algorithm, FCM with DE (FCM-DE), FCM with Self Adaptive DE (FCM-SADE).

Keywords: Differential Evolution, Fuzzy Clustering, Global Optimization, Evolutionary Algorithm.

1 Introduction

Fuzzy Clustering [1] plays a vital role in the fields of statistics and pattern recognition. It finds extensive applications in machine learning, data mining, pattern recognition and image segmentation. In fuzzy C-means (FCM) algorithm, to account for the fuzziness present in a data set, each data sample is assumed to belong to every cluster with a specific degree, with the constraint that the sum of its membership degrees in all the clusters equals to 1. In spite of being an effective algorithm FCM sometimes produces shoddier results because of random initialization. Consequently, FCM can get trapped in local minima when started with poor initialization.

In order to avoid the problems associated with poor initialization generally the procedure is run with different initializations hoping that some runs will lead to global optimal solution [2]. Thus many researchers have formulated the clustering task of FCM as an optimization process and have used various metaheuristics methods for

* Corresponding author.

S.C. Satapathy et al. (Eds.): Proc. of Int. Conf. on Front. of Intell. Comput., AISC 199, pp. 469–475.
DOI: 10.1007/978-3-642-35314-7_53

the solution of the same. Some of the noteworthy metaheuristics that have been used in conjunction with FCM are simulated annealing (SA) [3], genetic algorithms (GA) [4], tabu search [5], particle swarm optimization (PSO) [6] and differential evolution (DE) [7] [8].

DE [9] is considered as one of the reliable, accurate, robust and fast optimization technique that has been successfully applied to solve a variety of numerical optimization problems. The values of the control parameters of DE are to be tuned by the user for each problem which is a time consuming task. A properly designed Adaptive or self-adaptive parameter control can enhance the robustness of a metaheuristics algorithm by dynamically adapting the parameters to the characteristic of different fitness landscapes. The adaptive and self-adaptive DE algorithms have shown faster and more reliable convergence performance than the classic DE algorithms without parameter control for many benchmark problems [10] [11]. However, self-adaptation schemes usually make the programming fairly complex and run the risk of increasing the number of function evaluations.

This paper introduces a new version of DE where the control parameters are adapted depending on the fitness of the population pool. The new version is called fitness-based adaptive differential evolution (FBADE). The main idea behind this adaptation mechanism is that the search-agents (DE-vectors) placed near to the optimum have small mutation step-size and during crossover, it passes more genetic information to its offspring for better exploitation. For agents that are far away from the optimum, more perturbation is performed so that during DE-type crossover, the offspring inherits lesser genetic information from the parent to facilitate exploration of alternate regions quickly.

The rest paper is organized as follows. Section 2 briefly introduces the fuzzy clustering algorithm. Classical DE and the self-adaptive DE (SADE) are briefly explained in Section 3. The proposed FBADE approach is described in Section 4. The test benchmark functions, parameter settings, simulation results and discussions are presented in Section 5. Finally, the work is concluded and summarized in Section 6.

2 Fuzzy Clustering Algorithm

Fuzzy clustering is formulated as a non-linear optimization problem as follows:

$$minimize\ z(U,v) = \sum_{i=1}^{cc} \sum_{k=1}^{n} (\mu_{ik})^m \left|\left|x_k - v_i\right|\right|^2 \tag{1}$$

$$Subject\ to\ \sum_{i=0}^{cc} \mu_{ik} = 1, 1 \le k \le n, 0 \le \mu_{ik} \le 1, for\ i = 1, \ldots, cc; k = 1, \ldots, n$$

where, v_i is the ith Cluster Center, $x_k$is the kth data sample, m is the degree of fuzziness μ_{ik} is the membership value of kth sample in the ith cluster, U is the membership matrix, cc is the number of clusters, n is the total number of sample points and z is the value of objective function .

3 Differential Evolution and Self Adaptive Differential Evolution

Originally proposed by Price and Storn [9], DE is a stochastic, population based optimization method. Because of the simple, fast and robust nature of DE, it has found widespread applications [12].

The ith individual (parameter vector) of the population at generation (time) t is a D-dimensional vector containing a set of D optimization parameters:

$$\overrightarrow{Z_i}(t) = [Z_{i,1}(t), Z_{i,2}(t), \dots \dots, Z_{i,D}(t)] \tag{2}$$

In each generation to change the population members $\overrightarrow{Z_i}(t)$ (say), a *donor* vector $\overrightarrow{Y_i}(t)$ is created. It is the method of creating this donor vector that distinguishes the various DE schemes. In one of the earliest variants of DE, now called DE/rand/1 scheme, to create $\overrightarrow{Y_i}(t)$ for each i^{th}member, three other parameter vectors (say the r_1, r_2, and r_3-th vectors such that $r_1, r_2, r_3 \in [1, NP]$ and $r_1 \neq r_2 \neq r_3$) are chosen at random from the current population. The donor vector $\overrightarrow{Y_i}(t)$ is then obtained multiplying a scalar number F with the difference of any two of the three. The process for the jth component of the ith vector may be expressed as,

$$\overrightarrow{Y_{i,j}}(t) = Z_{r1,j}(t) + F.\left(Z_{r2,j}(t) - Z_{r3,j}(t)\right) \tag{3}$$

A 'binomial' crossover operation takes place to increase the potential diversity of the population. The binomial crossover is performed on each of the D variables whenever a randomly picked number between 0 and 1 is within the Cr value. In this case the number of parameters inherited from the mutant has a (nearly) binomial distribution. Thus for each target vector $\overrightarrow{Z_i}(t)$, a trial vector $\overrightarrow{R_i}(t)$is created in the following fashion:

$$R_{i,j}(t) = \begin{cases} Y_{i,j}(t) & if\ rand_j(0,1) \leq Cr\ or\ j = rn(i) \\ Z_{i,j}(t) & if\ rand_j(0,1) > Cr\ or\ j \neq rn(i) \end{cases} \tag{4}$$

For $j = 1, 2 \dots D$ and $rand_j(0,1) \in [0,1]$ is the jth evaluation of a uniform random number generator. $rn(i) \in [1, 2, \dots \dots, D]$ is a randomly chosen index to ensures that $\overrightarrow{R_i}(t)$ gets at least one component from $\overrightarrow{Z_i}(t)$. Finally 'selection' is performed in order to determine which one between the target vector and trial vector will survive in the next generation i.e. at time $t = t + 1$. If the trial vector yields a better value of the fitness function, it replaces its target vector in the next generation; otherwise the parent is retained in the population:

$$\overrightarrow{Z_i}(t+1) = \begin{cases} \overrightarrow{R_i}(t) & if\ \ f\left(\overrightarrow{R_i}(t)\right) \leq f\left(\overrightarrow{Z_i}(t)\right) \\ \overrightarrow{Z_i}(t) & if\ \ f\left(\overrightarrow{R_i}(t)\right) > f\left(\overrightarrow{Z_i}(t)\right) \end{cases} \tag{5}$$

where $f(.)$ is the function to be minimized.

Selection of Control Parameters in DE is very crucial. There have been several approaches for adaptation and self-adaptation of these parameters. Brest et al. [11] discussed an efficient technique for adapting control parameters associated with DE. JADE [10] has been implemented with a new mutation strategy "DE/current-to-pbest" with optional external archive and updating control parameters in an adaptive manner. The DE/current-to-pbest in JADE is a generalization of the classic "DE/current-to-best". Here, the optional archive operation utilizes historical data to provide information of progress direction. In Self-adaptive DE (SaDE) [13], both trial vector generation strategies and their associated control parameter values are gradually self-adapted by learning from their previous experiences in generating promising solutions. Salman et al. [14] have investigated a self-adaptive version of DE, called SDE which was tested on nine benchmark functions where it generally outperformed other well-known and other adaptive versions of DE.

DE has been used along with FCM [8] [15] and has been tested on a range of datasets. Results indicate that DE based FCM algorithms are robust in obtaining the optimal number of clusters and can be treated as viable alternatives to the FCM when stability and accuracy are essential criteria.

4 Fitness-Based Adaptive Differential Evolution

Storn has suggested that the DE control parameters are to be adjusted as $F \in [0.5, 1]$, $CR \in [0.8, 1]$ and $NP = 10D$ [9]. Here an approach which is based on the fitness of the population pool has been adapted that was originally proposed for the control of parameters of GA [16]. This method for adapting of the control parameters depends on the fitness of the population pool. Here we have based the strategy on the *DE/rand/1/bin* scheme. Each vector was extended with its own *F* and *CR* values and the control parameters were self-adjusted in every generation for each individual according to the scheme:

$$F_{i,G+1} = \begin{cases} k_1 & if \;\; f_{i,G} \leq \bar{f}_G \\ k_2 \dfrac{f_{max,G+1} - f_{i,G}}{f_{max,G} - \bar{f}_G} & \text{otherwise} \end{cases}$$

$$Cr_{i,G+1} = \begin{cases} k_3 & if \;\; f_{i,G} \leq \bar{f}_G \\ k_4 \dfrac{f_{max,G+1} - f_{i,G}}{f_{max,G} - \bar{f}_G} & \text{otherwise} \end{cases} \tag{6}$$

where, $f_{max,G+1}$ and $f_{max,G}$ are the maximum fitness values of generation $G+1$ and G respectively, $f_{i,G}$ is the fitness of the ith individual of Gth generation, $\bar{f}_G$ is the average fitness of the generation G. $F_{i,G+1}$ and $Cr_{i,G+1}$ are the values of F and Cr of the ith individual in the generation $G+1$.

The values of $k_1, k_2, k_3, k_4 \leq 1.0$.

Therefore by using the fitness based adaptive DE algorithm the user does not have to adjust the values of F and Cr parameters while the time complexity does not increase.

5 Results and Discussion

The algorithms were implemented in MATLAB 2010 on a Pentium Core2Duo machine with 2.20 Ghz and 1 GB RAM under Windows XP environment. Performance of the algorithms was compared on 3 natural datasets Iris, Wine, Motorcycle and 2 synthetic datasets Data1 and Data2 [17]. The table-1 lists the different aspects of the datasets that have been used in the paper. The table-2 lists the parameters settings that are used for different algorithms. The degree of fuzziness m was fixed at 2.

Results are reported as the mean of the objective functions of 50 independent runs each run containing 1000 generations. Detailed results of best, worst, mean objective function value and standard deviation are tabulated in the table-3. The best values of each have been marked in bold letters. From this table we see that the FBADE algorithm has better performance in almost all the datasets.

Table 1. The Test Datasets

Data Set	n	d	cc
Iris data	150	4	3
Wine data	178	13	3
Motorcycle	133	3	3
Data set 1	100	2	2
Data set 2	200	2	4

Table 2. Parameter Settings of different algorithms

FCM-DE [8]	**FCM-SDE [14]**	**FCM-FBADE**
$F = 0.6$, $Cr = 0.9$	$F_i = N[0.5, 0.15]$, $Cr_i = N[0.5, 0.15]$	$K_1 = 0.6, K_2 = 0.9$ $K_3 = 0.6, K_4 = 0.9$
$NP = 70$	$NP = 70$	$NP = 70$
$MAXGEN = 1000$	$MAXGEN = 1000$	$MAXGEN = 1000$

Table 3. Comparison of fuzzy clustering between FCM, FCM-DE [8], FCM-SDE [14] and FCM-FBADE algorithms

Test Data Set	Algorithm	Best	Worst	Average	Standard Deviation
Iris	FCM	60.575958	105.873104	61.26987	4.901618
	FCDE	60.505956	60.581776	0.000479	0.016508
	FC-SDE	60.515521	60.570125	60.535252	0.030288
	FC-FBADE	**60.505710**	**60.505766**	**60.505733**	**4.0193e-005**
Wine	FCM	1796082.76	2698550.23	1805107.43	90019.71
	FCDE	1796082.76	**1796103.25**	**1796083.37**	2.273523
	FC-SDE	1796125.93	1796483.29	1796245.05	206.321
	FC-FBADE	**1796076.26.**	1796323.25	1796157.37	**2.255310**
Motor Cycle	FCM	4.8670087 e+004	4.8670087 e+004	4.8670087 e+004	3.652541 e-003
	FCDE	4.8670084 e+004	4.8670085 e+004	4.8670084 e+004	7.605277 e-004
	FC-SDE	4.8670086 e+004	4.8670086 e+004	4.8670084 e+004	2.471007 e-006
	FC-FBADE	**4.8670084 e+004**	**4.8670084 e+004**	**4.8670084 e+004**	**0**
Data1	FCM	48.96972	59.74185	48.98811	0.020438
	FCDE	48.90943	48.90943	48.90943	2.004859 e-014
	FC-SDE	48.90963	48.90983	48.90963	8.702335 e-014
	FC-FBADE	**48.90943**	**48.90943**	**48.90943**	**1.409567 e-014**
Data2	FCM	1.29965 e+002	1.30121 e+002	1.29988 e+002	0.085212
	FCDE	1.29942 e+002	1.29972 e+002	1.29948 e+002	0.013702
	FC-SDE	1.29942 e+002	1.30104 e+002	1.30027 e+002	0.081197
	FC-FBADE	**1.29940 e+002**	**1.29942 e+002**	**1.29941 e+002**	**0.003254**

6 Conclusion

The performance of DE is sensitive to the choice of control parameters and finding the right combination of values for these parameters for each problem is a time consuming task. This paper investigated an adaptive version of DE, called FBADE. The approach was tested on 5 clustering datasets where it generally outperformed fuzzy clustering algorithms using DE and SDE. Thus we can infer that for fuzzy clustering problems, FBADE is an attractive choice because of the fact that it does not require any parameter tuning.

References

[1] Bezdek, J.C.: Fuzzy Mathematics in Pattern Classification, Ph. D. thesis, Center for Applied Mathematics, Cornell University (1973)

[2] Kuncheva, L.I., Bezdek, J.C.: Selection of cluster prototypes from data by a genetic algorithm. In: Proc. 5th European Congress on Intelligent Techniques and Soft Computing (EUFIT), Aachen, Germany, vol. 18, pp. 1683–1688 (1997)

[3] Sun, L.X., Danzer, K.: Fuzzy cluster analysis by simulate annealing. Journal of Chemometrics 10, 325–342 (1996)

[4] Hall, L.O., Ozyurt, I.B., Bezdek, J.C.: Clustering with a genetically optimized approach. IEEE Transactions on Evolutionary Computation 3, 103–112 (1999)

[5] Al-Sultan, K.S., Fedjki, C.A.: A tabu search-based algorithm for the fuzzy clustering problem. Pattern Recognition 30, 2023–2030 (1997)

[6] Runkler, T.A., Katz, C.: Fuzzy Clustering by Particle Swarm Optimization. In: IEEE International Conference on Fuzzy Systems, pp. 601–608 (2006)

[7] Das, S., Sil, S.: Kernel-induced fuzzy clustering of image pixels with an improved differential evolution algorithm. Information Sciences 180(8), 1237–1256 (2010)

[8] Kao, Y., Lin, J., Huang, S.: Fuzzy Clustering by Differential Evolution. Intelligent Systems Design and Application (1), 246–250 (2008)

[9] Storn, R., Price, K.: Differential evolution a simple and efficient heuristic for global optimization over continuous spaces. Journal of Global Optimization 11(4), 341–359 (1997)

[10] Zhang, J., Sanderson, A.C.: JADE: Adaptive Differential Evolution with Optional External Archive. IEEE Transactions on Evolutionary Computation 13(5), 945–958 (2009)

[11] Brest, J., Greiner, S., Boskovic, B., Mernik, M., Zumer, V.: Self-Adapting Control Parameters in Differential Evolution: A Comparative Study on numerical benchmark problems. IEEE Transactions on Evolutionary Computation 10(6), 646–657 (2006)

[12] Das, S., Suganthan, P.N.: Differential Evolution: A Survey of the State-of-the-Art. IEEE Transactions on Evolutionary Computation 15(1), 4–31 (2011)

[13] Qin, A.K., Huang, V.L., Suganthan, P.N.: Differential Evolution Algorithm with strategy adaptation for Global Numerical Optimization. IEEE Transactions on Evolutionary Computation 13(2) (2009)

[14] Salman, A., Engelbrecht, A.P., Omran, M.G.H.: Empirical analysis of self-adaptive differential evolution. European Journal of Operational Research 183, 785–804 (2007)

[15] Ravi, V., Aggarwal, N., Chauhan, N.: Differential Evolution Based Fuzzy Clustering. In: Panigrahi, B.K., Das, S., Suganthan, P.N., Dash, S.S. (eds.) SEMCCO 2010. LNCS, vol. 6466, pp. 38–45. Springer, Heidelberg (2010)

[16] Srinivas, M., Patnaik, L.M.: Adaptive probabilities of crossover and mutation in genetic algorithms. IEEE Transactions on Systems, Man, and Cybernetics - Part A: Systems and Humans 24(4), 656–667 (1994)

[17] Blake, C.L., Merz, C.J.: UCI repository of machine learning databases. University of California, Department of Information and Computer Science, Irvine (1998), `http://www.ics.uci.edu/mlearn/MLRepository.html`

A Study of Roulette Wheel and Elite Selection on GA to Solve Job Shop Scheduling

Sandhya Pasala, Balla Nandana Kumar, and Suresh Chandra Satapathy

ANITS, Vhskapatnam,
Swarnandhra College of Engineering and Technology, Narsapur
{sandhya.cse40,nandankumar007}@gmail.com,
sureshsatapathy@ieee.org

Abstract. Usage of Genetic algorithm to solve NP hard problems like job shop scheduling yields remarkable results. The choice of crossover and mutation parameters however effect the GA performance and still the selection off - springs plays a major role in tuning the GA performance and has remarkable significance in controlling early convergence or local convergence. In this paper we tried to study the results of roulette wheel and elite selection process on a linear chromosome structure.

Keywords: Genetic Algorithm, Job Shop Scheduling, Roulette wheel, Elite Selection.

1 Introduction

The Job-shop Scheduling Problem (JSP) is one of the most difficult problems, as it is classified as an NP-complete one (Carlier and Chretienne, 1988; Garey and Johnson, 1979). In many cases, the combination of goals and resources exponentially increases the search space, and thus the generation of consistently good scheduling is particularly difficult because we have a very large combinatorial search space and precedence constraints between operations. Exact methods such as the branch and bound method and dynamic programming take considerable computing time if an optimum solution exists. In order to overcome this difficulty, it is more sensible to obtain a good solution near the optimal one. Stochastic search techniques such as evolutionary algorithms can be used to find a good solution. They have been successfully used in combinatorial optimization, e.g. in wire routing, transportation problems, scheduling problems,

Several problems in various industrial environments are combinatorial. This is the case for numerous scheduling and planning problems. Generally, it is extremely difficult to solve this type of problems in their general form. Scheduling can be defined as a problem of finding an optimal sequence to execute a finite set of operations satisfying most of the constraints. The problem so formulated is extremely difficult to solve, as it comprises several concurrent goals and several resources which must be allocated to lead to our goals, which are to maximize the utilization of individuals and/or machines and to minimize the time required to complete the entire process being scheduled.

S.C. Satapathy et al. (Eds.): Proc. of Int. Conf. on Front. of Intell. Comput., AISC 199, pp. 477–485.
DOI: 10.1007/978-3-642-35314-7_54

1.1 Description of Job-Shop Scheduling

The task of production scheduling consists in the temporal planning of the processing of a given set of orders. The processing of an order corresponds to the production of a particular product. It is accomplished by the execution of a set of operations in a predefined sequence on certain resources, subject to several constraints. The result of scheduling is a schedule showing the temporal assignment of operations of orders to the resources to be used. In this study, we consider a flexible job shop problem. Each operation can be preformed by some machines with different processing times, so that the problem is known to be NP hard. The difficulty is to find a good assignment of an operation to a machine in order to obtain a schedule which minimizes the total elapsed time (makespan).

1.2 The Problem Formulation

Flexible Job Shop Problem can be described as follows:

1. A set of jobs, J = {J1, J2, ℵ, Jn}. The number of $ jobs in this set is n.
2. A set of machines, M = {M1, M2, ℵ, Mm}. The number of machines in this set is m.
3. Every job consists of a sequence of operations. J_i= {O_{i1}, O_{i2}, … , O_{ini} }. O_{ij} means that this operation is the j-th operation in J_i. The number of operations in J_i is n_i.
4. Every operation O_{ij} can be processed in a machine subset of M. We called M_{ij} is the machine set of operation O_{ij}.

Table 1. Processing time of the operations on different machines.

	M_1	M_2	M_3	M_4	M_5
$O_{1,1}$	1	8	3	7	5
$O_{2,1}$	3	5	2	6	4
$O_{3,1}$	6	7	1	4	3
$O_{1,2}$	1	4	5	3	8
$O_{2,2}$	2	8	4	9	3
$O_{3,2}$	9	5	1	2	4
$O_{1,3}$	1	8	9	3	2
$O_{2,3}$	5	9	2	5	3

2 The Genetic Algorithm

The genetic algorithm is the process of simulating biological survival on to a randomly generated solution space tending towards finding the optimum solution. Each a solution in the solution space is mapped to a mathematical structure and this structure is called a chromosome. The below are the stages of this algorithm.

1. **Selection of the chromosome structure**
   The problem has to be translated into a chromosome representation. Each gene of the chromosome corresponds to a decision variable of the problem. According to problem complexity, the chromosome structure can be either conventional (a binary string) or not (reals, a series or a sequence of orders, a parallel form, etc.).
   In this paper, we use the task sequencing list representation proposed by Kacem et al [9]. The gene of an individual is formed by a triple (i, j, k). It means that the operation Oij is processed on machine Rk. And the length of individual is the number of operations in all jobs. As an example, we can have an indivudual, which shows in Fig 1.

(1,1,3)	(1,2,1)	(2,1,3)	(2,2,2)	(3,1,4)	(1,3,1)	(3,2,1)	(2,3,4)

Fig. 1. The coding of an individual

2. **Initialization of the population of chromosomes**
   The initial population can be generated at random if the problem structure allows it.
   In this paper, we get the in-degree of all operations. Make a set A. Make a set B as empty set. Put the operation that its in degree is 0 into A , If set A is empty, the process will exit and the sequence of operations has been recorded in list L. Otherwise, select an operation O_{ij} from set A. Get the machine M_k which handles operation O_{ij}. Set B to empty Put all operations in set A and assigned to Mk into B. Select an operation from B with roulette wheel method. If O_{ab} is an operation in set B..
3. **Perform genetic operations on chromosomes**
   Some operators are introduced in genetic algorithms to produce a solution. Among them there are two categories of operators: crossover and mutation.

   a. **Crossover**
      Here, the Precedence Preserving Order-based crossover (POX) [10] strategy is adopted.

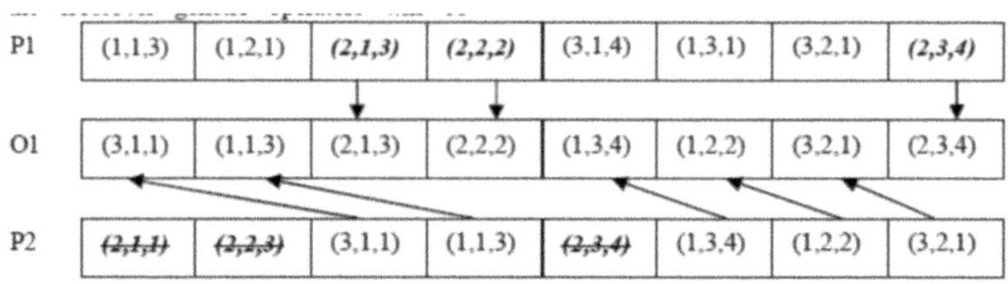

Fig. 2. Precedence preserving order-based crossover

For an easy statement, we call the first parent P1, the second parent P2, the first offspring O1 and the second offspring O2. The process of POX strategy is described as follows: Select an operation Oij for P1. Oij belongs to Ji. Copy all the operations in Ji of P1 to O1 and complete O1 with the remaining operations, in the same order as they appear in P2. Fig 2 shows an example of POX. Supposed that the gene (2, 2, 2) is selected. And the operation of gene (2, 2, 2) is belonged to J2. So, all the operations in J2 will be copied to O1. These operations will be deleted in P2. And put all the operations remained in P2 into O1.

b. Mutation

Here, we take the Precedence Preserving Shift mutation (PPS) strategy [11]. PPS selects an operation from a single parent chromosome and moves it into another position, taking care of the precedence constraints for that operation [10]. Fig 3 shows an example of PPS.

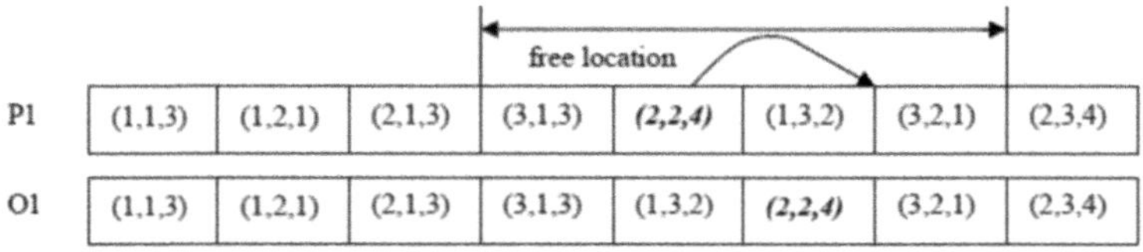

Fig. 3. Precedence preserving shift mutation

4. Evaluation chromosomes

During evolutionary generation, an evaluating system is set up to assess the chromosomes and to select those chromosomes that are fit enough for the next generation.

3 The Roulette Wheel Method

The basic part of the selection process is to stochastically select from one generation to create the basis of the next generation. The requirement is that the fittest individuals have a greater chance of survival than weaker ones. This replicates nature in that fitter individuals will tend to have a better probability of survival and will go forward to form the mating pool for the next generation. Weaker individuals are not without a chance. In nature such individuals may have genetic coding that may prove useful to future generations.

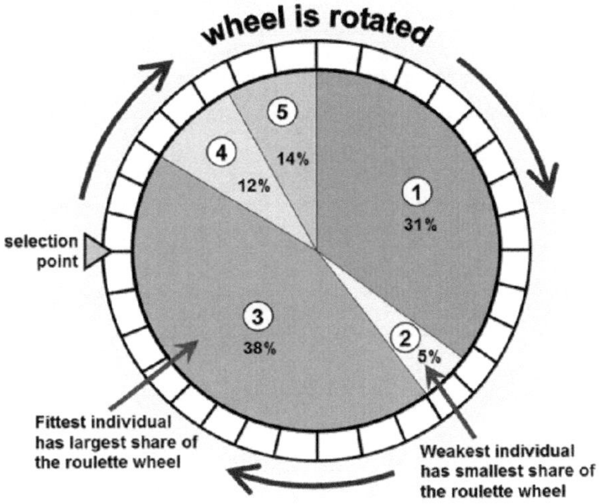

4 The Elite Method

The generations with no change in highest-scoring (elite) chromosome (GensNoChange) is the second termination criterion which is the number of generations that may pass with no change in the elite chromosome before that elite chromosome will be returned as the search answer.

5 Results and Conclusion

With the below set of parameters for the genetic algorithm we have deduced the following results as tabulated and represented by the apprprate graph.

Crossover rate = 75%,
Mutation rate = 5%,
Number of generations = 5000.

Table 2. Generation number giving the best maksepan

Run number	Generation number	Makespan
1	2265	7
2	2136	7
3	1965	7
4	1936	7
5	2035	7
6	**1853**	7
7	1896	7
8	1906	7
9	1885	7
10	1869	7

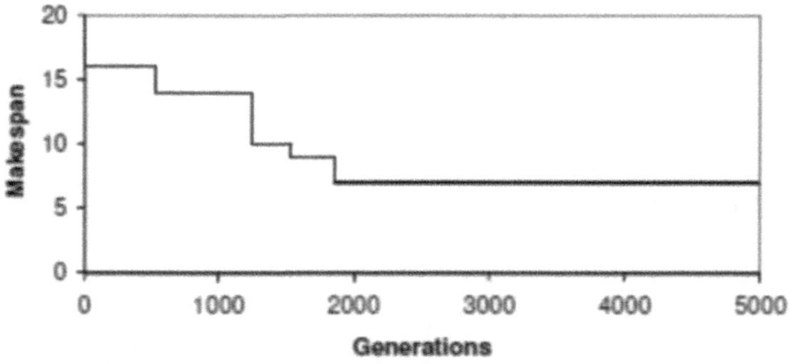

Fig. 4. Graph representing the decrease of schedule cost

Herewith, we conclude that the inclusion of elite and roulette wheel in the selection process of the off springs to be subjected to the crossover and mutation genetic operation, we could identify a large enhancement in the schedule cost reduction. We also identified that the off spring quality is better in this scenario.

References

[1] Essafi, I., Mati, Y., Dauzere-Peres, S.: A genetic local search algorithm for minimizing total weighted tardiness in the job-shop scheduling problem. Computers & Operations Research 35, 2599–2616 (2008)

[2] Jain, A.S., Meeran, S.: Deterministic job-shop scheduling: past, present and future. European Journal of Operational Research 113(2), 390–434 (1999)

[3] Volta, R., Della Croce, G., Tadei, F.: A genetic algorithm for the job shop scheduling problem. Computers and Operations Research 22, 15–24 (1995)

[4] Xing, L.-N., Chen, Y.-W., Yang, K.-W.: Multiobjective flexible job shop schedule: Design and evaluation by simulation modeling. Applied Soft Computing 9, 362–376 (2009)
[5] Van Laarhoven, P.J.M., Aarts, E.H.L., Lenstra, J.K.: Job shop scheduling by simulated annealing. Computers and Operations Research 40(1), 113–125 (1992)
[6] Waligora, G.: Tabu search for discrete-continuous scheduling problems with heuristic continuous resource allocation. European Journal of Operational Research 193, 849–856 (2009)
[7] Vilcot, G., Billaut, J.-C.: A tabu search and a genetic algorithm for solving a bicriteria general job shop scheduling problem. European Journal of Operational Research 190, 398–411 (2008)
[8] Lourenco, H.R.: Job-shop scheduling: computational study of local search and large-step optimization methods. European Journal of Operational Research 83, 347–364 (1995)
[9] Kacem, I., Hammadi, S., Borne, P.: Approach by localization and multiobjective evolutionary optimization for flexible job-shop scheduling problems. IEEE Transactions on Systems, Man, and Cybernetics, Part C 32(1), 1–13 (2002)
[10] Pezzella, F., Morganti, G., Ciaschetti, G.: A genetic algorithm for the Flexible Job-shop Scheduling Problem. Computers & Operations Research 35, 3202–3212 (2008)
[11] Lee, K.M., Yamakawa, T., Lee, K.M.: A genetic algorithm for general machine scheduling problems. International Journal of Knowledge-Based Electronic 2, 60–66 (1998)

Appendix

Data Source Selection

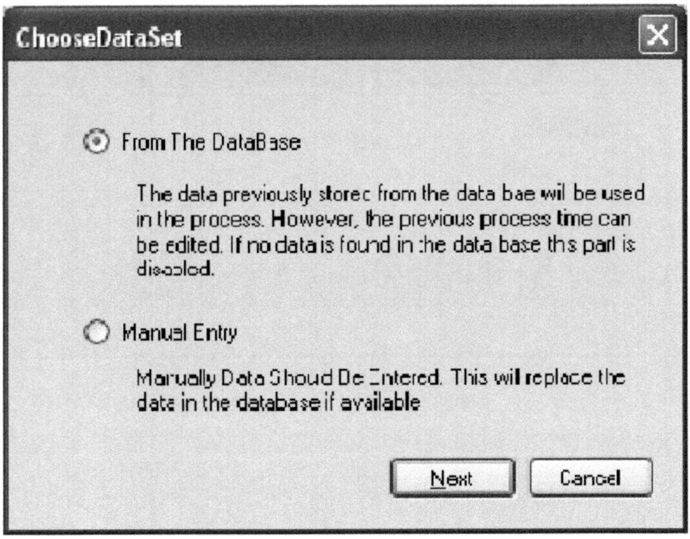

Machines and jobs Selection

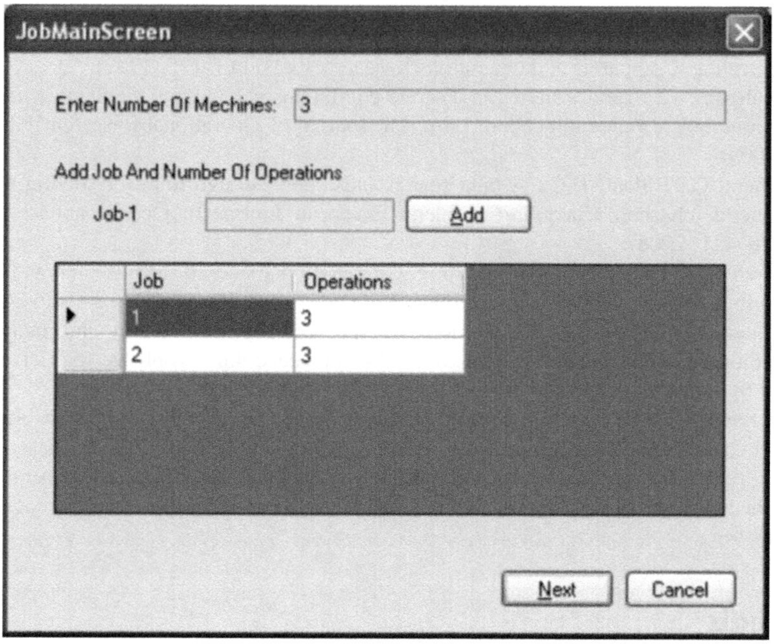

Process Time

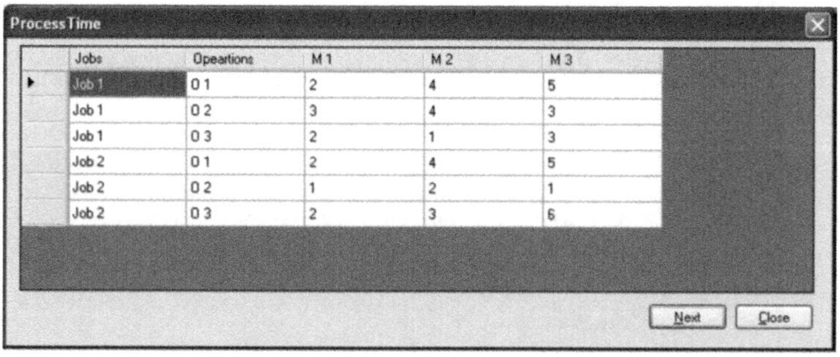

Experimentation Console

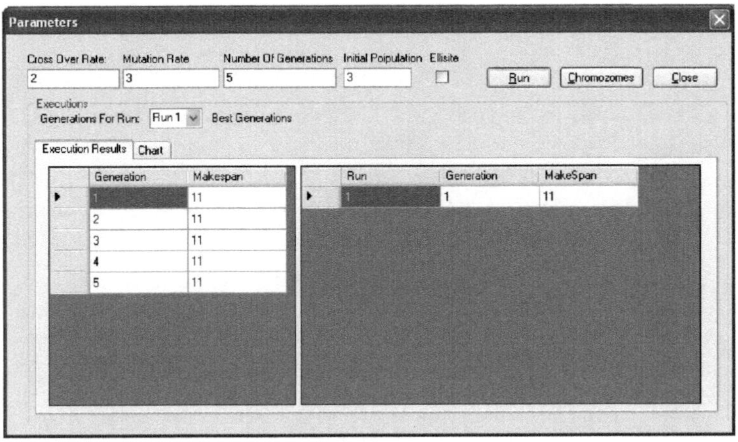

The Graph

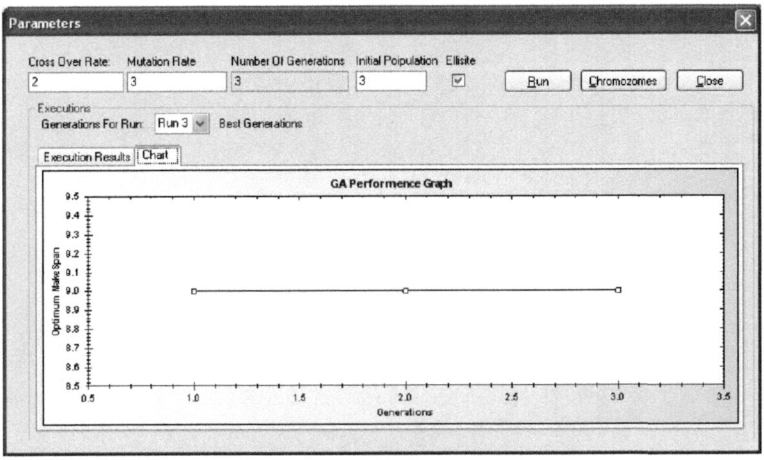

Teaching Learning Based Optimized Mathematical Model for Data Classification Problems

Polinati Vinod Babu[1], Suresh Chandra Satapathy[2], Mohan Krishna Samantula[3], P.K. Patra[4], and Bhabendra Narayan Biswal[5]

[1] Swarnaandhra College of Engineering and Technology, Narsapur, India
[2] MIEEE, ANITS, Visakhapatnam, India
[3] MIEEE, GITAM University, Visakhapatnam, India
[4] CET, Bhubaneswar, India
[5] Bhubaneswar Engineering College, Bhubaneswar, India
vinodbabusir@gmail.com, {sureshsatapathy,smkrishna}@ieee.org, {hodcse,bhabendra_biswal}@yahoo.co.in

Abstract. This paper presents application of yet one more optimization technique based on evolutionary computation approach to locate the optimal values of the coefficients of terms of polynomial equations which is developed to classify the unknown dataset. A recent optimization technique known as Teaching Learning Based Optimization (TLBO) is used here for optimizing the coefficients of polynomial terms for classifying many bench mark datasets. The original mathematical model for classification problems are developed using Polynomial neural network (PNN) and then the coefficients of the terms of polynomials are optimized separately with Least mean square (LSE) and TLBO approach. The comparisons of the two approaches are suitably presented with classification accuracies for many bench mark datasets. The results reveal that TLBO optimized polynomials are performing better than LSE- optimized polynomial, herein known as PNN models for all investigated datasets.

Keywords: Polynomial Neural Network, Group Methods of Data Handling, TLBO.

1 Introduction

Neural networks, Decision Trees, SVM etc are used extensively for pattern recognition [1-9]. Suitable mathematical model for classification problems have been attracted researchers to investigate deep into the subject. Group Method of data handling (GMDH) based Polynomial Neural Network (PNN) is a popular approach for evolving a short-term polynomial equation for data classification. This mathematical model is developed using the features of the data sets as input to the PNN. Number of terms in the polynomial equation, degree of the polynomial, number of features in each term, the degree of the polynomials, the number of terms in the polynomial equations and number and type of features are determined while the model is developed using PNN technique. Mishra et al [10] have investigated the approach of PNN with different real world data sets. Although this approach is a

S.C. Satapathy et al. (Eds.): Proc. of Int. Conf. on Front. of Intell. Comput., AISC 199, pp. 487–496.
DOI: 10.1007/978-3-642-35314-7_55

suitable one but it suffers from the classification performance accuracy. In this paper we suggest a suitable approach of developing mathematical models in terms of polynomial equations using Teaching Learning Based Optimization (TLBO)[11] techniques which is comparatively less complex than PNN providing competitive performance. The TLBO is based on the principle teaching learning process in a class room scenario. The degree of polynomials, number of terms in the equation and the variables in the equation (i.e. features) are randomly chosen in suitable ranges for developing the model using TLBO technique. Our derived polynomial equations using TLBO are found to be performing better compared to PNN approach. For datasets like Iris, Diabetes, Pima, Iono and Balance scale.

The section 2 describes the PNN approach and the motivation for our proposed model. The basic TLBO is discussed in section 3. Section 4 and section 5 describe our model and simulation results respectively. Finally conclusion and further enhancements are given in the section 6.

2 GMDH-Type Polynomial Neural Network Model

The GMDH is a category of inductive self-organization data driven approach. Relationship between input –output variables can be approximated by Volterra functional series, the discrete form of which is Kolmogorov-Gabor Polynomial [14]

$$y = C_0 + \sum_{k1} C_{k1} x_{k1} + \sum_{k1k2} C_{k1k2} x_{k1} x_{k2} + \sum_{k1k2k3} C_{k1k2k3} x_{k1} x_{k2} x_{k3} + \ldots.. \quad (1)$$

where C_k denotes the coefficients or weights of the Kolmogorov-Gabor polynomial & x vector is the input variables. A new algorithm called GMDH is developed by Ivakhnenko [15-16] which is a form of Kolmogorov-Gabor polynomial. He proved that a second order polynomial i.e.

$$y = a_0 + a_1 x_i + a_2 x_j + a_3 x_i x_j + a_4 x_i^2 + a_5 x_j^2 \quad (2)$$

which takes only two input variables at a time and can reconstruct the complete Kolmogorov-Gabor polynomial through an iterative procedure. The GMDH-type Polynomial Neural Networks are multilayered model consisting of the neurons/active units /Partial Descriptions (PDs) whose transfer function is a short- term polynomial described in equation (2).

The details of the model developed by PNN and least square estimation technique are explained below.

Let the input and output data for training is represented in the following manner

$$\begin{bmatrix} x_{11} & x_{12} & \cdots & x_{1m} & y_1 \\ x_{21} & x_{22} & \cdots & x_{2m} & y_2 \\ . & . & . & . & . \\ x_{n1} & x_{n2} & \cdots & x_{nm} & y_n \end{bmatrix}$$

In general, it is expressed as

$$(X_i, y_i) = (x_{1i}, x_{2i}, \ldots, x_{mi}, y_i)$$

where i =1, 2, 3, … ,n.

The input and output relationship of the above data by PNN algorithm can be described in the following manner:

$$y = f(x_1, x_2, x_3, \ldots, x_m)$$

where m is the number of features in the dataset.

The architecture of a PNN with four input features is shown in "Fig. 1"

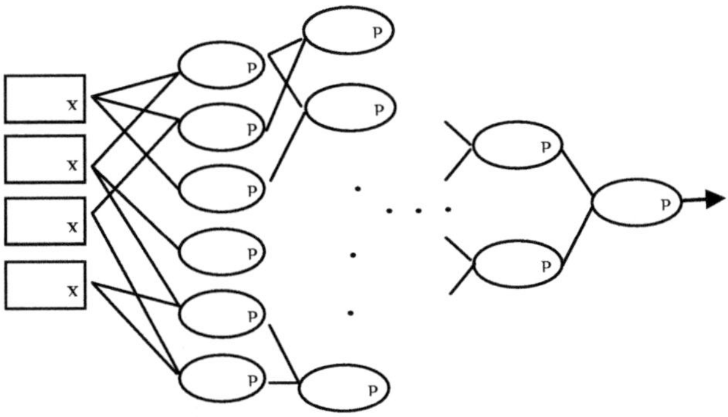

Fig. 1. Basic PNN Model

Number of PDs, K in each layer depends on the number of input features M as below

$$K = \overset{M}{\underset{2}{C}} = M(M-1)/2$$

The input index of features (p,q) to each PD, may be generated using the following algorithm

1. Let layer is l.
2. Let k=1,
3. for i =1 to m-1
4. for j = i+1 to m
5.   Then PD_k^1 will receive input from the features
6.   p=i; & q=j;
7.   k=k+1;
8.   end for
9 end for

Let us consider the equations for the first PD of layer1, which receives input from feature 1 and 2.

$$\begin{bmatrix} d_1 = y_1 - (c_{11} + c_{12}x_{11} + c_{13}x_{12} + c_{14}x_{11}x_{12} + c_{15}x_{11}^2 + c_{16}x_{12}^2 \\ d_2 = y_2 - (c_{11} + c_{12}x_{21} + c_{13}x_{22} + c_{14}x_{21}x_{22} + c_{15}x_{21}^2 + c_{16}x_{22}^2 \\ \cdots\cdots\cdots\cdots\cdots\cdots\cdots\cdots\cdots\cdots \\ d_n = y_n - (c_{11} + c_{12}x_{n1} + c_{13}x_{n2} + c_{14}x_{n1}x_{n2} + c_{15}x_{n1}^2 + c_{16}x_{n2}^2 \end{bmatrix}$$

where the d vector is the error estimation between the target and the obtained outputs.

This equation in general may be written as

$$d_i = y_i - (c_{j1} + c_{j2}x_{ip} + c_{j3}x_{iq} + c_{j4}x_{ip}x_{iq} + c_{j5}x_{ip}^2 + c_{j6}x_{iq}^2$$

where

(i) i=1, 2, ..., n.
(ii) j=1, 2, ..., k
(iii) k=m(m-1)/2

The equations for the least square are

$$\begin{aligned} \Pi &= d_1^2 + d_2^2 + \ldots + d_n^2 \\ &= \sum_{i=1}^{n} d_i^2 \\ &= \sum_{i=1}^{n} \left[y_i - (c_{j1} + c_{j2}x_{ip} + c_{j3}x_{iq} + c_{j4}x_{ip}x_{iq} + c_{j5}x_{ip}^2 + c_{j6}x_{iq}^2) \right]^2 \end{aligned}$$

To minimize the error, we get the first derivatives of Π in terms of all the unknown variables (i.e. the coefficients).

$$\begin{bmatrix} \dfrac{\partial \Pi}{\partial c_{j1}} = 2\sum_{i=1}^{n} \left[y_i - (c_{j1} + c_{j2}x_{ip} + c_{j3}x_{iq} + c_{j4}x_{ip}x_{iq} + c_{j5}x_{ip}^2 + c_{j6}x_{iq}^2 \right] = 0 \\ \dfrac{\partial \Pi}{\partial c_{j2}} = 2\sum_{i=1}^{n} x_{ip} \left[y_i - (c_{j1} + c_{j2}x_{ip} + c_{j3}x_{iq} + c_{j4}x_{ip}x_{iq} + c_{j5}x_{ip}^2 + c_{j6}x_{iq}^2 \right] = 0 \\ \dfrac{\partial \Pi}{\partial c_{j3}} = 2\sum_{i=1}^{n} x_{iq} \left[y_i - (c_{j1} + c_{j2}x_{ip} + c_{j3}x_{iq} + c_{j4}x_{ip}x_{iq} + c_{j5}x_{ip}^2 + c_{j6}x_{iq}^2 \right] = 0 \\ \dfrac{\partial \Pi}{\partial c_{j4}} = 2\sum_{i=1}^{n} x_{ip}x_{iq} \left[y_i - (c_{j1} + c_{j2}x_{ip} + c_{j3}x_{iq} + c_{j4}x_{ip}x_{iq} + c_{j5}x_{ip}^2 + c_{j6}x_{iq}^2 \right] = 0 \\ \dfrac{\partial \Pi}{\partial c_{j5}} = 2\sum_{i=1}^{n} x_{ip}^2 \left[y_i - (c_{j1} + c_{j2}x_{ip} + c_{j3}x_{iq} + c_{j4}x_{ip}x_{iq} + c_{j5}x_{ip}^2 + c_{j6}x_{iq}^2 \right] = 0 \\ \dfrac{\partial \Pi}{\partial c_{j6}} = 2\sum_{i=1}^{n} x_{iq}^2 \left[y_i - (c_{j1} + c_{j2}x_{ip} + c_{j3}x_{iq} + c_{j4}x_{ip}x_{iq} + c_{j5}x_{ip}^2 + c_{j6}x_{iq}^2 \right] = 0 \end{bmatrix}$$

On expanding the above equations, we get

We know that,

$$\begin{aligned} XA &= Y \\ X^T XA &= X^T Y \\ A &= \left(X^T X\right)^{-1} X^T Y \end{aligned}$$

Here

$$X = \begin{bmatrix} \sum_{i=1}^{n} 1 + \sum_{i=1}^{n} x_{ip} + \sum_{i=1}^{n} x_{iq} + \sum_{i=1}^{n} x_{ip}x_{iq} + \sum_{i=1}^{n} x_{ip}^2 + \sum_{i=1}^{n} x_{iq}^2 \\ \sum_{i=1}^{n} x_{ip} + \sum_{i=1}^{n} x_{ip}^2 + \sum_{i=1}^{n} x_{ip}x_{iq} + \sum_{i=1}^{n} x_{ip}^2 x_{iq} + \sum_{i=1}^{n} x_{ip}^3 + \sum_{i=1}^{n} x_{ip}x_{iq}^2 \\ \sum_{i=1}^{n} x_{iq} + \sum_{i=1}^{n} x_{ip}x_{iq} + \sum_{i=1}^{n} x_{iq}^2 + \sum_{i=1}^{n} x_{ip}x_{iq}^2 + \sum_{i=1}^{n} x_{ip}^2 x_{iq} + \sum_{i=1}^{n} x_{iq}^3 \\ \sum_{i=1}^{n} x_{ip}x_{iq} + \sum_{i=1}^{n} x_{ip}^2 x_{iq} + \sum_{i=1}^{n} x_{ip}x_{iq}^2 + \sum_{i=1}^{n} x_{ip}^2 x_{iq}^2 + \sum_{i=1}^{n} x_{ip}^3 x_{iq} + \sum_{i=1}^{n} x_{ip}x_{iq}^3 \\ \sum_{i=1}^{n} x_{ip}^2 + \sum_{i=1}^{n} x_{ip}^3 + \sum_{i=1}^{n} x_{ip}^2 x_{iq} + \sum_{i=1}^{n} x_{ip}^3 x_{iq} + \sum_{i=1}^{n} x_{ip}^4 + \sum_{i=1}^{n} x_{ip}^2 x_{iq}^2 \\ \sum_{i=1}^{n} x_{iq}^2 + \sum_{i=1}^{n} x_{ip}x_{iq}^2 + \sum_{i=1}^{n} x_{iq}^3 + \sum_{i=1}^{n} x_{ip}x_{iq}^3 + \sum_{i=1}^{n} x_{ip}^2 x_{iq}^2 + \sum_{i=1}^{n} x_{iq}^4 \end{bmatrix}$$

$$A = \begin{bmatrix} c_{j1} \\ c_{j2} \\ c_{j3} \\ c_{j4} \\ c_{j5} \\ c_{j6} \end{bmatrix}, \text{ and } \quad Y = \begin{bmatrix} \sum_{i=1}^{n} 1 y_i \\ \sum_{i=1}^{n} x_{ip} y_i \\ \sum_{i=1}^{n} x_{iq} y_i \\ \sum_{i=1}^{n} x_{ip} x_{iq} y_i \\ \sum_{i=1}^{n} x_{ip}^2 y_i \\ \sum_{i=1}^{n} x_{iq}^2 y_i \end{bmatrix}$$

After obtaining the values of the coefficients with the testing dataset, we estimate the target

$$\hat{y}_i = (c_{j1} + c_{j2}x_{ip} + c_{j3}x_{iq} + c_{j4}x_{ip}x_{iq} + c_{j5}x_{ip}^2 + c_{j6}x_{iq}^2)$$

If the error level is not up to our desired value, we construct next layer of PNN by taking the output of the previous layer i.e. z_i, which is the input to next layer

$z_j = (c_{j1} + c_{j2}x_{ip} + c_{j3}x_{iq} + c_{j4}x_{ip}x_{iq} + c_{j5}x_{ip}^2 + c_{j6}x)$ and the process is repeated until the error decreases.

From the simulation of [10] we found that as the PNN layers grow the number of PDs in each layer also grows. Hence, pruning of the PDs are necessary to limit the computational complexity. As an example, if there are 10 features in the dataset, then it requires generation of 20000 PDs at layer 6th even after pruning. From the simulation we have seen that even though the classification error decreases up to 10th layer, however the classification accuracy is not competitively improved compared to 3rd/4th layer. These have been the input ideas to develop an alternate approach of data classification compared to PNN and our proposed model is based on our simulation experiences of [10].

3 Teaching Learning Based Optimization

This optimization method is based on the effect of the influence of a teacher on the output of learners in a class. It is a population based method and like other population based methods it uses a population of solutions to proceed to the global solution. A group of learners constitute the population in TLBO. In any optimization algorithms there are numbers of different design variables. The different design variables in TLBO are analogous to different subjects offered to learners and the learners' result is analogous to the 'fitness', as in other population-based optimization techniques. As the teacher is considered the most learned person in the society, the best solution so far is analogous to Teacher in TLBO. The process of TLBO is divided into two parts. The first part consists of the 'Teacher Phase' and the second part consists of the 'Learner Phase'. The 'Teacher Phase' means learning from the teacher and the 'Learner Phase' means learning through the interaction between learners. In the sub-sections below we briefly discuss the implementation of TLBO.

3.1 Initialization

Following are the notations used for describing the TLBO:

N: number of learners in a class i.e. "class size"
D: number of courses offered to the learners
$MAXIT$: maximum number of allowable iterations

The population X is randomly initialized by a search space bounded by matrix of N rows and D columns. The jth parameter of the ith learner is assigned values randomly using the equation

$$x_{(i,j)}^{0} = x_j^{min} + rand \times (x_j^{max} - x_j^{min}) \tag{3}$$

where *rand* represents a uniformly distributed random variable within the range (0, 1), x_j^{min} and x_j^{max} represent the minimum and maximum value for jth parameter. The parameters of ith learner for the generation g are given by

$$X_{(i)}^{g} = [x_{(i,1)}^{g}, x_{(i,2)}^{g}, x_{(i,3)}^{g}, \dots\dots, x_{(i,j)}^{g}, \dots\dots, x_{(i,D)}^{g}] \tag{4}$$

3.2 Teacher Phase

The mean parameter M^g of each subject of the learners in the class at generation g is given as

$$M^g = [m_1^g, m_2^g, \dots\dots, m_j^g, \dots\dots, m_D^g] \tag{5}$$

The learner with the minimum objective function value is considered as the teacher $X_{Teacher}^{g}$ for respective iteration. The Teacher phase makes the algorithm proceed by shifting the mean of the learners towards its teacher. To obtain a new set of improved learners a random weighted differential vector is formed from the current mean and the desired mean parameters and added to the existing population of learners.

$$Xnew_{(i)}^{g} = X_{(i)}^{g} + rand \times (X_{Teacher}^{g} - T_F M^g) \tag{6}$$

Where T_F is a teaching factor which is randomly taken at each iteration to be either 1 or 2 . If $Xnew_{(i)}^{g}$ is found to be a superior learner than $X_{(i)}^{g}$ in generation g, than it replaces inferior learner $X_{(i)}^{g}$ in the matrix.

3.3 Learner Phase

In this phase the interaction of learners with one another takes place. The process of mutual interaction tends to increase the knowledge of the learner. The random interaction among learners improves his or her knowledge. For a given learner $X_{(i)}^{g}$, another learner $X_{(r)}^{g}$ is randomly selected$(i \neq r)$. The ith parameter of the matrix $Xnew$ in the learner phase is given as

$$Xnew_{(i)}^{g} = \begin{cases} X_{(i)}^{g} + rand \times \left(X_{(i)}^{g} - X_{(r)}^{g}\right) & if\ f\left(X_{(i)}^{g}\right) < f\left(X_{(r)}^{g}\right) \\ X_{(i)}^{g} + rand \times \left(X_{(r)}^{g} - X_{(i)}^{g}\right) & oterwise \end{cases} \tag{7}$$

3.4 Algorithm Termination

The algorithm is terminated after $MAXIT$ iterations are completed.

Details of TLBO can be refereed in [11].

4 TLBO Optimized Polynomial: Our Proposed Approach

In our proposed approach we apply TLBO technique to evolve few polynomial equations to classify the data set. Finally, we choose a suitable polynomial equation as our model for the data set under investigation which gives better classification accuracy.

The polynomial equation considered in our approach can be expressed in below given form

$$y = C_0 + \sum_{i=1}^{n} C_i \prod_{j=1}^{p} x_r^q \tag{8}$$

where

n is the number of polynomial terms chosen randomly from a suitable range

p is the number of features in each term chosen randomly from the given set of features for the dataset under consideration.

r is the index of feature a random integer value between 1 and number of features in the respective dataset.

q is the degree of the feature, a random integer value chosen from a suitable range.

Our proposed model is a mimic of the PNN model. So for obtaining the suitable ranges of terms like n, p and q, we have extensively analyzed the PNN model developed [10].

5 Simulation and Result

Following are the datasets used for developing classifier mathematical model using TLBO technique. A brief description of the properties of these dataset is presented in Table 1.

Table 1. Data sets

#	Datasets	Cases	Classes	Attributes
1	Iris	150	3	4
2	Diabetes	768	2	8
3	Pima	699	2	8
4	Iono	351	2	34
5	Balance Scale	625	3	4

The entire dataset is first grouped into two parts. First part is for training and the second part is for testing. The model is developed with the training dataset and then it is tested for accuracy with the testing dataset. Usually two third of dataset is taken for training. The care is taken to have uniform samples of data in each group. Several models are developed using the training dataset. Using TLBO the coefficients of the polynomial equations are optimized. If a model gives better performance over the previous best model, then it is preserved for the future reference. We have obtained competitive classification accuracy for our data sets with 100 simulations using our proposed TLBO approach.

The percentage of correct classification for each data set using the PNN model[10] and our proposed model(TLBO Model)discussed in this paper is presented in the Table 2.

Table 2. Classification Accuracy

Dataset	**% of correct classification**	
	PNN	TLBO model
Iris	98.69	99.33
Diabetes	77.34	85.54
Pima	41.79	81.32
Iono	35.89	86.55
Balance Scale	58.24	78.01

The mathematical models two datasets are illustrated below as samples for PNN [10] and TLBO models.

Mathematical Model for Iris data set:

1. Developed by PNN

$$PD_2^1 = [-0.96751,-0.25663,0.65004,-0.13994, 0.044319,0.10067]*\text{poly}(x_1, x_3),$$

$$PD_3^1 = [1.633,-1.0407,1.4825,-0.051333, 0.09108,-0.067678]*\text{poly}(x_1, x_4);$$

$$PD_1^1 = [-1.7572,1.4898,-2.4317,0.3308,-0.15092, -0.018686]*\text{poly}(x_1, x_2);$$

$$PD_{16}^2 = [2.0965,1.4134,-0.84284,0.046201, -0.57289,0.079276]*\text{poly}(PD_2^1, x_3);$$

$$PD_{28}^2 = [2.2683,1.7305,-1.2497,-0.22651, 0.15639,0.15237]*\text{poly}(PD_3^1, x_2);$$

$$PD_{38}^2 = [-1.8284,-0.58075,0.42213,0.081458, -0.25631,0.015491]*\text{poly}(PD_1^1, x_3);$$

$$PD_{496}^3 = [-0.054529,0.54087,0.45923,-1.1876, 0.6309,0.60812]*\text{poly}(PD_{16}^2, PD_{28}^2);$$

$$PD_{557}^3 = [-0.043133,0.59432,0.41454,-0.75286, 0.39684,0.3952]*\text{poly}(PD_{28}^2, PD_{38}^2);$$

$$y = [0.0057053,1.158,-0.15925,5.3642, -2.208,-3.156]*\text{poly}(PD_{496}^3, PD_{557}^3);$$

2. Developed By TLBO_Polynomial

$$\begin{aligned} y_{iris} &= 1.142\,e^{-009} * x_1^7 * x_3^4 \\ &- 6.3452e^{-007} * x_2^4 * x_3^2 * x_4^3 \\ &- 4.753e^{-006} * x_2^7 \\ &- 8.9229e^{-009} * x_4^{11} * x_2^4 \\ &- 9.2988e^{-011} * x_4^8 * x_1^7 \\ &+ 0.23424 * x_3^2 \\ &- 0.000376 * x_1^4 * x_4^2 \\ &+ 0.076236 * x_3^2 \\ &+ 4.38812e^{-007} * x_4^4 * x_1^5 \\ &- 2.1123 \end{aligned}$$

6 Conclusion and Future Enhancement

This paper presents a new approach of developing mathematical models for classification problems using Polynomial neural Networks (PNN) optimized with Teaching learning Based optimization (TLBO) technique. The mathematical models suggested using TLBO approach is computationally less expensive compared to the model derived using PNN. The TLBO approach is able to provide good results with very few features as compared to PNN model. We have compared our result with the PNN model for five real world data sets. In all cases TLBO model outperforms PNN model with respect to classification accuracy.

References

1. Breitman, L., Friedman, J.H., Olshen, R.A., Stone, C.J.: Classification and Regression tress. Wadsworth, Belmont (1984)
2. Buntine, W.L.: Learning classification trees. Statistics and Computing 2, 63–73 (1992)
3. Cover, T.M., Hart, P.E.: Nearest neighbour pattern classification. IEEE Trans. on Information Theory 13, 21–27 (1967)
4. Duda, R.O., Hart, P.E.: Pattern Classification and Scene Analysis. Wiley, Newyork (1973)
5. Hanson, R., Stutz, J., Cheeseman, P.: Bayesian classification with correlation and inheritance. In: Proceedings of the 12th International Joint Conference on Artificial Intelligence, Sydney, Australia, vol. 2, pp. 692–698. Morgan Kaufmann (1992)
6. Michie, D., et al.: Machine Learning, Neural and Statistical Classification, Ellis Horwood (1994)
7. Richard, M.D., Lippmann, R.P.: Neural network classifiers estimate Bayesian a-posterior probabilities. Neural Computation 3, 461–483 (1991)
8. Tsoi, A.C., et al.: Comparison of three classification Techniques, CART, C4.5 and multilayer perceptrons. In: Advances in Neural Information Processing Systems, vol. 3, pp. 963–969 (1991)
9. Adem, J., Gochet, W.: Mathematical Program based heuristics for improving LP-generated classifiers for the multi-class supervised classification problem. European Journal of Operation Research (1994)
10. Misra, B.B., Satapathy, S.C., Biswal, B.N., Dash, P.K., Panda, G.: Pattern Classification using Polynomial Neural Network. Accepted for presentation at IEEE International Conferences on Cybernetics & Intelligent Systems (CIS) and Robotics, Automation & Mechatronics (RAM), CIS-RAM 2006 (2006)
11. Rao, R.V., Savsani, V.J., Vakharia, D.P.: Teaching–learning-based optimization: A novel method for constrained mechanical design optimization problems. Computer-Aided Design 43, 303–315 (2011)

Power Loss Minimization by the Placement of DG in Distribution System Using PSO

D. Sattianadan[1], M. Sudhakaran[2], S.S. Dash[1], K. Vijayakumar[1], and Bishnupriya Biswal[1]

[1] SRM University, Chennai, India
[2] Pondicherry Engg. College, Pondicherry, India
{sattia.nadan,bishnupriya06}@gmail.com,
{karan_mahalingam,munu_dash_2k}@yahoo.com,
kvijay_srm@rediffmail.com

Abstract. Power loss minimization is an important aspect in distribution System where the load variation is more compared other systems. There are different methods to minimize the power loss like DG placement, capacitor placement, load balancing etc. Among those methods DG placement was much beneficial because it is directly related to real power loss. This paper is based on power loss minimization by the placement of distributed generators (DG) in distribution system. The location of DG is found with the help of voltage stability index (VSI) and DG size is varied in small steps and corresponding power loss is calculated by running the power flow and the result obtained is verified by Particle Swarm Optimization Technic. The simulation study is carried out on a 33 bus Distribution System by considering different load models.

Keywords: Distributed generation, Voltage stability index, Particle Swarm Optimization.

1 Introduction

In the modern world, day by day the load demand increases rapidly due to Industrial and Domestic needs. On the other hand the conventional energy sources are decreasing rapidly. In this case we need an alternative method to meet the load demand, distributed generation is meant for that. It has huge potential benefits about which this paper is concerned.

The distributed generation has been defined by many researchers [1,2], but in general a distributed generation is nothing but a small generator which is connected at the consumer terminal. Placement of DG is an important factor because improper location may leads voltage instability and power loss. The Newton Rapson load flow method used in [3]. This method reduces the power loss and the cost factor very effectively, but the conventional method of load flow analysis was not applicable for distribution system because of its high R/X ratio, large value of resistance and reactance of the line and radial structure of the distribution system.

S.C. Satapathy et al. (Eds.): Proc. of Int. Conf. on Front. of Intell. Comput., AISC 199, pp. 497–504.
DOI: 10.1007/978-3-642-35314-7_56

Tuba Gozel used loss sensitivity factor for determination of the optimal size and location of DG to minimize total power loss [4]. Andrew used Linear Programming Technique for placement of DG with multiple constraints [5]. Mallikarjuna used Simulated Annealing for determining the optimal location and size of DG units in a microgrid, given the network configuration and heat and power requirements at various load points [6]. Krueasuk used PSO to find optimal location and size of DG [7]. Lalitha used fuzzy approach to find optimal DG localization [8]. Hughifam used multi-objective function to minimize cost of energy losses, Investment cost of DG and Operation and maintenance cost [9]. Ochoa minimized real power loss and simple phase short circuit level [10]. Celli used multi objective approach, based on the non-dominated sorbing Genetic Algorithm has been adopted to solve the optimal placement of different types of generation simultaneously. He saved the energy in the form of greenhouse gas emission reduction[11]. Vinoth Kumar addressed minimizing the multi objective index using genetic algorithm for the optimal Placement of DG[12].

This paper minimizes the Power loss by the Placement of optimal size of DG and is organized as follows: Section-2: Defines the Objective Function and The load flow analysis of distribution system by Power flow is a crucial part of power system design procedures, and it is categorized into transmission power flow and distribution power flow. The distribution networks commonly have some special features such as: Unbalanced loads and unbalanced operation; being radial with sometimes weakly-meshed topology; and high resistance to reactance R/X ratios. Due to these features the conventional load flow like gauss sedial and Newton Rapson fail to solve. Hence we need a special method to solve the load flow on Distribution Systems, here network Topological based load flow has been considered for load flow Analysis [13,14]. Section-3: Candidate Bus Selection by using VSI and Load Modelling has been discussed. Section-4: Explains PSO Technic. Section-5: Test Results and Discussion Section 6: Concludes the paper..

2 Objective Function

The objective of the present optimization problem is to minimize the Distribution network power loss

$$\text{Min.}f_1 = \sum_{b=1}^{N_b} (I_b)^2 \cdot R_b \tag{1}$$

Where,

N_b- Total number of branches in the given radial distribution system

b - Branch number

I_b - Branch current in branch b

R_b- Resistance of branch b

2.1 Load Flow Analysis for Radial Distribution System

The simple distribution system shown in Fig.1 will be used as an example. The power injections can be converted into the equivalent current injections using Eq. (2)

$$I_i = (P_i + Q_i / V_i)^* \tag{2}$$

And a set of equations can be written by applying Kirchhoff's Current Law (KCL) to the distribution network. Then, the branch currents can be formulated as a function of the equivalent current injections. For example, the branch currents *B*5, *B*3 and *B*1 can be expressed as,

$$\begin{aligned} B_5 &= I_6 \\ B_3 &= I_4 + I_5 \\ B_1 &= I_2 + I_3 + I_4 + I_5 + I_6 \end{aligned} \tag{3}$$

The Bus-Injection to Branch-Current (***BIBC***) can be obtained by using above equations. The BCBV matrix is responsible for the relations between the branch currents and bus voltages. The corresponding variation of the bus voltages, which is generated by the variation of the branch currents, can be found directly by using the BCBV

$$V_2 = V_1 - B_1 Z_{12} \tag{4}$$

$$V_3 = V_2 - B_2 Z_{23} \tag{5}$$

$$V_4 = V_3 - B_3 Z_{34} \tag{6}$$

By using equ (4) and equ (5), The voltage of Bus 4 can be rewritten as,

$$V_4 = V_1 - B_1 Z_{12} - B_2 Z_{23} - B_3 Z_{34} \tag{7}$$

2.2 Algorithm for Distribution System Load Flow

A brief idea of how bus voltages can be obtained for a radial system is given below.

1. .Input data.
2. Form the *BIBC* matrix.
3. Form the *BCBV* matrix.
4. Form the *DLF* matrix.
5. Iteration $k = 0$.
6. Iteration $k = k + 1$.
7. Solve the equations iteratively and update voltages $I_i^k = (P_i + Q_i / V_i)$
   $[\Delta V^{k+1}] = [DLF]\,[I^k]$
   If $I_i^{k+1} - I_i^k$ >tolerance, go to step(6)
   else print result.

3 Candidate Bus Selection Using Voltage Stability Index

A system experiences a state of voltage instability when there is a progressive or uncontrollable drop in voltage magnitude following a disturbance, increase in load demand or change in operating condition. It is usually identified by an index called voltage stability index of all the nodes in radial distribution system [4].

$$VSI(n_2) = V_1^4 - 4[P_2R_1 + Q_2X_1] V_1^4 - 4[P_2X_1 - Q_2R_1]^2 \tag{8}$$

Nodes with minimum voltage instability, in different laterals of the distributed system are chosen as the candidate location for placement of distributed generators. Following steps are involved in optimal siting of the distributed generator

a) Performed load flow to calculate the bus voltage magnitudes and total network power loss in the RDS.
b) Compute the Voltage Stability Index (VSI).
c) Select the buses with the highest priority(First, Second) and place DG.
d) Run the power flow program again and find losses of power system.
e) Check the voltage profile limitation at each bus of the system.
f) Change the size of DG to small step and calculate loss by running load flow.
g) Find the bus which lead to the lowest power system losses.

3.1 Load Modelling

A balanced load that can be represented either as constant power, constant current or constant impedance load has been considered here. The general expression of load is given below

$$P(m) = P_n [a_1 + a_2V(m) + a_3V(m)^2] \tag{9}$$

$$Q(m) = Q_n [b_1 + b_2V(m) + b_3V(m)^2] \tag{10}$$

Where,

P_n, Q_n - Nominal real and reactive power respectively
$V(m)$ – Voltage at node m

For all the loads, equation (9) and equation (10) are modeled as

$$a1 + a2 + a3 = 1.0 \tag{11}$$

$$b1 + b2 + b3 = 1.0 \tag{12}$$

For Constant Power (CP) load $a_1 = b_1 = 1$ and $a_i = b_i = 0$ for i=2, 3. For Constant Current (CI) load $a_2 = b_2 = 1$ and $a_i = b_i = 0$ for i= 1, 3. For Constant Impedance (CZ) load $a_3 = b_3 = 1$ and $a_i = b_i = 0$ for i=1, 2

4 Particle Swarm Optimization Technique for Optimal Sizing and Placement of DG

In this paper, a PSO technique is used identify the sizes and placement of the DG for minimizing the Power loss. The method has been developed through a simulation of simplified social models. The features of the method are as follows [15][16][17]. The method is based on research on swarms such as fish schooling and bird flocking. It is based on a simple concept. It works in two steps, which are calculating the particle velocity and updating its position. Therefore, the computation time is short, and it requires little memory. PSO is basically developed through simulation of bird flocking in two-dimensional space. Each agent tries to modify its position using the following information:

- the current position vector Si = [Sxi, Syi],
- the current velocity vector vi = [vxi, vyi],
- the distance between the current position and pbest, introduced as (pbesti – Si), and
- the distance between the current position and gbest, introduced as (gbest – Si).

This modification can be represented by the concept of velocity.

The velocity of each agent can be modified by the following equation:

$V^{k+I} = wV^{k}_{i} + c_1 rand * (pbest-s^{k}_{i}) + c_2 rand * (gbest-s^{k}_{i})$

Where,

v^k is the velocity of agent i at iteration k,

w is the adaptive inertia weight linearly adapted to decrease from wmax = 0.9 to wmin = 0.04, such that w = wmax – [(wmax – wmin)/number of iterations]∗ current iteration number, cj are the accelerating coefficients within the range [0,4], which are conventionally set to a fixed value of 2, rand is random number between 0 and 1,

$s^k i$ is the current position of agent i at iteration k,

pbesti is the pbest of agent i, and gbest is the gbest of the group.

Using the above equation, a certain velocity, which gradually gets close to pbest and gbest, can be calculated. The current position (searching point in the solution space) can be modified by the following equation:

$$S^{k+1}_{,i} = S^{k} + V^{k+1}_{,i}$$

5 Test Result and Discussion

To analysis the effect of DG on the Distribution system, the 33Bus system has been considered here. The effect of DG with different load models has been shown in Fig.2. From the fig it was clear that the optimal size of DG is 0.14MW for reduced loss and the optimal location has been selected by using VSI, Which is at location 18. The same test system has been tested with PSO. The test result shown in Fig.3. From the fig.3 it was found that the best location is 32 for minimizing power loss at 209.802kW the corresponding DG size is 0.2MW. Table.1 gives the information about the comparative analysis of power loss reduction. The effect on Bus Voltages by the Placement of DG for the test system has been shown in fig.4. From the fig.4.it has been found that PSO gives the better voltage profile improvement compare to VSI and Base case.

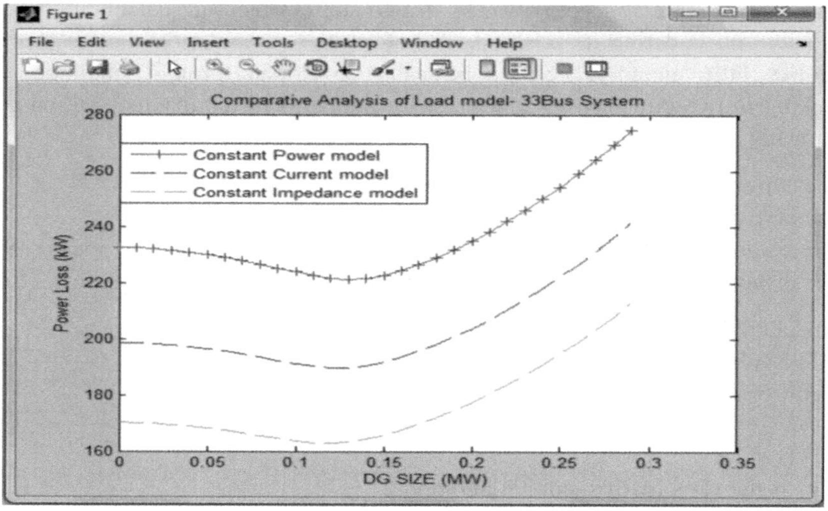

Fig. 1. Comparative Analysis of Load model- 33Bus System

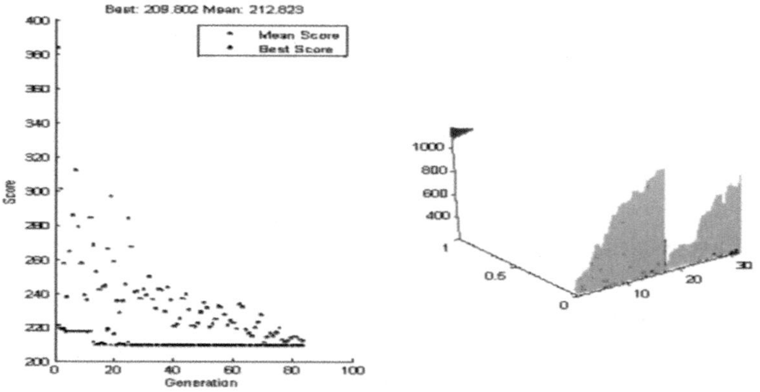

Fig. 2. Output of PSO- 33Bus System

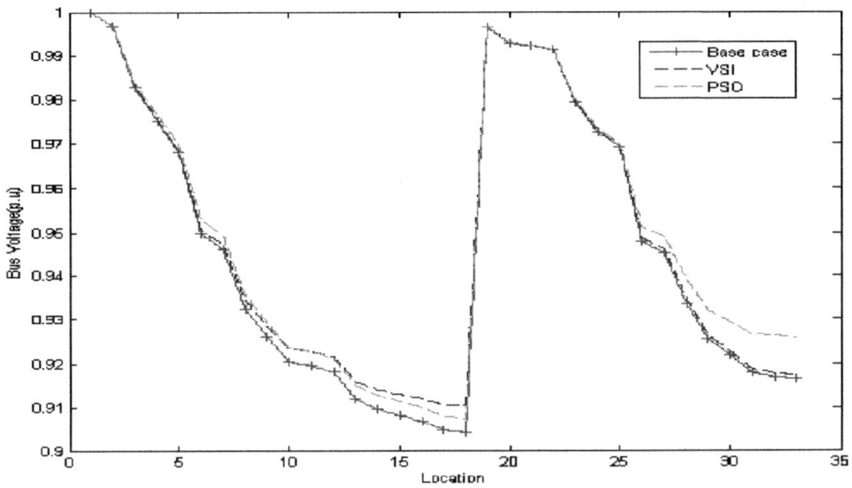

Fig. 3. Comparative Analysis of Bus Voltages- 33Bus System

Table 1. Comparative Analysis of Power Loss Reduction

Test case	33 bus system	
Approach	VSI	PSO
Optimal location	18	32
Optimal Size of DG in MW	0.14	0.2
Loss in kW	221.2841	209.802

6 Conclusion

In this paper the optimal location of DG is obtained using VSI and the size of the DG is obtained by trial and error method to minimize the real losses. PSO is used simultaneously to find the location and size both to minimize the losses. Even through the size of DG obtained by PSO is high, for this size the loss obtained by VSI method is greater than the loss obtained by PSO, so PSO gives better performance when compared to other method.

References

1. Ackermann, Anderson, G., Sooder, L.S.: Distributed generation: A generation. Electrical Power Systems Research 57 (2001)
2. El-Khattam, W., Salama, M.M.A.: Distributed generations technologies, definitions and benefits. Electrical Power Systems Research, 119–128 (2004)
3. Ghosh, S.: Optimal sizing and placement of distributed generation in a network system. Electrical Power and Energy Systems, 849–856 (2010)
4. Gozel, T., Hakan Hocaoglu, M.: An analytical method for the sizing and siting of distributed generators in radial systems. Electrical Power Systems Research, 912–918 (2009)
5. Keane, A., O'Malley, M.: Optimal Allocation of Embedded Generation on Distribution Networks. IEEE Transactions on Power Systems 20(3), 1640–1646 (2005)
6. Vallem, M.R., Mitra, J.: Siting and Sizing of Distributed Generation for Optimal Microgrid Architecture. In: IEEE Conference, pp. 611–616 (2005)
7. Wichit, K., Weerakorn, O.: Optimal Placement of Distributed Generation Using Particle Swarm Optimization. M. Tech Thesis. AIT, Thailand (2005)
8. Lalitha, M.P., Reddy, V.C., Usha, V., Reddy, N.S.: Application of fuzzy and PSO for DG placement for minimum loss in radial distribution system. ARPN Journal of Engineering and Applied Sciences 5(4), 32–37 (2010)
9. Haghifam, M.-R., Falaghi, H., Malik, O.P.: Risk-based distributed generation placement. IET Gener. Transm. Distrib. 2(2), 252–260 (2008)
10. Ochoa, L.F., Padilha, F.A., Harrison, G.P.: Evaluating distributed time-varying generation through a multiobjective index. IEEE Trans. Power Delivery 23(2), 1132–1138 (2008)
11. Celli, G., Ghiani, E., Mocci, S., Pilo, F.: A multiobjective evolutionary algorithm for the sizing and sitting of distributed generation. IEEE Trans. 20(2), 750–757 (2005)
12. Vinothkumar, K., Selvan, M.P.: Impact of DG Model and Load Model on Placement of Multiple DGs in Distribution System. In: IEEE Conference, pp. 508–513 (2010)
13. Teng, J.H.: A Network-Topology Based Three Phase Load Flow for Distribution Systems. Proceedings of National Science Council ROC (A) 24(4), 259–264 (2000)
14. Teng, J.H.: A Direct Approach for Distribution System Load Flow Solutions. IEEE Transaction on Power Delivery 18(3), 882–887 (2003)
15. Kennedy, J., Eberhart, R.: Particle swarm optimization. In: Proc. IEEE Neural Networks Conf., Piscataway, NJ, pp. 1942–(1948)
16. Shi, Y., Eberhart, R.: A modified particle swarm optimizer. In: Proc. IEEE Int. Conf. on Evolutionary Computation, pp. 69–73 (1998)
17. Sattianadan, D., Sudhakaran, M.: Optimal Placement of Capacitor in radial Distribution System Using PSO. In: Second International Conference on Sustainable Energy and Intelligent System (SEISCON 2011), pp. 326–331 (2011)

Particle Swarm Approach for Identification of Unstable Processes

V. Rajinikanth[1] and K. Latha[2]

[1] Department of Electronics and Instrumentation Engg, St.Joseph's College of Engineering, Jeppiaar Nagar, Chennai, Tamilnadu, India
[2] Department of Instrumentation Engineering, M.I.T. Campus, Anna University, Chennai, Tamilnadu, India

Abstract. In this paper, a step response based closed loop system identification procedure for a class of time delayed unstable chemical process loops using Particle Swarm Optimization algorithm is proposed. A novel objective function is developed using the time domain specification data to guide the PSO algorithm. The step response based identification is a simple closed loop test with a Proportional (P) controller. The PSO algorithm finds the best possible values for the process model parameters such as process gain (K), process time constant (τ), and the closed loop delay (θ_c). The method is tested on a class of unstable process models in the presence and absence of measurement noise. The performance of the proposed PSO based identification procedure is compared with the classical identification scheme existing in the literature. The results evident that, the proposed method helps to accomplish a better transfer function model with considerably reduced model mismatch.

Keywords: Unstable system, Particle Swarm, Identification, Step response, Model validation.

1 Introduction

In process industries, the real time chemical process loops such as jacketed CSTR, bioreactor, exothermic stirred reactors with back mixing, and polymerization reactor are to be operated in unstable steady state due to the economical and the safety reasons [1]. In order to design a controller, it is necessary to develop an approximated process model around the operating region.

System identification is the preliminary practice, widely executed in the field of process control to develop mathematical models from experimental data. Closed loop step test and relay tuning are the two most popular closed loop identification schemes widely employed in industries to identify the approximated process model for unstable systems. The practical application of relay based system identification is limited by delay time/process time constant (θ/τ) ratio. An unstable process under relay feedback produce limit cycle when θ/τ ratio < 0.693 [2]. Moreover the relay based identification cannot provide an appropriate response when the process time of the real time system is very large (Ex: Cell growth in bioreactor). Hence, closed loop step test can be used as an alternate for relay based identification.

S.C. Satapathy et al. (Eds.): Proc. of Int. Conf. on Front. of Intell. Comput., AISC 199, pp. 505–513.
DOI: 10.1007/978-3-642-35314-7_57

The existing identification procedures proposed for unstable system requires cumbersome computations. To reduce the computation time and also to increase the model accuracy, it is necessary to developed heuristic algorithm based system identification technique. The nature inspired algorithm based system identification process is initially discussed by Shin et.al [3]. They discussed about the real-coded genetic algorithm based system identification procedure for stable processes. From the recent literature, it is observed that the PSO algorithm based optimization process have emerged as a powerful tool for finding the solutions for a variety of complex engineering problems [4-7]. In PSO, the number of parameters to be assigned is very few and the number of iteration required by the search is small compared to other nature inspired algorithms.

In this paper, a 'P' controller based closed loop system identification is implemented to attain an approximated First Order Plus Time Delayed (FOPTD) model for a class of unstable processes. To evaluate the performance of the proposed method, a simulation study is carried out using a class of unstable system models. Through a comparative study with the classical system identification techniques existing in the literature, the performance of the identified model is validated.

2 Particle Swarm Optimization (PSO)

Particle Swarm Optimization (PSO) technique, developed by Kennedy and Eberhart [8], is a population based heuristic optimization technique and is widely applied in various engineering problems due to its high computational efficiency [4-7]. In this, a group of bird is initialized with arbitrary positions 'S_i' and velocities 'V_i'. At early searching stage, each bird in the swarm is scattered randomly throughout the 'D' dimensional search space. With the supervision of the Objective Function (OF), own flying experience and their companions flying experience, each particle in the swarm dynamically adjust their flying position and velocity. During the optimization search, each particle remembers its best position attained so far, and also obtains the global best information achieved by any particle in the population.

The search operation is mathematically described by the following equations;

$$V_{i,D}^{t+1} = W.V_{i,D}^{t} + C_1.R_1.(P_{i,D}^{t} - S_{i,D}^{t}) + C_2.R_2.(G_{i,D}^{t} - S_{i,D}^{t}) \quad (1)$$

$$W = W_{max} - \frac{W_{max} - W_{min}}{iter_{max}}.iter \quad (2)$$

$$V_{i,D}^{t+1} = \Psi \left| V_{i,D}^{t} + C_1.R_1.(P_{i,D}^{t} - S_{i,D}^{t}) + C_2.R_2.(G_{i,D}^{t} - S_{i,D}^{t}) \right| \quad (3)$$

$$\Psi = \frac{2}{\left| 2 - \varphi - \sqrt{\varphi^2 - 4\varphi} \right|}\text{; Where } \varphi = C_1 + C_2, \varphi > 4 \quad (4)$$

$$S_{i,D}^{t+1} = S_{i,D}^{t} + V_{i,D}^{t+1} \quad (5)$$

Where: W = inertia weight; $V_{i,D}^{t}$ = current velocity of the particle; $S_{i,D}^{t}$ = current position of the particle; R_1, R_2 are the random numbers in the range 0-1; C_1, C_2 are the cognitive and global learning rate respectively; $V_{i,D}^{t+1}$ = updated velocity; $S_{i,D}^{t+1}$ = updated position; W_{max} = maximum iteration number; W_{min} = minimum iteration number; *iter* = current iteration; $iter_{max}$ = maximum iteration number; Ψ = constriction factor; and i = 1,2, … N, = particles.

Eqn 1 represents the velocity update equation for the PSO. In this Eqn, the updated velocity depends on the inertia weight '*W*'. From Eqn 2, it is noted that, the inertia weight '*W*' requires the additional parameters such as W_{min}, W_{max}, *iter*, and $iter_{max}$. In the literature, there is no guide line to assign the value for these parameters. Due to the above reason, in this study, we considered Eqn 3 for velocity update [5]. The updated velocity depends mainly on the constriction factor 'Ψ', and its value can be easily assigned as in Eqn 4. Eqn 5 shows the position update for the PSO algorithm, and it depends on the current position of the 'i^{th}' particle and the updated velocity of the 'i^{th}' particle in the 'D' dimensional search space.

3 Closed Loop System Identification Procedure

The proportional (P) controller based system identification technique is very simple and the parameter to be adjusted is only the proportional gain 'K_p' [14, 15]. It is a closed loop test and it can be used for the unstable system having the 'θ/τ' up to 0.8. The previous study reports that, when the identification procedure is performed with a stable under-damped like process response, it is possible to minimize the mismatch between the original process and the identified FOPTD model [9].

The closed loop structure of the feedback control scheme proposed is shown in Fig 1. Where $G_p(s)$ is the process transfer function to be identified, $G_{c1}(s)$ is the 'P' controller.

.

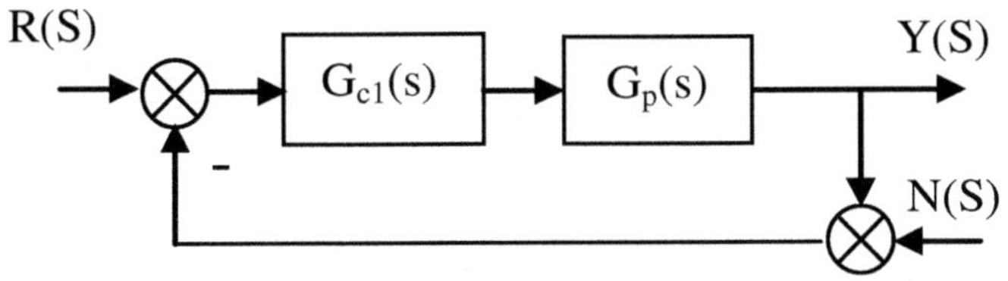

Fig. 1. Closed loop control system

During the system identification procedure, the proportional controller gain 'Kp' is tuned manually by the operator. In the proposed work, the identification procedure is performed in the presence and absence of the measurement noise term 'N(s).

The key steps in 'P' controller based system identification are as follows;

Step.1 Consider the closed loop system with 'K_p' only,
Step.2 Excite the system with an unity step signal 'R(s)',
Step.3 Manually adjust the value of 'K_p' until the closed loop response is similar to an under damped second order system
Step.4 Calculate the values of Y_p, Y_v, Δt and Y_∞ using PSO
Step.5 Find the FOPTD model of the system using Equations 6 - 13,
Step.6 Validate the model.

Fig.2 depicts the under damped second order like response of the inner loop for both the noisy and the noise free response for the unstable system.

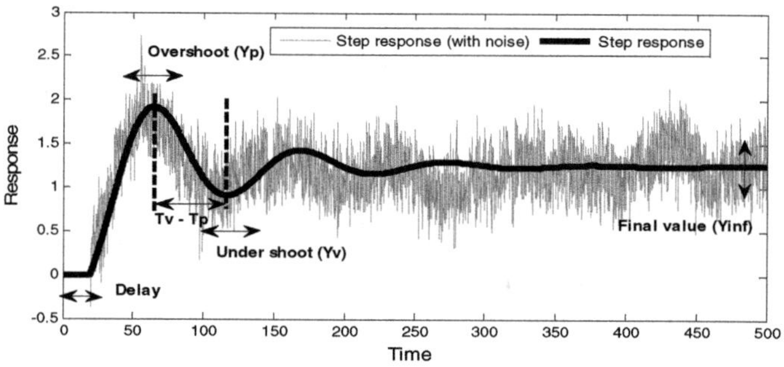

Fig. 2. Close loop response for system identification

In order to get the noisy response, a band-limited white noise with a noise power of 0.25 and a mean value of 'zero' is introduced along with the feedback signal. The parameters of the process model are identified by considering the following equations discussed by Padmasree and Chidambaram [10];

$$P_1 = \sqrt{(1-\xi^2)(K_k - 1)} \tag{6}$$

$$P_2 = \xi\sqrt{(K_k - 1)} + \sqrt{(K_k + 1) + (\xi^2 (K_k - 1)} \tag{7}$$

$$K_k = K\ K_p \tag{8}$$

$$\xi = \frac{-ln(V)}{\sqrt{\pi^2 + \{ln(V)\}^2}} \tag{9}$$

$$V = \frac{Y_\infty - Y_V}{Y_P - Y_\infty} \tag{10}$$

Where: P_1, P_2, V = variables, ξ = damping ratio, K_k = closed loop gain.

The process parameters can be found by considering the following mathematical relations;

$$\text{Process gain} \quad = K = \frac{Y_\infty}{K_p(Y_\infty - 1)} \qquad (11)$$

$$\text{Process time constant} \quad = \tau = \frac{(\Delta t \ P_1 \ P_2)}{\pi} \qquad (12)$$

$$\text{Closed loop delay} \quad = \theta = \frac{(2 \Delta t \ P_1)}{(P_2 \ \pi)} \qquad (13)$$

$$\text{The identified FOPDT unstable process model will be} = G_p(s) = \frac{K \ e^{-\theta s}}{\tau s - 1} \qquad (14)$$

A comparative study is then executed between the identified model, original system, and the model developed using the classical system identification procedures existing in the literature in order to find the model with best fit.

4 PSO Based System Identification

In this paper, we proposed a time domain based novel Objective Function (OF), to support the PSO based system identification search. Fig 3 shows the time response for the closed loop system.

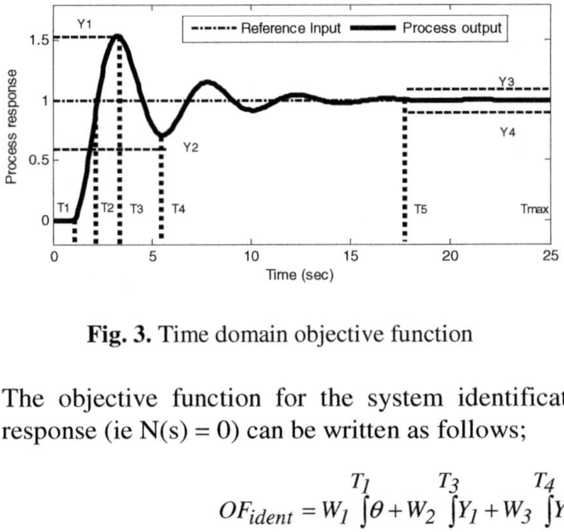

Fig. 3. Time domain objective function

Where; T1 - process delay, T2 = rise time, T3 - peak time (Tp), T4 – valley time (Tv), T5 - settling time, Tmax = time to reach Yinf (Y∞), Y- desired process output (reference signal), Y1 - peak overshoot, Y2 - peak undershoot, Y3 – maximum limit for Y; Y4 – minimum limit for Y.

The objective function for the system identification procedure for the noise free response (ie N(s) = 0) can be written as follows;

$$OF_{ident} = W_1 \int_0^{T_1} \theta + W_2 \int_{T_1}^{T_3} Y_1 + W_3 \int_{T_3}^{T_4} Y_2 + W_4 \int_{T_4}^{T_{max}} Y_\infty \qquad (15)$$

Where W_1 to W_4 are the weighting parameters (real numbers), which supports the optimization of the parameters such as delay (θ), over shoot (Y_p), under shoot (Y_v), the time difference between overshoot and undershoot (Tv - Tp) , and the final steady state value of process response ($Y\infty$) (Fig 2).

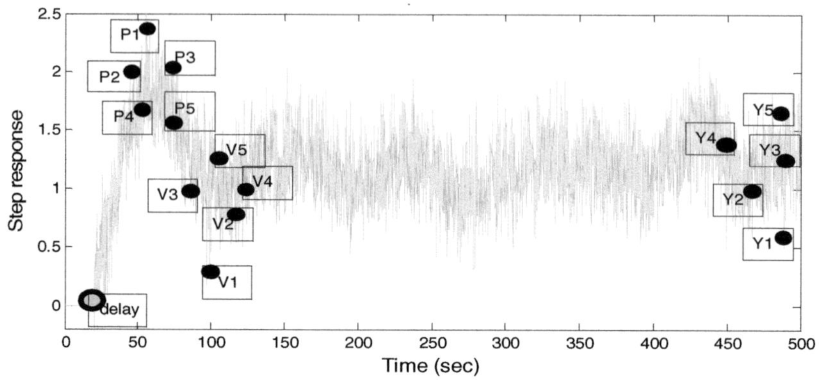

Fig. 4. Multiple point identification for the response with noise

- The PSO algorithm parameters are assigned with the following values;
  The dimension of the search (D) = 3 (ie. K, τ, and θ_c), number of swarm (N) = 20, number of swarm steps = 20, the cognitive learning rate (C_1) = the global learning rate (C_2) = 2.1 (ie. C1+C2= $\varphi > 4$), total number of iterations during the search = N* number of swarm steps.
- The maximum simulation time is fixed based on the delay time 'θ' present in the process ;
  $\theta < 1$: The simulation time (T_{max}) is fixed as 100 sec
  $\theta \leq 20$: The simulation time is set as 500 sec

The PSO algorithm calculates the FOPDT process model using Eqn. 6 to 13. If the process response has the measurement noise (ie N(s) = 0.25), then the system identification procedure will be very complex. In this case, single point identification proposed for noise free data may fails to provide an approximated process model. Hence, multi point identification as presented in Fig 4 should be used to obtain the better model from the experimental data.

As discussed earlier, the mean value for noise = zero. For noisy data, the PSO algorithm is allowed to search five best possible values for peak point, valley point,

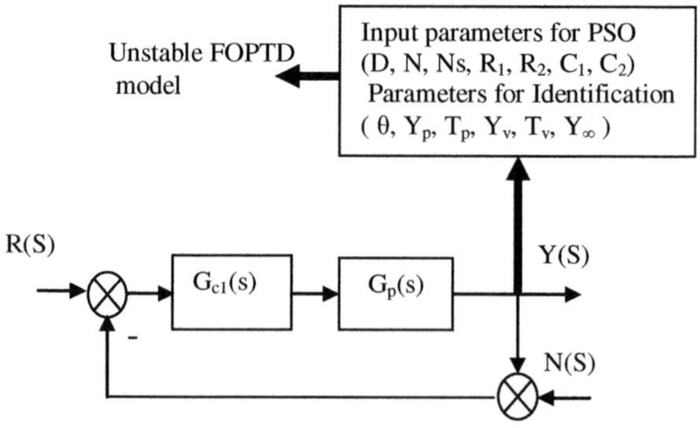

Fig. 5. PSO based System Identification practice

and final steady state value as in Fig 4. The average of the above values is considered to identify the model. The peak value Y_p = [(P1+P2+P3+P4+P5)/5]; Y_v = [(V1+V2+V3+V4+V5)/5]; and Y_∞ = [(Y_1+Y2+Y3+Y4+Y5)/5].

A relative analysis between the identified model and the original process is carried to calculate the model mismatch. The identified model shows smaller contradiction with the original system and the existing models in the literature. The identified model is then adopted to design the controller.

5 Results and Discussions

To study the performance of the PSO based system identification practice, three examples are considered from the literature. The following simulation study demonstrates the competence of the proposed method.

Example 1: Let us take an unstable second order process with one unstable pole as stated in Eqn 16[11].

$$G_p(s) = \frac{1e^{-s}}{(2s-1)(0.5s+1)} \tag{16}$$

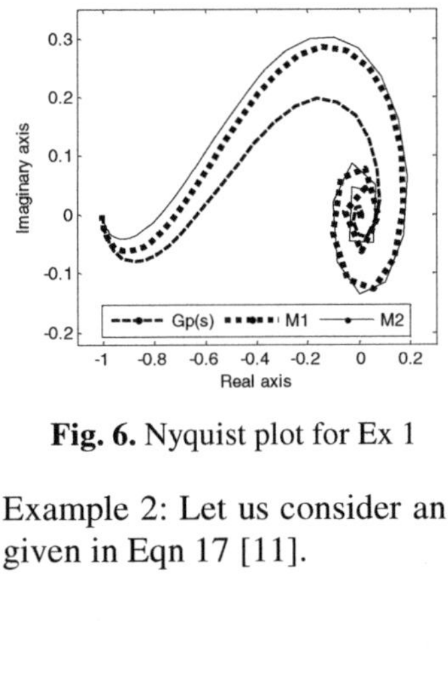

Fig. 6. Nyquist plot for Ex 1

Initially the controller gain 'K_p' is continuously adjusted manually until the system exhibit an under damped response as depicted in Fig 3. The system identification procedure for the above process model is proposed using PSO algorithm for N(s) = 0, and N(s) = 0.25. The parameters obtained during the identification study are presented in Table 1 and the identified process models are presented in Table 2. Fig 6 depicts the Nyquist plots for the original and identified model. The model 'M1' (noise free) has best fit with Gp(s) compared to 'M2' (noisy).

Example 2: Let us consider an unstable third order process with one unstable pole as given in Eqn 17 [11].

$$G_p(s) = \frac{1e^{-0.5s}}{(5s-1)(0.5s+1)(2s+1)} \tag{17}$$

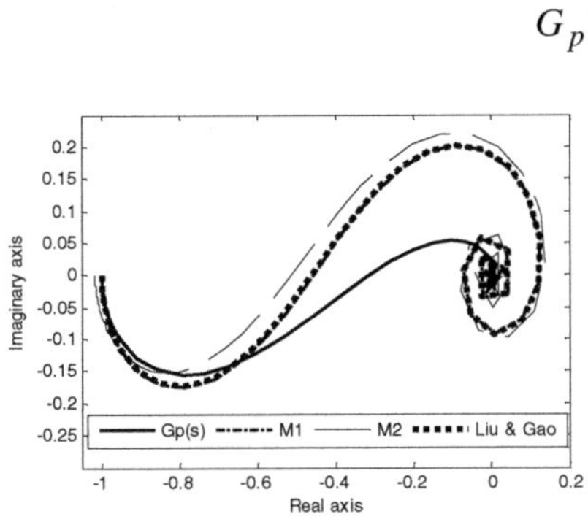

Fig. 7. Nyquist plot for Ex 2

For the above process, 'P' controller based identification is executed with K_p=2 and the PSO based identified values are tabulated in Table 1 and Table 2. Liu and Gao [12] proposed a relay feedback test for the above process model, and the identified values are: K=1.0001, τ = 5.7663 and θ =3.2821. From Fig 7, it is observed that, the identified FOPTD model 'M2' is very close with the model developed by Liu and Gao.

Table 1. PSO based identified values and the modeling parameters

	N(s)	K_p	Values from process response				Calculated values				
			Yp	Yv	Y∞	Δt	V	ξ	K1	P1	P2
Ex.1	0	1.2	8.859	4.64	5.99	9.29	0.471	0.233	1.200	0.435	1.591
	0.25		8.605	4.38	6.06	8.96	0.660	0.131	1.197	0.441	1.542
Ex.2	0	2.0	3.259	1.20	2.00	10.66	0.635	0.143	2.000	0.989	1.880
	0.25		2.831	1.27	1.97	10.59	0.813	0.066	2.031	1.013	1.809
Ex.3	0	2.5	2.014	0.70	1.22	20.01	0.636	0.142	5.762	2.160	2.929
	0.25		2.106	0.72	1.27	20.38	0.658	0.132	4.704	1.907	2.656

Table 2. Identified FOPTD model parameters and its fit with the original process

	N(s)	K	τ	θ_c	Model fit (%)
Ex.1	0	1.0031	2.0486	1.6176	93.17
	0.25	0.9982	1.9380	1.6304	90.20
Ex.2	0	1.0000	6.3164	3.5712	87.25
	0.25	1.0155	6.1781	3.7758	91.88
Ex.3	0	2.3048	40.776	9.3898	96.11
	0.25	1.8815	32.868	9.3186	91.84

Example 3: Continuous Stirred Tank Reactor (CSTR) with nonideal mixing considered by Liou and Yu-Shu [13] has the following transfer function model:

$$G_p(s) = \frac{2.22(1+11.133s)}{(98.3s-1)} e^{-20s} \tag{18}$$

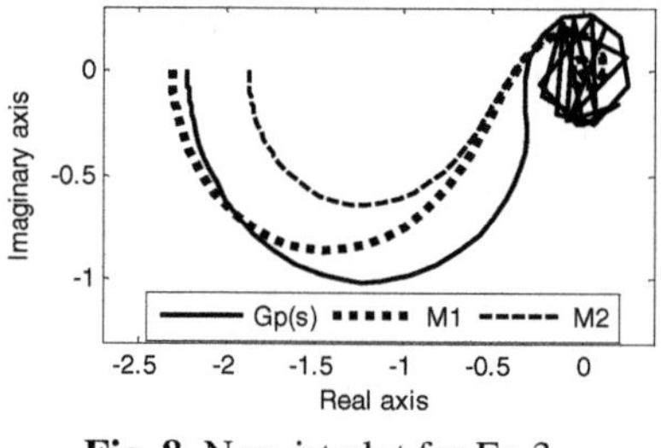

Fig. 8. Nyquist plot for Ex 3

This system has one stable zero and an unstable pole. As stated earlier, the system identification task is executed with K_p=2.5 and the PSO based identified values are presented in Table 1 and Table 2. From the Nyquist plot (Fig.8), the observation is that, model 'M1' has the best fit with the Gp(s) compared to 'M2'.

6 Conclusions

In this paper, PSO based system identification procedure is proposed for unstable system using a proportional controller. It is a closed loop test, provides a first order plus time delayed model, widely considered in model based controller tuning practice. A novel time domain based objective function is developed to guide the PSO search. The identification analysis is tested on a varied class of unstable process models in the presence and absence of measurement noise. The proposed identification procedure shows a robust performance on the process data with measurement noise. The simulation result illustrates that, the discussed method helps to attain a satisfactory process model with considerably reduced model mismatch.

References

1. Panda Ramesh, C.: Synthesis of PID controller for unstable and integrating processes. Chem. Eng. Sci. 64(12), 2807–2816 (2009)
2. Padhy, P.K., Majhi, S.: Relay based PI-PD design for stable and unstable FOPDT processes. Comput. Chem. Eng. 30, 790–796 (2006)
3. Shin, G.-W., Song, Y.-J., Lee, T.-B., Choi, H.-K.: Genetic Algorithm for Identification of Time Delay Systems from Step Responses. IJCAS 5(1), 79–85 (2007)
4. Zamani, M., Sadati, N., Ghartemani, M.K.: Design of an Hα PID Controller Using Particle Swarm Optimization. IJCAS 7(2), 273–280 (2009)
5. Pillay, N., Govender, P.: PSO Tuned PI/PID Controller for Open-Loop Unstable Processes with Time Delay. In: EPIA 2011, pp. 223–237 (2011)
6. Khalid, M., Luo, Q., Duan, H.: A cultural algorithm based particle swarm optimization approach to linear brushless DC motor PID controller. Sci. Res. and Essays 7(3), 318–326 (2012)
7. Illoul, R., Loudini, M., Selatnia, A.: Particle Swarm Optimization of a Fuzzy Regulator for an Absorption Packed Column. The Medit. J. of Meas. and Cont. 7(1), 174–182 (2011)
8. Kennedy, J., Eberhart, R.C.: Particle swarm optimization. In: Proceedings of IEEE International Conference on Neural Networks, pp. 1942–1948 (1995)
9. Rajinikanth, V., Latha, K.: Identification and Control of Unstable Biochemical Reactor 1(1), 106–111 (2010)
10. Padma Sree, R., Chidambaram, M.: Control of Unstable Systems. Narosa Publishing House, India (2006)
11. Chen, C.-C., Huang, H.-P., Liaw, H.-J.: Set-Point Weighted PID Controller Tuning for Time-Delayed Unstable Processes. Ind. Eng. Chem. Res. 47(18), 6983–6990 (2008)
12. Liu, T., Gao, F.: Identification of integrating and unstable process from relay feedback. Comput. Chem. Eng. 32, 3038–3056 (2008)
13. Liou, C.-T., Yu-Shu, C.: The effect of nonideal mixing on input multiplicity in a CSTR. Chem. Eng. Sci. 46(8), 2113–2116 (1991)
14. Ananth, I., Chidambaram, M.: Closed loop identification of transfer function model for unstable systems. J. Franklin Inst. 336, 1055–1061 (1999)
15. Pramod, S., Chidambaram, M.: Closed loop identification of transfer function model for unstable bioreactors for tuning PID controllers. Bioprocess Engineering 22, 185–188 (2000)

Fast Convergence in Function Optimization Using Modified Velocity Updating in PSO Algorithm

Nanda Dulal Jana[1], Tapas Si[2], and Jaya Sil[3]

[1] Dept. of Information Technology
National Institute of Technology, Durgapur, West Bengal, India
[2] Dept. of Computer Science & Engineering
BankuraUnnayani institute of Engineering, Bankura, West Bengal, India
[3] Dept. of Computer Science & Technology
Bengal Engineering & Science University, Howrah, West Bengal, India
{nanda.jana,c2.tapas}@gmail.com, js@cs.becs.ac.in

Abstract. In this paper, a new version of Particle Swarm Optimization (PSO) Algorithm has been proposed where the velocity update equation of PSO has been modified. A new term is added withthe original velocity update equation by calculating difference between the global best of swarm and local best of particles. The proposed method is applied on eight well known benchmark problems and experimental results are compared with the standard PSO (SPSO). From the experimental results, it has been observed that the newly proposed PSO algorithm outperforms the SPSO in terms of convergence, speed and quality.

1 Introduction

The Particle Swarm Optimization (PSO) is a population based global optimization technique, inspired by the social behavior of bird flocking and fish schooling [1][2]. The PSO algorithm maintains a swarm of particles, called individuals where each particle (individuals) represents a candidate solution. Particles follow a very simple behavior: emulate the success of neighboring particles and own success achieved. The position of particle is therefore influenced by the best particle in a neighborhood, as well as the best solution found by the particle. Initially PSO was designed for continuous optimization problem but shown its capability of handling non-differentiable, discontinuous and multimodal objective functions and has gained increasing popularity in recent years due to its ability to efficiently and effectively tackle several real-world applications [3][4].

Several variations [5][6][7][8][9][10][11] has been proposed to improve the performance and the convergence behavior of PSO algorithms. One class of variations include the modification of velocity update equation in PSO. In T. Ziyu and Z. Dingxue [5] was proposed a new version of PSO without the velocity of the previous iteration and a novel selection of acceleration coefficients was introduced in the algorithm. In [6], cognitive component and social component was replaced by two terms of the linear combination of global best of swarm and personal best of particle.

S.C. Satapathy et al. (Eds.): Proc. of Int. Conf. on Front. of Intell. Comput., AISC 199, pp. 515–521.
DOI: 10.1007/978-3-642-35314-7_58

In this paper, we have proposed a new PSO algorithm, in which velocity update equation has been modified by adding the difference between the global best of swarm and the personal best of particle. Initial experimental results in a benchmark set consisting of eight difficult high dimensional benchmark functions demonstrate that this is a promising approach.

The rest of the paper is organized as follows: Section 2 describes the basic operations of the standard PSO. In section 3, we propose the new PSO algorithm, while in section 4, experimental analysis and results are presented. The paper concludes with a short discussion and some pointers for future work.

2 The Standard Particle Swarm Optimization (SPSO)

The beauty of Particle Swarm Optimization lies in its simplicity and ease of applicability. It is a kind of algorithm to search for the best solution by simulating the movement offlocking birds. It uses a swarm of individual called particles. Each particle has its own position and velocity to move around the search space. The coordinates of each particle represent a possible solution associated with two vectors- the position vector and the velocity vector. Particles have memory and each particle keep track of previous best position and corresponding fitness. The previous best value is called as *pbest*. It also has another value called *gbest*, which is the best value of all the particles *pbest* in the swarm.

Consider a D-dimensional function $f(x)$, want to be optimized

$$Minimize\ f(x),\ where\ f: R^D \rightarrow R$$

The position vector and velocity vector of the *ith* particle is represented by $X_i = (x_{i1}, x_{i2}, \ldots\ldots, x_{iD})$ and $V_i = (v_{i1}, v_{i2}, \ldots., v_{iD})$, where i denote the swarm size. A swarm consists of a number of particles that proceed (fly) through the search space towards the optimal solution. Each particle update its position based on its own best exploration, overall best swarm exploration and its previous velocity vector according to the following equations:

$$V_i(t+1) = V_i(t) + c_1 r_1(pbest_i(t) - X_i(t)) + c_2 r_2(gbest_i(t) - X_i(t)) \quad (1)$$

$$X_i(t+1) = X_i(t) + V_i(t+1) \quad (2)$$

Where $c_1$and $c_2$are two positive acceleration coefficients, r_1 and r_2 are uniformly distributed random numbers in [0, 1],$X_i = (x_{i1}, x_{i2}, \ldots\ldots, x_{iD})$ is the current position of the *ith* particle.$pbest_i = (x_{i1}^{pbest}, x_{i2}^{pbest}, \ldots\ldots, x_{iD}^{pbest})$is the best position of the *ith* particle achieved based on its own experience. $gbest_i = (x_1^{gbest}, x_2^{gbest}, \ldots\ldots, x_D^{gbest})$is the position of the best particle based on the overall swarms experience. T is the iteration counter.A constant, maximum velocity (V_{max}) is used to limit the velocities of the particles and improve the resolution of the search space. Shi and Eberhart [12][13] proposed to use an "inertia weight" factor ω, in order to overcome the premature convergence of PSO. The resulting velocity update equation (1) becomes:

$$V_i(t+1) = \omega V_i(t) + c_1 r_1(pbest_i(t) - X_i(t)) + c_2 r_2(gbest_i(t) - X_i(t)) \quad (3)$$

In this paper, the inertia weight version of PSO is regarded as the SPSO.

3 The Proposed Algorithm

The motivation behind designing the new PSO algorithm is to accelerate its convergence rate, success rate and exploration capability in finding global optimum solution. In this approach, the difference between *gbest* and *pbest*is scaled while multiplyingby a constant c_3 and a random number r_3. This term is added with the velocity update equation of SPSO, expressed by (3). The new velocity update equation is defined by equation (4).

$$V_i(t+1)=\omega V_i(t) + c_1r_1(pbest_i(t) - X_i(t))+ c_2r_2(gbest_i(t)- X_i(t)) + c_3r_3(gbest_i(t) -pbest_i(t)) \quad (4)$$

In equation (4), the first term represents current velocity of the particle and can be thought of as momentum term. The second term is responsible for attraction of particles at the current position towards the positive direction of its own best position (*pbest*) i.e. cognitive component. The third term is responsible for the attraction of particles at current position towards the positive direction of the global best position (*gbest*) i.e. social component. The fourth term is responsible for the distance between *gbest* and *pbest*which implies how far-away the *pbest* position is from the *gbest* position. The values of $(gbest_i- pbest_i)$ are decreasing over the increasing number of iterations (t). Finally,$(gbest_i–pbest_i)\cong 0$ because the *pbest* closely reached to the *gbest* of the swarm.

Moreover, the velocity of the SPSO may be a small value, if both $(pbest - X_i)$ and $(gbest - X_i)$ are small enough. In such situations, exploration capability of SPSO has been lost at some generation. At the early stage of evolution process, $(gbest_i) - pbest_i)$ preventssuch situation to arrive(exploration capability) in the swarm. But, the loss of diversity for $(gbest_i - pbest_i)$ is typically occurred in the latter stage of evolution process because the value of $(gbest_i - pbest_i)$reaches close to zero.

The proposed algorithm is presented below.

Input: Randomly initialized position and velocity of the particles: $X_i(0)$ and $V_i(0)$.
Output: Position of the approximate global optima X^*

while maximum number of iteration *(t)* or minimum error criterion is not attained

fori= 1 to N (Swarm size)
for j= 1 to D (Dimension of the problem)
Velocity update equation
$V_i(t+1) = \omega V_i(t) + c_1r_1(pbest_i(t) - X_i(t)) + c_2r_2(gbest_i(t) - X_i(t)) + c_3r_3(gbest_i(t) - pbest_i(t))$

Position update equation
$X_i(t+1) = X_i(t) + V_i(t+1)$
end for j
Evaluate fitness of the update position of the particles
Update previous pbest and gbest information, if needed
end for i
end while

4 Results and Discussions

This section compares the performance of the proposed PSO algorithm with the SPSO algorithm. To verify the effectiveness of the proposed approach we have used eight widely known high dimensional benchmark functions with different characteristics [14].The four functions (f_1 – f_4) are high dimensional and scalable benchmark functions. The remaining four functions (f_5 – f_8) are high dimensional and multimodal functions, where the number of local minima increases exponentially with their dimensionality. A brief description of the functions is provided in Table 1. More specifically, D denotes the dimension of the problem, search space is the optimization range box and objective function value is the global minimized value.

The SPSO and the proposed PSO algorithm are coded in MatLab2010b and developed on AMD FX-8150 Eight-Core machine with 4GB RAM under Windows 7 platform. Fifty independent runs with different seed for the generation of random are taken. However, the same seed is used for generating the initial swarm for SPSO and proposed PSO algorithm. We consider*best- run-error* values for function optimization, which is calculatedusing following inequality:

$$|f^*(x) - f(x)| < 0.001 \tag{5}$$

Where $f^*(x)$ is the best known objective function and $f(x)$ is the best objective function found in the corresponding run. Right side of the equation (5)denotes the threshold value representing precision in the objective function. Themaximum number of function evaluations are fixed,say 1, 00,000. The swarm size and dimension are fixed to 40 and 30, respectively. The inertia weight ω is 0.72984 and the acceleration coefficients for SPSO and proposed PSO are set to c_1= c_2 = 1.49445 experimentally.

The quality of solution obtained is measured by the best, mean and standard deviation of the objective function values out of fifty runs. The performance of solution is measured by the mean and standard deviation of the total number of function evaluations out of fifty runs. In Table 2 and Table 3, best-run-error values and number of function evaluations of the SPSO and the proposed PSO are tabulated, considering c_3=c_1 in the proposed algorithm. For c_3= 1, the best-run-error values and number of function evaluations of the SPSO and the proposed PSO are tabulated in Table 4 and Table 5, respectively. The convergence graph for function f8 is given in Fig. 1.

From Table 2, it has been observed that the proposed PSO gives better result than SPSO for the functions f1, f5, f7, f8. Also in Table 3, the proposed PSO shown better performance compared with SPSO. In observing Table 4 and Table 5, it has been concluded that the quality of solution and performance obtained by our proposed PSO is better or equal to the SPSO. Analysis of the results obtained by the proposed PSO algorithm outperforms on multimodal functions than unimodal functions using SPSO.

Table 1. A brief description of the benchmark function set

Test Function	S	fmin
$f_1(x)=\sum_{i=1}^{D} x_i^2$	[-100,100]	0
$f_2(x)=\sum_{i=1}^{D}(\sum_{j=1}^{i} x_j)^2$	[-100,100]	0
$f_3(x)=\sum_{i=1}^{D}(10^6)^{\frac{i-1}{n-1}} x_i^2$	[-100,100]	0
$f_4(x)=[100(x_{i+1}-{x_i}^2)^2-(1-x_i\ )^2]$	[-100,100]	0
$f_5(x)=-x_i * \sin(\sqrt{\lvert x_i \rvert})$	[-500,500]	-12569.50
$f_6(x)=\sum_{i=1}^{D}\frac{x_i^2}{4000}-\prod_{i=1}^{D}\cos(\frac{x_i}{\sqrt{i}})+1$	[-600,600]	0
$f_7(x)=-20 * \exp\left(-0.2 * \sqrt{\frac{1}{D}\sum_{i=1}^{D} x_i^2}\right) - \exp\left(\frac{1}{D}\sum_{i=1}^{D}\cos(2\pi x_i)\right)+20+e$	[-32,32]	0
$f_8(x)=\sum_{i=1}^{D}[x_i-10\cos(2\pi x_i)+10]$	[-5.12,5.12]	0

Table 2. Best-run-error values achieved by SPSO and Proposed PSO (for c_3=c_1)

Test#	**SPSO**		**Proposed PSO**	
	Mean	Std. Dev.	Mean	Std. Dev.
f1	9.36e-04	**0.0014**	**0.0014**	**0.0014**
f2	**2.84e-04**	**2.64e-04**	3.03e-04	2.77e-04
f3	**9.21e-04**	**7.26e-05**	0.2796	0.3461
f4	**35.5684**	**45.0933**	354.9286	470.18
f5	7.09e+03	597.031	**4.23e+03**	**544.55**
f6	**0.0212**	**0.0223**	0.0232	0.0271
f7	1.5119	0.9891	**0.0244**	**0.0698**
f8	33.76	8.9137	**23.3976**	**8.7277**

Table 3. Number of function evaluations for SPSO and Proposed PSO (for c_3=c_1)

Test#	**SPSO**		**Proposed PSO**	
	Mean	Std. Dev.	Mean	Std. Dev.
f1	**13256**	**1.21E+003**	9.38E+004	6.34E+003
f2	**1.88E+003**	**1.01E+003**	3.62E+003	4.35E+003
f3	2.24E+004	2.32E+003	**100000**	**0**
f4	**100000**	**0.00E+000**	**100000**	**0**
f5	**100000**	**0.00E+000**	**100000**	**0**
f6	**7.76E+004**	**3.83E+004**	9.96E+004	1.30E+003
f7	8.54E+004	3.16E+004	**100000**	**0**
f8	100000	0.00E+000	**100000**	**0**

Table 4. Best-run-error values achieved by SPSO and Proposed PSO (for c_3=1)

Test#	SPSO		Proposed PSO	
	Mean	Std. Dev.	Mean	Std. Dev.
f1	9.36e-004	0.0014	**9.44e-04**	**6.11e-05**
f2	**2.84e-04**	**2.64e-04**	3.71e-04	3.28e-04
f3	**9.21e-04**	**7.26e-05**	9.22e-04	**7.26e-05**
f4	**35.5684**	45.0933	50.0599	**40.155**
f5	**7.09e+03**	**597.031**	5.47e+03	636.716
f6	0.0212	0.0223	**0.0157**	**0.0177**
f7	1.5119	0.9891	**0.0241**	**0.1632**
f8	**33.7688**	**8.9137**	37.4701	13.0016

Table 5. Number of function evaluations for SPSO and Proposed PSO (for c_3= 1)

Test#	SPSO		Proposed PSO	
	Mean	Std. Dev.	Mean	Std. Dev.
f1	**13256**	**1.21E+003**	3.67E+04	2.42E+03
f2	1.88E+003	**1.01E+003**	**2656**	2.00E+03
f3	**2.24E+004**	**2.32E+003**	5.51E+04	3.92E+03
f4	**100000**	**0.00E+000**	**100000**	**0.00E+000**
f5	**100000**	**0.00E+000**	**100000**	**0.00E+000**
f6	**7.76E+004**	3.83E+004	8.18E+04	**2.81E+04**
f7	8.54E+004	**3.16E+004**	**5.05E+04**	8.09E+03
f8	**100000**	**0.00E+000**	**100000**	**0.00E+000**

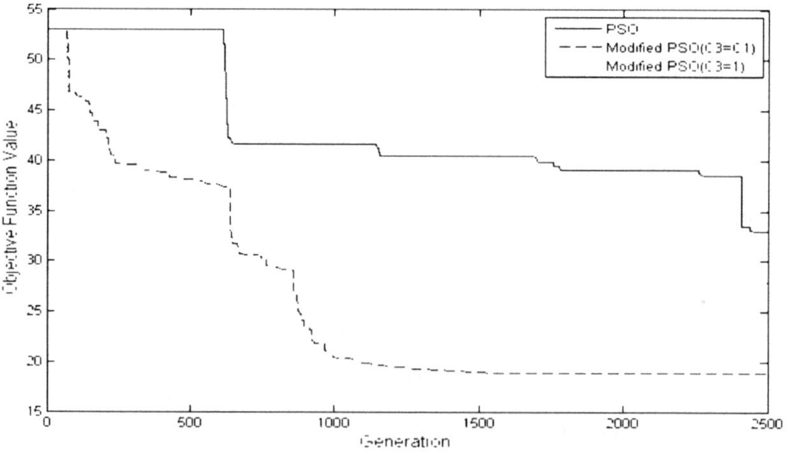

Fig. 1. Convergence graph for function f8

5 Conclusions

In the present study, a new PSO algorithm has been proposed for function optimization. It is based on the basic change in the velocity updating equation. One more term, $(gbest_i - pbest_i)$ is added to the velocity update equation of SPSO. It is tested on eight high dimensional benchmark functions. It is shown that the proposed PSO algorithm outperforms SPSO in terms of efficiency, accuracy and effectiveness. Particularly for multimodal functions, proposed PSO algorithm superior than SPSO. In this paper, we are using only two values ($c_3 = c_1$ and $c_3 = 1$) of the parameter c_3. Therefore, the effective changes of this parameter are notexplored. In a future study parameters fine tuning may be carried out for better performance. Also the application of proposed PSO to the real world problems would be interesting as a future research.

References

1. Eberhart, R.C., Kennedy, J.: A New Optimizer Using Particle Swarm Theory. In: International Symposium on Micromachine and Human Science, pp. 39–43 (1995)
2. Kennedy, J., Eberhart, R.C.: Particle Swarm optimization. In: IEEE International Joint Conference on Neural Networks, pp. 1942–1948. IEEE Press (1995)
3. Clerc, M.: Particle Swarm Optimization. ISTE Publishing Company (2006)
4. Englelbrecht, A.: Computational Intelligence: An Introduction. Halsted Press (2002)
5. Ziyu, T., Dingxue, Z.: A Modified particle Swarm Optimization with an Adaptive acceleration coefficients. In: Asia-Paciffic Conference on Information Processing (2009)
6. Deep, K., Bansal, J.C.: Mean Particle Swarm Optimization for function optimization. International Journal of Computational Intelligence Studies 1(1), 72–92 (2009)
7. Zhan, Z.-H., Zhang, J., Li, Y., Chung, H.S.-H.: Adaptive particle swarm optimization. IEEE Transactions on Systems, Man, and Cybernetics, 1362–1381 (2009)
8. Xinchao, Z.: A perturbed particle swarm algorithm for numerical optimization. Applied Soft Computing, 119–124 (2010)
9. Chen, M.-R., Li, X., Zhang, X., Lu, Y.-Z.: A novel particle swarm optimizer hybridized with external optimization. Applied Soft Computing, 367–373 (2010)
10. Pedersen, M.E.H.: Tuning & Simplifying Heuristically Optimization, Ph.D. thesis, school of Engineering Science, University of Southampton, England (2010)
11. Singh, N., Singh, S.B.: One Half Global Best Position Particle Swarm Optimization Algorithm. International Journal of Scientific & Engineering Research 2(8), 1–10 (2012)
12. Shi, Y., Eberhart, R.C.: A modified particle swarm optimizer. In: Proceedings of the IEEE international Conference on Evolutionary Computation, pp. 69–73 (1998)
13. Shi, Y., Eberhart, R.C.: Parameter Selection in particle swarm Optimization. In: 7th Annual Conference on Evolutionary Programming, San Diego, USA (1998)
14. Yao, X., Liu, Y., Lin, G.: Evolutionary programming made faster. IEEE Transactions on Evolutionary Computation 3, 82–102 (1999)

Concurrency Aware Dynamic Scheduler for Virtualized Environment

Pratik Shinde, Avinash Tomar, Komal Shah, Shipra Kalra, and D.A. Kulkarni

Pune Vidyarthi Griha's Collge of Engineering and Technology, Pune-09
{pracshi,avinash.tomar1405,shkomal19,shipra91kalra}@gmail.com, dineshakulkarni@yahoo.com

Abstract. Virtual Machines (VMs) running on a Symmetric Multiprocessors (SMP) are called SMP VMs. Operating System primitives can be easily applied directly to the virtualized environment. However, virtualization disrupts the basis of spinlock synchronization in the guest operating system. In non-virtualized environment lock-holder thread never gets pre-empted assuring that lock-waiter thread never gets scheduled before lock-holder. In virtualized environment scheduler is completely unaware of the association between Virtual CPUs (VCPUs). Scheduler may schedule VCPU executing lock-waiter thread before lock-holder VCPU or it may pre-empt lock-holder VCPU. This violates the most basic primitive of Operating System and causes problems like VCPU stacking due to which VCPUs of the same domain are stacked on the run queues of the same physical processor. To address this problem, the solution of co-scheduling is proposed. In co-scheduling all the VCPUs of a domain are scheduled at the same time or none of them is scheduled. This approach suffered from the drawbacks like CPU fragmentation and priority inversion. Our solution proposes a dynamic scheduler that is completely aware of the concurrency of the domain. Rather than co-scheduling every VCPU of the domain, we are relaxing the constraint on co-scheduling. Our solution proposes co-scheduling only if concurrency degree θ of domain is greater than threshold value δ.

Keywords: Virtual Machine, Virtual CPU, Synchronous scheduling, VCPU stacking, Priority inversion.

1 Introduction

Virtual Machine Monitor (VMM) is a software layer between hardware and operating system (OS). VMM allows the multiple OSs to run simultaneously on the same hardware platform[1]. Virtualization can be thought as a step towards green technology where multiple servers not having too much load can share hardware. Suppose there are two servers: a mail server and a web server both are having moderate load, without virtualization we must have two separate computers one

S.C. Satapathy et al. (Eds.): Proc. of Int. Conf. on Front. of Intell. Comput., AISC 199, pp. 523–533.
DOI: 10.1007/978-3-642-35314-7_59

running mail server and another running web server. Virtualization allows both servers to exist on the same machine, each server is encapsulated in a guest OS. Each guest OS is having an illusion that it is the only OS running on the hardware. It is the job of VMM to maintain balance between all the domains, to look after their requests and to schedule each domain. By using virtualization technique, two separate servers can be consolidated on a single machine diminishing the need of extra hardware.

In cloud computing we need lots of different operating environments. The hypervisor becomes an ideal delivery mechanism by showing the same application on lots of different systems. Because hypervisors can load multiple operating systems, they are a very practical way of getting things virtualized quickly and efficiently. Various VMM are available in market like Xen, VMWare ESXI and Virtual Box. Xen is an open source VMM can be setup on the Linux platform.

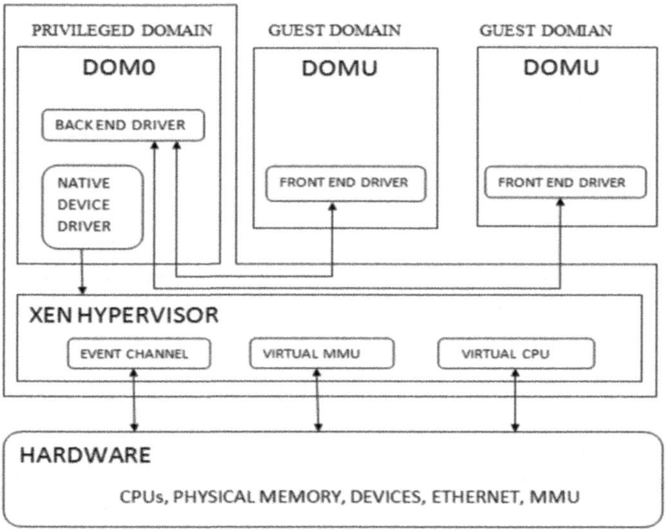

Fig. 1. Xen Architecture

Fig.1 gives architecture of Xen. There are multiple Operating Systems running as an application on the abstraction layer provided by Xen VMM. Each Operating System is called Domain or Guest OS or Virtual Machine (VM). There is a special domain called Dom0 which is more privileged than other domains. Other domains are called DomU (Unprivileged Domain). Xen uses split device driver architecture in which Dom0 contains the backend device drivers and all DomUs contain only front ends of the device drivers [13]. Xen provides resources such as CPU, memory and I/O to the domain as per predefined policy. In Xen system there are three levels of scheduling in between user level thread and CPU:

1) User space threading library mapping the user space threads to guest kernel threads.
2) Guest kernel mapping threads to VCPUs.
3) Hypervisor mapping VCPUs to physical CPUs.

Hypervisor scheduler lying at the bottom of this stack needs to be predictable [1]. The above layers make the assumptions on the behaviour of the underlying scheduling and will make highly suboptimal scheduling decisions if these assumptions are invalid. This leads to unpredictable behaviour of processes in domains. Generally VMs with multiple VCPUs are seen as a symmetric multiprocessing system by the guest kernel. As a rule of thumb, any domain can't have more VCPUs than actual number of physical processors [7]. So any domain will always have number of VCPU equal to or less than number of physical CPUs. When workload in VM is a non concurrent application, then Xen schedules the VCPUs asynchronously. This method simplifies the implementation of CPU scheduling and is adopted in most of the VMM along with Xen. However when workload in the domain is concurrent and requires synchronization, this method gives the poor performance. In OS semaphores or spinlocks are used to avoid the contention between competing processes [2]. The same rule is applied to Virtualized Environment (VE) also. But VE disrupts the normal behaviour of the OS policy that lock-holder thread never get pre-empted. As a result, spinlocks may have longer waiting time [9]. Our solution is to provide the balance scheduler which is aware of the degree of concurrency of each domain.

1.1 Lock Holder Pre-emption Problem

Operating System schedules multiple threads of a concurrent application on the different CPUs to fully utilize the support of multiple CPUs. To allow the communication between these concurrent threads Operating System provides two kind of lock primitives: semaphores and spinlocks. Spinlocks are basically used in SMP systems and they are held for very short period of time. Operating System primitive says "kernel pre-emption is disabled in every critical region protected by spinlock" [3].

Synchronization Latency is the time taken by lock-waiter thread to success fully acquire lock. Current scheduler in Xen allows VCPU to be scheduled on any physical CPU. As shown in Fig.2, let VCPU1 and VCPU2 be VCPUs of the same domain. During time slice T0, VCPU1 is holding a spinlock and VCPU2 is waiting for the same lock. Credit Scheduler which is unaware of this situation, pre-empts VCPU1 (Lock Holder) and VCPU2 (Lock Waiter) when their time slice is over. During next time slice T1, it may select VCPU2 which is lock waiter VCPU to execute on physical CPU1 and any other job on CPU2. As this VCPU is waiting for the lock to be release from VCPU1 so it just keeps spinning and wastes CPU cycles. It gets pre-empted and at next time slice T2 VCPU1 is scheduled. Now VCPU1 executes and releases the lock. This lock can be used by VCPU2 in next time slice T3.

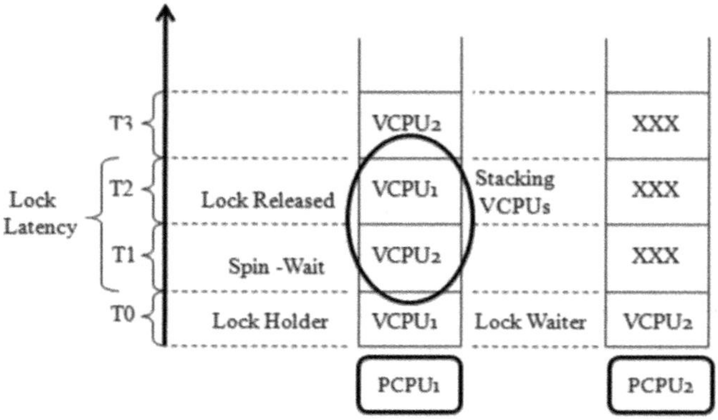

Fig. 2. VCPU Stacking

Latency, which is experienced by the job executing on VCPU2 is: Total_Time_To_Acqire_Lock+Time_To_Execute.

= (T0+T1) +T3. As VCPU2 has to wait for T0+T1 time. This problem may lead to severe problem like kernel panic [4]. The inter processor operations performed by the guest kernel uses spinlocks and are expected to complete within reasonable period of time. It may timeout if such operations take unreasonably long time [7].

1.2 Guest Unaware of the Uniform Progress of CPU

Also asynchronous scheduling of VCPUs of a domain violates another assumption of guest that "All processors it manages are making progress at the same rate". In virtual environment VCPU may be either pre-empted, running or in blocked state. With asynchronous scheduling the VCPU may be schedule at any time, breaking guest Operating Systems assumption [5]. This confuses the guest operating system and it may lead to malfunction of the guest kernel. For example, watchdog timer is expecting a response from its sibling VCPU in specified time. If sibling VCPU doesn't respond, it may crash [16].

2 Co-scheduling

To overcome above problem, previous work [4],[7] proposes co-scheduling. In this scenario, either all VCPUs of the domain are scheduled at the same time or none of the VCPU is scheduled. So execution of the domain is deferred until we find a time slice at which required numbers of physical CPUs are idle.

This approach removes VCPU stacking problem as spinlock released by VCPU1 can be used by VCPU2 instantly. Also this approach fulfils the guest assumption that all VCPUs are making equal progress because we are scheduling all VCPUs of the domain simultaneously. But this approach suffers from the serious drawbacks like CPU fragmentation and priority inversion [7].

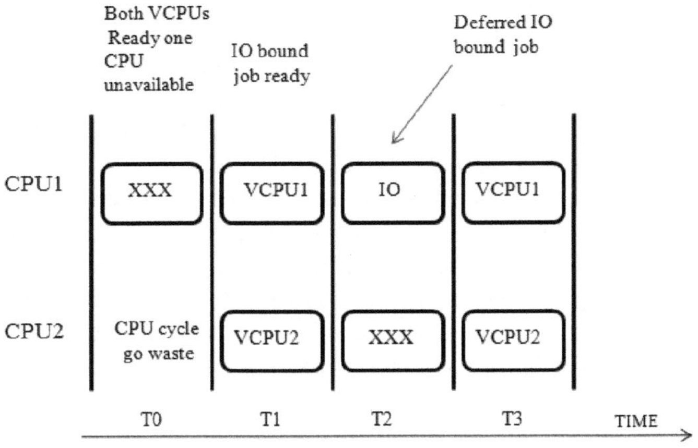

Fig. 3. CPU Fragmentation and Priority Inversion

2.1 CPU Fragmentation

As shown in Fig.3, with co-scheduling approach VCPU1 and VCPU2 cannot be scheduled at time slice T0 although both becomes runnable at T0 because there is only one CPU idle at T0. If there is no any other job ready to schedule on physical CPU2, one cpu cycle on CPU2 goes waste. This is called CPU fragmentation which reduces CPU utilization and delays the VCPU execution.

2.2 Priority Inversion

An I/O bound job is given a priority to run whenever it is ready. As shown in Fig. 3, although an I/O job is ready at T1, the I/O job has to wait until T2 because the scheduler already has assigned the slot T2 to both the CPUs. This problem adversely affects I/O bound jobs. Latency time for the I/O job increases. This is called priority inversion.

3 Proposed Design

Asynchronous scheduling and co-scheduling both have their own pros and cons. Our solution is the midway between these two approaches. We dynamically change the scheduling policy depending upon the type of load present on the domain. If the domain is executing the jobs like inter processor communication which require spinlock synchronization then we are forcing VCPUs to schedule simultaneously i.e. co-scheduling and if jobs on the domain doesn't require spinlock synchronization then we are allowing asynchronous scheduling.

As shown in Fig.4, Consider a guest Operating System is having four VCPUs. Let, a process P executing on guest spawns four concurrent threads T1', T2', T3' and T4' these threads are mapped to the kernel threads T1, T2, T3 and T4 respectively by the guest Operating System kernel and scheduled on the set of VCPUs(VCPU1,VCPU2,VCPU3 and VCPU4) assigned to the guest. Xen receives the set of these VCPUs to schedule on actual physical CPUs. Our work proposes the optimization in between the mapping of these VCPUs to the physical CPUs. When hypervisor scheduler receives the set of VCPUs to schedule, it calculates the θ (theta) for corresponding set. For every domain, θ is an absolute value, representing number of jobs executing on that domain which require spinlock synchronization. It is the characteristic of the set of VCPUs not of the single VCPU. All the VCPUs in the set have same value of θ.

3.1 VCPU Monitor

VCPU monitor inspects the set of the VCPUs arriving for the scheduling. It obtains the information regarding the concurrency of the threads scheduled on those VCPUs by making the contact with the process control block (PCB) where the information regarding each thread is stored. VCPUs which are executing the threads of the same process and whose resources are dependent on each other may need to be synchronized Fig 4. As per this information, VCPU monitor calculates the degree of the concurrency θ for each set of VCPUs it receives. It then prepare the data structure where the VCPUs need to be synchronized are regarded as the siblings of each other.

3.2 Decision Switch

When the dispatcher selects the VCPU for scheduling, control is taken by the Decision Switch. It obtains the information of the degree of concurrency of VCPU simply from the data structure prepared by the VCPU monitor.

Threshold, δ (delta), represents the upper bound for the concurrency of the domain up to which co-scheduling would be an overhead.

Value of the δ is static and is passed as a boot parameter to the scheduler. Decision switch obtains the value θ by inspecting the set of VCPUs currently selected. It compares the value of θ with δ. If θ is greater than or equal to δ, it indicates that domain is executing the jobs which require spinlock synchronizations.

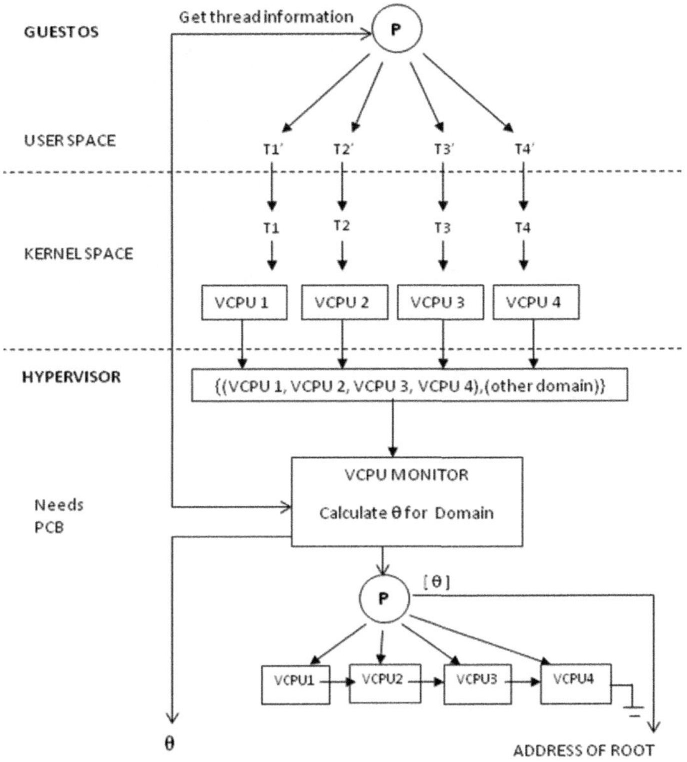

Fig. 4. VCPU Monitor

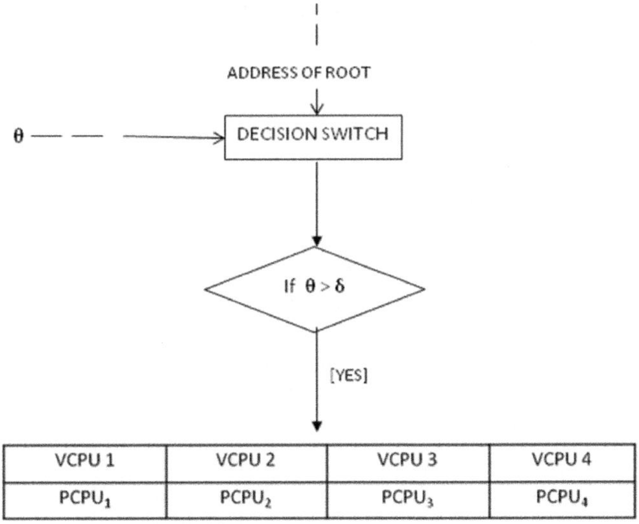

Fig. 5. Decision Switch

So it would be beneficial if we co-schedule all other sibling VCPUs of the set. From the data structure prepared by VCPU monitor, Decision Switch comes to know the sibling VCPUs of the currently selected VCPU. It then distributes the VCPUs in the run queues of the different physical CPUs. As shown in Fig.6, VCPU1, VCPU2, VCPU3 and VCPU4 are distributed in the run queue of physical processors PCPU1, PCPU2, PCPU3 and PCPU4. Distribution is necessary because these VCPUs are to be co-scheduled. Decision Switch prepares the data structure which gives the clear cut idea of which VCPU is in the run queue of which physical CPU. Distribution is done by dynamically setting the affinity of VCPU to corresponding physical CPU [10].

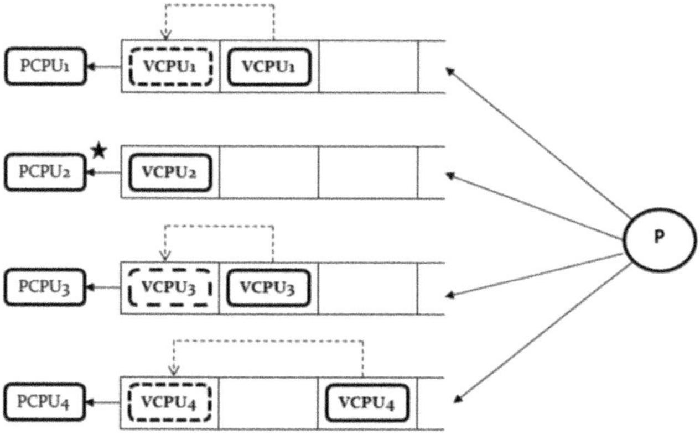

Fig. 6. Boosting VCPU

If value of θ is less than δ, it indicates that domain is not executing the jobs which require spinlock so no need to co-schedule. This approach just ensures that VCPU of the domain are distributed on different physical CPUs.

3.3 Co-scheduling

Most of the code of parallel threads can be executed without the need of synchronization. Synchronization is needed only in the critical section. So that any VCPU holds Spin-lock and value of θ is greater than δ, then we are co-scheduling the siblings of VCPUs using BOOSTing mechanism provided by Credit Scheduler. BOOSTing increases the priority of the VCPU temporarily. As shown in Fig 6, let the value of θ for set of VCPUs {VCPU1,VCPU2,VCPU3,VCPU4} is greater than δ. So VCPUs of the domain P, are distributed in the run queues of the different physical CPUs. As soon as VCPU2 holds the spinlock, all other siblings of VCPU2 i.e. VCPU1, VCPU3 and VCPU4 are BOOSTed. So that their priority is increased temporarily and they are executed simultaneously. This is the strict co-scheduling. This strict co-scheduling will continue until any VCPU belonging to the set is holding a spinlock. When none of the sibling VCPU holds a spinlock, we are relaxing the co-scheduling.

4 Advantages

This design has the following advantages over current scheduler

1. Design is flexible as it provides both co-scheduling and asynchronous scheduling options.
2. VCPU stacking problem for the spinlock synchronization is minimized by co-scheduling the VCPUs.
3. Spinlock waiting time is reduced so that VCPUs executing inter processor operations are expected to complete without timeout within reasonable period of time.
4. Co-scheduling is done only if domain require spinlock synchronization, so that overhead of co-scheduling, for the domain which are not executing such jobs is reduced.
5. CPU fragmentation and Priority inversion is minimized as synchronization of VCPUs is achieved only within critical sections.

5 Results

Chart shows that without our patch, as the size of the critical section increases waiting time of the thread on the spinlock increases exponentially. With our patch this growth is linear. There is almost 26.46% decrease in involuntary waiting time of VCPU.

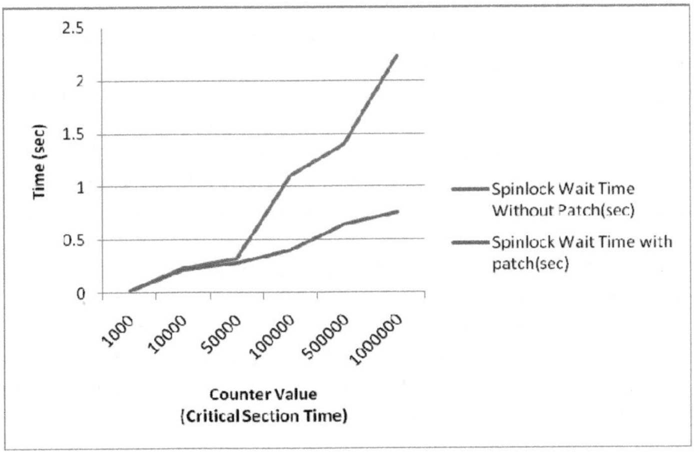

6 Conclusion and Future Work

Virtualization has a negative impact on synchronization in guest operating system and this issue is especially serious, when concurrent workloads such as multithreaded programs run in the virtualized environment. We propose a dynamic scheduling method to mitigate the problem by avoiding unnecessary overhead for co-scheduling and performance degradation of concurrent workloads while keeping the performance of non-concurrent workloads on virtual machine.

Currently, the authors are working on implementation of the proposed design. Future work can be done in mitigating the problem which arises due to the asynchronous progress of the VCPUs when scheduler decides not to do co-scheduling.

Acknowledgement. We would like to thank Dr. George Dunlap, Xen scheduler expert and author of credit2 scheduler of Xen, for providing valuable guidance, introducing us to various challenges in the project, encouragement and support.

References

[1] Chisnall, D.: The Definitive Guide to the Xen Hypervisor. Prentice Hall (2008)
[2] Stallings, W.: Operating Systems. Prentice Hall
[3] Bovet, D.P.: Understanding the Linux Kernel, 3rd edn. O'Reilly
[4] VMware, Inc., Co-scheduling SMP VMs in VMware ESX Server, `http://communities.vmware.com/docs/DOC-4960`
[5] VMware, Inc., VMware vSphere 4:The CPU scheduler in VMware ESX 4, white paper (September 2010), `http://www.vmware.com/pdf/perf-vsphere-cpu_scheduler.pdf`
[6] Yu, Y., Wangy, Y., Guo, H., He, X.: Hybrid Co-Scheduling Optimizations for Concurrent Applications in Virtualized Environments. In: Sixth IEEE International Conference on Networking, Architecture, and Storage (2011)
[7] Sukwong, O., Kim, H.S.: Is co-scheduling too expensive for SMP VMs. In: EuroSys 2011. ACM (April 2011)
[8] Cherkasova, L., Gupta, D., Vahdat, A.: Comparison of the Three CPU Schedulers in Xen. In: Proceedings on 30th IEEE Conference on Cloud
[9] Weng, C., Wang, Z., Li, M., Lu, X.: The hybrid scheduling framework for virtual machine systems. In: Proceedings of the 2009 ACM SIGPLAN/SIGOPS International Conference on Virtual Execution Environments (VEE), pp. 111–120 (2009)
[10] Li, Z., Bai, Y., Zhang, H., Ma, Y.: Affinity-aware Dynamic Pinning Scheduling for Virtual Machines. In: IEEE International Conference on Cloud Computing Technology and Science (2010)
[11] Weng, C., Liu, Q., Yu, L., Liu, M.: Dynamic Adaptive Scheduling for Virtual Machines. In: HPDC 2011, June 8-11 (2011)
[12] Wells, P.M., Chakraborty, K., Sohi, G.S.: Hardware Support For Spin Management in Overcommitted Virtual Machines. In: PACT 2006, Sepetember 16-20, pp. 16–20 (2006)
[13] Chen, H., Jin, H., Hu, K., Huang, J.: Dynamic Switching-Frequency Scaling: Scheduling Overcommitted Domains in Xen. In: 39th International Conference on Parallel Processing, September 13-16 (2010)
[14] Xen, `http://www.xen.org`
[15] Milenkovic, M.: Operating Systems Concepts and Design. Tata Mcgraw-Hill
[16] VMWare, `http://www.vmware.com`
[17] Xen documentation, `http://wiki.xensource.com`
[18] Intel Corporation, `http://software.intel.com/en-us/articles/multiple-approaches-to-multithreaded-applications/`

[19] Feitelson, D.G., Rudolph, L.: Distrubuted Hierarchical Control For Parallel Processing (May 1990)
[20] Braham, P., Dragovich, B., Fraser, K., Hand, S., Harris, T., Ho, A., Neugebauer, R.: xen and art of virtualization. In: Proc. of 19th ACM Symposium on Operating Systems Principles (SOSP 2003), pp. 164–177 (October 2003)
[21] Love, R.: Linux Kernel Development. Addison-Wesley Professional (2010)
[22] Credit Scheduler, http://wiki.xensource.com/xenwiki/CreditScheduler

Performance Analysis of Network Layer Security Attack in WiMAX System

Rakesh Kumar Jha[1] and Suresh Limkar[2]

[1] Department of Electronics and Communication Engineering, SVNIT Surat, India
[2] Department of Computer Engineering, AISSMS's IOIT, Pune, India
{Jharakesh.45,sureshlimkar}@gmail.com

Abstract. In the last few years there has been significant growth in the wireless communications. Security issues are prime concern in wireless communication networks. In this paper analyzed and observed the performance one of the most challenging attack i.e Black hole attacks in WiMAX-WLAN interface network. Black hole attack is very important for NGN (Next Generation Network) because this attack is very much possible in WiMAX-WLAN interface network with high impact with fewer efforts by intruder nodes. In our case study intruder node associated with less buffer size and it is moving with defined trajectory from router WiMAX-WLAN converter. In this observation there are three possibility of black hole attack has been studied i.e on the basis of less buffer size, medium buffer size and finally default buffer size. This attack is effected the performance of entire network like decrease the throughput and increase the packet dropped or delay. In last phase observed that the AP (Access Point) is highly sensitive for black hole attack with respect to buffer size.

Keywords: WiMAX, WLAN, WiMAX, WLAN, Black Hole attack, NGN, Security, Throughput, Delay, OPNET Modeler.

1 Introduction [1]

We have taken the specification required in black hole attack on the basis of Irshad Ullah thesis. In the case of protocol based flooding attack, the malicious node reply will be received by the requesting node before the reception of reply from actual node; hence a malicious and forged route is created. When this route is establish, now it's up to the node whether to drop all the packets or forward it to the unknown address. External attacks physically stay outside of the network and deny access to network traffic or creating congestion in network or by disrupting the entire network. External attack can become a kind of internal attack when it take control of internal malicious node and control it to attack other nodes in WiMAX-WLAN interface network. External black hole attack can be summarized in following facts. Malicious node detects the active route and notes the destination address. Malicious node sends a Route Reply Packet (RREP) including the destination address field spoofed to an unknown destination address. Hop count value is set to lowest values and the sequence number is set to the highest value. Malicious node send RREP to the nearest

S.C. Satapathy et al. (Eds.): Proc. of Int. Conf. on Front. of Intell. Comput., AISC 199, pp. 535–542.
DOI: 10.1007/978-3-642-35314-7_60

available node which belongs to the active route. This can also be send directly to the data source node if route is available.

The RREP received by the nearest available node to the malicious node will relayed via the established inverse route to the data of source node. The new information received in the route reply will allow the source node to update its routing table. New route selected by source node for selecting data. The malicious node will drop now all the data to which it belong in the route [1].

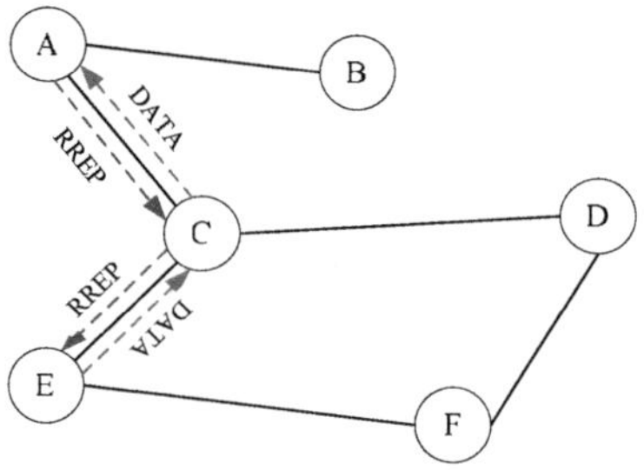

Fig. 1. Black hole attack specification [1]

From above figure we observed that the malicious node “A” first detect the active route in between the sender “E” and destination node “D”. The malicious node “A” then send the RREP which contains the spoofed destination address including small hop count and large sequence number than normal to node “C”. This node “C” forwards this RREP to the sender node “E”. Now this route is used by the sender to send the data and in this way data will arrive from the many malicious nodes. These data will then be dropped. In this way sender and destination node will be in no position any more to communicate in state of black hole attack [1].

In last phase we had compared the performance analysis with four possible conditions.

This paper is organized as follows; Section 2 illustrate the WiMAX Network Model implementation, in section 3 parameters set for the BS and SS according to the IEEE 802.16 standards and buffer condition for all four scenarios, in section 4 simulation results to investigate the black hole attacks in all possible conditions, in section 5 result analysis and finally in section 6 describes future scope.

2 Network Model for WiMAX-WLAN Interface

The WiMAX Base Station (BS) may access maximum number of nodes depending on its capacity described in IEEE 802.16 standards; here only one WiMAX node is taken which is covered by one Base Station (BS). There are two Access Points (APs) in the Subnet which are covered by two Base Stations (BSs) each. The Access Points (APs) used in a subnet is not only a regular Access Points (APs) used in ADHOC Network in Wi-Fi environment, but it is one type of router which takes WiMAX packets from Base Station (BS) and converts it to Wi-Fi packets and route to the WLAN clients, the Access Points (APs) works as a WiMAX clients.

Similarly each Access Points (APs) may contain maximum number of WLAN clients depending on its capacity, in my case each Access Points (APs) consists of seven (7) WLAN nodes. As shown in the Figure of a Subnet the upper Access Point (AP) is in connection with Base Station (BS) WiMAX_BS_B via WiMAX link (Radio link) and the lower Access Point (AP) is in connection with WiMAX_BS_A via the same WiMAX link (Radio link). Again from the specification of IEEE 802.16 the Access Points (APs) are within the coverage area of Base Station (BS) (which is 30 kms practically). The WLAN clients are placed in a circular fashion which surrounds their respective Access Points (APs).

In our proposed case study we have compared the result with 4 scenarios: First scenario: It is consists of the WiMAX-WLAN subnet without malicious node (or intruder node). Second scenario: Two malicious nodes are included in WiMAX-WLAN subnet with less buffer size of about 64Kb. Third scenario is similar to scenario 2; only the buffer size of malicious nodes is somewhat more to about 256Kb. Forth scenario: The buffer size of the malicious nodes is same as the actual WLAN client, which is 1Mb.

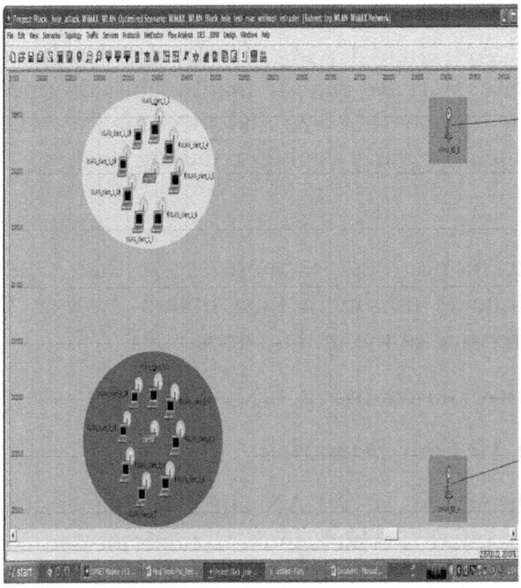

Fig. 2. Network model without Black hole Attack

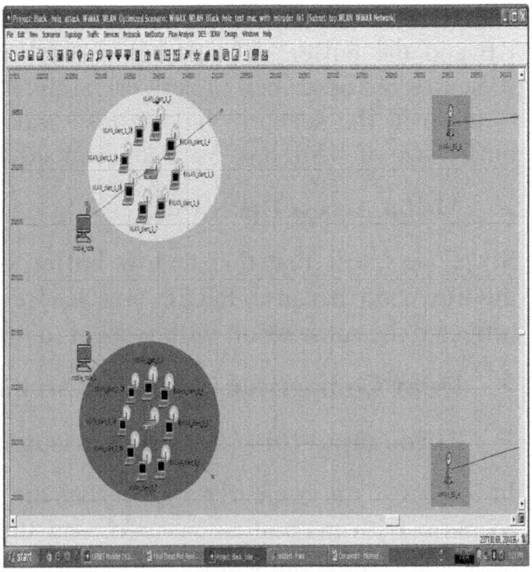

Fig. 3. Network model with Black hole Attack

3 Simulation Parameter

In this network model 4 scenarios are made, in which first scenario consists of the WiMAX-WLAN subnet without malicious node (or intruder node).In the second scenario two malicious nodes are included in WiMAX-WLAN subnet with less buffer size of about 64Kb The third scenario is similar to scenario 2; only the buffer size of malicious nodes is somewhat more to about 256Kb In forth scenario the buffer size of the malicious nodes is same as the actual WLAN client, which is 1Mb.

Table 1. Buffer Size of Client for different Scenarios

	Scenario 1	Scenario 2	Scenario 3	Scenario 4
Buffer Size	1024 Kb	**64 Kb**	**256 Kb**	**1024 Kb**

4 Result Analysis

In this simulation we have observed the performance of the network behavior in four different scenarios for five minutes. In first phase analyzed the result of WLAN client and in second phase observed the performance of black hole attack on WiMAX client.

4.1 Nodes in Ap_0 (Lower Network- From Figure 2 And Figure 3)

- **Data Dropped (buffer overflow) (bits/sec) for mobile node 1**

Mobile_node_1 is a malicious node which is creating a black hole attack in lower network near AP_0. The above figure shows the results of data dropped due to buffer overflow in malicious node mobile_node_1. This figure shows the results for three scenarios, which are, less buffer size having 64 Kb, medium buffer size having 256 Kb and the buffer size same as actual WLAN client having size of 1 Mb. These scenarios are described in network architecture. Here we can see that the data dropped is more in the scenario having less and medium buffer size and the scenario with buffer size of 1 Mb has less data dropped which is clearly visible in the Fig. 4.

- **Media Access Delay (Sec)**

Since we know that due to low buffer size mobile node is more active than other mobile node because RREP and RREP time is minimum than others. In Fig. 5. reflected the same result with respect to buffer size of the mobile node.

- **Delay Comparison in all Four Scenarios** (shown in Fig. 6.)

- **Throughput (bits/sec) Comparison in All Four Scenarios**

In the Fig. 7. it is clearly shown that the throughput at WLAN client is more in the scenario with no malicious node, and the scenarios which include malicious nodes have less throughput level

- **Data Dropped (Packet/Sec)**

In this Fig. 8. the data dropped related to WiMAX environment is shown. It is clearly visible that when there is no malicious node in the subnet than the packets dropped is less as compared to the scenarios in which the malicious node is present.

- **Data Dropped (Buffer Overflow) (packets/sec)**

Data dropped is more in the scenario with malicious node having less buffer size. We have moreover concluded that black hole attack is possible in WiMAX –WLAN interface network and our performance analysis given an idea about packet dropped, delay and throughput at mobile node (malicious node), AP and highly effected client. Our approach is one of the major attacks in this environment and this attack may be possible in different way. It is shown in Fig. 9.

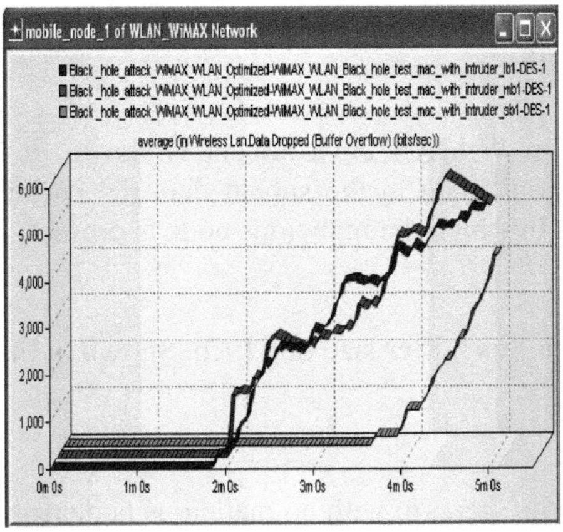

Fig. 4. WLAN dropped at mobile_node_1

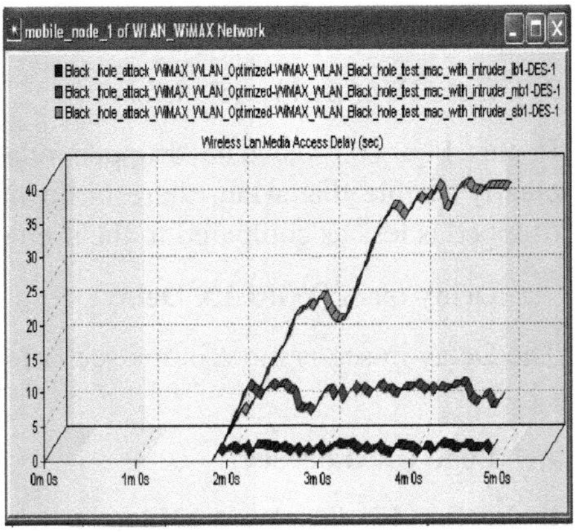

Fig. 5. WLAN MAC Delay

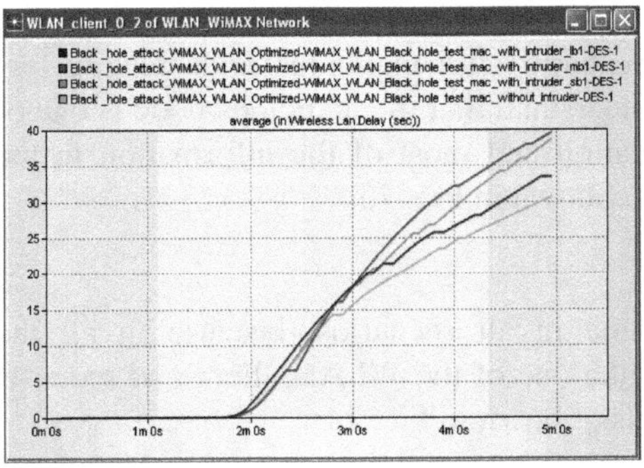

Fig. 6. WLAN delay in all four scenarios

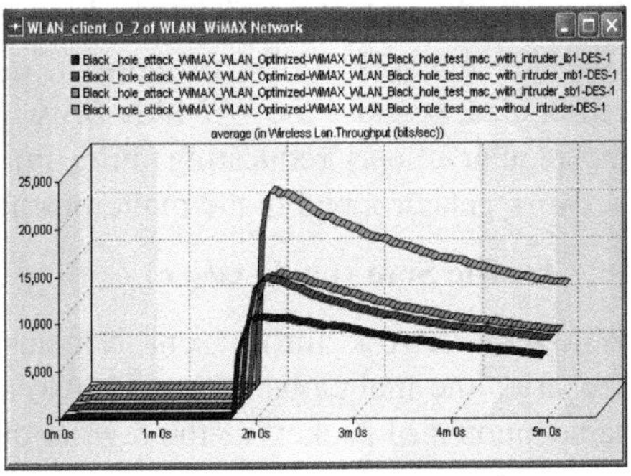

Fig. 7. WLAN Client throughput

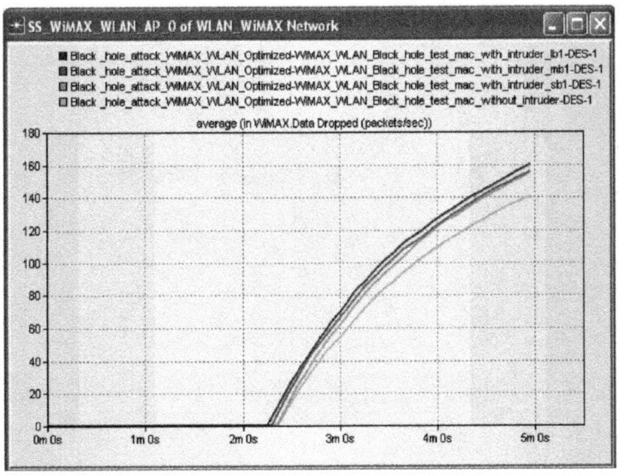

Fig. 8. WiMAX _WLAN AP_0 data dropped

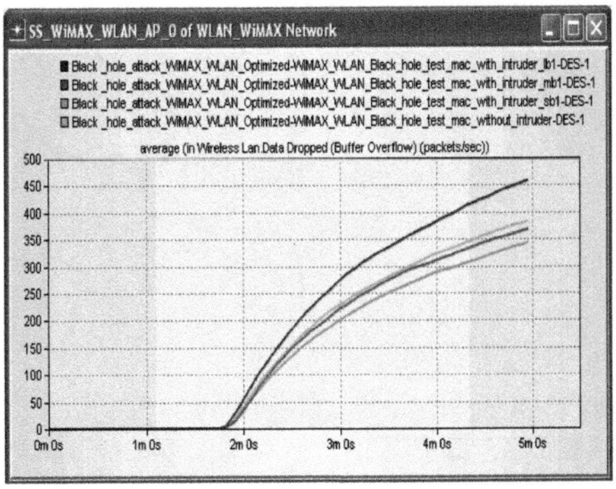

Fig. 9. Data Dropped in AP_0

4.2 WiMAX Related Results

➢ **Data Dropped (packets/sec)**

In this Fig. 10. The data dropped related to WiMAX environment is shown. It is clearly visible that when there is no malicious node in the subnet than the packets dropped is less as compared to the scenarios in which the malicious node is present.

➢ **Delay (sec) (WiMAX Delay)**

The Delay in sec is more in the scenario with less buffer size of 64 Kb. Shown in Fig. 11.

➢ **Load (packets/sec)**

In this Fig. 12. The WiMAX load is less in the scenario with no malicious node and it is almost same in other three scenarios where malicious node is present, this load is more due to the access amount of request packets are sent by the malicious nodes.

➢ **Traffic Received (packets/sec)**

WiMAX network traffic is more in the scenario with malicious node having less buffer size, because the WiMAX BS will understand that the malicious node is one of the regular clients requesting more information and most of the information in the network gets dropped in the malicious node. Shown in Fig. 13.

➢ **Traffic Sent (packets/sec)**

WiMAX network traffic sent remains same in all scenarios, because in all the scenarios the malicious node will also act like one of the WLAN client and requests same amount of packets as the regular clients.. Shown in Fig. 14.

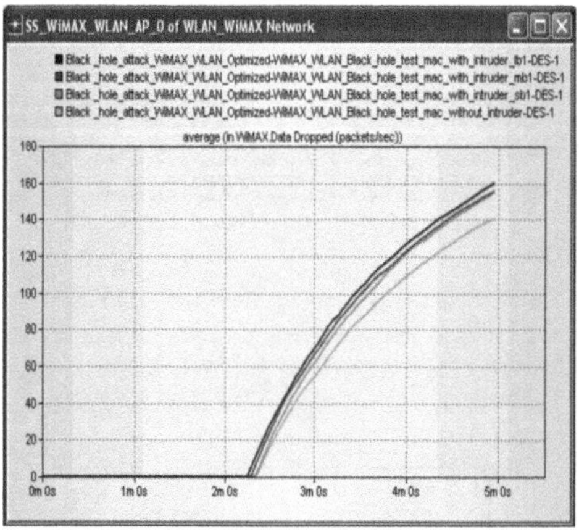

Fig. 10. WiMAX _WLAN AP_0 data dropped

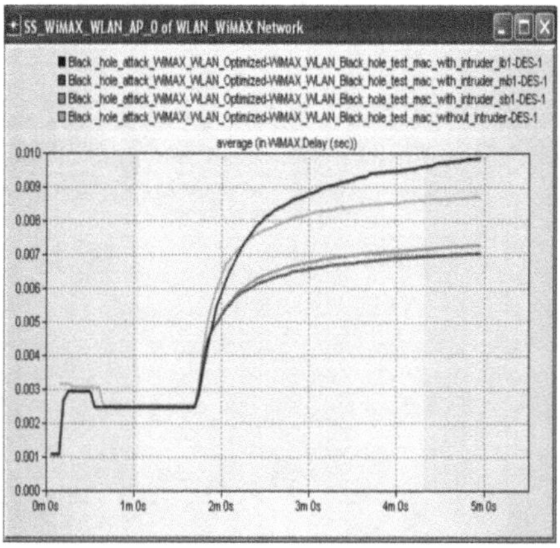

Fig. 11. WiMAX _WLAN AP_0 Delay

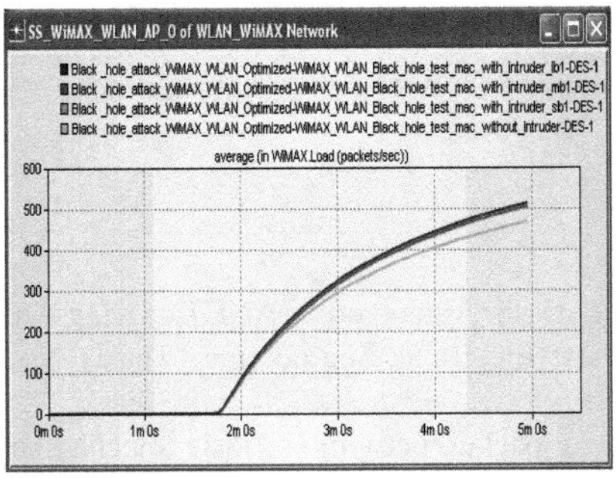

Fig. 12. WiMAX _WLAN AP_0 Load

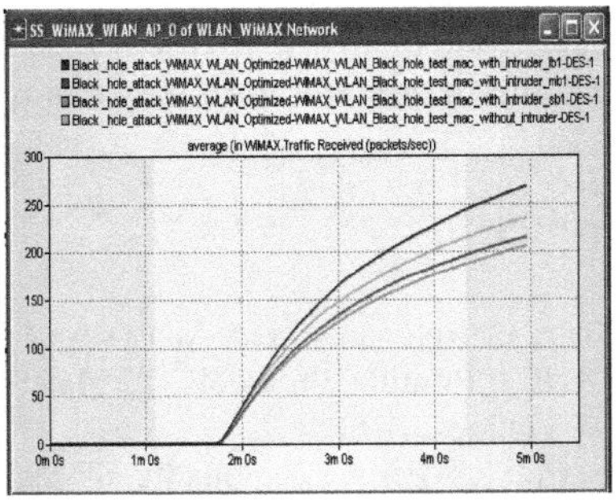

Fig. 13. WiMAX _WLAN AP_0 Traffic Received

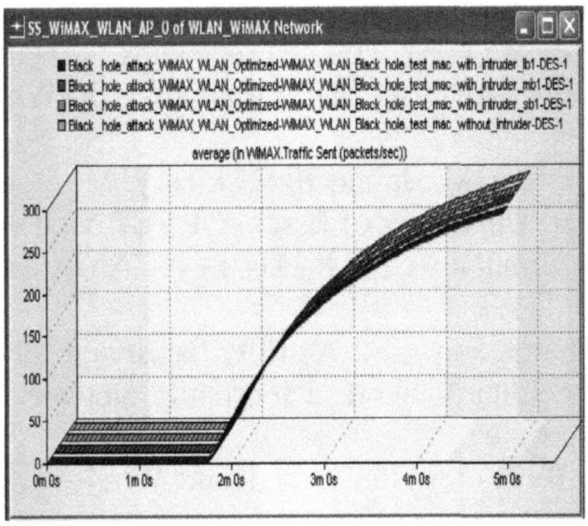

Fig. 14. WiMAX _WLAN AP_0 Traffic Sent

5 Conclusion

On the basis of simulation result observed that black hole attack is possible in WiMAX-WLAN interface network because buffer size has been played an important role in our proposed scenarios. This attack is highly sensitive attack as far as NGN network because enemy can destroy or degrade the network performance with minimum effort and very less expenditure. Our performance analysis has given proposal about the packet dropped, delay and throughput at mobile node (malicious node), AP and highly effected client in the presence of malicious node.

6 Future Scope

In future scenario there is a scope to analyze the black hole attack in WiMAX-WLAN interface network with different routing protocol like DSR, TORA and AODV.As far as for future security expects we can compare this attack with misbehavior node attack, Band-width attack, wormhole attack and Sybil attacks. Data base of AP can be train such that if any malicious node will come with low buffer size they can be detected and in next iteration they can be removed from the original data base.

References

1. Ullah, I., Rehman, S.U.: Analysis of Black Hole Attack on MANETs Using Different MANET Routing Protocols, A Mater Thesis, Electrical Engineering, Thesis No. MEE 10:62 (June 2010)
2. Guerin, R., Peris, V.: Quality of Service in packet networks: Basic mechanisms and directions. Computer Networks 31(3), 169–189 (1999)
3. Lu, J., Ma, M.: Cross-layer QoS support framework and holistic opportunistic scheduling for QoS in single carrier WiMAX system. Journal of Network and Computer Applications 34(2), 765–773 (2011)
4. Sayenko, A., Alanen, O., Karhula, J., Hämäläinen, T.: Ensuring the QoS requirements in 802.16 scheduling. In: MSWiM 2006: Proceedings of the 9th ACM International Symposium on Modeling Analysis and Simulation of Wireless and Mobile Systems (October 2006)
5. Lucio, G.F., Paredes-Farrera, M., Jammeh, E., Fleury, M., Reed, M.J.: Electronic Systems Engineering Department, University of Essex, OPNET Modeler and Ns-2: Comparing the Accuracy Of Network Simulators for Packet-Level Analysis using a Network Test bed (2005)
6. Saad, W., Dawy, Z., Sharafeddine, S.: A utility-based algorithm for joint uplink/downlink scheduling in wireless cellular networks., Journal of Network and Computer Applications (In Press, Corrected Proof)
7. Nuaymi, L.: WiMAX Technology for Broadband Wireless Access. John Wiley & Sons, France (2007)
8. Ozcelik, I.: Interconnection of CAN segments through IEEE 802.16 wireless MAN. Journal of Network and Computer Applications 31(4), 879–890 (2008)
9. IEEE. Standard 802.16-2004, Part 16: Air interface for fixed broadband wireless access systems (June 2004)
10. IEEE. Standard 802.16-2005, Part 16: Air interface for fixed and mobile broadband wireless access systems (December 2005)
11. Andrews, J.G., Ghosh, A., Muhammad, R.: Fundamentals of WiMAX Understanding Broadband Wireless Networking, pp. 1–63, 271–292 (2007)
12. Yan Zhang, H.-H.C.: Mobile WiMAX: Toward Broadband Wireless. Auerbach Publication, New York (2008)
13. Ahmedi, S.: Introduction to Mobile WiMAX Radio Access Technology: PHY and MAC Architecture. Intel Corporation
14. Jha, R., Dalal, U.: WiMAX System Simulation and Performance Analysis under the Influence of Jamming. Wireless Engineering and Technology(WET) Journal by Scientific Research 1(1), 20–26 (2010)
15. Jha, R., Dala, U.D.: A Journey on WiMAX and Its Security Issues. International Journal of Computer Science and Information Technologies 1(4), 256–263 (2010)

An Optimized Approach to Minimize Broadcast in Communication of Self Organized Wireless Networks

Neeta Shirsat and Pravin Game

Computer Engineering Department,
Pune Institute Of Computer Technology,
Pune-411043, India
{neeta.shirsat,pravingame}@gmail.com

Abstract. This paper proposes the strategy for effective connections in backbone of self organized wireless networks with role based approach. Various applications like disaster management, home monitoring and office automation, shows increasing demand for wireless networks. Nodes in a wireless and ad-hoc networks are free to move. Each node plays the efficient role for formation of backbone with local interaction. In this approach four roles are identified: Agent, Leader, Willingness to act as a Gateway and Gateway. Each node is playing one of the roles and backbone reconfiguration is performed with changes in environment. 'Willingness to act as gateway' node avoids the problem of duplicate gateways and unnecessary broadcast. Thus forming an efficient backbone provides good resource conservation property. Number of links of MST and proposed strategy are compared for performance analysis. As compared to the MST, proposed algorithm shows near solution for network connections. This approach utilizes resources in optimized way. In case of failure of original gateway on path, other appropriate device plays the role of original gateway.

Keywords: wireless sensor network, emergent behavior, clustering, self organization, wireless devices networks, wireless communication.

1 Introduction

A Wireless network is on-the-fly network formed by wireless devices like cell phones, PDAs, sensors etc. Because of the hardware and energy limitations, wireless networks needs extra mechanism to organize themselves in self organization than wired network.

In role based self organization every device need to perform certain task. In proposed strategy various roles are identified like agent, willingness to act as a gateway, gateway and leader. Network inconsistency due to duplicate gateway on single path increases number of communication links. Network inconsistency is removed by proposing new role: willingness to act as gateway. Every device will run IMPROVED self organization algorithm and clusters are formed [1]. Every cluster

S.C. Satapathy et al. (Eds.): Proc. of Int. Conf. on Front. of Intell. Comput., AISC 199, pp. 543–549.
DOI: 10.1007/978-3-642-35314-7_61

will have a cluster head called as Leader. Leaders can communicate with each other through gateways and all other members will act as an agent. When more than one gateway present on a single path, network inconsistency is identified. In proposed approach when original gateway is not available, duplicate gateway role can be modified to be a willing node to act as gateway.

A minimum connection in backbone of self organized network is proposed here. MST (Minimum Spanning Tree) is formed ideally for minimum connections. As compared to the MST, proposed algorithm shows near solution for network connections. This approach utilizes resources in optimized way as broadcast is minimized with effective connections. In case of failure of original gateway on path, other appropriate device plays the role of original gateway.

The remainder of the paper is organized as follows: In section 2 the related work is overviewed. In Section 3 describes the IMPROVED algorithm for various roles assignment and the environment. In Section 4 simulation results are shown. Conclusions that can be drawn are covered in section 5.

2 Related Work

Pure flooding allows each node to receive broadcast packets [2]. This requires maximum connections among the nodes and resources are not utilized in proper way. Energy-aware self-organization algorithms for small WSNs [4], allow deploying a WSN solution in monitoring contexts without a base station or central nodes. Sensors are self-organized in a chain and alternate between sleep and active mode where the sleep periods are longer than the activity periods. Effective creation of backbone is not considered. IDSQ ALGORITHM for Wireless sensor networks described in [5], consist of three components mainly: the sensor nodes, sensing object and the observer. The mutual cooperation between them, each sensor node has a small processors, some data need to be addressed was sent to the node summary, then through the multi-hop routing data about monitoring on the perceived object will be sent to the gateway, and finally by the gateway to data within the entire region is transferred to the remote center to manipulate. Hence this algorithm has constraints on roles and network reconfiguration.

Self organization proposes no need of any manual intervention and central control. Network topology changes with time and local interaction leads to global behavior. Various topology control approaches are designed and proposed like a Multi-Point Relay (MPR) based approach [7], a connected dominating set (CDS) based approach [10] and a cluster based approach [6]. But this strategy requires the knowledge of network in advance and constrained on equal transmission ranges [9]. Many cluster based approaches are proposed with self organization. Cluster based algorithm for self organization with various roles like member, leader and gateway is proposed [6] with variable transmission ranges. Each node performs some role like member, leader or gateway and they form backbone depending on local interaction. But backbone connection can have duplicate gateway for one path which leads increase in broadcast and extra energy consumption as multiple gateways can exist on single path [6]. This scenario is shown in figure 1.

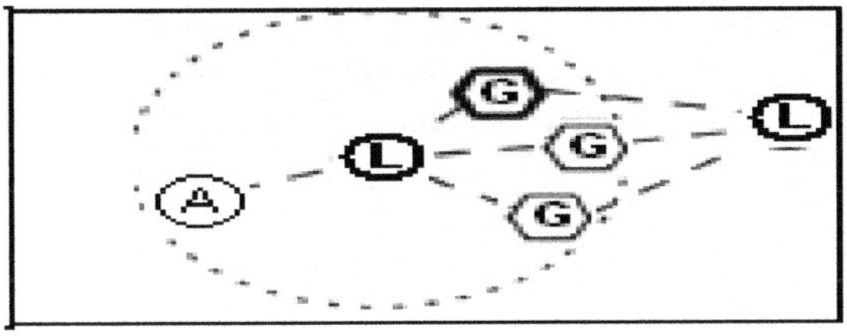

Fig. 1. Unnecessary Broadcasting

This paper proposes role based approach for effective connections in backbone in self organized wireless network. It tries to minimize the unnecessary broadcast with minimum connections, in turn optimal use of resources. Performance analysis is done by comparing number of links with MST.

3 Role Based Self Organization with IMPROVED Algorithm

In proposed strategy four different roles are identified with the help of IMPROVED Self Organization Algorithm. Firstly Leader Election algorithm is run to form clusters with cluster head. Then for cluster communication gateway role is identified. Cluster members are called as agents. Duplicate gateway is assigned with role Willingness to act as Gateway. Following algorithm describes the IMPROVED Role Based self-organization algorithm [1].

Algorithm 1 IMPROVED Role Based Self –Organization Algorithm

```
if NodeExist≠ 0
    if NodeLeaderNum= 0 then
        ROLE<=LEADER;
    else if ROLE =Leader then
        leaderElection();
    else if Node LeaderNum=1 then
        ROLE<= AGENT
    else
        ROLE<=GATEWAY
            if NodeGatewayNum >1 then
                ROLE<=WillingGateway
            end if
else
    ROLE<=ANY
end if
```

Algorithm 1 describes the assignment of various roles. As duplicate gateway problem is solved by identifying new role, minimum connections required to form backbone.

Figure 2 shows effective formation of backbone with proper role assignment [1]. When more than one gateway present on single path, only one gateway remains active and others will show willingness to become as gateway but perform function like an agent.

Fig. 2. Effective Formation of Backbone with proper role assignment

FDA is shown in figure 3.

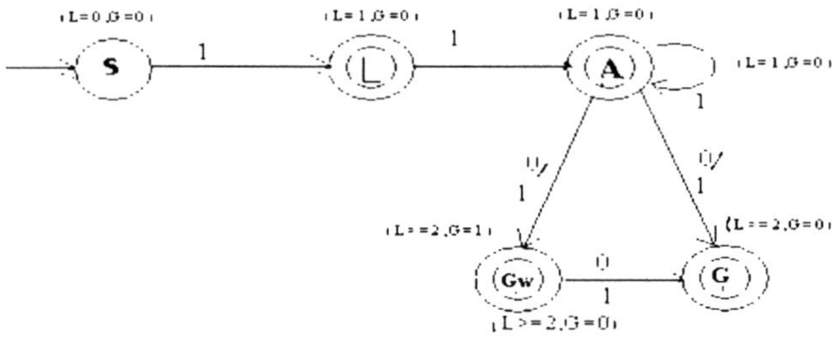

Fig. 3. DFA for Role Change

4 Simulation Results

The behavior and performance of proposed IMPROVED algorithm has been analyzed and simulated using NS-2 version 2.33. NS-2 is an event oriented network research simulator. In this section sample of simulation tests are shown.

I Scenario

A scenario of wireless sensor network has to be organized. It consists of 100 mobile nodes distributed within an environment of 100 x 100 meters.

Initial configuration of all nodes is same. Following are the configurations required:

- The network interface is 802.15.4
- The initial energy of every node is 2 joule.
- The maximum transmission range is 15 meters.
- The IMPROVED algorithm is run on every node.

II Running NS-2

The stabilizing time is considered as 10 second. At instant 25, the weight is estimated and at instant 30, clustering is done. Here various nodes are identified like agent, leader, gateway and willingness to act as gateway. Proper role assignment is shown in figure 4.

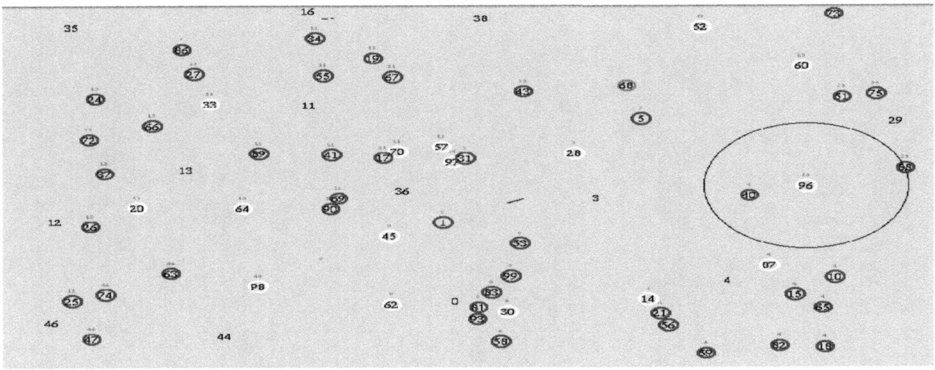

Fig. 4. Roles are assigned with each node

III Analyzing and Comparing Number of Connections with MST

The Minimum communication links generated by MST for 50 nodes is shown in figure 5 [2].

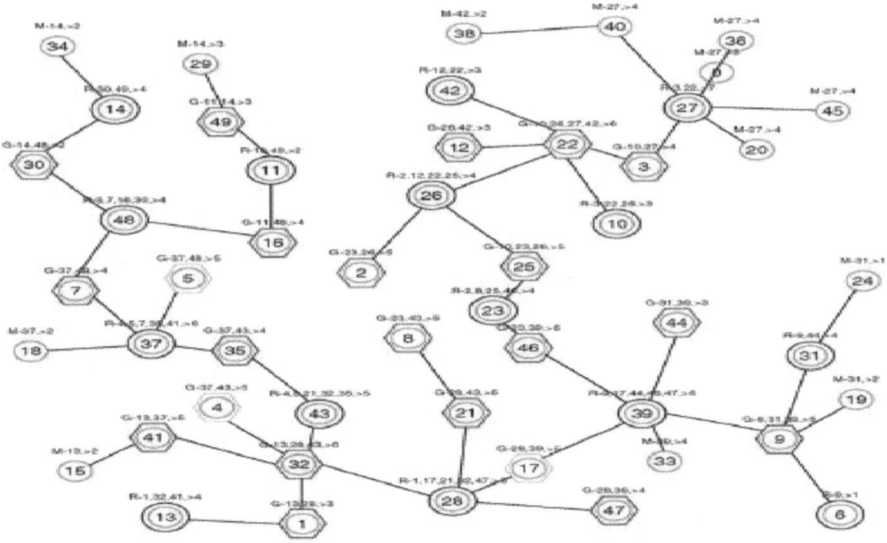

Fig. 5. Number of communication links generated by MST

Figure 6 shows connections formed by IMPROVED role based self organization algorithm for 50 nodes.

Fig. 6. Number of communication links generated by IMPROVED algorithm

Following table gives comparative analysis of number of links with MST [2].

Table 1. Comparitive analysis of MST and IMPROVED Algorithm

NS2 Environment			No of links		Performance	
Range	Agents	Dimensions	MST	IMPROVED Algorithm	Inactive Links	Proximity %
15 Mts	50	100x100	49	54	3	96%
15 Mts	60	100x100	59	69	3	89%
15 Mts	70	100x100	69	78	3	92%
15 Mts	80	100x100	79	90	8	96%
15 Mts	100	100x100	100	121	12	92%

Table 1 indicates that simulation results are near to MST and IMPROVED algorithm gives better performance.

5 Conclusion

Efficient use of resource is very important in case of wireless sensor network. Unnecessary broadcast due to duplicate gateway increases number of network connections and energy consumption. Proposed approach identifies roles efficiently and minimum connection backbone is formed. Simulation results have shown that simulation is near to MST. Thus role based IMPROVED algorithm forms efficient backbone and helps in minimizing network connections and in turn avoids unnecessary broadcast.

References

1. Shirsat, N., Game, P.: Role Based Approach for Effective Connections in Backbone of Self Organized Wireless Networks. In: Satapathy, S.C., Avadhani, P.S., Abraham, A. (eds.) Proceedings of the InConINDIA 2012. AISC, vol. 132, pp. 763–768. Springer, Heidelberg (2012)
2. Prehofer, C., Bettstetter, C.: Self organization in communication networks: Principles and design paradigms. IEEE Communication Magazine 43(7), 78–85 (2005)
3. Orfanus, D., Heimfarth, T., Janacik, P.: An Approach for Systematic Design of Emergent Self-Organization in Wireless Sensor Networks. In: Future Computing, Service Computation, Cognitive, Adaptive, Content, Patterns, Computation World 2009, November 15-20, pp. 92–98 (2009)
4. Kacimi, R., Dhaou, R., Beylot, A.-L.: Energy-Aware Self-Organization Algorithms for Wireless Sensor Networks. In: Global Telecommunications Conference, IEEE GLOBECOM 2008, November 30-December 4, pp. 1–5. IEEE (2008)
5. Yun, B., Song-Bo, J., Li, X.: Self-Organized Algorithm Simulation for Wireless Sensor Networks. In: 2009 Second International Symposium on Information Science and Engineering (ISISE), December 26-28, pp. 523–526 (2009)
6. Olascuaga-Cabrera, J.G., Lopez-Mellado, E., Ramos-Corchado, F.: Self-organization of mobile devices networks. In: Proc. IEEE Int. Conf. on Systems of Systems Engineering, pp. 1–6 (2009)
7. Liang, O., Ekercioglu, Y.A.S., Mani, N.: Gateway multipoint relays-an mpr-based broadcast algorithm for ad hoc networks. In: Proc.10th IEEE Singapore Int. Conf. Communication Systems (ICCS), pp. 1–6 (2006)
8. Zatout, Y., Campo, E., Llibre, J.-F.: WSN-HM: Energy-efficient Wireless Sensor Network for home monitoring. In: 2009 5th International Conference on Intelligent Sensors, Sensor Networks and Information Processing (ISSNIP), December 7-10, pp. 367–372 (2009)
9. Correia, L.H., Macedo, D.F., dos Santos, A.L., Loureiro, A.A., Nogueira, J.M.S.: Transmission power control techniques for wireless sensor networks. Comput. Netw. 51(17), 4765–4779 (2007)
10. Funke, S., Kesselman, A., Meyer, M.S.U.: A simple improved distributed algorithm for minimum CDS in unit disk graphs. In: Proc. IEEE Int. Conf. Wireless Mobile Computing, Networking, Communications (WiMob), vol. 2, pp. 220–223 (August 2005)
11. Nieberg, T., Hurink, J.: Wireless communication graphs. In: Proc. 2004 Intelligent Sensors, Sensor Networks, Information Processing Conf., pp. 367–372 (December 2004)
12. DARPA, The network simulator -ns-2 (1989), http://www.isi.edu/nsnam/ns/

Randomized Approach for Block Cipher Encryption

Srinivasan Nagaraj[1], D.S.V.P. Raju[2], and Kishore Bhamidipati[3]

[1] Dept. of CSE, GMRIT, Rajam
[2] Andhra University, Visakhapatnam.
[3] Dept. of CSE, MIT, Manipal
{sri.mtech04,kishore.gmr}@gmail.com

Abstract. In cryptography, a **substitution block cipher** is a method of encryption by which units of plain text are replaced with cipher text according to a regular system. The receiver deciphers the text by performing an inverse substitution. If the cipher operates on single blocks, it is termed as **simple substitution block cipher**. The proposed method that considers a random matrix which is chosen from the length of the given input text. Here the input text is stored in the matrices which are selected with minimum cost. There on the series of operations are performed and cipher text is produced, in decryption inverse matrix is calculated and the operations are performed in order to generate the plain text. As we are using random matrix and then the inverse matrix calculation is performed which adds more security for cracking the key. Thus, the cipher text obtained is very difficult to be broken without knowing the key, which provides high security.

Keywords: key length, random matrix, minimum cost, encryption, decryption.

1 Introduction

1.1 Cryptography

There are many aspects to security and many applications, ranging from secure commerce and payments to private communications and protecting passwords. One essential aspect for secure communications is that of cryptography, which is the focus of this paper. But it is important to note that while cryptography is necessary for secure communications, it is not sufficient by itself. The reader is advised, that the topics covered in this paper only describe the first of many steps necessary for better security in any number of situations. There are some specific security requirements, including:

- ***Authentication.*** The process of proving one's identity. (The primary forms of host-to-host authentication on the Internet today are name-based or address-based, both of which are notoriously weak.)
- ***Privacy/confidentiality.*** Ensuring that no one can read the message except the intended receiver.

S.C. Satapathy et al. (Eds.): Proc. of Int. Conf. on Front. of Intell. Comput., AISC 199, pp. 551–558.
DOI: 10.1007/978-3-642-35314-7_62

- ***Integrity.*** Assuring the receiver that the received message has not been altered in any way from the original.
- ***Non-repudiation.*** A mechanism to prove that the sender really sent this message.

1.2 Block Ciphers

In cryptography, a **block cipher** is a symmetric key cipher operating on fixed-length groups of bits, called *blocks*, with an unvarying transformation. A block cipher encryption algorithm might take (for example) a 128-bit block of plaintext as input, and output a corresponding 128-bit block of cipher text. The exact transformation is controlled using a second input the secret key. Decryption is similar: the decryption algorithm takes, in this example, a 128-bit block of cipher text together with the secret key, and yields the original 128-bit block of plain text. A message longer than the block size (128 bits in the above example) can still be encrypted with a block cipher by breaking the message into blocks and encrypting each block individually. To overcome this issue, modes of operation are used to make encryption probabilistic. Some modes of operation, despite the fact that their underlying implementation is a block cipher, allow the encryption of individual bits.

1.2.3 Types of Block Ciphers

There are different types of the block cipher which all aim to increase the security of encrypted text. These methods are described in detail below.

a. Integrated Block Cipher. One simple way to increase the effectiveness of the encryption algorithm is to repeatedly apply the same algorithm, each iteration encrypting the message further. The number of rounds selected is usually determined by the computational power available and the security level desired.

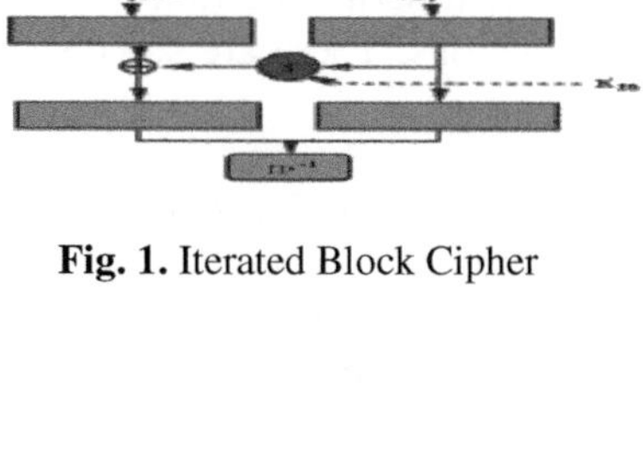

Fig. 1. Iterated Block Cipher

b. Cipher Block Chaining. Cipher block chaining is another common way to increase the effectiveness of block ciphers. In this method, the resulting cipher text is not only a function of the key and the encryption algorithm but also depends on the blocks which were previously encoded. The cipher text of an encoded block is XORed with the next plaintext block which is to be encoded. This means that to decode a particular block of cipher text, one also has to know the block that was encoded prior to it. The initial block is encrypted using an initialization vector which is randomly generated. The major drawback of this method is the same requirement. It implies that in order to decode a block, we must know all blocks prior. While this may seem beneficial, it can be hazardous when the cipher text blocks are transferred to a different user or workstation. If a single block of cipher text is lost, all subsequent blocks can no longer be decoded and must be transmitted again.

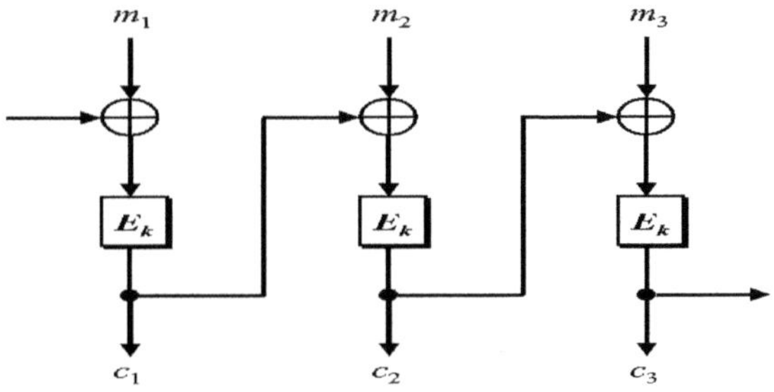

Fig. 2. Cipher Block Chaining

2 Existing System

Several well-known algorithms such as substitution techniques, transposition techniques, RSA, Deffi-Hellmen, DES, Triple DES Etc. algorithms can be used to perform this encryption and decryption. But there is no guarantee that Cipher text generated by them is safe from the intruders. Each technique has its own advantages and disadvantages. In this situation, we propose a new technique that gives support for the cipher text which is created by conventional encryption algorithm. Using this technique, one extra feature is added to the conventional encryption technique. In this we generated the random keys used to generate randomized cipher text and original cipher text.

The following drawbacks are to be eliminated:

- In some cryptographic systems, data can be amended or manipulated by unauthorized persons.
- These need more time for processing.

- These may be considered insecure for many applications.
- If we forget our key file then there is almost no chance of recovering the data.

3 Proposed System

The algorithm considers a random matrix which is chosen from the length of the given input text. Here the input text is stored in the matrix which is selected with minimum cost. There on the series of operations are performed and cipher text is produced, in decryption inverse matrix is calculated and the operations are performed in order to generate the plain text. As we are using random matrix and then the inverse matrix calculation is performed which adds more security for cracking the key.

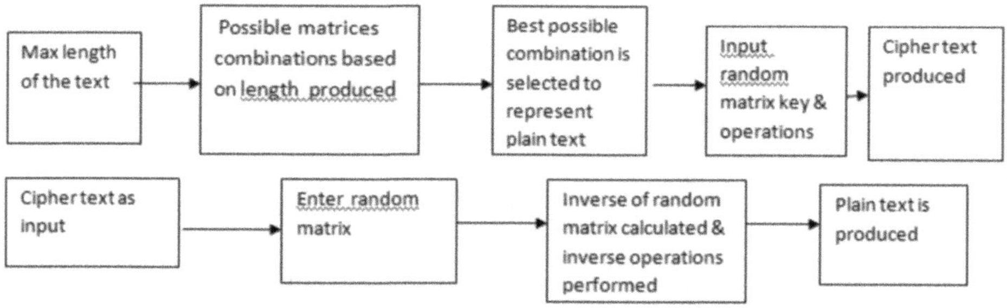

Fig. 3. Encryption and Decryption process

It has the following **advantages**:

Many of the cryptographic procedures involve using either any one of the algorithm specified and undergoes series of operations to produce the cipher text.

- By performing above operations we can say that cipher text produced is secured in all aspects as production of cipher text involves series of operations, and the key which was used in encryption process is highly difficult.
- Easy to implement and understand the concepts which was followed in this cryptography process.

4 Implementation of the Algorithm

4.1 Encryption Process

Step1. Enter the plain text.

hi how are you?

Step2. The length of the plain text is calculated and the possible matrices are taken to fit the plain text.

The length of the string is 15
Possible matrices are
1*15
3*5
5*3
15*1

Step3. The best fit matrix among the sequence of matrices is selected (depending on the cost of matrix)

Among the matrices generated the matrix with minimum cost is 3 * 5

Step4. Now the ASCII values of the plain text are stored in the best fit matrix as shown below

$$\begin{pmatrix} 104 & 105 & 32 & 104 & 111 \\ 119 & 32 & 97 & 114 & 101 \\ 32 & 121 & 111 & 117 & 63 \end{pmatrix} = af[gd][hd]$$

Step5. Depending on the best fit matrix the key matrix is selected for operations and we need to input the values for the key matrix. Key matrix is: 3 * 3

$$\begin{pmatrix} 1 & 5 & 3 \\ 2 & 11 & 8 \\ 4 & 24 & 21 \end{pmatrix} = a[gd][hd]$$

Step6. Now to get the cipher text in printable characters we have to subtract the values with 32 as af[gd][hd] - 32

$$\begin{pmatrix} 72 & 73 & 0 & 72 & 79 \\ 87 & 0 & 65 & 82 & 69 \\ 0 & 89 & 79 & 85 & 31 \end{pmatrix} = af[gd][hd]$$

Step7. Now multiply the matrix af[gd][hd] with the key matrix a[gd][hd]

$$\begin{pmatrix} 507 & 340 & 562 & 737 & 517 \\ 1101 & 858 & 1347 & 1726 & 1165 \\ 2376 & 2161 & 3219 & 4041 & 2623 \end{pmatrix} = af[l][m]$$

Step8. Now we perform operation to get the text accordingly {(af[l][m] % 95) + 32}

$$\begin{pmatrix} 64 & 87 & 119 & 104 & 74 \\ 88 & 35 & 49 & 48 & 57 \\ 33 & 103 & 116 & 83 & 90 \end{pmatrix}$$

Step9. Convert the above obtained values which are basically ASCII to normal text and the text obtained is cipher text.

@WwhJX#109!gtSZ

4.2 Decryption Process

Step1. Enter the cipher text.

@WwhJX#109!gtSZ

Step2. The length of the plain text is calculated and the possible matrices are taken to fit the plain text.

The length of the string is 15
Possible matrices are
1*15

3*5
5*3
15*1

Step3. The best fit matrix among the sequence of matrices is selected (depending on the cost of matrix).

Among the matrices generated the matrix with minimum cost is 3 * 5

Step4. Now the ASCII values of the plain text are stored in the best fit matrix as shown below

$$\begin{pmatrix} 64 & 87 & 119 & 104 & 74 \\ 88 & 35 & 49 & 48 & 57 \\ 33 & 103 & 116 & 83 & 90 \end{pmatrix}$$

Step5. Depending on the best fit matrix the key matrix is selected for operations and we need to input the values for the key matrix. Key matrix is: 3 * 3

$$\begin{pmatrix} 1 & 5 & 3 \\ 2 & 11 & 8 \\ 4 & 24 & 21 \end{pmatrix} = a[gd][hd]$$

Step6. Now for the decryption process the inverse of the matrix is calculated and to the inverse matrix the operations are performed accordingly.

$$\begin{pmatrix} 39 & -33 & 7 \\ -10 & 9 & -2 \\ 4 & -4 & 1 \end{pmatrix}$$

Step7. Now to get the cipher text in printable characters we have to subtract the values with 32. af[gd][hd]-32

$$\begin{pmatrix} 32 & 55 & 87 & 72 & 42 \\ 56 & 3 & 17 & 16 & 25 \\ 1 & 71 & 84 & 51 & 58 \end{pmatrix} = af[gd][hd]$$

Step8. Now multiply the matrix af[gd][hd] with the key matrix a[gd][hd]

$$\begin{pmatrix} -593 & 2543 & 3420 & 2637 & 1219 \\ 182 & -665 & -885 & -678 & -311 \\ -95 & 279 & 364 & 275 & 126 \end{pmatrix} = af[l][m]$$

Step9. Now we perform operation to get the text accordingly {(af[l][m] % 95) + 32}

$$\begin{pmatrix} 9 & 105 & 32 & 104 & 111 \\ 119 & 32 & 2 & 19 & 6 \\ 32 & 121 & 111 & 117 & 63 \end{pmatrix} = af[l][m]$$

Step10. Now to get the text in printable range for the values which are less than 32 add 32 to af[l][m]

$$\begin{pmatrix} 104 & 105 & 32 & 104 & 111 \\ 119 & 32 & 97 & 114 & 101 \\ 32 & 121 & 111 & 117 & 63 \end{pmatrix} = af[l][m]$$

Step11. Convert the above obtained values which are basically ASCII to normal text and the text obtained is plain text.

hi how are you?

5 Results

Description: Select Matrix

Fig. 4. Possible matrix for given length of text

Description: Inputting Key matrix

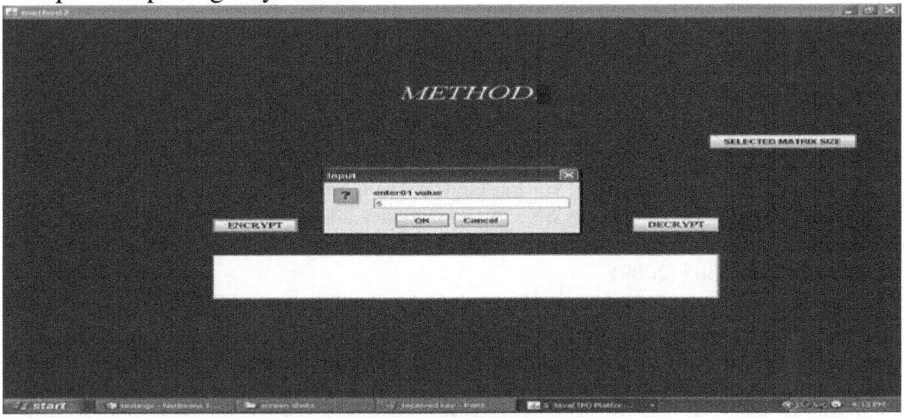

Fig. 5. Enter the matrix key

Description: Encryption

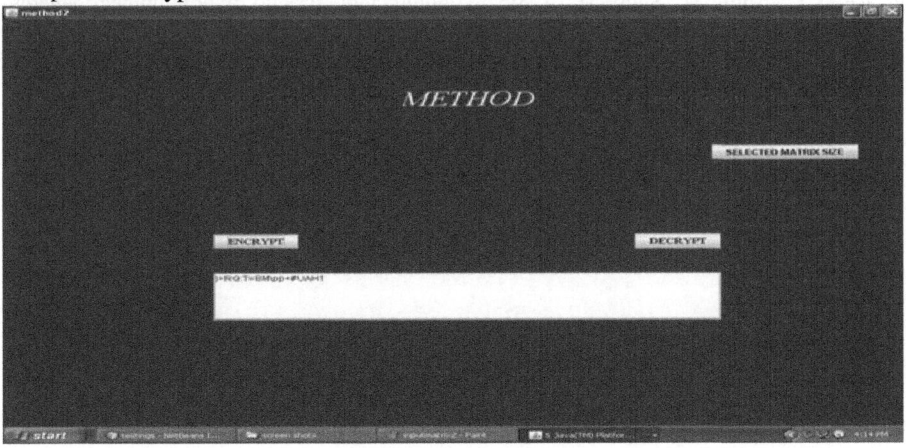

Fig. 6. Encryption process

6 Conclusion

Sending the data securely is a major important task now-a-days. There are many mechanisms which are satisfying the purpose. But still there are lot many chances of cracking the code which is sent to the receiver. In this proposed method the plain text is generated to cipher text based on printable ranges only and this forms the difficulty in identifying the key. This algorithm provides almost equal security at low computational overhead. And also the given algorithm is free from differential and linear crypto analysis, which makes it suitable in data encryption.

This ultimately benefits for encryption process as the key which is to be used is generated from combinations. We therefore assessed the generation of cipher text in more secured manner by overcoming the faults of identifying the key by 3rd party users.

References

1. Introduction to Cryptography–Ranjan Bose. Tata Mc-Grew–Hill Publisher Ltd. (2001)
2. Koblitz, N.: A course in number theory and Cryptography. Springer-Verlag, New York, Inc. (1994)
3. Nalani, N., Raghavendra Rao, G.: Cryptanalysis of Simplified Data Encryption Standard via Optimisation Heuristics. IJCSNS 6(1B) (January 2006)
4. Simmons, S.: Algebric Crypto analysis of Simplified AES. Proquest Science Journals 33(4), 305 (2009)
5. Ravi, S., Knight, K.: Attacking Letter Substitution Ciphers with Integer Programming. Proquest Science Journals 33(4), 321 (2009)
6. Kumar, A., Kumar, A.: Development of New Cryptographic Construct using Palmprint Based Fuzzyvoult. EURASIP Journal on Adv. In Signal Processing 21, 234–238 (2009)
7. Wang, B., Wu, Q., Hu, Y.: A Knapsack Based Probabilistic Encryption Scheme (March 2007), `http://www.citeseer.ist.psu.edu`
8. Bluekrypt 2009: Cryptographic Key length Recommendations
9. Blum, L., Blum, M., Shub, M.: A simple unpredictable pseudo random number generator. SIAM J. Compute. 15(2), 364–383 (1986)
10. Brics: Universally comparable notions of key exchange and secure channels. LNCS. Springer, Berlin (March 2004)

Steganography Based Visual Cryptography (SBVC)

Ritesh Mukherjee[1] and Nabin Ghoshal[2]

[1] Centre for Development of Advanced Computing,
Plot –E2/1, Block-GP, Sector-V, Kolkata – 700091, West Bengal, India
[2] Dept. of Engineering and Technological Studies,
University of Kalyani, Kalyani, Nadia-741235, West Bengal, India
mukherjee.ritesh@gmail.com, nabin_ghoshal@yahoo.co.in

Abstract. A digital signature performs the function of conventional handwritten signatures for authentication of documents, data integrity and non-repudiation. Most of the conventional digital signature schemes are based on complex mathematical computations to generate the keys and verify signatures. In late nineteen century comparatively less computation based scheme known as visual cryptography (VC) is introduced. VC has high security and requires comparatively less complex computations. In this paper, we consider a new digital signature scheme, based on the concept of Visual Secret Sharing associated to XOR-based non-expansion visual cryptography systems. Our proposed scheme is towards enhancement of security of born digital and digitized document by generating visual shares and embedding them within a cover image using steganographic techniques rather than complicated mathematical computations. The proposed scheme could be applied for ownership protection, copy control, authentication of digital media etc.

Keywords: VC, Steganography, MSE, PSNR, IF, RSA, DS.

1 Introduction

Increasing popularity of office automation is steadily changing the nature of conventional official communication from paper based to digitise one. In this transition the most important aspect is to ensure authenticity of born digital as well as digitized documents. Cryptography is one of the most powerful arrangements in this connection. Digital signature (DS) is the one of the effective technique to ensure integrity and authenticity issues of information security domain and capable of ensuring non-repudiation. DS is a verification method requires two keys: private key and a public key for verification of the authenticity of the message. The goal of DS is to verify that whether the message has been modified or tempered in between after it was signed and to ensure the confidence of the receiver that the message was send by the expected sender. The theory of DS algorithm was introduced by Diffie and Hellman. However, the first practical system was the RSA digital signature scheme developed by Rivest [6] and subsequent DS schemes such as ElGamal signature [7, 8] undeniable signature [9] and others were proposed. The major disadvantages of conventional cryptographic techniques including digital signature scheme suffers

S.C. Satapathy et al. (Eds.): Proc. of Int. Conf. on Front. of Intell. Comput., AISC 199, pp. 559–566.
DOI: 10.1007/978-3-642-35314-7_63

from its inherent complexity and requirement of computation power in all underlying steps. The Visual Cryptography system proposed by Naor and Shamir [1] was to overcome these issues without compromising the security. The main idea of this arrangement was to split an image into two random visual shares, which separately reveal no information on the original image but the same can be reconstructed using those two shares only. But, visible encrypted messages trigger a suspicion for any human being. Steganography [2] is a unique concept in which hides data inside other messages, images or files. Combination of cryptography and Steganography will be the right choice to generate innocent looking shares to avoid suspicion of the attacker. It is obvious that while cryptography protects the messages, steganography distinctly safeguards the communicating parties with protection of authenticity and non-repudiation, thus shielding the document from illegal hacking. Figure 1 shows the entire SBVC process.

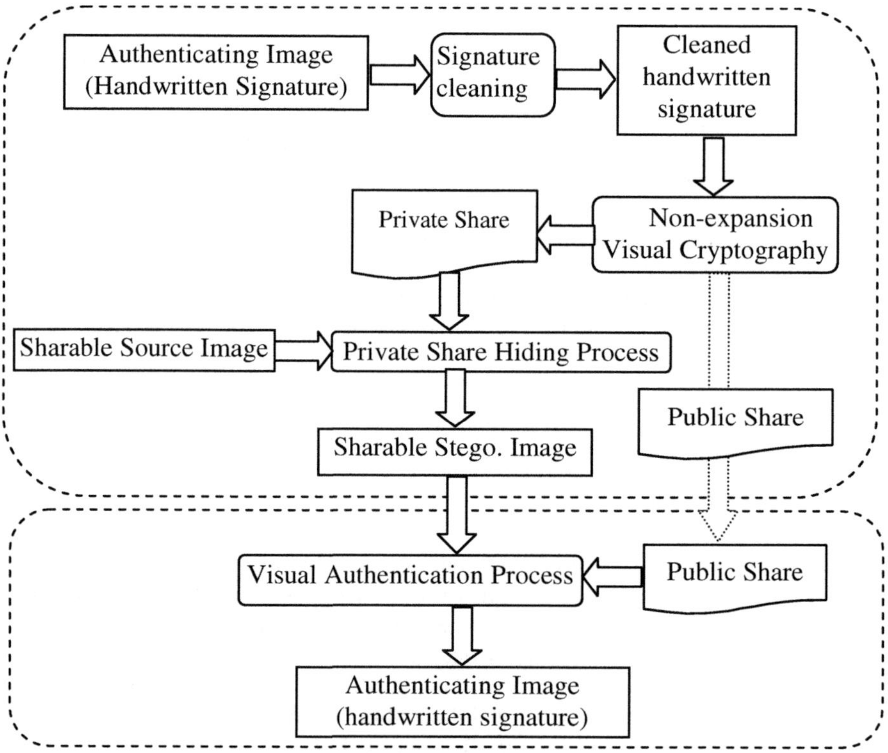

Fig. 1. The process Processes of SBVC

2 The Technique

In steganography based visual cryptography (SBVC) approach, we have proposed a novel technique to overcome these disadvantages of digital signature without compromising the benefits towards integrity, authentication and non-repudiation.

Fig 1 describes the main block of processes of the proposed SBVC solution. In our work clean digitized signature of the sender is used for creation of public and private shares. Digital handwritten signature cleaning is applied on each pixel of the born digital or digitized handwritten signature to obtain clearer gray-image with exact white and black pixels. Clarity enhancement is performed by applying a predefined threshold based classification of pixel values. In this approach all the pixels of the digitized signature are checked in row major order and modified with 255 or 0 based on the threshold comparison result. Cleaned digital signature is used to create private and public share.

2.1 Share Creation

Share creation is applied on each pixel of the cleaned digital handwritten signature image to obtain two shares out of it. Independently both the shares will have no resembles with the original image but jointly will be able to produce source image using XOR operation over each pixel values. The technique of visual share creation is explained below. Figure 2 shows the pixel description under non expandable VC scheme. Construction of two of two shares using non expandable VC scheme is represented in figure 3.

Fig. 2. Pixel description under non expandable VC scheme

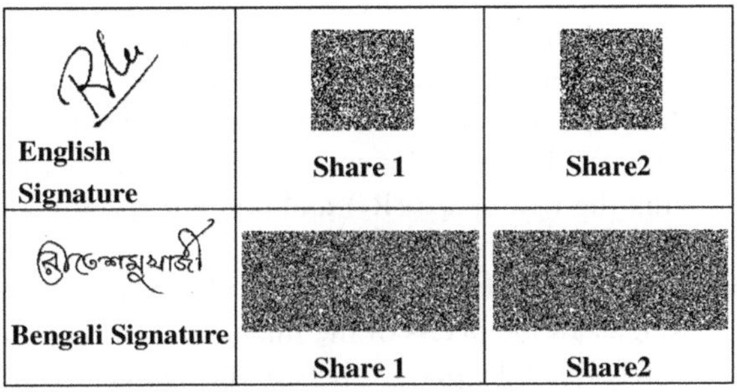

Fig. 3. Construction of two-out-of-two share using non expandable VC scheme

Input: Cleaned digital image of handwritten signature of m x n dimension.
Output: Two share image of same size (m x n) with black and white pixels.
Method: Visual share creation is performed only in the pixel values of input image and generates two pixel values to be written on the output shares. The algorithm is as follows:

1. Collect properties like image type, dimensions and maximum intensity of the input image and define the property set of the share images accordingly.
2. Repeat the following steps until all pixels have been read from the source image as well as all the pixels manipulated with extended visual cryptographic methods,
   2.1 Read the pixel values of the source image matrix in row major order.
   2.2 Determine the output pixel set {{(255,255), (0, 0)}, {(255, 0), (0,255)}} depending on the input pixel value in {255, 0}.

2.3 Decide applicable pixel pair for output share based on a modulo 9 of random function (range 100) with threshold 4 and publish the pixel values in the output shares accordingly.

3. Stop.

2.2 Sharable Stego. Image Creation

The proposed scheme uses gray scale image as the input to be embedded with another gray-scale image. Here, sharable source image is cover image with p x q dimension and either of the visual shares created from the cleaned digital handwritten signature of m x n dimension. The proposed technique embeds the visual share in the gray-scale cover image of dimension p x q, such that p>= m and q>=n.

Input: Two gray-scale digital images, first one to be used as cover image and second one is the visual share to be incorporated for sharing.
Output: Sharable digital cover image embedding visual share.
Method: Encoding is performed only in the pixel values of input images and generates pixel values to be written on the output share. The algorithm is as follows:

1. Collect properties like image type, dimensions and maximum intensity of the cover image and define the property set of the output image accordingly.
2. Read the dimension information of the cover image as well as visual share. Suppose dimension of cover image is (p, q) and visual share is (m, n).
3. Find the row space (R_S) and column space (C_S). Here R_S = round (p / m) and C_S =round (q / n).
4. Repeat the following steps until all pixels have been read from the source image as well as all the pixels of the image to be incorporated on the cover image,
   4.1. Read the pixel values of the source image matrix in row major order.
   4.2. Keep track of row position (R_p) and column position (C_p) of the pixel.
   4.3. Position modulo (P_m) = $\{(R_p \bmod R_S) + (R_p \bmod R_S)\}$ $\forall$ $R_p \in \{0, (m \times R_S)\}$ and $C_p \in \{0,(n \times C_S)\}$.
   4.4. If P_m = 0 than replace the bit value of {pixel position (P_p) modulo n} position with authenticating bit representative of the visual share. Modulo n will ensure the modification of image byte will takes place on the least n-1 bits of the cover image. Here n is a number within 1 to 3.
   4.5. Write the output pixel in the output visual share.
5. Stop.

2.3 Visual Authentication Process

During this process, the embedded image (Sharable Stego. Image) and the public share have been taken as the input data. The process of extracting the embedded message/image is reverse of the embedding process with the same traversing order of image bytes. The scheme uses gray scale images as the input to be overlapped to generate output image. The detailed steps of Authenticating Image extraction are as follows.

Input: Two gray-scale images (Sharable Stego. Image and public share image).
Output: Authenticating image consisting visually recognizable signature.
Method: Authenticating image extraction is performed only in the pixel values of input images and generates pixel values to be written on the output. The algorithm is as follows:

1. Collect properties like image type, dimensions and maximum intensity of the cover image and define the property set of the output image accordingly.
2. Read the dimension information of the sharable Stego. Image as well as visual share. Suppose dimension of sharable Stego. image is (p, q) and public share is (m, n).
3. Find the row space (R_S) and column space (C_S). Here R_S = round (p / m) and C_S = round (q / n).
4. Repeat the following steps until all pixels have been read from sharable Stego. Image as well as public share image,
   4.1. Read the pixel values of the Stego. image matrix in row major order.
   4.2. Keep track of row position (R_p) and column position (C_p) of the pixel.
   4.3. Calculate position modulo (P_m) = $\{(R_p \bmod R_S) + (R_p \bmod R_S)\}$ $\forall$ $R_p \in \{0, (m \times R_S)\}$ and $C_p \in \{0, (n \times C_S)\}$.
   4.4. If P_m = 0 than retrieve the bit value of {pixel position (P_p) modulo n} position where authenticating bit representative of the private share is already exists. Modulo n will ensure the exact position of the modified bit. Here n is a number within 1 to 3, same as sharable Stego. image creation process.
   4.5. If the retrieve bit value is 0 then treat comparable pixel value as 0 else 255.
   4.6. Read next pixel value of the public share image in row major order.
   4.7. Apply XOR operation on the extracted pixel values and write the same in the authenticating image.
5. Stop.

3 Result Comparison and Analysis

SBVC approach is combination of two different arrangement namely visual cryptography and steganography. There are various parameters to compare the performance of visual cryptography schemes, like pixel expansion and contrast etc. It can be easily observed that as pixel expansion increases, we got more number of black sub pixels for a white pixel in the reconstructed image and hence it corresponds to loss in resolution and expansion of share size. SBVC creates shares of cleaned signature to ensure maximum contrast. Share creation involves randomization to overcome time variant share generation problem. Use of XOR in SBVC ensured 0 expansions leads to loss less reconstruction of authenticating image. Comparison of major contributing parameters between SBVC and some existing approaches is described in table 1.

Table 1. Comparison of SBVC with other existing Visual Cryptography approaches

Author or Approach	Naor and Shamir [1] 1995	Rijmen and Preneel [11], 1996	Verhuel and Tilborg [12], 1997	Zhang et al [13], 2008	Tsai et al [14] 2009	SBVC
Pixel expansion	255	4	Variable	1	9	0
No of secret image	1	1	1	1	1	1
Share type (Rectangle)	Yes	Yes	Yes	Yes	Yes	Yes
Meaningful share	No	No	No	No	Yes	Yes
Multi pixel encoding	No	No	No	Yes	Yes	No

Source Image Name	Sharable Source Image	Experience with 90 X 90 pixel		Experience with 256X111 pixel	
		Authenticating Signature Image	Sharable Stego. Image	Authenticating Signature Image	Sharable Stego. Image
Fig. 2a. Baboon	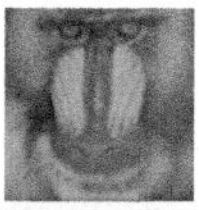		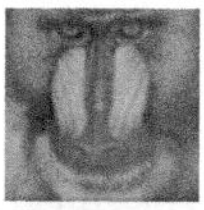	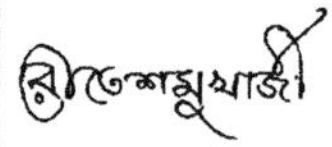	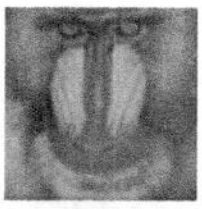
Fig. 2b. Lenna				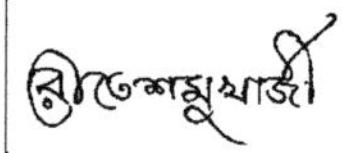	
Fig. 2c. Splash				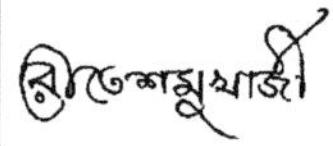	

Fig. 4. Sharable Stego. Images using SBVC

Fig 4 is showing some experiment results with different sizes of Authenticating Signature Image. In the entire cases sharable source images are taken from benchmark images of size 512 X 512. In this process, the source handwritten signature is gray image of 90 X 90 for the first and 256 X 111 for the second scenario. As a result visual shares created are of two different sizes for two different types of handwritten signature. In the proposed process the capacity of 512 X 512 gray cover images is 2097152 bit but in the mentioned scenarios SBVC will use either 8100 bit or 28416 bit only, which will produce sharable stego images with almost no visually recognizable information. The parameters used in the study are mainly Visual Interpretation, Mean Square Error (MSE), Peak Signal to Noise Ratio (PSNR), Image Fidelity (IF)

Table 2. MSE, PSNR and IF calculation after embedding different size of secrete data

Images	MSE		PSNR		IF	
Sizes	90 X 90	256 X 111	90 X 90	256 X 111	90 X 90	256 X 111
Baboon	0.038673	0.054741	61.435636	59.926639	0.999997	0.999997
lenna	0.038738	0.053802	61.937302	60.510678	0.999997	0.999996
peppers	0.038475	0.054276	61.382766	59.888517	0.999997	0.999996
airplane	0.038917	0.053825	61.408306	59.999888	0.999998	0.999998
Oakland	0.038811	0.054508	60.596131	59.121030	0.999997	0.999997
sailboat	0.038513	0.054363	61.820213	60.323253	0.999998	0.999997
SanDiego	0.039314	0.053539	61.364258	60.023034	0.999998	0.999997
Splash	0.038414	0.053874	61.867230	60.398255	0.999997	0.999995
Tiffany	0.038330	0.054458	62.295406	60.770137	0.999999	0.999998
Woodlad Hills	0.038105	0.054234	61.648419	60.115560	0.999998	0.999997
Average	**0.038629**	**0.054162**	**61.57557**	**60.1077**	**0.999998**	**0.999997**

Table 2 is showing analytical information pertaining to the embedding private visual share of two different sizes like 90 X 90 and 256 X 111 on different benchmark images of 512 X 512. Embedding public visual share for the first scenario is showing higher average PSNR values like 61.57557dB with lower MSE like 0.038629 and average IF value is 0.999998 which is very close to 1. In second scenario the average PSNR is lower than the first one but still the result is encouraging in terms of higher PSNR value like 60.1077 dB with lower MSE like 0.054162 and average IF is 0.999997, which is also very close to 1. From the comparison it is clear that with respect of required capacity of secret data embedding in SBVC approach PSNR is more, it yields better image fidelity. This analysis depicts integration of less noise and minimum distortion which ensures complete recovery of handwritten signature to ensure visual authentication

4 Conclusion

Proposed technique ensures integrity and authenticity factors for any grayscale image. In this algorithm it is very difficult to extract authenticating message using statistical method because of the random position. Moreover, if the image has been extracted it will be useless unless it overlapped with counter share. Hence, the proposed technique is quite secured from most of the likely attacks and the beauty of the technique is the simplicity.

Acknowledgement. The authors express their deep sense of gratitude to the faculty members of the Dept. of Engineering and Technological Studies, University of Kalyani, West Bengal, India, where the work has been carried out.

References

1. Naor, M., Shamir, A.: Visual Cryptography. In: De Santis, A. (ed.) EUROCRYPT 1994. LNCS, vol. 950, pp. 1–12. Springer, Heidelberg (1995)
2. Radhakrishnan, R., Kharrazi, M., Menon, N.: Data Masking: A new approach for steganography. Journal of VLSI Signal Processing 41, 293–303 (2005)
3. Naor, M., Pinkas, B.: Visual Authentication and Identification. In: Kaliski Jr., B.S. (ed.) CRYPTO 1997. LNCS, vol. 1294, pp. 322–336. Springer, Heidelberg (1997)
4. Amin, P., Lue, N., Subbalakshmi, K.: Statistically secure digital image data hiding. In: IEEE Multimedia Signal Processing MMSP 2005, Shanghai, China (October 2005)
5. Pavan, S., Gangadharpalli, S., Sridhar, V.: Multivariate entropy detector based hybrid image registration algorithm. In: IEEE Int. Conf. on Acoustics, Speech and Signal Processing, Philadelphia, Pennsylvania, USA, pp. 18–23 (March 2005)
6. Rivest, R., Shamir, A., Adleman, L.: A Method for Obtaining Digital Signatures and Public-Key Cryptosystems. Communications of the ACM 21(2), 120–126 (1978)
7. Laih, C.S., Chen, K.Y.: Generating visible RSA public keys for PKI. Int. J. Secur. 2(2), 103–109 (2004)
8. El Gamal: A public key cryptosystem and a signature scheme based on discrete logarithms. IEEE Trans. Inform. Theory IT 31(4), 469–472 (1985)
9. Alia, M., Samsudin, A.: A new Digital Signature Scheme Based on Mandelbrot and Julia Fract Seta. Americal Journal of Applied Science, AJAS 4(11), 850–858 (2007)
10. Al-Hamami, A.H., Al-Ani, S.A.: A New Approach for Authentication Technique. Journal of Computer Science 1(1), 103–106 (2005) ISSN 1549-3636
11. Rijmen, V., Preneel, B.: Efficient colour visual encryption for shared colors of Benetton. In: Eurocrypto 1996, Rump Session, Berlin (1996)
12. Verhuel, E.R., Tilborg, V.: Construction and Properties of k out of n visual secret sharing schemes. Designs, Codes and Cryptography 11, 179–196 (1997)
13. Zhang, H., Wang, X., Cao, W., Huang, Y.: Visual Cryptographic for General Access Structure by Multi-pixel Encoding with Variable Block Size. In: International Symposium on Knowledge Acquisition and Modeling (2008)
14. Tsai, D.-S., Horng, G., Chen, T.-H., Huang, Y.-T.: A Novel Secret Image Sharing Scheme For True-Color Images With Size Constraint. Information Sciences 179, 3247–3254 (2009)

A Novel Approach to Find an Optimal Path in MANET Using Reverse Reactive Routing Algorithm

Bhabani S. Gouda[1], Ashish K. Dass[2], and K. Lakshmi Narayana[3]

Department of Computer Science & Engg.
National Institute of Science & Technology, Berhampur
bhabani012@rediffmail.com, ashishkumardass@yahoo.co.in,
lakshmi2912@hotmail.com

Abstract. In Mobile Ad hoc Network, reactive protocols don't preserve routing information at the network node level, if there is no communication between the nodes. Reactive protocol determines a route to a specific destination when a particular packet is intends to send. We proposed a reverse reactive routing based route discovery approach, which is used to find an optimal route to the destination with lower overhead than flooding based reverse route discovery. Here we showed how the reverse reactive routing protocol performs to find an optimal path. The optimal path is obtained through three steps, which is reverse route calculation in route request (RREQ), reverse route calculation in route reply (RREP) and reverse route calculation in route error (RERR). Experiments have been carried out using network simulator (NS2) and the obtained results are performed better than reactive routing protocol (AODV).

Keywords: Mobile Ad-Hoc Networks, Reactive Routing Protocol, NS2, Route Discovery, RAODV.

1 Introduction

In the next generation of wireless communication systems, there will be a drastic need for the rapid deployment of independent mobile users for rescue operations, disaster relief, and military operations. Such type of network scenarios cannot rely on centralized connectivity, and can be conceive as applications of Mobile Ad Hoc Networks. The design of network protocols for these networks is a complex issue. Regardless of these applications, MANETs need efficient distributed algorithm to determine network organization, link scheduling, and routing. However, determining viable routing paths and delivering messages in a decentralized environment where network topology fluctuates is not a well-defined problem. Mobile ad-hoc networks are self-organizing and self configuration of multi-hop wireless networks, where to interpret the network changes dynamically due to mobility of nodes [1]. The reactive routing protocol algorithm creates routes between nodes on request of source nodes with network flexibility to allow nodes to enter and leave the network at any point of time. The newly created routes remain active only as long as data packets are travelling along the paths from the source to the destination. A routing procedure is

S.C. Satapathy et al. (Eds.): Proc. of Int. Conf. on Front. of Intell. Comput., AISC 199, pp. 567–574.
DOI: 10.1007/978-3-642-35314-7_64

always needed to find an optimal path to send the packets between the source and the destination [2]. Therefore the requirements of the protocol for wireless networks are path (source, destination), hop count and sequence number.

1.1 Reactive Routing Protocol

Reactive routing protocol is an on-demand routing protocol for mobile ad-hoc networks, which uses routing tables to store routing information. During routing, all route information maintained in tables for unicast as well as for multicast routes. These routing tables hold information like destination address, next-hop address, hop-count, destination sequence number and life time. Instead of keeping static route information from one node to every other node, a reactive routing protocol like AODV discovers the route as and when required and these routes are maintained as long as necessary. The protocol comprises of three main functions like route discovery, route establishment and route maintenance. In routing protocol, on request of source node, route discovery function is responsible for the discovery of new routes; route establishment function is responsible for detection of the link of discovered routes. Finally route maintenance function is responsible for detection of the link failures and repair of an existing route. Reactive routing protocols, such as the AODV [3] nodes have four types of message to communicate between each other: Route Request, Route Reply, Route Error and Hello messages with a key feature that it don't required to distribute routing information and to keep up the routing information about the failure links [4]. During packet transmission, every intermediate node in the discovery route create routing table to store the information regarding neighbour node and the destination node information. The routing table information updated for every packet transmission during the message transmission. When communication between two nodes completes, nodes discard all these routing and neighbour information.

2 Related Work

Numerous frameworks have been proposed in mobile Ad-hoc network for performance-based routing protocol. This framework uses the concept of reverse reactive routing to find an optimal path between source and destination.

Khan et al. [5] conclude that when the MANET setup for a small amount of time, then AODV is better because of low initial packet loss. DSR is not prefers because of its packet loss. On the other hand if we have to use the MANET for a longer duration then we can use both protocols, because after sometimes both have the same behavior. AODV is having very good packet receiving ratio in comparison with DSR. At the end, they concluded that the combined performance of both AODV and DSR routing protocol could be the best solution for routing in MANET. In [6], Barakovic et al. compared performances of three routing protocols: DSDV, AODV and DSR. They analyzed these routings with different load and mobility scenarios with Network Simulator version 2 (NS-2). In [7] Performance of AODV, TORA and DSDV protocols is evaluated under both CBR and TCP traffic pattern. Extensive Simulation is done using NS-2. Simulation results show that Reactive protocols perform better in terms of packet delivery ratio and average end-to-end delay.

3 Routing Protocol

Mobile ad-hoc networks, also well-known as short-term networks, are autonomous systems of mobile nodes forming network in the absence of centralized access point. Absence of fixed infrastructure poses several types of challenges for this type of networking. Among these challenges routing is one of them. Routing protocols of mobile ad-hoc network lean to need different approaches from existing protocols, since most of the existing Internet protocols were proposed to support routing in a network with fixed structure. The proposed routing protocol for find an optimal path in MANET using the following route discovery approaches.

3.1 Route Discovery

In Mobile ad-hoc network each node will create a reverse route table when it receives a RREQ (route request), the RREQ is discards if it has already been processed. It records and indicates the route to the source node; otherwise the source address and the broadcast ID from RREQ resolve is there buffered to prevent it from being processed again. Furthermore, each node will calculate the distance every time, and most importantly, this distance is the key reason to choose the shortest path from the source node. Initially, when a node receives RREQ, it will create a reverse route entry which indicates the next hop (forwarding the RREQ) of the source node and calculate the distance between the next hop node and the source node. Second, each node will also make the similar decision when it receives RREQ and update reverse route table or discard RREQ [8].

Once an intermediate node receives a RREQ, the node sets up a reverse route entry for the source node in its reverse route table. Reverse route entry consists of <Source IP address, Source seq. number, number of hops to source node, Destination IP address, Destination seq. number>.

The reverse route reply (RREP) message is created such that the source of the packet appears to be the requested destination and the destination of the packet is the source. The reverse route reply packet updates the routing table based on the number of hopes in the path from the source to destination. This kind of update information takes care about the multiple paths between the networks. The reply packet is routed by the trail content of the route request packet. The intermediate nodes on the return path automatically discover the requested node.

We have used the similar calculation mechanism to find the optimal path in forwarding RREP. The simply difference is that the distance we calculate in RREP is from the node forwarding RREP to the destination node.

3.2 Route Failure Handling

Route error (RERR) has been done by the data packets during the transmission. If the link fails between sources to destination, it will be detected by the periodic sending of 'hello' packets. If the hello reply message is not obtained within the specific timeout period (2sec) then that nodes neighbor route entry is deleted from the routing table. Therefore during the next route request process alternate route or newer path is generated. If all the route entries in the destination column reached the minimum pheromone then that destination field is removed from the routing table thinking that node has moved to the different location. Hence every node will maintain path for the source to destination node which are dynamically implicated in the process of routing.

4 Optimal Path Finding Approach

We study the problem of selecting an optimal route in terms of transition probability and link available time. Finally we calculate optimal path between source and destination node by three steps, which execute and forwarding RREQ (route request) packets, RREP (route reply) packet and RRER (route error) packets. Experiments have been carried out using NS2 as network simulator ware and results encouraging.

4.1 Computation of Reverse Route in RREQ

In mobile ad hoc network route request is responsible for generating route between source and destination. In this process RREQ is as follows:

Source node create RREQ packet and broadcast through the neighbors in the increasing order of cost value.

Every node, when it receives RREQ packet, it does the following:
if current addr = dest_addr
> *Construct the REPLY packet and copies the trail content of RREQ packet, send REPLY packet to node prevHop from which it has received RREQ and update the routing table according to the hop count*

else
> *Add the current node addr into the trail field and forward the RREQ packet to the nearest neighbor nodes*

At any node, when it receives REPLY packet it does the following:
if current addr = dest_addr
> *update the routing table according to the hop count and retrieves the packet from the data queue and insert into regular queue.*

else
> *update the routing table according to the hop count and forward the packet to the neighbor following the trail content*

4.2 Computation of Reverse Route in RREP

Route reply is responsible for the transmitting of data packet. It receives RREQ packet, sent reply packet to prevHop. In this process RREP is as follows:

At any Destination node receives RREQ packet and generate Reply packet, broadcast through the neighbors in the increasing order of cost value.

Every node, when it receives RREP packet, it does the following:
if current addr = dest_addr
Construct the Acknowledge and copies the trail content of RREP packet and start transmitting the data
else
route the route reply message to the sender.
else if route the route reply message to the sender, when it receives reply packet is unicasted

4.3 Computation of Reverse Route in RERR

Route maintenance module is responsible for the maintenance of the transmission of packet and generated path during the discovery phase. In this process RERR is as follows:

Every node, when it receives data packet from prevHop, it does the following:
if current addr = dest_addr
extort data and set type=ack in data packet , eliminate the data content and send acknowledge packet to prev_id
else
Obtain routing values of all links using neighbor table information, calculate probability for all nodes in neighbor routing table and send packets to that link which has highest probability.

decompose the table whenever accessed. If the routing value = 0.1 for every destination then delete the destination entry from the specific routing table.

5 Performance Evaluation

We have performed simulations to evaluate several performance metrics of our schemes. First, we would like to see how obtained optimal path of route discovered by reverse route calculation reduced. Then we compare our schemes with DSR in terms of packet delivery ratio, routing overhead and end-to-end delay.

5.1 Simulation Environment

To evaluate and compare the effectiveness of these routing protocols with existing proposed models [9], we performed extensive simulations in NS2. Each simulation is carried out under a constant mobility. The simulation parameters are listed in table-1

Table 1. Simulation Parameters

Parameter	Value	Description
Simulator	NS-2	Simulator tool
Simulation Time	30 sec	Maximum execution time
Simulation Area	400m *450m	Physical boundary of the network
Number of nodes	10	Nodes participating in the network
Transmission Range	85 m	Frequency of the node
Maximum speed	5 m/s	Speed of nodes
CBR Flows	18	Constant Bit Rate link used
Data payload	2000 bytes	Packet size
Sending rate	5 packets/sec	Maximum number of sending packets
Movement Model	Random Waypoint	Network connection

5.2 Results and Analysis

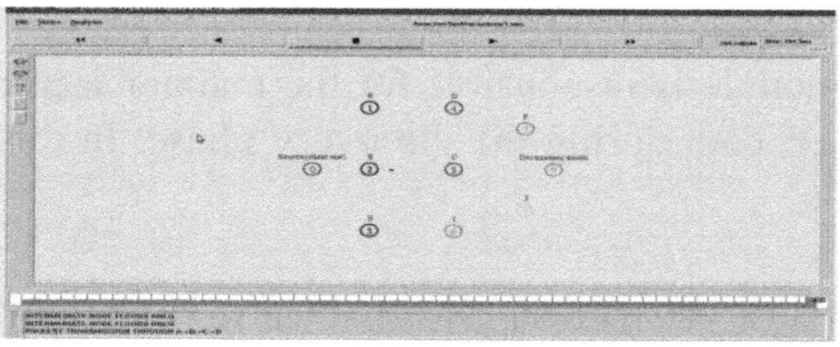

Fig. 1. Structure of Mobile Ad hoc Network

No. of Nodes vs. Packet Drop

A packet is dropped in two cases: the buffer is full when the packet needs to be buffered and the time that the packet has been buffered exceeds the limit. Packet dropping was observed for several nodes and varied the nodes each time and the dropped was counted at *destination node during* entire simulation period.

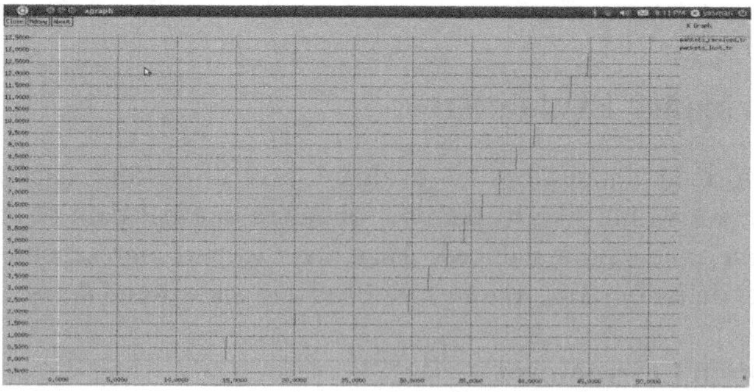

Fig. 2. Packet Lost variation

End-to-End Delay vs. Packet Delivery Ratio

It shows significant dependence on route stability, thus its packet received rate is lower. Although, the amount of packet received is inversely proportional to propagation delay, DSR has the best performance than AODV and RAODV.

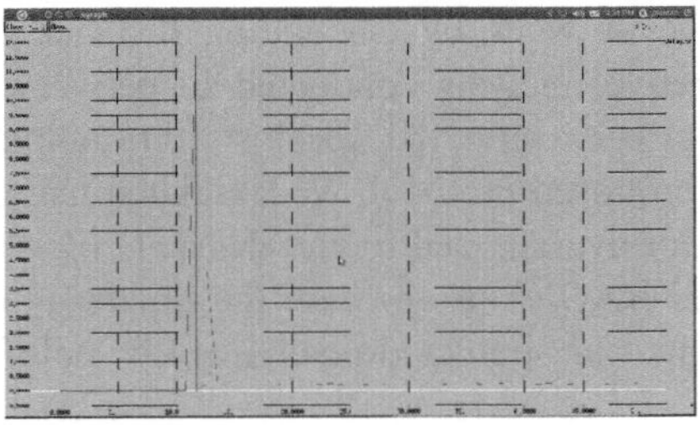

Fig. 3(a). End to End delay variation

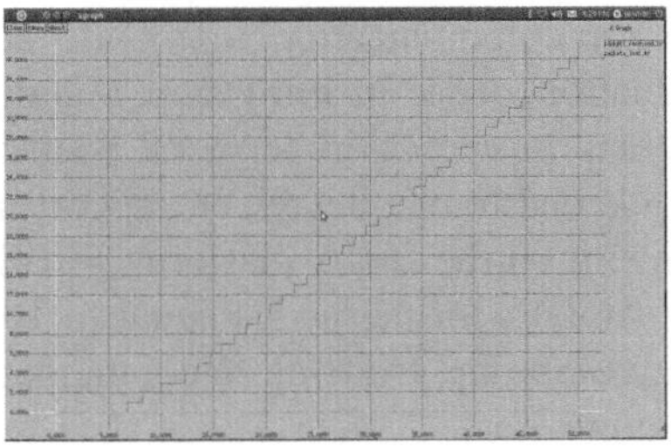

Fig. 3(b). Packet delivery variation

Throughput vs. Simulation Time

Throughput was gained at destination node against various dimension of networks and varied the simulation time uniformly for each protocol whose measure was as in fig 4.Throughput is the average rate of successful message delivery over a communication channel. This data may be delivered over a physical or logical link, or pass through a certain network node.

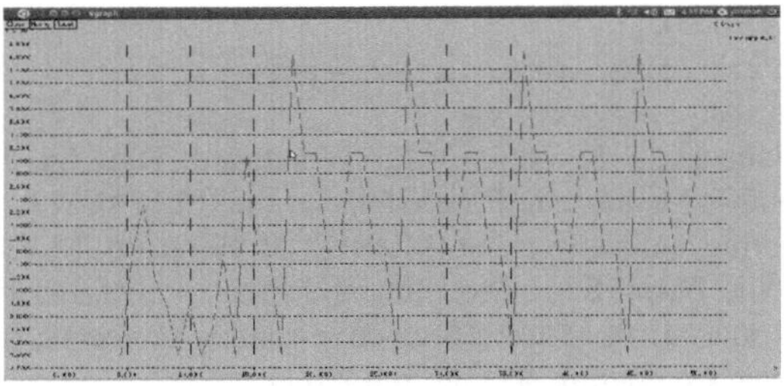

Fig. 4. Throughput variation

Path Optimality

The ratio between the numbers of hops of the shortest path to the number of hops in the actual path taken by the packets.

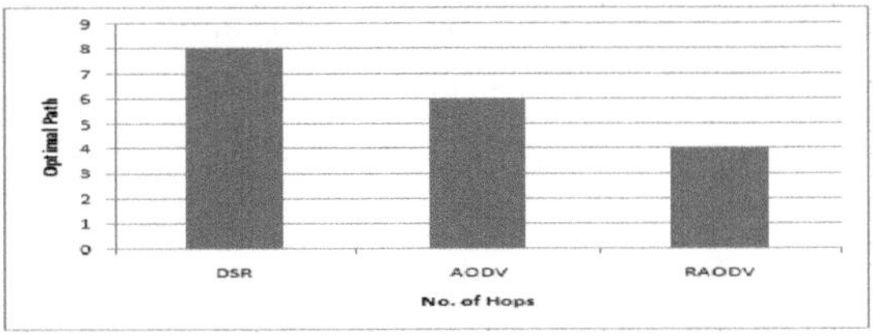

Fig. 5. Shortest path vs. Optimal Path

6 Conclusion

This study was conducted to propose a reactive routing protocol, consists of three steps to find the optimal path. Initially, we calculate the shortest path to the source node and create reverse route table. In second, we filter these paths to obtain optimal path for communication in mobile ad-hoc network by calculating distance to the destination node. In third step a comparative analysis conducted in between three different protocols in term of packet delivery ratio, packet lost, throughput and average end to end delay. To support the proposed protocol, we simulated using NS2 simulator on the Linux platform. Finally, for average end to end delay, DSR is lower than AODV, for the nodes equal to 10 and RAODV to increase the reliability of the reactive routing protocol. We anticipate that our simulated results can be helpful for the future work.

References

1. Li, J., Kameda, H., Li, K.: Optimal Dynamic Mobility Management for PCS Networks. IEEE/ACM Transactions on Networking 8(3) (June 2000)
2. Carofiglio, G., Chiasserini, C.-F., Garetto, M., Leonardi, E.: Route stability in MANETs under the Random Direction Mobility Model. IEEE Transactions on Mobile Computing 8(9) (September 2009)
3. Perkins, C., Belding-Royer, E., Das, S.: Ad hoc On-Demand Distance Vector (AODV) Routing (February 2003), `http://www.ietf.org/internet-drafts/draftietf-manet-aodv-13.txt`
4. Basagni, S., Chlamtac, I., Syrotiuk, V.R., Woodward, B.A.: A distance routing effect algorithm for mobility (dream). In: Proceedings of the IEEE/ACM International Conference on Mobile Computing and Networking (MOBICOM 1998), pp. 76–84 (1998)
5. Khan, J., Hyder, S.I., Fakar, S.M.: Modeling and simulation of dynamic intermediate nodes nnd performance analysis in MANETS reactive routing protocols. International Journal of Grid and Distributed Computing 4(1) (March 2011)
6. Barakovic, S., Kasapovic, S., Barakovic, J.: Comparison of MANET routing protocols in different traffic and mobility models. Telfor Journal 2(1) (2010)
7. Akshatha, P.S., Khurana, N., Rathi, A.: Optimal Path For Mobile Ad-Hoc Networks Using Reactive Routing Protocol. International Journal of Advances in Engineering & Technology, IJAET (2011) ISSN: 2231-1963
8. Kaur, S.: Performance Comparison of DSR and AODV Routing Protocols with Efficient Mobility Model in Mobile Ad-Hoc Network. IJCST 2(2) (June 2011)
9. NS-2, The ns Manual (formally known as NS Documentation) `http://www.isi.edu/nsnam/ns/doc`

Lossless Audio Steganography in Spatial Domain (LASSD)

Dipankar Pal[1], Anirban Goswami[2], and Nabin Ghoshal[3]

[1] Dept. of Computer Science and Engineering, Techno India, EM 4/1 Salt Lake, Sec-V, Kolkata-700091
[2] Dept. of Information Technology, Techno India, EM 4/1 Salt Lake, Sec-V, Kolkata-700091
[3] Dept. of Engineering and Technological Studies, University of Kalyani, Kalyani, Nadia-741235,West Bengal, India
mail2dpal@yahoo.com,
an_gos@yahoo.com,
nabin_ghoshal@yahoo.co.in

Abstract. In our proposed work we put an effort to make the technique of standard Least Significant Bit (LSB) coding more secured to embed secret information within an audio file. The prime focus here is to transmit any secret message using an audio signal which can only be retrieved by the intended recipient, while keeping the original characteristics of the carrier audio signal unaltered. In order to achieve the objective, a hash function has been devised to generate pseudorandom positions for insertion and extraction of the secret data bits. In the process of embedding sample amplitude values are read from the source audio and secret data bits are embedded in each of the sample values sequentially at pseudorandom positions. In the process of extraction the reverse technique is applied. Experimental results, both objective and subjective, reveal enhanced performance in terms of imperceptibility and security of the proposed technique.

Keywords: Audio Steganography, Hash Function, HAS, MSE, SNR, PSNR, MOS.

1 Introduction

Data transmission in public communication system is not secure because of interception and improper manipulation by eavesdropper. Steganography is one of the popular methods to achieve secret communication between sender and receiver by hiding message in any form of cover media such as an audio, video or image. Almost all digital file formats can be used for steganography, but image and audio files [2, 3, 9] are most suitable because of their high degree of redundancy. So, digital audio steganography [5, 6] has emerged as a prominent source of data hiding across novel telecommunication technologies such as covered voice-over-IP, audio conferencing, etc.

The particular importance of hiding secret data in audio files results from the prevailing presence of audio signals as information vectors in our human society. In fact, availability and popularity of audio files make them eligible to carry hidden secret information. Data hiding in audio files is especially challenging because of the sensitivity of the Human Auditory System (HAS). However, HAS still tolerates common

S.C. Satapathy et al. (Eds.): Proc. of Int. Conf. on Front. of Intell. Comput., AISC 199, pp. 575–582.
DOI: 10.1007/978-3-642-35314-7_65

alterations in small differential ranges. For example, loud sounds tend to mask out quiet sounds. Additionally, there are some common environmental distortions which may be ignored by listeners in most cases. These properties have led researchers to explore the utilization of audio signals as carriers to hide secret data.

Least significant bit encoding technique in audio samples in time domain [10] is one of the simplest algorithms with very high data rate. But improvement of watermark robustness due to increase in depth of the used LSB layer is restrained by perceptual transparency bound, which is the fourth LSB layer for the standard LSB coding algorithm.

In this paper we propose a method that hides information (image, text or audio) in a cover audio which is able to shift the limit of transparent data hiding in audio signals from standard LSB technique [7, 8, 11] to embedding in pseudorandom positions [4]. Embedding / Extraction positions are calculated using a self devised hash function to incorporate more security. In addition, objective as well as subjective (listening) tests are performed and the perceptual quality of the steganographic audio signal is found to be high.

Fig. 1, shown below, illustrates the overall insertion and extraction techniques followed by sec. 2 which explains the insertion and extraction algorithms of LASSD in detail. The experimental results based on MSE, SNR (in dB), PSNR (in dB) and MOS grade are discussed in sec. 3. Finally the conclusion is drawn in sec. 4.

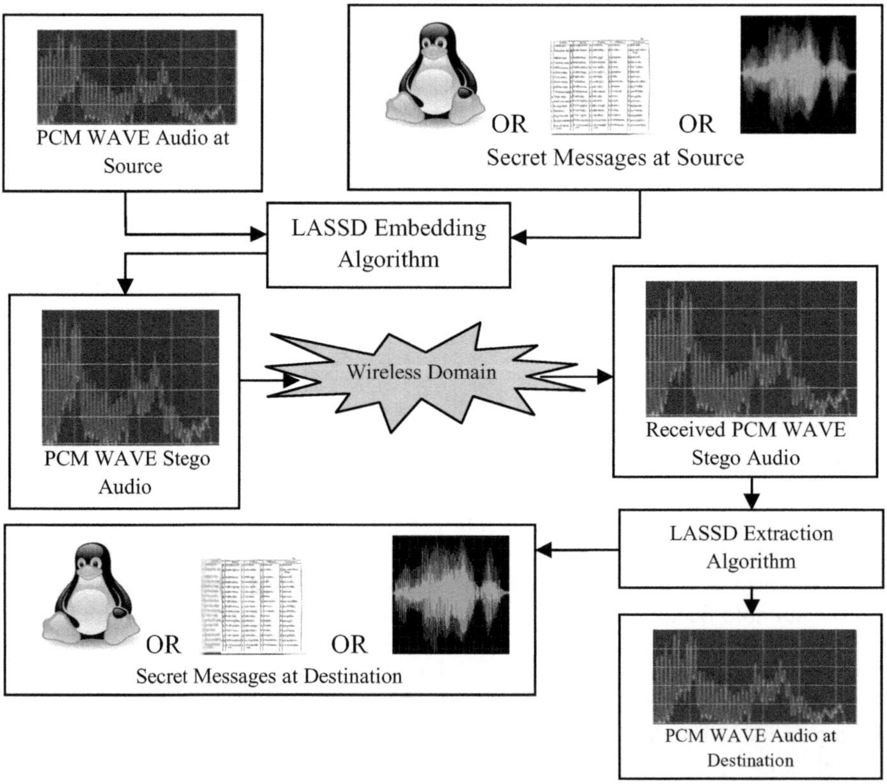

Fig. 1. Illustration of Embedding and Extraction Techniques of LASSD

2 The Technique

Embedding of secret message bits are done at pseudorandom positions (0 – 3), determined by a variable epos. A new hash function has been devised to generate the value for epos. Insertion is performed in a number of sample amplitude values depending on the volume of the secret message.

In case of extraction, the embedded message is retrieved from the sample amplitude values of the steganographic audio. The process of retrieval works by extracting one bit each from the embedded sample values sequentially and bytes are formed to reproduce the embedded message. At the time of extraction, the location of each hidden secret bit is derived with the help of the same hash function. The algorithms for insertion and extraction are vividly explained in subsections 2.1 and 2.2 respectively followed by the explanation of generation of pseudorandom position in subsection 2.3.

2.1 Embedding Algorithm

Input: A PCM WAVE source audio and a secret data file (image, text or audio).
Output: A steganographic PCM WAVE audio.
Steps:

1. Read the header information (RIFF, FMT and DATA) from the source audio and write into the output audio.
2. Read a byte of secret data.
   - 2.1 Read a sample amplitude value from the source audio.
   - 2.2 Generate a pseudorandom position (0-3) using a self devised hash function (sec 2.3).
   - 2.3 Embed a bit of the secret data byte into the sample value at the position obtained in the previous step.
   - 2.4 Write the modified sample value into the output audio.
   - 2.5 Repeat steps 2.1 to 2.4 until all the bits of a byte have been embedded.
3. Apply step 2 for all the bytes of the secret message.
4. Stop.

2.2 Extraction Algorithm

Input: A steganographic PCM WAVE audio.
Output: The embedded data file (image, text or audio).
Steps:

1. Read a sample amplitude value from the input audio.
2. Generate a pseudorandom position (0 - 3) using the same hash function (sec 2.3).
3. Extract a bit from the sample value from the position obtained in the previous step.

4. Repeat steps 2 to 3 to extract 8 consecutive bits to form a byte of the secret data.
5. Write the extracted byte obtained in step 4 into the output file.
6. Repeat steps 1 to 5 until all the bytes of the secret data are extracted.
7. Stop.

2.3 Generation of Pseudorandom Position

The mechanism for generating a pseudorandom value is formulated by a self devised hash function. It is represented as epos = h(x, y), where epos is a simple variable, x represents a Boolean value (either 0 or 1) and y is an integer which varies from 0 to 7 in value. The algorithm to produce a pseudorandom value in epos is described below,

Steps:

1. Consider three bits from right (starting from LSB) of y.
2. Prefix x with y, which now transforms into $xy_2y_1y_0$ (e.g. 1011).
3. Let $z = x\,y_2 \oplus y_1y_0$.
4. Define a queue of size N to hold previous N values of z.
5. Now implement the following steps to determine the final value of z,
   If the Queue is not full, then
   - 5.1 Insert the present value of z into the queue and return the value to epos.

     Else
   - 5.2 Compare the present value of z with all the existing values in the queue.
   - 5.3 If the present value of z differs at a single instance then
     - 5.3.1 Return the value of z immediately to epos.

       Else
     - 5.3.2 Modify the present value of z by the following expression

       If x = 0, then

       $z = z_1$ (complement of z_0)

       Else

       $z =$ (complement of z_1) z_0
     - 5.3.3 Insert this value of z into the queue and also return it to epos.
6. Stop.

The queue, defined in step 4 above, avoids consecutive repetitions of the same random value for more than N times.

3 Experimental Results and Analysis

The effectiveness of our proposed work has been tested by three objective measures, namely Mean Squared Error (MSE), Signal to Noise Ratio (SNR) in dB and Peak

Signal to Noise Ratio (PSNR) in dB as discussed below and the results are shown in table 2.

3.1 Objective Measures

1. Mean Squared Error (MSE) is defined as: MSE = $\frac{1}{N}\sum(|e|-|\bar{e}|)^2$, where N represents the total number of samples in the original audio, e corresponds to the original audio sample and ē corresponds to the stego audio sample.
2. Signal to Noise Ratio (SNR) is represented as: SNR = $10 \cdot \log_{10}(E(x) / MSE)$ (in dB), where E(x) = $\frac{1}{N}\sum(|x|)^2$ is denoted as the energy of the original audio and x corresponds to the original audio sample.
3. Peak Signal to Noise Ratio (PSNR) is represented as: PSNR = $10 \cdot \log_{10}((2^b - 1)^2/ MSE)$ (in dB), where b is the bit depth of the original audio signal.

3.2 Subjective Measure

Although objective measures are considered as most common procedure for measuring the noise difference between the original and stego audio signals, they do not prove to be authentic when it comes to considering the specific characteristics of HAS. So Mean Opinion Score (MOS), shown as a grading scale in table 1, has been carried out as the subjective measure. The MOS is the arithmetic mean of all the individual scores, and can range from 1 (worst) to 5 (best). The main purpose of MOS is to evaluate the quality of human voice at the point of termination on any type of phone connection. Scores of four or five are usually considered stable and within industry standards, and thus acceptable.

Table 1. MOS Grading Scale

MOS Grade	1	2	3	4	5
Description	Very Annoying	Annoying	Slightly Annoying	Perceptible, Not Annoying	Imperceptible

3.3 Results

The proposed scheme has been experimented with a variety of CD quality stereo audio signals. The signals are sampled at a rate of 44.1 KHz with 16 bit resolution. Fig. 2 shows some sample audio signals (both original and stego) involved in the experimentation. The original audio signals are: Fig 2a – Adeline.wav (28 MB), Fig. 2d – Fur Elise.wav (25.5 MB) and Fig 2g – My Heart.wav (54.8 MB). The secret messages that were embedded are: Fig 2b - Melbourne.jpg (344064 bytes), Fig 2e - Text.txt (499845 bytes) and Fig 2h -Train Drive By.mp3 (3149952 bytes). The corresponding stego audio signals are shown in fig. 2c, 2f and 2i.

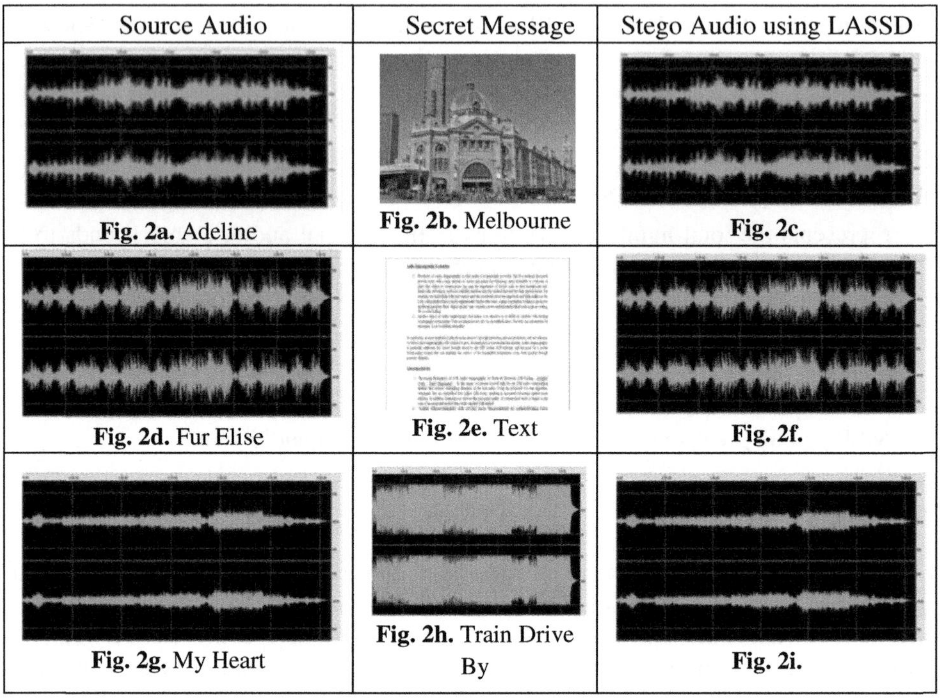

Source Audio | Secret Message | Stego Audio using LASSD

Fig. 2a. Adeline | **Fig. 2b.** Melbourne | **Fig. 2c.**

Fig. 2d. Fur Elise | **Fig. 2e.** Text | **Fig. 2f.**

Fig. 2g. My Heart | **Fig. 2h.** Train Drive By | **Fig. 2i.**

Fig. 2. Visual Interpretation of Audio Steganography using LASSD

More than 15 listeners were involved in the listening test to measure the subjective MOS [1] grade of the testing audio signals (Table 2). Ten pairs of signals were given for tests – each pair consisted of one original and the corresponding stego audio. The listeners were requested to pinpoint the differences between the two signals within a pair and the average grade for each pair is taken as the final grade for the respective pair.

Table 2. Metric values of Audio signals in LASSD

Source Audio	Size (MB)	Message Type	Embedded (Bytes)	MSE	SNR (dB)	PSNR (dB)	MOS
Adeline	28	Image	344064	2.0427	67.2617	87.2069	5
Lizzie	26.3	Image	344064	2.1754	63.1783	86.9334	5
Fur Elise	25.5	Text	499845	3.3779	67.1595	85.0224	5
My Heart	54.8	Audio	3149952	9.6332	55.8010	80.4712	5
Average	**33.65**	–	**4337925**	**4.3073**	**63.3501**	**84.9085**	**5**

4 Conclusion

LASSD is proposed to utilize audio signals to transmit any secret message which can only be retrieved by the intended recipient. The imperceptibility of the carrier audio signals is also maintained and is assessed by both objective and subjective metrics. Authenticity has been incorporated by embedding the data bits of the secret message (image, text or audio) in pseudorandom positions of the carrier audio samples. Emphasis has been given to make the steganographic audio imperceptible on account of noise accruement. Also, the embedded message using this algorithm is very difficult to perceive due to the dynamic insertion position of the authenticating message bits in the original audio. Hence, this technique is very much effective for safeguarding the secret messages from any potential hacking attempt.

Acknowledgement. The authors express their deep sense of gratitude to the faculty members of the Dept. of Engineering and Technological Studies, University of Kalyani, West Bengal, India, where the work has been carried out.

References

1. Dhavale, S.V., Deodhar, R.S., Patnaik, L.M.: High Capacity Lossless Semi-fragile Audio Watermarking in the Time Domain. In: Wyld, D.C., Zizka, J., Nagamalai, D. (eds.) Advances in Computer Science, Engg. & Appl. AISC, vol. 167, pp. 843–852. Springer, Heidelberg (2012)
2. Bhattacharyya, S., Sanyal, G.: Audio Steganalysis of LSB Audio Using Moments And Multiple Regression Model. International Journal of Advances in Engineering & Technology, IJAET 3(1), 145–160 (2012)
3. Djebbar, F., Ayad, B., Abed-Meraim, K., Hamam, H.: A view on latest audio steganography. In: 7th IEEE Internationl Conference on Innovations in Information Technology, Abu Dhabi, UAE (2011)
4. Zamani, M., Ahmad, R.B., Manaf, A.B.A., Zeki, A.M.: An Approach to Improve the Robustness of Substitution Techniques of Audio Steganography. In: Proc. IEEE International Conference on Computer Science and Information Technology, ICCSIT, pp. 5–9 (2009)
5. Kekre, H.B., Archana, A.A.: Information hiding using LSB technique with increased capacity. International Journal of Cryptography and Security 1(2) (October 2008)
6. Bandyopadhyay, S.K., Bhattacharyya, D., Ganguly, D., Mukherjee, S., Das, P.: A Tutorial Review on Steganography, `http://www.jiit.ac.in/jiit/ic3/IC3_2008/IC3-2008/APP2_21.pdf`
7. Sridevi, R., Damodaram, A., Narasimham, S.V.L.: Efficient Method of Audio Steganography By Modified LSB Algorithm And Strong Encryption Key With Enhanced Security. Journal of Theoretical and Applied Information Technology (2005)
8. Cvejic, N., Seppanen, T.: Increasing Robustness of LSB Audio Steganography using a novel embedding method. Proc. IEEE Int. Conf Info. Tech.: Coding and Computing 2, 533–537 (2004)

9. Bandyopadhyay, S.K., Datta, B.: Higher LSB Layer Based Audio Steganography Technique. International Journal of Electronics & Communication Technology 2(4) (October - December 2011) ISSN: 2230-7109, ISSN: 2230-9543
10. Bassia, P., Pitas, I., Nikolaidis Robust, N.: audio watermarking in the time domain. IEEE Transactions on Multimedia 3(2), 232–241 (2001)
11. Cvejic, N., Seppänen, T.: Reduced distortion bit-modification for LSB audio steganography. In: ICSP Proceedings. IEEE (2004)

Genetic Algorithm Based Method for Analyzing Reliability State of Wireless Sensor Network

Vipin Pal[1], Girdhari Singh[2], and R.P. Yadav[3]

[1] Department of Electronics and Communication Engineering, Malaviya National Institute of Technology Jaipur, Jaipur-302017, India
vipinrwr@yahoo.com

[2] Department of Computer Engineering, Malaviya National Institute of Technology Jaipur, Jaipur, India
girdharisingh@rediffmail.com

[3] Vice-Chancellor, Rajasthan Technical University Kota, Kota-302017, India
rp_yadav@yahoo.com

Abstract. Wireless sensor networks are application specific and consist of large number of sensor nodes deployed in a harsh environment/area. Sensor nodes are deployed randomly to monitor or sense the area of interest. Wireless sensor networks have a wide range of application such as military surveillance, environment monitoring, agriculture, health monitoring and many more. Reliability of a wireless sensor network is defined in terms of area covered by sensor nodes and redundancy in sensed data. Redundancy in data is caused by overlapping in sensed area of nodes. Redundancy is required to gather high quality of information. Sensor nodes have limited battery power and harsh deployed environment makes it quite impossible to recharge or replace the battery of nodes. Energy of nodes is consumed in sensing, computing and communicating data. Due to the non-uniform energy consumption of nodes in field, nodes start dying over the time. As nodes start dying, area is not completely sensed and redundancy of data also decreases. That makes the network unreliable. Hence the gathered data from the network is also unreliable. So it is necessary to find when network is in unreliable state so that proper action can be taken. In this paper, we have proposed a genetic algorithm based method to find whether wireless sensor network is reliable or unreliable. Our proposed method finds the optimal minimal number of nodes that can sense the whole area with minimum desired redundancy in data to have good quality of information gathered. Minimum number of nodes and minimum required redundancy for a network are application dependent. We have experimented with different number of topologies and different parameters. Our result show that for a network to be in reliable state at least 48% to 52% of the initial nodes (random) should be active to sense the complete area with 20% to 30% overlapping the sensed area respectively.

Keywords: Wireless Sensor Network, Redundancy, Reliability, Genetic Algorithm.

1 Introduction

Wireless sensor networks[1, 2] are application specific and consist of large number of sensor nodes deployed in a harsh environment/area. Wireless sensor networks have a

S.C. Satapathy et al. (Eds.): Proc. of Int. Conf. on Front. of Intell. Comput., AISC 199, pp. 583–591.
DOI: 10.1007/978-3-642-35314-7_66

wide range of application for an example military surveillance [3], environment monitoring [4], agriculture [5], health monitoring [6], automotive [7], industry [8], critical information infrastructure protection [9] and many more. Wireless sensor networks are classified in following two classes according to their applications - Data Gathering and Event Driven. In a data gathering network, nodes continuously sense area and send data with a fixed interval to base station. In an event-driven wireless sensor network, nodes send data only when an event is triggered.

A sensor node consumes energy in sensing, computing and communicating the data. Communication between the nodes is the most energy consuming process. In most of the application, the harsh environment makes it quit impossible to recharge or replace the battery of sensor nodes. So the on-board battery power of nodes derives the lifetime of overall network. Energy of nodes should be consume very efficiently and economically to prolong the life of network. So energy efficiency is the prime design issue for protocol design of wireless sensor networks.

In a wireless sensor network, nodes sense the area and send data to base station. Quality of data gathered by a wireless sensor network is very important. High quality of data can be gathered only from a reliable wireless sensor network. Reliability of wireless sensor network can be defined in terms of area sensed by all nodes and redundancy of data gathered. Redundancy of data is caused by the overlapped sensed area of nodes. As there is dense deployment of sensor nodes in field so there is overlapping in sensed area of nodes. A network can be considered as a reliable network if it senses the almost whole area of interest and produces the desired minimum redundancy in sense data for accurate information about the phenomenon.

Even after implementation of energy efficient algorithms, nodes go out of battery power. Sensor nodes consume inconsistent energy in field and die randomly. As the number of nodes start decreasing in the field, reliability of network in terms of covered (sensed) area and redundancy in the sense data also starts decreasing. Data gathered at that time is of low quality. So, it is necessary to find state of wireless sensor network, whether it is reliable or unreliable.

Finding the minimum number of nodes (random) that covers the almost whole area with minimum desired overlapped area is an NP-HARD issue [10]. Genetic algorithm [11] can be used in many NP-HARD problems like optimization and Traveling Salesman Problem (TSP). In this paper, the problem of finding the minimum number of nodes (random) that covers the almost whole area with desire minimum redundancy according to application of network is optimized by genetic algorithm (GA).

The rest of the paper is organized as: section 2 define the problem description, section 3 gives description of genetic algorithm, section 4 describes the experimental results and usefulness of the procedure, and section 5 has conclusion.

2 Problem Description

Wireless sensor networks are application specific networks so the performance of the network should stick to the requirements of application. Wireless Sensor networks should be enough reliable to provide high quality information about the phenomenon. The two main requirements of wireless sensor networks to be considered as reliable are:

1. Almost each point is covered (i.e. all area is sensed)
2. Minimum desired overlapped area (i.e. amount of redundancy in sense data)

Sensor nodes are deployed in a spatial region and hence can be considered as points in two dimensional planes. If the positions of the sensors are pre-engineered, then the minimum number of active nodes that meet the above stated requirements of a reliable network can be calculated as:

$$\frac{((Area\,of\,Field)+(Minimum\,Overlapped\,Area))}{Area\,Covered\,by\,a\,Single\,Sensor} \tag{1}$$

But in most of the application, sensor nodes are deployed randomly not by a pre-engineered procedure to engineer the location of nodes. Communication between the nodes is the main source of energy dissipation of nodes that heavily depends upon the distance between two. Energy consumption of nodes is not uniform that makes sensors nodes to die randomly in field. Number of alive nodes in field starts decreasing. The reliability of network also decreases as the decreased number of nodes will not provide enough redundant data. Minimum desired redundancy in data depends upon application of network. Over the time, the nodes will not be able to sense the complete area along with redundancy below desired. The gathered data is not of good quality and does not provide correct information about phenomenon. Now the network can be considered in unreliable state.

Hence it is necessary to find, when the network will transit from a reliable state to unreliable state. For this, find the minimum number of nodes (random) that makes the network reliable. Genetic algorithms are used to optimize the stated problem.

3 Genetic Algorithm

Dynamic nature of the network makes the stated problem more complex. In the other hand, genetic algorithm is very flexible in solving such dynamic problems. In this paper, genetic algorithm is applied in a way to find constraint on minimum number of nodes alive in the field while fulfilling the requirements.

GA maintains a population of chromosomes and each chromosome represents a solution. Each chromosome is evaluated to calculate the fitness according to a function that defines the problem. Genetic transformations, crossover and mutation, are applied to selected chromosomes. A new population is generated with the combination of chromosomes with better fitness in current population and new chromosomes generated by genetic transformation. After several generations, the algorithm converges to the best solution.

3.1 Genetic Algorithm for Problem

Population. The initial population consists of randomly generated set of chromosomes. Binary representation is used for chromosome definition and each bit corresponds to one sensor node. So, the length of each chromosome is equal to initial number of nodes in field. Presence of a node in field constitute a "1" in chromosome otherwise a "0".

Fitness Function. Survival of a chromosome depends upon its fitness. Fitness of each individual in population is evaluated. Problem stated in the paper has three parameters for calculating the fitness of chromosomes.

1. Number of active nodes should be minimum.
2. Nodes present in the field should cover the whole area.
3. Sensed area of nodes should overlap at least to desired percentage.

If number of active nodes are less than the minimum that means the network is not reliable any more. Network either not covers the whole area or data produced is not having the minimum redundancy to make data reliable.

So the fitness function for the problem consists of three objectives and is represented as:

$$Fitness = (n, covered\,area, overlapped\,area) \tag{2}$$

After scaling the fitness function:

$$Fitness\,Value = \left\{ \begin{array}{l} \left\{ \frac{100 - No.\,of\,Active\,Nodes}{100} + \frac{Total\,Covered\,Area}{100} \right\} \\ -\left\{ \frac{(\pm(Overlapped\,Area - Minimum\,Overlapped\,Area}{Minimum\,Overlapped\,Area} \right\} \end{array} \right.$$

Fitness scaling is applied to fitness for better selection procedure and to avoid premature convergence and slow finishing.

Selection. Selection is the process of choosing two parents from the population for crossing. The purpose of the selection process in a genetic algorithm is to give more reproductive chances to those population members that are better fit. The selection procedure may be implemented in a number of ways like Roulette Wheel selection, Tournament selection, Boltzmann selection, Rank selection, Random selection, etc. In this work Roulette Wheel selection procedure is applied to select chromosomes for generating new population. The procedure creates a biased roulette wheel in which the slots are sized in proportion to the fitness of chromosomes. Selection procedure is random but the chance of being selected is proportional to fitness of chromosomes. The chromosomes with higher fitness values are more likely to be selected as the chromosomes of population in the next generation.

Due to random nature of selection procedure, best member of the population may fail to live on to the next generation. The elitist strategy fixes this loss by copying the best member of each generation into the next generation. It improves performance of genetic algorithm.

Crossover Operator. In this paper, one-point crossover method is used. The crossover operation takes place between two chromosomes with probability specified by crossover rate. These two chromosomes exchange portions that are separated by the crossover point. The following is an example of one point crossover:

Chromosome 1	0	1	1	1	0	**0**	**1**	**1**	**1**	**0**	**1**	**0**
Chromosome 2	1	0	1	0	1	**1**	**0**	**0**	**1**	**0**	**1**	**0**

After crossover, two offspring are created as below:

Offspring 1	0	1	1	1	0	**1**	**0**	**0**	**1**	**0**	**1**	**0**
Offspring 2	1	0	1	0	1	**0**	**1**	**1**	**1**	**0**	**1**	**0**

Mutation Operator. The mutation operator is applied to each bit of a chromosome with a probability of mutation rate. After mutation, a bit that was "0" changes to "1" and vice versa.

Before Mutation	0	1	1	**1**	0	0	1	1	1	0	1	0
After Mutation	0	1	1	**0**	0	0	1	1	1	0	1	0

4 Results

The purpose of work is to optimize the problem explained in section 2 using genetic algorithm. Sample sensor networks are generated with different number of nodes deployed in different dimensions. Sensor nodes are deployed randomly and each node knows about its exact location. The node density of all sample networks is same. All nodes are stationary once deployed in the field. Parameters used are listed in Table 1.

Table 1. Parameters

Parameters	Values
Number of Nodes (N)	25,100,150
Sensing Range of Nodes	10m
Network Dimensions	50mX50m,100mX100m,125mX125m
Size of Population	N
Length of Chromosome	N
Selection Type	Roulette Wheel
Crossover Rate	0.65-0.75
Mutation Rate	0.01-0.02

For the work minimum acceptable overlapping in the sensed area is varied from 20% to 30% of the whole area.

4.1 For Minimum 20% Overlapped Area

For 50mX50m with 25 Nodes. There is an increase in the best fitness value of the current population over generations. After 170 generations the best fitness value is constant .i.e. genetic algorithm applied optimize the problem.

As the fitness function is increasing the value of best fitness value while the number of active nodes is decreasing to an optimal value. The results show that the solution for the stated problem converges at 12 nodes (random) for 50mX50m with 25 nodes i.e. 48% of the initial nodes in field.

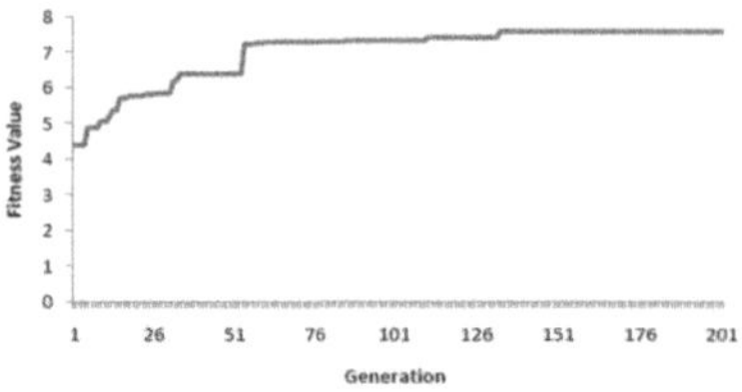

Fig. 1. Best Fitness Value Vs Generation

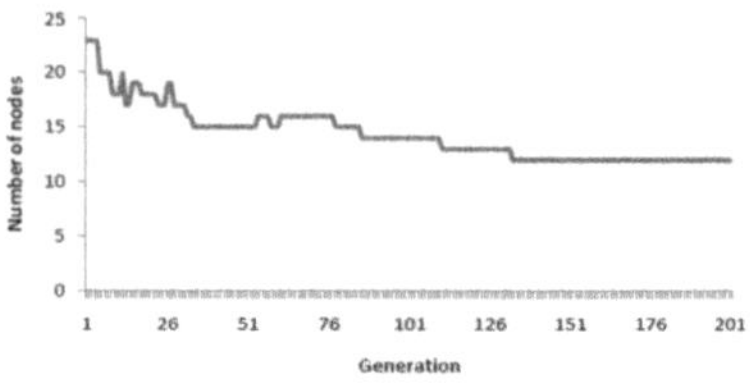

Fig. 2. Active Nodes Vs Generation

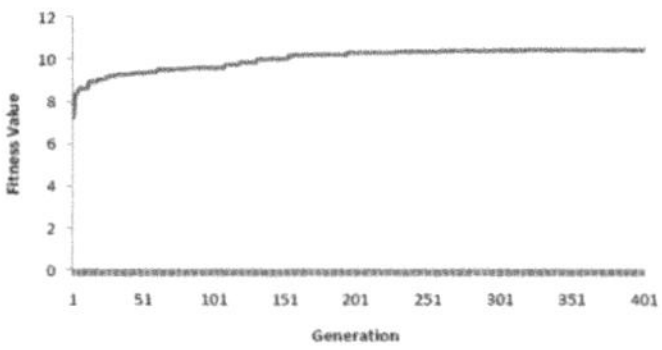

Fig. 3. Best Fitness Value Vs Generation

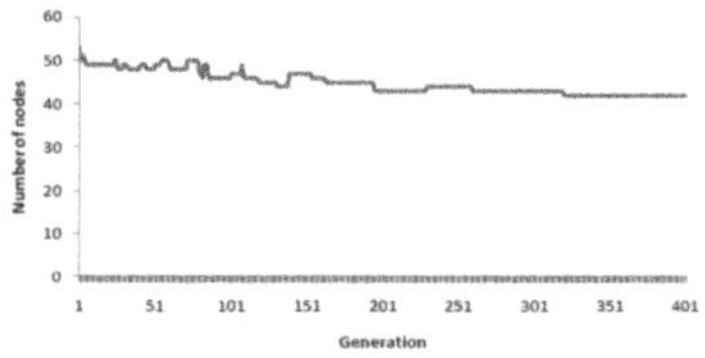

Fig. 4. Active Nodes Vs Generation

For 100m X 100m Area with 100 Nodes. Figure 3 shows that the fitness function in (2) increases the best fitness value of population over generations. Genetic algorithm is having a constant best fitness value after 250 generations.

Figure 4 shows that the number of active nodes is decreasing and the solution for the stated problem converges at 42 nodes (random) for 100m X 100m with 100 nodes i.e. 42% of the initial nodes in field.

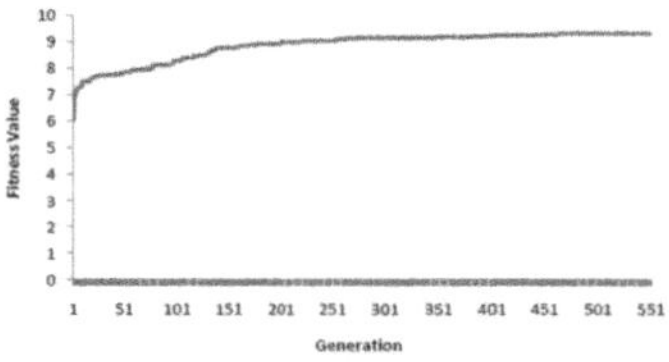

Fig. 5. Best Fitness Value Vs Generation

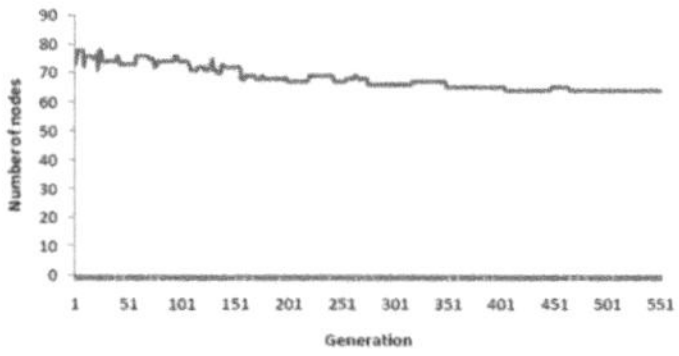

Fig. 6. Active Nodes Vs Generation

For 125mX125m Area with 150 Nodes. Figure 5 shows the increase in the best fitness value over generations. After 300 generations the increase is slight and after 470 generation it converges.

Figure 6 shows that the solution for the stated problem converges at 64 nodes (random) for 125mX125m with 150 nodes i.e.43% of the initial nodes in field.

4.2 For Minimum 30% Overlapped Area

For 50mX50m Area with 25 Nodes. There is an increase in the best fitness value of the current population over generations. After 140 generations the best fitness value is constant .i.e. genetic algorithm applied optimize the problem.

As the fitness function is increasing the value of best fitness value while the number of active nodes is decreasing to an optimal value. The results show that the solution for the stated problem converges at 13 nodes (random) for 50mX50m with 25 nodes i.e. 52% of the initial nodes in field.

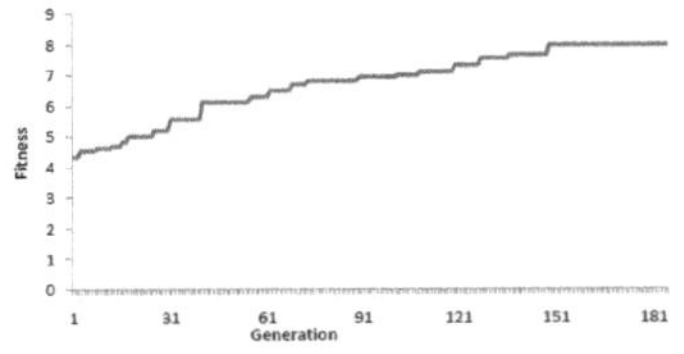

Fig. 7. Best Fitness Value Vs Generation

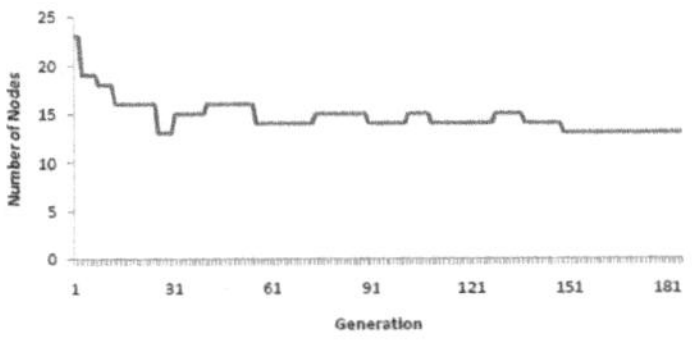

Fig. 8. Active Nodes Vs Generation

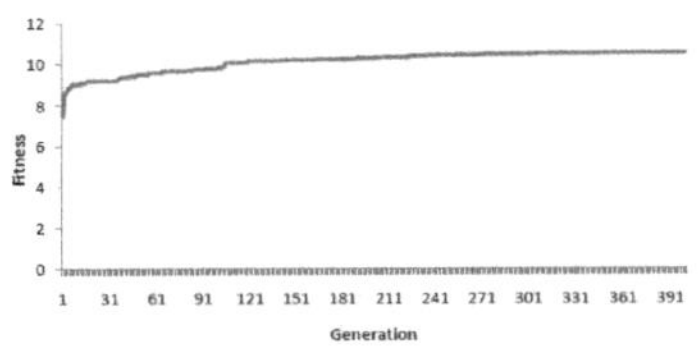

Fig. 9. Best Fitness Value Vs Generation

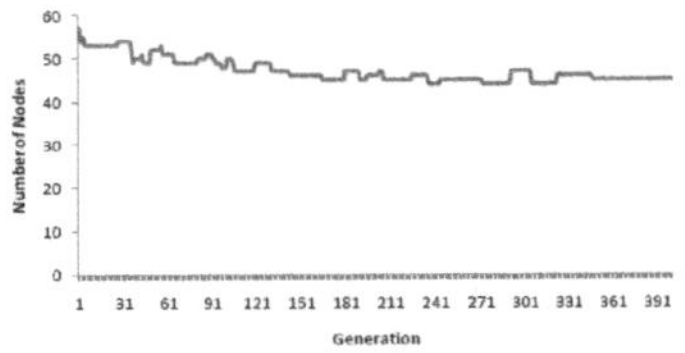

Fig. 10. Active Nodes Vs Generation

For 100mX100m with 100 Nodes. Figure 9 shows increases in the best fitness value of population over generations. Genetic algorithm is having a constant best fitness value after 340 generations.

Figure 10 shows that the number of active nodes is decreasing with the generations and the solution for the stated problem converges at 45 nodes (random) for 100mX100m with 100 nodes i.e. 45% of the initial nodes in field.

For 125mX125m Area with 150 Nodes. Figure 11 shows the increase in the best fitness value over generations. After 300 generations the increase is slight and after 450 generation it converges.

The results show that the solution for the stated problem converges at 70 nodes (random) for 125X125m2 with 150 nodes i.e. 47% of the initial nodes in field.

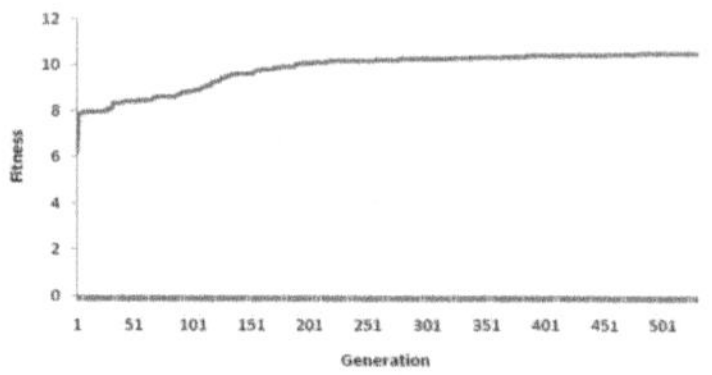

Fig. 11. Best Fitness Value Vs Generation

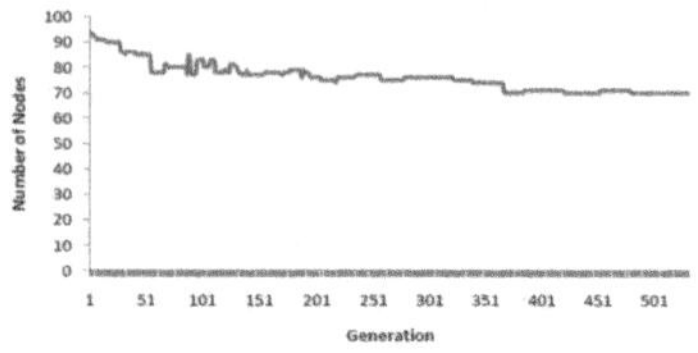

Fig. 12. Active Nodes Vs Generation

Table 2. Result

Number of Nodes	**Area**	Minimum No. of Nodes for 20% Overlapped area		Minimum No. of Nodes for 30% Overlapped area	
		From Eq. (1)	**Random**	**From Eq. (1)**	**Random**
25	50 X 50	10	12	11	13
100	100 X 100	39	42	41	45
150	125 X 125	60	64	65	70

Hence the above results show that 48% to 52% of the initial nodes (random) are required to complete the requirements of sensing the whole area with minimum overlapping in sensed area from 20% to 30%.

4.3 Application

The sensor nodes send their information about location to base station. Size of the data is very small, so it does not consumes much energy. Base station performs the optimization. Network knows in advance the minimum number of active nodes that will complete the requirements of the application. Hence the energy efficiency issue of the applied approach is not much affected.

In sleep/wake-up scheduling [12] approaches, if the selected active nodes are less than the calculated minimum active nodes, the result produced by the active nodes will be unreliable. The scheduling approaches should ensure that the number of selected nodes for a particular time slice should be higher than the acceptance level.

Reliability of clustering approach [13, 14] highly depends upon the covered area and redundancy in data. The network should be left with acceptable number of active nodes otherwise should inform the base station.

5 Conclusion

In a wireless sensor network, reliability of a network heavily depends upon the sensed area and the redundancy in data. Due to dynamic nature of wireless sensor network nodes die randomly in field. Node death in field causes decrease in redundancy of data as well as reduction in sensed area. Data gathered can be considered of high quality. The work of this paper is to find when network transit from reliable state to unreliable state with node deaths over the time. The work of paper optimizes the problem to find the minimum number of active nodes that will sense almost complete area with minimum acceptable redundancy in data. Genetic algorithm based approach is applied to fond the state of network. The optimal solution for the problem results that until 48% to 52% of initial nodes (random) network can be considered reliable for 20% to 30% overlapping respectively with complete sense area. The procedure consumes very less energy and hence does not affect the energy efficiency issue of the network.

References

[1] Akyildiz, I.F., Su, W., Sankarasubramaniam, Y., Cayirci, E.: Wireless sensor networks: a survey. Computer Networks 38(4), 393 (2002)
[2] Estrin, D., Govindan, R., Heidemann, J.S., Kumar, S.: Next century challenges: Scalable coordination in sensor networks. In: MOBICOM, pp. 263–270 (1999)
[3] Arora, A., Dutta, P., Bapat, S., Kulathumani, V., Zhang, H., Naik, V., Mittal, V., Cao, H., Demirbas, M., Gouda, M., Choi, Y., Herman, T., Kulkarni, S., Arumugam, U., Nesterenko, M., Vora, A., Miyashita, M.: A line in the sand: a wireless sensor network for target detection, classification, and tracking. Comput. Netw. 46(5), 605–634 (2004)
[4] Mainwaring, A., Culler, D., Polastre, J., Szewczyk, R., Anderson, J.: Wireless sensor networks for habitat monitoring. In: Proceedings of the 1st ACM International Workshop on Wireless Sensor Networks and Applications, WSNA 2002, pp. 88–97. ACM, New York (2002)
[5] Selavo, L., Wood, A.D., Cao, Q., Sookoor, T.I., Liu, H., Srinivasan, A., Wu, Y., Kang, W., Stankovic, J.A., Young, D., Porter, J.: Luster: wireless sensor network for environmental research. In: SenSys., pp. 103–116 (2007)
[6] Milenković, A., Otto, C., Jovanov, E.: Wireless sensor networks for personal health monitoring: Issues and an implementation. Comput. Commun. 29(13-14), 2521–2533 (2006)
[7] Tavares, J., Velez, F.J., Ferro, J.M.: Application of wireless sensor networks to automobiles. Measurement Science Review 8(3), 65–70 (2008)
[8] Flammini, A., Ferrari, P., Marioli, D., Sisinni, E., Taroni, A.: Wired and wireless sensor networks for industrial applications. Microelectron. J. 40(9), 1322–1336 (2009)
[9] Roman, R., Alcaraz, C., Lopez, J.: The role of wireless sensor networks in the area of critical information infrastructure protection. Inf. Secur. Tech. Rep. 12(1), 24–31 (2007)
[10] Cormen, T.H., Stein, C., Rivest, R.L., Leiserson, C.E.: Introduction to Algorithms, 2nd edn. McGraw-Hill Higher Education (2001)
[11] Mitchell, M.: An Introduction to Genetic Algorithms. MIT Press, Cambridge (1998)
[12] Keshavarzian, A., Lee, H., Venkatraman, L.: Wakeup scheduling in wireless sensor networks. In: Proceedings of the 7th ACM International Symposium on Mobile Ad Hoc Networking and Computing. MobiHoc 2006, pp. 322–333. ACM, New York (2006)
[13] Younis, O., Krunz, M., Ramasubramanian, S.: Node clustering in wireless sensor networks: Recent developments and deployment challenges. IEEE Network Magazine 20, 20–25 (2006)
[14] Liu, J.-S., Lin, C.-H.R.: Energy-efficiency clustering protocol in wireless sensor networks. Ad Hoc Networks 3(3), 371–388 (2005)

A Survey on Fault Management Techniques in Distributed Computing

Selvani Deepthi Kavila[1], G.S.V. Prasada Raju[2], Suresh Chandra Satapathy[1], Alekhya Machiraju[1], G.V.L. Kinnera[1], and K. Rasly[1]

[1] Department of Computer Science and Engineering,
Anil Neerukonda Institute of Technology and Sciences, Sangivalasa-531162, Visakhapatnam, Andhra Pradesh, India
{selvanideepthi14,sureshsatapathy,alu.1492,
kinnera.anits,raslykusumanchi}@gmail.com
[2] Department of Computer Science,
SDE, Andhra Univeristy, Visakhapatnam, India
gsvprajudr9@yahoo.co.in

Abstract. Now-a-days with the rapid increase in distributed computing systems faults are equally enhancing in scales in spite of many fault detection techniques proposed. Designing and implementing distributed computing systems is challenging due to their ever- increasing scales and the complexity. A faulty distributed system due to any reason during executing its processes can cause some damages. A fault management system helps the distributed systems by detecting malfunctions, errors or faults etc., We investigated different techniques of fault tolerance used in real time distributed system. The main concentration is on types of faults, fault detection techniques and their recovery techniques used. Link failure, resource failure or any other failure is to be detected and rectified for working the system accurately without any disturbances. The fault management applications are hereby enabled to determine the root cause of distributed systems failure automatically. In order to aspect faults detection in distributed systems we propose to combine proactive and reactive techniques in an expert system for managing the faults.

Keywords: Distributed computing systems, Fault Management, Reactive and Proactive.

1 Introduction

A distributed system can be much larger and more powerful with the combined capabilities of the distributed components, than combinations of stand-alone systems. But it's not easy - for a distributed system to be useful, it must be reliable. This is a difficult goal to achieve because of the complexity of the interactions between simultaneously running components. Distributed computing is a method of computer processing in which different parts of a program are run simultaneously on two or more computers that are communicating with each other over a network. Generally it can be referred to segment or parallel computing .Mostly it is referred to parallel

S.C. Satapathy et al. (Eds.): Proc. of Int. Conf. on Front. of Intell. Comput., AISC 199, pp. 593–602.
DOI: 10.1007/978-3-642-35314-7_67

computing because from the definition we can observe that program run simultaneously on two or more processors that are part of the same computer. While both the processing's required program to be segmented to run simultaneously. Some of the examples are telnet, network file system, network printer etc., ATM (cash machine).Distributed computing is a natural result of using networks to enable computers to communicate efficiently. The goal of distributed computing is to make such a network work as a single computer.

Fault is defined as an error present in the internal states of components of system or in the design of a system. Error is nothing but the part of the system that is incorrect. Failure is the state where the system is deviated from its behavior specified in its specification. Faults underwent in distributed computing are primarily of software, Depending on system resources and time factor they are categorized as transient and permanent. Apart from these misleading return values, hanging processes, hardware/software/network outages, misbehaving machines, over commitment of resources are also some of the faults that are experienced in distributed systems [4].Faults can be classified in number of ways which can be differentiated depending on several factors they are Network faults, Physical faults, Processor faults, Process faults, Service expiry fault. Basing on the system resources and time, faults can be categorized into Transient faults, intermittent faults and Permanent faults.

In reactive approaches, routing tables are computed only if needed. This allows constructing viable routes that take the current network status into account. It also reduces the overhead required to store and maintain routing tables but at the expense of slower response times.

In proactive approaches, routing tables are pre-computed and maintained at all times regardless of whether a source needs them or not. This separation between route computation and data delivery allows more powerful techniques to be used for route optimization and can significantly reduce the delivery time. Since reactive and proactive techniques complement each other, a better approach may be combining them to gain merits of both i.e., some nodes may use reactive technique while others use a proactive technique. Another way is to run a proactive technique first. Then, when there is a need, a reactive technique is called.

In our paper we deal with the introduction to distributed system in section 1, section 2 shows the related work of existing fault detection and fault tolerant system and we finally conclude and future scope in section3.

2 Related Work

2.1 Existing Fault Detection System in Distributed Computing Systems

The key challenge in building reliable distributed computing is to detect the faults. The aim of the fault detection mechanisms is to identify the faulty components, so that they can be separated and repaired accordingly. Where the component that caused the fault stops performing the sequence of steps .hence these general faults will have a considerable impact on the practical system and hence due to this problem it would be useful to apply fault detection in a wider extent of faults. Hardware faults are to be considered in a wider class in order to achieve the reliability of the distributed computing. Hardware can fail in many ways but because of the enhancement of the density of the integrated circuits and also the improvement of the electronic packaging these hardware faults cause less damage and hence the hardware reliability was much more improved. Since the reliability of the hardware is improved enormously it reflected the decrease in the heat production and also the power consumption in smaller circuits, reduction in off chip connections and wiring and manufacturing techniques are also improved [38].

Globus offers a software infrastructure that enables applications to handle distributed heterogeneous computing resources as a single virtual machine [18]. The Globus Heart Beat Monitor [30, 21] uses a generic failure detection service that enables applications to detect both host or network failure through a process of heartbeat missing, e.g. 'the task died without un-registering' notification message.

Condor-G [13] uses an 'ad hoc' failure detection mechanism and uses 'periodic polling to generic Grid server' to detect some specific types of failures e.g. host and or network failures. Condor-G can't detect task crash failures or user-defined exceptions, as is the case in Legion. Condor-G uses Retry on the same machine for fault tolerance in Grid environment [30].

Armen aghasaryan and Eric Fabre addressed the problem of the alarm correlation in large distributed systems by using concurrence of events in order to specify and also simplify the state estimation in faulty systems by a technique called Petri nets and their causality semantics [7].

Paul stelling and Ian foster proposed a wide variety of techniques for detecting and correcting faults. The implementation in a particular context may be difficult .A modular approach considering the nature of the grid services a specific fault detection service was proposed. This service used well known techniques based on the unreliable fault detectors to detect and report component failure [21].

The problems of the failure detection and consensus in asynchronous systems in which processes may crash and recover ,and links may lose messages a new failure detector that are particularly suitable to crash recover model were developed by Marcos Kawazoe Aguilera , Wei Chen. Many consensus algorithms to tolerate the link failures particularly efficient i.e. those with no failures or fail detector mistakes were developed [16].

Felix c freiling and rachid guerraoui presented a failure detector which is a fundamental abstraction in distributed computing and failure detectors are the building blocks to simplify the design of reliable distributed algorithms and illustrated how failure detectors can factor out timing assumptions to detect failures in distributed agreement algorithms and also addressed the weakest failure detector question and illustrated how failure detectors can be used to classify problems [8].

Yo-Ping Huang and Chih-Hsin Huang integrated three different modeling techniques, i.e., genetic algorithm, fuzzy logic, and grey theory to become a practical model for prediction purposes. The grey system is used to predict the next output from an unknown plant. Since the prediction error is inevitable, a fuzzy controller is designed to learn how to compensate for the output from the grey system.

The roughly determined fuzzy rule base is then tuned by the genetic algorithms. The results show that the proposed technique outperforms the conventional grey systems. Also, the proposed model demonstrates its simplicity in modeling, its applicability to real-world prediction problem, and its extension ability for future intelligent control [37].

Dong tian and kaigui wu combined adaptive heartbeat mechanism with fuzzy grey prediction algorithm and presented a novel implementation of failure detector. The main parts of the implementation are adaptive grey prediction layer and adaptive fuzzy rule-based classification layer. The former layer employs a GM (1, 1) unified-dimensional new message model, only needs a small volume of sample data, to predict heartbeat arrival time dynamically. Then, they predict value and the message loss rate in specific period are act as input variations for the latter layer to decide failure/non-failure. Furthermore, algorithms of how to predict arrival time and how to construct adaptive fuzzy rule-based classification system are also presented [6].

Ahmad shukri mohd nor and Mustafa mat deris proposed a most popular technique that used in detecting fault is heartbeat mechanism where it monitors the system resources consistently and in a very short interval. However, this technique has its weaknesses as it requires a period of times to detect the faulty node and therefore delaying the recovery actions to be taken [1].

2.2 Existing Fault Tolerance and Recovery Techniques

Many Fault Tolerance techniques such as Retry, Replication, Message logging, Check-Pointing, Scheduling are available in Distributed computing.

(a) Retry:
Retry is the simplest failure recovery technique in which we hope that whatever is the cause of failures, the effect will not be encountered in subsequent retries [30].

(b) Replication:
In replication based technique we have replicas of a task running on different machines and as long as not all replicated tasks crash (i.e. host crash etc), chances are that the task execution would succeed [30].

(c) Message Logging:
In message logging all participating nodes log incoming messages to stable storage and when a failure is encountered than these message logs are used to compute a consistent global state. Algorithms that take this approach can be further classified into those that use pessimistic and those that use optimistic message logging [17].

(d) Check-Pointing:
Check-pointing is relatively more popular fault tolerant approach used in distributed systems, where the state of the application is stored periodically on reliable and stable storage, normally a hard disk etc. In case of problem during execution, i.e. after crash etc, the application is restarted from the last checkpoint rather than from the beginning [19].

(e) Scheduling:

Scheduling is also one of the methods to tolerate fault from distributed system [36, 14]. It is used to overcome the drawback of check-pointing in distributed environment. It is categorized as time-sharing scheduling, space sharing scheduling, and hybrid combination of both. Scheduling is used for load balancing as well as fault tolerance in distributed system on the basis of space or time sharing [34], [31]. There are three approaches of scheduling such as space, time, and hybrid.

Space scheduling is used to tolerate permanent or hardware type of fault from a system. The Primary-Backup approach is applied in space redundancy, Time redundancy is used when there is intermittent type of fault in the system and Hybrid redundancy is used when both are required [16].

Net Solve [20] is a client/server application designed to solve computational science problems in a distributed environment. The Net solve system is composed of loosely coupled distributed systems, connected via LAN or WAN. Net solve uses a generic heartbeat mechanism for failure detection and uses Retry on another available machine for fault tolerance [30].

The Transaction processing which is a basic application of the distributed computing should maintain the fault tolerance protocols in order to make the transactions atomic and hence standard protocols such as BFT [22], BFTDC[35] handle the fault tolerance to the greater level , also a novel proactive based agreement which identifies the tentative faults in the system, and optimized reactive view change mechanism was also proposed by Poonam Saini1 and Awadhesh Kumar Singh to improve the failure resiliency with a less execution overhead [22].

Miguel Castro proposed the techniques that tolerate the byzantine faults (BFT) provide a potential problem of malicious attacks, operator mistakes, and software errors because the technique makes no assumptions about the behavior of faulty processes. It is a first byzantine fault tolerant, state machine replication algorithm that is safe in asynchronous systems such as internet. It integrates the mechanisms of to oppose against byzantine faulty clients and it recovers replicas proactively [17].

Wending Zhao proposed that byzantine fault tolerant distributed commit protocol (BFTDC) can endure the byzantine faults at the coordinator replicas and malicious faults at the participants. It is the enhanced technique of the traditional two-phase commit protocol where all the byzantine faults are addressed but not only benign faults [35].

Goyer, p.momtahan, and B.selic proposed a fault tolerant pattern for distributed computing systems which is suitable for a class of distributed applications that is characterized by the star like topology which is commonly used in practice. Where the system consists of a set of distributed agents, if an agent fails and recovers, it can restore its local state information from the controller. One interesting aspect to this topology is that it is not necessary for the controller to monitor its agents. A recovering agent needs merely to contact its controller to get its state information. This eliminates costly polling [11].

ISIS system [B. & J. 1987, B. 1985] aimed at providing user-transparent fault-tolerance in a general purpose distributed computing environment. And assumes nodes are fail silent (by crashing) they Provide a tool-kit that allow a programmer to build a distributed application that is made fault tolerant by replication of code and data.

Vilgot claesson and Neeraj suri presented a novel, efficient, and low-cost start-up and restart synchronization approach for TDMA environments. This approach utilizes information about a node's message length that forms a unique sequence to achieve synchronization such that communication overhead can be avoided and also presented a fault-tolerant initial synchronization protocol with a bounded start-up time. The protocol avoids start-up collisions by deterministically postponing retries after a collision [33].

Richard Ekwall and Andre Schiper published many atomic broadcast algorithms in the last twenty years. Token based algorithms represent a large class of these algorithms. All these rely on a group membership service and none of them uses unreliable failure detectors directly. This is the first token based atomic broad cast algorithm that uses an unreliable failure detector instead of group membership services. It requires a system size that is quadratic in the number of supported failures [26].

Sergio Gorender and Michel raynal presented an adaptive programming model for fault-tolerant distributed computing, which provides upper-layer applications with process state information according to the current system synchrony. The underlying system model is hybrid, composed by a synchronous part (where there are time bounds on processing speed and message delay) and an asynchronous part (where there is no time bound). However, such a composition can vary over time, and, in particular, the system may become totally asynchronous or totally synchronous. Moreover, processes are not required to share the same view of the system synchrony at a given time [29].

Cao Huaihu, Zhu Jianming proposed an adaptive replica creation algorithm with fault tolerance in the distributed storage network. This algorithm also maintains a rational replica number, not only satisfying the user anticipant availability, improving access efficiency and balancing overload, but also reducing bandwidth requirement, maintaining the system's stability, providing users with the satisfaction of Quality of service (QOS)[4].

Hamid Mushtaq, Zaid Al-Ars and Koen Bertels applied online fault tolerance techniques to reduce the probability of failures of distributed systems. These techniques need to be efficient as they execute concurrently with applications running on such systems. Fault tolerance is classified into four different steps which are proactive fault management, error detection, fault diagnosis and recovery [12].

Chi-Hsiang Yeh proposed the robust middleware approach to transparent fault tolerance in parallel and distributed systems. The proposed approach inserts a robust middleware between algorithms/programs and system architecture/hardware. With the robust middleware, hardware faults are transparent to algorithms/programs so that ordinary algorithms/programs developed for fault-free networks can run on faulty parallel/distributed systems without modifications. Moreover, the robust middleware automatically adds fault tolerance capability to ordinary algorithms/programs so that no hardware redundancy or reconfiguration capability is required and no assumption is made about the availability of a complete sub network (at a lower dimension or smaller size).Even a nomadic agent multithreaded programming as a novel fault-aware programming paradigm that is independent of network topologies and fault patterns. Nomadic agent multithreaded programming is adaptive to fault/traffic/workload patterns, and can take advantages of various components of the robust middleware, including the fault tolerance features and multiple embeddings, without relying on specialized robust algorithms[5].

LA-MPI -Los Alamos Message Passing Interface [23, 24] is a network-fault-tolerant implementation of MPI designed for terascale clusters. The two important goals in LA-MPI are fault tolerance and performance. In LA-MPI process fault tolerance has completely been ignored and it only focuses on network relevant fault tolerance.

FT-MPI -Fault Tolerance Message Passing Interface is another try of handling problems in MPI. Current fault tolerance and recovery strategies in FT-MPI [25] can regularly take checkpoints during a workflow step, that is normally a scientific application, and when a failure is encountered, the application is restarted from the last checkpoint [15].

S. Chakravorty et al. in [28] presents a fault tolerance solution for parallel applications that proactively migrates execution from processors where failure is imminent. The drawback in this idea is that it works on the assumption that failures are predictable

Table 1. shows the existing fault detection and tolerant techniques and their Performance in Distributed System

Table 1. Fault detection and fault tolerance techniques and the performance of Distributed system

System	**Fault detection technique**	**Faults detected**	**Fault tolerance techniques**	**Proactive/reactive**	**Disadvantage**
GLOBUS[18]	Heart beat monitor	Host and network failure	Resubmit the failed job	Reactive	Can't handle user defined exceptions
LA-MPI [23,24]	Checks unacknowledged list at specified intervals	Network related failure	Sender side retransmission	Reactive	Appropriate only for low error rate environments
Condor-G[13]	polling	Host, crash and network crash	Retry on same machine	Reactive	Use of Condor client interfaces on top of GLOBUS
FT-MPI[25]	Discovered by run time environment	Process failures, system crash	Check pointing and roll back verse replication	Proactive	Can't handle fault-tolerance on the application level.
S. Chakravorty et al.[28]	Environmental monitoring, event logging, parallel job monitoring and resource monitoring	Imminent faults	Process virtualization and dynamic task migration	proactive	Can't be applicable for unpredictable faults
Net Solve [20]	Generic Heart beat mechanism	Host crash, Network failure, k crash	Retry on another available machine	Reactive	Doesn't support diverse failure recovery mechanism
Miguel Castro	polling	Byzantine Fault	Byzantine fault tolerance algorithm (BFT) -state machine replication algorithm	Proactive	Inefficient state transfer and retransmission strategies

3 Conclusion and Future Scope

As the scale of distributed computing continues to grow, the reliability of the system becomes more crucial and failure prediction and correction mechanisms play a major role in increasing the system reliability. The distributed computing is very flexible by reducing workload, communicative which enhance human-to human communication. We address reliability issue by developing an efficient fault management system for Distributed systems. Various techniques have been proposed for detecting faults in distributed computing. Once the faults are detected, one may diagnose the system to track the root cause.

Till now various systems based on reactive and proactive techniques are separately developed for diagnosis of the system, we are now going to implement them in Expert system and we are combining both the reactive and proactive techniques for fault management in the distributed environment so that we prove that these techniques are reliable, working efficiently and properly in the real time scenario and are feasible.

Acknowledgement. We thank AICTE for the approval of our research work and encouraging us to develop an efficient system so forth.

References

[1] Shukri, A., Noor, M., Deris, M.M.: Dynamic hybrid fault detection methodology. Journal of Computing 3(6) (June 2011)

[2] Girault, A., Kalla, H., Sorel, Y.: Transient Processor/Bus Fault Tolerance For Embedded Systems with hybrid redundancy and data fragmentation

[3] Sistla, A.P., Welch, J.L.: Efficient distributed recovery using message logging. In: Proceedings of the Eighth Annual ACM Symposium on Principles of Distributed Computing, pp. 223–238 (June 1989)

[4] Cao, H., Zhu, J.: An Adaptive Replicas Creation Algorithm with Fault Tolerance in the Distributed Storage Network. IEEE (2008)

[5] Yeh, C.-H.: The robust middleware approach for transparent and systematic fault tolerance in parallel and distributed systems. In: International Conference on Parallel Processing, Proceedings (2003)

[6] Tian, D., Wu, K., Li, X.: A Novel Adaptive Failure Detector for Distributed Systems. In: Proceedings of the 2008 International Conference on Networking, Architecture, and Storage (2008) ISBN: 978-0-7695-3187-8

[7] "Fault detection and diagnosis in distributed systems: An approach by partial stochastic Petri nets", `http://resources.metapress.com/pdfpreview.axd?code=n6251v28705h7551&size=largest`

[8] Freiling, F.C., Guerraoui, R., Kuznetsov, P.: The failure detector abstraction (published in 2011)

[9] Fischer, M., Lynch, N., Paterson, M.: Impossibility of Distributed Consensus with One Faulty Process. Journal of the ACM 32(2), 374–382 (1985)

[10] Kola, G., Kosar, T., Livny, M.: "Faults in large distributed systems and what we can do about them" (2005), http://citeseerx.ist.psu.edu/viewdoc/summary?doi=10.1.1.108.725

[11] Goyer, P., Momtahan, P., Selic, B.: A Fault-Tolerant Strategy for Hierarchical Control in Distributed Computer Systems. In: Proc. 20th IEEE Symp. on Fault-Tolerant Computing Systems (FTCS20), pp. 290–297. IEEE CS Press (1990)

[12] Mushtaq, H., Al-Ars, Z., Bertels, K.: Survey of fault tolerance techniques for shared memory Multicore/multiprocessor systems. In: IEEE 6th International Design and Test Workshop, IDT (2011)

[13] Frey, J., Tannenbaum, T., Foster, I., Livny, M., Tuecke, S.: Condor-G: A Computation Management Agent for Multi-Institutional Grids. Cluster Computing 5(3) (2002)

[14] Abawajy, J.H.: Fault-Tolerant Scheduling Policy for Grid Computing Systems. In: Proceedings of the IEEE 18th International Parallel and Distributed Processing Symposium (IPDPS 2004),

[15] Kandaswamy, G., Mandal, A., Reed, D.: Fault Tolerance and Recovery of Scientific Workflows on Computational Grids. In: 8th IEEE International Symp. on Cluster Computing and the Grid, CCGRID 2008 (2008)

[16] Aguilera, M.K., Chen, W., Toueg, S.: Failure Detection and Consensus in the Crash-Recovery Model. In: Kutten, S. (ed.) DISC 1998. LNCS, vol. 1499, pp. 231–245. Springer, Heidelberg (1998)

[17] Castro, M.: Practical Byzantine Fault Tolerance and Proactive Recovery. Microsoft research (2002)

[18] Affan, M., Ansari, M.A.: Distributed Fault Management for Computational Grids. In: Fifth International Conference on Grid and Cooperative Computing, GCC 2006, pp. 363–368 (2006)

[19] Hussain, N., Ansari, M.A., Yasin, M.M.: Fault Tolerance using Parallel Shadow Image Servers (PSIS). In: Grid Based Computing Environment, IEEE–ICET, November 13-14 (2006)

[20] Jalote, P.: Fault Tolerance in Distributed Systems (1994) ISBN: 0-13-301367-7

[21] Stelling, P., Foster, I.: A Fault detection service for wide area distributed computations. The aerospace Cooperation, El Segundo, CA 90245-4691-USA

[22] Sainil, P., Singh, A.K.: Two New Protocols for Fault Tolerant Agreement., Department of Computer Engineering, National Institute of Technology, Kurukshetra, India, International Journal of Distributed and Parallel Systems (IJDPS) 2(1) (January 2011)

[23] Graham, R.L., Choi, S.-E., Daniel, D.J., Desai, N.N., Minnich, R.G., Rasmussen, C.E., Risinger, L.D., Sukalski, M.W.: A network-failure-tolerant message passing system for terascale clusters. In: Proceedings of the 16th International Conference on Supercomputing, pp. 77–83. ACM Press (2002)

[24] Aulwes, R.T., Daniel, D.J., Desai, N.N., Graham, R.L., Risinger, L.D., Sukalski, M.W.: LA-MPI: The design and implementation of a network-fault-tolerant MPI for terascale clusters. Technical Report LA-UR- 03-0939, Los Alamos National Laboratory (2003)

[25] Wolski, R., Pjesivac-Grbovic, N., London, K., Dongarra, J.: Extending the MPI Specification for Process Fault Tolerance on High Performance Computing Systems. ICS, Heidelberg (June 2004)

[26] Ekwall, R., Schiper, A.: A Fault-Tolerant Token-Based Atomic Broadcast Algorithm. Journal: IEEE Transactions on Dependable and Secure Computing - TDSC 8(5), 625–639 (2011)

[27] Chakravorty, S., Kalé, L.V.: A Fault Tolerance Protocol with Fast Fault Recovery. IEEE (2007)

[28] Chakravorty, S., Mendes, C.L., Kalé, L.V.: Proactive Fault Tolerance in MPI Applications Via Task Migration. In: Robert, Y., Parashar, M., Badrinath, R., Prasanna, V.K. (eds.) HiPC 2006. LNCS, vol. 4297, pp. 485–496. Springer, Heidelberg (2006)

[29] Gorender, S., Raynal, M.: An Adaptive Programming Model for Fault-Tolerant Distributed Computing. IEEE Transactions on Dependable and Secure Computing 4(1) (January-March 2007)

[30] Hwang, S., Kesselman, C.: A Flexible Framework for Fault Tolerance in the Grid. Journal of Grid Computing 1, 251–272 (2003)

[31] Krishnan, S., Gannon, D.: Checkpoint and restart for distributed components in XCAT3. In: Proceedings of the Fifth IEEE/ACM International Workshop on Grid Comp., GRID (2004)

[32] Haeberlen, A., Kuznetsov, P.: The Fault Detection Problem. In: Abdelzaher, T., Raynal, M., Santoro, N. (eds.) OPODIS 2009. LNCS, vol. 5923, pp. 99–114. Springer, Heidelberg (2009), `http://resources.metapress.com/pdf-preview.axd?Code=m131704524825417&size=largest`

[33] Claesson, V., Lönn, H., Suri, N.: Efficient TDMA Synchronization for Distributed Embedded Systems. In: Proc. 20th Symp. Reliable Distributed Systems, pp. 198–201 (October 2001)

[34] De Florio, V., Blondia, C.: A Survey of Linguistic Structures for Application-Level Fault Tolerance. ACM Computing Surveys 40(2), Article 6 (April 2008)

[35] Zhao, W.: A Byzantine Fault Tolerant Distributed Commit Protocol, Department of Electrical and Computer Engineering, Cleveland State University, 2121 Euclid Ave, Cleveland, OH 44115

[36] Luo, W., Yang, F., Tu, G., Pang, L., Qin, X.: TERCOS: A Novel Technique for Exploiting redundancies in Fault-Tolerant and Real-Time Distributed Systems. In: 13th IEEE International Conference on Embedded and Real-Time Computing Systems and Applications, RTCSA 2007 (2007)

[37] Huang, Y.-P., Huang, C.-H.: A Genetic-based fuzzy grey prediction model. IEEE (October 1995)

[38] "How and why computer systems fail", `http://resources.metapress.com/pdf-preview.axd?code=u3rk8052h46284uu&size=largest`

An Eminent Approach of Fault Management Using Proactive and Reactive Techniques in Distributed Computing

Selvani Deepthi Kavila[1], G.S.V. Prasada Raju[2], Suresh Chandra Satapathy[1], P.B.P. Niharika[1], S. Geetha[1], and S. Praveen Kumar[1]

[1] Department of Computer Science and Engineering,
Anil Neerukonda Institute of Technology and Sciences, Sangivalasa-531162, Visakhapatnam, Andhra Pradesh, India
{selvanideepthi14,sureshsatapathy,niharika.pavani, sangana.geetha,praveen8514}@gmail.com
[2] Department of Computer Science,
SDE, Andhra Univeristy, Visakhapatnam, India
gsvprajudr9@yahoo.co.in

Abstract. Many of the existing distributed systems perform remote monitoring, and even generate alarms when fault occurs. But they fail in finding the exact location of the fault, or even the automatic execution of appropriate fault recovery actions. Hence for large distributed systems with many computing nodes it may be rather time-consuming and difficult to resolve the problems in a short time by an exhaustive search in order to find the cause for failure. At the time of failure many problems like loss or delay of fault messages may occur. Also a failure may result in a number of unreliable alarms. A good sophisticated design for fault management should work out efficiently whenever there is a redundant and incomplete data. So we proposed a system comprising of several components where we have a fault detection engine in which various techniques have been proposed for detecting faults in distributed computing. Once the faults are detected, one may diagnose the system to track the root cause. For diagnosis the system we are going to implement that In Expert system and we are combining both the reactive and proactive techniques for fault management.

Keywords: Distributed Systems, Distributed computing, Fault Management, Expert System, Reactive, Proactive.

1 Introduction

1.1 Distributed Systems

Distributed system is an information processing system that contains a number of independent computers that co-operate with one another over a communication network in order to achieve specific objective.

S.C. Satapathy et al. (Eds.): Proc. of Int. Conf. on Front. of Intell. Comput., AISC 199, pp. 603–611.
DOI: 10.1007/978-3-642-35314-7_68

1.2 Distributed Computing

Distributed computing is a field that studies distributed system where we have number of systems or computing nodes and we need to compute the exact path from source node to destination there by making the computing process faster. In distributed computing, each and every processor has its private memory and the information is communicated through the processors by passing messages. The computing is done in the distributed systems.

Distributed computing is a method of computer processing in which different parts of a program are run simultaneously on two or more computers that are communicating with each other over a network. Distributed computing is also a typical processing where various modules of a program execute at the same time on multiple processors which are a part of the same computer. Distributed computing also makes sure that the program division considers different environments in which different modules of the program will be executed.[13]

1.3 Fault Management

In distributed computing, fault management is set of functions or rules that detect and correct faults in a communication network, adapts to the environmental changes and also can maintain and examine errors, accepting errors and acting on error detection, tracing and also identifying error, making certain tests for correcting faults, reporting the error conditions.

In our paper we deal with the introduction to distributed systems and distributed computing in section 1, section 2 discusses already existing systems, section 3 depicts our proposed architecture, section 4 explains briefly all the components of our model, section 5 describes the working of our system and we finally conclude and future work with section 6.

2 Existing Systems

There is a large body of research that aims to reactively recover systems as fast and efficiently as possible ([11],[4], [5]). Early work on this field (e.g., [11]) advocates the execution of some kind of recovery action (in general, restart the monitored process) after a fault is detected. More recently, the recovery oriented computing project proposed a set of methods to make recovery actions more efficient and less costly in terms of time [4][1].

Several techniques were developed in this project, either to detect failures, restart the minimum set of system components to put the system back to correct operation and to undo some operator configuration errors. Other works like [4] try to diagnose system faults through several monitors and evaluate which is the best set of recovery actions that must be taken in order to recover the system as fast as possible. These works do not consider Byzantine faults or security-compromised components and also do not rely on redundancy to ensure that the system stays available during recoveries, the main problems addressed in the present work [1].

S. Chakravorty et al. in [9] presents a fault tolerance solution for parallel applications that proactively migrates execution from processors where failure is imminent. The drawback in this idea is that it works on the assumption that failures are predictable.

3 Proposed System

We actually referred many research papers and found out that each system has some disadvantages regarding security issues, transmission problems etc., as mentioned in the previous section. In this paper we are trying to overcome some of the demerits of the already existing systems.

The description of the system and its components are explained in the next section.

4 Components of the System

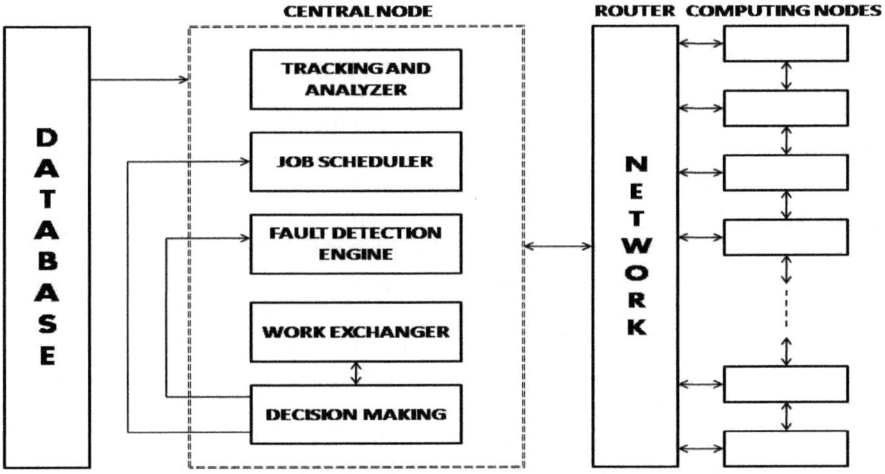

Fig. 1. Proposed architecture

4.1 Database

The database that we use in our system is distributed database is not a common processing unit such as C.P.U where all the storage devices are attached. It may be stored in multiple computers located in the same physical location, or may be dispersed over a network of interconnected computers. Unlike parallel systems, where the processors are tightly coupled and constitute a single database system, a distributed database system consists of loosely coupled sites that have no physical components. Collections of data can be distributed across multiple physical locations. A distributed database can reside on network servers on the Internet, on corporate intranets or extranets, or on other company networks. The distribution is transparent where the users interact with the system as if it were one logical system.

4.2 Tracking and Analyzer

Tracking: It involves the basic criteria of tracking down the paths or the steps which the specified job has to perform. Tracking involves thorough investigation of the job and its whereabouts. It forms a major role as if it is not done in a proper way then there is a chance of deadlock arising.

Analysis: It follows the tracking procedure and it specifies in brief the activities and the resources that are necessary for the job to be performed successfully.

4.3 Job Scheduler

Job scheduler is used to schedule the jobs that are to be handled by the system. In general, job scheduling is composed of at least two inter-dependent steps: the allocation of processes to workstations and the scheduling of the processes over time (time-sharing). When submission of a job to the system is done, job placement will be done. Along with the job submission, a description of the attributes of the job is also submitted to the system in order to specify the resource requirement, such as memory size requirement, expected CPU time, deadline time, etc.

In the meantime, the system always maintains an information table, either distributed or centralized, to record the current resource status of each workstation. Then it matches work to the most suitable set of workstations to meet the requirement of the job. This job is then decomposed into small components, i.e. processes, which are distributed to those assigned workstations. On each individual workstation, the local scheduler allocates some time-slices to the process based on some policies so as to achieve the scheduling requirement, such as response time and fairness. These decomposed components may require synchronization among themselves.

Various schemes are used to decide which particular job to run. Parameters that might be considered include:

- Job priority
- Compute resource availability
- License key if job is using licensed software
- Execution time allocated to user
- Number of simultaneous jobs allowed for a user
- Estimated execution time
- Elapsed execution time
- Availability of peripheral devices
- Occurrence of prescribed event.

4.4 Fault Detection Engine

Fault detection engine is the main component of our system. It detects the faults and recovers them using various techniques. It is necessary for the system to be able to detect the faults and take the appropriate actions to avoid further degradations of the service. Due to increasing demands for access to information, today's distributed systems have grown both in size and complexity. Distributed systems require systems to be highly reliable and easily available. In the future, distributed systems will get even larger, more complex and error prone. Hence, effective fault detection engine is a key to maintain the high availability and reliability in distributed systems.

The current fault management systems are not sufficient to solve the challenges in the Distributed Computing Environment. The existing systems are mostly based on the reactive systems. So in our research work we combine both reactive and proactive techniques to deal with the fault management problems in distributed computing [12].

4.4.1 Fault Tolerant Approaches

4.4.1.1 Reactive approach. In reactive approaches, routing tables are computed only if needed. This allows constructing viable routes that takes the current network status into account. It also reduces the overhead required to store and maintain routing tables but at the expense of slower response times.

4.4.1.2 Proactive approach. In proactive approaches, routing tables are pre-computed and maintained at all times regardless of whether a source needs them or not. This separation between route computation and data delivery allows more powerful techniques to be used for route optimization and can significantly reduce the delivery time. Since reactive and proactive techniques complement each other, a better approach may be combining them to gain merits of both i.e., some nodes may use reactive technique while others use a proactive technique. Another way is to run a proactive technique first. Then, when there is a need, a reactive technique is called.

4.4.2 Types of Faults

Faults in distributed systems may result from two types of causes. They are Software faults and Hardware faults.

a. Software faults:

In some situations, the running environment of a distributed system is fully trusted, i.e. all the participating nodes are trusted. In this scenario, faults are derived from bugs in the system design, implementation or configurations. We refer to this type of faults as software errors.

b. Hardware faults:

On the other hand, users may have full confidence of the system design and implementation, and the faults manifest themselves as malicious behaviors, as a result of part of the nodes being com- promised by adversaries. Compared to software errors, malicious behaviors are more difficult to be detected, as the compromised nodes may cheat about their behaviors and collude with each other. Moreover, the nodes not only behave erroneously, but also fail to behave consistently when interacting with multiple other peers [7].

4.4.3 Fault Detection Techniques

According to how the expected behavior of a distributed system is encoded, different techniques are applicable for fault detection. Based on that, the mechanisms can be classified into three categories. They are invariant checking, Reference implementation and Model checking.

4.4.1.1 Invariant Checking. The recorded data of system state are checked against the properties using online assertions or offline analysis. To facilitate users expressing the properties, in some fault management systems, declarative domain-specific languages are provided [12].

4.4.1.2 Reference Implementation. Given a reference implementation, users are able to detect faults by comparing the behavior of the actual system and the one of the reference implementation. Once all non-deterministic events, such as read/write of files and send/receive of messages, are recorded, the reference implementation's behavior exhibited in deterministic replay should be identical to the one observed in the actual system [12].

4.4.1.3 Model checking. In model checking, the behavior of a distributed system is modeled as a state machine. Starting from an initial state, a model checker performs exhaustive search by systematically enumerating all possible execution paths. The target system is steered to follow each of the execution paths to check whether it behaves correctly[12].

4.4.1.4 Fault recovery approach. In order to detect and recover the faults effectively in our systems we use one of the expert systems technique i.e., Rule-based reasoning [3].

Rule-Based Reasoning

Most expert systems use rule-based representation of their knowledge-base. In the rule based approach, the general knowledge of a certain area is contained in a set of rules and the specific knowledge, relevant for a particular situation, is constituted of facts, expressed through assertions and stored in a database [2].

There are two operation modes in a rule-based system. One is the forward mode which departs from an initial state and constructs a sequence of steps that leads to the solution of the problem ("goal"). When it comes to a fault diagnosis system, the rules would be applied to a database containing all the alarms received, until a termination condition involving one fault is reached. The other is the backward mode, which starts from a configuration corresponding to the solution of the problem and constructs a sequence of steps that leads to a configuration corresponding to the initial state. The same set of rules may be used for the two operation modes.

The rules are of the form:

Left Hand Side (LHS) ==> Right Hand Side (RHS).

The LHS is a set of conditions which must be matched in working storage for the rule to be executed. The RHS has the actions to be performed if the L.H.S is matched.

The knowledge engineer programs in Oops by creating rules for the particular application. The syntax of the rules is:

rule<rule id>:
[<N>: <condition>, ...]
==>
[<action>, ...].

Where:

rule id – A unique identification of the rule;
N - Optional identification for the condition;
condition - A rule that should match the working storage;
action – the action to be performed.

4.5 Work Exchanger

Work exchanging can also be termed as Symmetry breaking. Sometimes some nodes need to be selected to orchestrate the computation (and the communication). This is typically achieved by a technique called symmetry breaking. Whenever a fault is detected in a computing node the work exchanger help in exchanging the job to the other computing node thereby decreases the delay in completing the job and increasing the efficiency of the system.

4.6 Decision Maker

The decision maker is able to communicate with fault detection engine and job scheduler such that whenever it knows that a fault is detected by a fault detection engine it contacts job scheduler in order to schedule the job to the other computing node and so it makes a decision based on the fault and the scheduling done by the job scheduler.

4.7 Router

Router is one of the communicating devices used for transmitting data along the networks. It is connected at least to two or more networks and they might be two LAN's or WAN's. It helps in providing connectivity between two or more networks. It maintains a table which stores the information about the best route from the source to transmit the packet to the destination.

5 System Working

The main objective of our work is to discover useful information and forecast the failure occurrence in the large systems. Based on this failure prediction some decisions will be made in order to maintain the reliability of the system so that it optimizes application and the system performance.

Once the data or information that is to be transferred through the systems is ready in the database then we retrieve the data and start processing it. The router helps in tabulating the best route from the source to destination. The jobs are scheduled with the help of job scheduler to the destination systems or computing node. If the destination node fails to receive the data then the acknowledgement is sent to the server and fault detection engine helps in detecting the fault during the transmission process. With the help of decision maker, it tracks the fault and analyzes it and with the help of decision maker if it finds that the fault handling is time taking then with the help of work exchanger, the job scheduler will schedule to the other computing node which is efficient enough to proceed the job.

6 Conclusion and Future Work

We have gone through various proposed techniques and found out the disadvantages and so we want to implement a system and we will overcome the security issues by allowing constrained access for all the users. We make sure that we keep our system as simple as possible and use a dedicated database. . This paper proposed the combination of proactive and reactive recovery in order to increase the efficiency of the system so that the fault can be detected and tolerated easily and then we proposed expert system concept in order to recover the fault and we make sure that we implement the proposed system in the most efficient way so that the faults can be easily detected and recovered using our proposed system in the distributed systems using distributed computing. The only disadvantage with our system will be mostly the task of work exchanger as to exchange the work to which host and we make sure that we develop it in an efficient way so that we can overcome this disadvantage.

Acknowledgement. We thank AICTE for the approval of our research work and encouraging us to develop an efficient system so forth.

References

[1] Bessani, A.N., Correia, M., Neves, N.F., Verissimo, P., Sousa, P.: Highly Available Intrusion-Tolerant Services with Proactive-Reactive Recovery. IEEE Transaction on Parallel and Distributed Systems 21(4), 452–465 (2010)

[2] Paoli, A.: Fault Detection and Fault Tolerant Control for Distributed Systems-A general Frame work. Ph.D Thesis, University of Bologna-XVI Ciclo, A.A (2000–2003)

[3] Abraham, A.: Rule based expert system. In: Sydenham, P.H., Thorn, R. (eds.) Handbook of Measuring System Design, John Wiley & Sons, Ltd., UK (2005) ISBN: 0-470-02143-8

[4] Patterson, D., Brown, A., Broadwell, P., Candea, G., Chen, M., Cutler, J., Enriquez, P., Fox, A., Kiciman, E., Merzbacher, M., Oppenheimer, D., Sastry, N., Tetzlaff, W., Traupman, J., Treuhaft, N.: Recovery Oriented Computing (ROC): Motivation, Definition, Techniques and Case Studies. Technical Report UCB/CSD TR 02-1175, Computer Science Dept., Univ. of California at Berkeley (March 2002)

[5] Joshi, K.R., Hiltunen, M., Sanders, W.H., Schlichting, R.: Automatic Model-Driven Recovery in Distributed Systems. In: Proc. 24th IEEE Symp. Reliable Distributed Systems (SRDS 2005), pp. 26–38 (October 2005)

[6] Blanke, M., Kinnaert, M., Lunze, J.: Diagnosis and fault-tolerant control. Springer (2003)

[7] Blanke, M., Zamanabadi, R.I., Bogh, S.A.: Fault tolerant control systems: a holistic view. Control Engineering Practice 5(5) (1997)

[8] Blanke, M.: Aims and means in the evolution of fault tolerant control. In: Proceedings of the European Science Foundation COSY Workshop, Rome (1995)

[9] Chakravorty, S., Kalé, L.V.: A Fault Tolerance Protocol with Fast Fault Recovery. IEEE (2007)

[10] Bogh, S.A.: Fault Tolerant Control Systems - a Development Method and Real-Life Case Study. PhD thesis, Aalborg University, Department of Control Engineering (December 1997)

[11] Huang, Y., Kintala, C.M.R.: Software Implemented Fault Tolerance: Technologies and Experience. In: Proc. 23rd Int'l Symp. Fault Tolerant Computing (FTCS-23), pp. 2–9 (June 1993)

[12] Zhou, W.: Fault Management in Distributed Systems, university of Pennsylvania, Department of CIS,Technical Report (January 05, 2010)

[13] `http://en.wikipedia.org/wiki/Distributed_computing`

Hybrid Cluster Based Routing Protocol for Free-Space Optical Mobile Ad hoc Networks (FSO/RF MANET)

D. Kishore Kumar[1], Y.S.S.R. Murthy[2], and G. Venkateswara Rao[1]

[1] Assistant Professor, Gitam Institute of Technology, GITAM University
{Kishore_dasari,Vrgurrala}@yahoo.com
[2] Professor, Shri Vishnu Engineering College for Women, Bhimavaram
yssrmoorthy@gmail.com

Abstract. The existing routing protocols in mobile ad hoc network cannot be utilized for Free-Space Optical Mobile Ad hoc Networks (FSO/RF MANET) owing to its disjoint characteristics which include accommodating directionality, accuracy in routing information, memory, reduced overhead and delay. In this paper, a hybrid cluster based routing protocol for FSO/RF MANET is proposed. In this technique, a network connector is defined which connects the hybrid networks. The cluster formation is based on Neighborhood Discovery Algorithm. In each cluster, the node which is nearer to the network connector is elected as cluster head (CH). Each network connector gathers information of all cluster members and their CH and builds a routing table using network connector discovery algorithm. When the source node of a network wants to send data to destination node in another network, routing is performed through the network connectors.

Keywords: Free-Space Optical, Cluster Head, Neighborhood Discovery, Routing, Network connector.

1 Introduction

1.1 Free-Space Optical (FSO)

A fibreless, laser-driven technology upholding high bandwidth inclusive of installation connections for last-mile and campus scenarios is termed as Free-Space Optics (FSO). With the help of low powered lasers or LED's, the light pulses are transmitted in a small conical shaped beam through the atmosphere. [1] In order to set up a network link, an optical transceiver is positioned on each side of a transmission path. The transmitter emits a modulated IR signal which is characteristically an infra red (IR) or LED. The high throughput of FSO technology is its main advantage [2]. FSO technology is simple and analogous to fiber optics and makes use of optical transmitters and receivers. However it does not contain fiber cables. Though FSO systems can act up to several kilometers distance, there must be adequate power and glimpse among source and destination [3]. In FSO, the data can travel in both directions at the same time and hence it is considered as full duplex unit [1]. FSO can

S.C. Satapathy et al. (Eds.): Proc. of Int. Conf. on Front. of Intell. Comput., AISC 199, pp. 613–620.
DOI: 10.1007/978-3-642-35314-7_69

act as substitute for fiber based MAN solutions [1]. The transmitter/ receiver link configurations such as directed, diffuse and hybrid can be attained using FSO unit [4].

FSO is concentrated mainly for high altitude application such as space communications and building-top metro-area communications in order to attain high-speed wireless point-to-point communications [5].

FSO has certain issues in spite of offering high throughput which are described as follows.The fiber strands cannot be tapped without shattering the strands as they are highly thin which in turn will result in shutdown of the link [1]. The main limitations of FSO technology is that it necessitates optical inks for upholding line of sight (LOS) [2]. The climatic changes such as fog and severe weather causes harmful effects on FSO performance [1].

1.2 Routing in Free-Space Optical MANET

FSO technology is capable of offering enhanced per-node throughput for mobile ad hoc network (MANET). [4] FSO-MANETS can be feasible using "optical antennas". [6] According to the following two principles, FSO structures can be designed in MANET [5].

At the time of routing, FSO topology carries entire traffic in the network. The RF topology is utilized to offer the required backup, as the FSO links are vulnerable to environmental events that include fog, snow, clouds etc. The hybrid nodes help in interconnecting RF and FSO backbone domain space [2]. The protocol should take features of FSO, RF and MANET into consideration for FSO/RF MANET network. The conventional routing protocol in FSO MANET forwards the unicast packets utilizing directional antenna. But the broadcast packets such as hello packets keep flooding out at every interface to wrap full spread. [9]

1.3 Proposed Solution

Owing to certain disjoint features of FSO and MANET, the existing routing protocols in MANET cannot be used for FSO MANET. The issues concerned with traditional routing protocol are as follows.

- The routing protocols such as DSDV and AODV make use of reverse path technique but do not take unidirectional links into account [8]. Hence these protocols cannot be directly applied to FSO MANET.
- The protocols namely DSR [12], ZRP [13], or SRL [14] considers unidirectionality by detecting unidirectional links, and subsequently offering a bi-directional abstraction for such links. [8]
- Through probabilistic routing, the routing information decays with time and possess reduced accuracy. [9]

In order to solve routing problems in RF ad-hoc domains, hierarchical state routing (HSR) protocol is used. It is based on a cluster and logical sub network. The source node transmits the data to the destination node through indirect path which results in delay and overhead. In addition, the scheme for clustering and cluster head selection is not specified. To rectify these problems, a hybrid cluster based routing protocol for FSO/RF MANET is proposed in this paper.

2 Hybrid Cluster Based Routing Protocol

The proposed technique consists of three phases namely.

- Cluster Formation,
- Network Connector Discovery
- FSO/RF MANET Routing.

2.1 Phase 1 Cluster Formation

Let N_i represent the nodes deployed in the network, where i = 1, 2, 3, ….., n
Let N_{nei} represents the neighbor nodes.
Let C_i represents the clusters
Let CH_i be the cluster head.
Let $n(C_i)$ be the number of nodes in the cluster
Let NC_i represent the network connectors that connects hybrid networks and it has virtual link with all cluster member via CHs.
Let Th represents the threshold for nodes.

Also we assume that each node maintains a neighbor table that includes the ID of the neighboring nodes and link status.

The steps involved in cluster formation are as follows.

Step 1
Each NC_i broadcast HELLO message to all N_{nei} using Neighbor Discovery Algorithm (NDA).

$$N_i \xrightarrow{Hello} N_{nei}$$

The HELLO message will include the sender nodes ID, sender nodes status (say energy) and neighboring node status. The format of HELLO message is as follows.

Table 1. Format of HELLO Message

Sender node ID	Node status	Neighboring Table	
		N_{nei} ID	Node status

Step 2
When any N_{ei} receives the HELLO message, it verifies whether the message possess its ID using the following condition.

If HELLO message contains N_{nei}'s ID
Then
N_{ei} will join NC_i to form the cluster.
End if

Step 3

The process in step 2 is iterated until a pre-defined threshold of nodes join the cluster.

If n (C_i) = Th
Then
    Pause step 2.
End if

Step 4

When any N_{nei} receives more than one HELLO message, it will join with nearer NC_i with maximum energy. Currently, the hybrid network comprises of many clusters.

Step 5

In each cluster, the node, which is nearer to NC_i can be elected as cluster head (CH_i).

Fig 1 demonstrates the broadcast of HELLO message by the network connectors. The network connector node NC_3 broadcast the HELLO message to its neighbor nodes N_{ne1}, N_{ne2}, N_{ne4}, N_{ne5}, N_{ne7}, and N_{ne10}. Similarly NC_8, NC_{11}, NC_{13} and NC_{14} broadcast the HELLO message to its neighbor nodes (N_{ne5}, N_{ne7}, N_{ne9}, N_{ne11}, N_{ne13}, and N_{ne15}), (N_{ne7}, N_{ne12}, N_{ne13}, and N_{ne14}), (N_{ne8}, N_{ne11}, N_{ne14}, N_{ne15}, $N_{ne16,}$ and N_{ne17}) and (N_{ne12}, N_{ne13}, N_{ne17}, N_{ne18}, N_{ne19}, and N_{ne20}).

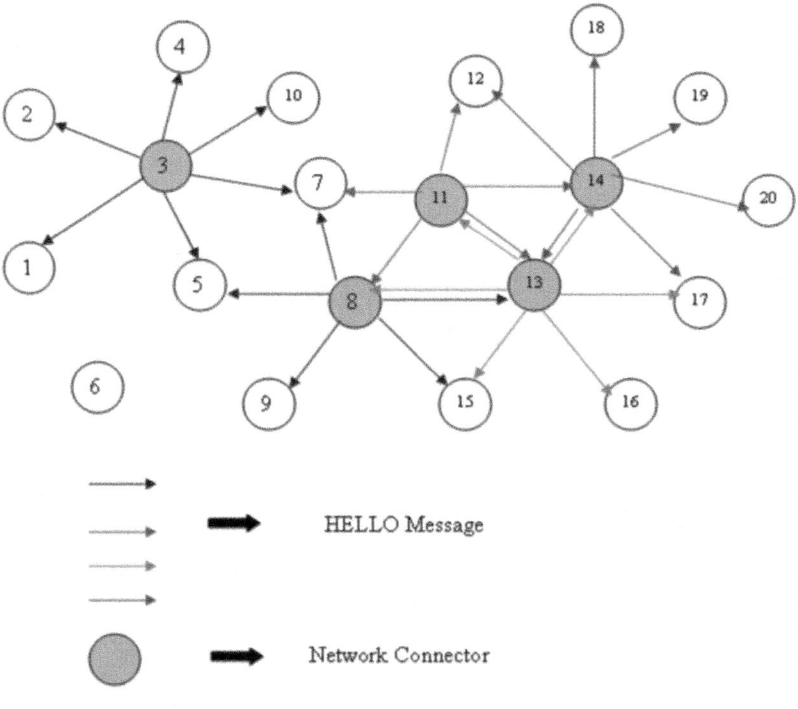

Fig. 1. Broadcast of Hello messages

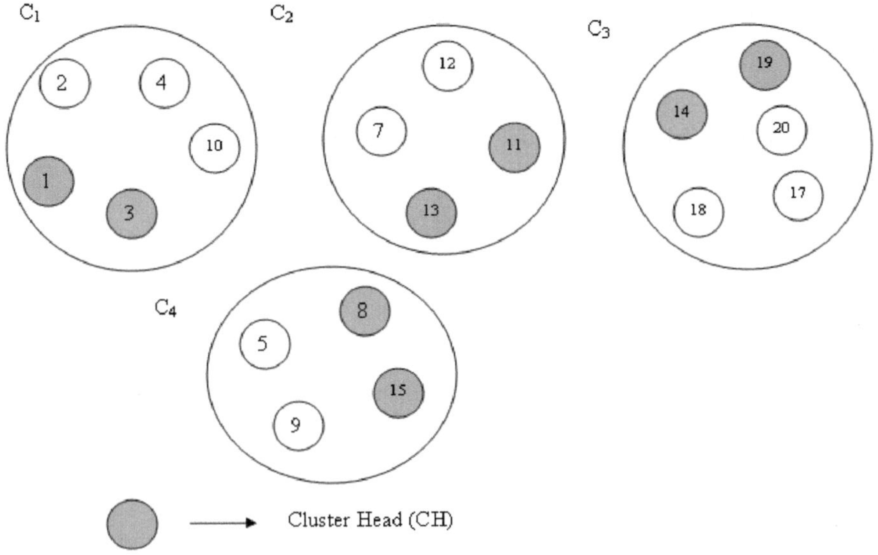

Fig. 2. Cluster Formation

Fig 2 shows the formation of clusters C_1, C_2, C_3 and C_4. The nodes N_1, N_{13}, N_{15}, and N_{19} which are near to NC_3, NC_{11}, NC_8, and NC_{14} are selected as CHs.

2.2 Phase 2 Network Connector Discovery Algorithm (NCDA)

Let *ID(NC)* represent the identity of NC.
Let *Sq(ID)* represents the sequence number of NC.
Let CM_i be the cluster members
Let CH_i be the cluster heads.

Consider that there are n clusters $\{C_1, C_2, C_3 \ldots., C_n\}$. Each NC should contain the routing table that includes the information about all the cluster members and their respective cluster heads. The routing table is constructed with the help of network connector discovery algorithm (NCDA). The steps involved in NCDA are as follows.

Step 1
Each NC_i sends network connector discovery (NCD) message to all other NC_i in the network through CH_i.

The format of NCD message is as follows.

Table 2. Format of NCD message

ID(NC)	Sq(NC)	CM_i	CH_i

Step 2
Any NC_i upon receiving NCD message builds its routing table with the information in NCD.

Step 3
The process involved in step 1 and 2 is iterated until NC_i receives NCD message from CH_n i.e. last CH.

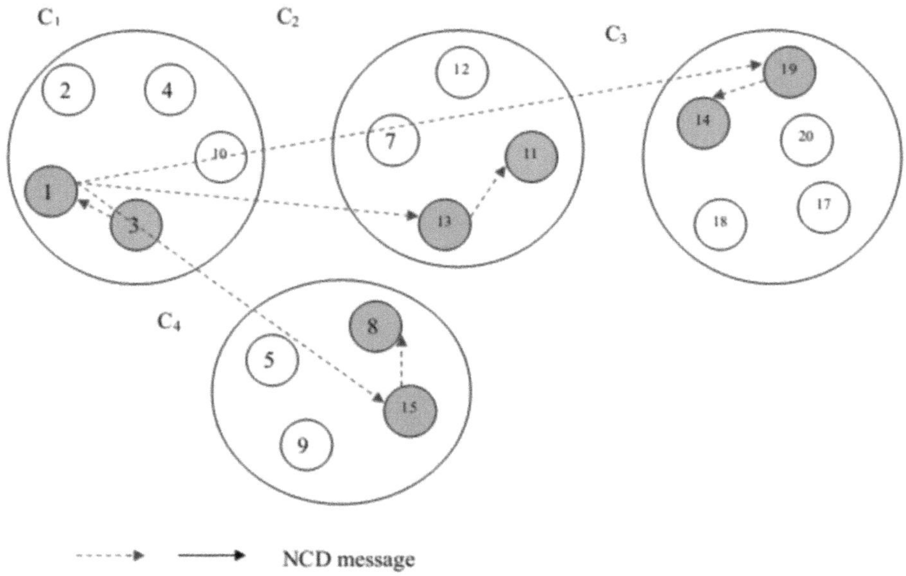

Fig. 3. Network Connector Discovery

Consider Fig 3. NC_3 transmits NCD message to its CH_1. CH_1 broadcast the NCD message to CH_{13}, CH_{15} and CH_{19}. These CHs transmits its NCD to its NC such as NC_{11}, NC_8 and NC_{14} respectively. All these network connectors build its routing table with information about the C_1 obtained from the NCD message. This process is iterated until NC_3 receives NCD message from last CH.

2.3 Phase-3 FSO/RF MANET Routing

Let S and D be source and destination node respectively.
Let ID_D be the identity of destination node
Let Ψ represent the set of NC_i in the hybrid network.
Let DNC be the destination NC.

When S wants to transmit the data to D in another network, S initially transmits the data along with the D's identity (ID_D) to its respective NC_i via its CH_i. This NC_i is considered as home NC (HNC)

$$S \xrightarrow{[ID_D][DATA]} CH_i \xrightarrow{[ID_D][DATA]} HNC$$

HNC verifies its local routing table for ID_D.

If HNC does not contain ID_D

Then

$$HNC \xrightarrow{[REQ(ID_D)]} \Psi$$

End if

If any NC_i within Ψ contains ID_D

Then

Particular NC_i becomes DNC

$$DNC \xrightarrow{[Info(D)]} HNC$$

$$HNC \xrightarrow{[data]} DNC \xrightarrow{[data]} D$$

End if

The NC first checks in the local routing table for the destination node id. If it is not found, broadcast a request to other NCs. The NC which contains the corresponding destination node, send the details to the home NC. The home NC then sends the data to the destination NC, which in turn sends it to the destination node by using its routing table information.

3 Conclusion

In this paper, a hybrid cluster based routing protocol for FSO/RF MANET proposed. The protocol consists of a network connector which connects the hybrid networks. The cluster formation phase, the node which is nearer to the network connector is elected as cluster head (CH) in each cluster. Each network connector gathers information of all cluster members and their CH and builds a routing table using network connector discovery algorithm. When the source node of a network wants to send data to destination node in another network, routing protocol is invoked through the network connectors. The proposed routing protocol is simulated in network simulator (NS-2) [15]. Simulation results have shown that the proposed routing protocol minimizes the delay and overhead, while enhancing the end-to-end throughput.

References

1. Sans Institute Infosec, Free-Space Optics: A Viable, Secure Last-Mile Solution?, http://www.sans.org/.../ free-space-optics-viable-secure-last-mil
2. Bilgi, M., Yuksel, M.: Multi-Element Free-Space-Optical Spherical Structures with Intermittent Connectivity Patterns. In: IEEE INFOCOM (2008)
3. Derenick, J., Thorne, C., Spletzer, J.: On the Deployment of a Hybrid Free-space Optic/Radio Frequency (FSO/RF) Mobile Ad-hoc Network. IEEE Intelligent Robot and Systems (2005)
4. Umar, A.: Ten Emerging Wireless Networks: UWB, FSO, MANET and Flash OFDM, http://www.amjadumar.com
5. Nakhkoob, B., Bilgi, M., Yuksel, M., Hella, M.: Multi Transceiver Optical Wireless Spherical Structures for MANETs. IEEE Journals on Selected Areas in Communications (2009)
6. Derenick, J., Thorne, C., Spletzer, J.: Hybrid Free-Space Optics/Radio Frequency (FSO/RF) Networks for Mobile Robot Teams. Multi Robot Systems (2005)
7. Cheng, B.-N., Yuksel, M., Shivkumar, Kalyanaraman: Using Directionality in Mobile Routing. In: IEEE International Conference on Mobile Adhoc and Sensor Systems (2008)
8. Okorafor, U.N., Kundur, D.: Efficient Routing Protocols for a Free Space Optical Sensor Network. In: IEEE International Conference on Mobile Adhoc and Sensor Network (2005)
9. Kashyap, A., Rawat, A., Shayman, M.: Integrated Backup Topology Control and Routing of Obscured Traffic in Hybrid RF/FSO Networks. In: IEEE Globe COM (2006)
10. http://www.wikipedia.org
11. Hu, Z., Verma, P., Sluss Jr., J.: Routing in Degree-constrained FSO Mesh Networks. International Journal of Hybrid Information Technology (2009)
12. Johnson, D.B., Maltz, D.A.: Dynamic source routing in ad hoc wireless networks. Mobile Computing 353 (1996)
13. Pearlman, M.R., Haas, Z.J., Samar, P.: The zone routing protocol (zrp) for ad hoc networks. Internet Draft - Mobile Ad Hoc NETworking (MANET) Working Group of the Internet Engineering Task Force, IETF (2001)
14. Ramasubramanian, V., Chandra, R., Mosse, D.: Providing a bidirectional abstraction for unidirectional ad hoc networks. In: IEEE INFOCOM, pp. 1258–1267 (2002)
15. Network Simulator, http://www.isi.edu/nsnam/ns

DCT Based Image Authentication with Partial Noise Reduction (DCTIAPNR)

Anirban Goswami[1], Dipankar Pal[2], and Nabin Ghoshal[3]

[1] Dept. of Information Technology, Techno India, EM 4/1 Salt Lake, Sec-V, Kolkata-700091
[2] Dept. of Computer Science and Engineering, Techno India, EM 4/1 Salt Lake, Sec-V, Kolkata-700091
[3] Dept. of Engineering and Technological Studies, University of Kalyani, Kalyani, Nadia-741235,West Bengal, India
an_gos@yahoo.com,
mail2dpal@yahoo.com,
nabin_ghoshal@yahoo.co.in

Abstract. The concept, proposed here, may prove to be an innovative derivation in the field of frequency domain steganography for grayscale images. Sub image blocks, each of size 2x2 in row major order are taken from the carrier image and are passed through the process of Discrete Cosine Transform (DCT) to obtain the corresponding frequency components. In the process of embedding a single bit of secret message/image is inserted into the real part of the frequency component of 2nd, 3rd & 4th carrier image bytes of each block. The level of data obscurity scales up due to the application of a self devised hash function. It generates pseudorandom positions which control embedding and extraction of secret bits in and from the frequency components. In addition an adjustment is made on each embedded frequency component to optimally reduce the noise effect due to embedding. The results after experimenting with the proposed technique prove statistically improved performance compared to other exiting techniques.

Keywords: Image Authentication**,** Steganography, DCT, IDCT, MSE, PSNR, IF, SSIM.

1 Introduction

Steganography [1, 8] is defined as a unique concept for communicating sensitive data between the sender and the receiver to elude one's discerning vision. Normally image, article or other text are treated as secret data. Generally visible encrypted messages trigger a query or suspicion for any human being, so it is obvious that while cryptography protects the message content, steganography distinctly safeguards both the message and the communicating parties.

Today as internet is flooded with a huge number of websites and online piracy appears to be not too difficult, copyright [4, 5] protection is a matter of great concern. Digital watermarking [2, 3] helps users to utilise content legally, with the addition of

S.C. Satapathy et al. (Eds.): Proc. of Int. Conf. on Front. of Intell. Comput., AISC 199, pp. 621–629.
DOI: 10.1007/978-3-642-35314-7_70

security, thus shielding the document from illegal hacking. In case of invisible watermarking which can be noted as a further refinement, only an authorised person can extract the watermark using some mathematical calculations. This may claim to provide for more secure and robust usage than visible watermarking. In general there is a reciprocation between the watermark embedding [6, 7] strength and its quality. Notching up of robustness, demands strong and increased embedding, culminating in a visual degradation and tampering with secret image/message. The proposed scheme DCTIAPNR emphasises on enhanced security by the creation of a hash function which generates a pseudorandom position for embedding and extraction of secret bits.

The flow diagram in Fig. 1 depicts the overall insertion and extraction processes. Two dimensional Discrete Cosine Transform and Inverse Discrete Cosine Transform have been utilised by the proposed technique and are represented in sub-section 1.1 and 1.2 respectively. Sec.2 illustrates the insertion and extraction algorithms of DCTIAPNR. The experimental results of the proposed technique have been presented in terms of SSIM, MSE, and PSNR in dB and IF in sec.3 followed by conclusion in sec.4.

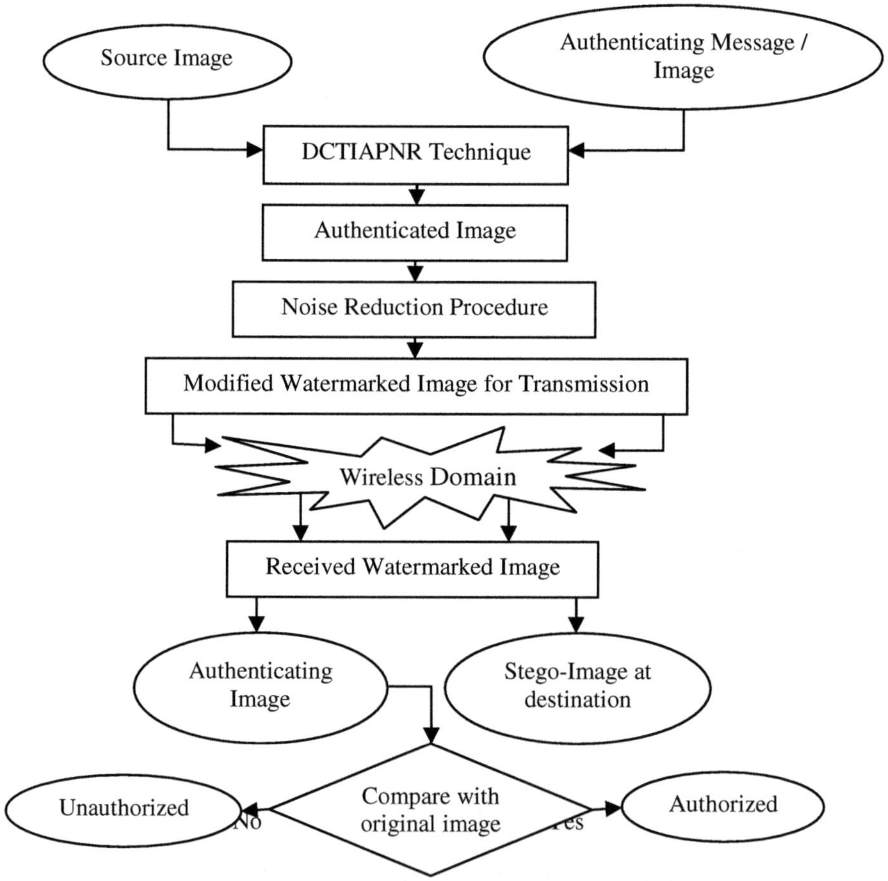

Fig. 1. The process of embedding and extraction of secret data using DCTIAPNR

1.1 Two Dimensional Discrete Cosine Transform

Two-dimensional DCT, implemented on M x N matrix is represented as follows:

$B_{pq}= \alpha_p\alpha_q\sum_{m=0}^{M-1}\sum_{n=0}^{N-1} A_{mn}\cos\frac{\pi(2m+1)p}{2M}\cos\frac{\pi(2n+1)q}{2N}$, where $0 \leq p \leq M-1$ and $0 \leq q \leq N-1$. The terms α_p and α_q are represented as,

$$\alpha_p = \begin{cases} \frac{1}{\sqrt{M}}, & p = 0 \\ \sqrt{2}/M, & 1 \leq p \leq M-1 \end{cases} \qquad \alpha_q = \begin{cases} \frac{1}{\sqrt{N}}, & q = 0 \\ \sqrt{2}/N, & 1 \leq q \leq N-1 \end{cases}$$

The values B_{pq} are called the DCT coefficients of spatial value A_{mn}. After DCT, the frequency components {W, X, Y, Z} for four spatial values {a, b, c, d} taken as a block of size 2x2 from the source image are represented as: DCT(a) = ½ (a + b + c + d) = W, DCT(b) = ½ (a – b + c – d) = X, DCT(c) = ½ (a + b – c – d) = Y, DCT(d) = ½ (a – b – c + d) = Z.

1.2 Two Dimensional Inverse Discrete Cosine Transform

IDCT is actually invertible DCT transform, and is shown as,

$A_{mn}=\sum_{p=0}^{M-1}\sum_{q=0}^{N-1} \alpha_p\alpha_q B_{pq}\cos\frac{\pi(2m+1)p}{2M}\cos\frac{\pi(2n+1)q}{2N}$, where $0 \leq m \leq M-1$ and $0 \leq n \leq N-1$. The terms α_p and α_q are represented as,

$$\alpha_p = \begin{cases} \frac{1}{\sqrt{M}}, & p = 0 \\ \sqrt{2}/M, & 1 \leq p \leq M-1 \end{cases} \qquad \alpha_q = \begin{cases} \frac{1}{\sqrt{N}}, & q = 0 \\ \sqrt{2}/N, & 1 \leq q \leq N-1 \end{cases}$$

The corresponding IDCT values are $DCT^{-1}(W)$ = ½ (W + X + Y + Z), $DCT^{-1}(X)$ = ½ (W – X + Y –Z), $DCT^{-1}(Y)$ = ½ (W + X –Y –Z), $DCT^{-1}(Z)$ = ½ (W – X – Y + Z).

2 The Technique

Discrete Cosine Transform is applied on each block (size 2x2) taken from the carrier image to obtain the frequency values of the pixels. Embedding of secret message bits are done at pseudorandom positions (0 - 3), defined by a variable ipos, generated by executing a self devised hash function. Insertion is performed in each block at 2nd, 3rd and 4th frequency coefficients only. Immediately after embedding, an adjustment is made only on the modified frequency components to reduce the noise effect due to embedding at positions other than LSB. A minor modification is also made only in the 1st component (DC coefficient) of each block to maintain the pixel values positive and non-fractional in spatial domain.

In case of extraction, the authenticating message/image is extracted from the frequency components of embedded image pixels. The frequency values are obtained after performing DCT operation on the current block of size 2 x 2, derived from the embedded image. The process of retrieval works by extracting one bit each from the

2nd, 3rd and 4^{th}pixel of each sub image block. The location of extraction of the bit is derived with the help of the same hash function.

The techniques of insertion and extraction are vividly explained below.

2.1 Insertion Algorithm

Input: A source image and authenticating message/image.
Output: An embedded image.
Method: Embedding of secret bits is performed only in the integer part of the frequency value. The fractional part is unaltered and is re-added with the embedded integer part. The algorithm is as follows:

1. Read the header information (image type, dimensions and maximum intensity) from source image and write the same into the output image.
2. Repeat the following steps until all pixels have been read from the source image as well as all the pixels/bits of authenticating image/message are embedded,
   - 2.1 Take a 2x2 block of pixels from the source image matrix in row major order.
   - 2.2 Apply Discrete Cosine Transform on the current block of pixels.
   - 2.3 Generate ipos (0 - 3) using the hash function.
   - 2.4 Read the authenticating message/image (i.e. secret data).
   - 2.5 Embed the authenticating bits (LSB to MSB order) in the integer part of each frequency component at the position earmarked by the variable ipos.
   - 2.6 Adjust frequency components (as applicable) to reduce the noise effect due to embedding.
   - 2.7 Apply Inverse Discrete Cosine Transform (IDCT) on current block of pixels to obtain the corresponding value in spatial domain.
   - 2.8 Apply readjustment procedure. (see below)
   - 2.9 Write the modified block into the output image in row major order.
3. Stop.

Readjustment procedure: After embedding operation, the following problems may occur when the frequency values are reverted back to spatial values using IDCT:

1. Due to any change in quantum value of the frequency components, corresponding negative spatial values may occur. This may be alleviated by incrementing gradually the value at the 1st position of the current block.
2. Fractional pixel values may be obtained due to the multiplication operation by½ in the expression of DCT (sec 1.1). This is eliminated by changing the value of the sum obtained after DCT operation to an even number.

2.2 Extraction Algorithm

Input: An embedded image.
Output: An authenticating message/image.

Method: Extraction process involves only the integer part of each frequency component while the floating point part remains untouched. The algorithm is as follows:

1. Read the image type and maximum intensity from the embedded image and write into the output image.
2. Repeat the following steps until each and every pixel have been read from the input image,
   - 2.1. Take 2x2 blocks of pixels from the embedded image matrix in row major order and perform DCT on the current block.
   - 2.2. Generate ipos (0 - 3) using the hash function.
   - 2.3. Extract the embedded bit from the integer part of the frequency values from the position as specified by ipos.
   - 2.4. A byte is constructed by combining 8 consecutive extracted bits, representing a gray level intensity value and is written in the output image.
   - 2.5. Repeat the steps from 2.1 to 2.4 until all pixels of the authenticating message/image have been retrieved from the received embedded image.
3. Stop.

2.3 The Hash Function and the Noise Reduction Mechanism

The value of ipos, used in embedding and extraction algorithms above, has been calculated by implementing a hash function which conceives two parameters, each 32 bit integer and produces a pseudorandom value between 0 and 3. The function is defined as, ipos = ((x AND 03H) AND ((y AND 0cH) SHR 2)), where x and y are dynamic parameters. After that the mean value M of the current block is computed and the generated value of ipos is further modified depending on the following condition: if (M < T) then ipos is either 0 or 1, else ipos is between 0 to 3, where T is a threshold value between 64 and 192. The resulting value is used as the position for embedding or extraction.

An additional approach undertaken for noise reduction is that of modifying the unaltered bits on the right (i.e. towards LSB) and/or left (i.e. towards MSB) of the target frequency component with respect to ipos. If the existing bit at ipos is changed (i.e. 1 becomes 0 or vice versa) then only the process is applied.

3 Result Comparison and Analysis

Now we discuss the results and make a comparative study of the proposed technique DCTIAPNR with other existing watermarking methods such as DCT, QFT and Spatio-Chromatic DFT-based. The parameters used in the comparative study are mainly Visual Interpretation, Mean Square Error (MSE), Peak Signal-to-Noise Ratio (PSNR), Image Fidelity (IF) and Structural Similarity Index Metric (SSIM). After applying the proposed technique on more than fifty PGM grayscale images it has been experienced that the algorithm can overcome any type of attack like visual or statistical. Experiments were done on systems with hardware and software configurations like, a processor with 2.00 GHz clock speed, primary memory capacity of 1 GB or higher, Unix/Linux OS and Gimp (GNU Image Manipulation Program) application installed on it. Some statistical and mathematical analysis are done, experimenting on a variety of carrier images, out of which 'tiffany', 'splash', 'kaya' and 'sailboat' are shown in fig 2a, 2b, 2c and 2d respectively. The dimension of each such carrier grayscale image is 512x512 and that of the authenticating grayscale image 'Earth' (Fig. 2q) is 150x150. The resultant embedded grayscale images are shown in Fig 2e, 2f, 2g and 2h respectively. 2i, 2j, 2k, and 2l are magnified versions of the source images and 2m, 2n, 2o, and 2p are magnified versions of the embedded images respectively. The visual interpretation is shown in fig. 2 and it can be observed that there is hardly any difference between the source and embedded images in terms of HVS.

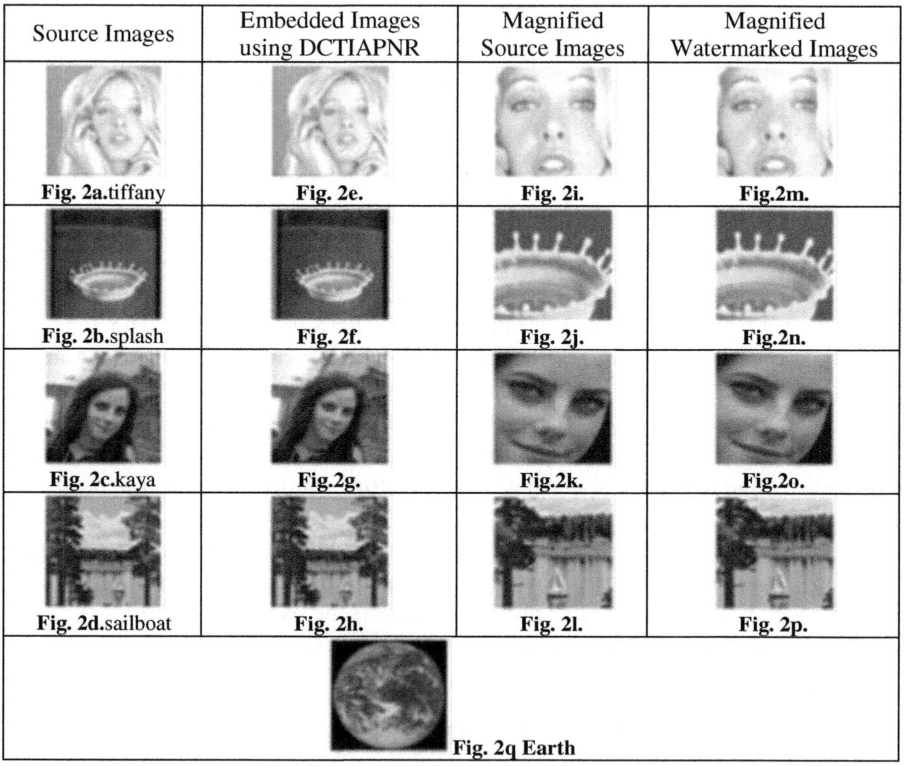

Fig. 2. Visual Interpretation of embedded images using DCTIAPNR

The quality of each steganographic image has been adjudged by applying the concept of PSNR and SSIM and their values computed on single level embedding of authenticating data byte (EL=0) have been shown in Table 1. The comparative study between DCTIAPNR, Reversible Data Hiding Based on Block Median Preservation (RDHBBMP) [12] and Image Authentication Technique Based on DCT (IATDCT)

[13] depicts enhancement in terms of hiding capacity of secret data and PSNR in dB in our proposed scheme as shown exclusively in Table 2. The average improvement of embedding secret data in the proposed scheme is 19857 bits more than RDHBBMP and 2904 bits more than IATDCT along with the increase of 1.86 dB and 0.66 dB respectively of PSNR in EL=0 which mean low rate of bit-error.

Table 3 shows better PSNR values than the existing techniques like DCT-based [9], QFT-based [10], and SCDFT-based [11] watermarking in frequency domain. In DCT based watermarking, the regions are selectively chosen that do not generate visible distortion for embedding, thus decreasing the authenticating data size. In QFT based watermarking compensation mark allows the watermark to be undetected even if the strength of it is high. For low compression factor it cannot completely recover the embedded message. Capacities of the existing techniques are 3840 bytes each and the PSNR values are 30.1024 dB, 30.9283 dB, and 30.4046 dB in SCDFT, QFT, and DCT, but the capacity of DCTIAPNR is 22500 bytes and PSNR is 47.320626 which is fully recoverable. 18660 bytes more of secret data embedding are possible in DCTIAPNR technique than the existing techniques with an average of 17 dB more PSNR value.

Table 1. Capacities and PSNR values of DCTIAPNR

Source Images	Capacity (Byte)	MSE	PSNR	IF	SSIM
Tiffany	22500	1.335735	46.873600	0.999646	0.998991
Splash	22500	1.278393	47.064159	0.999729	0.999781
Kaya	22500	1.324913	46.908932	0.999268	0.999869
Sailboat	22500	0.932178	48.435814	0.999768	0.999903
Average	**22500**	**1.217804**	**47.320626**	**0.999602**	**0.999636**

Table 2. Results and comparison in capacities and PSNR of RDHBBMP, IATDCT & DCTIAPNR

Test images	Indicator	EL=0	EL=0	EL=0
		RDHBBMP	**IATDCT**	**DCTIAPNR**
Lenna	C(bits)	26,465	54,896	57,800
	PSNR	49.68	51.16	52.16
Baboon	C(bits)	36,221	54,896	57,800
	PSNR	49.80	50.91	51.21

Table 3. Results and comparison in capacities and PSNR of DCTIAPNR and DCT, QFT, SCDFT [12]

Technique	Capacity (bytes)	PSNR in dB
SCDFT	3840	30.1024
QFT	3840	30.9283
DCT	3840	30.4046
DCTIAPNR	**22500**	**47.320626**

4 Conclusion

DCTIAPNR has been proposed for increasing the security of data hiding as compared to other existing algorithms. Authenticity is incorporated by embedding secret data in specific carrier image bytes at pseudorandom positions. As compared to RDHBBMP and IATDCT, the proposed algorithm is applicable for any type of grayscale image with increased security and authenticity. Also acute emphasis has been given on visible noise reduction to avoid any suspicion. The embedded image in this algorithm is very difficult to detect due to pseudorandom insertion position of the authenticating message/image bits in the carrier image. Hence, the proposed technique can defeat any possible hacking attempt.

Acknowledgement. The authors express their deep sense of gratitude to the faculty members of the Dept. of Engineering and Technological Studies, University of Kalyani, West Bengal, India, where the work has been carried out.

References

1. Radhakrishnan, R., Kharrazi, M., Menon, N.: Data Masking: A new approach for steganography. Journal of VLSI Signal Processing 41, 293–303 (2005)
2. EL-Emam, N.N.: Hiding a large Amount of data with High Security Using Steganography Algorithm. Journal of Computer Science 3(4), 223–232 (2007) ISSN 1549-3636
3. Amin, P., Lue, N., Subbalakshmi, K.: Statistically secure digital image data hiding. In: IEEE Multimedia Signal Processing, MMSP 2005, Shanghai, China, pp. 1–4 (October 2005)
4. Pavan, S., Gangadharpalli, S., Sridhar, V.: Multivariate entropy detector based hybrid image registration algorithm. In: IEEE Int. Conf. on Acoustics, Speech and Signal Processing, Philadelphia, Pennsylvania, USA, pp. 18–23 (March 2005)
5. Al-Hamami, A.H., Al-Ani, S.A.: A New Approach for Authentication Technique. Journal of computer Science 1(1), 103–106 (2005) ISSN 1549-3636
6. Ker, A.: Steganalysis of Embedding in Two Least-Significant Bits. IEEE Transaction on Information Forensics and Security 2(1), 46–54 (2008) ISSN 1556-6013
7. Yang, C., Liu, F., Luo, X., Liu, B.: Steganalysis Frameworks of Embedding in Multiple Least Significant Bits. IEEE Transaction on Information Forensics and Security 3(4), 662–672 (2008) ISSN 1556-6013
8. Wu, H.C., Wu, N.I., Tsai, C.S., Hwang, M.S.: Image steganographic scheme based on pisel-value differencing and LSB replacement methods. Proc. Inst. Elect. Eng., Vis. Images Signal Processing 152(5), 611–615 (2005)
9. Ahmidi, N., Safabkhsh, R.: A novel DCT-based approach for secure color image watermarking. In: Proc. Int. Conf. Information Technology: Coding and Computing, vol. 2, pp. 709–713 (April 2004)
10. Bas, P., Biham, N.L., Chassery, J.: Color watermarking using quaternion Fourier transformation. In: Proc. ICASSP, Hong Kong, China, pp. 521–524 (June 2003)

11. Tsui, T.T., Zhang, X.-P., Androutsos, D.: Color Image Watermarking Using Multidimensional Fourier Transfomation. IEEE Trans. on Info. Forensics and Security 3(1), 16–28 (2008)
12. Luo, H., Yu, F.-X., Chen, H., Huang, Z.-L., Li, H., Wang, P.-H.: Reversible data hiding based on block median preservation. Information Sciences 181, 308–328 (2011)
13. Ghosal, N., Goswami, A., Mondal, J.K., Pal, D.: Image Authentication Technique Based on DCT (IATDCT). In: Wyld, D.C., Zizka, J., Nagamalai, D. (eds.) Advances in Computer Science, Engg. & Appl. AISC, vol. 167, pp. 863–871. Springer, Heidelberg (2012)

Adaptive Steganography for Image Authentication Based on Chromatic Property (ASIACP)

Nabin Ghoshal[1], Anirban Goswami[2], and H.S. Lallie[3]

[1] Dept. of Engineering and Technological Studies, University Of Kalyani, Kalyani, Nadia-741235
[2] Dept of Information Technology, Techno India, EM 4/1 Salt lake, Sec-v, Kolkata-700091
[3] International Digital Laboratory (WMG), University of Warwick, Coventry, CV4 7AL
nabin_ghoshal@yahoo.co.in,
an_gos@yahoo.com,
h.s.lallie@warwick.ac.uk

Abstract. This paper presents a new adaptive data hiding method in colour images using complement value (CV) of higher order three bits (b7, b6 and b5) of each colour image byte to achieve large embedding capacity and imperceptible stego-images. The technique exploits the complement value (CV) of each colour image byte to estimate the number of bits to be embedded into the image byte. Image bytes located in the edge areas are embedded by k-bit LSB substitution technique with a large value of k in deep colored area than that of the image bytes located in the light colored areas. The range of complement values is adaptively divided into lower level and higher level respectively. An image byte is embedded by the k-bit LSB substitution technique. The value of k is adaptive and is decided by the complement value. In order to keep the fidelity of the embedded image at the same level of the source image, a re-adjustment phase termed as handle is implemented. The experimental results obtained are compared with the existing studies of Wu et al's and of Yang et al.'s LSB replacement method based on pixel-value differencing (PVD) in gray images. It proves that the proposed algorithm is capable to hide more volume of data while retaining better image quality.

Keywords: Steganography, Complement Value (CV), colour image.

1 Introduction

Digital images are transmitted over popular communication channels such as the Internet. For secured communication, image authentication techniques have gained more attention due to their importance for a large number of multimedia applications. A number of approaches have been proposed, which includes conventional cryptography, fragile and semi-fragile watermarking and digital signatures. Digital watermarking is the process of hiding [1] the watermark imperceptibly in the content. Steganography [3, 5] has become an important technique for proving image authentication and identification. Data hiding primarily refers to a digital watermark which is a piece of information hidden in a multimedia content, in such a way that it

S.C. Satapathy et al. (Eds.): Proc. of Int. Conf. on Front. of Intell. Comput., AISC 199, pp. 631–638.
DOI: 10.1007/978-3-642-35314-7_71

is imperceptible to a human observer, but easily detected by a computer. The principal advantage is that the watermark is inseparable from the content. Ownership verification [2, 4] and authentication [6, 7] is the major task for military people, research institutes, and scientists.

Information security and image authentication has become very important in order to protect digital image documents from unauthorized access. In steganographic applications, the hidden data may be a secret message, hologram or video. Data hiding represents a useful alternative to constructing a hypermedia document or image, which is more difficult to manipulate.

Prior research into this area has focused on:

1. Identifying bits that can be used for Steganography**:** Nameer N. EL-Emam [7] used entropy based technique for detecting the suitable areas in the image where data can be embedded with minimum distortion.
2. Methods of implementing the Steganography**:** Ker [8] and C. Yang [9] presented a general structural steganalysis framework for embedding in two or more LSB's. H. C. Wu [10] and Cheng-Hsing Yang [11] constructed a method of LSB replacement into the edge areas using pixel value differencing (PVD) where PVD was used to distinguish between the edge and smooth areas.

The aim of this paper is to present an algorithm that would facilitate color image authentication using a data hiding procedure which embeds the data adaptively by considering the concept of human vision, with features of high capacity and low distortion. The Adaptive Steganography for Image Authentication based on Chromatic Property (ASIACP) method emphasises on:

1. Information and image protection against unauthorized access
2. Inserting large amounts of messages/image data into the source image for image identification
3. Transmitting secure messages within the image

In our technique the tolerance levels of deep colored edge areas, light colored edge areas and smooth colored areas are incorporated and it can embed large volumes of secret data whilst maintaining the high quality of stego-images. All image bytes are embedded through the LSB substitution method with different numbers of secret bits, the number of secret bits is decided by the complement value (CV) of the three higher order bits of each image byte. The embedding is not a direct LSB substitution, but embeds at even and odd positions in each image byte alternatively among the lower parts of image bytes to give added protection against attack. In order to increase the quality of stego-images, and to ensure proper decoding, a handle is proposed to re-adjust the pixel values.

The rest of this paper is structured as follows. Section 2 of the paper discusses the proposed technique. Results, comparison and analysis are given in section 3 and conclusion is drawn in section 4.

2 The Technique

The proposed embedding scheme utilizes Adaptive LSB substitution based on the technique of CV (complement value) of three higher order bits ($b_7b_6b_5$) of each image byte. The complement value determines that edge areas may embed a larger number of authenticating bits than the smooth areas. For any image pixel, each image byte is embedded using *k*-bits LSB substitution, where the value of *k* is decided by the CV of each image byte.

Usually, the deep colored edge areas can tolerate more changes than the smooth areas and light colored edge areas. The value of *k* will be large in deep colored area than in the light colored area because the intensity value of a deep colored pixel is less than that of a light colored pixel. In the case of deep colored pixels, higher order bits are mostly zeroes, so, the CV of the higher three bits tend towards 7, high value of k.

In ASIACP the value of *k* is divided into two levels namely lower and higher levels. A lower level means tolerable level (allowed to change) and is defined as region $R_l$with l-h values and the higher level means protected levels (not allowed to change) and is defined as region R_2 with l-h values. The division of ranges are, range $R_l = [0, 4]$ and range $R_2 = [5, 7]$. The lower level consists of image bytes with the CVs *k* falling into the region R_l; these will be embedded by *k* bits (which as we have already pointed out are calculated using the LSB substitution technique). The higher levels consists of image bytes with the CVs *k* falling into R_2 and set the value of $k = 4$ i.e. 4 bits of authenticating message/image can be embedded in each byte of source image. As a result the higher order bits remain unchanged.

Here the range of l-h values means that the CV of an image byte falling in R_l or in R_2, which indicates that the entire embedding process will be done by l-bits to h-bits authenticating data i.e. any one cover image byte can be authenticated by any number of secret bits which belong to the least value l and highest value h i.e. $l \leq k \leq h$.

2.1 The Fidelity Handle

In order to improve the fidelity of the stego-images, a handle has been applied to the embedded image byte. The purpose of this handle is to reduce the square error between the original pixel and embedded pixel by increasing or decreasing the most-significant-bit (MSB) by 1. The MSB i.e. unaltered part is increased or decreased by 1 if the embedded image byte is decreased or increased by an amount of 2^k. The handle is applicable only to those stego-image bytes where after applying the handle the bits $b_7b_6b_5$ remain the same. In order to extract authenticating data properly, the CVs before and after embedding must be the same and should belong to the same region. The embedding and extraction algorithms of our approach are described in subsections 2.1.1 and 2.1.2 respectively.

2.1.1 Procedure for Embedding

Source images are represented as 24 bit RGB colour components. The proposed ASIACP technique embeds the authenticating message/image $AI_{p,q}$ along with the size of authenticating message/image (16 bits) for the purpose of authenticating the source image $SI_{m,n}$ of size *m* x *n* bytes.

The first step in ASIACP is to read an image byte in row major order and calculate the CV of higher order three bits. Authenticating message/image bits are inserted between 1st to 4th positions starting from LSB of the byte. To enhance the security of authentication process the proposed ASIACP uses an even and odd position embedding strategy alternatively for consecutive image bytes in such a way that, if the value of k is 1 then secret bit will be embedded at LSB of an image byte and if the values of k = 2, 3 or 4, the secrets bits are embedded in 2nd, 3rd and 4th bit positions starting from LSB respectively. The embedding will start from even or odd position alternatively and during this process if adequate positions are not available to replace k bits in even or odd positions of LSB part, unaltered bit positions from LSB are used. The detailed embedding steps are as follows.

1. Obtain the size of the authenticating message/image (16 bits representation).
2. For each source image byte (Pi) do
   - 2.1 Calculate the complement value CV_i for upper three bits of image bytes i.e. CV of ($b_7b_6b_5$)), where CV_i=complement of MSB part of P_i.
   - 2.2 Find the region where CV_i belongs: If CV_i belongs to R_1(i.e. has a value of 0 to 4) $k = CV_i$. If CV_i belongs to R_2(i.e. has a value of 5 to 6) $k = 4$
   - 2.3 Calculate K and embed k bits
     - 2.3.1 Calculate K which is the decimal value of original k bits from LSB
     - 2.3.2 Embed k bits secret bits into P_i by k-bit LSB substitution. The decimal value of k secret bits from LSB is K_1
     - 2.3.3 EP_i = the embedded image byte of P_i.
   - 2.4 Execute handle on EP_i by calculating the revised complement value CV_i' (complement of $b_7b_6b_5$). The handle is used as follows.

$$EP_i = \begin{cases} (EP_i + 2^k), & \text{If } CV_i = CV_i' \text{ and if } -(2^k-1) \le (K_1 - K) \le -2^{k-1} \\ (EP_i - 2^k), & \text{If } CV_i = CV_i' \text{ and if } (2^k-1) \ge (K_1 - K) \ge 2^{k-1} \end{cases}$$

   If either of the above relations is not satisfied the handle is not to be executed on EP_i.
3. Repeat step 2 for the whole authenticating message/image content and along with the size of the authenticating data.
4. Stop.

2.1.2 Procedure for Extraction

During decoding process, the embedded image has been taken as the input data and the authenticating message/image and the size of secret data are extracted from the embedded image. The process of extracting the embedded message/image is same as the embedding process with the same traversing order of image bytes. The detailed steps for extraction are as follows.

1. Read embedded source image byte in row major order
2. For each embedded image byte do

2.1 Calculate the CV on embedded image byte ECV_i for upper three bits of each image byte (i.e. CV of ($b_7b_6b_5$)), say EP_i, using ECV_i= complement of MSB part of EP_i.
2.2 Find the region that ECV_i belong to. Let k = ECV_i, if ECV_i belongs to R_1 otherwise $k = 4$ (i.e. high value of R_1) if CV_i belongs to R2.
2.3 Extract the k-bits of authenticating message/image from embedded image byte in even and odd positions alternatively, in the same manner as the secret data was embedded.
2.4 Replace extracted bit positions in each embedded image byte with '1'.
2.5 For each 8 (eight) bits of extracted data, construct one alphabet/one primary (R/G/B) colour image byte. Generate pixel of authenticating data using the composition of R, G and B.

3 Repeat steps 1 and 2 to complete decoding as per the size of the authenticating message/image.
4 Stop.

3 Experimental Results, Comparison and Analysis

In this section we present some experiments to demonstrate the performance of proposed adaptive data embedding approach. The comparative study has been made on several images using the proposed ASIACP technique. Fifteen cover colour images with size 512 x 512 are used in the ASIACP for experiment and two of them are shown Fig. 1(b) Baboon and 1(c) Peppers. Here the colour image 'Earth' shown in Fig. 1(a) is used to authenticate carrier source images. Authentication is done in various ways such as 0-3 bits, 0-4 bits, 2-3 bits, and 2-4 bits embedding process. Here 0-3 bits embedding process means the embedding capacity of each source image is from 0 bit to 3 bits depends on CVs. Similarly 0-4 bits, 2-3 bits, and 2-4 bits embedding process means the embedding capacity of each source image byte are from 0 bit to 4 bits, from 2 bits to 3 bits and, from 2 bits to 4 bits respectively depends on CVs. The mechanism is defined as *l-h* (*l*-least number of bits, *h*- highest number of bits) process, where k number of bits may be inserted in each image byte and the range of k is defined as $l \leq k \leq h$.

Experimental results of stego-images with 0-3 bits and 0-4 bits embedding are shown in Fig. 2(a, b) and Fig.2(c, d) respectively. In other instances stego-images with 2-3 bits and 2-4 bits embedding are shown in Fig. 3(a, b) and Fig. 3(c, d) respectively. Fig. 4(a, b) and 4(c) show the extracted authorized and authenticating images at destination. To measure visual quality of the authenticated images with respect to source images we use peak signal-to-noise ratio (PSNR). Different experimental results using various *l-h* ranges are given in Table 1. Table 1 show that a noticeable amount of secret data embedding is done with higher PSNR values using ASIACP. In 0-3 and 2-3 bits embedding process, the average embedding capacities are 234220 and 254415 bytes with high PSNR values 42.83 and 42.34 respectively. But in 0-4 and 2-4 bits embedding process, the average PSNR values are slightly decreased (though it is higher than the existing methods) for large amount of data embedding. Here the amount of average hidden data is 274921 and 298116 bytes with PSNR values 38.33 and 38.06 respectively.

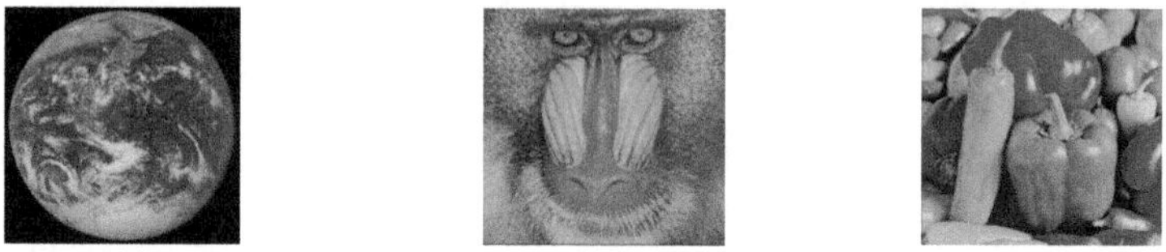

(a) (b) (c)

Fig. 1. Authenticating image (a) Earth, and cover images (b) Baboon. (c) Peppers

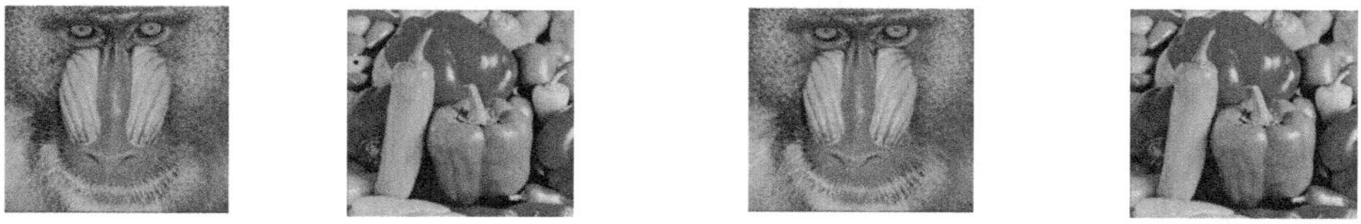

(a) (b) (c) (d)

Fig. 2. Stego-images generated by ASIACP with 0-3 bits embedding (a), Baboon (b), Peppers and with 0-4 bits embedding (c) Baboon (d) Peppers

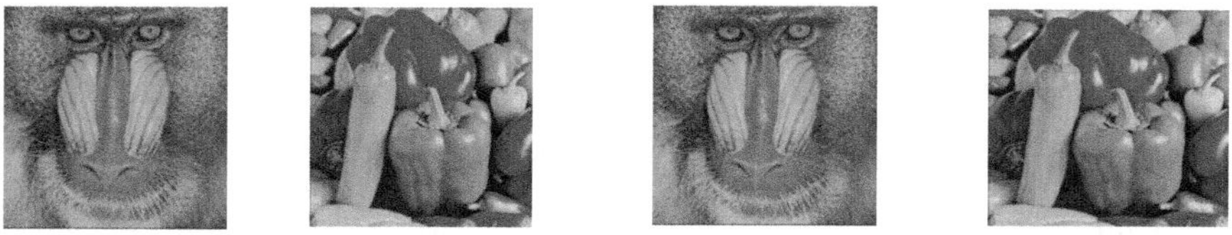

(a) (b) (c) (d)

Fig. 3. Stego-images generated by ASIACP with 2-3 bits embedding (a) Baboon (b) Peppers and with 2-4 bits embedding (c) Baboon (d) Peppers

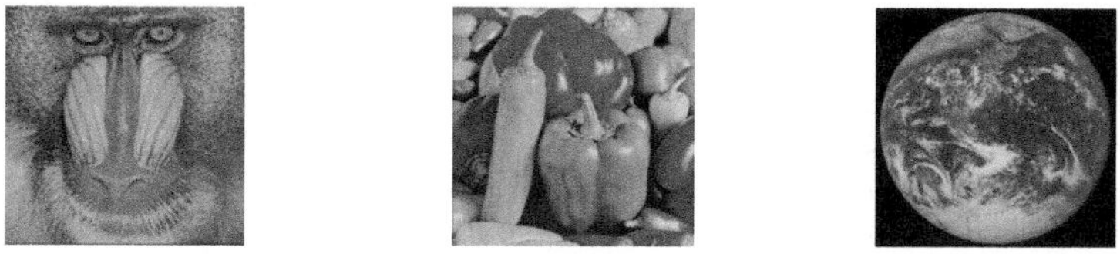

(a) (b) (c)

Fig. 4. Carrier images and one extracted authenticating image after extraction at receiver end by ASIACP (a) Baboon (b) Peppers and (c) Earth

Table 2 shows the comparisons between ASIACP, Wu et al.'s and Yang et al.'s method where ASIACP uses a 0-4 bits and 2-4 bits embedding and Yang et al.'s used 3-4 bits embedding. Though the change of pixel intensity in colour image is more sensitive than the change of pixel in gray image, the ASIACP embeds more or a similar amount of secret data than the existing methods but with higher PSNR values.

Another important difference between this proposal and the research of Wu et al. and Yang et al. is that their methods were developed and experimented on gray images, where as ASIACP method is implemented and experimented on colour images. From Table 2, it is clear that the capacity of secret data embedding in ASIACP is more and PSNR is also high, which yields a better image fidelity. Using 0-4 bits and 2-4 bits insertion, we can obtain an increase in insertion of 2,429 bytes and 31,326 bytes respectively using ASIACP than Wu et al.'s method. The PSNR value increases by 2.69 dB and 2.52 dB on average than the Wu et al.'s method. The amount of data embedded in ASIACP using 0-4 bits insertion is similar to Yangs et al.'s method, but the PSNR value scales to by 1.70 dB and in 2-4 bits insertion, ASIACP embeds 1,200 more bytes and the PSNR value increases by 1.52 dB.

Table 1. Experimental Results of Capacity and PSNRs in dB using different l-h values

Cover Colour Images	0-3		0-4		2-3		2-4	
	Capacity (Bytes)	PSNR (dB)	Capacity (Bytes)	PSNR (dB)	Capacity (Bytes)	PSNR (dB)	Capacity (Bytes)	PSNR (dB)
Peppers	248348	41.551084	305099	36.094550	264245	41.465189	320996	36.064536
Lena	237815	42.026793	292561	36.558530	264222	41.797940	318967	36.455448
Baboon	244976	41.825842	297563	36.654319	265498	41.706988	318085	36.491314
Sailboat	230536	42.309488	281505	37.010748	258008	42.064236	318978	36.915914
Tiffany	225583	45.521403	246988	44.531878	236915	43.397080	258320	42.827864
Splash	229604	41.911328	292563	36.267049	264717	41.613059	327676	36.142381
Airplane	233869	45.486718	257883	41.458166	248056	44.329064	292070	40.993031
Woodland	231855	42.390342	263252	37.916675	251626	42.259575	283025	37.905372
San Diego	207693	43.272028	230255	39.154955	237677	42.857563	260239	39.085132
Oakland	251924	42.017711	281534	37.675779	253183	41.889990	282793	37.694783
Average	234220	42.831274	274921	38.332265	254415	42.338068	298116	38.057578

Table 2. Comparisons of results of ASIACP with WU et al.'s and Yang et al.'s methods

Cover Colour Images	Wu et al.'s results		Yang et al.'s results (3-4)		0-4 in ASIACP		2-4 in ASIACP	
	Capacity (bytes)	PSNR (dB)	Capacity (bytes)	PSNR (dB)	Capacity (bytes)	PSNR (dB)	Capacity (bytes)	PSNR (dB)
Peppers	96281	35.34	102923	37.17	305099	36.094550	320996	36.064536
Lena	95755	36.16	104667	36.28	292561	36.558530	318967	36.455448
Baboon	89731	32.63	114501	33.01	297563	36.654319	318085	36.491314
Sailboat	94596	33.62	100865	36.43	281505	37.010748	318978	36.915914
Airplane	97790	36.60	101229	36.39	257883	41.458166	292070	40.993031
Average	94831	34.87	104873	35.86	286922	37.555263	315819	37.384049

4 Conclusions

The proposed technique is a novel attempt to implement image authentication in spatial domain using adaptive data hiding method to embed secret data into RGB colour images without making a perceptible distortion. Image bytes located in deep coloured edge areas are embedded by k-bits LSB substitution method in even and odd positions in each image byte alternatively in successive bytes with a large value of k than that of image bytes located in smooth areas. The CV approach is used to distinguish between deep coloured areas and smooth colour areas. ASIACP may hide huge amount of data in the form of text message/image. The proposed algorithm shows better results than Wu et al.'s method and Yang et al.'s method, which may produce better authenticated images. The l-h bits insertion may yields higher capacity and higher PSNR.

References

1. Amin, P., Lue, N., Subbalakshmi, K.: Statistically secure digital image data hiding. In: IEEE Multimedia Signal Processing, MMSP 2005, Shanghai, China, pp. 1–4 (October 2005)
2. Al-Hamami, A.H., Al-Ani, S.A.: A New Approach for Authentication Technique. Journal of computer Science 1(1), 103–106 (2005) ISSN 1549-3636
3. Ghoshal, N., Mandal, J.K.: A Novel Technique for Image Authentication in Frequency Domain using Discrete Fourier Transformation Technique (IAFDDFTT). Malaysian Journal of Computer Science 21(1), 24–32 (2008) ISSN 0127-9094
4. Ghoshal, N., Sarkar, A., Chakraborty, D., Ghosh, S., Mandal, J.K.: Masking based Data Hiding and Image Authentication Technique (MDHIAT). In: Proceedings of 16th International Conference of IEEE on Advanced Computing and Communications, ADCOM 2008, December 14-17, pp. 119–122. Anna University, Chennai (2008) ISBN: 978-1-4244-2962-2, `http://ieeexplore.ieee.org/xpls/abs_all.jsp?arnumber=4760437&tag=1`
5. Radhakrishnan, R., Kharrazi, M., Menon, N.: Data Masking: A new approach for steganography. Journal of VLSI Signal Processing 41, 293–303 (2005)
6. Pavan, S., Gangadharpalli, S., Sridhar, V.: Multivariate entropy detector based hybrid image registration algorithm. In: IEEE Int. Conf. on Acoustics, Speech and Signal Processing, Philadelphia, Pennsylvania, USA, pp. 18–23 (March 2005)
7. EL-Emam, N.N.: Hiding a large Amount of data with High Security Using Steganography Algorithm. Journal of Computer Science 3(4), 223–232 (2007) ISSN 1549-3636
8. Ker, A.: Steganalysis of Embedding in Two Least-Significant Bits. IEEE Transaction on Information Forensics and Security 2(1), 46–54 (2008) ISSN 1556-6013
9. Yang, C., Liu, F., Luo, X., Liu, B.: Steganalysis Frameworks of Embedding in Multiple Least Significant Bits. IEEE Transaction on Information Forensics and Security 3(4), 662–672 (2008) ISSN 1556-6013
10. Wu, H.C., Wu, N.I., Tsai, C.S., Hwang, M.S.: Image steganographic scheme based on pixel-value differencing and LSB replacement methods. Proc. Inst. Elect. Eng., Vis. Images Signal Process. 152(5), 611–615 (2005)
11. Yang, C.H., Weng, C.Y., Wang, S.J., Sun, H.M.: Adaptive Data Hiding in edge areas of Images With Spatial LSB Domain Systems. IEEE Transaction on Information Forensics and Security 3(3), 488–497 (2008) ISSN 1556-6013
12. Weber, A.G.: Theuscsipi image database. Signal and Image Processing Institute at the University of Southern California (October 1997), `http://sipi.usc.edu/services/database/Database.html`

Performance Analysis of JPEG Algorithm on the Basis of Quantization Tables

Mumtaz Ahmad Khan[1], Qamar Alam[2], and Mohd Sadiq[3]

[1] Section of Electrical Engineering, University Polytechnic, Faculty of Engineering and Technology, Jamia Millia Islamia, A Central University, New Delhi-110025, India
makhan2@jmi.ac.in

[2] Department of Computer Science, Institute of Management Studies, Roorkee, U.K., India
alamqamar786@yahoo.com

[3] Department of Computer Engineering
National Institute of Technology, Kurukshetra-136119, Haryana, India
sadiq.jmi@gmail.com,
msq_delhi@yahoo.co.in

Abstract. Image compression techniques are used to reduce the storage and transmission costs. Joint Photographic Experts Group (JPEG) is one of the most popular compression standards in the field of still image compression. In JPEG technique, an input image is decomposed using the DCT, quantized using quantization matrix and further compressed by using entropy encoding. Therefore, the objective of this paper is to carry out the performance analysis of JPEG on the basis of quantization tables. In this paper, we have employed the nelson algorithm for the generation of quantization table; and an attempt has been made for the identification of the best quantization table, because for digital image processing (DIP), it is necessary to discover a new quantization tables to achieve better image quality than the obtained by the JPEG standard. Research has shown that quantization tables used during JPEG compression can also be used to separate images that have been processed by software from those that have not been processed; and it is also used to remove JPEG artefacts or for JPEG recompression.

Keywords: JPEG, DCT, Quantization tables.

1 Introduction

JPEG stands for Joint Photographic Experts Group. It was found in 1987 by a joint effort of the Photographic Experts Group (PEG: a branch of ISO) and CCITT to produce an image compression standard for transmitting graphics image data for digital communication networks [1]. It issued the first JPEG standard in 1992, which was approved in September 1992 as ITU-T Recommendation T.81 [9] and in 1994 as ISO/IEC 10918-1. Compression of digital signal can be implemented either in software or in hardware. JPEG standard is one of the most popular and comprehensive still frame compression standards.

S.C. Satapathy et al. (Eds.): Proc. of Int. Conf. on Front. of Intell. Comput., AISC 199, pp. 639–647.
DOI: 10.1007/978-3-642-35314-7_72

Compression algorithms are basically of two types: (a) lossless, and (b) lossy. Lossless algorithms recover the original picture perfectly. Lossy algorithms are those which are able to recover a similar kind of original picture where in the changes are not visible to the human eye. The lossy techniques tend to give better compression ratio. Therefore, they are more often applied to image and video compression than lossless techniques. JPEG is basically a lossy method of compression commonly used for digital photography. Table-1 represents the systematic difference between lossless compression techniques and lossy compression techniques.

Table 1. Difference between lossless and lossy compression techniques

S. No.	Lossless Compression Techniques	Lossy Compression Techniques
1	Original Image can be perfectly recovered from the compressed image.	Decompressed image is not exactly identical to the original image.
2	Quality of image is important rather than compression ratio.	Compression ratio is high.
3	Examples: Run length encoding, Huffman Encoding, Lempel-Ziv-Welch Coding and Area Coding.	Vector Quantization, Fractal Coding, Block Truncation Coding, Sub-band Coding, Transformation Coding.

JPEG is commonly used on the World Wide Web for the storage and transmission of images. It works best on photographs and paintings of natural scenery where there is gradual change in the tone and color. JPEG is not suitable for line drawings where there is sudden change in the intensity values. JPEG handles only still images, but there is a related standard called MPEG for motion pictures. Lossless JPEG also exists created by the same ISO standard is a completely different method that really is lossless. However, it doesn't compress nearly as well as baseline JPEG; it typically can compress full-color data by around 2:1[2]. It does not provide useful compression of palette-color images or low-bit-depth images.

Lossless JPEG has never been popular. In fact, no common applications support it and it is now largely obsolete. Since the baseline JPEG is a lossy mode of compression and it cannot not be used for scientific or medical images where exact reconstruction is required. It basically defines the codec to compress an image into stream of bytes and again decompress it back to image format. It discards useless data during the encoding process. It works on the principle of eliminating image information which is not visible to the human eye. It has been observed that slight changes in color are not noticed by the human eye but small minute changes in the intensity are well noticed. Therefore, the JPEG algorithm is stricter towards intensity and less careful towards color. The end user can easily tune in the quality of the JPEG encoder as per requirement. It can be adjusted to produce very high compression ratios with very low picture quality but still suitable for many applications. On the other hand it can also produce very high picture quality with very low compression rates. The best known lossless compression methods can compress data about 2:1 on average. JPEG can typically achieve 10:1 to 20:1 compression without visible loss, bringing the effective storage requirement down to 1 to 2 bits/pixel. 30:1 to 50:1 compression is possible with small to moderate defects, while for very-low-quality purposes such as previews or archive indexes, 100:1 compression is quite feasible [3,13].

The paper is organized as follows: Section 2 presents the steps that we have employed in JPEG. Performance analysis of JPEG is given in section 3. Finally, we conclude the paper in section 4.

2 Steps Involved in JPEG

This section presents the steps which are involved in JPEG. In literature [6, 7, 8, 10, 11, 14], we have identified JPEG and JPEG 2000 algorithm. The basic differences between these two algorithms are: (i) In JPEG the source image is divided into 8X8 blocks; and each block is transformed by using Discrete Cosine transform (DCT). JPEG 2000 is based on Discrete Wavelet Transform (DWT), which is applied on image tiles. DWT tiles are decomposed into different decomposition levels. Block diagram of JPEG encoder is given in figure-1

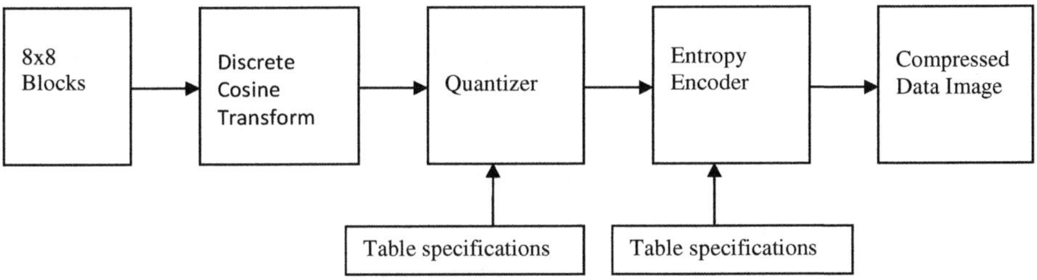

Fig. 1. Block Diagram of a Sequential JPEG Encoder (adopted from [12, 13])

2.1 Discrete Cosine Transform

Image samples are grouped into blocks of 8x8. They are further shifted from the range of unsigned integers [0, 2^P-1] to the range of signed integers [-2^{P-1}, 2^{P-1}-1] and are send for DCT transformation. Let the horizontal index is u and the vertical index is v then the DCT coefficient at coordinates (u, v) can be written as

$$G_{u,v} = \sum_{x=0}^{7}\sum_{y=0}^{7} \alpha(u)\alpha(v) g_{x,y} \cos\left[\frac{\pi}{8}\left(x+\frac{1}{2}\right)u\right]\cos\left[\frac{\pi}{8}\left(y+\frac{1}{2}\right)v\right] \quad (1)$$

Where (i) u is the horizontal spatial frequency, for the integers 0<= u<8 and (ii) v is the vertical spatial frequency, for the integers 0<= v <8 .

$$\alpha(u) = \begin{cases} \sqrt{\frac{1}{8}}, & \text{if } u = 0 \\ \sqrt{\frac{2}{8}}, & \text{otherwise} \end{cases}$$

- $g_{x,y}$ is the pixel value at coordinates (x,y)
- $G_{u,v}$ is the DCT coefficient at coordinates(u, v)

The DCT is Fourier related transform but uses only real numbers. DCT is mainly used for compression of images as it has the property of energy compaction and constraints the signal information to be concentrated in a few low-frequency components of the DCT. Each sample of the 8x8 block consists of 64 point discrete signals which are a function of two dimensions x and y. The output of the DCT consists of 64 unique two-dimensional (2D) "spatial frequencies''. The leftmost coefficient is the DC coefficient and the remaining 63 are known as AC coefficients.

2.2 Quantization

DCT is a lossless transformation which doesn't compress the image. The purpose of quantization is to achieve compression by discarding the less significant information. Quantization is a many to one mapping [3] and therefore introduces losses. Quantization error [4] is the fundamental reason for making the compression lossy. For performing quantization a standard quantization matrix has been provided by JPEG [5] as shown in table 2.

Table 2. Standard Quantization Matrix

16	11	10	16	24	40	51	61
12	12	14	19	26	58	60	55
14	13	16	24	40	57	69	56
14	17	22	29	51	87	80	62
18	22	37	56	68	109	103	77
24	35	55	64	81	104	113	92
49	64	78	87	10	121	120	101
72	92	95	98	11	100	103	99

Quantization is performed by dividing each DCT coefficient by its corresponding value in the quantization table. The result is further rounded off to its nearest integer. The function to implement it is given below:

$$F\,[u, v] = \text{round}\,(F\,[u, v] \,/\, Q\,[u, v]) \tag{2}$$

Here Q [u, v] is the quantum value taken from the quantization matrix.

2.3 Zigzag Scan

Instead of compressing the coefficients in the horizontal or vertical direction, JPEG algorithm moves diagonally in a zigzag sequence [6] as shown in Figure 2 .The quantized values are reordered and placed in a zigzag pattern; and thereafter, it is mapped into a 1 x 64 vector. This ordering helps in entropy encoding because it places the low (more likely to be non-zero) frequency coefficients prior to the high (more likely to be zero) frequency coefficients.

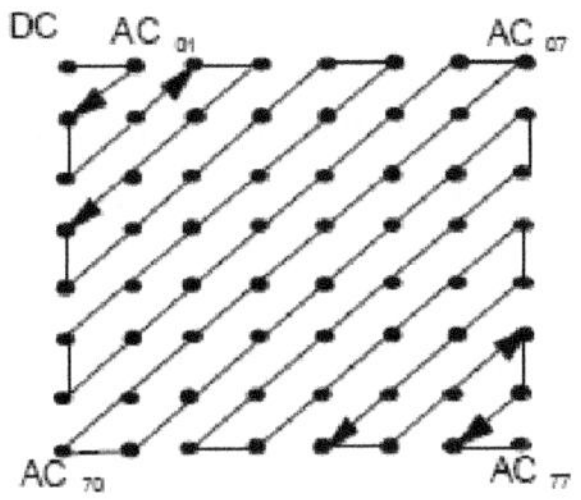

Fig. 2.

2.4 Entropy Encoding

The final step achieves additional compression by encoding the DCT coefficients more compactly on their statistical behaviour. Two entropy encoding methods have been proposed by the JPEG standard – Huffman encoding [7] and arithmetic encoding [8]. Arithmetic encoding tends to give better compression rates. But the baseline codec sticks to Huffman encoding as it is faster to implement and is not covered by patents which require heavy royalty licenses. For Huffman encoding it is required that tables be specified by the application for both compression and decompression. The same table can be used on both the ends.

3 Performance Analysis

This section presents the performance analysis of JPEG on the basis of quantization tables. The objective of the quantization table is to discard the information which is not visually significant. In literature [3, 6, 13], we have identified various methods for the generation of quantization tables like: nelson algorithm, existing standard JPEG quantization tables and so on. Quantization tables has great impact during the compression, for example, quantization tables are used to identify the origin of digital images and it is the only methods for conducting the digital ballistics. The idea of using quantization tables for digital ballistics was first proposed by Farid [18]. Bauschke H Z [17] explain that JPEG image compression method is employed in a large number of intensive applications; and in some of these applications, requnatization is required when the amount of compression needed is unknown in advance. As we know that, everybody would like to work with original images when requnatizing, since JPEG compression is lossy. Therefore, requnatizing an already quantized image can lead to seemingly unpredictable behaviour and unwanted artifacts. Fan Z. [15] work is based on the identification of Bitmap Compression History. There work is based on whether the image has ever been compressed using the JPEG standard. On the basis of the literature [14, 15, 16, 17, 18], we can say that, how much important it is, to analyse the quantization table. Therefore, during the compression, it is important to identify the best quantization table. An attempt has been made, to identify the best quantization table. In this paper, we have generated the quantization table using nelson algorithm, and results are given in the following table i.e., table 3. The basic measure for the performance of a compression algorithm is Compression Ratio and it can be defined as:

$$Cr = \textit{Original Image Size} \,/\, \textit{Encoded Picture Size} \tag{3}$$

One measure for the quality of the picture, proposed by Wallace [3], is the number of bits per pixel in the compressed image (N_b) which is defined as the total number of bits in the compressed image divided by the total number of pixels.

$$N_b = \textit{Encoded number of Bits/ Number of Pixels} \tag{4}$$

The DCT-based encoders generally follow the following levels of picture quality for the calculated N_b [3].

- 0.25-0.5 bits/pixel: moderate to good quality, sufficient for some applications;
- 0.5-0.75 bits/pixel: good to very good quality, sufficient for many applications;
- 0.75-1/5 bits/pixel: excellent quality, sufficient for most applications;
- 1.5-2.0 bits/pixel: usually indistinguishable from the original, sufficient for the most demanding

For the performance analysis of the JPEG algorithm, we have employed the image of cameraman.tif; and we have applied the JPEG algorithm and Nelsons algorithm. We have summarized the results of our analysis in table-4. It contains the information about the size of the image , size after compression , compression ratio and the number of bits per pixel (N_b) for different values of quality parameters like QP= 2, 5, 10, 15, 20, 25.

On the basis of the results of the above table, we can say that, as we are increasing the value of the QP, i.e., from 2 to 25, we are achieving the good compression ratio. Results of the Nelson algorithm indicates that quantization table generated from the QP=25, is better than QP=20 and so on. So from this way, we can identify the best quantization table.

Table 3. Generated values of quantization matrix on the basis of different values of QP

Value of Quality Parameter (QP)	Generated values of the quantization matrix							
	7	9	11	13	15	17	19	21
	9	11	13	15	17	19	21	23
	11	13	15	17	19	21	23	25
QP=2	13	15	17	19	21	23	25	27
	15	17	19	21	23	25	27	29
	17	19	21	23	25	27	29	31
	19	21	23	25	27	29	31	33
	21	23	25	27	29	31	33	35
	16	21	26	31	36	41	46	51
	21	26	31	36	41	46	51	56
	26	31	36	41	46	51	56	61
QP=5	31	36	41	46	51	56	61	66
	36	41	46	51	56	61	66	71
	41	46	51	56	61	66	71	76
	46	51	56	61	66	71	76	81
	51	56	61	66	71	76	81	86
	31	41	51	61	71	81	91	101
	41	51	61	71	81	91	101	111
	51	61	71	81	91	101	111	121
QP=10	61	71	81	91	101	111	121	131
	71	81	91	101	111	121	131	141
	81	91	101	111	121	131	141	151
	91	101	111	121	131	141	151	161
	101	111	121	131	141	151	161	171
	46	61	76	91	106	121	136	151
	61	76	91	106	121	136	151	166
	76	91	106	121	136	151	166	181
QP= 15	91	106	121	136	151	166	181	196
	106	121	136	151	166	181	196	211
	121	136	151	166	181	196	211	226
	136	151	166	181	196	211	126	241
	151	166	181	196	211	226	241	256
	61	81	101	121	141	161	181	201
	81	101	121	141	141	161	181	201
	101	121	141	161	181	201	221	241
QP=20	121	141	161	181	201	221	241	261
	141	161	181	201	221	241	261	281
	161	181	201	221	241	261	281	301
	181	201	221	241	261	281	301	321
	201	221	241	261	281	31	321	341
	76	101	126	151	176	201	226	251
	101	126	151	176	201	226	251	276
	126	151	176	201	226	251	276	301
QP=25	151	176	201	226	251	276	301	326
	176	201	226	251	276	301	326	351
	201	226	251	276	301	326	351	376
	226	251	276	301	326	351	376	401
	51	276	301	326	351	376	401	426

Fig. 3. Cameraman.tif

Table 4. Results of baseline and nelson algorithm

QP	Results of baseline JPEG algorithm				Results of nelson algorithm			
	Size of the Image	**Size after compression**	**Compression ratio**	**Nb**	**Size of the Image**	**Size after compression**	**Compression ratio**	**Nb**
2	65536	5194	12.61	0.6340	65536	11320	5.7894	1.3818
5	65536	2924	22.4131	0.3569	65536	6474	10.1230	0.7903
10	65536	1772	36.9842	0.2163	65536	4034	16.2459	0.4924
15	65536	1368	47.9064	0.1670	65536	3046	21.5154	0.3718
20	65536	1140	57.4877	0.1392	65536	2488	26.3408	0.3037
25	65536	998	65.6673	0.1218	65536	2096	31.2672	0.25529

4 Conclusion(s)

This paper presents the performance analysis of JPEG algorithm on the basis of quantization tables. Quantization tables play an important role during image compression. There are some cases, in which it is important to identify the bitmap compression history using JPEG detection and Quantizer estimation; and to identify the best quantization table. In this paper, we have employed the nelson algorithm for the generation of quantization table; and as result we have explain a method for the identification of the best quantization table. For digital image processing (DIP), it is necessary to discover a new quantization tables to achieve better image quality than the obtained by the JPEG standard. Research has shown [14, 15, 16, 17] that quantization tables used during JPEG compression can also be used to separate images that have been processed by software from those that have not been processed. Therefore, the identification of best quantization table can greatly reduce the number of images an examiner must consider during an investigation. Identification of Quantization table can also be used to remove JPEG artefacts or for JPEG recompression.

References

[1] Standardization of Group 3 Facsimile apparatus for document transmission. CCITT Recommendations, Fascicle VII.2, Recommendation T.4 (1980)

[2] http://www.fileformat.info/mirror/egff/ch09_06.html

[3] Wallace, G.K.: The JPEG still picture compression standard. IEEE Transactions on Consumer Electronics 38(1), 18–34 (1992)

[4] Gersho, A., Gray, R.M.: Vector Quantization and Signal Compression. Kluwer Academic Publishers, Boston (1992)

[5] Encoding Parameters of Digital Television for Studios. CCIR Recommendations, Recommendation 601 (1982)

[6] Nelson, M., Gailly, J.-L.: The Data Compression Book. M&T Books, New York (1996)

[7] Hudson, G.P., Yasuda, H., Sebestyén, I.: The international Standardization of a Still Picture Compression Technique. In: Proceedings of the IEEE Global Telecommunications Conference, pp. 1016–1021. IEEE Communications Society (1988)

[8] Pennebaker, W.B., et al.: Arithmetic Coding. IBM J. Res. Dev. 32(6), 717–774 (1988)

[9] T.81: Information Technology - Digital Compression and Coding of Continuous-Tone Still Images - Requirements and Guidelines, http://www.itu.int/rec/T-REC-T.81

[10] Dewan, M.A.A., Islam, R., Sharif, M.A., Islam, M.A.: An Approach to Improve JPEG for Lossy Still Image Compression. Computer Science & Engineering Discipline, Khulna University, Khulna 9208, Bangladesh

[11] Sadiq, M.: Implementation of VQ Techniques, M.Tech Project, Department of Computer Engineering, Aligarh Muslim University (AMU), Aligarh, U.P., India (December 2004)

[12] Sadiq, M.: Study of the JPEG Algorithm and its Implementation using DCT Based Encoder, M.Tech Dissertation, Department of Computer Engineering, Aligarh Muslim University (AMU), Aligarh, U.P., India (June 2005)

[13] Ansari, F.J., Sadiq, M., Ali, A.: A Comparison of the Baseline and Nelsons Algorithm for the JPEG Image Compression Encoder. In: INDIACom 2012, Delhi, India (2012)

[14] Richardo, L.: Processing JPEG-Compressed Images and Documents. IEEE Transactions on Image Processing 7(12), 1661–1672 (1998)

[15] Fan, Z., Richardo, L.: Identification of Bitmap Compression History: JPEG Detection and Quantizer Estimation. IEEE Transactions on Image Processing 12(2), 230–235 (2003)

[16] Kornblum, J.D.: Using JPEG Quantization Tables to Identify Imagery Processed by Software. Digital Investigation, S21–S25 (2008)

[17] Bauschke, H.Z., et al.: Recompression of JPEG Images by Requnatization. IEEE Transactions on Image Processing 12(7), 843–849 (2003)

[18] Hany, F.: Digital image ballistics from JPEG quantization. Technical Report TR 2006-583, Department of Computer Science, Dartmouth College (2006)

Zone Centroid Distance and Standard Deviation Based Feature Matrix for Odia Handwritten Character Recognition

Debananda Padhi[1] and Debabrata Senapati[2]

[1] Department of MCA,
Purushottam Institute of Engineering & Technology,
Rourkela, Odisha, India
debananda.padhi106@gmail.com
[2] Department of CA,
ITER, SOA University, Bhubaneswar, Odisha, India
debabratasenapati@gmail.com

Abstract. Optical character recognition (OCR) is a type of document image analysis where scanned digital image that contains either machine printed or handwritten script input into an OCR software engine and translating it into an editable machine readable digital text format. In this paper we designed a novel and robust two stage recognition system for Odia handwritten characters as well as we prepare a standard deviation and zone centroid average distance based feature matrix for more accuracy while training and testing the Neural Network. The OHCR System is based on the algorithm of feed forward BPNN in two stage to perform the optimum feature extraction and recognition. The Odia characters are classified into four groups according to similarity of their shapes and features. The system uses ANN in two stages, having different parameters, the first stage classifies the characters into similar groups and in the second stage individual characters are recognized.

Keywords: Zone, ANN, centroid, Character Recognition, Morphological analysis, Standard deviation.

1 Introduction

The biggest challenge in the field of image processing is to recognize documents both in printed and handwritten format. Optical Character Recognition (OCR) is a type of document image analysis where scanned digital image that contains either machine printed or handwritten script input into an OCR software engine and translating it into an editable machine readable digital text format. Development of OCRs for Indian script is an active area of research today. We are making an attempt to develop the Hand written Character recognition system for Odia language, which is the official language of Odisha[8]. Odia language present great challenges to an OCR designer due to the large number of letters in the alphabet, the sophisticated ways in which they combine, and the complicated graphemes they result in.

S.C. Satapathy et al. (Eds.): Proc. of Int. Conf. on Front. of Intell. Comput., AISC 199, pp. 649–658.
DOI: 10.1007/978-3-642-35314-7_73

There are lots of application areas where, OCR can help. Major areas are described below:

1. Preserve old documents in electronics format.
2. Save document images within limited space.
3. Help visually impaired persons to read the content on the document.

In a broad sense Document Image Processing consists of 4 steps namely

1. Preprocessing
2. Feature Extraction and Classification
3. Recognition
4. Post Processing

1.1 About Odia Script

Odia is one of the scheduled languages of India. It is the principal language of communication in the state of Odisha, spoken by over 23 million people comprising 84% of population (1991 Census). It is the official language of the state. Odia belongs to the Eastern group of Indo-Aryan language family and has evolved around 10th century AD. It is the southernmost Indo-Aryan language placed at the boundary of Dravidian family of languages along with some Munda group of languages belonging to Austro-Asiatic family of languages[8]. The modern Odia script consists of simple and complex characters. There are 12 vowels, 3 vowel modifiers, 37 simple consonants, 10 numerical digits and about 159 composite characters (juktas) in Odia alphabets. One of the major characteristics of Odia elementary characters is that most of their upper one third is circular and a subset of them has a vertical straight line at their rightmost part. The conjuncts have quite complex shapes. The matras are comparatively small in size[8]. In writing a text document all elementary characters and some matras fall along a base line. Different matras take relative positions with consonant characters like before or after them, or upper or lower to the base line, and sometimes at the upper-right or lower-right corners. The matras sometimes get touched with common characters, and more than one modifier combined forming a composite modifier.

In this paper we present a generic approach to Character Recognition System, based on the two stage classification and recognition system using Back Propagated Neural Network. In 1st stage of classification the zone centroid average distance and average angle based feature matrix is used and in 2nd stage i.e., coarse level of classification the standard deviation values, zone centroid average distance & average angle based feature value matrix is prepared for actual classification of the Odia character set. The structure of this paper is as follows: in Section 2 about proposed OHCR system is discussed, in section 3 algorithm for feature matrix of two stage classification & recognition System is discussed, in section 4 ANN design of two stage classification for training and testing, & results showing percentage of recognition is given.

2 Proposed Handwritten Character Recognition System

The proposed handwritten Odia Character Recognition system is based on the algorithm of Feed forward Back Propagated Neural Network for two stage classification and recognition(figure 2). The handwritten Odia characters are classified according to similarity of their shapes and features from the data set collected from different person's handwritten text. Then the feature matrix for classification is prepared and given as input to the ANN and the target values are prepared according to it. We prepared 12 feature values for first level classification and 45 feature values for coarse level classification of each individual character.

Fig. 1. Odia character sets with some modifiers and numerical digits

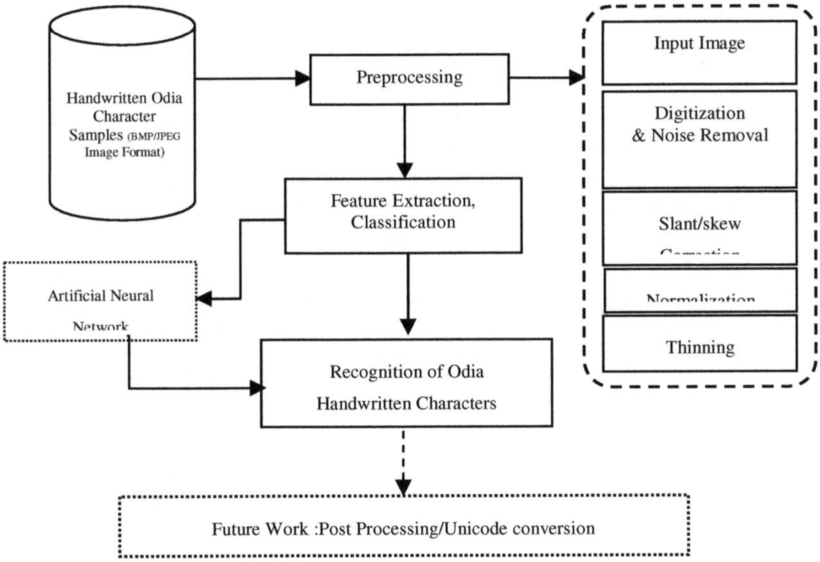

Fig. 2. Block Diagram of Proposed ANN Based Odia Handwritten Character Recognition System

3 Feature Matrix for Two Stage Classification

The Odia characters are classified according to its appearance as follows:

Class1 : having vertical bar on right most part of the character
Class2 : Having upper big circle and a small diagonal line(signature tail) at right lower part of the character.
Class3 : Having upper half big circle and recursive circle at left most part of the character.
Class4 : Having recursive circle at left and right most part of the character.
For the above classification we divide the cropped character into 6 zones i.e., 3 zone row wise and 3 zones column wise, from which the feature vector is prepared for the first stage classification. For the second stage classification the cropped image is divided into 9 equal zones and the feature matrix is prepared taking consideration of each zone.

3.1 Algorithm for First Level Classification

Input : the cropped image
Output: 12 feature values of the character.
Step 1: check whether the cropped image is binary or not if not then binarize it.
Step 2: convert the image into skeleton form (thinning)
Step 3: calculate the Centroid value of the image.
Step 4 : 1st divide the image into 3 equal part row wise.
Step 5 : 2nd divide the image into 3 equal part column wise.
Step 6: compute the distance between the pixel value present in each zone row wise of the character and the centroid of the image.
Step 7: repeat step 6 for all the pixel of the character present in the zone.
Step 8 : find the average distance of the zone. (1 feature)
Step 9 : repeat step 6-8 for each zone row wise and column wise (total 6 feature)
Step 10 : compute angle between image centroid and pixel present in the zone.
Step 11 : compute average angle of each zone .(total 6 feature)
finally we got 6 average distance and 6 average angle from the image centroid to the different zone which are actually the different part of the character.

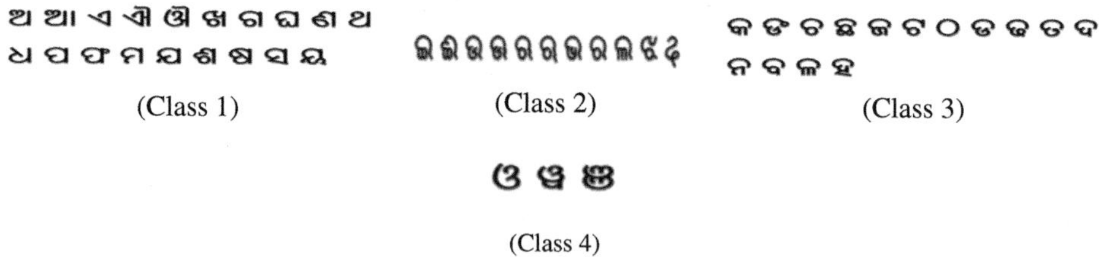

Fig. 3. Odia character sets showing different classes

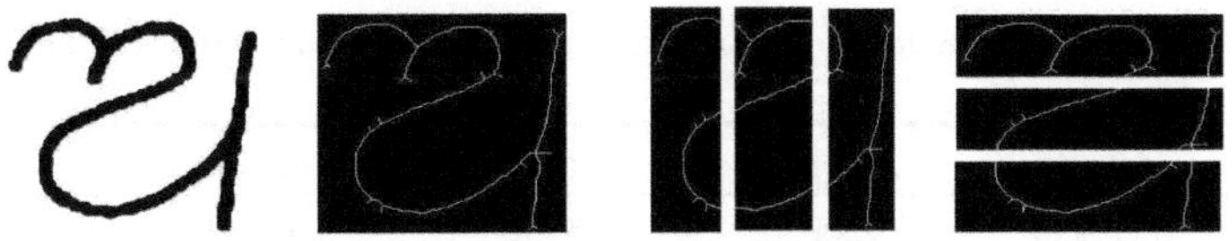

Fig. 4. Handwritten character (@), its thinned form and row, column zones

If coordinates of two point is X = (x_1 y_1) and Y = (x_2 y_2) then Distance(D) between X and Y is calculated using the function as:

$$D = sqrt((x_1-x_2)^2+(y_1-y_2)^2))$$

And the angle (A) between two points X and Y is calculated as:

$$A = atan2(y_2-y_1, x_2-x_1)$$

3.2 Algorithm for Second Level (Coarse) Classification

Input : Classified cropped character (class1,2,3,or 4)
Output :Detail feature matrix of the character for recognition
Step 1: input the image and check it for binarization, if not then binarize it.
Step 2: Skeletonize the image to get the thinned image or single pixel lined character for process.
Step3 : Compute the centroid of the thinned image.
Step 4: Divide the image into 9 equal zones.
Step 5: Compute the distance between image centroid and pixel value of thinned character in the zone.
Step 6: Repeat step 5 for all the pixels in the zone.
Step 7: Repeat step 5 and step 6 for all the 9 zones.
Step 8: Compute the average distance of all the 9 zones. (9 features)
Step 9: Compute the standard deviation of all 9 zones (9 features)
Step 10: As in step 5 compute the angle between image centroid and pixel of the zone.
Step 11: Repeat step 10 for all the pixels in the zone.
Step 12: Repeat step 10 and step 11 for 9 zones.
Step 13 : find the average angle of all the 9 zones (9 features).
Step 14: compute the centroid value of a zone.
Step 15: compute the average distance from zone centroid to pixels present in the same zone.
Step 16 : repeat step 15 for all 9 zones (9 features);
Step 17 : Compute the average angle from zone centroid to pixels present in the same zone.
Step 18: repeat step 17 for all 9 zones (9 features).

So finally we got 9 standard deviation values, 9 average distance values from image centroid, 9 average angle values from image centroid, 9 average distance values from zone centroid,9 average angle values from zone centroid. Total of 9x5=45 feature values for a single character.

Table 1. Feature values of 4 super classes, taking one example of each class

	Zones	Class1 (ଘ)	Class2(ଲ)	Class3(ଢ)	Class4(ଓ)
Average Distance	Zone1	0.4000	0.3336	0.4656	0.7564
	Zone2	0.4348	1.3980	0.7949	1.9039
	Zone3	2.4412	0.9413	0.5105	0.4638
	Zone4	0.2365	0.2576	0.3673	0.4690
	Zone5	1.2034	1.2333	1.4431	1.4431
	Zone6	0.3821	0.7515	0.4521	10.6878
Average Angle	Zone1	0.0194	0.0177	0.0175	0.0198
	Zone2	0.0137	0.0174	0.0135	0.0176
	Zone3	0.0395	0.3336	0.0132	0.0153
	Zone4	0.0395	0.0333	0.0132	0.0153
	Zone5	-0.0040	0.0056	-0.0061	-0.0084
	Zone6	0.0227	0.0226	0.0242	0.0242

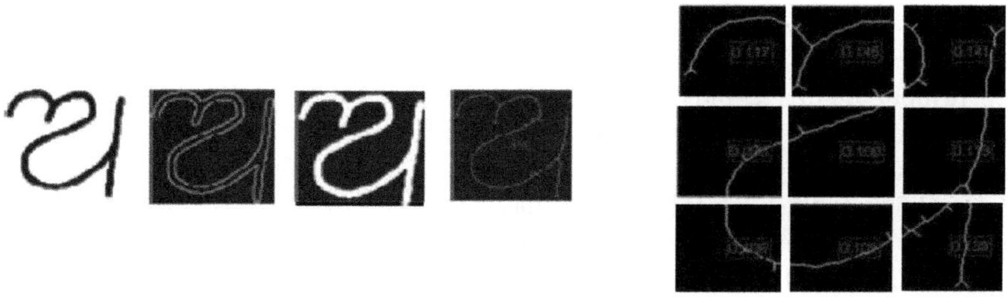

Fig. 5. The Character (A) and its 9 equal zones showing standard deviation values

Table 2. 45 Feature values of single character (A) of class 1 type

Image zones	Standard Deviation	Average distance from image centroid	Average distance from zone centroid	Average angle from image centroid	Average angle from zone centroid
Zone 1	0.1173	0.4666	1.1268	0.0235	0.0108
Zone 2	0.0977	0.5862	1.4144	0.0373	0.0147
Zone 3	0.1060	0.5159	0.9396	-0.0303	0.0110
Zone 4	0.1454	0.3511	0.8457	0.0174	0.0081
Zone 5	0.1001	0.6511	1.7985	0.0263	0.0140
Zone 6	0.1088	0.4691	1.1279	0.0302	0.0118
Zone 7	0.1409	0.4748	1.4830	-0.0104	0.0053
Zone 8	0.1126	0.4303	1.3352	0.0217	0.0107
Zone 9	0.1387	0.5448	2.3215	0.0232	0.0142

From the feature values we are calculated the four target values of four classes for the first level ANN according to the average values maximum-minimum range as for the class1 type character, only the right most column zone values is required, for class2 type character lower row zone as well as upper row zone values, for class3 type only upper row zone values and for class4 type both left and right column zone are required so we are not taking consideration for the middle zone. After getting the target values for training the ANN is designed as given in figure 7.

4 ANN Design for Two Stage Classification

In our proposed multistage recognition scheme we designed recognition process of two stage, in first stage it classifies the Odia character set into 4 super classes using the 1st phase of feature set as described in section 2. It is classified as shown in figure 3, the classes are divided into four groups and class 1 having 20 sub classes, class 2 having 10 subclass , class 3 having 15 subclass and class 4 have only 3 subclass as total of 48 characters.

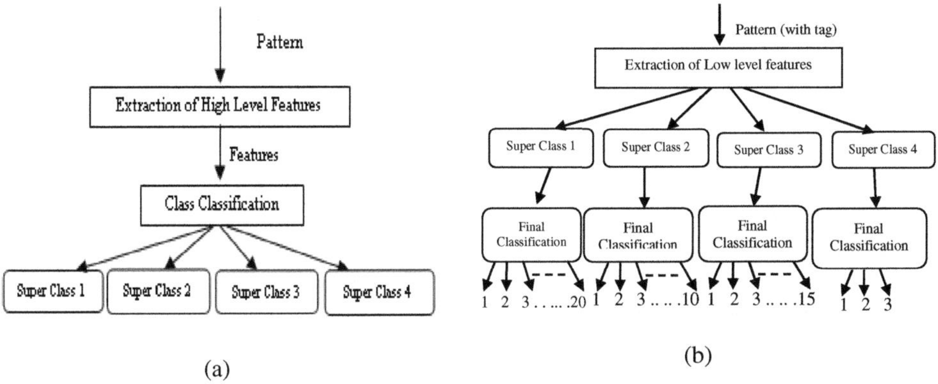

Fig. 6. Multistage recognition scheme (a) Stage I. (b) Stage II

Stage I: In this stage, the experts are trained individually on the training samples from the chosen classes individually.

Stage II: In this stage, the already trained experts are exposed individually to further training samples from the chosen classes. In stage one the classes of the Odia character are decide according to its shape and the network is trained as having input neurons 12 and targets are 4 values as shown in figure 7 (a) (b).

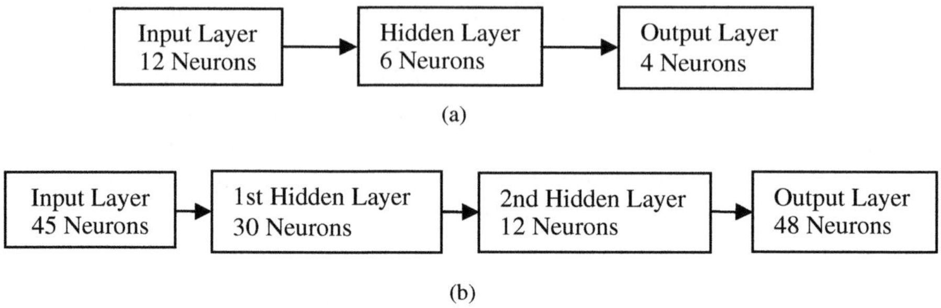

Fig. 7. ANN design for stage I (a) & II (b) training & testing

Table 3. Main confusion pairs of Odia characters

Confusing Character pairs		% of confusion overall
D	E	0.59 %
B	C	0.23 %
F	G	0.19 %
M	N	0.27 %
Z	H	0.25 %
T	V	0.11 %
`	$	0.45 %
i	I	0.32 %

Table 4. Classification rate for 4 classes based on stage I of ANN

Class	Class1(20 characters)	Class2 (10 Characters)	Class 3(15 Characters)	Class 4(3 characters)
Class 1	**90.2**	0.0	4.32	0.0
Class 2	0.0	**92.4**	3.45	0.0
Class 3	3.06	0.0	**89.7**	2.12
Class 4	0.51	0.0	0.0	**95.36**

Table 5. Recognition Rate based on 2nd Phase NN Training of class 2 having 10 characters

Characters	B	C	D	E	F	G	b	e	M	T
B	**88.2**	0.34	0.0	0.0	0.1	0.1	0.0	0.0	0.3	0.0
C	0.2	**84.3**	0.0	0.0	0.0	0.0	0.0	0.0	0.2	0.0
D	0.01	0.0	**91.2**	0.12	0.0	0.0	0.0	0.0	0.0	0.0
E	0.0	0.0	0.1	**92.1**	0.0	0.0	0.0	0.0	0.0	0.0
F	0.0	0.0	0.0	0.0	**89.3**	0.72	0.0	0.12	0.0	0.0
F	0.0	0.0	0.0	0.0	0.98	**87.3**	0.0	0.15	0.0	0.0
b	0.0	0.0	0.0	0.0	0.03	0.01	**82.1**	0.0	0.0	0.0
e	0.0	0.0	0.0	0.0	0.21	0.31	0.0	**94.2**	0.0	0.0
m	0.11	0.0	0.0	0.0	0.0	0.0	0.0	0.0	**89.9**	0.0
T	0.0	0.0	0.0	0.0	0.0	0.0	0.0	0.12	0.0	**93.3**

Table 6. Recognition Rate based on 2nd Phase Neural network Training of class 4 having 3 characters

Characters	J	g	U
J	**89.9**	0.63	0.32
g	0.65	**90.3**	0.42
U	0.05	0.02	**91.3**

There are several pairs of characters having similarity in shapes in the same class for which ANN may get confused while recognizing it. The main confusion pairs of Odia characters are shown in table 3. The overall classification rate in stage I neural network training is given in table 4 and due to lack of space it is not possible to give all the relevant tables of showing recognition rate of coarse level stage II training and testing.

5 Conclusion and Future Work

Various methods are implemented for recognition of Odia printed documents character recognition, segmentation, skew/slant corrections etc, but still now for handwritten Odia characters very few steps has been taken, and few number of papers are published particularly for Odia handwritten characters. We are presented a new method of Odia handwritten character recognition using a unique and robust combination of artificial neural networks in two stage classification and recognition. On programming and testing the modules a very high efficiency has been noted. More work has to be done on Juktakhyara (compound characters) feature extraction as we are done only on simple characters. We are also proposed to optimize the recognition rate using genetic algorithm for better accuracy.

Acknowledgements. To complete this work, we have got valuable suggestions and guidance from different experts of this field. The Book Digital Image Processing (Third Edition) by Rafael C. Gonzalez and Richard E. Woods is very helpful to start research in this field. Also the book "Data Mining, Concepts and Techniques" (2nd edition), by Han & Kamber is very useful for Neural Network and data mining approaches. We also thankful to all the authors of the reference section, it is their research work from which we got the valuable idea for this work.

References

1. Rejean, P., Srihari Sargur, N.: On-line and Off-line Handwriting Recognition: A comprehensive survey. IEEE Transactions on Pattern Analysis and Machine Intelligence 22(1), 63–84 (2000)

2. Prema, K.V., Subba, R.N.V.: Two-tier architecture for unconstrained handwritten character recognition. Sadhna 27, Part 5, 585–594 (2002)
3. Tripathy, N., Pal, U.: Handwriting segmentation of constrained Oriya text. Sadhna 31, Part 6, 755–769 (2006)
4. Rajashekararadhya, S.V., Vanaja Ranjan, P.: A Novel Zone Based Feature Extraction Algorithm for Handwritten Num Recognition of Four Indian Scripts. Digital Technology Journal 2, 41–51 (2009) ISSN 1802-5811
5. Heutte, L., Paquet, T., et al.: A structural statistical feature based vector for handwritten character recognition. Pattern Recognition Letters 19, 629–641 (1998)
6. Wang, X., Ding, X., Liu, C.: Gabor filters-based feature extraction for character recognition. Pattern Recognition Society (2004)
7. Pal, U., Roy, P.P.: Multi-oriented and curved text lines extraction from Indian documents. IEEE Trans. on Systems, Man and Cybernetics-Part B 34, 1676–1684 (2004)
8. Mohanty, S., Behera, H.K.: A complete OCR Development System for Oriya Script. In: Proceedings of SIMPLE 2004, IIT Kharagpur (2004)
9. Pal, U., Wakabayashi, T., Kimura, F.: A System for Off-line Oriya Handwritten Character Recognition using Curvature Feature. IEEE (2007) 0-7695-3068-0/07
10. Ren, J.: Multi-order Standard Deviation Based Distance Metrics and its Application in Handwritten Chinese Character Recognition. In: 18th International conference on Pattern Recognition (ICPR 2006). IEEE (2006)
11. Liu, H., Ding, X.: Handwritten Character Recognition Using Gradient Feature and Quadratic Classifier with Multiple Discrimination Schemes. In: Proceedings of the 2005 Eighth International Conference on Document Analysis and Recognition (ICDAR 2005). IEEE (2005)
12. Blumenstein, M., Verma, B., Basli, H.: A Novel Feature Extraction Technique for the Recognition of Segmented Handwritten Characters. In: Proceedings of the Seventh International Conference on Document Analysis and Recognition, ICDAR 2003. IEEE (2003) 0-7695-1960-1/03

Multiresolution Scene Based Digital Video Watermarking

T. Geetamma[1] and K. Padma Raju[2]

[1] Dept. of ECE, GMRIT,
A.P, India
[2] Dept. of ECE, JNTU College of Engineering,
Kakinada, A.P, India
geeta.tummalapalli@gmail.com,
padmaraju_k@yahoo.com

Abstract. The recent progress in the digital multimedia technologies has offered many facilities in the transmission, reproduction and manipulation of data. However, this advancement has also brought the challenge such as copyright protection for content Providers. Digital watermarking is one of the best solutions for copyright protection of multimedia data. This paper present a copyright protection method, a video watermarking technique based on the video scene segmentation and 3-level wavelet transform in which watermarks are embedded into the corresponding decomposed video. Multiresoluted watermarks are used for embedding which are obtained by pyramidal decomposition using Gaussian and laplacian pyramids. This method does not affect the perceptual quality, maintains statistical invisibility and increases the level of security to the content. The main concept is categorized to four steps watermark preprocessing, video pre-processing, embedding process and extraction process.

Keywords: Disrete wavelet Transform, Multiresolution , Digital Watermaking, Video.

1 Introduction

One of the reasons for the rapid development in research in digital watermarking is the need to find a solution for protecting intellectual properties of digital material. In order to embed watermark information in host data, watermark embedding techniques apply minor modifications to the host data in a perceptually invisible manner, where the modifications are related to the watermark information. The watermark information can be retrieved afterwards from the watermarked data by detecting the presence of these modifications. Most of the watermarking techniques proposed till date are applicable to gray scale images, but can be easily extended to color images by watermarking luminance component. Most of the existing watermarking algorithms can be classified according to the following criteria. The selection of locations where the watermark is embedded. The domain in which algorithm operates. For example, an algorithm can modify the image in the spatial domain directly to embed the watermark or it can transform the image in to other domains like DCT [11], DFT [9], DWT

S.C. Satapathy et al. (Eds.): Proc. of Int. Conf. on Front. of Intell. Comput., AISC 199, pp. 659–666.
DOI: 10.1007/978-3-642-35314-7_74

[10] and fractal. The rest of this paper is organized as follows: section2 will describe about proposed video watermarking technique, section3 will give implementation and results,section4 will give conclusions and future scope.

2 Video Watermarking Using DWT

2.1 Proposed Video Watermarking Method

With the rapid growth of the Internet and multimedia systems in distributed environments, it is easier for digital data owners to transfer multimedia documents across the Internet. Therefore, there is an increase in concern over copyright protection of digital contents[12]-[15]. Traditionally, encryption and control access techniques were employed to protect the ownership of media. These techniques, however, do not protect against unauthorized copying after the media have been successfully transmitted and decrypted. Recently, watermark techniques are utilized to maintain the copyright. Video watermarking introduces a number of issues not present in image watermarking. Due to a large amount of data and inherent redundancies between frames, video signals are highly susceptible to piracy attacks, including frame averaging, frame dropping, frame swapping, statistical analysis,etc. this problem can be overcome by applying scene change detections and scrambled watermarks in a video. The scheme is robust against different attacks that are possible.

2.2 Video Watermarking Scheme

The watermarking scheme is based on embedding multiresoluted watermark images into corresponding resolution of the decomposed video. In this scheme, a video is taken as the input, and then a watermark is decomposed into different resolutions which are embedded in corresponding resolutions of different frames in the original video. As applying a fixed image watermark to each frame in the video leads to the problem of maintaining statistical and perceptual invisibility. This scheme employs independent watermarks for successive frames in a original video. However, applying independent watermarks to each frame also presents a problem if regions in each video frame remain little or no motion frame after frame. These motionless regions may be statistically compared or averaged to remove the independent watermarks; consequently, a watermark of different resolutions is used within each frame[7][8]. With these mechanisms, the proposed method is robust against the attacks of frame dropping, averaging, swapping, and statistical analysis.

The proposed video watermarking scheme is categorized into four main steps

- Watermark preprocess,
- Video preprocess,
- Watermark embedding,
- Watermark detection.

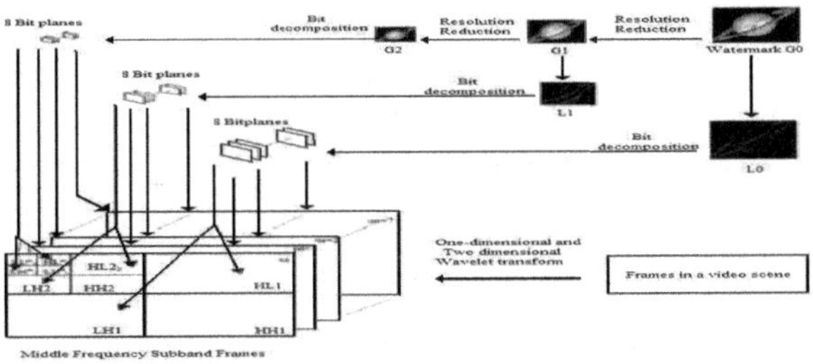

Fig. 1. Block diagram of video watermarking scheme

2.3 Watermark Preprocess

In watermark preprocess this paper considered a 2-D gray-level digital image (64x64) as a watermark, which is a visually recognizable pattern. In order to embed the watermark invisibly, the watermark information should adapt itself to the detail of original video. Hence, decomposed the watermark into a multiresolution hierarchical structure of images L0, L1 and G2 by the resolution-reduction method[2][6] as shown in fig.4. where G0 (64x64), the watermark image, is level0 of the Gaussian pyramids, G1 (32x32) is level1 of the Gaussian pyramids, G2(16x16) is level2 of the Gaussian pyramids, G'1 (64x64) andG'2(32x32) are the Gaussian Pyramid interpolation results of G1(32x32) and G2 (16x16) respectively, L0(64x 64) is level 0 of the Lapacian pyramids, and L1(32x32) is level 1 of the Lapacian pyramids.

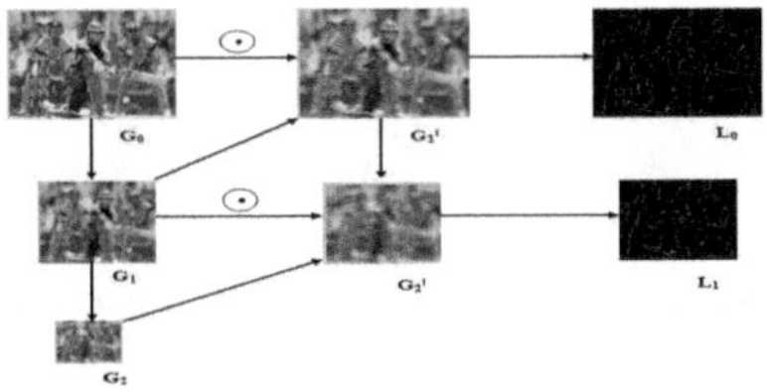

Fig. 2. Decomposition of original image

The watermark image is decomposed to multi resolutions by using the following image pyramids. Gaussian pyramids and Laplacian pyramids

Gaussian Pyramids

This process of Gaussian pyramid decomposition is to low-pass filter the original image G0 to obtain image G1 and say that G1 is a "reduced" version of G0 in that both resolution an sample density are decreased. In a similar way, this form $G2$ as a ' reduced version' of $G1$, and so on filtering is performed by a procedure equivalent to convolution with one of a family of local, symmetric weighting functions[5]. An important member of this family resembles the Gaussian probability distribution, so the sequences of images G0, G1, G2…….Gn are called as "Gaussian pyramids". Suppose the image is represented initially by the array G0 which contains *C* columns and R rows of pixels. Each pixel represents the light intensity at the corresponding image point by an integer *I* between 0 and K -1. This image becomes the bottom or zero level of the Gaussian pyramid. Pyramid level 1 contains image G1, which is a reduced or low-pass filtered version of G0. Each value within level 1 is computed as a weighted average of values in level 0 within a 5-by-5 window. Each value within level 2, representing G2, is then obtained from values within level 1 by applying the same pattern of weights. A graphical representation of this process in one dimension. The size of the weighting function is not critical. This has selected the 5-by-5 pattern because it provides adequate filtering at low computational cost. The level-to-level decomposition process is performed by the function REDUCE.

$$G_k = REDUCE \quad (G_{k-1}) \tag{1}$$

Where G_k is the kth level of Gaussian pyramid and G_{k-1} is the next level of gaussian pyramids.

The Generating Kernel: The 5-by-5 pattern of weights w is used to generate each pyramid array from its predecessor. This weighting pattern called the generating kernel should be same for all the levels of pyramids.
The one-dimensional, length 5, weighting function w is given in eq 6

$$\sum_{i=-2}^{2} w(i) = 1 \tag{2}$$

It is also symmetric

w(i) = w(-i) for I = 0, 1,2.

An additional constraint is called equal contribution. This stipulates that all nodes at a given level must contribute the same total weight (=1/4) to nodes at the next higher level. Let w(0) = 1, w(-1) = w(l) = b, and w(-2) = w(2) = c. In this case equal contribution requires that a + 2c = 2b. These three constraints are satisfied when the following conditions satisfies

w(0) = a
w(-1) = w(1) = ¼
w(-2) = w(2) = 1 /4 – a/2.

Gaussian Interpolation: In order to find the Gaussian interpolation results of an image we used an EXPAND function. Its effect is to expand an (M + 1)-by-(N + 1) array into a (2M + l)-by-(2N + 1) array by interpolating new node values between the given values. Thus, EXPAND applied to array G1 of the Gaussian pyramid would yield an array G1', which is the same size as G1. The function used to expand the lower level of pyramids is as shown in equation (7)

$$G_k' = EXPAND(G_{k-1}) \tag{3}$$

Where G_k' the interpolation is result of G_{k-1} Gaussian pyramid and G_{k-1} is the k-1 th level of Gaussian pyramid.

Laplacian Pyramids: The Laplacian pyramids are the sequence of error images L0, L1,…, LN. Each is the difference between a level of Gaussian pyramid and the interpolated image of next subsequent pyramid. Thus, for $0 < 1 < N$,

$$L_1 = g_l - EXPAND \quad (g_{l-1}) \tag{4}$$

Bit Planes Generation: After obtaining the laplacian pyramids from resolution reduction method, these pyramids are converted into bit planes. Bit-plane is the plane that one specific bit of every pixel create. For a gray scale image of 255 levels the number of bit planes obtained will be 8. Since the number of bits required to represent a pixel will be 8 bits. The first bit-plane is the least significant one (LSB) and most of the time is hardly related to the main shapes of the picture. On the other hand, the last bit-plane is the most significant one (MSB) and contains the main lines and edges of the picture.

2.4 Video Preprocess

The video which is to be watermarked is taken and frames from that video are read. All frames in the video are transformed to the wavelet domain. The general procedure is used for video decomposition into frames. The frames are decomposed into 3-level subband frames by separable two-dimensional (2-D)wavelet transform. It produces a low-frequency subband LL , and three series of high-frequency sub bands LH,HL,HH. According to the energy distribution, LL is the most important than LH,HL,HH . For different levels, the higher the level, the more important the sub bands. In this scheme, the watermark is only embedded in the middle frequency sub bands. If less than 3-levels is applied, the capacity of the scheme would be decreased. If larger than 3-levels is applied, the quality of the watermarked video is affected.

2.5 Watermark Embedding Algorithm in Video Stream

The decomposed watermarks are embedded to the decomposed video frames by changing position of some DWT coefficients with the following condition: If $W[j]=1$

$$Ch\ [i] = C\ [i] + eps$$

$$Cv\ [i] = C\ [i] + eps$$

Else if $W\ [j] = 0$

$$Ch\ [i] = C\ [i] - eps$$

$$Cv\ [i] = C\ [i] - eps$$

Where w[j] is the jth bit plane value of the decomposed watermark image. $Ch\ [i]$, $Cv\ [i]$ are the modified dwt coefficients which are HL and LH bands respectively. $C\ [i]$ is the original dwt coefficients of the of the corresponding sub bands and eps is a constant value which is a user choice to determine the value .Coefficients of LL (i.e. the low frequency sub-band) are not watermarked, as video energy is concentrated on lower frequency wavelet coefficient. If they are altered, it will affect perceptual quality. Only alter LL coefficients make the hidden mark not perceptible under domestic condition. Coefficients of HH (i.e. the high frequency sub-band) are also not watermarked. It can make the watermark survive MPEG lossy compression as lossy compression removes the details (i.e. the high frequency components) of the image. Inverse of the 2-DDWT and 1-D DWT By inversing the watermarked 2-D and 1-D DWT wavelet coefficient frames, this obtain the watermarked video.

2.6 Watermark Extraction Algorithm

The video is processed to detect the video watermark. In this step, scene changes are detected from the tested video i.e., with this algorithm. The extraction steps are as follows: The watermarked video is broken into scenes and each scene is decomposed into multiresolution temporal representation (a series of wavelet coefficient frames) by 2D-DWT along the temporal axis of the video. Each video frame is transformed to wavelet domain by 2D-DWT with 3 levels. Watermark is extracted from the frames by checking the magnitude of some DWT coefficients to extract the 8 bit-planes of images L0, L1 and G2.

The condition is shown as follow:

$$\begin{array}{ll} W[j]=1 & \text{if} \quad Ch[i] \geq C[i] \\ W[j]=0 & \text{otherwise} \end{array}$$

Where Ch[i] is the i-th DWT coefficient of the watermarked video frame, W[j] is the j-th pixel of a certain bit-planes. By composing these bit-planes it get the gray-level images G2, L1, and L0.

3 Implementation and Results

Fig. 3. Original image(64x64)

Fig. 4. RecoveredOriginal image(64x64), PSNR=49.835

Fig. 5. Watermark1 L1(64x64)

Fig. 6. Recovered Watermark1 RL1(64x64)

Fig. 7. Watermark2 L2(32x32)

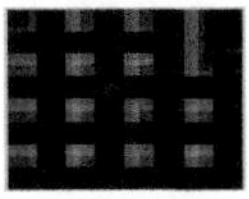

Fig. 8. Recovered Watermark2 RL2(32x32)

Fig. 9. Watermark3 G2(16x16)

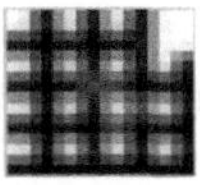

Fig. 10. Recovered Watermark3 RG2 (32x32)

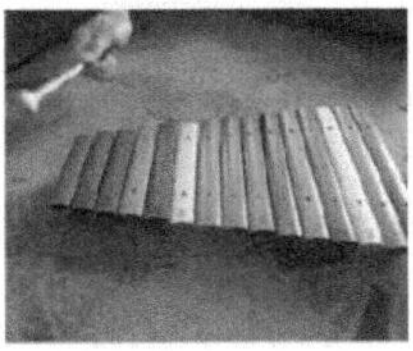

Fig. 11. Original frame1(f1)

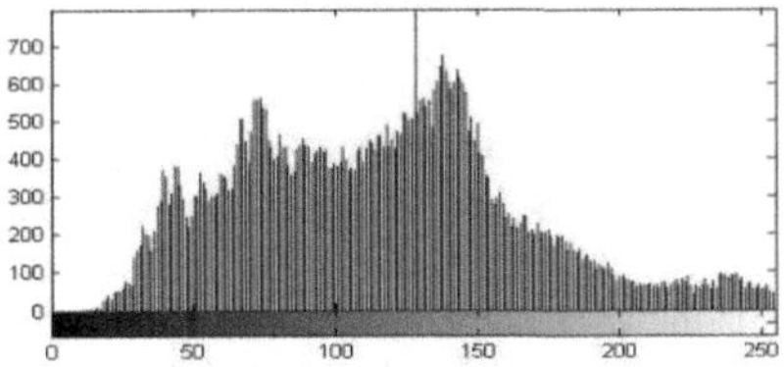

Fig. 12. Histogram of f1

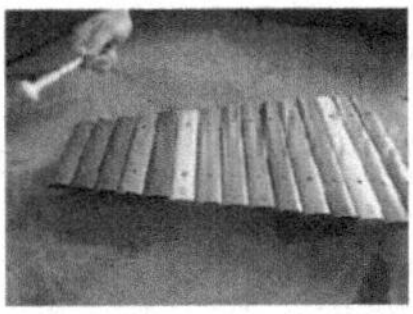

Fig. 13. Watermarked frame1(wf1)

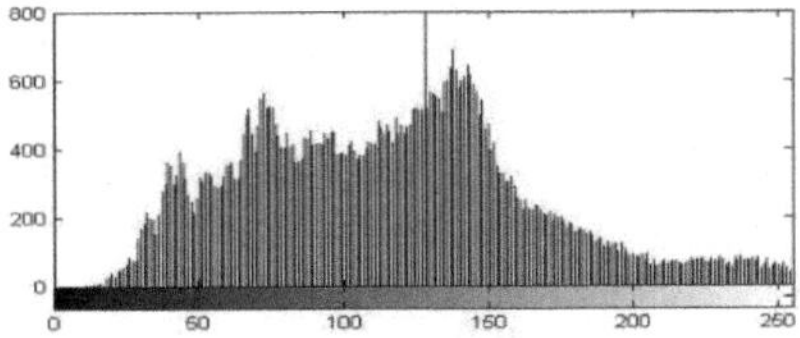

Fig. 14. Histogram of wf1

4 Conclusion and Future Scope

This paper presents a video watermarking technique that uses multiresoluted watermarks used for embedding. Multiresoluted watermarks are obtained by pyramidal decomposition using Gaussian and laplacian pyramids. For embedding watermarks, the video frames are subjected to wavelet decomposition and the wavelet coefficients of these frames are modified for embedding based on watermark bits. Proposed watermarking technique has following advantages: By using multiresoluted watermarks which are embedded in corresponding resolution of decomposed video will maintain the statistical invisibility. By embedding three different watermarks into the video the level of security is increased. The perceptual quality of the watermarked video is not changed because of embedding data in middle frequency sub bands.

Future Scope

For further development, this paper proposes the following improvements:

- This scheme can be enhanced by combining with audio watermarks
- This proposed watermarking scheme can further be associated with different applications to achieve a sophisticated system and the fidelity can be improved by applying genetic algorithm. By including the watermark detection devices in electronic devices like DVD players, cell phones the watermarked data cannot be used by unauthorized user.

References

1. Essaouabi, A., Ibnelhaj, E.: A 3D Wavelet based method for Digital Video Watermarking. In: IEEE Conference Paper (2009)
2. Niu, X., Sun, S.: A New Wavelet-Based Digital Watermarking for Video. In: 9th IEEE Digital Signal Processing. IEEE (2000)
3. Swanson, M.D., Zhu, B., Tewfik, A.H.: Multiresolution Scene-Based Video Watermarking Using Perceptual Models. IEEE Journal on Selected Areas in Communications 16(4), 540–550 (1998)
4. Zhuang, H.-Y., Li, Y., Wu, C.-K.: A blind spatial-temporal algorithm based on 3D wavelet for video watermarking. In: 2004 IEEE International Conference on Multimedia and Expo (ICME), pp. 1727–1730 (2004)
5. Burt, P.J., Adelson, E.H.: The Laplacian Pyramid as a Compact Image Code. IEEE Trans. on Communications 31(4), 532–540 (1983)
6. Burt, P.J., Adelson, E.H.: The Laplacian Pyramid as a Compact Image Code. IEEE Trans. on Communications 31(4), 532–540 (1983)
7. Barda, J.: Network security through access control and watermarking. EBU Technical Review (Autumn 1999)
8. Doërr, G., Dugelay, J.-L.: Video watermarking overview and challenges. In: Furht, B. (ed.) Handbook of Video Databases: Design and Applications. CRC Press (September 2003) ISBN :084937006X
9. Pereira, S., Pun, T.: Robust template matching for affine resistant image watermarks. IEEE Trans. Image Process. 9(6), 1123–1129 (2000)
10. Hong, I., Kim, I., Han, S.: A blind watermarking technique using wavelet transform. In: Proc. IEEE Int. Symp. Industrial Electronics, vol. 3, pp. 1946–1950 (2001)
11. Duan, F., King, I., Xu, L., Chan, L.: Intra-block algorithm for digital watermarking. In: Proc. IEEE 14th Int. Conf. Pattern Recognition, vol. 2, pp. 1589–1591 (August 1998)
12. Piva, A., Bartolini, F., Barni, M.: Managing copyright in open networks. IEEE Trans. Internet Computing 6(3), 18–26 (2002)
13. Lu, C., Yuan, H., Liao, M.: Multipurpose watermarking for image authentication and protection. IEEE Trans. Image Process. 10(10), 1579–1592 (2001)
14. Lu, C., Huang, S., Sze, C., Liao, H.Y.M.: Cocktail watermarking for digital image protection. IEEE Trans. Multimedia 2(6), 209–224 (2000)
15. Lee, J., Jung, S.: A survey of watermarking techniques applied to multimedia. In: Proc. 2001 IEEE Int. Symp. Industrial Electronics (ISIE), vol. 1, pp. 272–277 (2001)

Encryption Based Image Authentication Technique Using Random Circular Insertion and Adjustment (EIATRCIA)

Soumit Chowdhury[1,*], Dipankar Nag[1], Krishnendu Sadhu[1], and Nabin Ghoshal[2]

[1] Dept. of Computer Sc. & Engg., Govt. College of Engg. & Ceramic Technology, Kolkata, West Bengal, India
[2] Dept. of Engg. & Technological Studies, University of Kalyani, West Bengal, India
{joy_pinu,dipankar_cmpstudent,nabin_ghoshal}@yahoo.co.in, krishnendusadhu@gmail.com

Abstract. This paper demonstrates a dynamic encryption based image authentication technique in the spatial domain that hides one authenticating color image inside another carrier color image where the authenticated receiver can only extract this embedded secrete image using the secrete key(s). The EIATRCIA technique actually embeds the secrete image bits into the randomly generated bit positions of each carrier image pixel bytes using the encryption with dynamic secrete key(s). The embedding of four numbers of the secrete image bits into the sub-image blocks of size 2×2 can also be organized in a randomly generated circular list involving encryption associating secrete key(s) and stenographic key. Finally a delicate readjustment in the respective bits of the concerned bit-embedded pixel bytes can minimize the introduced distortion as well. This carrier image after embedding also avoids the visual distortion and the experimental result shows the robustness along with performance of this scheme while embedding.

Keywords: Steganography, Image Authentication, Digital Watermarking.

1 Introduction

Digital watermarking [6, 7] technique has evolved as a special case of Steganography [1], in which image authentication [2, 3] is done by hiding [4] secrete authenticating color image imperceptibly inside another carrier color image for maintaining the image authenticity. So the watermark secrete image exists in any legal copies and it helps the copyright [1] owner to identify who has an illegal copy, especially in multimedia applications. There is obviously a tradeoff between the watermark embedding strength (robustness), the quality (watermark invisibility) and security aspects, while considering visual degradation. This EIATRCIA scheme uses two secrete keys, where one key is the stenographic key of 1 byte long and the other key is

* Corresponding author.

S.C. Satapathy et al. (Eds.): Proc. of Int. Conf. on Front. of Intell. Comput., AISC 199, pp. 667–674.
DOI: 10.1007/978-3-642-35314-7_75

the 8 byte long secrete key, that are only available to the authenticated [10] users at the different nodes, with dynamic values for each communication. The 1 byte long stenographic key is encrypted with the 1 byte chunk of the secrete key, followed by the modulo operation with four, in order to find the starting pixel byte position in the sub-image block of size 2×2 for insertion of 4 nos. of secrete image bits into the concerned pixel bytes in a circular manner on a purely random basis. In addition the different combinations of the stenographic key can also be modulo with the number 4, for generating the random values between 0-3, and this two bit binary representation can be XOR-ed with the two MSB bits of the concerned pixel byte, for obtaining the secrete bit embedding position from the LSB. Finally a delicate readjustment in the concerned embedded pixel byte involving the four LSB bits except the bit position where embedding was made, will certainly reduce the noise distortion that may be introduced in the range of +8 to -8. This dynamic and random insertion of the secrete image bits will enhance the security aspects of the scheme as compared to the traditional LSB based embedding. Results of this EIATRCIA scheme is also compared with the existing DCT-based, QFT-based Spatiochromatic method, as well as DFT-based watermarking method [8], in terms of the visual quality, Mean Square Error (MSE), Pick Signal-to-Noise Ratio (PSNR) in dB and Image Fidelity (IF), for robustness [5] with imperceptibility aspects.

2 Technique Details

Color image files are considered as a two dimensional matrix, where $C_{m,n}$ is the matrix element representing the corresponding pixel value, with m and n indicating the row and column position of the matrix respectively. The two associated keys are 1 byte long Stenographic key, and the 8 byte long secrete key, denoted by 'stgk' and $sk_i=\{sk_1, sk_2,\ldots, sk_8\}$ respectively, that can be generated dynamically for each communication, with keys are available at both communicating ends. For maintaining less distortion, just 1 secrete data bit per pixel, is embedded, and hence the carrier image file size is at least 8 times larger than the secrete image file size.

2.1 Secrete Bit Insertion

Flag Compute. In order to dynamically specify the size of the secrete image, a flag is introduced within the carrier image, with the resolution involving the width (say w) and the height (say h) of the secrete image, for proper extraction at the receiver end.

n_l denotes the number of bits required to represent the maximum size of hidden file.

$$n_1 = \log_2({}^{wh}/_{8})\ . \tag{1}$$

Since (w×h) is the total number of pixel bytes present in the carrier color image, so (w×h)/8, can be the maximum size of the hidden secrete image file in terms of bytes.

Flag is the n bit representation of the size of the file to be hidden in bytes and it is inserted just before the secrete image bits into the sub image matrices of size 2×2.

$$n = \begin{cases} n_1, & \text{if } n_1 \% 4 = 0 \\ n_1 + 4 - (n_1 \% 4), & \text{else} . \end{cases} \quad (2)$$

Creating 2×2 Square Matrices. The whole carrier image is partitioned into 2×2 sub matrices, consisting of 4 pixels in each matrix, with 'w' and 'h' are even.

$$S = \{M : M = \begin{pmatrix} a_{2i,2j} & a_{2i,2j+1} \\ a_{2i+1,2j} & a_{2i+1,2j+1} \end{pmatrix} \text{ where } a \in C_{m,n}\} . \quad (3)$$

The whole secrete image file is a stream of bits and obviously 4 secrete bits can be hidden inside a single 2×2 matrix, with embedding of 1 bit per pixel byte.

Hiding Bits inside a 2×2 Matrix. The principle of generating the watermark embedded image '$I^{|}$' is, $\times\ k \times D \rightarrow I^{|}$, where 'I' is the original carrier image, 'k' is the combination of keys, which are also available to the receiver end and 'D' is the secrete image bits to be embedded within 'I'. Initially considering i = 1, for sk_i .

Step1. At first two variables, named as 'key' and 'ind', are defined, where

$$key = stgk \% 4 . \quad (4)$$

$$ind = (sk_i \oplus stgk) \% 4 . \quad (5)$$

$$stgk = bitwise\ reversal\ of\ (stgk + 3) \% 255 . \quad (6)$$

For example, let stgk = 125 and sk = {76, 65, 9, 34, 45, 78, 68, 7} so sk_1=76. Therefore, key = 125 % 4 = 1, ind = (76⊕125)%4 = $\{(01001100)_2 \oplus (01111101)_2\}\%4$ = $(00110001)_2$ % 4 = 49 % 4 = 1 and stgk = bitwise reversal of (125+3) % 255 = bitwise reversal of 128 = bitwise reversal of $(10000000)_2 = (00000001)_2$ = 1.

Step 2. The embedded matrix $M^{|}$ is generated using the transition function δ as,

$$M^{|} = \delta(M, D, key, ind) . \quad (7)$$

A. The elements of the matrix M is denoted as,

$$M = \begin{pmatrix} a_{0,0} & a_{0,1} \\ a_{1,0} & a_{1,1} \end{pmatrix} .$$

For example, $M = \begin{pmatrix} 34 & 37 \\ 38 & 40 \end{pmatrix} = \begin{pmatrix} 00100010 & 00100101 \\ 00100110 & 00101000 \end{pmatrix}$ and D = {1, 1, 0, 1}.

B. The 2 bit binary representation of the variable 'ind' indicates the starting pixel position (0-3) from where the secrete bit insertion starts in the 2×2 matrix and continuing in a circular way until the whole matrix is covered. If ind = 2 then from $a_{1,0}$ the data insertion starts followed by $a_{1,1}$, $a_{0,0}$ and $a_{0,1}$.

C. Now the first two binary bits (from the MSB position) of the present working matrix element 'a' are XOR-ed with the two binary bits of the variable 'key', and the resulting decimal value of two bit length is stored in a variable called 'pos' indicating the position where to embed the secrete bit from the LSB. Since pos ∈ {0, 1, 2, 3}, so the secrete image bit is embedded within the 4 LSB bits in each matrix element introducing a distortion from -8 to +8. The bit positions are numbered starting from the LSB with 0 index and consecutive increments towards the MSB part (0, 1, 2 …). For example, the XOR operation performed between the first two bits of $a_{0,1}$ and key yields $(01)_2$. Hence at position number 1 the secrete data bit '1' gets inserted in place of '0'. So $a_{0,1}$ now becomes $(00100111)_2$ and the changes in the pixel byte value is 2.

D. Here the bit-value at the position number indicated by 'pos' and the bit to be embedded are checked for equality. If both are same then no further changes are made, otherwise the secrete bit is embedded at the position indicated by 'pos'. Now if any changes have occurred in the pixel value, then step 3 is executed, otherwise the next secrete bit insertion will take place in the next available pixel value.

Step 3. Finally a delicate readjustment in the concerned embedded pixel byte is made in order to reduce the noise introduction due to embedding. If the secrete data bit value inserted as '0' in place of the bit value '1' in the pixel byte, then the value of the pixel byte needs to be increased by setting some 1's. So bits starting from the next position of 'pos' in the right direction, towards the LSB part are checked, with all the available '0's are changed to '1', otherwise if all the right hand side bits next to 'pos' are already 1, then the immediate left bit of 'pos' is checked. If it is '0' then it is made to '1', but if it is already '1' then no further changes are made. In case of secrete data bit value inserted as '1' in place of the bit value '0' in the pixel byte, then this same procedure applies by changing the possible 1's to 0's as discussed above. This adjustment is kept within the 4 LSB bits, and at first the right hand side is exploited for maintaining less distortion.

In the given example the secrete bit value '1' is inserted at the position number 1, in place of '0', so the next right position number 0 is checked and since '1' is found here, it is changed to '0', so that the pixel byte value now becomes as $(00100110)_2$, with a small distortion value of 1 is introduced in the pixel byte as compared to the original one. Hence after all the secrete data bit insertion the final 2x2 matrix looks as,

$$M^{|} = \begin{pmatrix} 34 & 38 \\ 38 & 40 \end{pmatrix}.$$

Step 4. Finally this secrete bit embedded matrix of size 2×2, denoted as $C^{|}_{m,n}$ is now written to an output image file, maintaining the same resolution and positions of the original carrier image. Now the next 2×2 matrix is taken with next circular value of 'i' as i = sk_i % 8 + 1. Again going to step 1 the embedded image $I^{|}$ is generated and sent.

2.2 Extraction

Creating 2×2 Square Matrices. We initially divide the whole watermarked image into sub matrices of size 2×2 as '$S^{|}$', and each element of the matrix is a pixel.

$$S^{|} = \{M^{|} : M^{|} = \begin{pmatrix} a_{2i,2j} & a_{2i,2j+1} \\ a_{2i+1,2j} & a_{2i+1,2j+1} \end{pmatrix} \; where \; a \in C^{|}_{m,n}\} . \tag{8}$$

Flag Compute. At First the value of the variables n_l and n are calculated using the equation (1) and (2) respectively, by obtaining the resolution of 'w' and 'h' from the carrier image. Variable n is the flag size in bits, and since each matrix hides 4 numbers of bits, so n/4 is the numbers of matrices for the flag just before the data bits.

Extracting Bits from a 2×2 Matrix. The principle for bit extraction is, $I^{|} \times k \to D$.

Step 1. Two variables 'key' and 'ind' are computed from equation (4) and (5).

Step 2. The extraction function with secrete image bit extraction steps are defined as,

$$D = \delta^{|}(M^{|}, key, ind) . \tag{9}$$

A. The elements of the matrix $M^{|}$ is denoted as,

$$M^{|} = \begin{pmatrix} a_{0,0} & a_{0,1} \\ a_{1,0} & a_{1,1} \end{pmatrix} .$$

B. Two bit binary representation of the variable 'ind' indicates the starting pixel position in the 2×2 matrix from where the data extraction starts and continues in a circular manner until the whole matrix is covered. For example, if ind = 2 then from the matrix element position $a_{1,0}$ the data extraction starts, followed by $a_{1,1}$, $a_{0,0}$ and $a_{0,1}$.

C. Now the first two MSB bits of the concerned matrix element and the two bit binary representation of the variable 'key' is XOR-ed, that results a decimal value between 0-3, as stored in the variable 'pos', indicating the position of the embedded secrete bit in the pixel byte, starting from the LSB position.

D. Using the above step C. all the 4 secrete bits are extracted from the 4 matrix elements of pixels, and these bits are written to the extracted output image file.

Step 3. Continue step 2 for extraction of the next four secrete image bits from the next available 2×2 matrix containing the pixels, until all the matrices are covered.

3 Results and Discussion

For testing purpose the ppm color images are taken, with carrier image resolution of 512×512, and the authenticating image resolution of 100×100. Using equation (1) and

(2) we get n = 16 bits as the flag size, that represents the maximum number of bytes that can be present in the 100×100 secrete image including the control information. After executing EIATRCIA scheme on different benchmark images with embedding of nearly 30,056 bytes, the results obtained at Table 1 and Table 2, shows the MSE, PSNR and IF values before and after adjustment of the pixel bytes. The comparative column chart obtained from Table 2, as shown in Fig. 3, is reflecting the PSNR values with and without adjustment of the pixel bytes. Finally, Table 3 shows the comparative study of this scheme with some existing techniques in spatial domain, and from the results it may be commented that this EIATRCIA image authentication process is relatively robust [8] and better in quality than the existing techniques.

Fig. 1. Sample benchmark images used for testing of EIATRCIA scheme [9]

Fig. 2. Visual interpretation of the carrier images after embedding of the authenticating images

Table 1. MSE, PSNR and IF values obtained, using EIATRCIA scheme, where Carrier image is same and different Authenticating images are present [5]

	Carrier Image (512×512)	Authenticating Image (100×100)	MSE	PSNR (dB)	IF
Without Adjustment	2101.ppm	2103.ppm	9.903061	38.173111	0.999819
		2107.ppm	9.890274	38.178719	0.999819
		2111.ppm	9.874451	38.185673	0.999820
		4104.ppm	9.886017	38.180592	0.999819
		4201.ppm	9.929237	38.161644	0.999819
With Adjustment	2101.ppm	2103.ppm	4.183071	41.915852	0.999924
		2107.ppm	4.174858	41.924385	0.999924
		2111.ppm	4.181526	41.917454	0.999924
		4104.ppm	4.201046	41.897228	0.999923
		4201.ppm	4.209713	41.888279	0.999923

Table 2. MSE, PSNR and IF values obtained, using EIATRCIA scheme, where Authenticating image is same and different Carrier images are present [5]

	Carrier Image (512×512)	Authenticating Image (100×100)	MSE	PSNR (dB)	IF
Without Adjustment	2103.ppm	2101.ppm	8.721058	38.725113	0.999851
	2107.ppm		9.459869	38.371952	0.999897
	2111.ppm		10.112103	38.082390	0.999815
	4104.ppm		9.834221	38.203403	0.999804
	4201.ppm		10.285126	38.008709	0.999802
With Adjustment	2103.ppm	2101.ppm	3.340279	42.892975	0.999943
	2107.ppm		4.807537	41.311577	0.999948
	2111.ppm		4.204620	41.893536	0.999923
	4104.ppm		4.140629	41.960140	0.999918
	4201.ppm		4.469086	41.628616	0.999914

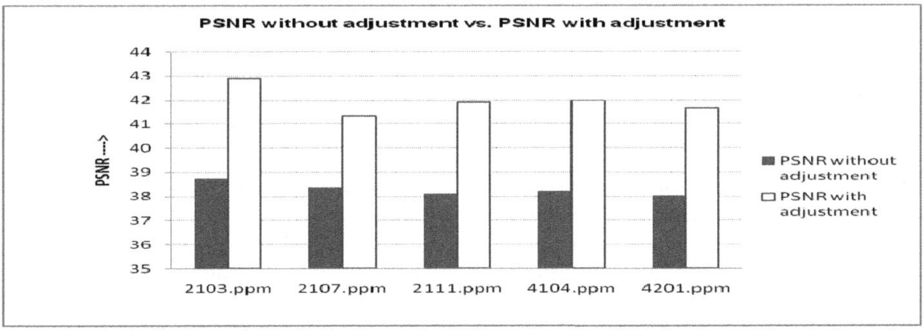

Fig. 3. PSNR value comparison for different carrier images, having same embedded Authenticating image, with and without adjustment of the pixel bytes

Table 3. Comparison of EIATRCIA (applying adjustment) with some existing schemes [8]

Technique	Capacity (bytes)	Average PSNR in dB
SCDFT	3840	30.10240
QFT	3840	30.92830
DCT	3840	30.40460
EIATRCIA technique	30056	41.93405

Acknowledgement. The authors express their deep sense of gratitude to the Dept. of Engineering and Technological Studies, University of Kalyani, Nadia, West Bengal, India, where this research work has been carried out.

References

1. Ghoshal, N., Mandal, J.K.: A Novel Technique for Image Authentication in Frequency Domain using Discrete Fourier Transformation Technique (IAFDDFTT). Malaysian Journal of Computer Science 21(1), 24–32 (2008) ISSN 0127-9094
2. Ghoshal, N., Mandal, J.K.: A Bit Level Image Authentication/Secrete Message Transmission Technique (BLIA/SMTT), Association for the Advancement of Modelling & Simulation Technique in Enterprises (AMSE). AMSE Journal of Signal Processing and Pattern Recognition 51(4), 1–13 (2008)
3. Ghoshal, N., Mandal, J.K., et al.: Masking based Data Hiding and Image Authentication Technique (MDHIAT). In: Proceedings of 16th International Conference of IEEE on Advanced Computing and Communications, ADCOM 2008, December 14-17, pp. 119–122. Anna University, Chennai (2008) ISBN: 978-1-4244-2962-2
4. EL-Emam, N.N.: Hiding a large Amount of data with High Security Using Steganography Algorithm. Journal of Computer Science 3(4), 223–232 (2007) ISSN 1549-3636
5. Kutter, M., Petitcolas, F.A.P.: A fair benchmark for image watermarking systems. In: Electronic Imaging 1999, Security and Watermarking of Multimedia Contents, Sans Jose, CA, USA, vol. 3657 (1999)
6. Ahmidi, N., Safabkhsh, R.: A novel DCT-based approach for secure color image watermarking. In: Proc. Int. Conf. Information Technology: Coding and Computing, vol. 2, pp. 709–713 (2004)
7. Bas, P., Biham, N.L., Chassery, J.: Color watermarking using Quaternion Fourier Transformation. In: Proc. ICASSP, Hong Kong, China, pp. 521–524 (2003)
8. Tsui, T.T., Zhang, X.-P., Androutsos, D.: Color Image Watermarking Using Multidimensional Fourier Transformation. IEEE Trans. on Info. Forensics and Security 3(1), 16–28 (2008)
9. Weber, A.G.: The usc-sipi image database. Signal and Image Processing Institute at the University of Southern California (October 1997), `http://sipi.usc.edu/services/database/Database.html`
10. Ghoshal, N., Mandal, J.K.: Discrete Fourier transform based Multimedia Color Image Authentication for Wireless Communication (DFTMCIAWC). In: 2nd International Conference on Wireless Communications, Vehicular Technology, Information Theory and Aerospace & Electronic Systems Technology, Le Royal Meridian Chennai, India (2011) ISBN: 978-1-4577-0787-2/11

Gödelization and SVD Based Image Watermarking under Wavelet Domain

P. Raja Mani[1] and D. Lalitha Bhaskari[2]

[1] Department of Computer Science Engineering,
Gitam University
Visakhapatnam, Andhra Pradesh, India
rajamani09@gmail.com
[2] Department of Computer Science and Systems Engineering,
AUCE(A), Andhra University
Visakhapatnam, Andhra Pradesh, India
lalithabhaskari@yahoo.co.in

Abstract. In this digital era, with the ever growing size of multimedia database and digital media, there is a need for designing of robust methods to enhance the security of multimedia data against various attacks. One such popular technique is digital watermarking, and in this paper two different approaches for multimedia data hiding and copyright protection are implemented. Gödelization and Singular Value Decomposition techniques under frequency domain are implemented and compared.

Keywords: Wavelets, Watermarking, Gödelization, Alphabetic Coding, SVD.

1 Introduction

Internet represents an insecure channel for exchanging information leading to a high risk of intrusion or fraud. We enunciate the need for watermarking to deter copyright protection, ownership and security of digital data. Watermarking systems were put forward as an efficient way of solving the security issues of the digital data. Digital watermarking is a general solution that can be used to identify illegal copying and ownership, authentication, or other applications by inserting information into the digital data in an imperceptible way. Robustness is one of the most basic requirements for invisible image watermarks. Usually digital watermark can be embedded either in spatial domain[1] or frequency domain[2], while both have their own advantages and disadvantages.

Frequency domain watermarking schemes are more robust to tampering and attacks than those in spatial domain. In addition discrete wavelet transform(DWT) has good time-frequency features and accurate matching of the human visual system(HVS). As cited in [3] among the available many watermarking methods in wavelet domain, Haar wavelet is efficient and so it has been adopted in this paper for the purpose of image watermarking.

S.C. Satapathy et al. (Eds.): Proc. of Int. Conf. on Front. of Intell. Comput., AISC 199, pp. 675–681.
DOI: 10.1007/978-3-642-35314-7_76

In the next section the basic concepts used in the approaches are explained. Section 3 focuses on the details of the embedding and extracting algorithms. Sections 4 and 5 show experimental results and observations. Section 6 deals with the conclusion and future work followed by references.

2 Understanding the Basic Concepts

2.1 Haar Wavelet Transform

In Discrete Wavelet Transformation domain, a 2-D digital image can be decomposed into its various resolutions based on the approximate weight (LL), and the detailed weights of the Horizontal (HL), vertical direction (LH), and diagonal direction (HH). Decomposition can be done at different DWT levels. Second Level Wavelet Decomposition along with the frequency bands are as shown in Fig 1(a) and 1(b). Low frequency bands will bring about stronger robustness [4], so the watermark is hidden in the LL2 band.

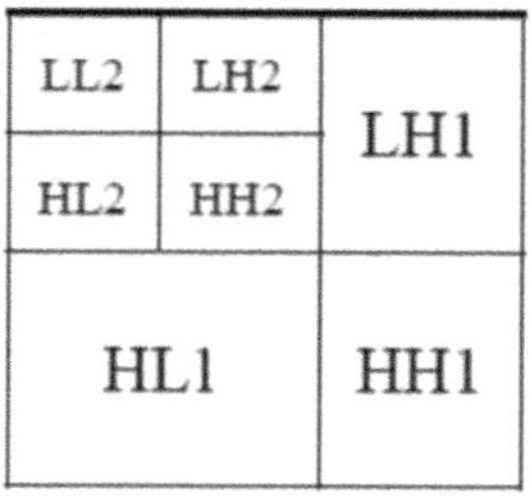

Fig. 1(a). Second level wavelet decomposition

Fig. 1(b). Second level wavelet decomposition for flower image

2.2 Singular Value Decomposition (SVD)

Singular value decomposition is used to approximate large, unmanageable matrices to smaller invertible square matrices. The application of the singular value decomposition to an image compresses without significant data loss[5]. It reduces the space required to store images. Suppose A is an n-by-p matrix, the SVD theorem states: $A_{nxp} = U_{nxn} S_{nxp} V^T_{pxp}$ where $U^T U = I_{nxn}$, $V^T V = I_{pxp}$ (i.e. U and V are orthogonal) where the columns of U are the left singular vectors, S has singular values and is diagonal, V^T has rows that are the right singular vectors. The SVD represents an expansion of the original data in a coordinate system where the covariance matrix is diagonal. Zhou et al.[6] has presented an analysis of the effects of geometric distortions on the singular values of an image, including transpose, flip, rotation, scaling, translation, etc. For majority of the attacks, the change in the largest singular value is very small. In this methodology, SVD is used for reducing the size of the watermark. This not only enhances the security of the watermark but also increases the data payload capacity.

2.3 Gödelization and Alphabetic Coding

Prime factorization theorem states that every positive integer greater than one can be factored into primes. Based on this theorem an encoding scheme was developed by the logician Kurt Gödel [7] to assign numbers to statements and formulas in an axiomatic system. Gödelization[8] as proposed by Lalitha Bhaskari et al., is a process of converting any positive integer greater than 1 into a sequence called Gödel Number Sequence(GNS). Gödel number sequence of 198 is GN(1,2,0,0,1). The sequence 1,2,0,0,1 can be encoded as $2^1 \times 3^2 \times 11^1$ as GN(0) =2, GN(1)=3, GN(2)=5 and so on [8]. We can hide the image using Gödel Numbering.

A gray scale image is a collection of pixels (intensity values) ranging from 0 to 255[9]. Each of these pixel is converted into a Gödel Number Sequence GNS[8].These sequences are stored in an array and each sequence is delimited by a special character. This is called the Gödel String of the image. To reduce the length, the Gödel String is again encoded using Alphabetic coding(AC) technique[8]. As we have a sequence of more 0's and 1's in the string, we represent 0's with 'A' , 1's with 'B' , 2's with 'C' and so on. If we encounter more than 3 same characters then we represent as the number of occurrences first and then the character. So the string 00001000$1200100000$000000010$ is encoded as 4AB3A$BC2AB5A$7ABA$. The length is reduced to 19 from 30. With AC technique the length is reduced as well as second level of security is also provided. After this the string obtained from AC technique is embedded into the decomposed image.

3 Proposed Methodology

The proposed approaches employ wavelet decomposition of the original image using Haar wavelet Transform, then in the first approach the watermark is encrypted using Singular Value Decomposition and in the second approach the watermark is encrypted using Gödelization and Alphabet Coding. Now the encrypted watermark is embedded into the LL2 band of the decomposed image and a transformed watermarked image is obtained. The wavelet image reconstruction is the inverse transform of the wavelet decomposition.

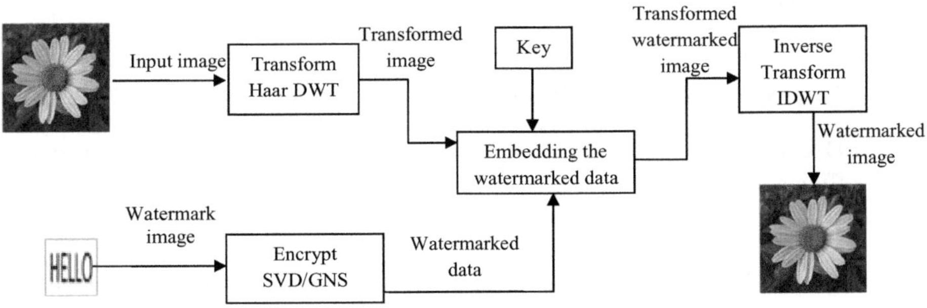

Fig. 2(a). Watermark Embedding

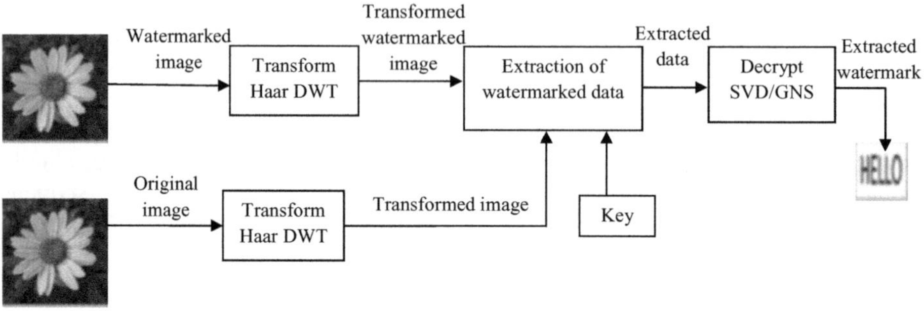

Fig. 2(b). Watermark Extraction

3.1 Watermark Embedding Procedure Using SVD

Step 1: Decompose Cover image using Haar wavelet upto 2 levels to get approximation coefficients, LL2.
Step 2: Read the watermark data.
Step 3: Encrypt the watermark using Singular Value Decomposition(SVD) technique.
Step 4: Generate the key.
Step 5: Embed the 'S' into LL2 band of the decomposed image from step 1using the generated key as positions.
Step 6: Apply inverse transformation to get the watermarked image.
Step 7: The left singular vectors U and right singular vectors V^T are transmitted securely using any public key encryption techniques.

3.2 Watermark Extraction Procedure Using SVD

Step 1: Decompose Cover and watermarked images using 'Haar' wavelet upto 2 levels to get approximation coefficients, LL2.
Step 2: Compute S from the approximation coefficients, LL2 of watermarked image and cover image using the key.
Step 3: Perform $U*S*V^T$ to recover the watermark.

3.3 Watermark Embedding Procedure Using GNS and AC

Step 1: Decompose Cover image using 'Haar' wavelet upto 2 levels to get approximation coefficients, LL2.
Step 2: Read the watermark i.e. the data (the intensity values of the pixels of the image).
Step 3: Starting from the first value, factorize each and every value into its primes.
Step 4: Store the prime numbers obtained from step 3 according to Gödel number sequencing.

Step 5: The Gödel number sequencing is again encoded with AC technique.
Step 6: Generate the key.
Step 7: Embed the string of characters in the positions of LL2 band indicated by the key. Now the transformed watermarked image is obtained.
Step 8: To get the watermarked image apply the inverse transform of the wavelet decomposition(IDWT) for 2 levels.

3.4 Watermark Extraction Procedure Using GNS and AC

Step 1: Decompose Cover and watermarked images using 'Haar' wavelet upto 2 levels to get approximation coefficients, LL2.
Step 2: Extract the string of characters from LL2 band of the watermarked image and cover image using key.
Step 3: Decode the string of characters with AC technique to obtain Gödel Number Sequence.
Step 4: The Gödel number sequencing is decoded to reconstruct the image.

4 Experimental Results

The SVD and Gödelization techniques are implemented on 25 standard database images of size 512x512 and watermark images of size 128x128 are considered, out of which 4 test cases for each technique are provided below in tables 1 and 2. Perceptual transparency is an evaluation metric which measures the performance. Perceptual transparency means the presence of watermark should not destroy the perceiving quality of the watermarked image and is measured by PSNR. Higher is the PSNR value, higher is the fidelity of the watermarked image. Figure 3 (a), 3 (c) shows that the cover image and the watermarked image are perceptually the same and there is no distortion in the extracted watermark as shown in figure 3 (b) and 3 (d).

Fig. 3(a). Cover Image

Fig. 3(b). Watermark

Fig. 3 (c). Watermarked Image

Fig. 3 (d). Extracted Watermark

Table 1. Experimental Results for standard images using SVD encryption technique

SVD				
Weighting Factor	0.01	0.006	0.01	0.01
PSNR of watermark	34.5116	33.1205	28.9373	31.6974
PSNR of cover image	45.0116	49.5508	45.0321	45.0203
Recovered Watermark				

Table 2. Experimental Results for standard images using GNS & AC technique

GNS & AC				
Weighting Factor	0.01	0.006	0.01	0.01
PSNR of Watermark	36.4526	36.1572	34.3649	33.9843
PSNR of cover image	46.1963	48.5629	47.2516	45.0526
Recovered Watermark				

5 Observations

Depending upon the results obtained, it can be observed that PSNR is high for the watermark encrypted with GNS and AC technique when compared to SVD technique. The storage space required for an image of size nxp using SVD technique is nxn+nxp+pxp bytes for the matrices U, S, V^T. Using GNS & AC it is nxpx35bits where each pixel can be represented by a maximum of 7 digits and after alphabetic coding each digit is represented by 5 bits. Even though the data bits to be embedded are more than SVD, because of its security and high PSNR, GNS proves to be more strong and efficient.

6 Conclusions and Future Work

The proposed methods perform encrypted watermark embedding into an image based on Discrete Wavelet Transform. The watermark is encrypted using two techniques, one with Singular value decomposition and another with Gödelization Technique and

Alphabet Coding. The comparative analysis of these two approaches show that Perceptual Transparency is high in Gödelization technique than in SVD technique and the watermark encrypted with Gödelization technique can be efficiently extracted when compared to SVD technique. But the storage space needed to implement SVD technique is less with respect to GNS & AC. The work can be further extended to overcome the limitation by improving the Alphabetic Coding technique to reduce the storage size. So any of these methods can be implemented depending on the requirement. GNS & AC can be used where Security is a high measure of priority. These methods can be further implemented for color images.

References

1. Schyndel, R.G.V., Tirkel, A.Z., Osborne, C.F.: A Digital Watermark. In: International Conference on Image Processing, vol. 2, pp. 86–90. IEEE, Austin (1994)
2. Li, L., Xu, H.-H., Chang, C.-C., Ma, Y.-Y.: A novel image watermarking in redistributed invariant wavelet domain. The Journal of Systems and Software 84, 923–929 (2011)
3. Liu, J.F., Huang, D.R., Hu, J.Q.: The orthogonal wavelet bases for digital watermarking. Journal of Electronics and Information Technology 25(4), 453–459 (2003)
4. Huang, D.R., Liu, J.F., Huang, J.W.: A DWT-based Image Watermarking Algorithm. In: Proc. of IEEE, ICME, Tokyo, Japan, vol. I, pp. 429–432 (2001)
5. Ogden, C.J., Huff, T.: The Singular Value Decomposition and It's Applications in Image Processing (December 1997)
6. Zhou, B., Chen, J.: A Geometric Distortion Resilient Image Watermarking Algorithm Based on SVD. Chinese Journal of Image and Graphics 9, 506–512 (2004)
7. Martin, J.: Introduction to Languages and the theory of Computation, 3rd edn., p. 462. TMH Publications
8. Lalitha Bhaskari, D., Avadhani, P.S., Damodaram, A.: A Combinatorial Approach for Information Hiding Using Steganography and Gödelization Techniques. International Journal of Systemics, Cybernetics and Informatics, 21–24 (2007) ISSN 0973-4864
9. Gonzalez, R.C., Woods, R.E.: Digital Image Processing, 2nd edn., pp. 50–51. Pearson Education
10. Lalitha Bhaskari, D., Damodaram, A., Avadhani, P.S.: Watermark Insertion Algorithm implementation using the auxiliary carry and LSB methods. In: Int. Conf. ICSCI 2006, India, pp. 666–668 (2006)
11. Kim, I., Hong, Han, S.S.: A Blind Watermarking Technique Using Wavelet Transform. In: ISIE, Pusan, Korea, pp. 1946–1950 (2001)
12. Tang, X., Wen, Q., Nian, G.: An Improved Robust Watermarking Technique in Wavelet Domain. In: Second International Conference on Multimedia and Information Technology, pp. 270–273 (2010)

Stereo Correspondence for Underwater Video Sequence Using Graph Cuts

C.J. Prabhakar and P.U. Praveen Kumar

Department of P.G. Studies and Research in Computer Science
Kuvempu University, Shankaraghatta - 577451
Karnataka, India
psajjan@yahoo.com, praveen577302@gmail.com

Abstract. In this paper, we introduce stereo correspondence method for underwater video sequence using Graph Cuts. The propagation property of light in the underwater causes variations in color information between two underwater video frames taken under same imaging conditions. To render the color values changed by the propagation property of light in the underwater environment, we use Markov Random Fields - Belief Propagation (MRF-BP) based approach for color correction. The conventional window-based correlation methods are often employed to estimate the disparity between the image pair, but these techniques are sensitive to illuminative variations, leads to fattening effect at the object boundaries and relatively lower performance in the featureless regions. Therefore, we employ energy minimization method such as Graph Cuts for the pair of color corrected underwater video frames to estimate disparity map. We compared and evaluated our approach qualitatively with well known window-based stereo correspondence techniques for the captured underwater video test frames. The experimental result reveals that our approach yields a visually suitable dense disparity map for the captured underwater video test frames compared to a window-based stereo correspondence techniques.

Keywords: Underwater video sequence, SSD, SAD, NCC, ZNCC, Graph cuts, Disparity Map.

1 Introduction

Underwater vision is one of the scientific fields of investigation for researchers. Autonomous Underwater Vehicles (AUVs) is usually employed to capture the data such as underwater mines, shipwrecks, coral reefs, pipelines and telecommunication cables from underwater environment [3]. Stereo correspondence technique helps us to obtain 3D information by finding the correct correspondence between a pair of stereo images captured from different point of views or at different time instance. Finding the accurate correspondence between a pair of stereo images is not an easy task in underwater environment due to absorption and scattering process of the light in water which degrades quality of underwater images. Forward scattering (randomly deviated light on its way from an object

S.C. Satapathy et al. (Eds.): Proc. of Int. Conf. on Front. of Intell. Comput., AISC 199, pp. 683–690.
DOI: 10.1007/978-3-642-35314-7_77

to the camera) generally leads to blur of the image features. On the other hand, backward scattering (the fraction of the light reflected by the water towards the camera before it actually reaches the objects in the scene) generally limits the contrast of the images, generating a characteristic veil that superimposes itself on the image and hides the scene. Absorption and scattering effects are not only due to the water itself but also to other components such as dissolved organic matter or small observable floating particles. The presence of the floating particles known as marine snow (highly variable in kind and concentration) increase absorption and scattering effects.

The amount of light is reduced when we go deeper, colors drop off one by one depending on their wavelengths. In contrast to light propagation in air, the single light rays are much more effected by the densely packed water molecules: they are attenuated and scattered, having a great effect on the image colors [9]. Colors of underwater images are dominated by a strong green or blue hue. This is due to the strong wavelength dependent attenuation of the different colors. In order to provide an accurate visualization of a scene, it is necessary to remove the green or blue hue and try to reconstruct the real colors of the object. In our approach, color-correction is done using MRF-BP based technique, which is modeled as a sample function of a stochastic process based on the Gibbs distribution, that is, as a Markov Random Field (MRF).

In this paper, we introduce stereo correspondence method to estimate the dense disparity map of the underwater video images. To render the color values which are changed by the propagation property of light in the underwater environment, we use MRF-BP approach, which distributes the color values equally in all regions and normalizes the color values. The color-corrected images are further used to find the correspondence points between two frames of the same scene captured in different viewpoint. We employ graph cuts based stereo correspondence method to estimate dense disparity map. To find the effectiveness, we compared and evaluated visually our approach with window based approaches such as Sum-of-Squared-Difference (SSD), Sum-of-Absolute-Difference (SAD), Normalized Cross-Correlation (NCC) and Zero-mean Normalized Cross-Correlation (ZNCC) [8]. The experimental result reveals that unlike window based approaches our approach based on graph cuts work well for underwater images.

The organization of the paper is as follows: Section 2 discuss the MRF-BP based color-correction for underwater images. In Section 3, we present our stereo correspondence approach to estimate the dense disparity map from underwater video images. The experimental results are presented in Section 4. Finally, Section 5 draws the conclusion.

2 MRF-BP Approach for Color Correction

The solution of the color correction problem can be defined as the minimum of an energy function. The idea on which our approach is based, is that an image can be modeled as a sample function of a stochastic process based on the Gibbs distribution, that is, as a Markov Random Field (MRF). We consider the color correction a task of assigning a color value to each pixel of the input image that best describes its surrounding structure using the captured image patches.

2.1 The Pairwise MRF Model

Denote the input color depleted image by $B = \{b_i\}$, $i = 1, ..., N$, where $N \in Z$ is the total number of pixels in the image and b_i is a triplet containing the RGB channels of pixel location i. We wish to estimate the color-corrected image $C = \{c_i\}$, $i = 1, ..., N$, where c_i replaces the value of pixel b_i with a color value.

A pairwise MRF model (also known as Markov network) is defined as a set of hidden nodes x_i representing local patches in the output image, and the observable nodes y_i representing local patches in the input test image. Each local patch is centered to pixel location i of the respective images.

Denoting the pairwise potentials between variables x_i and x_j by ψ_{ij} and the local evidence potentials associated with variables x_i and y_i by ϕ_i, the joint probability of the MRF model under variable instantiation $x = (x_1, ..., x_N)$ and $y = (y_1, ..., y_N)$, can be written [1] as:

$$P(x, y) = \frac{1}{Z} \prod_{ij} \psi_{ij}(x_i, x_j) \prod_{i} \phi_i(x_i, y_i), \tag{1}$$

where Z is the normalization constant. We wish to maximize $P(x, y)$, that is, we want to find the most likely state for all hidden nodes x_i, given all the evidence nodes y_i.

A color pixel value in C is synthesized by estimating the maximum a posteriori (MAP) solution of the MRF model using the captured Data set. The MAP solution of the MRF model is:

$$x_{MAP} = \arg\max_{x} P(x|y), \tag{2}$$

where

$$P(x|y) \alpha P(y|x) P(x) \alpha \prod_{i} \phi_i(x_i, y_i) \prod_{(i,j)} \psi_{ij}(x_i, x_j). \tag{3}$$

Calculating the conditional probabilities in an explicit form to infer the exact MAP in MRF models is intractable. We cannot efficiently represent or determine all the possible combinations between pixels with its associated neighborhoods. We compute a MAP estimate, by using a learning-based framework on pairwise MRFs, as proposed by [4], using belief propagation (BP).

The compatibility functions $\phi(x_i, y_i)$ and $\psi(x_i, x_j)$ are learned from the captured data set using the patch-based method in [4]. They are usually assumed to obey a Gaussian distribution to model Gaussian noise. The $\phi(x_i, y_i)$ compatibility function is defined as follows

$$\phi(x_i, y_i) = e^{\frac{-|y_i - y_{x_i}|^2}{2\sigma_i^2}}, \tag{4}$$

where x_i is a color-corrected patch candidate, y_{x_i} is the corresponding test image patch of x_i, and y_i is the patch in the input test image.

The image is divided so that the corresponding color-corrected patches overlap. If the overlapping pixels of two node states match, the compatibility between those states is high. We define $\psi(x_i, x_j)$ as:

$$\psi(x_i, x_j) = e^{\frac{-d_{ij}(x_i, x_j)}{2\sigma_i^2}}, \qquad (5)$$

where d_{ij} is the difference between neighborhoods i and j.

Images in the captured data set are pairs of small image regions of the greenish image with its corresponding color-corrected image, thus the compatibility functions depend on each particular input image.

2.2 MRF-MAP Inference Using BP

Belief propagation (BP) was originally introduced as an exact algorithm for tree-structured models [6], but it can also be applied for graphs with loops, in which case it becomes an approximate algorithm, leading often to good approximate and tractable solutions [10]. For MRFs, BP is an inference method to efficiently estimate Bayesian beliefs in the network by the way of iteratively passing messages between neighboring nodes.

The message send from node i to any of its adjacent nodes $j \in N(i)$ is

$$m_{ij}(x_j) = Z \sum_{x_i} \psi(x_i, x_j)\phi(x_i, y_i) \prod_{k \epsilon N(i) \backslash \{j\}} m_{ki}(x_i), \qquad (6)$$

where Z is the normalization constant. The maximum a posteriori scene patch for node i is:

$$x_{iMAP} = \arg\max_{x_i} \phi(x_i, y_i) \prod_{j \epsilon N(i)} m_{ji}(x_i). \qquad (7)$$

Candidate states for each patch are taken from the captured data set. For each greenish patch in the image, we search the training set for patches that best resemble the input. The color-corrected patches corresponding the best k patches are used as possible states for the hidden nodes.

3 Stereo Correspondence

Stereo correspondence is the process of finding corresponding points in two or more images. The crux of the stereo correspondence problem is how to deal with the inherent point ambiguity that results from the ambiguous local appearance of image points. If the local structures of neighboring image points are quite similar as in textureless or repetitive-textured regions, it may be very difficult to find their correspondences in other images without any proper global reasoning. To resolve the point ambiguity problem in stereo correspondence, many methods have been proposed during the last few decades.

The traditional window-based methods yield a dense disparity map by matching fixed square windows as a whole, relying on the assumption that nearby points within the support window usually have similar displacements. Therefore, depth variations will introduce errors in the calculation. This usually happens at depth discontinuities and weakly-texture regions. Whether the introduced error can be neglected or leads finally to the wrong decision depends on the similarity between the object, the occluded and visible part of the background, which is covered by the support window. The wrong decisions usually lead to the extending of objects horizontally and the fuzzy disparities in the weakly-textured areas.

Energy minimization methods try to overcome the problems found in conventional window-based methods. In these methods, the matching does not depend only on the neighbors of left and right images, but also depends on the matches of their neighbors. Hence, the match of a pixel influences the matches of its neighbor pixels. This influence is modeled by regularization constraints on the matches set [11]. Therefore, we employ the stereo correspondence algorithm, which is an energy optimization method. This method transforms the matching problem to a minimization of a global energy function. The minimization is achieved by finding out an optimal cut (of minimum cost) in a graph. When applied to minimize a global cost function in stereo vision, for each pixel, all possible disparities between minimum and maximum values is considered. Energy optimization method gives excellent results, performing better in textureless areas and near depth discontinuities.

Roy S. and Cox I.J. [7] were the first to use energy minimization approach in the context of multi-camera stereovision. Later, Boykov et al., [2] and Kolmogorov et al., [5] generalizes for many previous constructions and is easily applicable to vision problems that involve large numbers of labels, such as stereo, motion, image restoration, and scene reconstruction.

Let function f be the disparity function associated to each pixel of an image. We search labeling f that minimizes the energy. To define this energy function for f, a cost function is introduced, based on a photoconsistence criterion (similarity between intensities of a pixel $\mathbf{p}$ in the first image and the pixel $(\mathbf{p} + f_p)$ in the second image) called data term. A second term, called smoothness term, which penalizes discontinuities between neighborhood pixels. Thus, the energy can be written as:

$$E(f) = \sum_{p \epsilon P} D_p(f_p) + \sum_{\{p,q\} \epsilon N} V_{\{p,q\}}(f_p, f_q), \tag{8}$$

where term D_p is the data term and $V_{\{p,q\}}$ is the smoothness term penalty between adjacent pixels.

4 Experimental Results

The imaging of underwater objects is carried out in a small water body in which the object was kept at a depth of 5 feet from the surface level of the water. The image capturing setup consists of a waterproof camera which is Canon-D10 and objects (pipes and valves). The camera is moved around an object at a distance of 1 - 2 m to capture the video sequence. We have captured two video sequence in two different water conditions with turbidity levels. From each video sequence, we select two test frames for experimentation. Fig. 1(a), 1(b) (data set1) and Fig. 2(a), 2(b) (data set2) shows the underwater video test frames in two different water conditions. Since there is no benchmark database available for underwater images, the evaluation criteria such as Root Mean Square Error (RMSE) and Bad Pixel Map (BPM) cannot be applied to evaluate our approach. The evaluation is carried out visually using results of our method with results of a window-based stereo correspondence methods such as SSD, SAD, NCC and ZNCC.

In order to be able to correct the color of the images using MRF-BP method, the training data from the same underwater environment is needs to be gathered. We considered the neighboring frames of data set1 and data set2 as training data. The training data should be visually better than the input image where MRF-BP learns the compatibility functions required for color correction. Since our training data and input data (data set1 and data set2) are having same quality, we need to enhance the quality of training data compared to data set1 and data set2. The training data were enhanced by using a commercial software. These images are certainly much better, in terms of color and clarity, then the dataset1 and data set2, and they can be used to train our algorithm to color correct. Fig. 1(c), 1(d) (data set1) and Fig. 2(c), 2(d) (data set2) shows the result after color-correction using MRF-BP method. There are some factors that influence the quality of the results, such as the adequate amount of reliable information as an input and the statistical consistency of the images in the training set.

Fig. 3 and Fig. 4 shows the dense disparity maps obtained using SSD, SAD, NCC, ZNCC and proposed approach for color-corrected data set1 and data set2 respectively. Since we do not have ground truth data for this, we cannot measure the performance of our algorithm, however it can be seen that the resulting disparity map looks visually good. It is observed that our approach yields the visually suitable results compared to other window-based approaches. This is due to fact that, the graph cuts energy optimization method performing better in textureless areas and near depth discontinuities.

Fig. 1. Data Set1: Captured Underwater video frames (a) Frame 1 and (b) Frame 10; After color-correction (c) Frame 1 and (d) Frame 10

Fig. 2. Data Set2: Captured Underwater video frames (a) Frame 1 and (b) Frame 20 ; After color-correction (c) Frame 1 and (d) Frame 20

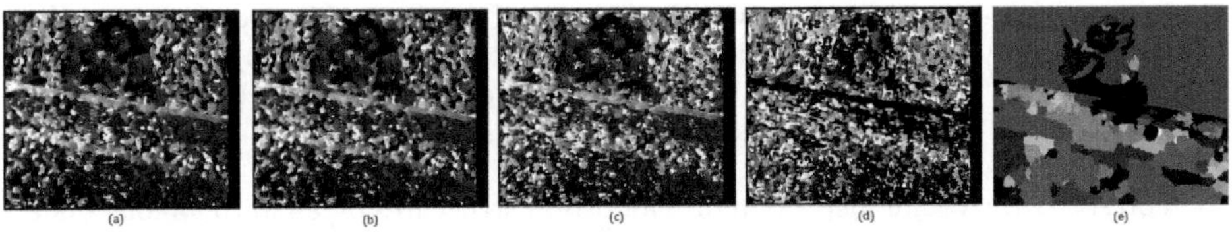

Fig. 3. Estimated Disparity Map for Data Set1 (a) SSD (b) SAD (c) NCC (d) ZNCC (e) Our Approach

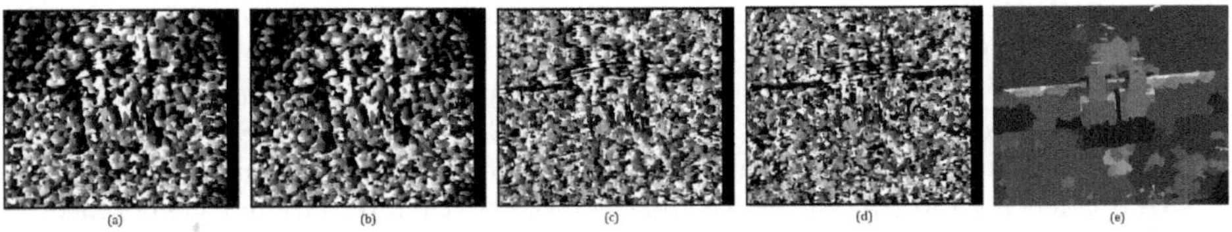

Fig. 4. Estimated Disparity Map for Data Set2 (a) SSD (b) SAD (c) NCC (d) ZNCC (e) Our Approach

5 Conclusion

In this paper, we introduce stereo correspondence approach to estimate dense disparity map for underwater video sequence. The experiments are conducted for underwater video frames, which are captured in two different water conditions. The video test frames are processed for color-correction using MRF-BP method. The color-corrected frames are used to estimate the dense disparity map based on Graph cuts. The experimental result shows that the our approach yields smooth disparity map compared to disparity maps of other window-based approaches such as SSD, SAD, NCC and ZNCC for underwater environment. Even though our approach exhibits a very high computational cost but delivers good results suitable for underwater applications. The obtained dense disparity map is evaluated visually because there is no benchmark database available for underwater images to evaluate quantitatively. The development of underwater image database could be one of the future research lines from which the underwater community would certainly beneficiate.

Acknowledgments. The authors would like to thank the anonymous reviewers for their valuable comments. This research was supported by Naval Research Board (Grant No.158/SC/08-09), DRDO, New Delhi, India.

References

1. Besag, J.: Spatial Interaction and the Statistical Analysis of Lattice Systems. Journal of the Royal Statistical Society. Series B (Methodological) 36(2), 192–236 (1974)
2. Boykov, Y., Veksler, O., Zabih, R.: Fast Approximate Energy Minimization via Graph Cuts. IEEE Transactions on Pattern Analysis and Machine Intelligence 23(11), 1222–1239 (2001)
3. Foresti, G.L.: Visual inspection of sea bottom structures by an autonomous underwater vehicle. IEEE Transactions on Systems, Man and Cybernetics, Part B 31, 691–705 (2001)
4. Freeman, W.T., Pasztor, E.C., Carmichael, O.T.: Learning Low-Level Vision. International Journal of Computer Vision 40(1), 25–47 (2000)
5. Kolmogorov, V., Zabih, R.: What Energy Functions Can Be Minimized via Graph Cuts? IEEE Transactions on Pattern Analysis and Machine Intelligence 26(2), 147–159 (2004)
6. Pearl, J.: Probabilistic Reasoning in Intelligent Systems: Networks of Plausible Inference. Morgan Kaufmann Publishers Inc., San Francisco (1988)
7. Roy, S., Cox, I.J.: A Maximum-Flow Formulation of the N-Camera Stereo Correspondence Problem. In: Proceedings of the Sixth International Conference on Computer Vision, pp. 492–499 (1998)
8. Scharstein, D., Szeliski, R.: A taxonomy and evaluation of dense two-frame stereo correspondence algorithms. International Journal of Computer Vision 47, 7–42 (2001)
9. Sedlazeck, A., Koser, K., Koch, R.: 3D Reconstruction Based on Underwater Video from ROV Kiel 6000 Considering Underwater Imaging Conditions. In: OCEANS 2009 - EUROPE, pp. 1–10 (2009)
10. Yedidia, J.S., Freeman, W.T., Weiss, Y.: Constructing Free-Energy Approximations and Generalized Belief Propagation Algorithms. IEEE Transactions on Information Theory 51(7), 2282–2312 (2005)
11. Zureiki, A., Devy, M., Chatila, R.: Stereo Matching and Graph Cuts. In: Bhatti, A. (ed.) Stereo Vision, pp. 349–372. I-Tech (2008)

Removal of High Density Impulse Noise from Digital Images and Image Quality Assessment

V. Anji Reddy and J. Vasudeva Rao

Department of Computer Science and Engineering,
GMR Institute of Technology, Rajam,
Srikakulam District, Andhra Pradesh (A.P)
anjisoftware@hotmail.com, jvr.vasu@gmail.com

Abstract. The digital images are corrupted by noise due to errors generated in camera sensors, analog to digital conversion and communication channels. Many types of noises may occur. In this paper we are concentrating on the impulse noise. Therefore it is necessary to remove impulse noise in order to provide the quality image. Filtering a noise image is one of the important methods in image processing. This paper presents a new method for impulse noise removal from digital images using iterative scheme. The performance of the iterative scheme is compared with other filters like SMF, DBA and VMF using image quality assessment metrics called peak signal to noise ratio, structural similarity (SSIM) index and region of index SSIM. The experimental results show the proposed algorithm can perform significantly better in terms of noise reduction by using image quality assessment algorithms PSNR.SSIM, ROI-SSIM.

Keywords: Impulse Noise, SMF, DBA, PSNR, SSIM, ROI-SSIM.

1 Introduction

Generally digital images may get the impulse noise during acquisition or transmission. The restoration of noise free images is carried out as a pre processing task in a wide range of image applications e.g.: In medical imaging and astronomical imaging applications. The impulse noise contains major two properties first one certain percentage of image pixels are corrupted with noise, the second of intensity of corrupted pixels is significantly different from noise free pixels [4]. There are two types of impulse noise models are used i.e. salt- and- pepper and random value noise. The noisy pixels contains with salt-and-pepper noise have two values – the minimum I_{min} and the maximum I_{max} value within the dynamic range [I_{min} , I_{max}]. However, Noisy pixels of image, corrupted with random-valued noise, have many random value from the dynamic range [I_{min} , I_{max}] . With the increase of noise density, in the image, numbers of Noisy pixels are increased. If the numbers of noisy pixels are greater than noise-free pixels then noise filtering become crucial. In this work to highlight the effectiveness of the proposed iterate scheme, we use only the salt-and-pepper noise [1,5, 4, 7].

There are number of filter schemes are available. In this paper we used the Standard Median Filter (SMF)[1] as the first scheme. The SMF based techniques replace

S.C. Satapathy et al. (Eds.): Proc. of Int. Conf. on Front. of Intell. Comput., AISC 199, pp. 691–698.
DOI: 10.1007/978-3-642-35314-7_78

every image pixel with the median value computed within the window without considering the status of (Noisy/Noise-free) pixels. Decision Based Algorithm (DBA)[2] is the second scheme. In this method the impulse noise pixels can take the maximum and minimum values in the dynamic range is (0, 255)[2]. If the value of the pixel processed is within the range, then it is an uncorrupted pixel and left unchanged. If the value does not lie within this range, then it is a noisy pixel and is replaced by the median value of the window or by its neighbourhood values. Vector Median Filter (VMF) [3] is the third scheme. In this method the vector pixels in a particular kernel or window are ordered based on a suitable distance measure. The sum of the distances between each vector pixel and the other vector pixels in the window is calculated. The distances are arranged in the ascending order and then the same ordering is associated with the vector pixels. The vector pixel with the smallest sum of distances is the vector median pixel [9].

Human Visual system cannot find the quality of the image after removing the noise from image. In this paper we assess the quality of image using three metrics. First Peak Signal to Noise Ratio (PSNR)[11], Structural Similarity (SSIM) Index[10], Region of Index-SSIM[10]. The following sections describe the proposed iterative scheme and the image quality assessment using these metrics.

2 Iterative Scheme

Here the iterative scheme is the proposed scheme to estimate the noise-free pixels with in a small neighbourhood. There are three steps to implement the iterative scheme[1].

- Construction of Detection Map
- Impulse Noise Filtering Method
- Update Noisy Image and Detection Map

2.1 Construction Of Detection Map

In this step, the detection map is constructed from the input noisy image *X*. In case of salt-and-pepper noise, the maximum and the minimum intensity values of the image dynamic range $[I_{min}, I_{max}]$ provide information about the Corrupted pixels that are used to detect the noisy pixels[1].

For 8-bit gray scale image, we assume 0 or 255 pixel values indicate the image pixel is corrupted with salt-and pepper noise. In case of noise-free images, very small number of pixels can have these two extreme values. Considering this assumption, we assign a binary value to each elements $d_{ij} \in D$ of the detection map *D*. The detection map is computed from the noisy image as follows

$$d_{i,j} = \begin{cases} 1, if \ x_{i,j} = I_{max} \\ 1 \ if \ x_{i,j} = I_{min} \\ 0, \ other \ wise \end{cases} \tag{1}$$

The entries of "1" and "0" in the detection map *D* represent the noisy and the noise-free pixels, respectively.

This map provides useful information about the noise intensity in the corrupted image. This information is used during filtering process. However, if the dynamic range of noise has more than two extreme values, the above simple detection method may not be accurate. In that case, the detection map can be developed using various derivative based detection methods.

2.2 Impule Noise Filtering Method

We use a small window W_{ij} neighbourhood of size 3x3 at each pixel location (i, j) of the noisy image X and the detection map D. We prefer to use small window because the larger size window may not be too efficient and effective. This is because the correlation between pixels decreases as pixels are separated apart. Moreover, the larger window may also remove the edges and fine image details. By applying small window, we obtain the noisy image patch $W_{i,j}^{x}$ and the detection map patch, $W_{i,j}^{d}$ in the form of 3X3 matrix form [1].

For each iteration, we count the number of noisy pixels in the detection map D. If the value of count K is a positive integer and the central pixel $x_{i,j}$ within 3x3 windows is noisy, and then an array R is populated with noise-free-pixels. The length of array, depending upon the noise density within the window, varies from zero to eight. The minimum length zero shows all pixels in the window are noisy, whereas the maximum length eight indicates all eight pixels are noise-free. To estimate the value of noisy pixel, we emphasize noise free pixels and a constraint of minimum three noise-free pixels ($M_f = 3$) in the array R. If this condition is satisfy, then we replace the central noisy pixel with the estimated value i.e.,

$$g_{i,j} = \begin{cases} e_s \ if \ d_{i,j} = 1 \vee L_R \geq M_f \\ x_{i,j}\ , otherwise \end{cases} \tag{2}$$

Where e_s is the estimated value for the noisy pixel, and L_R= length(R) is the length of the array R. Currently, we estimated the value of noisy pixels taking average from the noise-free pixels. Other statistical estimates can be employed. However, in case of noise-free pixels, we found average estimate better and computationally efficient. The value of e_S is computed as

$$e_S = \frac{1}{L_R} \sum_{i=1}^{L_R} R_i \tag{3}$$

If the condition of minimum noise free pixels is not satisfied (i.e., length($R<M_f$) then the central noisy pixel is left untreated. In other worlds, there must be at least three noise free pixels in the window to estimate the average value. In case of high noise density, there is a more possibility that the central noisy pixel we may not satisfy the constraint $L_R \geq M_f$.

Therefore, many noisy pixels are not estimated in early iterations.

2.3 Update Noisy Image and Detection Map

If the noisy pixel is replaced with the average estimated, then the detection map is also updated by changing the entries at the corresponding location in the detection map from "1" to "0" i.e.

$$d_{i,j} = \begin{cases} 0 \text{ if } d_{i,j} = 1 \vee L_R \geq M_f \\ d_{i,j} \quad \text{otherwise} \end{cases} \tag{4}$$

At the end of each iteration, we obtain a refined image *G* and updated detection map *D*. The number of entries in the detection map reduces. After a few iterations, depending upon the intensity of salt-and-pepper noise, all the entries in the detection map become zeros. The updating process terminated and we obtained a restored image *G*.

1. Taking initial noisy image X
2. Computation of initial detection map D
3. Compute the value of K that represent the noise- free pixels in D and assign X →G
4. Check If K = 0, output resorted images G & stop iteration process , otherwise do
   i. Check If central pixels X_{ij} is noisy , then do
   ii. Fill the array R with noise-free pixels
   iii. Check if $L_R \geq M_f$ do
   iv. Update d_{ij} and g_{ij} using average estimate
   v. Process each X_{ij} and get updated G and D
   vi. For next iteration ;assign G→X and go to step iii

3 Image Quality Assessment

After the removal of noise from digital images, by human visual system is unable to estimate the quality of the resulted image. This is the reason to assess the quality of the image by using quality metrics. So many metrics are available. But here we used three types of metrics called PSNR, SSIM and ROI-SSIM.

The peak signal to noise ratio (PSNR) is most commonly used as a measure of quality of reconstruction of digital images. The PSNR is applied to original image with noise and reconstructed image. A higher PSNR [11] would normally indicate that the reconstruction is of higher quality. It is most easily defined PSNR via the mean squared error (MSE) which for two m×n monochrome images

$$MSE = \frac{1}{M^2} \sum_{i-1}^{M} \sum_{j-1}^{M} (X'_{i,j} - G_{i,j})^2 \tag{5}$$

Where $X'_{I,J}$ is the original noise-free image, $G_{i,j}$ is the restored image, and M x M indicates the size of pixels of the original and the restored images.

$$PSNR = 10 \log_{10} \frac{(255)^2}{MSE} \tag{6}$$

The structural similarity (SSIM) index is a method for measuring the similarity between two images. The SSIM index is the measuring of image quality based on an initial uncompressed or distortion-free image as reference. SSIM is designed to improve on traditional methods like peak signal-to-noise ratio (PSNR) and mean squared error (MSE), which have proved to be inconsistent with human eye perception.

The structural similarity paradigm hypothesized that the comparison between a original image and a distorted image consists of three factors. They are luminance comparison, contrast comparison and structural comparison [10]. Their definitions are as follow:

$$\mathbf{l(x,y)} = \frac{2\mu_x\mu_y+C_1}{\mu_x^2+\mu_y^2+C_1} \tag{7}$$

$$\mathbf{c(x,y)} = \frac{2\sigma_x\sigma_y+C_2}{\sigma_x^2+\sigma_y^2+C_2} \tag{8}$$

$$\mathbf{s(x,y)} = \frac{\sigma_{xy}+C_3}{\sigma_x\ \sigma_y+C_3} \tag{9}$$

l(x, y) is the form of luminance comparison, c(x,y) is the form of contrast comparison, and s(x,y) is the form of structural comparison. x is reference image, y is distorted image.μ_x and μ_y are the means of x and y. σ_x and σ_y are their standard deviations. σ_{xy} is the covariance between x and y.They are estimated as follow

$$\sigma_{xy} = \frac{1}{N-1}\sum_{i=1}^{N}(X_i - \mu_X)(y_i - \mu_y) \tag{10}$$

$$\mu_x = \overline{X} = \frac{1}{N}\sum_{i=1}^{N}X_i \tag{11}$$

$$\mu_y = \overline{Y} = \frac{1}{N}\sum_{i=1}^{N}Y_i \tag{12}$$

$$\sigma_x = \left(\frac{1}{N-1}\sum_{i=1}^{N}(X_i - \mu_x)^2\right)^{\frac{1}{2}} \tag{13}$$

$$\sigma_y = \left(\frac{1}{N-1}\sum_{i=1}^{N}(Y_i - \mu_y)^2\right)^{\frac{1}{2}} \tag{14}$$

The SSIM index of each block is:

$$\text{SSIM(x,y)=l(x,y) c(x,y) s(x,y)} \tag{15}$$

Finally SSIM of entire image is

$$\text{MSSIM(x,y)}=\frac{1}{k}\sum_{j=1}^{k} SSIM(X_j, Y_j) \tag{16}$$

Where k =number of blocks

When humans observe any image, their focus will be only on particular region only. Usually, HVS is more sensitive to structural variation rather than luminance variation. To overcome this problem we go for Region Of Index SSIM (ROI-SSIM)[10]. First, an image is divided into blocks with size of 11×11 and the structural value σ of each block is calculated. Second, the structural value of current block is compared with its four-sided neighbors'. If the differences are smaller than a threshold, they would be merged up and the structural value of the newly merged block would be updated. Otherwise the block is kept as a region.

$$\sigma_x = \left(\frac{1}{N-1}\sum_{i=1}^{N}(X_i - \mu_x)^2\right)^{\frac{1}{2}} \tag{17}$$

We assume that the smaller each region is the more observers tend to focus on it while seeing an image. According to the dividing strategy mentioned above, we do some alternation on SSIM. Set each block in *i*-th region with a weigh ω_i

$$\omega_i = \frac{1}{N_i \, X \, M} (i = 1 \text{ to } M) \tag{18}$$

Where M is the number of divided regions in reference image, N_i is the number of blocks in each region. Therefore, the smaller each region is, the larger the weights of each block in the region. ROI-SSIM can be defined as follows

$$\text{ROI-SSIM}(x,y) = \sum_{j=1}^{N_i} \sum_{i=1}^{M} \omega SSIM(X_i, Y_j) \tag{19}$$

4 Experimental Results

In this section of the paper the experiment results are discussed. The experimentation is carried out on the LENA image which is shown in Fig.1. The results of the experiment performed on LENA image are shown in the Table 1 and Fig.2. Here the image is corrupted with various noises from 0.1 to 0.9. The LENA image is corrupted with salt-and-pepper noise as shown in Fig.1.

We used peak signal to noise ratio (PSNR) [1], SSIM and ROI-SSIM to assess the quality of image after applying the proposed approach. Table1 shows the PSNR values with different noise density values obtained using different algorithms. Table 2 and Table 3 show the SSIM and ROI-SSIM values with different noise density values obtained using different algorithms. Fig.3 describes performance level of the various algorithms SMF, DBA, VMF and Iterative Scheme using PSNR.

Fig. 1. Original LENA image and LENA image with noise

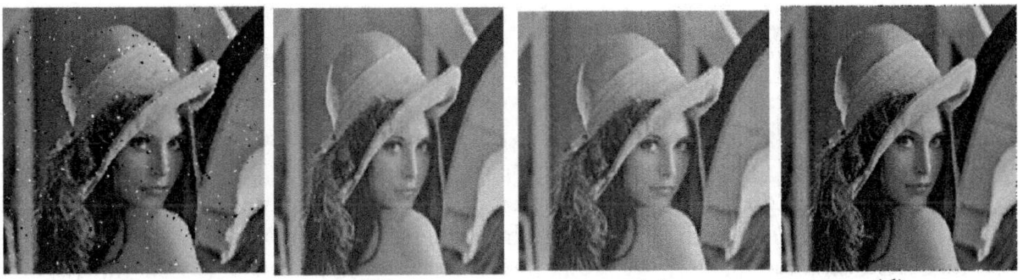

Fig. 2. Noise removal using (a) SMF (b) DBA (c) VMF (d) ITERATIVE

Table 1. Performance Comparison of LENA Image with Various Noises Densities using PSNR

Noise Density	Smf	Dba	Vmf	Iterative
0.1	29.5645	40.3689	37.7696	38.5165
0.2	27.8938	36.148	33.597	37.875
0.3	22.795	33.2256	28.856	36.1878
0.4	18.6153	30.9438	23.774	35.8546
0.5	14.7285	28.3112	20.1827	33.5786
0.6	12.0145	26.5689	17.3473	30.6201
0.7	9.0606	23.5004	14.9313	25.407
0.8	7.8882	20.8504	13.1824	23.0729
0.9	6.3577	17.5529	11.5224	19.055
mean	16.5465	28.6078	22.3514	31.1297

Table 2. Performance Comparison of LENA Image with Various Noises Densities using SSIM

noise density	smf	dba	vmf	iterative
0.1	0.9896	0.1058	0.9891	0.9979
0.2	0.9856	0.1061	0.9852	0.9956
0.3	0.9758	0.1059	0.9776	0.9928
0.4	0.9590	0.1059	0.9759	0.9897
0.5	0.9215	0.1051	0.9258	0.9854
0.6	0.8689	0.1042	0.8674	0.9807
0.7	0.7864	0.1038	0.7786	0.9743
0.8	0.6788	0.1005	0.6892	0.9599
0.9	0.5729	0.0911	0.5838	0.9323

Table 3. Performance Comparison of LENA Image with Various Noise Densities using ROI-SSIM

Noise Density	smf	dba	vmf	Iterative
0.1	0.9423	0.2054	0.9414	0.9946
0.2	0.9341	0.2024	0.9294	0.9832
0.3	0.8859	0.1981	0.8817	0.9653
0.4	0.8383	0.2164	0.8417	0.9851
0.5	0.7182	0.1797	0.8056	0.9403
0.6	0.6977	0.2093	0.7105	0.9172
0.7	0.5999	0.1570	0.5506	0.8793
0.8	0.5019	0.1569	0.5174	0.8586
0.9	0.4418	0.1230	0.3867	0.7380

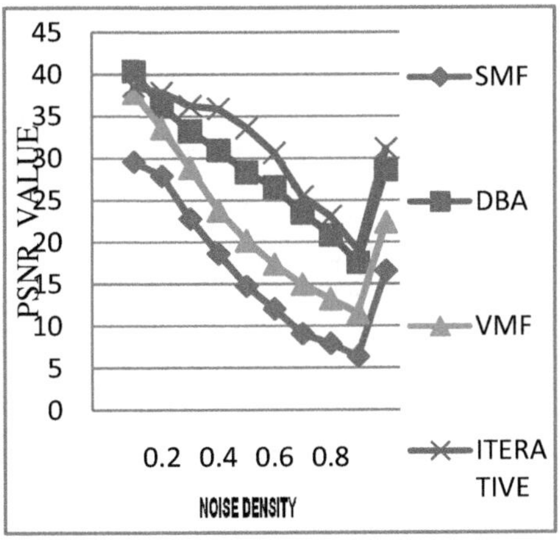

Fig. 3. Performance Comparison for the LENA image corrupted with various noise densities

5 Conclusion

In this paper the iterative scheme is discussed which is able to process noise pixels by using noise-free pixels. This scheme provides better performance when compared to other algorithms. As per the results the iterative scheme is capable of removing impulse noise from high density digital images. The proposed method is having good performance compared with other algorithms. The SMF, DBA and VMF[3] methods are used to compare the iterative scheme. Finally it can be stated that the impulse noise can be removed using iterative scheme effectively with high peak signal to noise ratio (PSNR) and better average of SSIM and ROI-SSIM.

References

1. Majid, A., Mahmood, M.: A Novel Technique for removal of high density impulse noise from digital images. IEEE (2010)
2. Srinivasan, K.S., Ebenzer, D.: A new and efficient decision-based algorithm for removal of high density impulse noises. IEEE (2007)
3. PremKumar, Harikiran, J., SaiChandana, B., RajeshKumar, P.: Performance evaluation of Image Fusion for Impulse Noise Reduction in Digital Images Using an Image Quality Assessment. IJCSI (2011)
4. Petrovic, N.I., Crnojevic, V.: Universal Impulse Noise Filter Based on Genetic Programming. IEEE Transactions on Image Processing 17, 1109–1120 (2008)
5. Zhengya, X., Hong Ren, W., Bin, Q., Xinghuo, Y.: Geometric Features-Based Filtering for Suppression of Impulse Noise in Color Images. IEEE Transactions on Image Processing 18, 1742–1759 (2009)
6. Hancheng, Y., Li, Z., Haixian, W.: An Efficient Procedure for Removing Random-Valued Impulse Noise in Images. IEEE Signal Processing Letters 15, 922–925 (2008)
7. Ghanekar, U., Singh, A.K., Pandey, R.: A Contrast Enhancement Based Filter for Removal of Random Valued Impulse Noise. IEEE Signal Processing Letters 17, 47–50 (2008)
8. Gouchol, P., Jyh-Charn, L., Nair, A.S.: Selective removal of impulse noise based on homogeneity level information. IEEE Transactions on Image Processing 12, 85–92 (2003)
9. Zhou, W., Zhang, D.: Progressive switching median filter for the removal of impulse noise from highly corrupted images. IEEE Transactions on Circuits and Systems II: Analog and Digital Signal Processing 46, 78–80 (1999)
10. Liu, Y., Sun, H., Di, Y., Zhou, Y.: A New Region of Intrest Based Image Quality Assessment Algorithm, 978-1-4244-3709-2/10 on 2010 IEEE
11. Hore, A., Ziou, D.: Image Quality metics:PSNR vx SSIM. In: IEEE International Conference on Pattern Recognition (2010)

A Differential Evolutionary Multilevel Segmentation of Near Infra-Red Images Using Renyi's Entropy

Soham Sarkar[1], Nayan Sen[1], Abhinava Kundu[1],
Swagatam Das[2], and Sheli Sinha Chaudhuri[3]

[1] Electronics and Communication Engineering Department, RCC Institute of Information Technology, Kolkata – 700015, India
[2] Electronics and Communication Sciences Unit, Indian Statistical Institute, Kolkata – 700108, India
[3] Electronics and Telecommunication Engineering Department, Jadavpur University, Kolkata – 700032, India

{sarkar.soham,nayansen90,animesh.k.rcc}@gmail.com,
swagatamdas19@yahoo.co.in,
shelism@rediffmail.com

Abstract. In recent years remote sensing image processing has got some intense attention of the researchers for its utility in land cover study, natural calamity detection, object tracking etc. In case of remote sensing image processing, the primal objective is to sub divide the image into more than one segment. In doing so, Multi-level thresholding based image segmentation techniques play an useful role in accomplishing this critical task. Endeavor of this paper is to focus on obtaining the optimal multiple threshold points from a LISS III Near Infra-Red (NIR) band by employing Renyi's Entropy. Moreover, a state-of-art meta-heuristics like Differential Evolution (DE) is incorporated to acquire optimal threshold values in reduced computational time with precision.

Keywords: Multilevel Image Segmentation, Remote Sensing Images, Renyi's Entropy, Differential Evolution, LISS-III, Near Infra-Red (NIR).

1 Introduction

Image segmentation is a method to distinguish between foreground and background. It forms the basis of computer vision, especially in areas related to feature extraction, identification etc. The segmented regions from the image are used as a basis for the machine Intelligence. In bi-level segmentation the image is segmented into two classes of objects and backgrounds, whereas in multilevel such as tri-level, quad-level thresholding the image is divided into multiple homogeneous regions based on intensity, resulting in higher number of components being extracted from the same image.

Multi-level entropy based techniques, which allows separation between different objects by considering them as dissimilar components in that image, can be used to successfully achieve those aforesaid objectives. Over the years Renyi's entropy has

S.C. Satapathy et al. (Eds.): Proc. of Int. Conf. on Front. of Intell. Comput., AISC 199, pp. 699–706.
DOI: 10.1007/978-3-642-35314-7_79

been applied to image segmentation paradigm for bi-level thresholding, which exhibits pleasing outcomes [1, 2]. Here, in this paper a Renyi's entropy based algorithms for multi-level thresholding is being proposed to determine optimal threshold values.

However, multi-level thresholding methods require significantly large computational time. Among the several alternative techniques that we find in literature, use of global optimization techniques like Genetic Algorithm (GA), Particle Swarm Optimization (PSO) etc. These are the most efficient and widely used approaches in order to find the maximum additive entropy between various classes [3]. In this paper Differential Evolution is used as it potentially outperforms other meta-heuristics in terms of convergence, speed, and accuracy, for solving numerous complex optimization demands [4, 5]. DE also exhibits superior performance when it is applied in the field of multi-level image segmentation [6].

The rest of the paper includes the basic concept of Renyi's entropy in Section 2. The Section 3 describes the proposed algorithm for multi-level image thresholding. A brief introduction of Differential Evolution (DE) is given in Section 4, whereas experimental results and comparative performance are presented in Section 5. Lastly the paper is concluded in Section 6.

2 Renyi's Entropy

For a system having a finite probability distribution of $P = (p_1, p_2, p_3 \dots p_n)$ where $p_i \geq 0$ where $i \in \{1, 2, \dots, n\}$ and $\sum_{i=1}^{n} p_i = 1$, the measure of uncertainty about the outcome of an experiment whose results have the above probability distribution is given by $H_S[P]$, as introduced by Shannon and defined by:

$$H_S[P] = \sum_{i=1}^{n} p_i \log_2\left(\frac{1}{p_i}\right) \tag{1}$$

Alfred Renyi [7] proposed a generalized form of Shannon's entropy for additively independent events, defined by:

$$H_\alpha[P] = \frac{1}{1-\alpha} \log_2\left(\sum_{i=1}^{n} {p_i}^{\alpha}\right) \tag{2}$$

Where $\alpha > 0$, and is known as the order of the entropy. Thus $H_\alpha[P]$ is known as the entropy of order α of the distribution P. It can be shown that Shannon entropy is a limiting case of Renyi entropy where $\alpha \to 1$,

$$\lim_{\alpha \to 1} \frac{1}{1-\alpha} \log_2\left(\sum_{i=1}^{n} {p_i}^{\alpha}\right) = \sum_{i=1}^{n} p_i \log_2\left(\frac{1}{p_i}\right) = H_s[P] \tag{3}$$

Where $H_s[P]$ is Shannon's entropy.

3 Multi-level Renyi's Entropy

In the case of image segmentation using multilevel thresholding, the 1-D histogram of the image is divided into several classes of pixels delimited by grayscale intensity values $(k_1, k_2, k_3 \ldots k_n)$ which are also known as '*threshold values*'. This formulate the basis of multilevel thresholding as each class of pixels is assigned a fixed threshold, and so segmenting the pictures into regions of varying grayscale intensity levels such as bi-level, tri-level, etc.

The corresponding probability of occurrence of pixels of grayscale intensity 'i' is defined as:

$p_i = \frac{n_i}{total\ number\ of\ pixels}$; n_i is the number of pixels at grayscale intensity "i"

This gives a finite probability distribution for the entire image as $P = (p_0, p_1, p_2 \ldots p_{255})$. Assuming a set of threshold values as $K = (k_1, k_2, k_3 \ldots k_n)$ .This set of threshold values divides the entire histogram into classes of pixels, whose probability distribution are as follows:

$$C_1 = (p_0, p_1, p_2 \ldots p_{k_1}) \text{ , } C_2 = (p_{k_1+1}, \ldots, p_{k_2}) \ \ldots \ C_{n+1} = (p_{k_n+1}, p_{k_n+2}, \ldots, p_{255})$$

The sum of the probabilities of a class or the total class probability is given by,

$$P(C_1) = \sum_{C_1} p_i = \sum_{i=0}^{k_1} p_i \text{ , } P(C_2) = \sum_{C_2} p_i = \sum_{i=k_1+1}^{k_2} p_i \text{ , } \ldots \text{ , } P(C_{n+1}) = \sum_{C_{n+1}} p_i = \sum_{i=k_n+1}^{255} p_i$$

Now the Renyi's entropy of each class is given by

$$H_\alpha[C_1] = \frac{1}{1-\alpha}\left[\ln \sum_{i=0}^{k_1} \left(\frac{p_i}{P(C_1)}\right)^\alpha\right]$$

$$H_\alpha[C_2] = \frac{1}{1-\alpha}\left[\ln \sum_{i=k_1+1}^{k_2} \left(\frac{p_i}{P(C_2)}\right)^\alpha\right]$$

$$H_\alpha[C_{n+1}] = \frac{1}{1-\alpha}\left[\ln \sum_{i=k_n+1}^{255} \left(\frac{p_i}{P(C_{n+1})}\right)^\alpha\right] \tag{4}$$

The probabilities of each class are normalized by dividing by the class probability so as to follow the postulate that $\sum_C p_i = 1$

The total Renyi entropy of the entire image, extended from Sahoo [1], is given as:

$$H_\alpha[I] = H_\alpha[C_1] + H_\alpha[C_2] + \cdots + H_\alpha[C_{n+1}] \tag{5}$$

Extending

$$H_{\alpha}[I] = \frac{1}{1-\alpha}\left[\ln \sum_{i=0}^{k_1} \left(\frac{p_i}{P(C_1)}\right)^{\alpha} + \ln \sum_{i=k_1+1}^{k_2} \left(\frac{p_i}{P(C_2)}\right)^{\alpha} + \cdots \right.$$
$$\left. + \ln \sum_{i=k_n+1}^{255} \left(\frac{p_i}{P(C_{n+1})}\right)^{\alpha}\right] \tag{6}$$

The optimum threshold value set $K^* = (k_1{}^*, k_2{}^*, k_3{}^* \ldots k_n{}^*)$ for which the entropy is maximum, is given as:

$$K_{\alpha}^* = Arg \max_{K^* \in L^n} \{H_{\alpha}[I]\} \tag{7}$$

For a prior chosen α the threshold set K^* is used to segment the image into n levels.In this paper DE is used to maximizing equation (7) and hence finding the optimal threshold values.

4 Differential Evolution (DE)

DE, a population-based global optimization algorithm, was proposed by Storn [4] in 1997. The ith individual (parameter vector) of the population at generation (time) t is a D-dimensional vector containing a set of D optimization parameters:

$$\overrightarrow{Z_i}(t) = [Z_{i,1}(t), Z_{i,2}(t), \ldots \ldots, Z_{i,D}(t)] \tag{8}$$

In each generation to change the population members $\overrightarrow{Z_i}(t)$ (say), a *donor* vector $\overrightarrow{Y_i}(t)$ is created. It is the method of creating this donor vector that distinguishes the various DE schemes. In one of the earliest variants of DE, now called DE/rand/1 scheme, to create $\overrightarrow{Y_i}(t)$ for each i^{th}member, three other parameter vectors (say the r_1, r_2, and r_3-th vectors such that $r_1, r_2, r_3 \in [1, NP]$ and $r_1 \neq r_2 \neq r_3$) are chosen at random from the current population. The donor vector $\overrightarrow{Y_i}(t)$ is then obtained multiplying a scalar number F with the difference of any two of the three. The process for the jth component of the ith vector may be expressed as,

$$\overrightarrow{Y_{i,j}}(t) = Z_{r1,j}(t) + F.\left(Z_{r2,j}(t) - Z_{r3,j}(t)\right) \tag{9}$$

A 'binomial' crossover operation takes place to increase the potential diversity of the population. The binomial crossover is performed on each of the D variables whenever a randomly picked number between 0 and 1 is within the Cr value. In this case the number of parameters inherited from the mutant has a (nearly) binomial distribution. Thus for each target vector $\overrightarrow{Z_i}(t)$, a trial vector $\overrightarrow{R_i}(t)$is created in the following fashion:

$$\begin{aligned} R_{i,j}(t) &= Y_{i,j}(t) \qquad if\ rand_j(0,1) \leq Cr\ or\ j = rn(i) \\ &= Z_{i,j}(t) \qquad if\ rand_j(0,1) > Cr\ or\ j \neq rn(i) \end{aligned} \tag{10}$$

For $j = 1, 2, \ldots, D$ and $rand_j(0,1) \in [0,1]$ is the jth evaluation of a uniform random number generator. $rn(i) \in [1, 2, \ldots \ldots, D]$ is a randomly chosen index to ensures that

$\overrightarrow{R_\iota}(t)$ gets at least one component from $\overrightarrow{Z_\iota}(t)$. Finally 'selection' is performed in order to determine which one between the target vector and trial vector will survive in the next generation i.e. at time$t = t + 1$. If the trial vector yields a better value of the fitness function, it replaces its target vector in the next generation; otherwise the parent is retained in the population:

$$\left.\begin{aligned} \overrightarrow{Z_\iota}(t+1) &= \overrightarrow{R_\iota}(t) \quad if\ f\left(\overrightarrow{R_\iota}(t)\right) \le f\left(\overrightarrow{Z_\iota}(t)\right) \\ &= \overrightarrow{Z_\iota}(t) \quad if\ f\left(\overrightarrow{R_\iota}(t)\right) > f\left(\overrightarrow{Z_\iota}(t)\right) \end{aligned}\right\} \tag{11}$$

where $f(.)$ is the function to be minimized.

5 Experimental Results

The simulations are performed with MATLAB R2011b in a workstation with Intel® Core™ i3 3.2 GHz processor. For testing and analysis, 3 LISS III Band 4 (Near Infrared Band 0.77 - 0.86 micron) images are used from the Bhuvan – NRSC Open EO Data Archive (http://bhuvan-noeda.nrsc.gov.in/download/).LISS III, a multispectral sensor with spatial resolution of 23.5 m, is boarded on Indian Resourcesat-1 satellite. The toposheet no. of the used images is NF45K06, NE44U16, and NF45T13 respectively.

As the RAW images are low contrast images, Contrast-limited adaptive histogram equalization (CLAHE) is applied to enhance the dynamic range of the image [8].Fig. 1 displays the NIR RAW images and their histogram, before and after the contrast enhancement technique is applied. Performance evolution metrics like Weighted Peak Signal to Noise Ratio (WPSNR) [9] and Mean Structural Similarity Index Measurement (MSSIM) [10] between original grayscale image and segmented grayscale image are also used to establish desired differences of results.

The segmented greyscale image is formed by using a generalized equation.

$$f_S(x,y) = \left\{k * \left\lfloor \frac{L-1}{n+1} \right\rfloor \right. \quad if\ (t_{k-1}, s_{k-1}) < f(x,y) \le (t_k, s_k), \tag{12}$$

where $k = 1, 2, \ldots, n+1$.

The performance of DE based method is compared with other efficient global optimization techniques like GA and PSO. In case of DE, the following parametric setup is used for all the test images: $Cr = 0.9,\ F = 0.5.$ Results are reported as the mean of the objective functions of 50 independent runs. Each run contains 500 generations. The value of α is set to 0.3 .Through detail analysis it is found that Renyi's entropy based multi-level segmentation performs efficiently for α's value below 0.4.

Table 1. displays the threshold values acquired for different levels using MRET-DE. WPSNR and MSSIM values along with mean objective function value (f_{mean}) and standard deviation (f_{std}), are displayed in Table 1. From this table it can be established that with the increase of levels WPSNR and MSSIM values were also increasing. This indicates better segmentation with the increase of level.

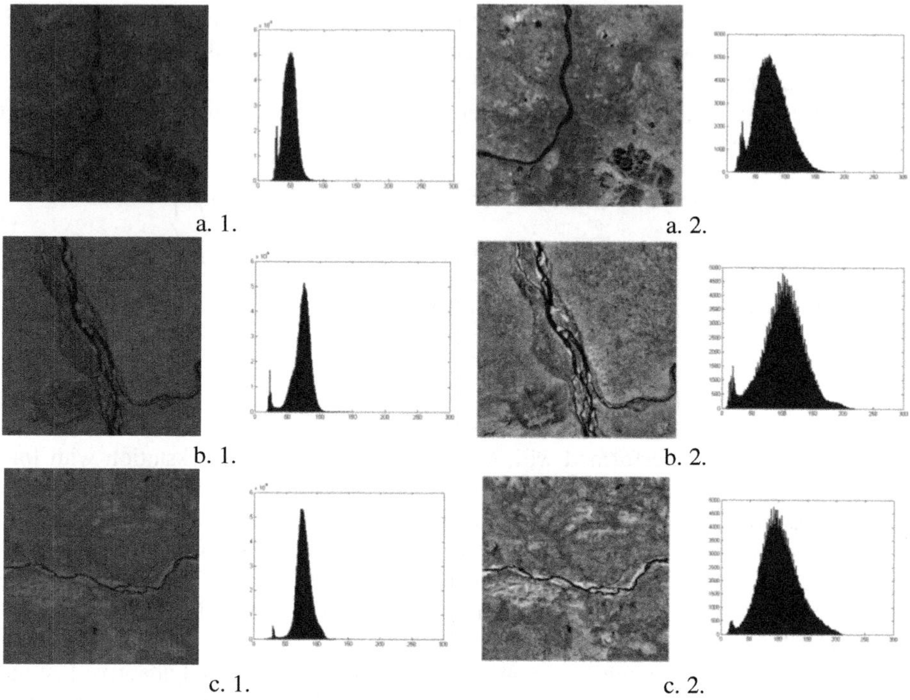

Fig. 1. Test Images and their histograms (a. 1.), (b. 1.), (c. 1.) RAW Image and its histogram, (a. 2.), (b. 2.), (c. 2.) After contrast enhancement

Fig. 2 displays the segmented gray level test images. Although in fig a. 3. the river is not distinguishable in all areas from its highly populated banks but when the number of levels are increased the river is easily separable. A change of colors indicated the density of the population in fig a. 4. and fig a.5. Similar results can be seen in fig b. 4. – b. 5 and in c. 3. – c. 5.However more than 5 level segmentation could be given as it is not supported in a gray scale map. The convergence plot of DE is shown in Fig. 3. for various levels of segmentations. It was clearly observable in those figures that DE achieved expected convergence with minimal iteration.

Table 1. Threshold values acquired by using DE and comparisons of WPSNR, MSSIM , mean objective function value (f_{mean}) and standard deviation (f_{std})

Image	L	Threshold values	WPSNR	MSSIM	f_{mean}	f_{std}
	2	69 124	22.4645	0.3573	11.7295	0.0
NF45K06	3	54 100 143	26.1600	0.5227	14.6716	0.0
	4	46 82 117 152	27.0943	0.6170	17.3553	0.0
	2	73 152	25.5338	0.4814	12.5515	0.0
NE44U16	3	58 113 164	27.0384	0.6176	15.6882	0.0
	4	49 90 132 171	27.4639	0.7030	18.5301	0.0
	2	67 141	22.7712	0.4656	12.3751	0.0
NF45T13	3	57 109 160	25.6147	0.6073	15.4783	0.0
	4	49 89 130 170	26.7285	0.6911	18.3026	0.0

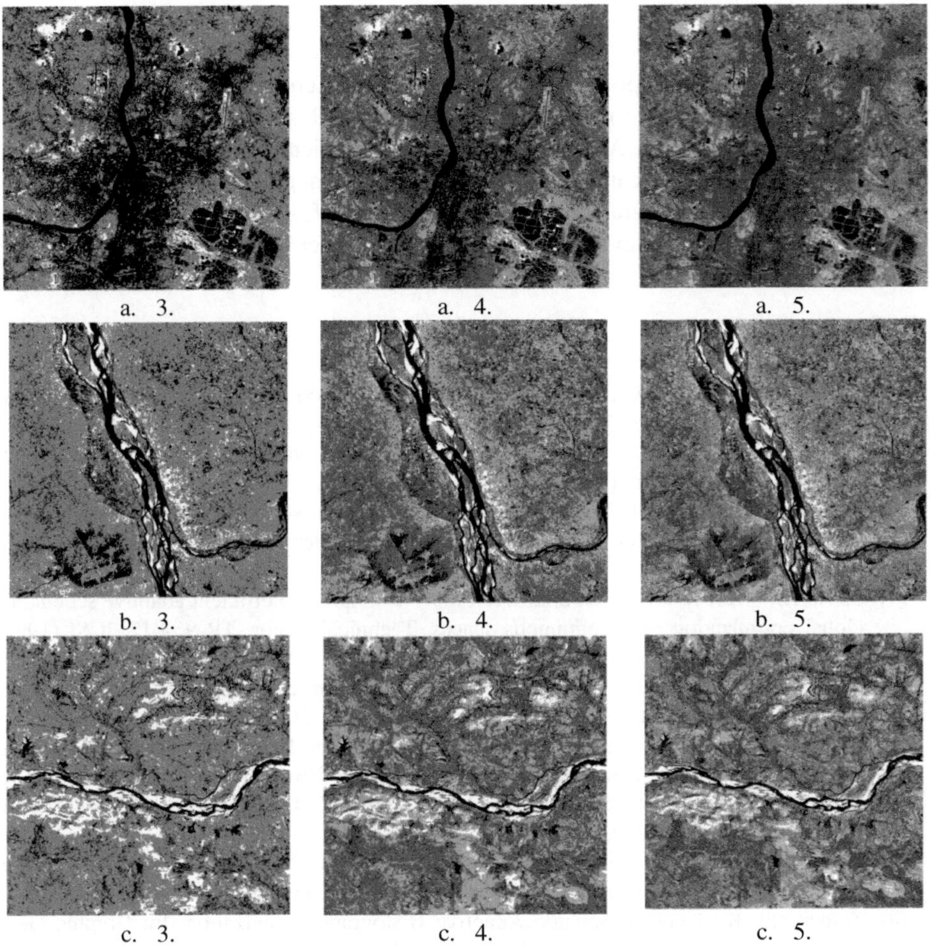

Fig. 2. Segmented images obtained by MRET-DE method (a. 3.), (b. 3.), (c. 3.) 3-level thresholding, (a. 4.), (b. 4.), (c. 4.) 4-level thresholding, (a. 5.), (b. 5.), (c. 5.) 5-level thresholding

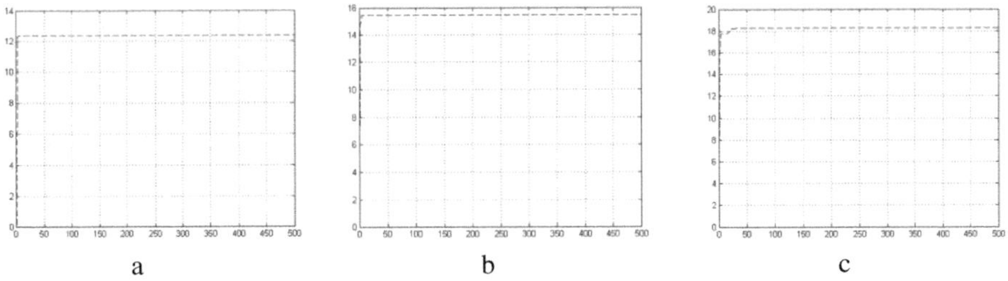

Fig. 3. Convergence plot of DE with 500 iterations (a) 3-level segmentation (b) 4 – level segmentation and (c) 5 – level segmentation

6 Conclusion

In this paper we have proposed a scheme based on differential evolution for multiple thresholding using MRE for NIR images of LISS III sensor. DE adds speed and accuracy to the algorithm. Although the results are typically encouraging, suggesting possibility of further researches on complex image segmentation and recognition problems, still 2-D histogram based approaches and fuzzy, rough set theory based approaches could also be implemented to achieve a better performance.

References

[1] Sahoo, P.K., Wilkins, C., Yeager, J.: Threshold selection using Renyi's entropy. Pattern Recognition 30, 71–84 (1997)

[2] Sahoo, P.K., Arora, G.: A thresholding method based on 2-D Renyi's entropy. Pattern Recognition, 1149–1161 (2004)

[3] Hammouchea, K., Diaf, M., Siarry, P.: A comparative study of various meta-heuristic techniques applied to the multilevel thresholding problem. Engineering Applications of Artificial Intelligence 23(5), 676–688 (2010)

[4] Storn, R., Price, K.V.: Differential Evolution - a simple and efficient adaptive scheme for global optimization over continuous spaces. Technical Report TR-95-012, ICSI (1995), `http://http.icsi.berkeley.edu/~storn/litera.html`

[5] Das, S., Suganthan, P.N.: Differential evolution – a survey of the state-of-the-art. IEEE Transactions on Evolutionary Computation 15(1), 4–31 (2011)

[6] Sarkar, S., Patra, G.R., Das, S.: A Differential Evolution Based Approach for Multilevel Image Segmentation Using Minimum Cross Entropy Thresholding. In: Panigrahi, B.K., Suganthan, P.N., Das, S., Satapathy, S.C. (eds.) SEMCCO 2011, Part I. LNCS, vol. 7076, pp. 51–58. Springer, Heidelberg (2011)

[7] Rényi, A.: On measures of information and entropy. In: Proceedings of the 4th Berkeley Symposium on Mathematics, Statistics and Probability, pp. 547–561 (1960)

[8] Zuiderveld, K.: Contrast Limited Adaptive Histograph Equalization. In: Graphic Gems IV, pp. 474–485. Academic Press Professional, San Diego (1994)

[9] Miyahara, M., Kotani, K., Algazi, V.R.: Objective picture quality scale (PQS) for image coding. IEEE Trans. on Communications 46(9), 1215–1226 (1998)

[10] Wang, Z., Bovik, A., Sheikh, H., Simoncelli, E.: Image quality assessment: From error visibility to structural similarity. IEEE Trans. Image Process. 13(4), 600–612 (2004)

An Experimental Based Study on Different Sources of Noise for High Dynamic Range Imaging

B. Ravi Kiran, M. Madhavi, M. Kranthi Kiran, and S.R. Mishra

Anil Neerukonda Institute of Technology and Sciences,
Sangivalasa, Bheemunipatnam[M], Visakhapatnam
{brk2007,madhavi.it25,soumyaranjanmishra.in}@gmail.com,
kranthikiranm@ieee.org

Abstract. In image processing, computer graphics, and photography, high dynamic range imaging (HDRI or simply HDR) is set of techniques that allow a grater dynamic range between the lightest and darkest areas of an image than current standard digital imaging techniques or photographic methods. By fusing several low dynamic range (LDR) images together we can get a high dynamic range image. In this process we need to fuse less noisy LDR images together in order to get more pleasing output images. A detailed study on different sources of noise is carried out on this paper. They include photon shot noise, read noise, pattern noise, pixel response non-uniformity (PRNU), quantization error and thermal noise. We used a Canon DSLR camera for experimental results.

Keywords: Noise, High Dynamic Range (HDR) Imaging.

1 Introduction

Image noise is the random variation of brightness or color information in images produced by the sensor and circuitry of digital camera. That is, in CCDs, the readout circuitry sits on top of the photosite and partially obscures them, so that some of the light falling on a sensor doesn't make it to the photosite to be detected. In CMOS sensors, each photosite's amplifier and related circuitry are adjacent to the photosite, directly on the sensor. Careful design and selection of sensors will reduce certain types of noise up to some extent, but we cannot reduce some sorts of noise, e.g. photon shot noise [1]. In this paper we have discussed the experimental way to know the sources of noise.

Natural scenes routinely produce HDR which in general not possible to get with a digital camera in a single capture. That means the scene exceeds the capability of the sensor. For example, if we want take a picture of a room from inside of that room and obviously we get some exterior vision also, through the window. In most cases, the exterior scene will be bright due to the illumination by the sun or by any light, while the interior illumination is far darker compared to the outside area. Hence, we can acquire an image, which shows properly exposed details in the room and a saturated window, in which all details of the exterior scene are lost. In other capture, the

S.C. Satapathy et al. (Eds.): Proc. of Int. Conf. on Front. of Intell. Comput., AISC 199, pp. 707–713.
DOI: 10.1007/978-3-642-35314-7_80

exterior scene is properly exposed, but all details of the room are lost in the underexposed (dark) areas of that image. To overcome this problem, several methods have been introduced that combine differently exposed images into one image of greater dynamic range.

Before discussing further about estimating the individual sources of noise, I would like to answer a basic question, "why high dynamic range imaging researcher needs to study about noise?". The answer is we have already discussed that, during the process of HDRI creation, we need to combine multiple exposures of the same scene to get a resultant HDR image. In this process if we combine noisy LDR images together, then the image will also be amplified and the resultant HDR image will look noisy. Knowing various types of noise, how various design choices in digital cameras can reduce noise and understanding how different choices of photographic exposure can help mitigate noise, we can definitely improve the results of HDRI experiments [3]. As we know the image noise of two types, that is, Chrominance and Luminance. Chrominance noise simply means color noise. Luminance noise is the non colored, i.e., brightness noise. These noises come because of the different sources of noise which we have discussed in Section 3.

2 Literature Review

The basic definition of noise is, the deviation from the original signal, coming to noise in digital image, it is the undesirable by-product of image capture [14]. For obtaining greater dynamic range we need to reduce the noise in the image. So for that purpose only we have to know about type and the source of that noise which we have discussed in section 3. The high dynamic range can be acquired by combining differently exposed images of same scene into one single image. These images can be combined by a weighted average into a single HDR image because of this weighted average the noise in the resulting HDR, will be less when compared to the noise in each single LDR image [4].

The natural images we see in open environment are analog signals which will be converted to digital signals by the sensor when we capture. In this conversion process only the random noise occurs. Random noise is due to the microscopic variations of the surface within one pixel, when the light is reflected from the surface, there will be random variations in phase and amplitude [13]. There is a measure to know the sharpness of the image that the camera is capable of. That is Transfer function of the camera (CTF). Even if we know the camera transfer function (CTF), we cannot get the image with high dynamic range we should know the optimal set of images to fuse and we need to reduce the noise in those images [4]. So after estimating the CTF we need to reduce the noise in the resultant HDR image. This indicates that we need to know the noise type and the process to reduce it.

Before the fusing of LDR images with different exposures, some people has proposed the other way to get wide dynamic range that is, to "increase the bit depth". But using which the cost and complexity of the device increases. Not only that, the dynamic range doesn't increase with the increase in the bit depth due to the noise in the image [2]. We used a Canon DSLR camera for the noise experiments and the experiments were performed as given in Emil Martinec's page [2]. We are able to estimate different sources of noise in the Canon DSLR camera model. They include photon shot noise, read noise, pattern noise, and pixel response non-uniformity (PRNU) of the Canon DSLR camera.

3 Sources of Noise

In this section we are going to explain different sources of noise.

3.1 Photon Shot Noise

Signal shot noise is fundamentally connected to the way photons spatially arrive on a detector. The average number of photons that strike a sensor is proportional to the intensity of light. Stream of photons that strike the sensor will have an average flux; although the scene and intensity of the light is fixed, there will be fluctuations around that average. This fluctuation in photon count is visible in images as photon shot noise, which follows the Poisson distribution. For example, if 10000 photons are collected on average, the typical fluctuation away from this average number of photons will be about $\sqrt{10000} = 100$. The counts will typically range from about 9900 to 10100. If instead, 100 photons are collected on average, then the counts will typically range from 90 to 110. Based on these observations we can conclude that if the number of photons collected is large then the effect of photon shot noise will be less and vice-versa [11],[3].

The distribution of the photon count is a Poisson distribution. In the Figure 1, ADU values are placed on the X-axis. In the Figure 1a, since we are measuring the raw values instead of photons the distribution does not possess the characteristics of the Poisson distribution. The reason for this is raw level in ADU of raw data is not the photon count; it is merely proportional to the photon count. In terms of ADU, if there are G photons/raw level then, the relation between mean and std dev is:

$$(\text{std dev})\_ADU*G = \sqrt{}mean\_ADU*G$$

The reason for the gaps in the Figure 1a is the faculty ADC in the camera. We repeated the same process and did not find any gaps in the Olympus E-450 (Figure 1b). if we observe the pixel values for the images captured by the camera (Canon dslr), we can see gaps at pixel values of 73, 78, 84, 90, 96, 101, 107, 113, 119, 124, 130, 136, 142 etc. if we observe the zero density bins carefully, we can see the second zero density bin at a distance of five ADUs from the first, third at a distance of six ADUs from the second, fourth at a distance of six ADUs from the third, and fourth at a distance of six ADUs from the fourth. In addition, the pattern repeats from the sixth zero density ADU bin. Since there are no under exposed pixels in the image, values below 73 are not present in the image statistics.

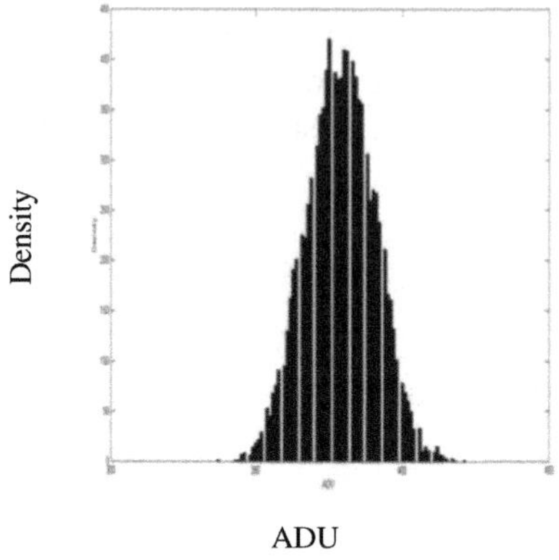

Fig. 1a. Histogram of a Photon shot noise by cropping a single uniform patch in a Macbeth color checker chart at ISO 100 for Canon dslr model.

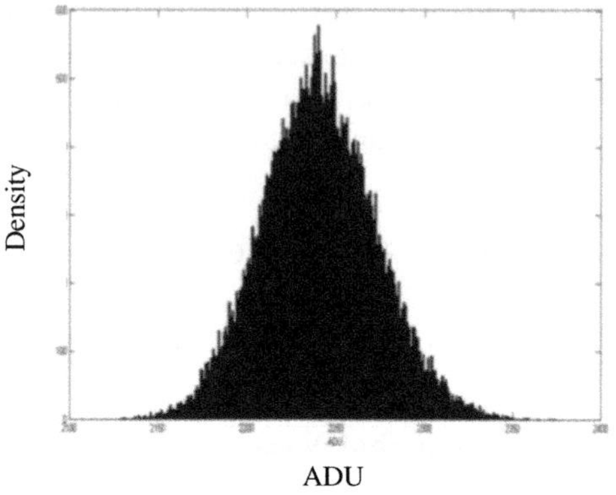

Fig. 1b. Histogram of the photon shot noise for Olympus E-450 model.

3.2 Read Noise

Theoretically, the raw data should be directly proportional to the photon count. However, in the real world this is not the case, the reason is read noise. The voltage fluctuations in the electronic components such as sensor element readout, ISO gain, and digitization, contribute to a deviation of the raw value from the ideal value proportional to the photon count. These fluctuations are the main cause of the sensor read noise [2],[6],[8],[9],[10]. To identify the read noise one can capture the image with lens cap on using the highest available shutter speed. Figure 2 is a combination of both read noise and pattern noise of the Canon dslr camera. The image in Figure 2 is captured with the lens cap on and the highest available shutter speed. Because of these settings, there are no photons captured and only electronic noise from reading the sensor remains. The image in Figure 2, Figure 3a, and 3b are cropped to

Fig. 2. Read noise of a Canon dslr at ISO 800, shutter speed 1/8000, F-stop f/8

Fig. 3a. A template created by averaging 20 black frames (exposure time 1/8000, F-stop f/8, ISO 800)

256x256. The original size images can be found in the link http://picasaweb.google.com/ravi0024.ntu/PatternNoise#. Figures_2, 3, 4 are obtained after converting them to 8-bit integer values 9values outside0-255 are mapped to the range of 0-255) and histogram equalization is applied.

Read noise can be isolated in a bias frame by subtracting one bias frame from another, or even better, by subtracting the pixel-by-pixel average of multiple bias frames smoothes out the random noise in a single bias frame getting you closer to the true bias of the pixel in the CCD.

3.3 Pattern Noise

Human eye is adapted to perceive patterns, this pattern noise or banding noise is more apparent than white noise. Even it comprises a smaller contribution to the overall noise. Pattern noise can have both fixed component which does not vary from image to image and variable component that, while not random from pixel to pixel, and is not same from image to image. Pixels response is linear in low light region and logarithmic in high light region. Previously a logarithmic pixel with reduced dark current has been introduced to get high dynamic range. But these pixels do not respond to the low light level scene properly. We need to remove the temporal noise before we correct the fixed pattern noise [5].

Now coming to experimental results of pattern noise, upon close observation of the Figure 2, we can see the vertical patterns in the image. To remove the pattern noise a template is created by averaging 20black frames. To remove the pattern noise we need to subtract the template from the Figure 2. the template is shown in Figure 3 and the Figure 3b is obtained by subtracting the template (Figure 3a from the Figure 2).

Fig. 3b. An image after removing the pattern noise which contains only read noise

3.4 Pixel Response Non-Uniformity(PRNU)

PRNU is a pixel-to-pixel variation in the sensor responsiveness and it is expected to be linear as a function of the exposure. Otherwise, PRNU is noise due to the in-homogeneity of pixels properties on the sensor. In a single input image if the read noise is R, photon shot noise is P, PRNU is W ignoring pattern noise and quantization error, the total noise is N then

$$N=\sqrt{R^2+P^2+W^2}.$$

If we take the difference of two test images then we can eliminate all the fixed noise and only read noise and random pattern noise will remain in the image. By using the above equation we can get $W^2=N^2-(R^2+P^2)$,which is nothing but the variance of single test input image with standard deviation square of read + shot noise subtracted. By plotting PU is about 0.2%.

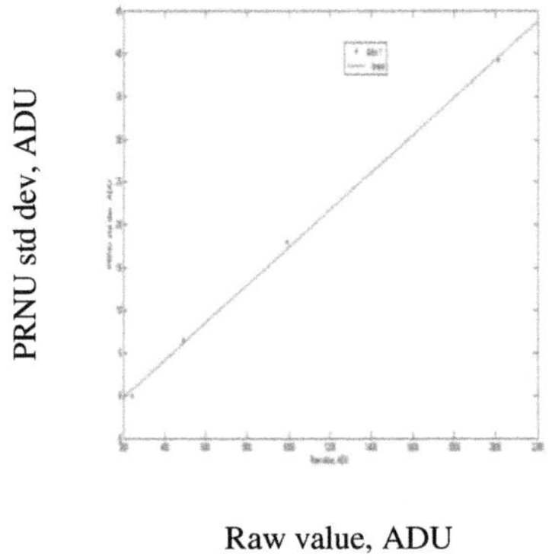

Fig. 4. Noise due to pixel response non-uniformity (PRNU) of a Canon dslr at ISO 100

3.5 Quantization Error

In image acquisition process, when the analog voltage signal from the sensor is digitized into a raw value, it is rounded to a nearby integer value. Due to this round-off, the raw value indicates the analog signal by a slight amount of deviation. The error introduced by digitization is called quantization error. Though it is having minor contribution to the image noise, we have to measure this noise also. We cannot oppose the quantization noise after the quantization is done. In general, quantization noise cannot be avoided. Previously, while analyzing the quantization noise people used to have an assumption that the Analog to Digital Converter (ADC) output is well matched with the input analog values. But where that assumption is not true [12]. So we have to study on it clearly and we need do experiments on this quantization error. This has been taken as our future work.

3.6 Thermal Noise

Thermal agitation of electrons in a sensel can liberate a few electrons. These thermal electrons are indistinguishable from the electrons freed by photon absorption, and thus cause a distortion of the photon count represented by the raw data. Thermal electrons are freed at a relatively constant rate per unit time, thus thermal noise increases with exposure time. The most common and conventional way to reduce the thermal noise is designing the pixel to minimize leakage current and cooling the sensor. On this also we need to study in depth and know any other the way to reduce this noise. This has been considered as of future work.

4 Future Work and Conclusions

In this paper we have experimentally studied some sources of noise available on Emil Martinec's website. There are other sources of noise which we have just given a brief discussion that is, thermal noise and quantization error. We need to work on the affect of these two noises on dynamic range of the image. That should be carried out as our future work.

We have done the experiments on sources of noise such as photon shot noise, read noise, pattern noise and pixel response non-uniformity. And also the methods to measure and reduce some of these noises. And we also need to modify the noise function for the high dynamic range images.

References

[1] Akyuz, A.O., Reinhard, E.: Noise reduction in high dynamic range imaging. Journal of Visual Communication and Image Representation 18(5), 366–376 (2007)

[2] Martinec, E.: Noise, Dynamic Range and Bit Depth in Digital SLRs. Internet (May 22, 2008), `http://theory.uchicago.edu/~ejm/pix/20d/tests/noise/` (February 15, 2012)

[3] Hasinoff, S.W., Durande, F., Freeman, W.T.: Noise-Optimal Capture for High Dynamic Range Photography. Massachusetts Institute of Technology Computer Science and Artificial Intelligence Laboratory

[4] Bell, A.À., Seiler, C., Kaftan, J.N., Aach, T.: Noise in high dynamic range imaging. In: ICIP 2008, pp. 561–564. IEEE (2008)

[5] Choubey, B., Collins, S.: Fixed pattern noise correction for wide dynamic range linear-logarithmic pixels. IEEE (2007)

[6] Robertson, M.A., Borman, S., Stevenson, R.L.: Estimation-theoretic approach to dynamic range improvement using multiple exposures. Journal of Electronic Imaging 12(2), 219–228 (2003)

[7] Liu, C., Szeliski, R., Kang, S.B., Zitnick, C.L., Freeman, W.T.: Automatic estimation and removal of noise from a single image. TPAMI 30(2), 299–314 (2008)

[8] Clark, R.N.: Digital camera sensor performance summary (2012), `http://www.clarkvision.com/imagedetail/digital.sensorperformance.summary/`

[9] Grossberg, M.D., Nayar, S.K.: High dynamic range from multiple images: Which exposures to combine? In: Workshop on Color and Photometric Methods in Comp. Vision, pp. 1–8 (2003)

[10] `http://www.cloudynights.com/item.php?item_id=2001`

[11] `http://spie.org/samples/PM170.pdf`

[12] Brajovic, V.: Brightness perception, dynamic range and noise unifying model for adaptive image sensors. In: CVPR 2004. IEEE (2004)

[13] Silwal, S., Wang, H., Maldonado, D.: Assessment of Randome –Noise Contamination in Digital Images via Testing on wavelet Coefficents

A New Contrast Measurement Index Based on Logarithmic Image Processing Model

Mridul Trivedi, Anupam Jaiswal, and Vikrant Bhateja

Deptt. of Electronics and Communication Engineering,
Shri Ramswaroop Memorial Group of Professional Colleges,
Faizabad Road, Lucknow-227105, (U.P.), India
{mridultrivedi1991,anupam4.srmc,bhateja.vikrant}@gmail.com

Abstract. With the introduction of more complex enhancement algorithms, there is a need for an effective method of enhancement measurement that can assess image quality in accordance with Human Visual System (HVS) characteristics. This paper presents a new quality index for measurement of contrast in digital images based on Logarithmic Image Processing (LIP) model. The proposed quality index evaluates the degree of contrast manipulation (provided by an enhancement algorithm) by considering the difference in the average gray level values in its foreground to that of background. The calculated statistical parameters for foreground and background regions are mathematically combined using the LIP operators to ensure processing of images from HVS point of view. The quality index is computed for different contrast manipulating algorithms which are applied to test images taken from standard MATLAB library as well as LIVE Database. Simulation results illustrate the precision and efficiency of the proposed index in comparison to other contrast evaluation methods proposed in literature.

Keywords: No-Reference, LIP Model, Contrast Measurement Index (CMI), Image Quality Assessment, Hadamard Transform.

1 Introduction

Image enhancement algorithms seek to enhance the apparent visual quality of an image or emphasize certain features based on the knowledge of source of degradation. The principle objective of an enhancement algorithm is to process an image such that the result is more suitable than the original image for a specific application. An important feature that influences image enhancement is contrast, as it is a perceptual measure that defines the difference between the perceived brightness [1]. Image quality assessment in digital domain is critical in all applications of image processing because when an image is transformed; the viewer is the ultimate judge of how well a transformation method works. Being a highly subjective process, visual evaluation of images by human becomes impractical for real time applications. Therefore, objective evaluation methods for digital images are preferred as they can dynamically monitor

S.C. Satapathy et al. (Eds.): Proc. of Int. Conf. on Front. of Intell. Comput., AISC 199, pp. 715–723.
DOI: 10.1007/978-3-642-35314-7_81

and adjust image quality, optimize parameter settings and can be used to benchmark various digital images processing systems [2]. Objective methods of quality evaluation are of three types: Full-Reference, Reduced-Reference, and No-Reference. In full-reference methods, a processed image is compared to its original one by calculating the difference between corresponding pixels in two images found at the same pixel locations. MSE (Mean Squared Error) [2] and SSIM (Structural Similarity) [3] are the two most commonly used full-reference signal fidelity measures. MSE exhibits weak performance and has been widely condemned for serious shortcomings as it does not confirm to the expected results when it is used to predict human perception of assessing image quality. On the other hand, SSIM cannot handle geometrical distortion in digital images which are non-structural in nature. To overcome the drawbacks of above methods, reduced-reference evaluation methods [4] were introduced which involve sending or supplying some amount of information such as statistical parameters about the reference images along with the distorted image that is useful in quality computation. Reduced-reference methods are widely criticized for not correlating well with perceived image quality. On the other hand, no-reference methods [5] do not require any reference images for assessment rather a blind evaluation is performed based on some characteristics of the given image. On the basis of assessment of image quality, these measures return an absolute value which is generally content dependent and calculated based on specific type of distortions. EME (Measure of Enhancement) and the EMEE (Measure of Enhancement by Entropy) are common no-reference methods for evaluation of image contrast proposed by Agaian *et al.* [5]. These methods evaluated the image quality by computing the ratio of local maximum and minimum pixels in small $k_1 \times k_2$ sized blocks and the results are averaged for the entire image. But, these methods rely on linear algorithms and can assess satisfactorily only when there is a large background with small single test object. LogAME and LogAMEE were the two modifications of the previous methods developed by incorporating a non-linear framework to the Michelson Contrast Law [5]. As these methods work on small blocks in an image so the results are affected by noise and steep edges in images [6]. A. K. Tripathi *et al.* in their work [7] proposed Histogram Flatness Measure (HFM) and Histogram Spread (HS) measures for evaluation of enhancement on the basis of statistical parameters of image histogram like geometric mean, quartile distance and range. However, it was concluded by the authors that both the measures worked satisfactorily well only for evaluation of histogram based enhancement methods and that too with reduce sensitivity. Hence, this paper presents a new quality index for measurement of contrast in the absence of any reference image. The proposed index is based on LIP model and evaluates the contrast by calculating mean gray values in two rectangular windows around the centre pixel called foreground and background. The quality index evaluates the contrast manipulated by enhancement algorithms on test images taken from standard MATLAB library as well as LIVE Database [8]. The remaining part of this paper is organized as follows: Section 2 describes the problem formulation in context to discussion of basic definition of contrast and LIP model. The simulation results and their discussion are given in section 3. Section 4 concludes the paper.

2 Problem Formulation

2.1 Background

Contrast has a great influence on the quality of an image in human visual perception as well as in image analysis. The definition of local contrast proposed by Morrow *et al.* [9] can be stated as:

$$C = \frac{m_f - m_b}{m_f + m_b} \tag{1}$$

where: m_f is the maximum luminance equivalent to the mean gray level value of a particular object in the image called the foreground. In the same context, m_b equals minimum luminance which is the mean gray level value of region surrounding that object, called the background. If the difference in the intensities between foreground and background is more than 2% then change in contrast cannot be properly distinguished by human eye. With this concept, at times the measurement of contrast using (1) yields poor sensitivity. Moreover, the variation of contrast provided by certain enhancement algorithms Histogram Equalization (HE) [9], Adaptive Histogram Equalization (CLAHE) [10], Unsharp Masking (UM) [11], Adjusting the Black to White Mapping (ABWM) [12], Morphological operations such as Bottom Hat filtering by square (BHS) and line (BHL) [13] structuring elements were not quantized sharply when evaluated using the above expression.

2.2 Logarithmic Image Processing (LIP) Model

The LIP model of Jourlin and Pinoli [14] is a mathematical framework that provides a specific set of non-linear algebraic and functional operations for the processing and analysis of pixel intensities. The LIP model has been proved to be physically justified by some important laws and characteristics of human brightness perception. This is designed to both maintain the pixels values inside the range as well as for more accurate processing of images from a HVS [15] point of view. The LIP operations are defined using gray tone functions expressed as $k\,(i, j)$ in (2).

$$k\,(i, j) = m - f\,(i, j) \tag{2}$$

where: $f(i, j)$ is the original image and m denotes the maximum value of the pixel in that image. $k\,(i, j)$ in (2) is the gray tone function used to generate negatives of the original images. Addition and subtraction using LIP operators can now be expressed in term of gray tone functions as follows:

$$k_1 \widetilde{\oplus} k_2 = k_1 + k_2 - \frac{k_1 k_2}{m} \tag{3}$$

$$k_1 \widetilde{\Theta} k_2 = m\,\frac{k_1 - k_2}{m - k_2} \tag{4}$$

where: $\widetilde{\oplus}$ and $\widetilde{\Theta}$ are operators for LIP addition and subtraction respectively. k_1 and k_2 represents the corresponding gray tone functions. Instead of processing the pixels with basic arithmetic operations, LIP arithmetic operators yields more robust functioning well within the acceptable range of image.

2.3 Proposed Quality Index

As discussed in section 2.1, the variation of contrast provided by certain enhancement algorithms [9]-[13] were not quantized sharply when evaluated using the expression of Morrow *et al.* [1] and the results are also not coherent as per HVS characteristics. Further, this contrast measurement approach is not versatile as the value of contrast varies with the selection of foreground and background regions. Hence, there is a need of a quality index which can effectively measure contrast in digital images overcoming the limitations discussed shortcoming as well as capable to discriminate between increasing and decreasing contrast in accordance to HVS. This paper proposes a contrast measurement index for assessment of various contrast manipulation algorithms, using the operators of LIP model. The degree of improvement in contrast provided by an enhancement algorithm can be adjudged by the fact that it should enhance the difference between the average gray level values lying in the foreground and background regions respectively. On this basis, the procedure for computation of proposed quality index for evaluation of contrast in an image (I) of size $r{\times}c$ (where r, $c \in$ odd numbers) is explained as under:

***Step 1*:** The centre $o(x, y)$ of the input image, I of size $r{\times}c$ is determined as $(r+c)/2$.
***Step 2*:** Taking $o(x, y)$ as centre, two concentric square windows (of sizes 3×3 and 5×5) are selected. The smaller window of size 3×3 can be referred to as the 'foreground' whereas the larger window of size 5×5 is called background as shown in fig. 1(a). Mean gray level values of these foreground and background regions can be computed and denoted as M_f and M_b respectively.
Step 3: The contrast within this region can be calculated as:

$$C_i = \ln \left| \left(\frac{M_f \tilde{\Theta} M_b}{M_f \tilde{\oplus} M_b} \right)_i \right| \tag{5}$$

where: $\tilde{\oplus}$ and $\tilde{\Theta}$ represents LIP addition and subtraction operators. C_i denotes to contrast evaluated in a sub-region of the input image during the first iteration.
***Step 4*:** Keeping the centre fixed at $o(x, y)$, the foreground and background window sizes are incremented by a factor of 2 and respective values of mean gray levels is computed for both foreground and background regions. Hence, during the second iteration, mean gray level values are calculated for foreground and background regions of window sizes 5×5 and 7×7 respectively as shown in figure 1(b). This can be stated as:

$$C_{i+1} = \ln \left| \left(\frac{M_f \tilde{\Theta} M_b}{M_f \tilde{\oplus} M_b} \right)_{i+1} \right| \tag{6}$$

***Step 5*:** The procedure is iteratively repeated for the entire image, by incrementing the foreground and background window sizes, each time by a factor of 2, till the time the background window size reaches r.

***Step 6*:** The contrast value computed during each iteration are averaged for all the iterations and multiplied by a factor α which is equivalent to the maximum of the Hadamard Transform [17] computed for the original image I. The Hadamard Transform is an orthogonal transformation technique that decomposes a signal into a set of basis function, which is a rectangular or square wave with values of +1 or -1. Hence, the proposed quality index known as Contrast Measurement Index (*CMI*) can be mathematically formulated as:

$$CMI = \frac{\alpha}{N}\sum_{i=1}^{N} \ln \left| \left(\frac{M_f \widetilde{\Theta} M_b}{M_f \widetilde{\oplus} M_b} \right)_i \right| \tag{7}$$

where: N is the total no. of iterations applied for contrast evaluation. Equation (7) yields an absolute value for the input image (I) which is a measure of its contrast. Higher values of *CMI,* characterizes better performance of the contrast enhancement algorithms. The range of *CMI* can be bounded between zero and infinity where zero indicates a totally black image and infinity for totally white image. The absorption and transmission of light follows a logarithmic relationship, both when processed by the human eye and when travelling through a medium [14] because of this it is natural to combine the statistical parameters of contrast measurement using logarithmic operators. This justifies the usage of LIP model to ensure HVS based contrast evaluation.

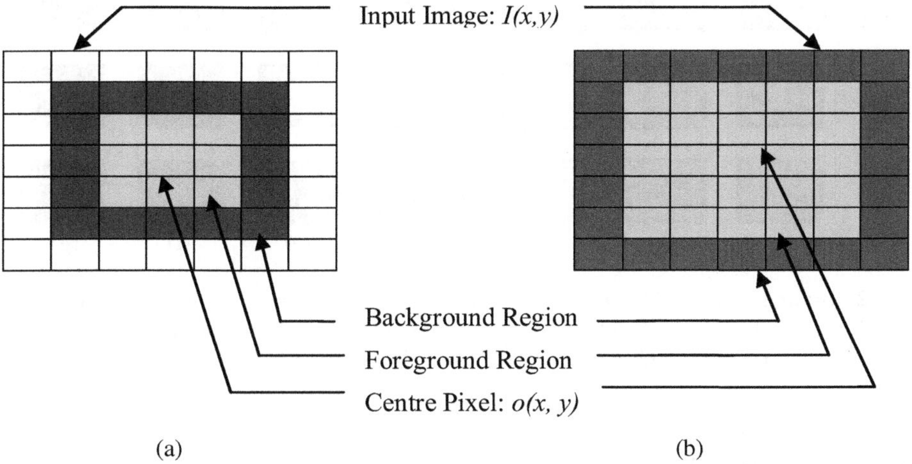

Fig. 1. Selection of foreground and background regions for contrast evaluation during the first two iterations. **(a)** Shows the 3×3 foreground and 5×5 background window selected about the centre pixel: *o(x,y)*. **(b)** Shows the increment in window sizes to previous ones, with 5×5 foreground and 7×7 background windows about the same centre pixel: *o(x,y)*.

3 Experimental Results

3.1 Test Images

To test the versatility of the proposed quality index number of experiments were included on two categories of test images: standard MATLAB images (1, 2 & 3) and Live database images (4, 5 & 6) as shown in fig. 2(a). The test images are initially processed by converting it from RGB to gray-scale and then normalized, prior to the application of contrast manipulation algorithms. Certain algorithms are applied for increasing the contrast such as Histogram Equalization (HE) [9], Adaptive Histogram Equalization (CLAHE) [10], Unsharp Masking (UM) [11] as given in fig. 2(b)-(d), whereas the contrast is decreased by adjusting the black to white mapping (ABWM) [12], applying Morphological operations such as Bottom Hat filtering with a square (BHS) and line structuring element (BHL) respectively [13] as shown in fig. 2(e)-(g).

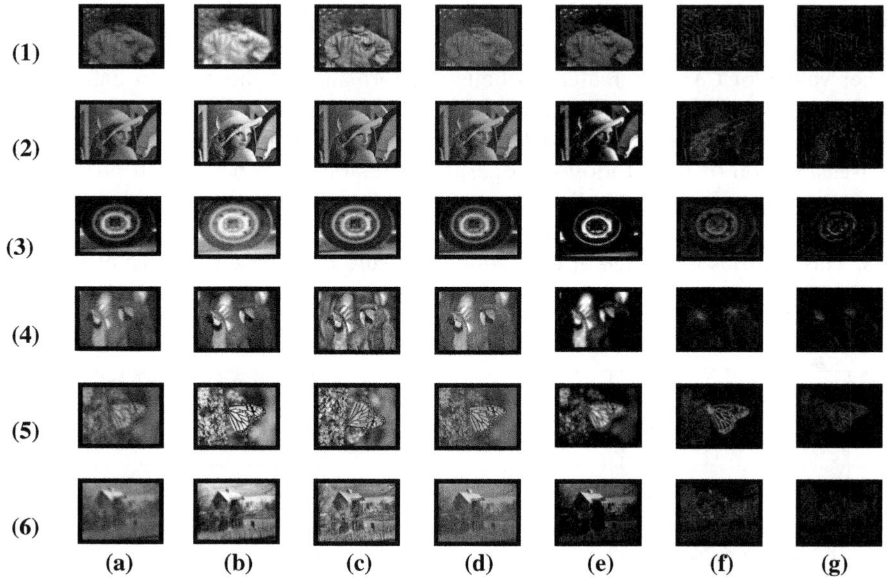

Fig. 2. Different Contrast Manipulation Algorithms Applied on Test images (1)-(6) **(a)** Original Test Images, (1)-(3): from MATLAB Library and (4)-(6): from LIVE Database. Images transformed by **(b)** HE [9], **(c)** CLAHE [10], **(d)** UM [11]. **(e)** ABWM [12], **(f)** BHS [13], **(g)** BHL [13].

3.2 Simulation Results

The proposed index *CMI* is calculated for all the test images in fig. 2 and tabulated under table 1. From the data in table 1, it can be interpreted that the value of *CMI* increases with respect to its value for the original image, upon application of increasing contrast algorithms [9-11]. Similarly, a decrease in value of *CMI* (in comparison to original image) is seen with decreasing contrast algorithms [12-13]. This illustrates that the proposed index is capable to discriminate between images of good and poor contrast without the knowledge of original reference images. The sensitivity of *CMI* is better as there is sharp change in its value with application of different contrast manipulating algorithms. HE [9] enhances the contrast better than other algorithms as it equalizes the histogram of the image uniformly, resulting in an image consisting of gray-levels with density, increasing the dynamic range of the image. This can be clearly verified by proposed index as it evaluates to a maximum *CMI* value for images transformed using HE [9] among various enhancement algorithms. CLAHE [10] uses a clip level to limit the local histogram such that the amount of contrast enhancement for each pixel can be limited. Hence, there is always a control on over-enhancement was performed with HE. Therefore, values of *CMI* for images transformed using CLAHE are less in magnitude in comparison to those with HE. For the purpose of comparisons, the contrast of the test images is also calculated using previously proposed well known measures of enhancements, namely EME [5] and logAME [6] and the values are given in table 2 and 3 respectively. It can be observed that, these evaluation methods proved to be non-satisfactory in discriminating good and poor contrast images. There is an increase in the values of both EME and logAME for images transformed with decreasing contrast algorithms [12-13]. Not only this, the sensitivity of logAME also comes out to be very low in comparison to the proposed index. UM [11] approach generates a high contrast image, by adding some portion of the original image to the blurred image. The obtained values of EME for images transformed using UM are not as stable as *CMI* values.

Table 1. Calculation of Proposed Qulaity Index (*CMI*) for different Contrast Manipulation Algorithms

Image No.	**Original Image**	**HE[9]**	**CLAHE[10]**	**UM[11]**	**ABWM[12]**	**BHS[13]**	**BHL[13]**
(1)	3.6549	6.8891	5.5572	8.6181	2.8602	1.3364	0.7536
(2)	14.3211	19.7244	19.7318	17.3581	13.1974	4.2173	2.1413
(3)	5.5989	9.8221	7.1893	5.8116	4.6944	3.6748	2.3887
(4)	16.9881	20.4317	17.1944	17.4777	14.3840	2.2711	1.4624
(5)	11.5987	15.3462	14.3949	19.5886	8.2491	4.0599	2.0596
(6)	13.2383	16.3650	15.5038	17.9331	7.5857	3.4218	2.3669

Table 2. Calculation of EME [5] for different Contrast Manipulation Algorithms

Image No.	Original Image	HE[9]	CLAHE[10]	UM[11]	ABWM[12]	BHS[13]	BHL[13]
(1)	11.5574	24.0704	10.4715	11.9591	91.0002	107.9761	117.3311
(2)	11.8921	26.0983	18.0461	51.1398	46.1724	107.0428	120.4132
(3)	15.5299	28.0337	29.3526	102.1440	56.5704	106.3981	116.6379
(4)	4.6756	13.2586	8.4120	20.2031	26.5801	101.0695	107.2062
(5)	8.2163	21.3697	13.7141	34.5570	56.2511	102.8437	111.1503
(6)	7.6399	23.4268	16.8241	45.5389	67.0442	106.6452	128.0127

Table 3. Calculation of l*ogAME* [6] for different Contrast Manipulation Algorithms

Image No.	Original Image	HE[9]	CLAHE[10]	UM[11]	ABWM[12]	BHS[13]	BHL[13]
(1)	0.0010	0.0010	0.0011	0.0012	0.0019	0.1056	0.0280
(2)	0.0718	0.0683	0.0665	0.0648	0.0441	0.0417	0.0348
(3)	0.0013	0.0013	0.0012	0.1571	0.0817	0.1559	0.0222
(4)	0.5302	0.0505	0.0505	0.0188	0.0718	0.0449	0.0167
(5)	0.0520	0.0491	0.0492	0.0197	0.0404	0.4088	0.0252
(6)	0.0511	0.0462	0.0463	0.0645	0.0427	0.0486	0.0171

4 Conclusion

Unlike the earlier proposed methods of contrast evaluation, *CMI* evaluates the local contrast by calculating the difference in the mean gray-level values of foreground and the background regions using the operators of LIP model. In addition, the defined criteria for selection of foreground and background regions make the method more robust for assessment of contrast. It can be inferred from the simulation results that *CMI* provides precise and improved evaluation of contrast based on HVS characteristics for images transformed with different enhancement algorithms. The proposed index uses a no-reference evaluation approach and is capable to effectively discriminate low and high contrast images. Hence, the proposed quality index can serve to evaluate and benchmark the performance of different contrast enhancement algorithms.

References

1. Morrow, W.M., et al.: Region-Based Contrast Enhancement of Mammograms. IEEE Transaction on Medical Imaging 11(2), 121–134 (1992)
2. Sheikh, H.R., Saber, M.F., Bovik, A.C.: A Statistical Evaluation of Recent Full Reference Image Quality Assessment Algorithm. IEEE Transactions on Image Processing 15(11), 3441–3452 (2006)
3. Wang, Z., Bovik, A.C., Sheikh, H.R., Simoncelli, E.P.: Image Quality Assessment from Error Visibility to Structural Similarity. IEEE Transactions on Image Processing 13(4), 600–612 (2004)
4. Wang, Z., Simoncelli, E.P.: Reduced-Reference Image Quality. In: Proceedings of International Symposium on Electronic Imaging, San Jose, CA, USA (2005)
5. Agaian, S.S., Panetta, K., Grigoryan, A.M.: Transform-based Image Enhancement Algorithms with Performance Measure. IEEE Transactions on Image Processing 10(3), 367–382 (2001)
6. Panetta, K., Wharton, E.J., Agaian, S.S.: Human Visual System based Image Enhancement and Logarithmic Contrast Measure. IEEE Transactions on Image Processing 38(1), 174–188 (2008)
7. Tripathi, A.K., Mukhopadhyay, S., Dhara, A.K.: Performance metrics for image contrast. In: IEEE Conference on Image Information Processing, Shimla, India (2011)
8. Sheikh, H.R., Wang, Z., Cormack, L., Bovik, A.C.: LIVE Image Quality Assessment Database Release 2, `http://live.ece.utexas.edu/research/quality`
9. Chen, S.D., Ramli, A.R.: Contrast Enhancement using Recursive Mean-Separate Histogram Equalization for Scalable Brightness Preservation. IEEE Transaction on Consumer Electronics 49(4), 1301–1309 (2003)
10. Zuiderveld, K.: Contrast Limited Adaptive Histogram Equalization. In: Graphic Gems IV, pp. 474–485. Academic Press Professional, San Diego (1994)
11. Panetta, K., Zhou, Y., Agaian, S.S., Jia, H.: Nonlinear Unsharp Masking for Mammogram Enhancement. IEEE Transaction on Information Technology in Biomedicine 15(6), 234–255 (2011)
12. Gonzalez, R.C., Woods, R.E.: Digital Image Processing. Addison-Wesley, Reading (2002)

13. Gao, X., Wang, Y., Li, X., Tao, D.: On combining Morphological Component analysis and Concentric Morphology Model for Mammographic Mass Detection. IEEE Transaction on Information Technology in Biomedicine 14(2), 266–273 (2010)
14. Jourlin, M., Pinoli, J.C.: Logarithmic Image Processing, The Mathematical and Physical Framework for the Representation and Processing of Transmitted Images. Advances in Imaging and Electron Physics 115(2), 129–196 (2001)
15. Panetta, K., Wharton, E.J., Agaian, S.S.: Human Visual System based Image Enhancement and Logarithmic Contrast Measure. IEEE Transaction on Image Processing 38(1), 174–188 (2008)
16. Beauchamp, K.G.: Applications of Walsh and Related Functions. Academic Press (2001)

A Novel Color Edge Detection Technique Using Hilbert Transform

Ankush Gupta, Ayush Ganguly, and Vikrant Bhateja

Deptt. of Electronics and Communication Engineering,
Shri Ramswaroop Memorial Group of Professional Colleges,
Faizabad Road, Lucknow-227105, (U.P.), India
{ankushh.guptaa,ayushnascarp,bhateja.vikrant}@gmail.com

Abstract. This paper presents a new technique for edge detection in color images, employing the concept of Hilbert transform. The color image at the input is initially transformed using a RGB color triangle, followed by the application of the proposed edge detection technique. Combination of bilateral filtering with Hilbert transform in the proposed technique makes it effective for edge detection in noisy environment. Computer simulations are performed on noise free as well as noisy images (corrupted with impulse noise), which portray marked improvement in edge detection in comparison to other techniques.

Keywords: Bilateral filtering, Edge detection, Hilbert transform, Pratt's figure of merit (PFOM), RGB color space.

1 Introduction

Edges in an image occur predominantly due to abrupt mutations or discontinuities in its physical characteristics such as brightness, geometry and reflectance of objects. Edge detection deals with the localization of significant variations of the gray level in the image and assists in identifying the physical phenomena that originated them. Thus, edge detection is an essential pre-processing step for operations like image segmentation, data compression, pattern recognition etc. [1]. Color edge detection is highly desirable over grayscale approaches because edges which exist at the boundary separating regions of different colors, cannot be detected in grayscale images if there is no change in intensity. In addition, some form of color edge detection is necessarily required to resolve the points that have not been considered while using grayscale edge detection [2]. Considerable contributions have been made so far on the development of different edge detection techniques for grayscale as well as color images and still there is a continuing research in this field. Conventional edge detection operators like Sobel and Roberts [3] were based on the concept of 'enhancement and thresholding' but lacked noise immunity. Marr and Hildreth [4] were the first to come up with concept of including Gaussian smoothing as a pre-processing step in edge detection. The authors proposed Laplacian of Gaussian (LoG) operator which also had some limitations such as: considerable effect of noise on the quality of generated edge map, detection of false edges etc. J. F. Canny [5] proposed another gradient based

S.C. Satapathy et al. (Eds.): Proc. of Int. Conf. on Front. of Intell. Comput., AISC 199, pp. 725–732.
DOI: 10.1007/978-3-642-35314-7_82

edge detector making use of Gaussian filter for smoothening of image and noise suppression. But at the same time, it degraded the edges by smoothening them as well. Moreover, this technique required manual resetting of upper and lower threshold levels when some modifications were done in the scene or illumination, making it quite impractical. M. Gudmundsson *et al.* [6] used genetic algorithm based on optimization for performing edge detection which proved to be successful in detecting thin, continuous and well localized edges. But, still it had certain short comings such as: big computational time complexity and low computational efficiency. The manuscript presented by S. Zahurul and S. Zahidul [7] employed improved Sobel operator for performing edge detection in knee osteoarthritis images, but the detected edges were not superior and the need of developing an enhanced refining operator had been cited. Z. Fengjing *et al.* [8] proposed a color edge detection technique based on the concept of triangle similarity using an improved Sobel operator. However, the technique proved to be inefficient for noisy images. In addition, the authors have not performed any quality assessment of the obtained edge map. S. Pie *et al.* in their work [9] introduced Generalized Radial Hilbert Transform for performing edge detection which showed higher noise immunity because of its longer impulse response. Reduction in impulse response improved the edge map quality but resulted in decrement in noise immunity. Thus, there was a tradeoff between: better performance level and noise robustness. N. Gopalyegani *et al.* presented a technique employing Hilbert matrix [10] and its inverse for performing edge detection which gave inadequate results for images corrupted with impulse noise. In this technique, the processing of images required manual setting of the orders of both the Hilbert matrix as well as its inverse which was not a versatile solution. Considering the edge detection techniques developed previously, it can be inferred that most of them focused primarily on grayscale images while only a few concentrated towards edge detection in color images. On the other hand, color edge detection techniques gave unsatisfactory results in case of noise corrupted images. Thus, as a remedy to this problem a novel color edge detection technique is proposed in this paper using Hilbert transform [11]. The technique uses the concept of RGB color triangle which helps in maintaining the originality of image by avoiding the transformation of one color space into another. Usage of bilateral filtering [12], [13] as a pre-cursor to Hilbert transform catalyzes the performance of edge detection in noisy environment as it preserves edges while smoothening the image. Computer simulations of the proposed edge detection technique are performed on both synthetic as well as real images and their performance is quantitatively evaluated using PFOM [14] and Reconstruction estimation function [15] respectively. The remaining part of this paper is structured as follows: a detailed explanation of the proposed edge detection technique has been given in section 2. The simulation results along with performance evaluation have been discussed in section 3. Conclusions on the basis of obtained results are drawn in section 4.

2 Proposed Edge Detection Technique

The proposed edge detection technique is procedurally divided into three main modules namely: Pre-processing, Bilateral filtering and finally edge enhancement as shown in the block diagram given in fig. 1. The pre-processing module is marked by the transformation of the input color image *p(x)* using RGB color triangle [8] resulting in a transformed image *q(x)*. The transformed image *q(x)* is given as input to the bilateral filter for necessary smoothening and noise suppression yielding the filtered image *r(x)*. This step is followed by application of Hilbert transform for edge enhancement giving *h(x)* as the resultant image. For further segmentation, thresholding is performed on enhanced edges yielding the final edge map *e(x)*.

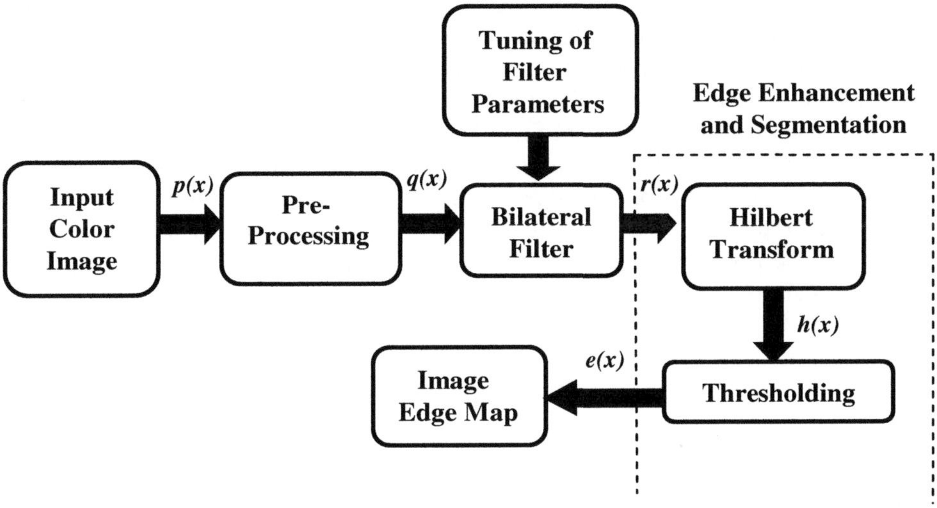

Fig. 1. Block Diagram of Proposed Edge Enhancement and Segmentation Technique

2.1 Pre-processing (Image Transformation Using RGB Color Triangle)

The pre-processing module consists of transformation of the input color image *p(x)*, by mapping each of its pixels to a RGB color triangle [8]. The sides of the RGB color triangle are assigned the R, G and B values of the corresponding pixels. Perimeter of the RGB scalene triangle is calculated using (1) and stored in a separate matrix. Thus, each pixel in the original color image is replaced by the perimeter of its corresponding RGB color triangle. The process is repeated for each and every pixel of the input image finally yielding the transformed image *q(x)*.

$$q(x) = aR + bG + cB \tag{1}$$

where: *a*, *b* and *c* are constants and have been assigned a value equal to (1/3). This pre-processing step avoids transformation of one color space into another and thereby assisted in maintaining the originality of image. The transformed image *q(x)* is then smoothened by using bilateral filtering.

2.2 Bilateral Filtering

Impulse noise, also known as Salt and Pepper noise is a special type of data dropout noise represented by a sprinkle of bright and dark spots just like salt and pepper granules. This noise usually corrupts the image during transmission, sensor faults or due to ac power interference. Edges represent points in an image where there is a sharp change of intensity from one value to another. Hence, edges are present in the high frequency structure of the image and are more susceptible to noise making edge detection unstable and inaccurate leading to detection of false edges or at times loss of true edges. Gaussian low pass filter and Median filters are commonly used for image noise suppression but during this process, these filters suppress the high frequency structure leading to the removal of finer details along with noise. In the present work, bilateral filtering [12], [13] is used as an edge preserving technique to catalyze the performance of proposed edge detector in noisy environments. In bilateral filtering, each pixel x in the input image $q(x)$ is replaced by the normalized weighted average of the pixels present in the spatial neighborhood $N(x)$ and can be expressed mathematically as,

$$r(x) = \frac{\sum_{a \in N(x)} W(x) q(a)}{\sum_{a \in N(x)} W(x)} \tag{2}$$

where: a signifies the nearby pixel location in $N(x)$ and $r(x)$ represents the image reconstructed after bilateral filtering. The weight $W(x)$ in (2) is calculated by the multiplication of two other weight components as given in (3).

$$W(x) = W_S(x).W_R(x) \tag{3}$$

where: $W_S(x)$ and $W_R(x)$ can be stated as:

$$W_S(x) = e^{-\frac{1}{2}\left(\frac{\|a-x\|}{\sigma_d}\right)^2} \tag{4}$$

$$W_R(x) = e^{-\frac{1}{2}\left(\frac{\|q(a)-q(x)\|}{\sigma_r}\right)^2} \tag{5}$$

In (4), $\|a\text{-}x\|$ represents the Euclidean distance between a and x and in (5), $\|q(a)\text{-}q(x)\|$ measures the intensity difference between two points a and x. The parameters σ_d and σ_r represent the geometric spread and the photometric spread respectively. The component, $W_s(x)$ in (3) performs filtering in spatial domain acting as *domain filter* whereas $W_R(x)$ performs filtering in intensity domain acting as *range filter.* The role of domain filter is mainly noise suppression and smoothening whereas range filter assists in preserving crisp edges. Thus, bilateral filter can be interpreted as a combination of these two filters acting simultaneously on the input image. The performance of this filter can be tuned with the help of spread parameters σ_d and σ_r respectively depending upon the noise content present in an image.

2.3 Hilbert Transform

G. L. Livadas and A. G. Constantinides [11] were the first to present the idea of performing edge detection using Hilbert transform. This transform can be considered as a filter which shifts phases of all the frequency components of its input signal by $-\pi/2$ radians. Hilbert transform of a single dimension signal, *r(x)* can be stated as:

$$h(x) = \frac{1}{\pi}\int_{-\infty}^{\infty} \frac{r(\tau)}{\tau - x} d\tau \tag{6}$$

Following algorithm is employed for computing Hilbert transform (in two-dimension) for digital images: firstly, the Fast Fourier Transform (FFT) of the input image *r(x)* is calculated row wise and the result is stored in a matrix *P(x)*. Next, a matrix *Q(x)* is created based on the following condition:

$$Q(i) = \begin{cases} 1; & \text{for } i=1 \text{ or } (N/2)+1. \\ 2; & i=2, 3, \ldots, (N/2). \\ 0; & i=(N/2)+2, \ldots, N. \end{cases} \tag{7}$$

where: *N* denotes the total number of pixels in the input image. The element-wise product of *P(x)* and *Q(x)* is calculated and stored in a separate matrix, whose inverse FFT returns *H(x)*. The same procedure is repeated column wise on *H(x)* to yield the Hilbert transformed image *h(x)* with enhanced edges. With the application of Hilbert transform, there is a clear demarcation between the edge pixels and the background in the resultant image. This results into appearance of dark and bright regions in the image with sharp boundaries resulting in edge enhancement. For segmentation of features, a suitable threshold selection is needed; this is achieved by calculating *T* as the average of the absolute values of pixel intensities in *h(x)* as given by equation (8).

$$T = \frac{1}{N}\sum_{x=1}^{N} |h(x)| \tag{8}$$

Finally, thresholding is performed on *h(x)* based on the condition given in (9) to generate the edge map *e(x)*.

$$[h(x) - h(x+1)] > T \tag{9}$$

3 Results and Discussion

3.1 Simulation Results

The performance evaluation of the proposed edge detection technique has been done by carrying out simulations for both synthetic as well as real images. Hence, in this paper, *PFOM* [14] has been used as a parameter for quality evaluation of synthetic (edge maps) images where as for edge maps of real images, reconstruction estimation function [15] is employed. Higher the value of *PFOM*, better is the quality of the obtained edge map whereas lower the value of Mean Squared Error (*MSE)*, better is the performance of edge detector for real images. A 'shapes' image (synthetic image) and a 'house' image (real image) are taken as test images for simulations in this work. Simulations are carried out on noise free as well as noisy versions (corrupted with impulse noise in the intensity range of 5-20%) of the above images. During pre-processing, each of the pixels in $p(x)$ is incremented by one and then transformed using (1). The pre-processed image is then given as input to a bilateral filter with σ_d ranging between 1 to 3 and σ_r varying from 10 to 120. The values of the spread parameters are tuned according to the noise content in the image. With the application of Hilbert transform and thresholding, final edge maps are generated. For the purpose of comparison, edge detection is performed on the above set of test images using Sobel operator [3], improved Sobel operator [8], generalized radial Hilbert transform [9] along with the proposed technique. Figure 2 shows the edge maps obtained by using different edge detection techniques for a 'shapes' image. Category-I in fig. 2 shows the simulations for noise free images, where it can be observed that the Sobel operator detects a double edge while this edge map gets degraded and dull in the presence of noise (as shown in category-II). Improved Sobel operator yields edges which are discontinuous in nature and the results are not favorable in the noisy environment. On the other hand, edge detection is good with generalized radial Hilbert transform are but the edge map is highly affected in presence of noise. However, it can be visualized that appreciable results are obtained with the proposed edge detection technique for both categories I and II of fig. 2(e). Table 1 enlists the *PFOM* values calculated for various edge detection techniques on the 'shapes' image which show relatively higher values for the proposed technique.

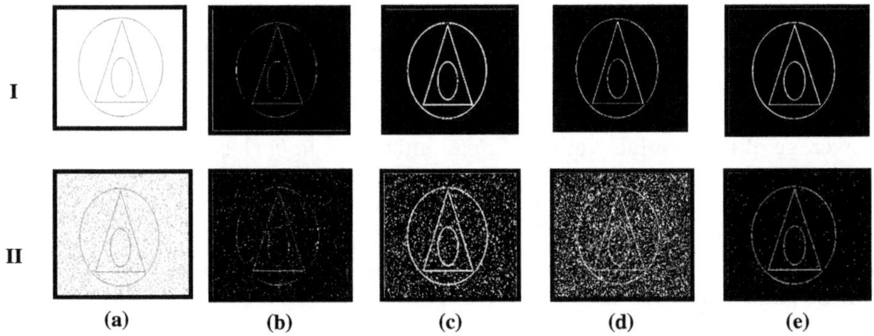

Fig. 2. **(a)** Original image: (I) Noise free (II) Contaminated with 10% impulse noise. Edge map obtained by: **(b)** Sobel Operator [3] **(c)** Improved Sobel Operator [8] **(d)** Generalized Radial Hilbert Transform [9] **(e)** Proposed Technique.

Table 1. PFOM values for different Edge Detection Techniques for the 'shapes' image

Amount of Noise Contamination	Sobel Operator [3]	Improved Sobel Operator [8]	Generalized Radial Hilbert Transform[9]	Proposed Technique
0% (Noise Free)	0.8020	0.8233	0.9117	0.9453
5%	0.6183	0.6211	0.5621	0.7312
10%	0.5543	0.5402	0.3819	0.6718
15%	0.4536	0.4784	0.2992	0.6107
20%	0.3832	0.4002	0.2211	0.5453

Similarly, fig. 3 shows simulation results for a 'house' image. The obtained edge maps clearly show the efficiency of the proposed technique over other edge detection techniques. For the improved Sobel operator, the detected edges are quite thick but significantly poor performance is seen in noisy images. Sobel operator and generalized radial Hilbert transform also appear to be unsuccessful in generating edge maps in noisy environment. Tabulation of *MSE* values (reconstruction estimation) are made in table 2 for edge maps of 'house' image. Obtained values of *MSE* are least for the edge maps generated with the proposed technique. In addition, there is a relative increment in *MSE* values with the increase in the noise intensity. But, even for a noise contamination of 20%, the obtained *MSE* with the proposed technique is least in comparison to other techniques.

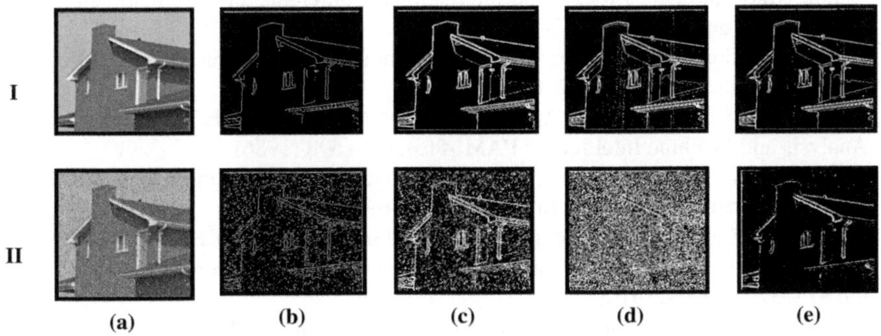

Fig. 3. (a) Original image: (I) Noise free (II) Contaminated with 10% impulse noise. Edge map obtained by: **(b)** Sobel Operator [3] **(c)** Improved Sobel Operator [8] **(d)** Generalized Radial Hilbert Transform [9] **(e)** Proposed Technique

Table 2. MSE values for different Edge Detection Techniques calculated for the 'house' image.

Amount of Noise Contamination	Sobel Operator [3]	Improved Sobel Operator [8]	Generalized Radial Hilbert Transform [9]	Proposed Technique
0% (Noise Free)	0.0161	0.0051	0.0051	0.0048
5%	0.0172	0.0067	0.0135	0.0051
10%	0.0194	0.0085	0.0205	0.0053
15%	0.0217	0.0105	0.0232	0.0064
20%	0.0263	0.0142	0.0309	0.0082

4 Conclusion

The color edge detection technique proposed in this paper provides favorable results for both synthetic as well as real images. The originality of the input color image has been maintained by avoiding color space transformation during pre-processing. Further, inclusion of bilateral filtering prior to performing edge detection using Hilbert transform makes the proposed technique highly promising for images corrupted with impulse noise. This has been validated by the scores of quality parameters calculated for the obtained edge maps.

References

1. Ziou, D., Tabbone, S.: Edge Detection Techniques: An Overview. International Journal of Pattern Recognition and Image Analysis 8, 537–559 (1998)
2. Novak, C.L., Shafer, S.A.: Color Edge Detection. In: Proc. of the DARPA Image Understanding Workshop, vol. 1, pp. 35–37 (1987)
3. Sharifi, M., Fathy, M., Mahmoudi, M.T.: A Classified and Comparative Study of Edge Detection Algorithms. In: Proc. of the International Conference on Information Technology: Coding and Computing, pp. 117–220 (2002)
4. Marr, D., Hildreth, E.C.: Theory of Edge Detection. Proc. of Royal Society of London, Series B, Biological Sciences 207, 187–217 (1980)
5. Canny, J.F.: A Computational Approach to Edge Detection. IEEE Transactions on Pattern Analysis and Machine Intelligence PAMI-8(6), 679–698 (1986)
6. Gudmundsson, M., et al.: Edge Detection in Medical Images Using a Genetic Algorithm. IEEE Transactions on Medical Imaging 17(3), 469–474 (1998)
7. Zahurul, S., Zahidul, S.: An Adept Edge Detection Algorithm for Human Knee Osteoarthritis Images. In: Proc. of International Conference on Signal Acquisition and Processing, pp. 375–379. IEEE (2010)
8. Fengjing, Z., et al.: Color Image Edge Detection Arithmetic Based on Color Space. In: Proc. of the International Conference on Computer Science and Electronics Engineering, pp. 217–220. IEEE (2012)
9. Pei, S., et al.: The Generalized Radial Hilbert Transform and its Applications to 2-D Edge Detection (any direction or specified direction). In: Proc. of the International Conference on Acoustics, Speech and Signal Processing, pp. 357–360 (2003)
10. Golpayegani, N., et al.: A Novel Algorithm for Edge Enhancement based on Hilbert Matrix. In: Proc. of 2nd International Conference on Computer Engineering and Technology, vol. 1, pp. 579–581 (2010)
11. Livadas, G.L., Constantinides, A.G.: Image Edge Detection and Segmentation Based on the Hilbert transform. In: Proc. of the International Conference on Acoustics, Speech and Signal Processing, pp. 1152–1155 (1988)
12. Tomasi, C., Manduchi, R.: Bilateral Filtering for Gray and Color Images. In: Proc. of the IEEE International Conference on Computer Vision, Bombay, India, pp. 839–846 (1998)
13. Elad, M.: On the Origin of Bilateral Filter and Ways to Improve it. IEEE Transactions on Image Processing 11(10) (2002)
14. Pratt, W.K.: Digital Image Processing, 2nd edn. Jhon Wiley and Sons, New York (1991)
15. Agaian, S.S., et al.: Boolean Derivatives with Application to Edge Detection for Imaging Systems. IEEE Transactions on Systems, Man and Cybernetics- Part B: Cybernetics 40(2), 371–382 (2010)

An Improved Algorithm for Noise Suppression and Baseline Correction of ECG Signals

Rishendra Verma, Rini Mehrotra, and Vikrant Bhateja

Department of Electronics and Communication Engineering,
Shri Ramswaroop Memorial Group of Professional Colleges,
Lucknow-227105(U.P.), India
{verma.rishendra90,rini684.mehrotra,bhateja.vikrant}@gmail.com

Abstract. Testing and analysis of Electrocardiogram (ECG) signals is one of the major requirements for clinical diagnosis of cardiovascular diseases and deciding future therapies. ECG being a weak non-stationary signal is often interfered by impulse noise as well as baseline drift. This paper presents an improved morphological algorithm for suppression of ailments posed by the above mentioned distortions using non-flat structuring element. Dimensions of the structuring element are optimally selected in a manner to achieve lower distortion rates. Simulation results show significant improvement in baseline correction and noise removal (yielding lower values of error indices and high signal to noise ratios) in comparison to other methods.

Keywords: baseline wandering, ECG, impulse noise, morphological filtering, non-flat structuring element.

1 Introduction

Electrocardiogram (ECG) is an interpretation of the electrical activity of the heart over a certain period of time, which is detected by electrodes attached to the heart and on the limbs. A single normal cycle of ECG represents the consecutive atrial and ventricular depolarization during every heartbeat which is associated with the peaks and troughs of ECG waveform [1-3]. ECG signals are frequently plagued by impulse noise in diverse forms. Power line interference of 50/60 Hz is a common artifact corrupting the raw ECG which appears as a sinusoidal wave. Another artifact is baseline wander, where the baseline (of ECG waveform) starts to drift up and down in a sinusoidal pattern due to respiration [4]. Therefore, ECG signal conditioning for providing baseline correction and noise elimination is a necessary pre-requisite for further analysis such as QRS detection and temporal alignment. Traditional methods for ECG signal conditioning include: high-pass filtering [5] and band-pass filtering [6]. Since baseline drift has relatively lower frequency; baseline correction is performed by high-pass filtering. Both of these filtering techniques possess sharp cut-off frequencies which often distort the signal. Morphological Filtering [7] is employed for baseline correction and denoising of ECG signals. Yan Sun *et al.* proposed MMF algorithm [8] which retained the characteristics of the ECG signal but could not

S.C. Satapathy et al. (Eds.): Proc. of Int. Conf. on Front. of Intell. Comput., AISC 199, pp. 733–739.
DOI: 10.1007/978-3-642-35314-7_83

denoise the signal effectively. In addition, the computational time of this algorithm was high. Wavelet based techniques [9-11] for ECG noise suppression requires numerous experiments for ruling out scales and thresholds. Chu *et al.* [12] used combination of morphological opening and closing operators for baseline correction which ultimately distorted the trait points in the signal which is not appropriate for ECG analysis. Liu *et al.* [13] proposed morphological filtering by using linear structuring element whose size depends on the sampling frequency. This approach successfully removed the distortion leading to baseline drift but was not capable in suppressing the noise. Also, other methods like FIR and IIR filters were also not appropriate in improving signal quality as additive noise has same frequency band as ECG signal. Yuan Gu *et al.* [14] used the morphological filters with flat structuring element but was able to reduce ST segment distortion only up to 50%. Hence, an improved algorithm for baseline correction and noise suppression is introduced using morphological filters. The proposed algorithm uses non-flat structuring element(s) which achieves low distortion and high noise suppression. Mean Squared Error (MSE), Percentage RMS Distortion (PRD) and Signal-to-noise ratio (SNR) are used as quality parameters to evaluate the performance of proposed algorithm in comparison to other methods. The rest of this paper is organized as follows. The proposed methodology is described in Section 2. Section 3 details the results and discussions and lastly, the conclusion is given under Section 4.

2 Proposed Algorithm

2.1 Morphological Filters

Background. Morphological filtering is a non-linear transformation technique primarily used for local modification of geometrical features of a signal. This shape information (of features) is extracted by using a structuring element (of appropriate dimensions) to operate on the input signal. Erosion, dilation, opening and closing are the common morphological operators used. Baseline wander removal and impulse noise suppression in the ECG signals can be attained by using various combinations of these operations. Erosion of a signal by structuring element is defined as moving local minima of the signal inside the structuring element or mask. Similarly dilation is defined as moving local maxima of the signal inside the structuring element. Therefore dilation enlarges the maxima of the signal while erosion enlarges the minima of the signal. Opening (erosion followed by dilation) by structuring element smoothes the signal from below by cutting down its peaks. Similarly, closing (dilation followed by erosion) by structuring element smoothes the signal from above by filling up its valleys. Hence, opening and closing operations can be used for detection of peaks and valleys in the signal [15].

Proposed Non-flat Structuring Element. As discussed above, a structuring element plays very significant role in extracting the characteristic information from the targeted signal. An important aspect in baseline removal and noise suppression is the selection of optimum size of the structuring element. Improper selection of structuring element may distort the adjacent wave in the ECG signal. Hence, the size of structuring element should be greater than the width of the characteristic wave [7]. Flat structuring elements often lead to distortion due to overlap of low frequency (ST) segment with the baseline wandering, thereby causing loss of relevant information [14]. Hence, a non-flat structuring element is proposed in this work as it improves the performance of morphological filtering in terms of smooth opening and closing of ECG signal. This enables proper extraction of characteristic wave for baseline correction and denoising without introducing any distortion in ECG waveform. A ball-shaped structuring element is used with the proposed morphological algorithm for performing baseline correction and noise filtering as shown in fig. 1(a) –(c). The dimensions of the structuring element are determined experimentally using the performance evaluation parameters.

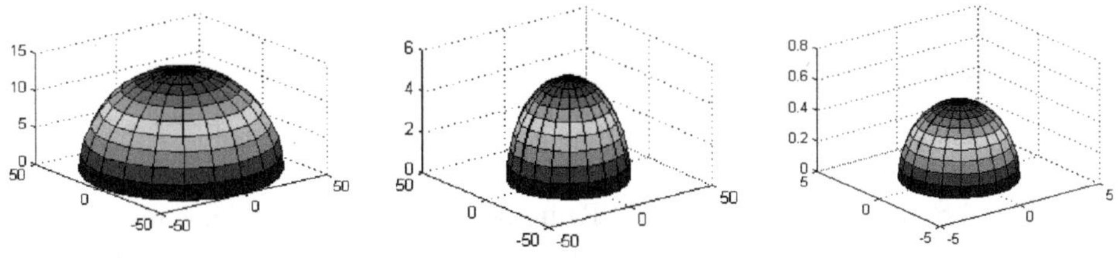

(a) (b) (c)

Fig. 1. Structuring Elements (se1-se3) for (a) Baseline Correction (se1) of radius=49, height=27; (b) Baseline Correction (se2) of radius=30, height=10; (c) Noise suppression (se3) of radius=3, height=1.

2.2 Proposed Algorithm for Baseline Correction

Baseline wandering being a low frequency artifact is removed by employing morphological operators that have both low pass and high pass filtering characteristics. Therefore, in the proposed algorithm, noisy baseline drifted signal is subjected to series of opening-closing operations. With a non-flat structuring element of required dimensions opening and closing operation(s) can be the thought of as non-linear filters which smoothes the contours of the input ECG signal. The size of the structuring element (se1>se2) used for opening operation is kept larger than that used for closing operation. This ensures proper extraction of characteristic wave and also helps in preprocessing of ECG signal. ECG Signal obtained after the first series of opening-closing is used to perform two sets of operations. In set-I, opening followed by closing is performed where as in set-II closing followed by opening is performed with the structuring elements mentioned in fig. 1. The signal obtained as output from I and II set of operations are averaged to produce the baseline wandering signal. It is then subtracted from noisy baseline drifted signal to yield the corrected baseline signal.

2.3 Proposed Algorithm for Noise Suppression

ECG signal is mainly corrupted by impulses (or spikes) i.e. very large positive or negative values of very short interval. During baseline correction, some of the impulse noise is removed. However, for further noise suppression, the signal obtained after baseline correction is made input to the proposed noise suppression algorithm. This involves morphological dilation followed by erosion in stage I and erosion followed by dilation in stage II using structuring element (se3). The structuring element is comparable to the width of impulse noise hence noise can be removed without sacrificing the signal quality. The results of stage I and stage II are averaged to get the denoised signal.

3 Results and Discussions

3.1 Simulation Results

The performance of proposed algorithm is evaluated with the help of three quality parameters: Mean Squared Error (MSE), Percentage Root mean square Difference (PRD) and Signal to Noise Ratio (SNR) in dB. MSE denotes the deviation of reconstructed signal from the original one. Hence lower the value of MSE better is the quality of the reconstructed signal. PRD is considered as a quality measure and is used to evaluate the reliability of the reconstructed signal.SNR is used to quantify the noise level in the ECG signal, thus higher the value of SNR, better is the amount of noise suppression [16]. For simulation purposes in the present work, ECG signals are adapted from internationally accepted MIT-BIH database that consists of recordings which are observed in clinical practice. The ECG signals (118 and 119 shown in fig. 2(a) and 3(a)) are taken from noise stress database section under MIT-BIH database [17]. Baseline wandering signal is added to the noisy signal to generate the baseline drifted noisy signal as shown in fig. 2(b) and 3(b) respectively. The distorted ECG signal is first processed through the proposed baseline correction algorithm using ball shaped structuring elements of radii 49 & 30 and heights 27 and 10 respectively. The baseline corrected signal (shown in fig. 2(c) and 3(c)) is then subjected to the proposed noise suppression algorithm with structuring element of radius 3 and height 1. The finally processed ECG signal is shown in fig. 2(d) and 3(d) respectively. Fig 2(c) and 3(c) shows the baseline corrected signal where baseline is adjusted to or brought down to zero reference. As it can be seen that characteristic wave of ECG are preserved therefore proposed baseline correction algorithm produces no distortion. Moreover, during baseline correction some amount of impulse noise is also reduced which occurs in the form of high peaks in the signal. It can be observed from fig. 2(d) and 3(d) that impulse noise is reduced by noise suppression algorithm without distorting the characteristic wave in the ECG signal.

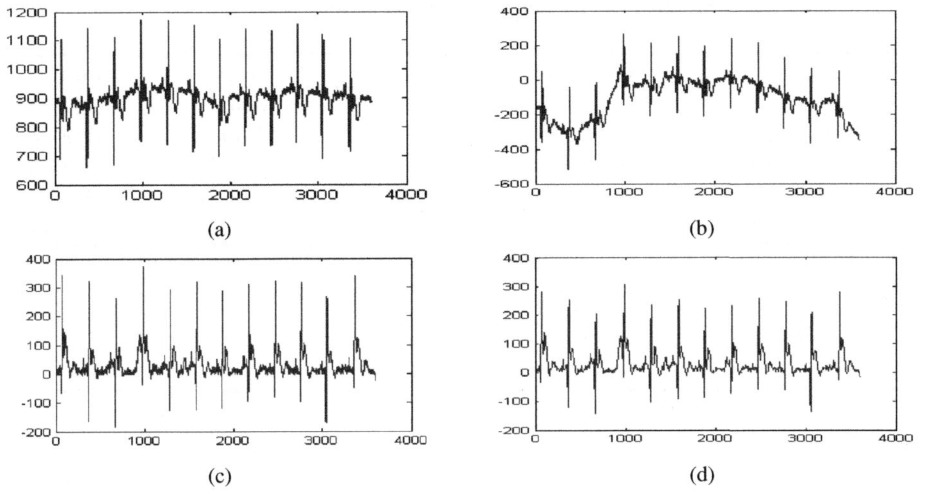

Fig. 2. (a) Original ECG signal (MIT BIH record 118) (b) Corrupted ECG signal (c) Baseline Corrected Signal (d) De-noised ECG signal

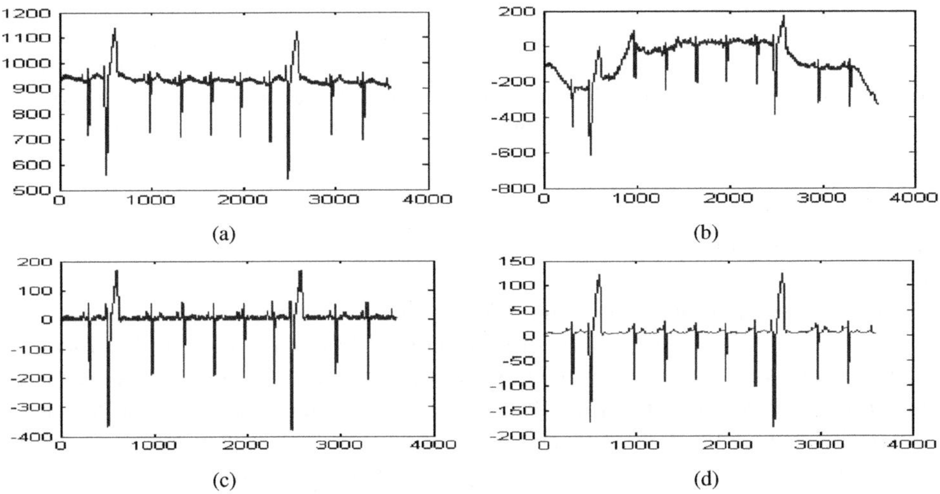

Fig. 3. (a) Original ECG signal (MIT BIH record 119) (b) corrupted ECG signal (c) baseline Corrected signal (d) De-noised signal

3.2 Comparison

Performance comparison of proposed algorithm is done with the results obtained from the morphological filter [13]. The results are evaluated with the help of above mentioned quality parameters and are tabulated in table 1 and 2.

Table 1. MSE and PRD for Performance Evaluation of Baseline Correction

Record#	Morphological Filter [13]		Proposed Algorithm	
	MSE	PRD%	MSE	PRD%
118e24[1](bw[2],noise1)	0.0025	0.0052	0.0006	0.0026
118e24(bw,noise2)	0.0069	0.0090	0.0025	0.0016
119e24[1](bw[2],noise1)	0.0250	0.0540	0.0057	0.0134
119e24(bw,noise2)	0.0056	0.0081	0.0015	0.0041

[1] noisy signals used here, [2] baseline wandering signal (corrupted with different noise intensities indicated as noise1 and noise2).

Table 2. Performance Evaluation of Noise Suppression using SNR(dB)

Record#	Original Signal	Corrupted Signal	Reconstructed Signal by morphological filter [13]	Reconstructed Signal by Proposed Algorithm
118e24	33.7604	24.9490	11.7832	41.3823
119e24	46.1549	20.5425	6.1426	27.0110

From above set of tabulations, it can be observed that in case of the proposed algorithm the values of MSE and PRD are lesser than those obtained by morphological filter [13], which shows that the reconstructed ECG signal contains minimum distortion. Moreover, higher values of SNR obtained in the present work depicts better noise suppression in comparison to existing methods.

4 Conclusion

In this paper, an improved algorithm using non-flat structuring element is proposed to remove baseline drift and impulse noise from the ECG signal. The non-flat structuring element possesses advantage over flat structuring element as it minimizes the distortion produced by flat structuring element during baseline line correction and noise removal. The results obtained in the present work gives lower values of error indices and shows significant improvement in SNR in comparison to other methods.

References

1. Dupre, A., Vincent, S., Iaizzo, P.: Basic ECG Theory, Recordings and Interpretation. In: Handbook of Cardiac Anatomy, Physiology, and Devices, pp. 191–201 (2005)
2. Gupta, R., Bera, J.N., Mitra, M.: Development of An Embedded System and MATLAB-Based GUI for Online Acquisition and Analysis of ECG Signal. Journal Measurement 43(9), 1119–1126 (2010)

3. Sayadi, O., Shamsollahi, M.B.: ECG Baseline Correction With Adaptive Bionic Wavelet Transform. In: Proc. of the 9th International Symposium on Signal Processing and Its Applications, Sharjah, pp. 1–4 (2007)
4. Kumar, Y., Malik, G.K.: Performance Analysis of Different Filters for Powerline Interface Reduction in ECG Signal. International Journal of Computer Applications (0975 – 8887) 3, 1–6 (2010)
5. Christov, I.I., Dotsinsky, I.A., Daskalov, I.K.: High-pass Filtering of ECG Signals Using QRS Elimination. Medical and Biological Engineering and Computing 30, 253–256 (1992)
6. Pei, S., Tseng, C.: Elimination of AC Interference in ECG using IIR Notch Filter with Transient Suppression. IEEE Transactions on Bio-Medical Engineering 42(11), 1128–1132 (1995)
7. Chu, C.-H.H., Delp, E.J.: Impulsive Noise Suppression and Background Normalization of Electrocardiogram Signal Using Morphological Operators. IEEE Transactions on Biomedical Engineering 36, 262–273 (1996)
8. Sun, Y., Chan, K.L., Krishnan, S.M.: ECG Signal Conditioning by Morphological Filtering. Computers in Biology and Medicine 32(6), 465–479 (2002)
9. Sayadi, O., Shamsollahi, M.: Multiadaptive Bionic Wavelet Transform: Application to ECG De-noising and Baseline Wandering Reduction. EURASIP Journal on Advances in Signal Processing (2007)
10. Alfaouri, M., Daqrouq, K.: ECG Signal De-noising by Wavelet Transform Thresholding. American Journal of Applied Sciences 5(3), 276–281 (2008)
11. Donoho, D.L.: De-noising by Soft-thresholding. IEEE Transactions on Information Theory 41(3), 612–627 (1995)
12. Chu, C.-H.H., Delp, E.J.: Electrocardiogram Signal Processing by Morphological Operators. In: Proc. of the Conference on Computers in Cardiology, Washington DC, pp. 153–156 (1988)
13. Liu, Z., Wang, J., Liu, B.: ECG Signal Denoising Based on Morphological Filtering. In: Proc. of the 5th International Conference on Bioinformatics and Biomedical Engineering, Wuhan, pp. 1–4 (2011)
14. Gu, Y., Zheng, G., Dai, M.: A Morphology Algorithm Based on 2-Dimensional Flat Structuring Element on ECG Baseline Wander Elimination. In: Proc. of the Conference on Computing in Cardiology, Hangzhou, pp. 817–820 (2011)
15. Maragos, P., Schafer, R.W.: Morphological filters part I and II. IEEE Transactions on Acoust. Speech Signal Process. 35, 1170–1184 (1987)
16. Mishra, A., Thakkar, F., Modi, C., Kher, R.: Comparative Analysis of Wavelet Basis Functions for ECG Signal Compression Through Compressive Sensing. International Journal of Computer Science and Telecommunications 3, 23–31 (2012)
17. Moody, G.B., Mark, R.G.: The Impact of the MIT-BIH Arrhythmia Database. IEEE Engineering in Medicine and Biology Magazine 20, 45–50 (2001)

A Reconstruction Based Measure for Assessment of Mammogram Edge-Maps

Vikrant Bhateja[1] and Swapna Devi[2]

[1] Deptt. of Electronics and Communication Engineering, SRMGPC, Lucknow (U.P.), India
bhateja.vikrant@gmail.com
[2] Deptt. of Electronics and Communication Engineering, NITTTR, Chandigarh, India
swapna_devi_p@yahoo.co.in

Abstract. Performance evaluation of for mammogram edge-maps is difficult because of the absence of reference image for comparison. This paper presents a novel approach for assessment of edge enhanced mammograms containing microcalcifications. It is a non-reference approach helpful in selection of most appropriate algorithm for edge enhancement of microcalcifications. Experiments results validate the efficiency of the proposed evaluation method in precise assessment of mammograms in accordance to human evaluation.

Keywords: Edge-maps, Interpolation, Mammograms, Morphological filter, Reconstruction estimation.

1 Introduction

Microcalcifications are small densities that appear as bright spots on mammograms. Direct interpretation of calcifications can be very subtle as they are often camouflaged by the dense fibro-glandular breast tissues. Their classification as benign or malignant requires accurate preservation of their morphological details; hence, edge detection of calcifications plays an important role in breast cancer detection at early stages in digital mammograms [1]. Conventional gradient based edge detectors (Sobel, Prewitt etc.) are known for their operational simplicity but cannot be directly made applicable to medical images, as they yield inaccurate results and lack sensitivity to noise. Laplacian edge detectors produces detection of only those edges possessing fixed characteristics in all the directions but fail to perform in the presence of noise. Canny edge detectors tend to improve the signal-to-noise ratio by smoothening the image, yielding better detection of edges in presence of noise. However, these operators fail to yield optimal results using a fixed operator [2]. A common approach for the calcification detection task performs localization of high spatial frequencies in the digital mammograms using wavelet transform where as non-wavelet based methods use the concept that calcifications possesses much higher intensity values than the surrounding breast tissues [3-4]. CLAHE [5] provides significantly sharp and defined edges but, these are coupled with the enhancement of other unwanted information. Unsharp Masking (UM) [6-7] did not produce satisfactory enhancement of edges in

S.C. Satapathy et al. (Eds.): Proc. of Int. Conf. on Front. of Intell. Comput., AISC 199, pp. 741–746.
DOI: 10.1007/978-3-642-35314-7_84

mammograms on account of the presence of some over shoots in the region of interest (ROI). Many techniques have been proposed in literature for quality evaluation of edge maps but their functionality is based on particular criteria and constraints; that limit their usage in a generalized mode. Objective evaluation methods for edge maps (of mammograms), can serve as important tool to dynamically monitor the quality independent of any viewing condition and would aid the radiologists in accurate diagnosis, thereby deciding future therapies and treatment patterns [8]. Pratt's Figure-of-Merit (PFOM) [9] can be used as an objective evaluation index for edge maps of synthetic images. They can be applied only when there is a complete knowledge of true edge locations but fail to assess for real images. On the other hand, Reconstruction based estimation [10] method is applied for evaluation of output edge maps for real images. These functions operate by deriving a reconstructed estimation of the original input image from the output edge map, based on the fact that edges contain the most important information in any image. The reliability of reconstruction estimation function is questionable as during assessments, the image reconstruction is mainly performed considering primarily the edge pixels. Direct application of Carlson's Reconstruction method [11] may not prove to be useful for assessment of mammographic edge maps, as improper reconstruction even for non-edge pixels may result in loss of diagnostically useful information. In mammograms, calcifications are either scattered throughout the region as tiny specks or occur in small clusters. Therefore, the quality of the reconstructed image during interpolation cannot be adjudged as a function of edge pixels density. Hence, this paper proposes a novel reconstruction estimation method for performance evaluation of edge detection algorithms for mammographic calcifications. It uses a non-reference approach which could aid in selection of most appropriate edge detection algorithm for calcifications. It is observed from the obtained results that evaluation done by the proposed reconstruction measure is consistent with the subjectivity of human assessment. The remainder of this paper is structured as follows: Section 2 details the proposed edge-map evaluation method. The results and discussions are given under section 3 and section 4 draws the conclusion.

2 Proposed Edge-Map Evaluation Method

The quality evaluation method for mammogram edge-maps operates in two phases. Phase-I deals with the detection of edge pixel information, while the reconstruction estimation and error assessment is performed during phase-II.

Phase-I: Detection of Edge Pixels

The detection of edge pixels in phase-I can be initiated by morphological dilation of the output edge map using a flat (diamond shaped) structuring element of size 3x3. The dilated edge map is logically multiplied with the edge enhanced region of interest (ROI) containing calcifications (prior to thresholding). The binary ROI obtained as the product is checked for the edge pixel locations. A new ROI is reconstructed by locating the corresponding pixels in edge map with magnitude unity (defined as edge points of calcifications). This reconstructed ROI contains the estimated edge pixel information. In case, the binary ROI returns a non-edge pixel, then the interpolation pixel value in the reconstructed ROI is determined using the reconstruction method in phase-II.

Phase-II: Reconstruction Using Interpolation Scheme

If the gray-level information in the binary edge-map, determines the edge pixels, then the remaining part of the ROI *f(x, y)* (containing the non-edge pixels) is reconstructed by interpolation using the minimum variation criteria [10] given in (1).

$$\iint \left(\frac{\partial f}{\partial x}\right)^2 + \left(\frac{\partial f}{\partial y}\right)^2 dxdy \tag{1}$$

The above variation measure can be converted to the discrete version using eq. (2). Evaluating the above variation criteria as a constrained optimization problem; the process starts by computing the values of f minimizing above equation subject to the condition that f attains particular values along the edges.

$$\sum_{x}\sum_{y}(f_{x,y} - f_{x,y-1})^2 + (f_{x,y} - f_{x-1,y})^2 \tag{2}$$

This can be further solved using the successive over-relaxation approach iteratively (defined in (3) & (4)) for finding the minima of the eq. (2).

$$f^{(i+1)}(x, y) = f^{(i)}(x, y) + \frac{\phi}{4}\Delta f^{(i)}(x, y) \tag{3}$$

where: $$\Delta f^{(i)}(x, y) = f^{(i+1)}(x-1, y) + f^{(i+1)}(x, y-1) + f^{(i)}(x+1, y) + f^{(i)}(x, y+1) - 4f^{(i)}(x, y) \tag{4}$$

Application of this approach tends to smoothen any of the unwanted gray-level variations or artifacts which might be introduced on account of incorrect pixels at non-edge points [10]. Lastly, the error assessment compares the reconstructed ROI [*f(x,y)*] with the original enhanced ROI [*g(x,y)*] to estimate the amount of error in the reconstructed ROI using Mean Absolute Error (*MAE*) defined in (5). Lower the error values, better is the quality reconstructed image. Hence, better is the performance of edge detection algorithm.

$$MAE = \frac{1}{MN}\sum_{i=1}^{M}\sum_{j=1}^{N}\left|g(i, j) - f(i, j)\right| \tag{5}$$

3 Results and Discussion

The original input mammographic images read from the Mammographic Image Analysis Society (MIAS) database [12] are normalized followed by extraction of ROI (containing calcifications) of size 256 x 256, to eliminate the unwanted portion of the image that consists of most of the background area. For performance evaluation, the output edge maps obtained from the following edge detection algorithms are used:

Co-ordinate Logic Filters (CLF) by J. Q. Domínguez *et al.* [13], is named as A-I; Another related work by Ch. Santhaiah *et al.* [14] used morphological gradient operator, is named A-II. Morphological top-hat and bottom hat Transformations proposed by Tomklav Stojic *et a.l* [15], is termed A-III. V. Bhateja and S. Devi proposed an edge detection algorithm [16] for microcalcifications using a two-stage morphological filter; is termed A-IV. These edge detection algorithms are applied to the enhanced ROI, which following by thresholding yields the output edge map, whose quality is evaluated using the proposed measure.

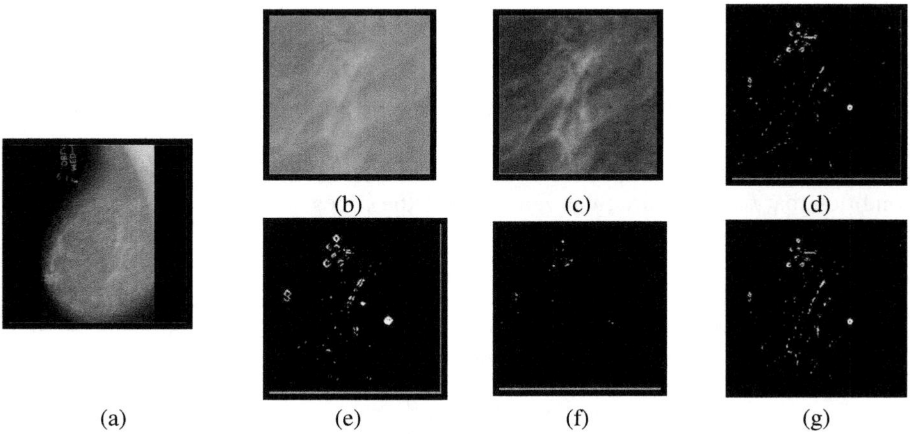

Fig. 1. (a) Original Mammogram (mdb231) containing calcifications embedded in a background of fatty-breast tissues. (b) Extracted ROI . (c) Enhanced ROI. Edge Maps obtained using: (d) A-I [13], (e) A-II [14], (e) A-III [15], (f) A-IV [16].

It can be observed in the edge maps of fig. 1(d) that some of the microcalcification clusters are missed owing to usage of small sized structuring element. In fig. 1(e) the edge map tend to miss some of the tiny scattered calcifications, on the other hand fig. 1(f) shows detection of very few calcifications. However, a reasonably good response can be seen in fig. 1(g) which shows recovery of both the micro-calcification clusters as well as the scattered ones. It can be seen in Table 1 that the values of *MAE* are least for A-IV (as in fig. 1(g)), validating its superior performance in comparison to other edge detectors. As visible in fig. 1(f), most of the calcifications are missed by the edge detector; hence A-III yields the highest error values among all four edge maps. Nearly comparable error values are obtained for A-I and A-II respectively. From the simulation results obtained and their corresponding evaluations, it can be ascertained that proposed reconstruction approach has proved to be a suitable evaluation method for edge maps, as it can return precise data to compare the performance of edge detectors. The proposed evaluation scheme quantifies that the performance of A-IV is significantly improved in comparison to other gradient based morphological edge detectors (A-I to III) proposed for detection of calcifications.

Table 1. Calculation of *MAE* for Mammographic Edge Detection Algorithms

Mammogram Ref. No.	A-I [13]	A-II [14]	A-III [15]	A-IV [16]
mdb231	0.0053	0.0043	0.0080	0.0023
mdb238	0.0043	0.0040	0.0075	0.0019
mdb211	0.0055	0.0026	0.0069	0.0011
mdb213	0.0063	0.0036	0.0072	0.0015
mdb219	0.0050	0.0034	0.0070	0.0012
mdb233	0.0061	0.0031	0.0071	0.0014
mdb223	0.0071	0.0057	0.0105	0.0015
mdb239	0.0106	0.0068	0.0140	0.0045
mdb240	0.0142	0.0064	0.0109	0.0048
mdb241	0.0076	0.0050	0.0110	0.0019

4 Conclusion

The advantage of the proposed evaluation method lies in the fact that the interpolation procedure used in performing reconstruction estimation helps in precise and effective generation of gray-scale enhanced images from the binary edge-maps. Hence, it can be concluded that the proposed evaluation measure will serve as an effective tool to benchmark the performance of various edge enhancement algorithms for detection of calcifications. This will further aid in extraction of features in mammographic images, improving diagnostic detection of breast cancer.

References

1. Rovere, G.Q., Warren, R., Benson, J.R.: Early Breast Cancer from Screening to Multidisciplinary Management, 2nd edn. Taylor & Francis Group, Florida (2006)
2. Sharifi, M., Fathy, M., Mahmoudi, M.T.: A Classified and Comparative Study of Edge Detection Algorithms. In: Proc. of the International Conference on Information Technology: Coding and Computing, ITCC 2002 (2002)
3. Strickland, R.N., Hahn II, H.: Wavelet Transform for Detecting Microcalcifications in Mammograms. IEEE Transactions in Medical Imaging 15(2), 218–229 (1996)
4. Dominguez, J.Q., Cortina-Januchs, M.G., Jevtić, A., Andina, D., Barron-Adame, J.M., Vega-Corona, A.: Combination of Nonlinear Filters and ANN for Detection of Microcalcifications in Digitized Mammography. In: Proc. of the International Conference on Systems, Man, and Cybernetics, USA, pp. 1516–1520 (2009)
5. Papadopoulos, A., Fotiadis, D.I., Costaridou, L.: Improvement of Microcalcification Cluster Detection in Mammography Utilizing Image Enhancement Techniques. Computers in Biology and Medicine 38(10), 1045–1055 (2008)
6. Wu, Z., Yuan, J., Lv, B., Zheng, X.: Digital Mammography Image Enhancement using Improved Unsharp Masking Approach. In: Proc. of 3rd International Congress on Image and Signal Processing (CISP), vol. 2, pp. 668–672 (2010)

7. Polosel, A., Ramponi, G., Mathews, V.J.: Image Enhancement via Adaptive Unsharp Masking. IEEE Transactions on Image Processing 9, 505–510 (2000)
8. Wang, Z., Bovik, A.C., Sheikh, H.R., Simoncelli, E.P.: Image Quality Assessment: from Error Visibility to Structural Similarity. IEEE Transactions on Image Processing 13(4), 600–612 (2004)
9. Trahanias, P., Venetsanopoulos, A.: Vector Order Statistics Operators as Color Edge Detectors. IEEE Transactions Systems, Man, Cybernetics B, Cybern. 26(1), 135–143 (1996)
10. Carlsson, S.: Sketch Based Coding of Gray Level Images. Signal Processing 15(1), 57–83 (1988)
11. Elder, J.: Are edges incomplete? International Journal on Comp. Vis. 34(2/3), 97–122 (1999)
12. Suckling, J., et al.: The Mammographic Image Analysis Society Mammogram Database. In: Proc. of 2nd International Workshop Digital Mammography, U.K, pp. 375–378 (1994)
13. Dominguez, J.Q., Sanchez-Garcia, M., Gozalez-Romo, M., Vega-Corona, A., Andina, D.: Feature Extraction using Coordinate Logic Filters and Artificial Neural Networks. In: 7th International Conference on Industrial Informatics (INDIN 2009), pp. 645–649 (2009)
14. Santhaiah, C., Babu, G.A., Rani, M.U.: Gray-level Morphological Operations for Image Segmentation and Tracking Edges on Medical Applications. International Journal of Computer Science and Network Security 9(7), 131–136 (2009)
15. Stojic, T., Reljin, B.: Enhancement of Microcalcifications in Digitized Mammograms: Multifractal and Mathematical Morphology Approach. FME Transactions 38(1), 1–9 (2010)
16. Bhateja, V., Devi, S.: A Novel Framework for Edge Detection of Microcalcifications using a Non-Linear Enhancement Operator and Morphological Filter. In: Proc. of 3rd International Conference on Electronics & Computer Technology (ICECT 2011), Kanyakumari, India, vol. 5, pp. 419–424 (2011)

Despeckling of SAR Images via an Improved Anisotropic Diffusion Algorithm

Anurag Gupta, Anubhav Tripathi, and Vikrant Bhateja

Deptt. of Electronics and Communication Engineering,
Shri Ramswaroop Memorial Group of Professional Colleges,
Faizabad Road, Lucknow-227105, (U.P.), India
{anuraggpt8,proabhi9,bhateja.vikrant}@gmail.com

Abstract. Synthetic Aperture Radar (SAR) is a powerful tool for producing high-resolution images but these images are highly contaminated with speckle noise. This paper proposes an improved Anisotropic Diffusion Algorithm for despeckling SAR images. The proposed algorithm is obtained by using a diffusion coefficient which consists of a combination of first and second order derivative operators. The spatial variation of this diffusion coefficient occurs in such a way that it prefers forward diffusion to backward diffusion resulting in improved structural details and edge preservation. The simulation results also show better computational efficiency in comparison to other denoising techniques.

Keywords: Speckle, SAR images, Diffusion coefficient, Multiplicative noise.

1 Introduction

Speckle is a kind of multiplicative noise that affects most of the coherent imaging systems. The presence of speckle noise in an imaging system reduces its resolution; especially for low contrast images such as Synthetic Aperture Radar (SAR) images. This creates problem in automatic processing of SAR images, used in various applications like crop monitoring, search and rescue operations, military target detection etc. Therefore, the suppression of speckle noise is an important consideration in the design of coherent imaging systems. Over the last few years, various despeckling techniques for SAR images have been proposed [1-3]. Among them, Anisotropic Diffusion (AD) filters [4] based on nonlinear heat diffusion equation surpass most others in terms of accuracy and robustness. These filters fall into the category of Partial Differential Equation (PDE) based image processing, originated from the work of Perona and Malik [5]. This method was capable of reducing the noise content of the image as well as enhancement of the boundary information within the data. However, this filter [5] introduced *blocky effects* and it blurs the edges with the number of iterations of the filter. Yu and Acton, therefore introduced an edge sensitive diffusion method, called Speckle Reducing Anisotropic Diffusion (SRAD) filter [6], which defined an instantaneous coefficient of variation to detect the edges in the noisy images. Aja-Fernández and Alberola-López [7] further developed this method by introducing a new AD filter,

S.C. Satapathy et al. (Eds.): Proc. of Int. Conf. on Front. of Intell. Comput., AISC 199, pp. 747–754.
DOI: 10.1007/978-3-642-35314-7_85

known as Detail Preserving Anisotropic Diffusion (DPAD) filter. Although, both the DPAD and SRAD methods enhanced the prominent edges during speckle filtering; they also resulted in blurring, thereby eradicating detailed features of the image. To enhance flow-like patterns, Weickert [8] developed Coherent Enhancing Diffusion (CED) filter in which the concept of structure tensor was introduced, which allowed the smoothing level to vary directionally. This concept was further utilized in Nonlinear Coherent Diffusion (NCD) technique [9] which implemented diffusion by discriminating between different levels of speckle and filtering only those regions which closely resembled an optimum level of speckle. Although, NCD yields high computational speed, robust parameter selection and texture preservation characteristics but it also introduced certain artifacts in the images. To overcome this drawback, Krissian [10] introduced Oriented Speckle Reducing Anisotropic Diffusion (OSRAD) filter which combined the matrix diffusion scheme and DPAD filter, for reducing speckle as well as preserving and enhancing the contours. But, the usage of diffusion matrix scheme requires heavy computational requirements [11]. Therefore, the present work proposes a novel AD algorithm incorporating a diffusion coefficient based on first and second order derivative operators. The proposed algorithm leads to efficient speckle suppression in homogeneous areas, thereby preserving edges and detailed features as well as lowering the unnecessary computational overhead. Peak Signal-to-Noise Ratio (*PSNR*), and Speckle Suppression Index (*SSI*) are used as quality parameters to evaluate the performance of proposed algorithm. The rest of the paper is organized as follows: Section 2 describes the proposed methodology; the quality parameters used for performance evaluation, experimental procedures and result analysis are explained under Section 3. Based on the analysis of obtained results, Section 4 draws the conclusion.

2 Proposed Methodology

2.1 Background

The AD filters use PDE based methods [13] to resolve an image in order to get expected results by removing the noise. The idea of using PDE based noise removal techniques can be explained as follows:-

Consider an image f_o contaminated with speckle noise ($\chi_{speckle}$) resulting in a noisy image f_n, such that:

$$f_n = f_o \cdot \chi_{speckle} \tag{1}$$

In order to regularize f_n, the variations posed by speckle $\chi_{speckle}$ has to be minimized which can be estimated by gradient norm of image:

$$\|\nabla f_n\| = \sqrt{f_{n_i}^{\,2} + f_{n_j}^{\,2}} \tag{2}$$

where: $\|\nabla f_n\|$ represents gradient norm of the noisy image. The variational problem of (2) is the minimization of the energy function, $E(f_n)$ as:

$$\min_{f_n:\Omega\to R} E(f_n) = \iint \|\nabla f_n\|^2 \, d\Omega \tag{3}$$

The necessary condition for minimizing $E(f_n)$ is given by Euler Lagrange equation:

$$\frac{\partial f}{\partial t} = c\nabla^2 f \tag{4}$$

Here, (4) is known as heat equation with the initial condition, $f = f_n$ at t = 0 and diffusion coefficient *c*. This equation can also be stated as:

$$\frac{\partial f}{\partial t} = div(c.\nabla f) \tag{5}$$

In terms of Continuity equation, the diffusion process can also be expressed as:

$$\frac{\partial f}{\partial t} = -div(J) \tag{6}$$

where: $J = -c.\nabla f$ is a flux created by concentration gradient ∇f and aims to overcome the gradient ∇f . Thus, the PDEs (5) and (6) for diffusion process can be classified on the basis of diffusion coefficient *c* in the two categories viz. Isotropic Diffusion equations, when *c* is a constant and Anisotropic Diffusion equations, when *c* is a function of gradient of image, i.e.

$$c = g(\| \nabla f \|) \tag{7}$$

Anisotropic Diffusion equations provide backward diffusion around transients and forward diffusion in smooth areas in favor of edge sharpening and noise removal [14].

2.2 Proposed Diffusion Coefficient

Perona & Malik replaced the classical isotropic diffusion (5), by introducing the concept of gradient in diffusion constant as shown in (7). By concept the Gradient operator serves to be an effective operator for detecting sharp edges as gradient of a scalar field is a vector field that points in the direction of greatest rate of increase of scalar field [5]. However, if the edges are not sharp i.e. pixel gray level do not change rapidly over space, then it produces very wide and blurred edges. In such cases, a laplacian operator proves to be more effective in comparison to a gradient operator. Laplacian is a second order derivative operator which has a zero crossing level in the middle of edges. Therefore, it can detect the edges more efficiently even when the edges are weak, by detecting the zero crossing level in the image [15]. Hence, a new diffusion coefficient with the combination of first order derivative (gradient) and second order derivative (laplacian) operators can be formulated as:-

$$c = g(\| \Delta f \|) \tag{8}$$

$$c = \left(\frac{1 + \| \Delta f \|}{1 + \| \nabla f \|} \right)^2 \tag{9}$$

where: Δ denotes laplacian operator and ∇ denotes gradient operator. The value of *c* evaluates to a real number ranging between 0 to 1.

2.3 Proposed Despeckling Algorithm

The algorithm proposed in this work is initiated by applying the solution of PDE mentioned in (5) to the noisy SAR images through several iterations until the diffusion gets saturated. Firstly, the gradient and laplacian operators are applied to the noisy input image (initialized as f^0 (x , y)) in order to calculate the proposed diffusion coefficient given by (9). Next, a 3x3 spatial mask centered at any pixel location f (i, j) is taken to calculate the directional derivatives of the central pixel in respective directions. This can be mathematically expressed as under:

$$\left.\begin{aligned}
\nabla_N f_{i,j}^n &= f_{i-1,j}^n - f_{i-1,j}^n \\
\nabla_{NE} f_{i,j}^n &= f_{i-1,j+1}^n - f_{i,j}^n \\
\nabla_{NW} f_{i,j}^n &= f_{i-1,j+1}^n - f_{i,j}^n \\
\nabla_E f_{i,j}^n &= f_{i,j+1}^n - f_{i,j}^n \\
\nabla_W f_{i,j}^n &= f_{i,j-1}^n - f_{i,j}^n \\
\nabla_{SE} f_{i,j}^n &= f_{i+1,j-1}^n - f_{i,j}^n \\
\nabla_{SW} f_{i,j}^n &= f_{i+1,j+1}^n - f_{i,j}^n \\
\nabla_S f_{i,j}^n &= f_{i+1,j}^n - f_{i,j}^n
\end{aligned}\right\} \tag{10}$$

where: n denotes n-th iteration and ∇_N denotes directional derivative in north direction. Similarily N, S, E, W, NE, NW, SE, SW mentioned as subscript with ∇_N denotes directional derivative in north, south, east, west, north-east, north-west, south-east, south-west directions respectively. Then, a simple numerical scheme [5] is used to discretize the solution of (5) which can be stated as:

$$f_{i,j}^{n+1} = f_{i,j}^n + \lambda \begin{bmatrix} c_N.\nabla_N f_{i,j}^n + c_{NE}.\nabla_{NE} f_{i,j}^n + c_{NW}.\nabla_{NW} f_{i,j}^n + c_E.\nabla_E f_{i,j}^n + c_W.\nabla_W f_{i,j}^n + \\ c_{SW}.\nabla_{SW} f_{i,j}^n + c_{SE}.\nabla_{SE} f_{i,j}^n + c_S.\nabla_S f_{i,j}^n \end{bmatrix} \tag{11}$$

where: $\lambda \in [0,1/4]$ and c_N is the diffusion coefficient for the North direction. Similarily, c_X denotes the diffusion coefficient in the respective X direction. This discrete version of the solution, when applied to the image results in a despeckled image (f^1 (x, y)) which is then compared with the original noisy image f^0 *(x, y)* and the resultant error is denoted as E_1:

$$E_1 = | f^1 (x, y) - f^0 (x, y) | \tag{12}$$

Then, the image f^0 *(x, y)* at the input is replaced by the resulting image, f^1 *(x, y)* and the entire process is repeated again such that the error computed is denoted as E_2. The difference between E_1 and E_2 is therefore:

$$\Delta E = E_1 - E_2 \tag{13}$$

Where: ΔE serves to define the stopping criterion. If $\Delta E \geq 0$, then the process is terminated else f^1 *(x, y)* is again replaced by the processed image f^2*(x, y)* and E_2 is stored in E_1 and the whole process is repeated until the error remains less than zero. In this way, the final image produced through several number of iterations is treated as the despeckled image.

3 Results and Discussion

3.1 Evaluation of Proposed Algorithm

Peak Signal-to-Noise Ratio (*PSNR*)

It is generally pre-assumed that higher the value of *PSNR* [12] better is the quality of restored image and is mathematically given as:

$$PSNR(dB) = 10\log_{10}\frac{(2^r - 1)}{MSE} \tag{14}$$

where: r is the number of bits and *MSE* is the Mean Squared Error. The term *MSE* is formulated as:

$$MSE = \frac{1}{MN}\sum_{i-1}^{M}\sum_{j-1}^{N}[f(i,j) - y(i,j)]^2 \tag{15}$$

where: $f(i\ ,\ j)$ denotes the original image, $y(i,\ j)$ denotes the despeckled image, i and j are the pixel position of the $M\ x\ N$ image. The advantage of using *PSNR* is its calculative simplicity and good mathematical convenience in terms of optimization.

Speckle Suppression Index (*SSI*)

It is an average measure of the amount of speckle present in the despeckled image as a whole when compared to the noisy image. So, the lower value of *SSI* signifies the better quality of the image. It is related to the ratio of the local deviation in pixel brightness to the mean pixel brightness averaged over the entire image. *SSI* is mathematically defined as:

$$SSI = \frac{1}{MN}\sum_{i-1}^{M}\sum_{j-1}^{N}\frac{\sigma(i,j)}{\mu(i,j)} \tag{16}$$

where: $\sigma(i,j)$ is the local deviation of the image pixels and $\mu(i,j)$ is the corresponding mean of the entire image. *SSI* is calculated for both noised and denoised image.

3.2 Experimental Procedures and Results

SAR input images (Crater) of size 256x256 is initially normalized to scale down the pixel intensity between the range 0 to 1. Speckle noise of varying intensities (ranging between $\sigma = 0.001$ to 0.1) is superimposed on the normalized SAR images to produce the noisy image. With the help of (9), the diffusion coefficient is calculated and is further used in diffusion process as discussed in section 2. The quality of the processed image is checked with the help of *PSNR* and *SSI* quality metrics where *PSNR* is used to calculate ΔE and thus, decides the stopping criterion. Therefore, the number of iterations required to achieve an optimum despeckling result depends on the value of *PSNR* of image at each iteration.

Fig. 2 shows the simulated results for Anisotropic Diffusion Algorithm of SRAD and DPAD along with those obtained by using proposed algorithm at various intensity levels of speckle noise. The speckle noise is significantly removed in the images obtained by using the proposed Anisotropic Diffusion Algorithm with minimum iteration [11]. The results are more reliable in terms of edge preservation and reduced speckle noise. Moreover, even at higher value of noise intensity such as 0.1 the results are quite relevant in comparison with those obtained from SRAD and DPAD.

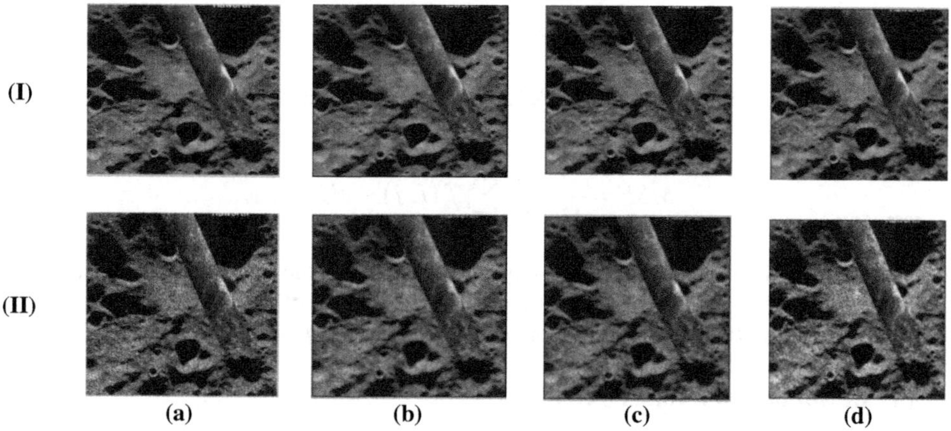

Fig. 1. (a) Original Image (Crater): **(I)** 0.01 speckle noise **(II)** 0.1 speckle noise. Despeckled SAR image by: **(b)** SRAD **(c)** DPAD **(d)** Proposed AD Algorithm.

3.3 Analysis

The analysis of AD filters is done on the basis of simulation results and corresponding values of *PSNR* and *SSI* obtained for a SAR image (crater). As discussed earlier, the *PSNR* values are required for quality assessment of restored images while *SSI* values measure the relative amount of speckle present in the denoised image when compared with the noisy image. Taking this under consideration, the results regarding to the values of *PSNR* and *SSI* are tabulated for DPAD, SRAD and for proposed AD algorithm at various speckle noise levels ranging from 0.001 to 0.1. The *PSNR* values for DPAD and SRAD in table 1 are high at low level noise and are decreasing with increase in noise level. The subsequent analysis of *PSNR* values for the proposed algorithm show high values at various noise levels and thus, the quality of images obtained through this algorithm can be considered to be better than the others. Similarly, the *SSI* values in table 2 are required to be as low as possible. The relative deviation in values for noisy and denoised SAR images for the proposed AD algorithm is very high in comparison with others, revealing the fact that the denoised image contains less amount of speckle noise. Therefore, this algorithm is efficient in preserving the edge details and despeckling the images while the earlier proposed filters like SRAD and DPAD exhibited these properties but with certain short comings of blurred edges and less filtered noise content. Moreover, the proposed algorithm also results in fast convergence of the filtering process as only 1 iteration is required to despeckle the noisy image having speckle intensity up to 0.1. However, for speckle intensity greater than 0.1, number of iterations required increases to 2 which is still much less that required for DPAD and SRAD filter (around 5-6 number of iterations).

Table 1. PSNR(dB) values for different AD Filters calculated for the SAR image

Noise Level (variance)	DPAD	SRAD	Proposed AD Algorithm
0.001	26.3596	26.4449	28.5470
0.01	25.7691	25.8422	27.8240
0.02	25.2080	25.2636	27.8141
0.04	24.5640	24.7008	25.9425
0.1	23.5550	23.7880	25.0727

Table 2. SSI values for different AD Filters calculated for the SAR image

Noise level (variance)	DPAD		SRAD		Proposed AD Algorithm	
	Noisy Image	Denoised Image	Noisy Image	Denoised Image	Noisy Image	Denoised Image
0.001	0.2023	0.1562	0.2023	0.1549	0.2023	0.1546
0.01	0.2299	0.1700	0.2299	0.1699	0.2299	0.1667
0.02	0.2548	0.1805	0.2548	0.1779	0.2548	0.1780
0.04	0.2970	0.1931	0.2970	0.1930	0.2970	0.1829
0.1	0.3926	0.1969	0.3926	0.1956	0.3926	0.1948

4 Conclusion

In this paper, an improved algorithm for AD filters is proposed along with a new coefficient of diffusion. The evaluation of proposed algorithm on the basis of quality metrics such as *PSNR* and *SSI* exhibits its relative dominance over other techniques in terms of preserving the edge details and despeckling of SAR images. Also, the requirement of less number of iterations for despeckling process proves the simplicity and effectiveness of the proposed diffusion coefficient in comparison to the other more complex methods like SRAD and DPAD.

References

1. Oliver, C.J.: Information from SAR Images. Journal of Applied Physics 24(5), 1493–1514 (1991)
2. Santosh, D.H.H., et al.: Efficiency Techniques for Denoising of Speckle and Highly Corrupted Impulse Noise Images. In: Proc. of the 3rd International Conference on Electronics and Computer Technology, vol. 3, pp. 253–257 (2011)

3. Lopes, A., Tauzin, R., Nezry, E.: Adaptive Speckle Filters and Scene Heterogenity. IEEE Transactions on Geoscience and Remote Sensing 28(6) (1990)
4. Loizou, C.P., et al.: Comparative Evaluation of Despeckle Filtering in Ultrasound Imaging of the Carotid Artery. IEEE Transactions on Ultrasonics, Ferroelectrics and Frequency 52(10), 1653–1669 (2005)
5. Perona, P., Malik, J.: Scale-Space and Edge Detection using Anisotropic Diffusion. IEEE Transactions on Pattern Analysis and Machine Intelligence 12(7), 629–639 (1990)
6. Yu, Y., Acton, S.: Speckle Reducing Anisotropic Diffusion. IEEE Transactions on Image Processing 11(11), 1260–1270 (2002)
7. Aja-Fernandez, S., Alberola-Lopez, C.: On the Estimation of the Coefficient of Variation for Anisotropic Diffusion Speckle Filtering. IEEE Transactions on Image Processing 15(9), 2694–2701 (2006)
8. Weickert, J.: Coherence-enhancing diffusion filtering. International Journal of Computer Vision 31(2-3), 111–127 (1999)
9. Abd-Elmoniem, K., Youssef, A.B., Kadah, Y.: Real-time Speckle Reduction and Coherence Enhancement in Ultrasound Imaging via Nonlinear Anisotropic Diffusion. IEEE Transactions on Biomedical Engineering 49(9), 997–1014 (2002)
10. Krissian, K., et al.: Oriented speckle reducing anisotropic diffusion. IEEE Transactions on Image Processing 16(5), 1412–1424 (2007)
11. Glavin, M., Jones, E.: Echocardiographic Speckle Reduction Comparison. IEEE Transactions on Ultrasonics, Ferroelectrics and Frequency Control 58(1), 82–101 (2011)
12. Wang, Z., et al.: Image Quality Assessment: From Error Visibility to Structural Similarity. IEEE Transactions on Image Processing 13(4), 600–612 (2004)
13. Narayan, S.K., Wahidabanu, R.S.D.: Despeckling of Medical Diagonostic Ultrasound Images via Laplacian Based Mixed PDE. In: IEEE International Conference on Communication and Control Computing Technologies (ICCCCT), pp. 525–530 (2010)
14. Kalaivani, S., Wahidabanu, R.S.D.: Condensed Anisotropic Diffusion for Speckle Reducton and Enhancement in Ultrasonography. EURASIP Journal on Image and Video Processing 12, 1687–5281 (2012)
15. Gonzalez, R., Woods, R.: Digital Image Processing, 3rd edn. Pearson Prentice Hall Press, New York (2009)

Two Stage Color Image Steganography Using DCT (TSCIS-DCT)

Anirban Goswami[1], Dipankar Pal[2], and Nabin Ghoshal[3]

[1] Dept. of Information Technology, Techno India,
EM 4/1 Salt Lake, Sec-V, Kolkata-700091
[2] Dept. of Computer Science and Engineering, Techno India,
EM 4/1 Salt Lake, Sec-V, Kolkata-700091
[3] Dept. of Engineering and Technological Studies,
University of Kalyani, Kalyani, Nadia-741235,West Bengal, India.
an_gos@yahoo.com, mail2dpal@yahoo.com,
nabin_ghoshal@yahoo.co.in

Abstract. In frequency domain steganography, use of color images for secret data hiding may prove to be a decisive innovation. The proposed concept uses two color images for hiding a color/gray authenticating message/image. The mathematical technique of Discrete Cosine Transform (DCT) is applied on each block of size 2x2 taken in row major order from three color planes (Red, Green & Blue) sequentially and from two carrier images alternatively. A single secret message/image bit is fabricated within the transformed real frequency component of each source image byte except the first frequency component of each mask. The first frequency component of each block is used for re-adjustment to maintain the quantum value positive, non-fractional in spatial domain and also to reduce the integrated noise due to embedding. The pseudorandom position of embedding and subsequent extraction is generated by a logical expression. Experimental results of this technique reveal more efficiency compared to other similar technique.

Keywords: Image Authentication, Digital Watermarking, Steganography, DCT, IDCT, MSE, PSNR, IF, SSIM.

1 Introduction

The popularity of Internet was one of the key factors in the process of Information Technology scaling heights with security of information being the prime concern. So a technique termed Cryptography was adopted for securing the secrecy of information and communication. But sometimes it is not only enough to keep the contents of a message secret, but also necessary to keep the existence of the message confidential [4, 5]. So the concept named Steganography [1] was conceived to emphasize invisible communication. Consequently steganography is mostly used today on computers where digital data is being the carrier and networks being the high speed delivery channels. In steganography many different carrier image file formats can be used, but digital images [2, 3] are the most popular because of their frequency format.

S.C. Satapathy et al. (Eds.): Proc. of Int. Conf. on Front. of Intell. Comput., AISC 199, pp. 755–763.
DOI: 10.1007/978-3-642-35314-7_86

An image file is a binary file containing a binary representation of the color or light intensity of each picture element (pixel). Images typically use either 8-bit or 24-bit color. A 24-bit color scheme uses 24 bits per pixel and provides a much better set of colors. In this case, each pixel is represented by three bytes, each byte representing the intensity of the three primary colors red, green, and blue (RGB) respectively.

The proposed watermarking [6] scheme uses two 24 bit color cover images to hide a secret color/gray image in distributive manner. Distribution of secret data is mathematically decided. Logical and reasonable distribution preserves robustness and imperceptibility of cover images. The embedding and extraction of secret bits are done based on a pseudorandom value [0.. 3] calculated by a logical expression.

Fig. 1 demonstrates the overall insertion and extraction processes. The proposed technique uses two dimensional Discrete Cosine Transform and Inverse Discrete Cosine Transform as presented in sec 1.1. Sec.2 details the technique of insertion and extraction processes of TSCIS-DCT. The experimental results based on SSIM, MSE, PSNR in dB and IF are explained in sec.3 followed by conclusion in sec.4.

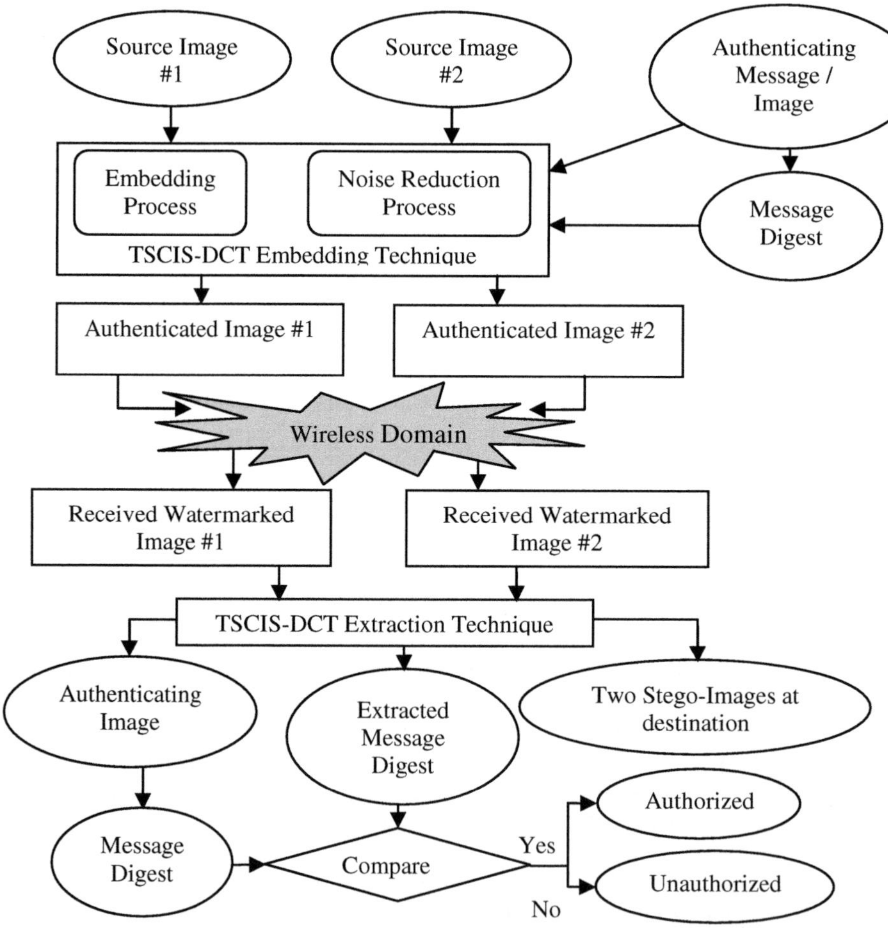

Fig. 1. Embedding and Extraction techniques of secret data using TSCIS-DCT

1.1 Two Dimensional Discrete Cosine and Inverse Discrete Cosine Transform

The representation of two-dimensional DCT implemented on M x N matrix is as follows:

$B_{pq} = \alpha_p \alpha_q \sum_{m=0}^{M-1} \sum_{n=0}^{N-1} A_{mn} \cos\frac{\pi(2m+1)p}{2M} \cos\frac{\pi(2n+1)q}{2N}$, where $0 \le p \le M-1$ and $0 \le q \le N-1$. The terms α_p and α_q are represented as,

$$\alpha_p = \begin{cases} \frac{1}{\sqrt{M}}, & p = 0 \\ \sqrt{2/M}, & 1 \le p \le M-1 \end{cases} \qquad \alpha_q = \begin{cases} \frac{1}{\sqrt{N}}, & q = 0 \\ \sqrt{2/N}, & 1 \le q \le N-1 \end{cases}$$

The DCT coefficient of spatial value A_{mn} is B_{pq}. The frequency components {W, X, Y, Z} obtained after performing DCT on four spatial values {a, b, c, d} taken as a block of size 2x2 from the source image are represented as: W = DCT(a) = ½ (a + b + c + d), X = DCT(b) = ½ (a – b + c – d), Y = DCT(c) = ½ (a + b – c – d) , Z = DCT(d) = ½ (a – b – c + d).

The invertible DCT transform, i.e. IDCT is expressed as,

$A_{mn} = \sum_{p=0}^{M-1} \sum_{q=0}^{N-1} \alpha_p \alpha_q B_{pq} \cos\frac{\pi(2m+1)p}{2M} \cos\frac{\pi(2n+1)q}{2N}$,where $0 \le m \le M-1$ and $0 \le n \le N-1$. The terms α_p and α_q are represented as,

$$\alpha_p = \begin{cases} \frac{1}{\sqrt{M}}, & p = 0 \\ \sqrt{2/M}, & 1 \le p \le M-1 \end{cases} \qquad \alpha_q = \begin{cases} \frac{1}{\sqrt{N}}, & q = 0 \\ \sqrt{2/N}, & 1 \le q \le N-1 \end{cases}$$

The corresponding IDCT values are $DCT^{-1}(W)$ = ½ (W + X + Y + Z), $DCT^{-1}(X)$ = ½ (W – X + Y –Z), $DCT^{-1}(Y)$ = ½ (W + X –Y –Z), $DCT^{-1}(Z)$ = ½ (W – X – Y + Z).

2 The Technique

Each color image is first split into three basic color planes, i.e. R, G & B. A sliding window block of size 2x2 is taken exclusively from these three planes in sequence and also from two cover images alternatively. DCT is applied on each such block to obtain the corresponding frequency values of the pixels. A generated pseudorandom number (0-3) defines the position for embedding of secret bits in 2nd, 3rd and 4th frequency coefficients of each block. A single bit is fabricated in the integer part of each frequency value while the fractional part remains unchanged. Noise generated due to embedding is reduced by making a minor re-adjustment on the modified frequency components. In the process of extraction, selection of blocks and application of DCT are done in the same fashion as mentioned above. Extraction is done by retrieving one bit each from the 2nd, 3rd and 4th pixel of each sub image block and the location of extraction is again decided by pseudorandom values. The following sub-sections explain the insertion and extraction algorithms in greater detail.

2.1 Insertion Algorithm

Input: Two color cover images and an authenticating color/gray message/image.
Output: Two color embedded images.
Steps:

1. Generate a message digest from the authenticating image.
2. Copy the header information of cover images into the output images.
3. Repeat the following steps until all pixels have been read from the two source images as well as the header information, the message digest and all the pixels of authenticating message/image are embedded,
   3.1 Take a 2x2 block of pixels from the cover image planes (related to Red, Green and Blue at a time and corresponding to two images alternatively) in row major order.
   3.2 Apply Discrete Cosine Transform on the current block of pixels.
   3.3 A pseudorandom value (0 - 3) is generated into a variable named ipos.
   3.4 Read the authenticating message/image (i.e. secret data).
   3.5 The authenticating message/image bits are embedded only in the integer part of each frequency component at the position defined by ipos.
   3.6 The noise effect due to embedding is reduced by adjusting some of the frequency components (refer to sec 2.3).
   3.7 To obtain the corresponding intensity value in spatial domain Inverse Discrete Cosine Transform is applied on the current block.
   3.8 Write the spatial block into the output image in row major order.
4. Stop.

Important Readjustments: The application of IDCT on the frequency values (obtained after the embedding operation) sometimes may generate erroneous results like:

1. Negative pixel values, which may be removed by subsequent increment of the value at the 1st position (DC coefficient) of the current block and re-applying the embedding procedure on that block.
2. Occurrence of fractional spatial values due to the expression of DCT being multiplied by 1/2 (sec 1.1). The problem may be winnowed out by changing the value of the sum obtained after the DCT operation to an even number.

2.2 Extraction Algorithm

Input: Two color watermarked images.
Output: An authenticating color/gray message/image.
Steps:

1. The following steps are repeated until the header information, the embedded message digest and all the pixels of the output image have been extracted from the received carrier images,
   1.1. Take a 2x2 block of pixels at a time from the embedded images (considering R, G & B planes separately and corresponding to two images alternately) in row major order.

1.2. Apply Discrete Cosine Transform on the current block of pixels.
1.3. A pseudorandom value (0 - 3) is generated into a variable termed ipos.
1.4. Consider the integer part of the frequency values to extract the embedded bit from the position specified by ipos.
1.5. Form the header information and the embedded message digest by combining the extracted bits.
1.6. After the formation of message digest, a byte is formed by blending consecutive 8 extracted bits. Three such bytes (for R, G, & B) of secret data together corresponds a pixel intensity value of the authenticating image.

2. Formulate a message digest from the extracted authenticating image.
3. Compare the extracted message digest with the formulated message digest to check the authenticity of the received images.
4. Stop.

2.3 Generation of ipos and the Noise Reduction

The value of the variable ipos is generated by the expression: ipos = (((x AND 0cH) SHR 2) XOR (y AND 03H)), where x & y varies in value. The mean value M of the current block is computed and the generated value of ipos is further modified depending on the following condition: if (M>T) then ipos = [0..3], else ipos = [0..2], where T could be any value between 32 and 128. The final value is used as the position for embedding or extraction.

The effect of resulting noise due to embedding is reduced by manipulating the unaltered bits of the embedded image byte only if the original image byte value and the resulting embedded image byte value are different. This is done by modifying the bits on the right (i.e. towards LSB) and/or left (i.e. towards MSB) of the target frequency component with respect to ipos.

3 Result Comparison and Analysis

A comparative study of the proposed technique TSCIS-DCT is done with other existing watermarking methods such as DCT-based [7], QFT-based [8], SCDFT-based [9], Z transform-based [11] and DFT based [12]. The study involves image quality assessment techniques like Visual Interpretation, Mean Square Error (MSE), Peak Signal-to-Noise Ratio (PSNR), Image Fidelity (IF) and Structural Similarity Index Metric (SSIM) as detailed below.

3.1 Image Comparison Metrices

1. Mean Square Error (MSE) is defined as:

$MSE = \frac{1}{(MXN)} \sum_{y=1}^{M} \sum_{x=1}^{N} [I(x,y) - I'(x,y)]^2$, where I(x, y) is the original image, I'(x, y) is the watermarked image and M, N are the dimension of the images.

2. Peak Signal To Noise Ratio (PSNR) is represented as: PSNR = 20 * $\log_{10}$ (255 / sqrt(MSE)), where 255 is the maximum intensity of the image.
3. Image Fidelity (IF) – Ability to discriminate between two images.

$$\text{IF} = 1 - \sum_{y=1}^{M}\sum_{x=1}^{N}\left[I(x,y)-I'(x,y)\right]^2 \Big/ \sum_{y=1}^{M}\sum_{x=1}^{N}[I(x,y)]^2$$

4. Structural Similarity Index Metric (SSIM) – This technique is sensitive to distortions that breaks down natural spatial correlation of an image, and expressed as l(f, g) x c(f, g) x s(f, g), where l(f, g) is luminance comparison function, c(f, g) is contrast comparison function and s(f, g) is structural comparison function. $l(f, g) = \frac{2\mu_f \mu_g + C1}{\mu_f^2 + \mu_g^2 + C1}$, $c(f, g) = \frac{2\sigma_f \sigma_g + C2}{\sigma_f^2 + \sigma_g^2 + C2}$, $s(f, g) = \frac{2\sigma_{fg} + C3}{\sigma_f \sigma_g + C3}$, where μ_f & μ_g are the mean, σ_f^2 & σ_g^2 are the variances, σ_{fg} is the covariance and f_{ij}, g_{ij} are the pixel values of the cover image (f) and steganographic image (g), each of size M X N respectively.

Source Images	Embedded Images using TSCIS-DCT	Magnified Source Images	Magnified Watermarked Images
 **Fig. 2a.**peppers	 **Fig. 2e.**	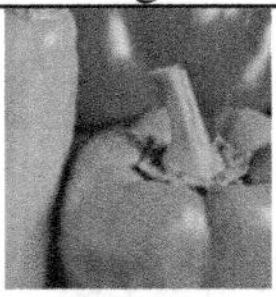 **Fig. 2i.**	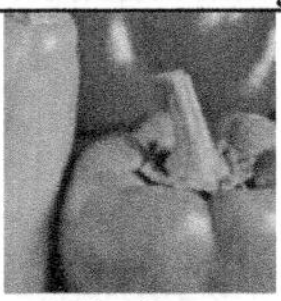 **Fig.2m.**
 **Fig. 2b.**sailboat	 **Fig. 2f.**	 **Fig. 2j.**	 **Fig.2n.**
 **Fig. 2c.**splash	 **Fig. 2g.**	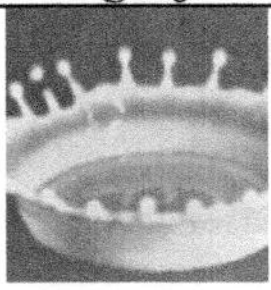 **Fig. 2k.**	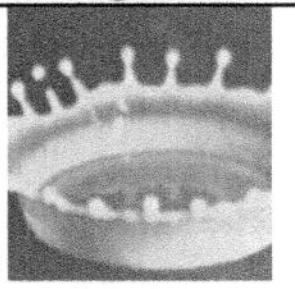 **Fig.2o.**
 **Fig. 2d.**tiffany	 **Fig. 2h.**	 **Fig. 2l.**	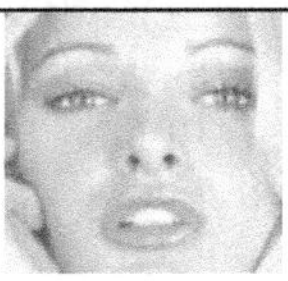 **Fig.2p.**
 **Fig. 2q Earth**			

Fig. 2. Visual Interpretation of embedded images using TSCIS-DCT

The proposed algorithm has been applied on many paired PPM images and it has been experienced that the algorithm can overcome any type of attack like visual or statistical. Some paired sample images like 'peppers', 'sailboat' and 'splash', 'tiffany' each of dimension 512x512 are shown in fig 2a, 2b, 2c and 2d respectively. The dimension of the authenticating image 'Earth' (Fig. 2q) is 210x210. 132300 bytes of secret data is fabricated within a pair of source images. The resultant embedded images are shown in Fig 2e, 2f, 2g and 2h respectively. 2i, 2j, 2k, and 2l are magnified source images and 2m, 2n, 2o, and 2p are magnified embedded images respectively. According to HVS, there is hardly any difference between the source and embedded images.

The quality of each steganographic image measured by MSE, PSNR, IF and SSIM is shown in Table 1. The comparative study between Reversible Data Hiding Based on Block Median Preservation (RDHBBMP) [10], A Steganographic Scheme for Color Image Authentication using Z-Transform (SSCIAZ) [11] and TSCIS-DCT shows scaling in terms of hiding capacity of secret data and PSNR in dB in our proposed scheme as shown in Table 2. The average improvement of embedding secret data in the proposed scheme is 185257 bits more than RDHBBMP and 151704 bits more than SSCIAZ along with the increase of 1.90 dB and 1.75 dB respectively of PSNR which mean low rate of bit-error.

Table 3 shows better PSNR values with much more capacity of embedding than the existing techniques like DCT-based, QFT-based, SCDFT-based and DFT-based (DFTMCIAWC [12]) watermarking schemes. Embedded data in SCDFT, QFT, and DCT is 3840 bytes each, in DFTMCIAWC is 73728 bytes and PSNR values are 30.10 dB, 30.93 dB, 30.40 dB and 44.65 dB respectively but that in TSCIS-DCT is 132300 bytes and PSNR is 45.59 dB which is fully recoverable.

Table 1. Capacities and Metric values of images inTSCIS-DCT

Source Images	Embedded (Bytes)	MSE	PSNR	IF	SSIM
Peppers	132300	1.763204	46.056034	0.998630	0.999863
Sailboat		1.733445	46.140722	0.999668	0.999913
Splash	132300	2.155649	45.183750	0.999281	0.999923
Tiffany		2.311454	44.980182	0.999225	0.999811
Average	**132300**	**1.990938**	**45.590172**	**0.999201**	**0.999877**

Table 2. Results and comparison in capacities and PSNR of RDHBBMP, SSCIAZ &TSCIS-DCT

Test images	Indicator	EL = 0	EL = 0	EL = 0
		RDHBBMP	**SSCIAZ**	**TSCIS-DCT**
Lenna	C(bits)	26,465	64,896	216600
	PSNR	49.68	49.89	52.20
Airplane	C(bits)	36,221	64,896	216600
	PSNR	49.80	49.87	51.07

Table 3. Comparisonbetween TSCIS-DCT and DCT, QFT, SCDFT & DFTMCIAWC

Technique	No. of Source images	Embedded (bytes)	PSNR (dB)
SCDFT	1	3840	30.10
QFT	1	3840	30.93
DCT	1	3840	30.40
DFTMCIAWC	1	73728	44.65
TSCIS-DCT	**2**	**132300**	**45.59**

4 Conclusion

TSCIS-DCT has been proposed for utilization of color images in increasing the security of data hiding. Authenticity is incorporated by embedding secret data in random positions of carrier image bytes. Attention has been imparted to avoid any visual suspicion due to accruement of noise. The embedded image in this algorithm is very difficult to detect due to dynamic insertion position of the authenticating message/image bits in the carrier image. Hence, this technique can safeguard the images from any possible hacking attempt.

Acknowledgement. The authors express their deep sense of gratitude to the faculty members of the Dept. of Engineering and Technological Studies, University of Kalyani, West Bengal, India, where the work has been carried out.

References

1. Wu, H.C., Wu, N.I., Tsai, C.S., Hwang, M.S.: Image steganographic scheme based on pisel-value differencing and LSB replacement methods. Proc. Inst. Elect. Eng., Vis. Images Signal Processing 152(5), 611–615 (2005)
2. EL-Emam, N.N.: Hiding a large Amount of data with High Security Using Steganography Algorithm. Journal of Computer Science 3(4), 223–232 (2007) ISSN 1549-3636
3. Amin, P., Lue, N., Subbalakshmi, K.: Statistically secure digital image data hiding. In: IEEE Multimedia Signal Processing MMSP 2005, Shanghai, China, pp. 1–4 (October 2005)
4. Pavan, S., Gangadharpalli, S., Sridhar, V.: Multivariate entropy detector based hybrid image registration algorithm. In: IEEE Int. Conf. on Acoustics, Speech and Signal Processing, Philadelphia, Pennsylvania, USA, pp. 18–23 (March 2005)
5. Al-Hamami, A.H., Al-Ani, S.A.: A New Approach for Authentication Technique. Journal of Computer Science 1(1), 103–106 (2005) ISSN 1549-3636
6. Ker, A.: Steganalysis of Embedding in Two Least-Significant Bits. IEEE Transaction on Information Forensics and Security 2(1), 46–54 (2008) ISSN 1556-6013
7. Ahmidi, N., Safabkhsh, R.: A novel DCT-based approach for secure color image watermarking. In: Proc. Int. Conf. Information Technology: Coding and Computing, vol. 2, pp. 709–713 (April 2004)

8. Bas, P., Biham, N.L., Chassery, J.: Color watermarking using quaternion Fourier transformation. In: Proc. ICASSP, Hong Kong, China, pp. 521–524 (June 2003)
9. Tsui, T.T., Zhang, X.–P., Androutsos, D.: Color Image Watermarking Using Multidimensional Fourier Transfomation. IEEE Trans. on Info. Forensics and Security 3(1), 16–28 (2008)
10. Luo, H., Yu, F.-X., Chen, H., Huang, Z.-L., Li, H., Wang, P.-H.: Reversible data hiding based on block median preservation. Information Sciences 181, 308–328 (2011)
11. Ghoshal, N., Chowdhury, S., Mandal, J.K.: A Steganographic Scheme for Color Image Authentication using Z-Transform (SSCIAZ). In: Advances in Intelligent Soft Computing, vis INDIA 2012 (2012) ISSN:1867-5662
12. Ghoshal, N., Mandal, J.K.: Discrete Fourier Transform based Multimedia Colour Image Authentication for Wireless Communication (DFTMCIAWC). In: 2nd International Conference on Wireless Communications, Vehicular Technology, Information Theory and Aerospace &Electronics Systems Technology, Wireless Vitae 2011, Chennai, India (2011) ISBN: 978-1-4577-0787-2/11

Author Index

GPSR Compliance

The European Union's (EU) General Product Safety Regulation (GPSR) is a set of rules that requires consumer products to be safe and our obligations to ensure this.

If you have any concerns about our products, you can contact us on ProductSafety@springernature.com

In case Publisher is established outside the EU, the EU authorized representative is:

Springer Nature Customer Service Center GmbH
Europaplatz 3
69115 Heidelberg, Germany

Batch number: 10755036

Printed by Printforce, the Netherlands